# Structural Dynamics

Master the principles of structural dynamics with this comprehensive and self-contained textbook, with key theoretical concepts explained through real-world engineering applications.

- The theory of natural modes of vibration, the finite element method, and the dynamic response of structures is balanced with practical applications to give students a thorough contextual understanding of the subject.
- Enhanced coverage of damping, rotating systems, and parametric excitation provides students with superior understanding of these essential topics.
- Examples and homework problems, closely linked to real-world applications, enrich and deepen student understanding.
- Curated mathematical appendices equip students with all the tools necessary to excel, without disrupting coverage of core topics.
- Containing all the material needed for a one- or two-semester course, and accompanied online by MATLAB/Python code, this authoritative textbook is the ideal introduction for graduate students in aerospace, mechanical, and civil engineering.

**Peretz P. Friedmann** is the François-Xavier Bagnoud Professor of Aerospace Engineering at the University of Michigan, Ann Arbor. He is a former Editor-in-Chief of the *AIAA Journal*, and was awarded the 2022 Reed Aeronautics Award – the highest honor bestowed by AIAA for notable achievements in aeronautics – as well as the Alexander Klemin Award, the highest honor bestowed by the VFS for advancing the field of vertical flight aeronautics. He is a Fellow of AIAA and an Honorary Fellow of VFS.

**George A. Lesieutre** is Associate Dean for Research and Professor of Aerospace Engineering at the Pennsylvania State University. He is Fellow of the AIAA and was awarded the SPIE Lifetime Achievement Award in Smart Structures and Materials.

**Daning Huang** is Assistant Professor of Aerospace Engineering at the Pennsylvania State University, where his research focuses on hypersonic aeroelasticity and aerothermoelasticity, multidisciplinary analysis and design, reduced-order modeling of nonlinear dynamical systems, and high-performance computing.

## Cambridge Aerospace Series

Editors: Wei Shyy and Vigor Yang

1. J. M. Rolfe and K. J. Staples (eds.): *Flight Simulation*
2. P. Berlin: *The Geostationary Applications Satellite*
3. M. J. T. Smith: *Aircraft Noise*
4. N. X. Vinh: *Flight Mechanics of High-Performance Aircraft*
5. W. A. Mair and D. L. Birdsall: *Aircraft Performance*
6. M. J. Abzug and E. E. Larrabee: *Airplane Stability and Control*
7. M. J. Sidi: *Spacecraft Dynamics and Control*
8. J. D. Anderson: *A History of Aerodynamics*
9. A. M. Cruise, J. A. Bowles, C. V. Goodall, and T. J. Patrick: *Principles of Space Instrument Design*
10. G. A. Khoury (ed.): *Airship Technology, Second Edition*
11. J. P. Fielding: *Introduction to Aircraft Design*
12. J. G. Leishman: *Principles of Helicopter Aerodynamics, Second Edition*
13. J. Katz and A. Plotkin: *Low-Speed Aerodynamics, Second Edition*
14. M. J. Abzug and E. E. Larrabee: *Airplane Stability and Control: A History of the Technologies that Made Aviation Possible, Second Edition*
15. D. H. Hodges and G. A. Pierce: *Introduction to Structural Dynamics and Aeroelasticity, Second Edition*
16. W. Fehse: *Automatic Rendezvous and Docking of Spacecraft*
17. R. D. Flack: *Fundamentals of Jet Propulsion with Applications*
18. E. A. Baskharone: *Principles of Turbomachinery in Air-Breathing Engines*
19. D. D. Knight: *Numerical Methods for High-Speed Flows*
20. C. A. Wagner, T. Hüttl, and P. Sagaut (eds.): *Large-Eddy Simulation for Acoustics*
21. D. D. Joseph, T. Funada, and J. Wang: *Potential Flows of Viscous and Viscoelastic Fluids*
22. W. Shyy, Y. Lian, H. Liu, J. Tang, and D. Viieru: *Aerodynamics of Low Reynolds Number Flyers*
23. J. H. Saleh: *Analyses for Durability and System Design Lifetime*
24. B. K. Donaldson: *Analysis of Aircraft Structures, Second Edition*
25. C. Segal: *The Scramjet Engine: Processes and Characteristics*
26. J. F. Doyle: *Guided Explorations of the Mechanics of Solids and Structures*
27. A. K. Kundu: *Aircraft Design*
28. M. I. Friswell, J. E. T. Penny, S. D. Garvey, and A. W. Lees: *Dynamics of Rotating Machines*
29. B. A. Conway (ed): *Spacecraft Trajectory Optimization*
30. R. J. Adrian and J. Westerweel: *Particle Image Velocimetry*
31. G. A. Flandro, H. M. McMahon, and R. L. Roach: *Basic Aerodynamics*
32. H. Babinsky and J. K. Harvey: *Shock Wave–Boundary-Layer Interactions*
33. C. K. W. Tam: *Computational Aeroacoustics: A Wave Number Approach*
34. A. Filippone: *Advanced Aircraft Flight Performance*
35. I. Chopra and J. Sirohi: *Smart Structures Theory*
36. W. Johnson: *Rotorcraft Aeromechanics vol. 3*
37. W. Shyy, H. Aono, C. K. Kang, and H. Liu: *An Introduction to Flapping Wing Aerodynamics*
38. T. C. Lieuwen and V. Yang: *Gas Turbine Emissions*
39. P. Kabamba and A. Girard: *Fundamentals of Aerospace Navigation and Guidance*
40. R. M. Cummings, W. H. Mason, S. A. Morton, and D. R. McDaniel: *Applied Computational Aerodynamics*

"Structural dynamics is one of the central topics in all of engineering, and especially as applied to aerospace and mechanical systems. The authors have provided a sound and novel treatment of this subject. While the standard topics are well treated, the discussion of Hamilton's principle and Lagrange's equations is particularly well done. Also, the authors introduce the concept of nonconservative forces and dynamic instability through the use of a follower force. Given the expertise and experience of the authors, it is not surprising that the inclusion of rotating systems (rotorcraft and turbomachinery) is a welcome addition, a topic too often neglected in other texts on structural dynamics. The stability of systems with periodic (in time) coefficients is nicely done, and the subject of damping is more extensively and insightfully treated than in many texts. Engineers in academe and industry will find this a useful and attractive treatment of the subject and a book they may well want to add to their library or course syllabus."

Earl Dowell, Duke University

"An authoritative overview of structural dynamics, from its fundamental principles to their application in aerospace and mechanical engineering. The book presents the material rigorously and comprehensively, including prerequisite concepts, an extensive collection of examples and problems, details on numerical methods, and advanced topics related to structural dynamics of rotating and periodic systems. It is an excellent resource for graduate students, educators, and professionals in the field."

Cristina Riso, Georgia Institute of Technology

"Professor Friedmann and his colleagues have written a must-read comprehensive book for students and engineers who want to enjoy the dynamics of flexible structures. It is the fruit of Professor Friedmann's decades of teaching and research experience. Read this book, and you will appreciate it as one of the best in the category."

Weihua Su, University of Alabama

"The authors have created an outstanding book that is highly recommended both for graduate-level instruction and as a reference text for structural dynamics researchers. I am not aware of any single reference text on structural dynamics that covers so many different important topics so thoroughly. Their approach combines the best of the classic theoretical reference texts along with modern computational approaches that are found in the commercially available analysis programs used by the aerospace industry."

John Kosmatka, University of California, San Diego

"The authors are well-known experts and active researchers in the subject area. They have provided a very thorough, comprehensive, and authoritative treatment of structural dynamics. I wish I could have had access to such an excellent book when I was a graduate student. Strongly recommend it to any students who are interested in this subject, including rotorcraft dynamics."

Wenbin Yu, Purdue University

# Structural Dynamics

## Theory and Applications to Aerospace and Mechanical Engineering

**Peretz P. Friedmann**
University of Michigan, Ann Arbor

**George A. Lesieutre**
Pennsylvania State University

**Daning Huang**
Pennsylvania State University

CAMBRIDGE
UNIVERSITY PRESS

Shaftesbury Road, Cambridge CB2 8EA, United Kingdom

One Liberty Plaza, 20th Floor, New York, NY 10006, USA

477 Williamstown Road, Port Melbourne, VIC 3207, Australia

314–321, 3rd Floor, Plot 3, Splendor Forum, Jasola District Centre, New Delhi – 110025, India

103 Penang Road, #05–06/07, Visioncrest Commercial, Singapore 238467

Cambridge University Press is part of Cambridge University Press & Assessment, a department of the University of Cambridge.

We share the University's mission to contribute to society through the pursuit of education, learning and research at the highest international levels of excellence.

www.cambridge.org
Information on this title: www.cambridge.org/highereducation/isbn/9781108842488

DOI: 10.1017/9781108909617

First published 2023

*A catalogue record for this publication is available from the British Library*

*A Cataloging-in-Publication data record for this book is available from the Library of Congress*

ISBN 978-1-108-84248-8 Hardback

Additional resources for this publication at www.cambridge.org/friedmann

This book is dedicated to the memory of my father Mauritiu Friedmann whose resourcefulness during World War II enabled the author to survive the Holocaust, and to my wife Esther, whose support during the last 58 years made everything possible.

PPF

To the memory of my parents, for love and prizing education;
To my students, for curiosity and questing; and
To Annie, my everything.

GAL

To Junyi, Guanle, Minjian, as well as future humans as a multiplanetary species.

DH

# Contents

Contents     xv

# Preface

This book addresses methods for the prediction of the dynamic response of structures, emphasizing practical applications in aerospace and mechanical engineering. It reviews and extends concepts from vibrations of mechanical systems and mechanics of structures to address free and forced response of large-scale structures, including numerical methods. It treats in detail several advanced topics that are critical to applications, such as damping, structural dynamics of rotating systems, and stability and response of periodic systems. Worked-out examples and problem sets enhance and augment the reader's understanding of the material covered.

Students who successfully complete a course based on this book will be able to:

- Describe the assumptions underlying models of the behavior of structural members such as rods, beams, and plates, including membrane loads;
- Develop equations of motion and boundary conditions for continuous structures using Newtonian or variational methods;
- Develop discretized equations of motion for structures using global approximate methods such as Rayleigh Ritz or Galerkin, as well as local approximate methods such as finite element analysis;
- Incorporate damping in equations of motion;
- Determine the undamped natural frequencies and mode shapes of structures by formulating and solving the structural dynamics eigenvalue problem;
- Determine the dynamic response of structures using the normal mode method, and perform dynamic response calculations in the time and frequency domains;
- Pursue advanced work in areas closely related to structural dynamics such as aeroelasticity, helicopter dynamics and aeromechanics, experimental modal analysis, passive vibration control, dynamic stability of structures, nonlinear structural dynamics, structural dynamics of rotating systems, structural acoustics, structural health monitoring, and active structural control.

The book is intended to be used as a textbook in a graduate-level one-semester or two-semester course in structural dynamics. Potential approaches for using the book in a one- or two-semester course are described next. The first chapter of the book focuses on the essential aspects of single-degree-of-freedom systems and it can be used in different ways.

In departments where the undergraduate program contains an introductory course in mechanical vibrations, it can be selected as a prerequisite for graduate-level structural dynamics course. The material provided in this chapter is considered a review and is assigned as a reading assignment. Only one lecture is devoted to the single-degree-of-freedom problem. Problems selected from the problem set given at the end of the chapter are used to test and enhance the students' understanding of the material. Another alternative depends on the instructor's assessment of the situation. If the majority of the class has had no exposure to an introductory course in mechanical vibrations, it is possible to spend up to three lectures on this chapter.

The road-map for a single-semester graduate-level course is described next. Chapter 2 has to be covered completely, followed by Chapter 3. In a one-semester course the instructor can reduce the coverage of Timoshenko beams, and skip the von Kármán plate theory. Chapter 4 is considered to be an important part of the course, but some topics could be omitted in a one-semester offering or treated as advanced topics at the end of the course. Such topics include finite element treatment of plates and Timoshenko beams as well as component mode synthesis. Chapter 5 is another important chapter; however, in a one-semester course this chapter can be treated with a degree of flexibility. For a one-semester course the numerical methods for solving the structural dynamics eigenvalue problem can be covered briefly due to the existence of numerical packages, as well as commercial structural codes that have such solvers embedded in them. A similar approach can be applied to Chapter 6. The instructor can select the material considered most important and not cover the material considered less important. Chapter 7 is an important chapter for dynamic response calculations. Both analytical and numerical response computations are covered in detail. The coverage of numerical response computations can be reduced substantially in a one-semester course. The last topic recommended for a one-semester course is a brief treatment rotating systems consisting of Sections 8.1 through 8.4. Such systems appear in many aerospace and mechanical engineering applications and students as well as practicing engineers have difficulty dealing with this class of problems.

The road-map for a two-semester graduate-level course in structural dynamics is described next. Items deleted from the coverage in the one-semester course should be covered in a two-semester course. This implies a complete coverage of Timoshenko beams as well as coverage of von Kármán plate theory. Chapter 4 needs to be covered in its entirety, but the topic of component mode synthesis could be deferred to follow the treatment of the natural modes of vibration of discrete systems in Chapter 5. The detailed discussion of numerical methods used in the solution of the structural dynamics eigenvalue problem presented in Chapter 5 is discretionary. A complete treatment of Chapter 6 is recommended. Again in Chapter 7 the instructor has flexibility in selecting the level of detail given to the treatment of numerical integration methods for response calculation. Chapter 8 is a unique feature of this book, and its complete coverage is recommended. Chapter 9 is an important topic for both aerospace as well as mechanical engineering applications and it is rarely clearly treated in books on structural dynamics. In a two-semester course we would also recommend adding a few lectures on random vibrations.

**Figure 0.1**
Illustration of the
book's structure

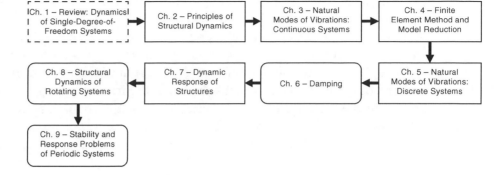

## Structure of the Book

- **Key to the graphical scheme of the book (Figure 0.1)**: Rectangular Boxes – material found in most books on structural dynamics; boxes with rounded corners – material unique to this particular book; box with a square dashed line – essentially review material.
- **A one-semester** graduate-level course in structural dynamics can be tailored from a judicious combination of the rectangular boxes, consisting of Chapters 1, 2, 3, 4, 5, and 7. Note the use of Chapter 1 is discretionary.
- **A two-semester** graduate-level course in structural dynamics can be tailored from a judicious combination of Chapters 1 through 9. The use of Chapter 1 is discretionary.

## About the Online Resources

Online resources for the book are available at www.cambridge.org/friedmann. These include a Solutions Manual for instructors, and the figures and tables provided in JPEG and PPT format for classroom use. MATLAB/Python code is provided for student use.

## Acknowledgments

The material presented in the book has evolved over a period that spans the 50 years that the first author has spent teaching structural dynamics at two institutions: the University of California, Los Angeles (UCLA, Department of Mechanical and Aerospace Engineering) and the University of Michigan (Department of Aerospace Engineering). At UCLA the author has alternated teaching structural dynamics with two colleagues: Dr. Stanley Dong and Dr. Oddvar Bendiksen. At the University of Michigan he has alternated with Dr. Carlos Cesnik. The author is grateful to these colleagues for influencing some of the material presented in the book. The author is also grateful to his PhD students Dr. Puneet Singh and Dr. Ryan Patterson for being his LaTeX consultants, together with Dr. A. Padthe, his

Post-doctoral Scholar. Finally, thanks are due to Carter Briggs, an undergraduate student, who has drawn many of the figures used in this book.

The second author took several courses from the first author at UCLA as a graduate student. At Penn State, he has taught much of the material in this text in a graduate-level structural dynamics course (and a follow-on course in vibration control) for 30 years. The structural dynamics course is cross-listed in three departments – aerospace engineering, mechanical engineering, and engineering mechanics – and serves as a prerequisite for courses such as experimental modal analysis, helicopter dynamics, and nonlinear vibrations. The author thanks his PhD student, Dr. Tianliang Yu, for assistance in creating the index for this book.

The third author took several courses from the first author at the University of Michigan as a graduate student. At Penn State, he has taught a portion of the material in this text in graduate-level courses such as aeroelasticity.

# 1     Review: Dynamics of Single-Degree-of-Freedom Systems

*This we shall obtain from the marvelous property of the pendulum, which is that it makes all its vibrations, large or small, in equal times.*
– Galileo Galilei, *Letter to Giovanni Battista Baliani (1639)*

The study of structural dynamics has to do with understanding and predicting the motion of structures that possess continuous distributions of stiffness and mass. A starting point for such study is a review of key concepts associated with the vibration of discrete systems – which possess *concentrated* stiffness and mass – as well as the static deformation of continuous structures.

The simplest discrete system is one that has only a single degree of freedom. Many of the concepts attendant to a single degree-of-freedom (SDOF) system are directly relevant to the principles and analysis techniques of structural dynamics.

## 1.1   Kinematics of Vibration

Vibration involves oscillatory motion, sometimes *periodic* or *harmonic*. *Kinematics* is the description of that motion.

### 1.1.1   An SDOF System

Consider the SDOF spring-mass system shown in Figure 1.1(a). It consists of a concentrated mass $m$ (weight, $W$) suspended from a rigid base via a massless spring $k$ and allowed to move along a line. A time-varying external force $P(t)$ may act on the mass. Figure 1.1(b) shows a version of this system in which gravity acts in the same direction as the motion.

The coordinate $u(t)$ describes the position of the mass relative to an equilibrium position. In an equilibrium position, the mass has zero velocity, and the internal force in the deformed spring balances any external static force(s). In Figure 1.1(a), no external force (such as gravity) acts on the mass, so the spring is unloaded in the equilibrium configuration. In Figure 1.1(b), however, gravity imposes a force $W = mg$ and, if the spring is linear with stiffness $k$ (units $F/L$), the associated static deflection of the spring is $u_{\text{st}} = W/k$. The dynamic motion of the mass, $u(t)$, is measured from this position. The external force $P(t)$

**Figure 1.1**
Single degree-of-freedom (SDOF) spring-mass system

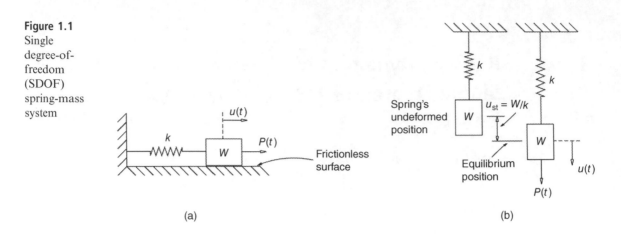

(a)                                        (b)

is positive in the same direction as $u(t)$. The two SDOF systems shown in Figure 1.1 are dynamically equivalent if the spring stiffness does not change with deflection.

If the mass is perturbed from its equilibrium position, the spring provides a *restoring* force that tends to move it back toward equilibrium. If the mass has some velocity, it maintains that velocity unless accelerated by forces acting on it. The interaction between the *inertia* of the mass and the restoring force of the spring results in natural oscillatory behavior.

A central problem in vibration analysis is the determination of the *response $u(t)$*, given $P(t)$, $m$, and $k$.

### 1.1.2  Periodic and Harmonic Motion

This text emphasizes linear systems – for which the principle of superposition holds. Many of the insights developed extend to nonlinear systems.

Some descriptions of periodic motion apply to both linear and nonlinear systems. A periodic motion is one that repeats itself in all respects in successive constant intervals of time, $T$, called the *period*, as illustrated in Figure 1.2. Its inverse $f = 1/T$ is the *frequency*, which is the number of repetitive cycles completed per unit time. Values for period and frequency are usually expressed in seconds and cycles per second (the SI unit: *hertz* (Hz)). Another common frequency measure is the *circular frequency*, $\omega = 2\pi f$ (units: radians per second). The circular frequency has the same units as angular velocity, which is useful in the treatment of harmonic motion.

**Harmonic Motion**

The simplest form of periodic motion is harmonic (sinusoidal) motion, in which the displacement response $u(t)$ has the general form:

$$u(t) = A\cos(\omega t + \phi)$$
$$u(t) = A\sin(\omega t + \pi/2 + \phi) = A\sin(\omega t + \psi),$$

(1.1.1)

where $A$ is the *amplitude* and $\phi$ and $\psi$ are *phase angles*, as illustrated in Figure 1.3.

**Figure 1.2**
Periodic motion

**Figure 1.3**
Harmonic motion

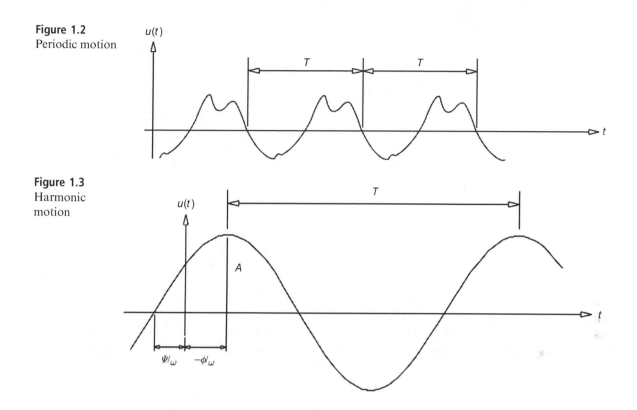

Time derivatives of a harmonic vibration (denoted by superior dots) are also harmonic:

Displacement: $u(t) = A \cos(\omega t + \phi)$.

Velocity: $\dot{u}(t) = -\omega A \sin(\omega t + \phi) = \omega A \cos(\omega t + \phi + \pi/2)$. $\qquad$ (1.1.2)

Acceleration: $\ddot{u}(t) = -\omega^2 A \cos(\omega t + \phi) = -\omega^2 A \sin(\omega t + \phi + \pi/2)$.

Figure 1.4 shows the relationships between displacement, velocity, and acceleration in harmonic vibration – represented as vectors of lengths $A$, $A\omega$, and $A\omega^2$ rotating counterclockwise with angular velocity $\omega$. The vectors rotate with phase differences of $\pi/2$ radians (90 degrees), with acceleration leading velocity, and velocity leading displacement. The corresponding time-varying scalar values are given by the projections of the vectors on the vertical axis, as time scrolls to the right.

Euler's formula provides another way to describe harmonic motion. Since $e^{i\theta} = \cos(\theta) + i \sin(\theta)$ (where $i = \sqrt{-1}$), harmonic vibration can be expressed in complex exponential form as:

$u(t) = A e^{i(\omega t + \phi)}$.

$\dot{u}(t) = (i\omega)A e^{i(\omega t + \phi)} = \omega A e^{i(\omega t + \phi)} e^{i\pi/2}$. $\qquad$ (1.1.3)

$\ddot{u}(t) = (i\omega)^2 A e^{i(\omega t + \phi)} = \omega^2 A e^{i(\omega t + \phi)} e^{i\pi}$.

**Figure 1.4**
Phase relation
between
displacement,
velocity, and
acceleration

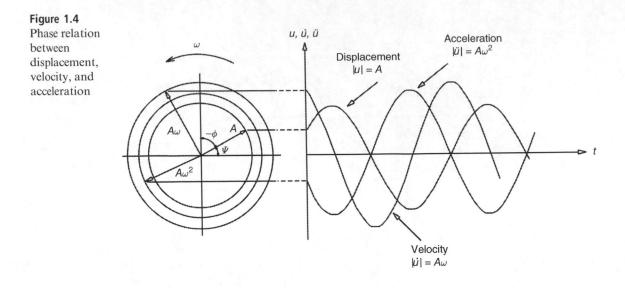

The corresponding time-varying values are given by the real parts of the expressions in Eq. (1.1.3). Note that multiplication by the imaginary unit $i = e^{i\pi/2}$ corresponds to a $\pi/2$ rotation in the complex plane, as evident in Eq. (1.1.3) as a 90-degree phase change from displacement to velocity and from velocity to acceleration.

## Superposition of Same-Frequency Harmonic Motions

Any number of harmonic motions sharing the *same* frequency $\omega$ but different amplitudes and phase angles can be combined into a single resultant harmonic motion at that frequency. Figure 1.5 shows the vectorial sum of two harmonic motions given in Eq. (1.1.4).

$$u_1(t) = A_1 \cos(\omega t + \phi_1).$$
$$u_2(t) = A_2 \cos(\omega t + \phi_2). \tag{1.1.4}$$

These can be combined to give

$$u(t) = A \cos(\omega t + \phi) = A \sin(\omega t + \psi), \tag{1.1.5}$$

where

$$A^2 = A_1^2 + A_2^2 + 2A_1 A_2 \cos(\phi_1 - \phi_2) \tag{1.1.6}$$

and

$$\tan(\phi) = -\frac{1}{\tan(\psi)}$$
$$= -\frac{A_1 \sin(\phi_1) + A_2 \sin(\phi_2)}{A_1 \cos(\phi_1) + A_2 \cos(\phi_2)}. \tag{1.1.7}$$

**Figure 1.5** Vector sum of two harmonic motions

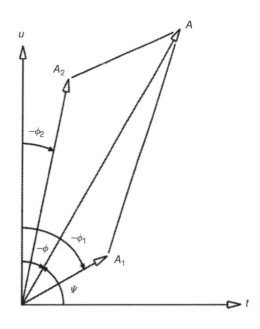

## Phase-Plane Representation

A useful tool for visualizing the time evolution of motion is the *phase plane*. The displacement and velocity of an SDOF system constitute its *state* and describe its configuration completely. A state is represented by a point on the phase plane and, as the motion evolves in time, the locus of state points traces a path called the *trajectory*.

A phase-plane plot shows the motion in a way that time does not appear explicitly and can reveal whether a given motion is asymptotically stable, marginally stable, or unstable. The dependence of the motion on its initial state can also be observed directly.

To illustrate the use of the phase plane, consider an undamped SDOF system. For harmonic motion, the displacement and velocity are given by

$$u(t) = A \cos(\omega t + \phi); \quad \dot{u}(t) = -\omega A \sin(\omega t + \phi). \tag{1.1.8}$$

Time can be eliminated from Eq. (1.1.8), yielding

$$(\dot{u}/\omega)^2 + u^2 = A^2, \tag{1.1.9}$$

which describes a circle in the phase plane with $u$ and $\dot{u}/\omega$ as the axes. As Figure 1.6(a) shows, the circular trajectory is generated by a clockwise rotation of radius vector $A$ at angular velocity $\omega$. Its initial condition is related to the phase angle $\phi$, i.e., $u(0) = u_0 = A \cos(\phi)$ and $\dot{u}(0) = v_0 = -A\omega \sin(\phi)$. Any point on the trajectory corresponds to an angle $\omega t$ measured clockwise from the initial position of the radius vector.

If the harmonic motion occurs about a displaced equilibrium position, as in a gravitational field, the total displacement differs from Eq. (1.1.8) by a static value $U_0$, i.e., $u(t) = U_0 + A\cos(\omega t + \phi)$. Then, the trajectory is described by

$$(\dot{u}/\omega)^2 + (u - U_0)^2 = A^2, \tag{1.1.10}$$

as shown in Figure 1.6(b).

**Figure 1.6**
Phase-plane representation of harmonic motion

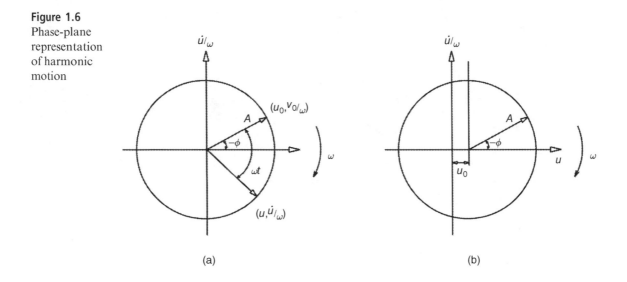

(a)                                   (b)

## 1.2    Free and Forced Response of Undamped SDOF Systems

### 1.2.1    Equation of Motion

Consider the system previously introduced in Figure 1.1(a). Establish a coordinate system with the origin at the block in its equilibrium position and let $u(t)$ denote its positive displacement from that position. Let $g$ (units: $L/T^2$) be the acceleration due to gravity. Figure 1.7 shows a free body diagram with all the forces acting on the block. (Since motion in the vertical direction is not allowed, the normal force $N$ equals $W$, the weight.) The governing equation of motion in the horizontal direction can be obtained using Newton's second law, i.e., $\sum F_x = ma_x$, or by D'Alembert's principle, i.e., $\sum F_x - ma_x = 0$. For $m = W/g$, the resulting equation of motion is

$$P(t) - ku(t) = m\ddot{u}(t) \quad \text{or} \quad m\ddot{u}(t) + ku(t) = P(t). \tag{1.2.1}$$

**Figure 1.7** Free body diagram of the SDOF system

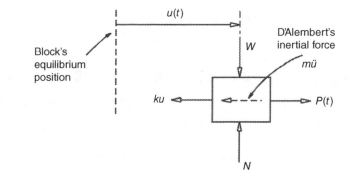

## 1.2.2  Free Vibration of Undamped SDOF Systems

For free vibration, $P(t) = 0$ and the motion of the mass proceeds under the influence of only elastic and "inertial" forces. Equation (1.2.1) reduces to

$$m\ddot{u}(t) + ku(t) = 0 \quad \text{or} \quad \ddot{u}(t) + \frac{k}{m}u(t) = 0. \tag{1.2.2}$$

Defining the *natural frequency of vibration*, $\omega_n$:

$$\omega_n^2 = \frac{k}{m}, \tag{1.2.3}$$

the unforced equation of motion may be written as:

$$\ddot{u}(t) + \omega_n^2 u(t) = 0. \tag{1.2.4}$$

The general solution of this homogeneous differential equation of motion corresponds to harmonic motion at the natural frequency, $\omega_n$:

$$u(t) = C_1 \sin(\omega_n t) + C_2 \cos(\omega_n t), \tag{1.2.5}$$

where $C_1, C_2$ are coefficients that can be determined from the initial conditions, i.e., the displacement and velocity at time $t = 0$:

$$u(0) = u_0 \quad \text{and} \quad \dot{u}(0) = v_0. \tag{1.2.6}$$

Evaluating $C_1$ and $C_2$ using the preceding conditions gives:

$$u(0) = u_0: \quad C_1 \sin(0) + C_2 \cos(0) = u_0 \rightarrow \quad C_2 = u_0.$$
$$\dot{u}(0) = v_0: \quad C_1 \omega_n \cos(0) - u_0 \omega_n \sin(0) = v_0 \rightarrow C_1 = v_0/\omega_n. \tag{1.2.7}$$

The undamped free vibration response is then:

$$u(t) = \frac{v_0}{\omega_n} \sin(\omega_n t) + u_0 \cos(\omega_n t) \quad \text{or} \quad u(t) = A \sin(\omega_n t + \psi), \tag{1.2.8}$$

where

$$A = \sqrt{u_0^2 + (v_0/\omega_n)^2} \quad \text{and} \quad \tan(\psi) = \frac{u_0}{v_0/\omega_n}. \tag{1.2.9}$$

### 1.2.3 Harmonically Forced Vibration of Undamped SDOF Systems

Consider a harmonic force of magnitude $P_0$ and frequency $\Omega$, such that $P(t) = P_0 \sin(\Omega t)$. The equation of motion, Eq. (1.2.1), takes the form:

$$\ddot{u}(t) + \omega_n^2 u(t) = \frac{P_0}{m} \sin(\Omega t) = \frac{P_0 \omega_n^2}{k} \sin(\Omega t). \tag{1.2.10}$$

The complete solution of this differential equation of motion consists of a homogeneous solution and a particular solution. The homogeneous solution has the same form as that found for free vibration, Eq. (1.2.5). Regarding the particular solution, since the system is linear it will respond at the same frequency as the forcing, with the possibility of a phase shift. In the absence of damping, the general solution has the form

$$u(t) = C_1 \sin(\omega_n t) + C_2 \cos(\omega_n t) + \frac{P_0}{k} \left\{ \frac{1}{1 - (\Omega/\omega_n)^2} \right\} \sin(\Omega t). \tag{1.2.11}$$

Evaluating the coefficients $C_1$ and $C_2$ assuming quiescent initial conditions, i.e., $u(0) = 0$ and $\dot{u}(0) = 0$, the solution is given by

$$u(t) = \frac{P_0}{k} \left\{ \frac{1}{1 - (\Omega/\omega_n)^2} \right\} \left\{ \sin(\Omega t) - \left( \frac{\Omega}{\omega_n} \right) \sin(\omega_n t) \right\}. \tag{1.2.12}$$

The terms in Eq. (1.2.12) have the following significances:

(1) $P_0/k$ – the response of the system to a static load $P_0$. Let $u_{st} = P_0/k$.
(2) $\frac{1}{1-(\Omega/\omega_n)^2}$ – a *magnification factor*, $A_d$, describing the multiplicative increase of the dynamic response $u(t)$ relative to the static response $u_{st}$.
(3) $\sin(\Omega t)$ – the *steady state* harmonic response associated with harmonic forcing at frequency $\Omega$ (and independent of the initial conditions).
(4) $(\Omega/\omega_n) \sin(\omega_n t)$ – the *transient* response, a free vibration term needed to satisfy the initial conditions. (If damping were present, this term would diminish with time, hence the descriptor *transient*.)

Let $D(t)$ denote the ratio of $u(t)$ to $u_{st} = P_0/k$, called the *Dynamic Response Factor*:

$$D(t) = \frac{u(t)}{u_{st}} = \left[ \frac{1}{1 - (\Omega/\omega_n)^2} \right] \left\{ \sin(\Omega t) - \left( \frac{\Omega}{\omega_n} \right) \sin(\omega_n t) \right\}. \tag{1.2.13}$$

The extreme (maximum and minimum) responses can be found by considering those times when $\dot{u}(t) = 0$.

### Resonance

When the forcing function frequency $\Omega$ coincides with the natural frequency $\omega_n$ of the system, *resonance* results. The dynamic load factor $D(t)$ is indeterminate, i.e., $D(t) = 0/0$ and

L'Hôpital's rule must be applied to determine the motion of the system. Differentiating the dynamic load factor $D(t)$ with respect to $\alpha = (\Omega/\omega_n)$ yields:

$$\lim_{a \to 1} \frac{\frac{d}{d\alpha}\{\sin(\alpha\omega_n t) - \alpha \sin(\omega_n t)\}}{\frac{d}{d\alpha}(1 - \alpha^2)} = \lim_{\alpha \to 1} \frac{\omega_n t \cos(\alpha\omega_n t) - \sin(\omega_n t)}{-2\alpha} \qquad (1.2.14)$$

$$= -\frac{1}{2}\{\omega_n t \cos(\omega t) - \sin(\omega_n t)\}.$$

Figure 1.8 shows the response versus time as well as in the phase plane. Note that the response grows unbounded with time and is no longer harmonic.

**Figure 1.8**
Resonance

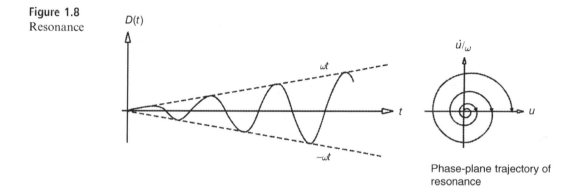

Phase-plane trajectory of resonance

## 1.2.4 Base Excitation

The SDOF system shown in Figure 1.1(a) consists of a mass $m$ suspended from a base via a spring $k$. This system is excited by an external force $P(t)$ acting on the mass. A different situation is sometimes encountered in practice, one in which the system is excited via *base motion*. Representative examples include vibration of buildings due to earthquakes, vibroacoustic loading of a spacecraft attached to a launch vehicle, and the vertical motion of the pilot's seat in a helicopter.

Consider the base-excited system shown in Figure 1.9. Let $u(t)$ denote the absolute motion of the mass (i.e. with respect to a fixed inertial coordinate system), $u_b(t)$ the motion of the base, and $u_r(t) = u(t) - u_b(t)$ the relative motion. The equation of motion can be formulated in terms of either the inertial or relative motions, $u(t)$ or $u_r(t)$.

**In Terms of Inertial Motion $u(t)$ and Base Displacement**

In this case, the spring force is given by $k(u - u_b)$ and the equation of motion is

$$m\ddot{u} + k(u - u_b) = 0 \quad \text{or} \quad \ddot{u} + \omega_n^2 u = \frac{k u_b}{m} = \frac{P_b(t)}{m}, \qquad (1.2.15)$$

**Figure 1.9** SDOF system with base motion

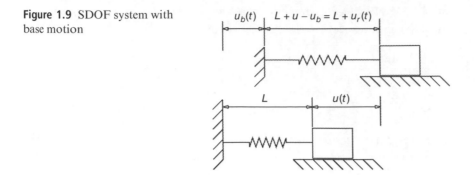

where $P_b(t)$ is the *effective dynamic load* based on the *displacement history* of the base. (Note that explicit dependence of $u(t)$, $u_b(t)$, and $\ddot{u}(t)$ on time has been omitted in this equation of motion.)

### In Terms of Relative Motion $u_r(t)$ and Base Acceleration

In practice, the relative motion is more readily observed than the absolute motion. An equation of motion based on the relative displacement takes the form

$$m(\ddot{u}_r + \ddot{u}_b) + ku_r = 0 \quad \text{or} \quad \ddot{u}_r + \omega_n^2 u_r = -\ddot{u}_b = \frac{P_{rb}(t)}{m}, \qquad (1.2.16)$$

where $P_{rb}(t)$ is the *effective dynamic load* given in terms of the *acceleration history* of the base.

## 1.3    Free and Forced Response of Damped SDOF Systems

A structure with damping converts energy associated with vibration into other forms, often gradually. Real vibrating structures all exhibit some level of damping. This energy dissipation can have many internal or external causes but common to all structures is material damping.

Three damping mechanisms are commonly considered for SDOF systems: *viscous* damping, *structural* or *hysteretic* damping, and *Coulomb* (friction) damping. Chapter 6 addresses the more specialized topics of hysteretic and Coulomb damping, while viscous damping is addressed here.

### 1.3.1    Equation of Motion with Viscous Damping

Viscous damping is the simplest damping model that provides essential energy dissipation. In this model, a linear damping force opposes and is proportional to the velocity of the mass. The magnitude of this damping force is given by

$$F_d(t) = c\dot{u}(t), \qquad (1.3.1)$$

where $c$ is the *viscous damping coefficient*, a positive parameter (units: $FT/L$).

Viscous damping is indicated by a dashpot in a system diagram, as shown in Figure 1.10(a). Considering the free body diagram in Figure 1.10(b), the equation of motion is found as:

$$m\ddot{u}(t) + c\dot{u}(t) + ku(t) = P(t) \quad \text{or} \quad \ddot{u} + \frac{c}{m}\dot{u} + \omega_n^2 u = \frac{P(t)}{m} = \frac{P(t)\,\omega_n^2}{k}. \tag{1.3.2}$$

This is the same as Eq. (1.2.1), with the addition of a damping term.

**Figure 1.10**
SDOF system with viscous damping

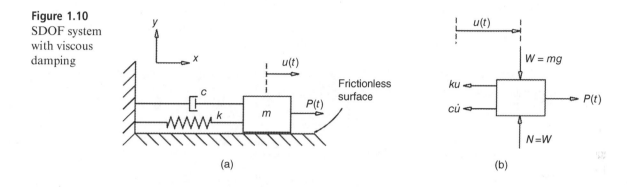

(a)          (b)

## 1.3.2  Free Vibration of Viscously Damped SDOF System

The free vibration of an SDOF system is the response in the absence of external forces. A *characteristic* response can be considered, as well as the response(s) associated with the initial displacement and velocity of the mass. These cases are addressed in turn.

### Unforced Characteristic Motion

Consider the homogeneous equation of motion, Eq. (1.3.2):

$$m\ddot{u}(t) + c\dot{u}(t) + ku(t) = 0. \tag{1.3.3}$$

This is a linear differential equation with constant coefficients, having a solution of the form:

$$u(t) = \bar{u}\, e^{st}. \tag{1.3.4}$$

Substituting this in the equation of motion yields:

$$(s^2 m + sc + k)\, \bar{u}\, e^{st} = 0. \tag{1.3.5}$$

For $\bar{u} \neq 0$ – the only possibility of physical interest – a quadratic equation in $s$ results:

$$s^2 + \left(\frac{c}{m}\right)s + \left(\frac{k}{m}\right) = 0. \tag{1.3.6}$$

Two general solutions to which can be found as:

$$s_{1,2} = -\frac{c}{2m} \pm \sqrt{\left(\frac{c}{2m}\right)^2 - \frac{k}{m}}. \tag{1.3.7}$$

These correspond to a characteristic *complex exponential* motion, $e^{st}$, that is an intrinsic property of the system – as distinguished from an actual response to a physical excitation or an initial condition.

Depending on the relative values of $m$, $c$, and $k$, these solutions represent different kinds of motions. Typically, damping is relatively "light." In that case, the characteristic response is oscillatory within a decaying exponential envelope. Another possibility is that the damping is very high, and the characteristic response corresponds to nonoscillatory exponential relaxation. These cases are addressed separately below.

***Underdamped ($c^2 < 4km$)***  In this case, the two roots of Eq. (1.3.7) comprise a complex conjugate pair, each having a negative real part:

$$s_{1,2} = -\frac{c}{2m} \pm i\sqrt{\frac{k}{m} - \left(\frac{c}{2m}\right)^2}. \tag{1.3.8}$$

And, in terms of alternate (modal) *frequency* and *damping* parameters,

$$\begin{aligned} s_{1,2} &= -\zeta\omega_n \pm i\omega_n\sqrt{1 - \zeta^2}, \\ &= -\zeta\omega_n \pm i\omega_d \end{aligned} \tag{1.3.9}$$

where

$$\omega_n^2 = \frac{k}{m}$$

is the *natural frequency* of vibration of the undamped SDOF system;

$$\zeta = \frac{c}{2}\sqrt{\frac{1}{km}} \tag{1.3.10}$$

is the (modal) *damping ratio*, with $0 \le \zeta < 1$ for an *underdamped* system; and the *damped natural frequency* is given by

$$\omega_d = \omega_n\sqrt{1 - \zeta^2}. \tag{1.3.11}$$

Figure 1.11 shows the location of these roots in the complex plane. As previously noted, these comprise a complex conjugate pair having the same negative real part and opposite complex parts. They are a distance $\omega_n$ from the origin, with complex parts of magnitude $\omega_d$. The common real part has magnitude $\zeta\omega_n$. Their location in the left-hand plane indicates decaying response and stability. The damping ratio $\zeta$ is equal to the sine of the angle between the imaginary axis and the line from the origin to the root having a positive complex part.

Typical values of $\zeta$ used in analyses of the lowest frequency vibrations of aerospace structures range from 0.04% for precision welded or bonded composite structures to 3% for non–precision bolted structures. Buildings and civil structures often have considerably higher damping.

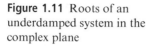

**Figure 1.11** Roots of an underdamped system in the complex plane

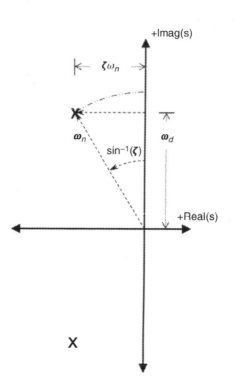

***Overdamped (c² > 4km)*** In this case $\zeta > 1$ and the two roots of Eq. (1.3.7) comprise distinct negative real roots:

$$s_1 = -\frac{c}{2m} - \sqrt{\left(\frac{c}{2m}\right)^2 - \frac{k}{m}} = -\omega_n(\zeta - \sqrt{\zeta^2 - 1})$$
$$s_2 = -\frac{c}{2m} + \sqrt{\left(\frac{c}{2m}\right)^2 - \frac{k}{m}} = -\omega_n(\zeta + \sqrt{\zeta^2 - 1})$$

(1.3.12)

The associated characteristic motion is nonoscillatory, corresponding to first-order exponential decay. The unforced free response to initial conditions consists of two decaying exponential functions:

$$u(t) = C_1 e^{-\omega\left(\zeta - \sqrt{\zeta^2 - 1}\right)t} + C_2 e^{-\omega\left(\zeta + \sqrt{\zeta^2 - 1}\right)t},$$

(1.3.13)

where $C_1$ and $C_2$ are found from the initial conditions as

$$C_1 = \frac{\omega\left(\zeta + \sqrt{\zeta^2 - 1}\right)u_0 + v_0}{2\omega\sqrt{\zeta^2 - 1}}; \quad C_2 = -\frac{\omega\left(\zeta - \sqrt{\zeta^2 - 1}\right)u_0 + v_0}{2\omega\sqrt{\zeta^2 - 1}}.$$

(1.3.14)

The descriptor *overdamped* is deserved for systems exhibiting nonoscillatory free response (damping ratio $\zeta > 1$). This level of damping in structures is rarely encountered in practice.

***Critically Damped*** ($c^2 = 4km$)    In this case, the two roots of Eq. (1.3.7) comprise repeated negative real roots:

$$s_1 = s_2 = -\frac{c}{2m} = -\omega_n. \tag{1.3.15}$$

A damping ratio $\zeta = 1$ is called *critical damping* and represents a transition from oscillatory to nonoscillatory motion. Since there are repeated roots, finding the unforced free response to initial conditions for a critically damped system requires the inclusion of an additional independent homogeneous solution, *e.g.*, $t\,e^{-\omega_n t}$. The solution in this case has the form:

$$u(t) = (C_1 + C_2 t)\,e^{-\omega t} \tag{1.3.16}$$

and, considering the initial conditions,

$$u(t) = [u_0 + \{(v_0/\omega) + u_0\}\,\omega t]\,e^{-\omega t}. \tag{1.3.17}$$

Figure 1.12 shows characteristic responses for underdamped, critically damped, and overdamped systems.

**Figure 1.12**
Motions of a
system with
viscous
damping

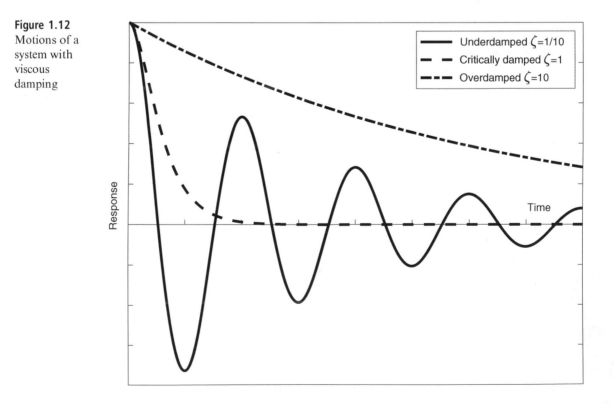

### Response to Initial Velocity

Consider a viscously damped SDOF system with *light* damping, the typical case of interest in structural dynamics. Noting that $c/m = 2\zeta\omega_n$, the unforced form of the equation of motion, Eq. (1.3.2), is

$$\ddot{u} + 2\zeta\omega_n\dot{u} + \omega_n^2 u = 0. \tag{1.3.18}$$

The general solution to this homogeneous equation of motion has the following form:

$$u(t) = e^{-\zeta\omega_n t}\,(C_1\cos\omega_d t + C_2\sin\omega_d t) \tag{1.3.19}$$

or, alternatively:

$$u(t) = A\,e^{-\zeta\omega_n t}\sin(\omega_d t + \Psi), \tag{1.3.20}$$

where the coefficients $C_1$ and $C_2$ (or equivalently, $A$ and $\Psi$) can be determined from the initial conditions $u(0) = u_0$ and $\dot{u}(0) = v_0$.

The free response associated with a nonzero initial velocity $v_0$ is given by

$$u(t) = v_0\left(\frac{1}{\omega_d}\right)e^{-\zeta\omega_n t}\sin(\omega_d t). \tag{1.3.21}$$

Figure 1.13 shows an illustrative response. The initial velocity is $\dot{u}(0) = v_0$; its oscillations follow a general sine response decaying within an exponential envelope; and, with light damping, the peak dynamic response is approximately $u_{peak} = v_0/\omega_n$.

**Figure 1.13**
Response to an initial velocity

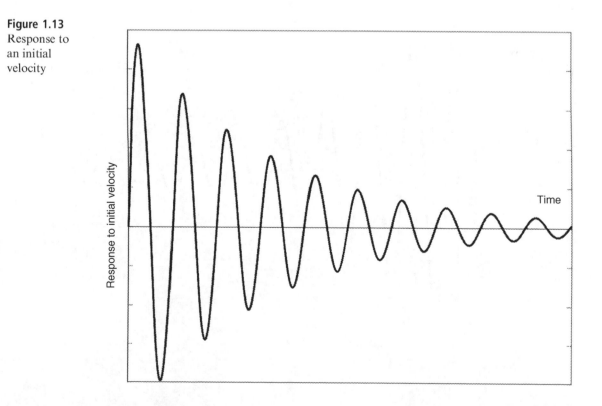

Response to initial velocity

Time

### Response to Initial Displacement

The free response associated with a nonzero initial displacement $u_0$ is given by

$$
\begin{aligned}
u(t) &= u_0 \left( \frac{\omega_n}{\omega_d} \right) e^{-\zeta \omega_n t} \cos\left( \omega_d t + \phi \right) \\
&= u_0 \left( \frac{\omega_n}{\omega_d} \right) e^{-\zeta \omega_n t} \sin\left( \omega_d t + \psi \right)
\end{aligned}
\quad , \tag{1.3.22}
$$

where

$$
\begin{aligned}
\phi &= \tan^{-1}\left( \frac{-\zeta}{\sqrt{1 - \zeta^2}} \right) = \tan^{-1}\left( \frac{-\zeta \omega_n}{\omega_d} \right) \\
\psi &= \tan^{-1}\left( \frac{\sqrt{1 - \zeta^2}}{\zeta} \right) = \tan^{-1}\left( \frac{\omega_d}{\zeta \omega_n} \right)
\end{aligned}
\quad .
$$

Figure 1.14 shows an illustrative response. The initial (and peak) displacement is $u(0) = u_0$ and its oscillations follow a general cosine response decaying within an exponential envelope.

**Figure 1.14**
Response to
an initial
displacement

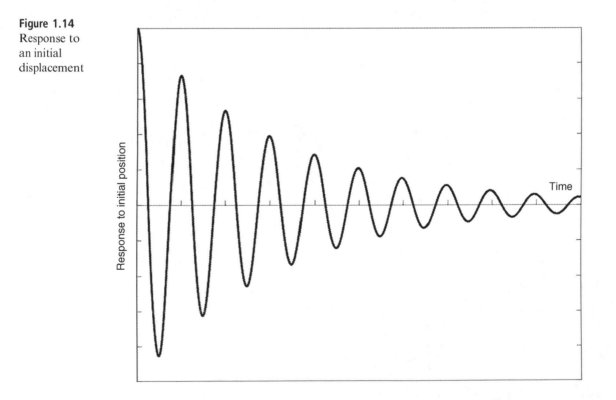

**Response to Combined Initial Conditions**

The preceding results can be combined to yield an expression for the response of a viscously damped SDOF system to general initial conditions:

$$u(t) = e^{-\zeta \omega_n t} \left( u_0 \cos \omega_d t + \frac{v_0 + \zeta \omega_n u_0}{\omega_d} \sin \omega_d t \right) \tag{1.3.23}$$

or, alternatively:

$$u(t) = A \, e^{-\zeta \omega_n t} \sin \left( \omega_d t + \Psi \right), \tag{1.3.24}$$

with

$$A - \sqrt{u_0^2 + \left( \frac{v_0 + u_0 \zeta \omega_n}{\omega_d} \right)^2}$$

and

$$\Psi = \tan^{-1} \left( \frac{u_0 \omega_d}{v_0 + u_0 \zeta \omega_n} \right).$$

## 1.3.3 Harmonically Forced Vibration of Viscously Damped SDOF Systems

The forced vibration response, $u(t)$, of an SDOF system is that part of the response specifically associated with external forcing, $P(t)$ – and not the initial conditions. Harmonic forcing is frequently encountered in practice due to the prevalence of rotating machinery and is of particular interest.

Consider a harmonic force of magnitude $P_0$ and frequency $\Omega$ such that $P(t) = P_0 e^{i\Omega t}$. The equation of motion (1.3.2) takes the form:

$$\ddot{u}(t) + 2\zeta \omega_n \dot{u}(t) + \omega_n^2 u(t) = \frac{P_0}{m} e^{i\Omega t} = \frac{P_0 \omega_n^2}{k} e^{i\Omega t}. \tag{1.3.25}$$

Neglecting for now the *transient response*, the *steady-state response* of this linear SDOF system must be harmonic at the same frequency as the forcing, possibly with a relative phase shift. That is,

$$u(t) = U(\Omega) e^{i(\Omega t + \phi(\Omega))}, \tag{1.3.26}$$

with $U(\Omega)$ and $\phi(\Omega)$ both real functions of $\Omega$, and the physical response the real part of the preceding expression. Substituting this solution form into Eq. (1.3.25) yields:

$$(-\Omega^2 + i2\zeta \omega_n \Omega + \omega_n^2) U(\Omega) e^{i\phi(\Omega)} = \frac{P_0}{m} = \frac{P_0 \omega_n^2}{k} \tag{1.3.27}$$

or

$$U(\Omega) \, e^{i\phi(\Omega)} = \frac{P_0}{m \left( (\omega_n^2 - \Omega^2) + i(2\zeta \omega_n \Omega) \right)}$$

$$= \frac{P_0 \omega_n^2}{k \left( (\omega_n^2 - \Omega^2) + i(2\zeta \omega_n \Omega) \right)}. \tag{1.3.28}$$

The term $e^{i\phi(\Omega)}$ is complex and has unit magnitude, so $U(\Omega)$ may be considered a real magnitude (within a sign change consistent with $\phi(\Omega)$). Then, the following expressions for $U(\Omega)$ and $\phi(\Omega)$ are found:

$$U(\Omega) = \frac{P_0}{k}\frac{1}{\sqrt{\left(1 - (\Omega/\omega_n)^2\right)^2 + (2\zeta(\Omega/\omega_n))^2}} \tag{1.3.29}$$

$$\tan(\phi(\Omega)) = -\frac{2\zeta(\Omega/\omega_n)}{1 - (\Omega/\omega_n)^2}. \tag{1.3.30}$$

Note that to find the total physical response to harmonic forcing "switched on" at time $t = 0$, the transient initial condition response must be included, and the (effective) initial conditions must account for the contribution of the steady-state response.

With $\alpha = (\Omega/\omega_n)$, the magnification factor $A_d$ is

$$A_d = \frac{1}{\sqrt{\left(1 - \alpha^2\right)^2 + (2\zeta\alpha)^2}}, \tag{1.3.31}$$

which, for $\zeta = 0$, reduces to the result previously found for undamped systems, Eq. (1.2.12). The dynamic response factor in this steady-state case is

$$D(t) = \frac{u(t)}{u_{st}} = \frac{\sin(\Omega t + \psi)}{\sqrt{\left(1 - \alpha^2\right)^2 + (2\zeta\alpha)^2}} = A_d \sin(\Omega t + \psi). \tag{1.3.32}$$

Differentiating Eq. (1.3.32) gives the velocity and acceleration:

$$\dot{D}(t) = A_v \cos(\Omega t + \psi) \quad \text{and} \quad \ddot{D}(t) = -A_a \sin(\Omega t + \psi), \tag{1.3.33}$$

where $A_v = \Omega A_d$ and $A_a = \Omega^2 A_d$. Figure 1.15 shows the magnification factors $A_d, A_v$, and $A_a$ vs $\alpha$. Figure 1.16 shows phase angle plots for various values of $\zeta$. Note that the phase changes more rapidly with frequency for lower values of damping.

## Frequency Response Function

The *frequency response function* (FRF) is a complex-valued function that describes the magnitude and phase of the output of a linear system relative to the input under conditions of steady-state harmonic response. The FRF is frequently used to assess the performance of vibration control approaches.

For the case of an SDOF system with viscous damping, the frequency response function can be found from Eq. (1.3.28) as:

$$H(\Omega) = \frac{U^*(\Omega)}{P(\Omega)} = \frac{1}{(k - \Omega^2 m) + i\,\Omega c} = \frac{1}{m\omega_n^2((1 - (\Omega/\omega_n)^2) + i\,2\zeta(\Omega/\omega_n))},$$
$$= \frac{1}{k((1 - (\Omega/\omega_n)^2) + i\,2\zeta(\Omega/\omega_n))} \tag{1.3.34}$$

**Figure 1.15**
Magnification
factors vs $\Omega/\omega_n$

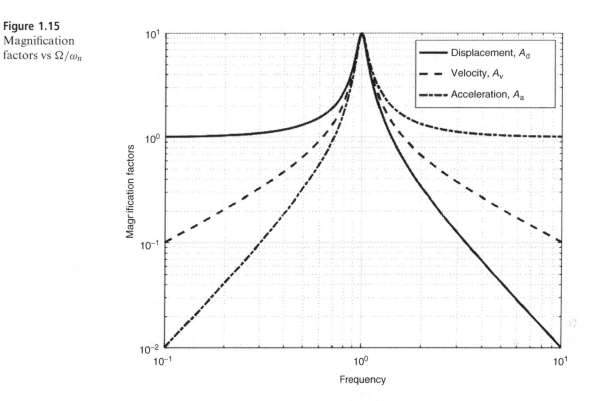

**Figure 1.16** Phase
angles vs $\Omega/\omega$

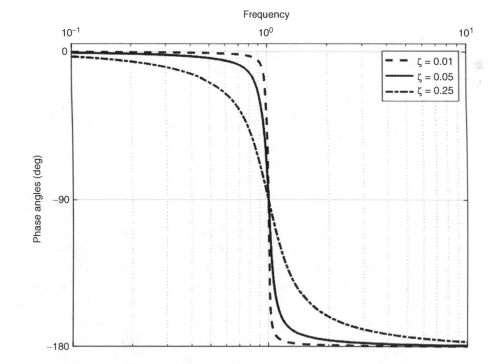

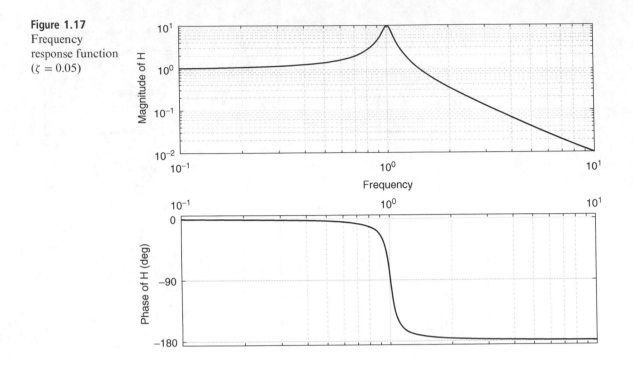

where $U^*(\Omega)$ is a complex function of $\Omega$. The magnitude, $|H(\Omega)|$, and phase, $\angle H(\Omega)$, of this frequency response function are:

$$|H(\Omega)| = \frac{1}{\sqrt{(k - \Omega^2 m)^2 + (\Omega c)^2}} = \frac{1}{k\sqrt{(1 - (\Omega/\omega_n)^2)^2 + (2\zeta(\Omega/\omega_n))^2}} \qquad (1.3.35)$$

$$\angle H(\Omega) = \tan^{-1}\left(\frac{-\Omega c}{k - \Omega^2 m}\right) = \tan^{-1}\left(\frac{-2\zeta(\Omega/\omega_n)}{1 - (\Omega/\omega_n)^2}\right). \qquad (1.3.36)$$

Frequency response functions are often visualized using a *Bode plot*. The magnitude is displayed on a log–log plot versus frequency, and the phase is displayed on a linear-log plot below it using the same frequency axis. Figure 1.17 shows a frequency response function for an SDOF system having unit stiffness ($k = 1$).

At low frequencies ($\Omega \ll \omega_n$), the response is very nearly independent of frequency, with the displacement in phase with the force and having a magnitude close to $1/k$, the static value. In this frequency range, the system is said to be "stiffness-controlled." At high frequencies ($\Omega \gg \omega_n$), the displacement magnitude drops rapidly with increasing frequency – with a slope of $-2$ on a log-log plot – and the displacement is out of phase with the force. In this frequency range, the system is said to be "mass-controlled." At an intermediate frequency ($\Omega \approx \omega_n$), the magnitude reaches a peak of approximately $\frac{1}{2\zeta k}$, and the phase transitions rapidly through nearly $-\pi$ radians ($-180$ degrees). Here, the system is said to be "damping-controlled."

### Energy Dissipation in Harmonic Forcing of an SDOF System with Viscous Damping

In steady-state vibration, the energy dissipation per cycle is balanced by the work input by the external force. The energy dissipation per cycle, $\Delta W$, is equal to the work done by the resistive damping force $F_d$ moving through its associated displacement or, equivalently, the integral of the dissipated power over one period of motion.

$$\Delta W = \int F_d \cdot du = \int_0^T F_d \cdot \dot{u}\, dt = \int_0^T c\dot{u}^2\, dt, \tag{1.3.37}$$

where $T = 2\pi/\Omega$. (Note that positive $\Delta W$ is the energy lost from (or input to) the system.) If the displacement and velocity in forced response are

$$u(t) = A \sin(\Omega t + \psi) \ \text{ and } \ \dot{u}(t) = \Omega A \cos(\Omega t + \psi), \tag{1.3.38}$$

then $\Delta W$ has the form:

$$\Delta W = c\Omega^2 A^2 \int_0^T \cos^2(\Omega t + \psi)\, dt = \pi c\Omega A^2. \tag{1.3.39}$$

Of particular interest is the energy dissipated at resonance, i.e., when $\Omega \approx \omega_n$. In that case,

$$\Delta W = \pi c\omega_n A^2 = 2\pi \zeta k A^2. \tag{1.3.40}$$

A graphical representation of the energy dissipated per cycle, a *hysteresis loop*, can be constructed by noting that the total internal force $F(t)$ consists of the elastic restoring force and the viscous damping force acting in parallel, i.e.,

$$F(t) = kA \sin(\Omega t + \psi) + c\Omega A \cos(\Omega t + \psi). \tag{1.3.41}$$

Combining $u(t)$ from Eq. (1.3.38) with $F(t)$ from Eq. (1.3.41), and eliminating time yields the following relation between displacement and internal force:

$$\left(\frac{u}{A}\right)^2 + \left(\frac{F - ku}{c\Omega A}\right)^2 = 1. \tag{1.3.42}$$

This relation describes an ellipse, as shown in Figure 1.18. The energy dissipated per cycle, $\Delta W$, is the area enclosed by this ellipse. In the limiting case of no damping ($c = 0$), the ellipse collapses into a straight line representative of lossless linear elasticity.

**Figure 1.18** Force–displacement hysteresis loop for viscous damping

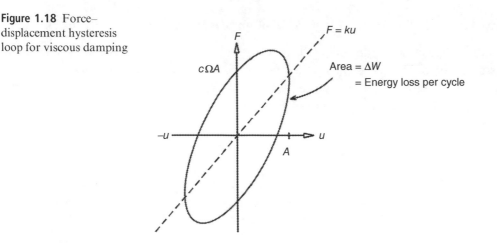

### 1.3.4 Response to Arbitrary Forcing

Previous sections address the response of undamped and damped SDOF systems to initial conditions and harmonic forcing. To completely evaluate the performance of a structural design, a design engineer must frequently consider the response to more general forcing. While a later chapter addresses the topic of *numerical integration* of equations of motion, this section reviews an analytical approach known as *Duhamel's integral*. This approach begins by considering the *impulse response* of an SDOF system.

#### Impulse Response

An *impulse* can be considered the limiting case of a force that acts over a very short time – such that its magnitude is indeterminate, but with finite impulse. The *unit impulse* function (also known as the *Dirac delta* function), $\delta(t)$, is zero everywhere except where its argument is zero, i.e., at $t = 0$, and is defined such that the associated impulse is

$$\lim_{\Delta t \to 0} \int_0^{0+\Delta t} \delta(t)\, \mathrm{d}t = 1. \tag{1.3.43}$$

Note that the function $\delta(t - \tau)$ is a unit impulse delayed to act at time $t = \tau$.

The *impulse response function*, $g(t)$, is the response of a quiescent SDOF system to a unit impulse. This can be found by recognizing that the force associated with the impulse instantaneously changes the velocity of the mass. This velocity change can be determined using the *impulse-momentum* equivalence principle, i.e., $m\, \mathrm{d}v = f\, \mathrm{d}t$. So, the initial velocity of the mass due to a unit impulse is $\dot{u}(0) = \frac{\int \delta(t)\mathrm{d}t}{m} = \frac{1}{m}$.

The system impulse response can then be found using the response to an initial velocity, Eq. (1.3.21), as:

$$g(t) = \left(\frac{1}{m\omega_d}\right) e^{-\zeta\omega_n t} \sin(\omega_d t). \tag{1.3.44}$$

### Duhamel's Integral

If an impulse $P(\tau)\mathrm{d}t$ acts at an arbitrary time $\tau$, it will change the velocity at that time by an amount $P(\tau)\mathrm{d}t/m$. As Figure 1.19 shows, the contribution of *this* impulse to the displacement at any later time $t$, $u(t, \tau)$, is given by

$$u(t, \tau) = P(\tau)g(t - \tau)\,\mathrm{d}t. \tag{1.3.45}$$

The total response of the system at some time $t$ can be considered to be the result of all prior impulses to the system. This requires considering the force-time history as a sequence of impulses, as shown in Figure 1.20, so that

$$
\begin{aligned}
u(t) &= \int_0^t P(\tau)g(t - \tau)\,\mathrm{d}\tau \\
&= \frac{1}{m\omega_d} \int_0^t P(\tau)e^{-\zeta\omega_n(t-\tau)} \sin(\omega_d(t - \tau))\,\mathrm{d}\tau
\end{aligned}
\tag{1.3.46}
$$

**Figure 1.19** Response of SDOF system to an impulse

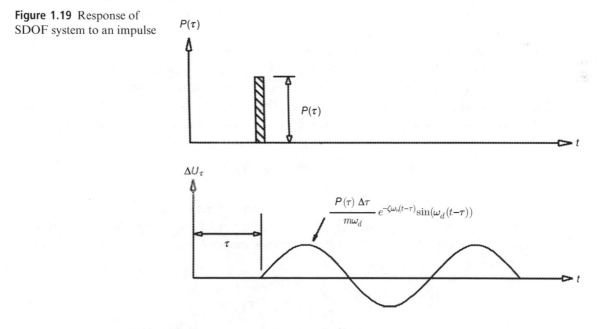

Equation (1.3.46) is known as *Duhamel's integral*. This *convolution* integral is especially useful for symbolically describing the response of an SDOF system to arbitrary forcing. As

**Figure 1.20**
Load history
as a sequence
of impulses

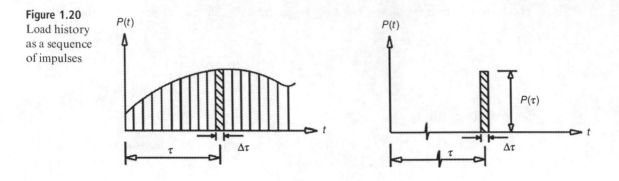

will be shown in a later chapter, this is essential for describing the response of a continuous structure to arbitrary forcing in terms of modal coordinates.

Sometimes, a special form of $P(t)$ permits explicit integration of this expression. If not, numerical quadrature of this expression is possible. In practice, however, as described in a later chapter, numerical "time-marching" schemes are usually preferred.

## Step Response

The response of a dynamical system to a suddenly applied nonperiodic force such as a rectangular step or pulse is of practical interest. Representative situations include aerodynamic gust loading, aircraft landing, spacecraft docking, and low-speed impacts. Certain features of responses to such loads indicate the suitability of a structural design. Such features include extremal responses, such as the maximum stresses and displacements.

A starting point for determining the response of a damped SDOF system to a rectangular pulse is the response to a step change in the forcing. Consider the sudden application of a constant load $P_0$ on a damped SDOF system that is initially at rest. As illustrated in Figure 1.21, this load can be expressed in terms of the Heaviside (unit) step function $H(t)$:

$$P(t) = P_0 H(t) \quad \text{where} \quad H(t) = \left\{ \begin{array}{ll} 0 & t < 0 \\ 1 & t \geq 0 \end{array} \right. . \tag{1.3.47}$$

Note that the function $H(t - \tau)$ is a unit step delayed to act at time $t = \tau$.

The *step response* of an SDOF system, $u_s(t)$, can be found using Duhamel's integral. (Note that the Heavyside step function is the time integral of the Dirac delta (unit impulse) function.)

$$
\begin{aligned}
u_s(t) &= \frac{P_0}{m\omega_d} \int_0^t e^{-\zeta\omega_n(t-\tau)} \sin(\omega_d(t-\tau)) \, d\tau \\
&= \frac{P_0}{k} \left[ 1 - \frac{e^{-\zeta\omega_n t}}{\sqrt{1-\zeta^2}} \cos(\omega_d t + \phi_s) \right] H(t)
\end{aligned}
\tag{1.3.48}
$$

**Figure 1.21** A step load

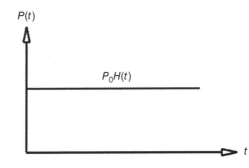

with $\tan\phi_s = -\zeta/\sqrt{1-\zeta^2}$. If the step load acts at some time $t_1$, the step response reflects that time delay:

$$u_s(t) = \frac{P_0}{k}\left[1 - \frac{e^{-\zeta\omega_n(t-t_1)}}{\sqrt{1-\zeta^2}}\cos\left(\omega_d(t-t_1)+\phi_s\right)\right]H(t-t_1). \qquad (1.3.49)$$

Figure 1.22 shows a representative step response function. When damping is present, the response eventually settles to the static displacement $P_0/k$. With light damping, the maximum dynamic response is very nearly twice the static deflection.

**Figure 1.22** Response of damped SDOF system to step load ($\zeta = 0.05$)

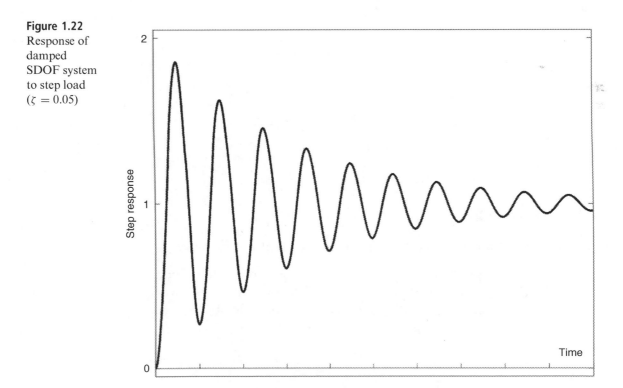

In the undamped case, the step response simplifies to

$$u_s(t) = \frac{P_0}{k}\left[1 - \cos\left(\omega_n t\right)\right] H(t). \tag{1.3.50}$$

The maximum dynamic response is precisely twice the static response, and the system oscillates indefinitely around the static deflection with $D(t) = 1 - \cos\left(\omega_n t\right)$.

**Rectangular Pulse**

Consider a rectangular pulse acting on an SDOF system that is initially at rest: it has amplitude $P_0$, begins at time $t_1$, and ends at time $t_2$, as shown in Figure 1.23. This loading can be considered to be the sum of two step functions, with the second having opposite sign and being delayed by the duration of the pulse:

$$P(t) = P_0\left(H(t - t_1) - H(t - t_2)\right) \tag{1.3.51}$$

$$P(t) = \begin{cases} 0 & t < t_1 \\ P_0 & t_1 \le t < t_2 \\ 0 & t \ge t_2 \end{cases}. \tag{1.3.52}$$

Since this is a linear system, the responses of the system to each step function can be found separately and summed to find the total response.

**Figure 1.23** A rectangular pulse

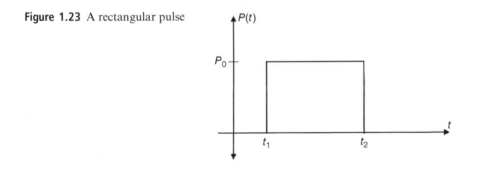

## 1.4    Periodic Response

### 1.4.1    Periodic Motion as a Superposition of Harmonic Motions

A periodic vibration can be expressed as a linear combination of harmonic vibrations using the *Fourier Series Expansion Theorem*. Given a periodic function $f(t)$ with period $T_p$ that satisfies the *Dirichlet conditions* (having to do with smoothness) – reasonable for vibratory mechanical systems – then $f(t)$ can be expressed as:

$$f(t) = \frac{a_0}{2} + \sum_{n=1}^{\infty} a_n \cos\left(2\pi nt/T_p\right) + \sum_{n=1}^{\infty} b_n \sin\left(2\pi nt/T_p\right), \tag{1.4.1}$$

where the coefficients $a_n$ and $b_n$ are given by the *Euler-Fourier* formulas:

$$a_n = \frac{2}{T_p} \int_0^{T_p} f(t) \cos (2\pi n t / T_p)\, dt;$$

$$b_n = \frac{2}{T_p} \int_0^{T_p} f(t) \sin (2\pi n t / T_p)\, dt; \quad n = 0, 1, 2, \ldots, \tag{1.4.2}$$

and the term $a_0/2$ is the mean value of $f(t)$ over $T_p$.

If $f(t)$ is continuous, the series in Eq. (1.4.1) converges uniformly. At a discontinuity in $f(t)$, say at $t_0$, it converges to the mean value of the function from both the left, $f(t_0^-)$, and the right, $f(t_0^+)$, i.e.,

$$f(t_0) \rightarrow \frac{1}{2} \left[ f(t_0^+) + f(t_0^-) \right]. \tag{1.4.3}$$

Some simplification in the coefficients, $a_n$ and $b_n$, is possible by decomposing $f(t)$ into two functions, $f_e(t)$ and $f_o(t)$, *even* and *odd* about $t = 0$,

$$f(t) = f_e(t) + f_o(t). \tag{1.4.4}$$

The cosine series in Eq. (1.4.1) represents $f_e(t)$, and the sine series, $f_o(t)$.

Occasionally, an alternative complex form for Fourier series is useful:

$$f(t) = \sum_{n=-\infty}^{n=+\infty} c_n e^{i(2\pi n t / T_p)}, \tag{1.4.5}$$

where the coefficients $c_n$ are defined by

$$\begin{aligned} c_n &= \frac{1}{T_p} \int_0^{T_p} f(t) e^{-i(2\pi n t / T_p)} dt \\ &= \frac{1}{T_p} \left[ \int_0^{T_p} f(t) \cos (2\pi n t / T_p) dt - i \int_0^{T_p} f(t) \sin (2\pi n t / T_p)\, dt \right]. \end{aligned} \tag{1.4.6}$$

The equivalence between Eqs. (1.4.1) and (1.4.5) can be established by means of Euler's formula, which shows that $c_n$ for $n > 0$ and $n < 0$ are complex conjugate pairs, i.e.,

$$c_{(n>0)} = (a_n - i b_n)/2 \ \text{ and } \ c_{(n<0)} = (a_n + i b_n)/2 = c_n^*. \tag{1.4.7}$$

Note also that $c_0 = a_0/2$ so that Eq. (1.4.5) can be recast into the form:

$$f(t) = c_0 + \sum_{n=1}^{\infty} \left\{ c_n e^{i(2\pi n t / T_p)} + c_n^* e^{-i(2\pi n t / T_p)} \right\}. \tag{1.4.8}$$

Fourier series are used in many applications, extending beyond the representation of periodic motions to include forcing and responses of all kinds.

### 1.4.2 Periodic Forced Vibration

Consider the case of a periodic forcing function such as that shown in Figure 1.24. The response of the system to such loading can be found using superposition – finding and

**Figure 1.24** Periodic nonharmonic forcing function

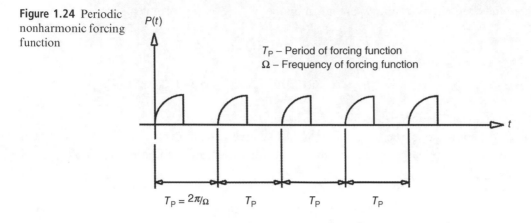

summing the responses of the linear system to the individual harmonic components of the forcing function. This requires the expansion of $P(t)$ in a Fourier series:

$$P(t) = a_0/2 + \sum_{n=1}^{N} a_n \cos\left(2\pi nt/T_p\right) + \sum_{n=1}^{N} b_n \sin\left(2\pi nt/T_p\right), \qquad (1.4.9)$$

where $T_p$ is the period of the forcing function. The coefficients $a_n$ and $b_n$ have the form:

$$\left.\begin{aligned}
a_0 &= \frac{2}{T_p} \int_0^{T_p} P(t)dt \\
a_n &= \frac{2}{T_p} \int_0^{T_p} P(t) \cos\left(\Omega_n t\right)dt \\
b_n &= \frac{2}{T_p} \int_0^{T_p} P(t) \sin\left(\Omega_n t\right)dt
\end{aligned}\right\}, \qquad (1.4.10)$$

where $n = 1, 2, \ldots$ and $\Omega_n = 2\pi n/T_p$.

The response $u(t)$ consists of a series of forced harmonic motions corresponding to each term in the series in Eq. (1.4.9), i.e., $a_0/2, a_n \cos\left(\Omega_n t\right)$, and $b_n \sin\left(\Omega_n t\right)$.

## BIBLIOGRAPHY

Balachandran, B. and Magrab, E. B. (2018). *Vibrations*. Cambridge University Press, 3rd edition.

Benaroya, H., Nagurka M. L., and Han, S. (2018). *Mechanical Vibration: Analysis, Uncertainties, and Control*. CRC Press, 4th edition.

Bishop, R. E. D. and Johnson, D. C. (2011). *The Mechanics of Vibration*. Cambridge University Press.

Den Hartog, J. P. (2008). *Mechanical Vibrations*. Crastre Press, 3rd edition.

Dimarogonas, A. D. (1996). *Vibration for Engineers*. Prentice Hall, 2nd edition.

Inman, D. J. (2014). *Engineering Vibration*. Pearson, 4th edition.

Palm III, W. J. (2006). *Mechanical Vibration*. Wiley.

Rao, S. S. (2017). *Mechanical Vibrations*. Pearson, Hoboken, 6th edition.

Sinha, A. (2010). *Vibration of Mechanical Systems*. Cambridge University Press.

Thomson, W. T. and Dahleh, M. D. (1997). *Theory of Vibration with Applications*. Pearson, 5th edition.

Tongue, B. H. (2001). *Principles of Vibration*. Oxford University Press, 2nd edition.

## PROBLEMS

1.  Rotating components are a common source of vibration in aerospace and mechanical systems. In some cases, additional devices can be used to counteract such vibration. Consider the following device, which consists of a point mass $m$ rotating counterclockwise about a fixed axis at a fixed frequency $\Omega$, restrained by a massless rigid rod of length $L$. At time $t = 0$, the mass is on the positive $x$-axis.

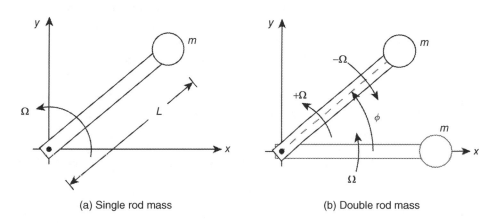

(a) Single rod mass                (b) Double rod mass

(a)  Develop expressions for:
  • The $x$ and $y$ coordinates of the mass as a function of time.
  • The $x$ and $y$ components of the velocity of the mass as a function of time.
  • The $x$ and $y$ components of the acceleration of the mass as a function of time.
  • The $x$ and $y$ components of the force on the axis as a function of time. What is the net force?

(b)  Now, consider the addition of a second mass at the end of a rod of the same length, rotating at the same frequency as the first (but perhaps in the opposite direction). At time $t = 0$, the second rod is at an initial phase angle $\phi$ relative to the $x$-axis.
  For rotation in the same direction:
  • What are the $x$ and $y$ components of the force on the axis as a function of time and the initial phase angle?
  • What is the maximum possible net force exerted on the axis? Minimum?
  For rotation in the opposite direction:
  • What are the $x$ and $y$ components of the force on the axis as a function of time and the initial phase angle?
  • What is the maximum possible net force exerted on the axis? Minimum?

2.  Aerospace vehicles typically comprise lightweight structures and mission-related mass. For dynamics analysis of some components it is reasonable to assume that the mass is concentrated in small regions, while the mass of the distributed stiffness is negligible by comparison. Consider the following cantilevered beam with tip mass:

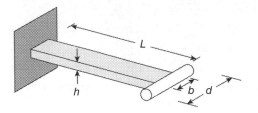

The beam is made from aluminum, with Young's modulus $E = 70$ GPa, length $L = 0.3$ m, width $b = 0.02$ m, and thickness $h = 0.003$ m. The tip mass is rod-shaped of length $d = 0.05$ m and mass $m_{tip} = 0.5$ kg.

(a) Estimate:
- the effective lateral (vertical) stiffness at the end of the beam (N/m) $\left(\text{Note: } w_L = \frac{F_{tip}L^3}{EI} \text{ and } I = \frac{bh^3}{12}\right)$;
- the effective axial stiffness at the end of the beam (N/m) $\left(\text{Note: } u_L = \frac{P_L L}{EA} \text{ and } A = bh\right)$;
- the effective torsional stiffness at the end of the beam (N m/rad) $\left(\text{Note: } \phi_L = \frac{T_L L}{GJ} \text{ and } J = \frac{bh^3}{3}\right)$; and
- the rotatory inertia of the mass (rod) about the $x$-axis (kg m$^2$).

(b) With this information, estimate the undamped natural frequency of the beam:
- in bending (rad/s) and (Hz);
- in extension (rad/s) and (Hz); and
- in torsion (rad/s) and (Hz).

3.  A vibration isolation system can be idealized as an SDOF mass-spring-damper system, as shown below.

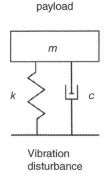

Isolated
payload

Vibration
disturbance

Experiments on this system yield the following information:
- The mass is measured to high accuracy as $m$ kg.
- The damped natural frequency is measured as $\omega_d$.
- The dynamic magnification factor is measured as $|H| = 1/200$.

Determine:
- The stiffness $k$ of the isolation system
- The modal damping ratio $\zeta$ of the system and the corresponding viscous damping coefficient $c$.
- The undamped natural frequency $\omega_n$ of the system.

4. A wing is modeled as an SDOF mass-spring-damper system. It has effective mass $m_w$ and stiffness $k_w$, and a modal damping ratio of $\zeta_w = 0.03$. In steady level flight, the wing deflects slightly under aerodynamic loads and its own weight. Consider this the reference position.

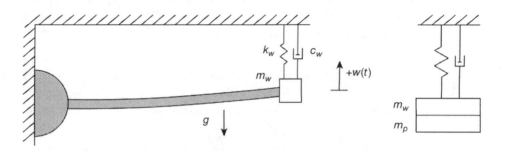

(a) A payload of mass $m_p$ is initially attached to the wing, which deflects slightly due to the additional weight. At time $t = 0$ the payload mass is released. Assume the aerodynamic loads do not change.
- Determine the lateral deflection of the wing $w(t)$ for $t \geq 0$.

(b) The wing (without payload mass) encounters a gust at time $t = 0$ that changes the aerodynamic loads. Idealize the gust load as an impulse of magnitude $\tilde{f}$.
- Determine the lateral deflection of the wing $w(t)$ for $t \geq 0$.

5. As shown in the figure below, a noncontact method is used to drive an SDOF mass-spring-damper system. The system has effective mass $m$ and stiffness $k$, and a modal damping ratio of $\zeta_w = 0.01$.

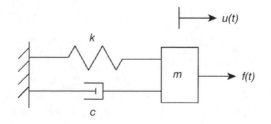

(a) Plot the time response $u(t)$ associated with release from an initial displacement, $u_0$.

(b) Plot the same response in the phase plane (velocity vs displacement). What is the long-time response?

(c) Plot the time response $u(t)$ associated with the imposition of an initial velocity, $v_0$.

(d) Plot the time response $u(t)$ associated with the imposition of a step force $f(t) = f_0 H(t)$ at $t = 0$.

(e) Plot the time response $u(t)$ associated with the imposition of a pulse of magnitude $f_0$. Consider three pulse durations:
   • Much shorter than $T_d$, the period associated with the damped natural frequency $\omega_d$;
   • Comparable to $T_d$; and
   • Much longer than $T_d$.

(f) Plot the time response $u(t)$ associated with the imposition of a pulse having constant total impulse $f_0 T_d$, where $T_d$ is the period associated with the damped natural frequency $\omega_d$. Consider three pulse durations:
   • Much shorter than $T_d$;
   • Comparable to $T_d$; and
   • Much longer than $T_d$.

6. The system in the preceding problem is driven harmonically. The drive frequency $\Omega$ can vary from 0.01 to 100 times the undamped natural frequency $\omega_n$ of the system.

(a) Develop an expression for the force-to-displacement frequency response function (FRF) of the system, $\frac{U}{F}(i\Omega)$.

(b) Plot the magnitude and phase of this FRF as a function of drive frequency. What is the peak response magnitude? What is the phase at resonance?

7. During flight, helicopter floor vibration is transmitted through seats, potentially subjecting crew members to high levels of whole-body vibration. The human body is especially sensitive to vibrations in the low frequency range between 0.5 Hz and 12 Hz, and those up to 80 Hz are of concern. Since typical rotors speeds are around 5 Hz, with two to five blades, considerable forcing exists in the range from 10 to 25 Hz.

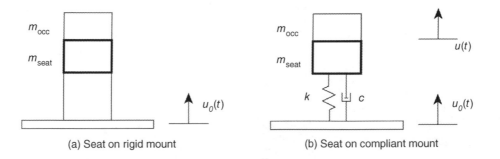

(a) Seat on rigid mount          (b) Seat on compliant mount

As shown in the figure, a passive seat mount consisting of a spring-damper system is one approach to reducing vibratory loads experienced by crew members. A nominal seat mass is 40 kg, and occupant mass ranges from 65 kg to 110 kg. Without the mount system, the maximum acceleration experienced at 10 Hz is 0.1 g.

Choose values for the stiffness $k$ and viscous damping $c$ of the mount that reduce this acceleration to no more than 0.005 g, while limiting seat excursions at any frequency to no more than 0.15 m.

8. During a *drop test* an airplane contacts the ground at a specified sink rate and the three tires all ideally stay in contact with the ground.

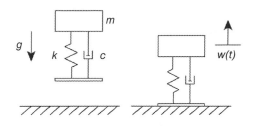

An airplane has a mass of 20,000 kg, and three vertical landing-gear struts (shock absorbers), each having stiffness $k = 2 \times 10^6$ N/m and damping $c = 4 \times 10^5$ N-s/m.

For an initial sink rate of $\dot{w}(t) = -6$ m/s, and assuming all the struts respond identically, determine:
- the eventual static compression of the landing gear struts and the associated loads;
- the maximum compression of the landing gear struts;
- the maximum load in each strut and when it occurs; and
- the maximum vertical acceleration of the airplane and when it occurs.

9. An open-loop positioning system aims to move the end effector of a flexible manipulator to a prescribed location $u = \Delta$ and come to rest as quickly as possible. The system has effective mass $m$ and stiffness $k$, and a modal damping ratio of $\zeta = 0.05$. The control inputs are step changes in the base location.

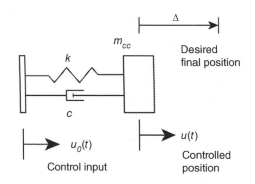

(a) One approach is to use a single step of size $\Delta$, but residual oscillations require time to settle.
- Determine the endpoint position response to a single step change $\Delta$ in the base location at time $t = 0$.
- The *response time* is the time after which the response remains within 5% of the target value. In this case, what is the response time?

(b) Another approach to positioning is to apply two such inputs sequentially. (The total response to two sequential step inputs is the sum of the responses to the individual steps.) Assume an initial step change $\Delta_1$ in the base location at time $t = t_1 = 0$.

• Considering the desired final position and velocity, identify potential good time(s) $t_2 > t_1$ at which to apply a second step change $\Delta_2$ in the base location. (Note that $\Delta_1 + \Delta_2 = \Delta$.) Discuss.

• Choose the time for the application of the second step position change that yields the desired endpoint position and velocity in the shortest time. Explain.

(c) Compare the two-step response time to that obtained for a single step.

10. A scanning mirror on an earth-observing satellite ideally executes a sawtooth pattern, tracking from side to side at a constant rate. The mirror is on a flexible mount and is driven by a superposition of harmonic moments, $M_i(t)$. The fundamental scanning frequency is $\Omega = \frac{1}{T}$ and, to achieve symmetric scanning, odd harmonics $(3\Omega, 5\Omega, \ldots)$ are also used.

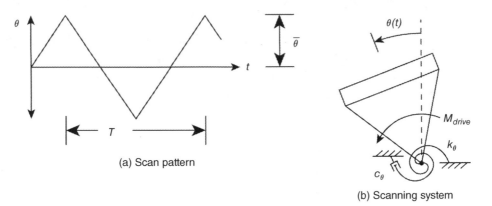

(a) Scan pattern

(b) Scanning system

(a) What are the harmonic components of a symmetric sawtooth signal having frequency $\Omega$ and amplitude $\bar{\theta}$?

(b) The mirror rotational inertia is $J_m$ kg-m$^2$. The rotational stiffness and damping of the mount are to be determined.

• Determine the angular position response $\theta(t)$ of the mirror to harmonic forcing of amplitude $M_f$ at some frequency $\Omega_f$.

• Consider the cases where $\Omega_f = \Omega, 3\Omega, 5\Omega, \ldots$ and find the associated angular responses $\theta_i(t)$.

• Consider the following combination of harmonic forcing: $M_1$ at $\Omega$ and $M_3$ at $3\Omega$, resulting in responses $\theta_1$ and $\theta_3$. By attempting to match the total response to the desired sawtooth response, while minimizing $M_1 + M_3$, find values for the rotational stiffness $k_\theta$ (and $\omega_n$) and damping $c_\theta$ (and $\zeta$) of the mount.

# 2     Principles of Structural Dynamics

## 2.1   Basic Concepts of Analytical Dynamics of Particles and Rigid Bodies

The concepts presented in this section are classical Newtonian dynamics that ignore relativity effects, and are easier to deal with than more modern versions of Newtonian dynamics, known as Kane's dynamics.

### 2.1.1   Newton's Laws

*Newton's First Law* states: Every body continues in its state of rest, or uniform motion in a straight line, unless compelled to change that state by forces acting on it.

*Newton's Second Law* states: The time rate of change of linear momentum of a body is proportional to the force acting on it and occurs in the direction in which the force acts.

*Newton's Third Law* states: To every action there is an equal and opposite reaction; that is, the mutual forces of two bodies acting on each other are equal and opposite in direction.

Newton's laws take a useful form when considering the motion of a particle subjected to a force that moves such that the force vector is equal to the time rate of change of the linear momentum vector $\mathbf{p}$

$$m\mathbf{V} = \mathbf{p}, \tag{2.1.1}$$

where $m$ is the particle mass and $\mathbf{V}$ is the velocity vector. Newton's second law states

$$\mathbf{F} = \frac{d}{dt}(m\mathbf{V}), \tag{2.1.2}$$

where $\mathbf{F}$ is the force vector. The motion is in an inertial reference frame, defined as a frame at rest or moving with uniform relative velocity relative to a fixed position (like a star). From Figure 2.1,

$$\mathbf{V} = \frac{d\mathbf{r}}{dt} = \dot{\mathbf{r}} \tag{2.1.3}$$

$$\mathbf{a} = \frac{d^2\mathbf{r}}{dt^2} = \ddot{\mathbf{r}}, \tag{2.1.4}$$

where $\mathbf{a}$ is the absolute acceleration vector and overdots indicate the derivative of the position vector $\mathbf{r}$ with respect to time. For the case of constant mass, Newton's second law, actually representing a combination of the first two laws, states

$$\mathbf{F} = m\mathbf{a}. \tag{2.1.5}$$

**Figure 2.1** Illustration of Newton's second law for a particle

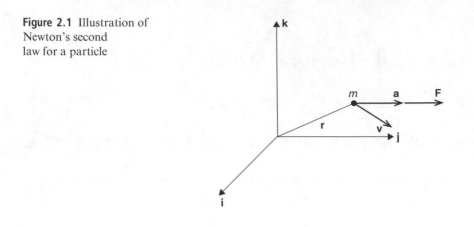

When $\mathbf{F}=0$, the statement represents the law of conservation of linear momentum.

*D'Alembert's Principle* consists of rewriting Newton's equation, Eq. (2.1.5), in the form

$$\mathbf{F} - m\mathbf{a} = 0, \tag{2.1.6}$$

which represents static equilibrium between two forces, the external force acting and the second term $m\mathbf{a}$, which represents an *inertia force*, sometimes called *reversed effective force*. D'Alembert's principle states that the laws of static equilibrium apply to a dynamical system if inertia forces are included by using a negative sign indicating that they act opposite to the acceleration vector $\mathbf{a}$. However, it is always important to distinguish between applied external forces and inertia forces.

### 2.1.2  Work and Kinetic Energy

The principles governing particle mechanics can be obtained from Newton's laws. One of the important principles is the principle of work and kinetic energy. Consider a particle having mass $m$ moving along a path under the action of an external force $\mathbf{F}$ from point $A$ to point $B$. The location of the points is determined by position vectors $\mathbf{r_A}$ and $\mathbf{r_B}$, as shown in Figure 2.2 respectively.

According to Newton's second law,

$$\mathbf{F} = m\ddot{\mathbf{r}}. \tag{2.1.7}$$

Evaluating the line integral of each side of the above equation between $A$ and $B$ yields

$$\int_A^B \mathbf{F}.d\mathbf{r} = \int_A^B m\ddot{\mathbf{r}}.d\mathbf{r}. \tag{2.1.8}$$

Note that the increment $d\mathbf{r}$ is taken along a tangent to the curve at each point, and furthermore

$$\ddot{\mathbf{r}}.d\mathbf{r} = \frac{1}{2}\frac{d}{dt}(\dot{\mathbf{r}}.\dot{\mathbf{r}})dt = \frac{1}{2}d(v^2), \tag{2.1.9}$$

**Figure 2.2** Motion of
a particle along a curved path

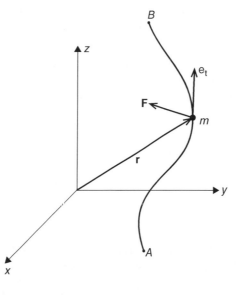

where it is assumed that infinitesimal changes in velocity and position occur during the same time interval, thus

$$\int_A^B m\ddot{\mathbf{r}}.d\mathbf{r} = \frac{m}{2}\int_A^B d(v^2) = \frac{m}{2}(v_B^2 - v_A^2), \tag{2.1.10}$$

where $v_A$ and $v_B$ are the particle velocities at points $A$ and $B$, respectively, along the path, thus

$$\int_A^B \mathbf{F}.d\mathbf{r} = \frac{1}{2}m(v_B^2 - v_A^2). \tag{2.1.11}$$

The term on the left of Eq. (2.2.11) is the work done by the force $\mathbf{F}$ and the term on the right is the kinetic energy $T$ of the particle relative to an inertial system

$$T = \frac{mv^2}{2}, \tag{2.1.12}$$

where $v$ is the speed of the particle, thus

$$W = T_B - T_A. \tag{2.1.13}$$

Note that the calculation of work and kinetic energy are dependent on the inertial system used as a reference frame and are independent of the path pursued. The general principle of work and kinetic energy is valid in any inertial reference frame. Next, consider Figure 2.2 again, and assume that the force $\mathbf{F}$ satisfies the following conditions: (a) it is a single-valued function of the position only (i.e. it is not a function of time) and (b) the line integral is a function of the end points only and is independent of the path followed between points $A$ and $B$. From property (a),

$$\int_A^B \mathbf{F}.d\mathbf{r} = -\int_B^A \mathbf{F}d\mathbf{r} \tag{2.1.14}$$

for the case when the integral on the right is taken in the reverse direction following the same path. However, due to property (b), it is true for any integration along any path connecting $A$ and $B$. A force having such properties is a *conservative* force, which implies that it forms a conservative force field. This property implies that the force is not dissipative, and the mechanical process is reversible. If the work $W$ is given by Eq. (2.1.13) that depends only on the location of the end points, then the integral is an exact differential

$$\mathbf{F}.d\mathbf{r} = -dV, \qquad (2.1.15)$$

where the minus sign has been chosen as a matter of convenience, thus

$$W = -\int_A^B \mathbf{F}.d\mathbf{r} = -\int_A^B \mathbf{F}dV = V_A - V_B. \qquad (2.1.16)$$

Equation (2.1.16) states that decrease in the potential energy $V$ associated with moving the particle from $A$ to $B$ is equal to the work done on the particle by a conservative force field. The potential energy is a single-valued scalar function of the position only. The sum of the potential energy and kinetic energy is known as the total energy $E$, thus

$$V_A + T_A = V_B + T_B = E. \qquad (2.1.17)$$

Equation (2.1.17) is a mathematical statement of the principle of conservation of mechanical energy.

## 2.2    Generalized Coordinates, Constraint Equations

The concept of generalized coordinates is of central importance in dynamics. Unlike the Cartesian coordinates, the generalized coordinates need not form the components of vectors, nor do they necessarily have the dimension of length. They are closely associated with the idea of "degrees-of-freedom," and to each degree of freedom there corresponds a generalized coordinate $q_i$. It is best understood by considering some typical examples:

(a) *Single particle*
Here, we have three degrees of freedom ($n = 3$), and we may choose the Cartesian coordinates $x, y, and\ z$ as the generalized coordinates $q_i$, as shown in Figure 2.3:

$$\{q\}^T = [x \quad y \quad z]. \qquad (2.2.1)$$

(b) *N particles*

$$\{q\}^T = [x_1 \quad y_1 \quad z_1 \quad x_2 \quad y_2 \quad z_2 \quad \cdots \quad x_N \quad y_N \quad z_N]. \qquad (2.2.2)$$

The number of degrees of freedom for the system of particles is obviously $n = 3N$ (Figure 2.3).

(c) *Rigid body*
A rigid body has six degrees of freedom ($n = 6$): three translational and three rotational as shown in Figure 2.5. In cases where one does not have to worry about finite (i.e. large) rotations, one can choose the coordinates as suggested in Figure 2.4, where the displacement

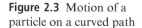
**Figure 2.3** Motion of a particle on a curved path

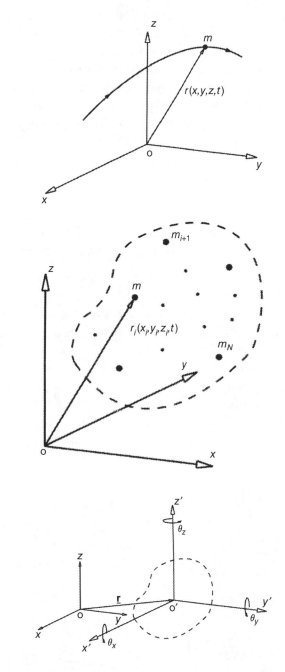

**Figure 2.4** Representation of a rigid body as a collection of particles

**Figure 2.5** Illustration of a rigid body with its six degrees of freedom

of an arbitrary point $O$ in the body is specified by the displacement vector $\mathbf{r}(x, y, z, t)$, and the rotations about the three coordinate axes are denoted by $\theta_x, \theta_y$, and $\theta_z$, respectively. A suitable set of generalized coordinates is then

$$\{q\}^T = [x \quad y \quad z \quad \theta_x \quad \theta_y \quad \theta_z].$$

(2.2.3)

For finite displacements, angular displacements are *noncommutative*, and it becomes necessary to use Euler angles, direction cosines, or some other suitable system, in order to account properly for finite rotations. Fortunately, many problems in engineering vibrations and dynamics of structures involve only small rotations about some mean equilibrium position, because the structure is restrained from large rigid-body rotations by kinematic constraints (boundary conditions). These problems can be formulated and analyzed without being concerned about the fact that angular displacements are not vectors and do not obey the transformation laws of vectors. Furthermore, infinitesimal rotations are so-called axial vectors, whose properties are very close to those of "genuine" vectors. Note that axial vectors differ from ordinary vectors in that they do not change sign upon a coordinate system inversion.

In many cases of practical interest, the rotation occurs about a *fixed* axis, as in the case of torsional vibration of rotating shafts. Here, the issue of noncommutative additions never arises, even in the presence of large angular rotations, and a single rotation angle $\theta$ can be used as a generalized coordinate.

(d) *Elastic body*

An elastic body has, by virtue of being composed of an infinite number of unconstrained particles, an infinite number of degrees of freedom ($n = \infty$). Adding a finite number of constraints that reflect boundary conditions or elastic/structural limits still leaves $n$ at infinity. Nevertheless, elastic structures are often modeled with only a finite number of degrees of freedom, by imposing *kinematic constraints* on the displacement field as shown in Figure 2.6 In practice, this is often accomplished by assuming that the displacement field can be approximated, to within the accuracy desired, by a finite number of *interpolation* or *shape functions*, each associated with a generalized coordinate.

**Figure 2.6**
Illustration of one-, two-, and three-dimensional constrained elastic bodies

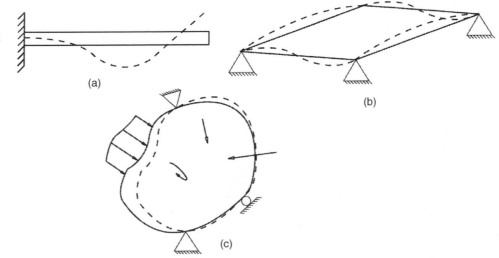

(a)

(b)

(c)

Examples of generalized coordinates for structures are provided next:

### 2.2.1 "Discrete" Coordinates

Consider an elastic body subjected to a set of forces $F_i$. Unless specified otherwise, the forces are considered to be "generalized" in the sense that moments are included in the $F_i$ s. Let $u_i$ denote displacements (or rotations) of the body at the point of application of $F_i$ *in the direction of $F_i$* as indicated in Figure 2.7

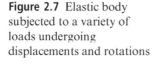

**Figure 2.7** Elastic body subjected to a variety of loads undergoing displacements and rotations

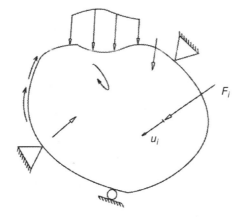

Consider now a set of *kinematically admissible* virtual displacements $\delta u_i$, that is, virtual displacements satisfying the kinematic boundary conditions. A virtual displacement is defined as an arbitrary infinitesimally small displacement that satisfies the kinematically admissible boundary conditions associated with the particular problem. It is also important to note that during a virtual displacement the applied loads on the system do not change. The virtual work done by the real forces $F_1$ is then

$$\delta W_E = \sum_{i=1}^{m} F_i \delta u_i = \sum_{j=1}^{n} Q_j \delta q_j, \qquad (2.2.4)$$

where the $Q_j$s are generalized forces associated with the set of generalized coordinates $q_j$, and Eq. (2.2.4) defines the generalized forces $Q_j$s. If the $u_i$s are linearly independent, as they are in this case, one can take $q_j = u_j$. But in the general case, where the original coordinates $u_i$ are either constrained or insufficient to model the structure, one needs to find a set of generalized coordinates $q_j$ **that are sufficient and linearly independent**. In the case of **nonholonomic**[1] systems, this procedure cannot be used and one is forced to work with constrained coordinates.

In the case where the $u_i$s associated with the applied forces are insufficient to model the structure properly, as is often the case in *dynamics* problems, one can add additional coordinates at other key locations of the structure and associate zero forces with these

---

[1] Constraints that must be expressed in terms of differentials of the coordinates and possibly time.

additional $u$-coordinates. Equation (2.2.4) can still be used to define the generalized forces $Q_j$ associated with any chosen generalized coordinates $q_j$.

For a linearly elastic (Hookean) material and small displacements, the force–displacement relations are linear:

$$F_i = k_{i1}u_1 + k_{i2}u_2 + \cdots + k_{in}u_n, \qquad (2.2.5)$$

where $k_{ij}$ are the appropriate elements of the stiffness matrix that govern the force–deflection relationship; these can be stated in matrix form as

$$\{F\} = [k]\{u\}. \qquad (2.2.6)$$

If the structure is restrained against rigid-body motions, so that the stiffness matrix $[k]$ is nonsingular, Eq. (2.2.6) can be inverted to give

$$\{u\} = [a]\{F\}, \qquad (2.2.7)$$

where

$$[a] = [k]^{-1} = \text{flexibility matrix}$$

$$(\det[k] \neq 0).$$

The $i$th row of Eq. (2.2.7) can be written as

$$u_i = a_{i1}F_1 + a_{i2}F_2 + \cdots + a_{in}F_n, \qquad (2.2.8)$$

where the $a_{ij}$ are the flexibility influence coefficients associated with the $i$th displacement $u_i$. Due to the Betti–Maxwell reciprocity theorem, both $[k]$ and $[a]$ are symmetric matrices, i.e., $k_{ji} = k_{ij}$ ; $a_{ji} = a_{ij}$.

## Examples

Three examples of discrete coordinates are shown in Figures 2.8 through 2.10. The first of these is a spring-mass damper system of the type studied extensively in undergraduate texts on mechanical vibrations and needs no further explanation here. The second example, Figure 2.9, is similar, except that the problem involves torsional vibrations of a shaft with two rigid disks, and the structure is a *continuous* system. The third example, shown in Figure 2.10, is a continuous frame structure, but it is again assumed that the structure can be modeled adequately by defining discrete coordinates, namely the displacements at the two joints of the frame.

**Figure 2.8** Discrete coordinates attached to each mass in this spring-mass system

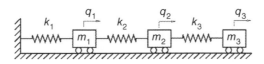

If we further assume, as is often done in analyzing slender beam structures, that the axial rigidity of a typical member of the frame is much greater than its transverse flexural rigidity ($EA/L >> EI/L^3$), it is reasonable to set

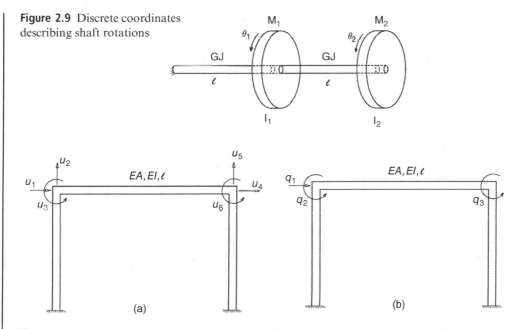

**Figure 2.9** Discrete coordinates describing shaft rotations

**Figure 2.10** Discrete coordinates for frame describing joint displacements and rotations

$$u_4 = u_1 \quad ; \quad u_2 = u_5 = 0$$

The reduced set of coordinates $q_1, q_2, q_3$ are indicated in part (b) of Figure 2.10. Equation (2.2.4) can then be used to define the generalized forces $Q_i$ and relate them to the $F_i$s.

In a *static* analysis of the last two structures, forces and/or moments can be associated with each discrete coordinate, using the basic force–deflection relationship of linear structures. The coordinates selected in the last two examples are sufficient for describing the deflected config-uration of the structure when subjected to a set of *static* loads at the coordinate locations, as defined in Figure 2.7. Thus, it is often possible to find a *finite* set of *generalized coordinates* that permits an "exact" static analysis to be carried out, without undue concern that the un-derlying continuous structure has an *infinite* number of degrees of freedom. The term "exact" here implies that the underlying structural theory used, e.g., Bernoulli–Euler beam theory, Kirchhoff–Love plate theory, etc., determines the solution. Obviously, one needs to know the location of the loads before selecting the coordinates.

For the case of *dynamics*, the situation is very different. Even if the loads act at the same locations as in the static examples, it is impossible to find a reduced set of generalized coor-dinates that are "exact" in the same sense as in the static case. As soon as the loads become functions of *time*, or one attempts to solve an initial-value problem or a free vibration prob-lem for the structure, the fact that the structure has an infinite number of degrees of freedom cannot be ignored. Fortunately, not all of these degrees of freedom are of equal importance in a given problem. This observation is useful when choosing the coordinates.

This reduction process of getting from $n = \infty$ to $n = N$, where $N$ is a finite number, always involves imposing *kinematic constraints* on the structure. That is, one makes certain assumptions about the global deflection behavior of the structure. In the case of the discrete coordinates illustrated in Figures 2.9 and 2.10, it is usually assumed that the structure deflects primarily in its *static* modes, so that the displacement distribution along the structural members in Figure 2.10 is of the same form as in the static case. For example, the twist distribution would be taken as linear functions of $x$ along the two shaft segments in Figure 2.9, whereas the flexural deflections along each frame member in Figure 2.10 would be assumed to be cubic functions of $x$.

Note that using these assumptions is equivalent to assuming that the deflected vibration modes of interest do not differ significantly from the corresponding static deflection shapes. Indeed, this would be true for the shaft with heavy disks and for the slender frame structure but might not be true for a more complex structure. For example, if one removes the two disks from the second example and considers only a uniform shaft, the static linear deflection shape becomes an inaccurate approximation to the dynamic deflection shapes (normal modes). In the latter case, the dynamic mode shapes are odd multiples of quarter-sine waves,

$$\phi_n(x) = \sin\left[(2n-1)\pi x/l\right] \quad ; \quad n = 1, 2, \ldots,$$

which are poorly approximated by a linear function.

## 2.2.2  "Distributed" Coordinates

Consider the simply supported beam shown. The examples below illustrate how so-called distributed coordinates can be used to model its dynamic displacement $w(x, t)$:

(a) *Fourier series*

$$w(x, t) = \sum_{n=1}^{\infty} a_n(t) \sin\left(\frac{n\pi x}{l}\right), \tag{2.2.9}$$

with generalized coordinates $(n = \infty)$

$$\{q\}^T = [a_1\ a_2\ a_3\ \ldots\ a_n\ \ldots]. \tag{2.2.10}$$

(b) *Truncated series*

$$w(x, t) = a_1(t) \sin\left(\frac{\pi x}{l}\right) + a_2(t) \sin\left(\frac{2\pi x}{l}\right) + a_3(t) \sin\left(\frac{3\pi x}{l}\right), \tag{2.2.11}$$

with generalized coordinates

$$\{q\}^T = [a_1\ a_2\ a_3] \quad (n = 3). \tag{2.2.12}$$

(c) *General expansion in terms of shape functions*

$$w(x, t) = q_1(t)\phi_1(x) + q_2(t)\phi_2(x) + \cdots q_N(t)\phi_N(x), \tag{2.2.13}$$

with generalized coordinates

$$\{q\}^T = [q_1\ q_2\ q_3\ \ldots\ q_N], \tag{2.2.14}$$

where $n = N$. Here, the shape functions $\phi_i(x)$ should satisfy the kinematic or geometric boundary conditions of the problem, which in the case of the simply supported beam under consideration:

$$\phi_i(0) = \phi_i(l) = 0 \ ; \ i = 1, 2, \ldots, N. \tag{2.2.15}$$

The requirement that the individual shape functions satisfy all the geometric boundary conditions can be relaxed, provided that the overall displacement field satisfies these conditions. It is possible to impose the geometric boundary conditions via *constraint equations*, using Lagrange multipliers. Note that the shape functions involved in examples (a) and (b) do satisfy the kinematic boundary conditions for the simply supported beam in Figure 2.11.

**Figure 2.11** Simply supported beam

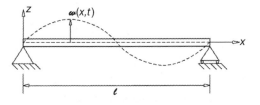

A more in-depth discussion of kinematic and natural boundary conditions will be presented in Section 2.7, when variational principles are discussed.

Distribution functions can also be used to form distribution vectors for structural models that are already in discrete form. Take the cantilever beam in Figure 2.12. This beam has been discretized into six elements and only the translational degrees of freedom are retained. The distribution vectors for this example may be constructed from the modes of vibration of a clamped-free beam:

**Figure 2.12** Distribution vectors for a cantilevered column

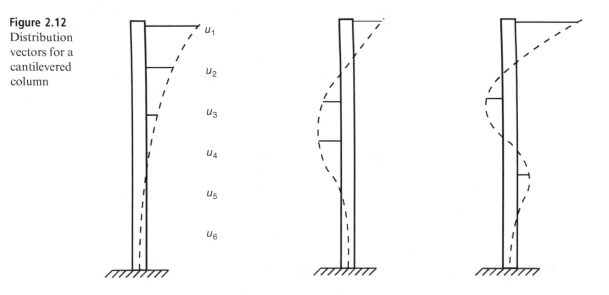

$$\phi_i = \cosh(\beta_i x) - \cos(\beta_i x) - \alpha_n(\sinh(\beta_i x) - \sin(\beta_i x)), \tag{2.2.16}$$

where $i = 1, 2, \ldots, n$ and $\beta_i$ are characteristic numbers.

Using the first three terms of Eq. (2.2.16), the displacements evaluated at the six coordinates may be written as:

$$
\begin{bmatrix} u_1 \\ u_2 \\ u_3 \\ u_4 \\ u_5 \\ u_6 \end{bmatrix}
=
\begin{bmatrix}
1.000 & 1.000 & 1.000 \\
0.752 & 0.550 & 0.167 \\
0.519 & 0.002 & -0.253 \\
0.313 & -0.257 & -0.217 \\
0.148 & -0.169 & 0.152 \\
0.039 & -0.050 & 0.065
\end{bmatrix}
\begin{bmatrix} a_1 \\ a_2 \\ a_3 \end{bmatrix}
\tag{2.2.17}
$$

or in abbreviated form:

$$\{u\} = [T]\{a\}.$$

Note that the matrix of distribution vectors [T] acts as a transformation matrix, which takes the original coordinates $\{u\}$ to the generalized coordinates $\{a\}$. If these three distribution vectors are deemed adequate to represent the response of the structure due to any given input to the original six discrete coordinates, then the model can be reduced to some combination of these three vectors. Again, there has been a constraint imposed on the original model by this process. In this case, one has a transformation from one discrete system to another but smaller discrete system.

$$
\{u\} =
\begin{bmatrix} u_1 \\ u_2 \\ u_3 \\ u_4 \\ u_5 \\ u_6 \end{bmatrix}
=
\begin{bmatrix} 1.000 \\ 0.752 \\ 0.519 \\ 0.313 \\ 0.148 \\ 0.039 \end{bmatrix} a_1
\; ; \;
\{u\} =
\begin{bmatrix} 1.000 \\ 0.550 \\ 0.002 \\ -0.253 \\ -0.164 \\ -0.050 \end{bmatrix} a_2
\; ; \;
\{u\} =
\begin{bmatrix} 1.000 \\ 0.167 \\ -0.253 \\ -0.217 \\ 0.152 \\ 0.065 \end{bmatrix} a_3.
$$

The reduction of the size of the structural model can be illustrated through the following example. Consider the extensional behavior of the bar system of variable cross section as shown in Figure 2.11. Five bar elements are used to model this system. The assembled stiffness matrix has the form:

$$
\begin{bmatrix} F_1 \\ F_2 \\ F_3 \\ F_4 \\ F_5 \end{bmatrix}
=
\frac{A_0 E}{l}
\begin{bmatrix}
2.2 & -1.1 & - & - & - \\
-1.1 & 2.0 & -0.9 & - & - \\
- & -0.9 & 1.8 & -0.9 & - \\
- & - & -0.9 & 2.0 & -1.1 \\
- & - & - & - & 1.1
\end{bmatrix}
\begin{bmatrix} u_1 \\ u_2 \\ u_3 \\ u_4 \\ u_5 \end{bmatrix}
\tag{2.2.18}
$$

or in abbreviated form:

$$\{F\} = [K]\{u\}.$$

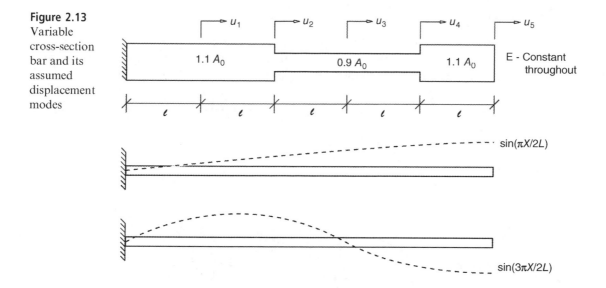

**Figure 2.13**
Variable cross-section bar and its assumed displacement modes

The strain energy of this system is given by

$$U = \frac{1}{2}\{u\}^T\{F\} = \frac{1}{2}\{u\}^T[K]\{u\}. \tag{2.2.19}$$

Based on the expected behavior of the system that it will be sufficiently accurate to use the first two modes of the natural vibration of a uniform bar that is clamped at one end as the distribution functions for the derivation of the distribution vectors for the system. The continuous distribution function for the displacements of the bar system is given by

$$u(x) = a_1 \sin(\pi x/2L) + a_2 \sin(3\pi x/2L). \tag{2.2.20}$$

Evaluating Eq. (2.2.20) at the discrete coordinate locations of the system yields the following transformation matrix:

$$\begin{bmatrix} u_1 \\ u_2 \\ u_3 \\ u_4 \\ u_5 \end{bmatrix} = \begin{bmatrix} 0.309 & 0.809 \\ 0.587 & 0.951 \\ 0.809 & 0.309 \\ 0.951 & -0.588 \\ 1.000 & -1.000 \end{bmatrix} \begin{bmatrix} a_1 \\ a_2 \end{bmatrix}. \tag{2.2.21}$$

Substituting Eq. (2.2.21) into the strain energy expression gives

$$U = \frac{1}{2}\{u\}^T[K]\{u\} = \frac{1}{2}\{a\}^T[\bar{K}]\{a\}, \tag{2.2.22}$$

where the generalized stiffness $[\bar{K}]$ is given by

$$[\bar{K}] = [T]^T[K][T] = \frac{A_0 E}{l} \begin{bmatrix} 0.2552 & 0.0533 \\ 0.0533 & 2.0239 \end{bmatrix}.$$

### Kinetic Energy

Consider a body of volume $V$ and mass density $\rho$, bounded by a surface $S$. If the velocity field is $\mathbf{v}(x, y, z, t)$ relative to the coordinate system $(x, y, z)$, the kinetic energy of the body in this system, shown in Figure 2.14, is by definition, the scalar $T$ given by

$$T = \frac{1}{2} \int_V \mathbf{v} \cdot \mathbf{v}\rho \, dV = \frac{1}{2} \int_V \dot{\mathbf{r}} \cdot \dot{\mathbf{r}}\rho \, dV. \qquad (2.2.23)$$

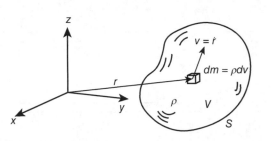

**Figure 2.14** Illustration of the kinetic energy associated with a body having volume V

Thus, the kinetic energy depends on the coordinate system or reference frame used in writing Eq. (2.2.23). Usually, one is interested in the kinetic energy with respect to inertial reference frames, as would be the case when using Lagrange's equations. However, it should be obvious from the definition of $T$ that it need not be the same in two different inertial systems. The difference is a constant independent of the generalized coordinates and velocities, which does not contribute terms to Lagrange's equations.

(a) *Discrete masses*
In this case, one can write

$$T = \frac{1}{2} \sum_i m_i \dot{\mathbf{r}}_i \cdot \dot{\mathbf{r}}_i = \frac{1}{2} m_i v_i^2. \qquad (2.2.24)$$

Consider next, kinetic energy expressions for some typical structural members.

(b) *Axial member (bar)*, shown in Figure 2.15

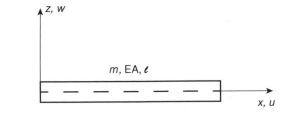

**Figure 2.15** Vibrating bar in the axial direction

Here, the only nonzero velocity is in the axial direction, and

$$T = \frac{1}{2} \int_V \rho \dot{u}^2 dV = \frac{1}{2} \int_0^l m \dot{u}^2 dx, \qquad (2.2.25)$$

where $m = \rho A$ is the mass per unit length of the bar.

(c) *Flexural member, Euler–Bernoulli beam*

If we consider symmetric bending in the $x$–$z$ plane, the velocity at a given point is

$$\dot{\mathbf{r}} - \dot{u}\hat{\mathbf{i}} + \dot{w}\hat{\mathbf{k}}$$

$$\dot{\mathbf{r}} \cdot \dot{\mathbf{r}} = \dot{u}^2 + \dot{w}^2.$$

The kinetic energy of the beam, using Eq. (2.2.23), is

$$T = \frac{1}{2} \int_V \rho A (\dot{u}^2 + \dot{w}^2) dV.$$

For a Bernoulli–Euler beam (i.e. neglecting shear deformations), the displacement of a point P in the cross section in the axial direction as a result of bending is, as shown in Figure 2.16,

**Figure 2.16**
Kinetic energy of a vibrating Euler–Bernoulli beam

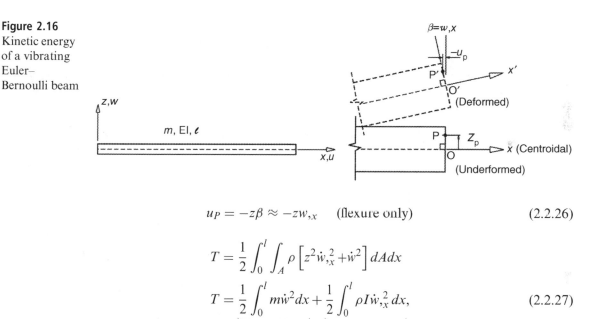

$$u_P = -z\beta \approx -zw_{,x} \quad \text{(flexure only)} \tag{2.2.26}$$

$$T = \frac{1}{2} \int_0^l \int_A \rho \left[ z^2 \dot{w}_{,x}^2 + \dot{w}^2 \right] dA dx$$

$$T = \frac{1}{2} \int_0^l m\dot{w}^2 dx + \underbrace{\frac{1}{2} \int_0^l \rho I \dot{w}_{,x}^2 dx}_{\text{rotational}}, \tag{2.2.27}$$

$$\underbrace{\phantom{\frac{1}{2} \int_0^l m\dot{w}^2 dx}}_{\text{translational}}$$

kinetic energy

where

$$I = \int_A z^2 dA$$

is the area moment of inertia about the $y$-axis. For slender beams, the rotational kinetic energy term is generally small and can be ignored; thus, the kinetic energy is

$$T = \frac{1}{2} \int_0^l m\dot{w}^2 dx.$$

(d) *Torsional member (shaft)*

Consider a simple circular shaft, for which Coulomb's theory of torsion yields the following

$$\dot{\mathbf{r}} = r\dot{\phi}\hat{e}_\theta$$

$$\dot{\mathbf{r}} \cdot \dot{\mathbf{r}} = r^2\dot{\phi}^2 = (y^2 + z^2)\dot{\phi}^2, \tag{2.2.28}$$

where $\phi(x, t)$ is the twist of the shaft at axial position $x$ at time $t$, and $\hat{e}_\theta$ is a unit vector in the tangential (i.e. perpendicular to the radius $r$ at a specified location) direction. In order to avoid confusion later on, we denote the total twist by $\phi$ rather than $\theta$, since $\theta$ is also used to denote the twist per unit length, $\theta = \phi,_x$.

The kinetic energy of the shaft is

$$T = \frac{1}{2}\int_0^l \rho J\dot{\phi}^2 dx,$$

where

$$J = \int_A r^2 dA = \int_A (y^2 + z^2)dA \tag{2.2.29}$$

is the polar (area) moment of inertia of the cross section. For noncircular cross sections, *warping* occurs and $u$ is different from zero.

**Figure 2.17**
Kinetic energy of a vibrating shaft

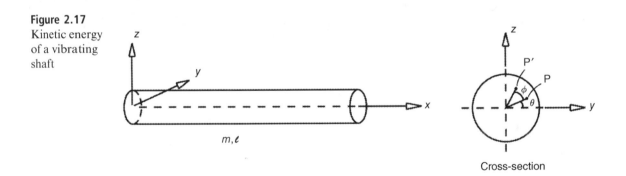

$m, \ell$

Cross-section

**Generalized Mass Matrix**   In an inertial reference frame, the kinetic energy $T$ can be written as a quadratic form

$$T = \frac{1}{2}\sum_i \sum_j m_{ij}\dot{q}_i\dot{q}_j = \frac{1}{2}\{\dot{q}\}^T [m] \{\dot{q}\}, \tag{2.2.30}$$

where $[m]$ is the generalized mass matrix. Let

$$u_i \; ; \; i = 1, 2, \ldots, m$$

represent an arbitrary set of coordinates. If the system is holonomic, we can always solve for the $u_i$s in terms of the generalized coordinates. In most cases of interest in structural dynamics the $u_i$s and the $q_i$s are related through a *linear* transformation. That is, a constant transformation matrix $[t]$ can be found such that

$$\{u\} = [t]\{q\}. \tag{2.2.31}$$

In the $u$-coordinates, the mass matrix is defined by the expression for $T$,

$$T = \frac{1}{2}\{\dot{u}\}^T [m]_u \{\dot{u}\}.$$  (2.2.32)

Assuming that the kinetic energy $T$ is the same in the two coordinate systems, Eqs. (2.2.31) and (2.2.32) yield

$$T = \frac{1}{2}\{\dot{q}\}^T [t]^T [m]_u [t] \{\dot{q}\}$$  (2.2.33)

or

$$T = \frac{1}{2}\{\dot{q}\}^T [m] \{\dot{q}\},$$  (2.2.34)

which implies that the generalized mass matrix is given by

$$[m] = [t]^T [m]_u [t].$$  (2.2.35)

If the structure is modeled using distributed coordinates, for example, as in a finite-element or Rayleigh–Ritz analysis, the generalized mass matrix is identified from the kinetic energy in the same way. Consider, as an example, the cantilever beam shown in Figure 2.18, and assume that the transverse bending deflection can be adequately described by an expansion of the type

$$w(x, t) = \sum_{i=1}^{N} q_i(t)\phi_i(x).$$

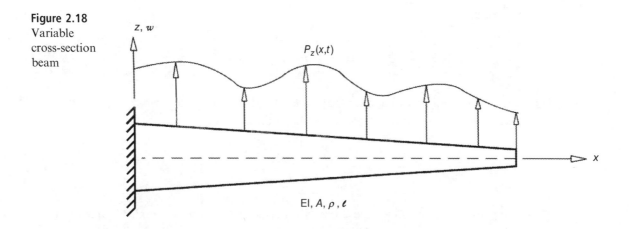

**Figure 2.18**
Variable cross-section beam

Assuming that the beam is slender, the kinetic energy is given by

$$T = \frac{1}{2}\int_0^l m\dot{w}^2 dx.$$  (2.2.36)

The velocity can be written as

$$\dot{w} = \sum_{i=1}^{N} \dot{q}_i(t)\phi_i(x) = \{\dot{q}\}^T \{\Phi\} = \{\Phi\}^T \{\dot{q}\}, \tag{2.2.37}$$

where

$$\{\Phi\}^T = [\phi_1 \ \phi_2 \ \dots \phi_N ]. \tag{2.2.38}$$

Thus,

$$\dot{w}^2 = \{\dot{q}\}^T \{\Phi\} \{\Phi\}^T \{\dot{q}\} \tag{2.2.39}$$

and the kinetic energy becomes

$$T = \frac{1}{2} \{\dot{q}\}^T \int_0^l m \{\Phi\} \{\Phi\}^T \, dx \, \{\dot{q}\} . \tag{2.2.40}$$

From the last expression, one concludes that the generalized $N \times N$ mass matrix for the beam is

$$[m] = \int_0^l m \{\Phi\} \{\Phi\}^T \, dx \tag{2.2.41}$$

with elements

$$m_{ij} = \int_0^l m\phi_i(x)\phi_j(x)dx. \tag{2.2.42}$$

The results from this example are easily generalized to two- and three-dimensional structures. Note that the generalized masses can be obtained from $T$

$$m_{ij} = \frac{\partial^2 T}{\partial \dot{q}_i \partial \dot{q}_j}. \tag{2.2.43}$$

As another example, consider the axial vibrations of a bar with linearly varying cross-sectional dimensions, shown in Figure 2.19. Assuming that the axial displacements of the bar can be represented by two generalized coordinates of the distributed type,

$$u(x, t) = a_1(t) \sin \frac{\pi x}{2L} + a_2(t) \sin \frac{3\pi x}{2L} \tag{2.2.44}$$

**Figure 2.19**
Axially loaded
bar with
varying cross
section

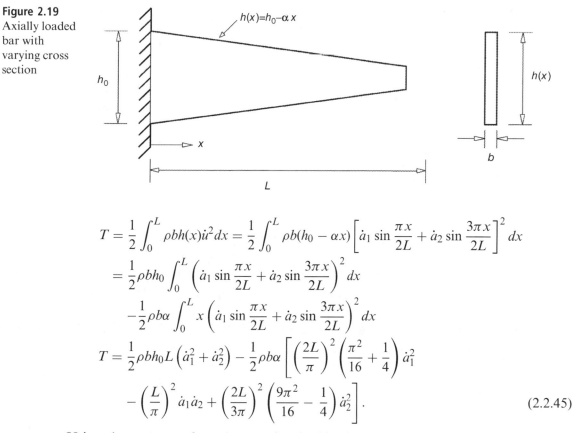

$$T = \frac{1}{2}\int_0^L \rho b h(x)\dot{u}^2\,dx = \frac{1}{2}\int_0^L \rho b(h_0 - \alpha x)\left[\dot{a}_1\sin\frac{\pi x}{2L} + \dot{a}_2\sin\frac{3\pi x}{2L}\right]^2 dx$$

$$= \frac{1}{2}\rho b h_0\int_0^L\left(\dot{a}_1\sin\frac{\pi x}{2L} + \dot{a}_2\sin\frac{3\pi x}{2L}\right)^2 dx$$

$$-\frac{1}{2}\rho b\alpha\int_0^L x\left(\dot{a}_1\sin\frac{\pi x}{2L} + \dot{a}_2\sin\frac{3\pi x}{2L}\right)^2 dx$$

$$T = \frac{1}{2}\rho b h_0 L\left(\dot{a}_1^2 + \dot{a}_2^2\right) - \frac{1}{2}\rho b\alpha\left[\left(\frac{2L}{\pi}\right)^2\left(\frac{\pi^2}{16} + \frac{1}{4}\right)\dot{a}_1^2\right.$$

$$\left.- \left(\frac{L}{\pi}\right)^2\dot{a}_1\dot{a}_2 + \left(\frac{2L}{3\pi}\right)^2\left(\frac{9\pi^2}{16} - \frac{1}{4}\right)\dot{a}_2^2\right]. \qquad (2.2.45)$$

Using the concept of a mass matrix, the kinetic energy for this system can also be written as

$$T = \frac{1}{2}\{\dot{a}\}^T[m]\{\dot{a}\}, \qquad (2.2.46)$$

where $\{\dot{a}\}^T = [\dot{a}_1 \quad \dot{a}_2]$ and $[m] = \begin{bmatrix} m_{11} & m_{12} \\ m_{21} & m_{22} \end{bmatrix}$ the matrix $[m]$ representing the inertia properties of the system is called the mass matrix, by virtue of its definition it is symmetric, that is, $m_{ij} = m_{ji}$. Expanding Eq. (2.2.46) one has

$$T = \frac{1}{2}\left(m_{11}\dot{a}_1^2 + 2m_{12}\dot{a}_1\dot{a}_2 + m_{22}\dot{a}_2^2\right). \qquad (2.2.47)$$

Comparing coefficients in Eqs. (2.2.45) and (2.2.47), the elements of the mass matrix are

$$m_{11} = \left[\rho b h_0 L - \rho b\alpha\left(\frac{2L}{\pi}\right)^2\left(\frac{\pi^2}{16} + \frac{1}{4}\right)\right]$$

$$m_{22} = \rho b h_0 L - \rho b\alpha\left[\left(\frac{2L}{3\pi}\right)^2\left(\frac{9\pi^2}{16} - \frac{1}{4}\right)\right] \qquad (2.2.48)$$

$$m_{12} = m_{21} = \rho b\alpha\left(\frac{L}{\pi}\right)^2.$$

Consider another example in which the system is described using discrete–discrete coordinates. The axial vibrations of the bar shown in Figure 2.20 are described by five coordinates $u_1, \ldots, u_5$.

**Figure 2.20**
Axially
vibrating bar
with varying
properties

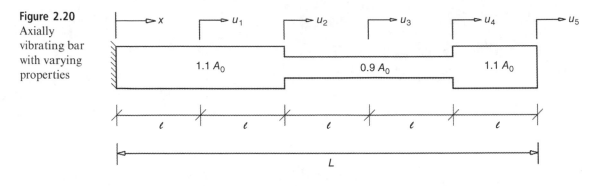

When using these discrete coordinates the inertial properties of the system are usually obtained by lumping the masses. Lumping of mass properties is a relatively ad hoc procedure which is dependent, somewhat, on the whims of the analyst. For this particular example, it is reasonable to select the mass located at a distance $1/2\ l$ behind and $1/2\ l$ in front of the coordinate location $u_i$ and lump it at the coordinate location. Using this procedure, the elements of the lumped mass matrix, based on the reference area $A_0$, are

$$m_{11} = 1.1\ \rho\ A_0 l$$
$$m_{22} = 1.0\ \rho\ A_0 l$$
$$m_{33} = 0.9\ \rho\ A_0 l$$
$$m_{44} = 1.0\ \rho\ A_0 l$$
$$m_{55} = 1.1\ \rho\ \frac{A_0 l}{2} = 0.55\ \rho\ A_0 l, \tag{2.2.49}$$

where $\rho$ is the mass/unit volume. The mass matrix is diagonal, that is, $m_{ij} = 0$ for $i \neq j$. This diagonal structure is one of the features of the lumped matrix. Using a transformation from this discrete set of coordinates to another set of two discrete generalized coordinates

$$u(x, t) = a_1 \sin \frac{\pi x}{2L} + a_2 \sin \frac{3\pi x}{2L}, \tag{2.2.50}$$

it is possible to reduce the size of the mass matrix in a manner similar to the reduction of the stiffness matrix in a previous example. Substituting the $x$-location of the $u_i$ coordinates into Eq. (2.2.50) one has

$$\{u\} = \begin{Bmatrix} u_1 \\ u_2 \\ u_3 \\ u_4 \\ u_5 \end{Bmatrix} = \begin{bmatrix} 0.309 & 0.809 \\ 0.587 & 0.951 \\ 0.809 & 0.309 \\ 0.951 & -0.588 \\ 1.000 & -1.000 \end{bmatrix} \begin{Bmatrix} a_1 \\ a_2 \end{Bmatrix} = [T]\{a\} \tag{2.2.51}$$

and the kinetic energy is given by

$$T_k = \frac{1}{2}\{\dot{u}\}^T[m]\{\dot{u}\} = \frac{1}{2}\{\dot{a}\}^T[T]^T[m][T]\{\dot{a}\} = \frac{1}{2}\{\dot{a}\}^T[\bar{m}]\{\dot{a}\},$$

where

$$[\bar{m}] = [T]^T[m][T] = \rho l A_0 \begin{bmatrix} 2.493 & -0.051 \\ -0.051 & 2.605 \end{bmatrix}, \tag{2.2.52}$$

where $[\bar{m}]$ is the reduced generalized mass matrix.

## D'Alembert's Principle

The principle has been briefly mentioned in Section 2.1.1. Starting with Newton's second law for a collection of particles, as shown in Figure 2.21, the force on the $i$th particle can be written as

$$\mathbf{F}_i = m_i \mathbf{a}_i^I = m_i\left(\ddot{\mathbf{R}}_0 + \ddot{\mathbf{r}}_{i_{xyz}}\right), \tag{2.2.53}$$

where the acceleration vector in the inertial reference frame $(X, Y, Z)$ is, for a system that is translating and rotating is given by elementary dynamics by

$$\ddot{\mathbf{R}}_i \equiv \mathbf{a}_i^I \equiv \ddot{\mathbf{R}}_0 + \ddot{\mathbf{r}}_{i_{xyz}} + \dot{\boldsymbol{\omega}} \times \mathbf{r}_{i_{xyz}} + 2\boldsymbol{\omega} \times \dot{\mathbf{r}}_{i_{xyz}} + \boldsymbol{\omega} \times (\boldsymbol{\omega} \times \mathbf{r}_{xyz})_i \tag{2.2.54}$$

i.e.

$$\mathbf{F}_i = m_i\left\{a_0 + \mathbf{a}_{i_{rel}} + \dot{\boldsymbol{\omega}} \times \mathbf{r}_i + 2\boldsymbol{\omega} \times \dot{\mathbf{r}}_i + \boldsymbol{\omega} \times (\boldsymbol{\omega} \times \mathbf{r}_i)\right\} \tag{2.2.55}$$

Write Eq. (2.2.55) as

$$\mathbf{F}_i - m_i \mathbf{a}_i^I = 0 \tag{2.2.56}$$

**Figure 2.21**
Geometry of the problem for obtaining D'Alembert's principle for a rigid body represented as a collection of particles

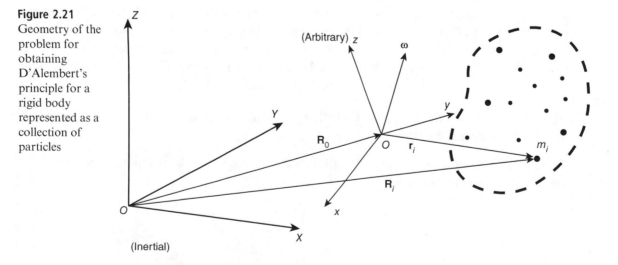

or simply

$$\mathbf{F}_i + \mathbf{F}_i^I = 0,$$

where $\mathbf{F}_i^I = -m_i \mathbf{a}_i^I$ is interpreted as an inertial force.

Sum over all particles:

$$\sum_i \left( \mathbf{F}_i + \mathbf{F}_i^I \right) = 0. \tag{2.2.57}$$

Apply the principle of virtual work to Eq. (2.2.57) (note that the equation as written looks like an equation of equilibrium from statics),

$$\delta W = \sum_i \left( \mathbf{F}_i + \mathbf{F}_i^I \right) \cdot \delta \mathbf{r}_i = 0. \tag{2.2.58}$$

In general, when constraint or interaction forces are present, one writes

$$\mathbf{F}_i = \mathbf{F}_i^A + \mathbf{f}_i, \tag{2.2.59}$$

where $\mathbf{F}_i^A$ are applied forces and $\mathbf{f}_i$ are constraint forces.

Substituting Eq. (2.2.59) into Eq. (2.2.58) yields

$$\delta W = \sum_i \left( \mathbf{F}_i^A + \mathbf{F}_i^I \right) \cdot \delta \mathbf{r}_i + \sum_i \mathbf{f}_i \cdot \delta \mathbf{r}_i = 0. \tag{2.2.60}$$

If one assumes a class of systems in which the constraints forces do no work, the last term in Eq. (2.2.60) is zero, and

$$\sum_i \left( \mathbf{F}_i^A - m_i \mathbf{a}_i^I \right) \cdot \delta \mathbf{r}_i = 0. \tag{2.2.61}$$

Equation (2.2.61) is D'Alembert's Principle, although some authors do not distinguish between Eqs. (2.2.58) and (2.2.61) in this respect. In its present form, the principle is not very useful for the purpose of deriving equations of motion, because the individual $\delta \mathbf{r}_i$s are not independent. The coefficients in Eq. (2.2.61) need not vanish independently of each other. D'Alembert's Principle is more useful if one finds a set of generalized coordinates and rewrites Eq. (2.2.61) in terms of these linearly independent coordinates. The results of such an operation are Lagrange's equations, which are treated in detail in Section 2.8.

Before doing so, an example where D'Alembert's Principle in its simplest form can be used to derive the equations of motions will be considered.

## Example: Direct Formulation of Equations of Motion for a Simple Structural Dynamic System

The equations of motion for any structural dynamic system can be formulated directly by applying Newton's second law

$$\mathbf{F}(t) = \frac{d}{dt} \left( m \frac{d\mathbf{r}}{dt} \right), \tag{2.2.62}$$

where

$\mathbf{F}(t)$ = applied force vector

$m$ = mass of the particle or the system

$\mathbf{r}$ = position vector of $m$ in an inertial reference system

for $m$ = constant Eq. (2.2.62) reduces to

$$\mathbf{F}(t) = m\frac{d^2\mathbf{r}}{dt^2} = m\ddot{\mathbf{r}}, \qquad (2.2.63)$$

where it is understood that $\ddot{\mathbf{r}}$ represents acceleration of the mass in an inertial reference or coordinate system.

Equation (2.2.63) can also be written as

$$\mathbf{F}(t) - m\ddot{\mathbf{r}}(t) = \mathbf{0}, \qquad (2.2.64)$$

where the second term in Eq. (2.2.64) is usually called an inertia force.

As mentioned earlier, D'Alembert's Principle states that inertial loads can be considered as regular quasi-static forces acting on the system, provided that the direction of the inertial force is taken as opposite of the corresponding acceleration vector. By using D'Alembert's principle one can conveniently write the equations of motion as equations of dynamic equilibrium.

Next, Newton's second law, Eq. (2.2.64), and D'Alembert's principle are combined to illustrate the direct formulation of the equations motion for a simple structural dynamic system shown below:

The system consists of a uniform beam having mass per unit length $m$ and stiffness $EI$. Three discrete, and fairly large, masses $M_1, M_2$, and $M_3$ are attached to the beam, which implies $M_1, M_2, M_3 \gg m\ell$. The motion of these discrete masses is represented by three discrete coordinates $u_1, u_2$, and $u_3$ as shown in Figure 2.22. For motion in the $u_1$ and $u_3$ degrees of freedom, damping is provided by two viscous type dampers shown, with damping coefficients $C_1$ and $C_2$, respectively.

**Figure 2.22** Example for a simple structural dynamic system

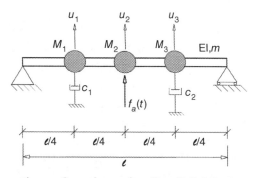

In formulating the equations of motion using Eq. (2.2.64) one recognizes that the force $\mathbf{F}(t)$ represents various types of forces acting on the system such as elastic forces, which are restraining forces opposing displacements, viscous damping forces which resist velocities and finally one has applied external loads. Thus,

$$\mathbf{F}(t) = \mathbf{F}_A(t) - \mathbf{F}_D(t) - \mathbf{F}_S(t), \qquad (2.2.65)$$

where the subscripts, $A$, $D$, and $S$ indicate applied, damping, and structural forces, respectively. Since there are three discrete coordinates, Eq. (2.2.65) can be rewritten as

$$\{F(t)\} = \{\mathbf{F}_A(t)\} - \{\mathbf{F}_D(t)\} - \{\mathbf{F}_S(t)\}. \tag{2.2.66}$$

Furthermore, since $M_1, M_2, M_3 >> ml$, the distributed mass of the beam yields negligible inertia forces. From the figure

$$\{\mathbf{F}_A(t)\} = \left\{ \begin{array}{c} 0 \\ f_A(t) \\ 0 \end{array} \right\} \tag{2.2.67}$$

$$\{\mathbf{F}_D(t)\} = \left[ \begin{array}{ccc} C_1 & 0 & 0 \\ 0 & 0 & 0 \\ 0 & 0 & C_2 \end{array} \right] \left\{ \begin{array}{c} \dot{u}_1 \\ \dot{u}_2 \\ \dot{u}_3 \end{array} \right\} = [C]\{\dot{u}\}. \tag{2.2.68}$$

Assuming that the stiffness properties of the beam are given by a stiffness matrix, proportional to $EI$, one has

$$\{\mathbf{F}_S(t)\} = EI \left[ \begin{array}{ccc} k_{11} & k_{12} & k_{13} \\ k_{12} & k_{22} & k_{23} \\ k_{13} & k_{23} & k_{33} \end{array} \right] \left\{ \begin{array}{c} u_1 \\ u_2 \\ u_3 \end{array} \right\} = [K] \left\{ \begin{array}{c} u_1 \\ u_2 \\ u_3 \end{array} \right\} \tag{2.2.69}$$

and the inertia loads are given by

$$\{\mathbf{F}_I(t)\} = - \left[ \begin{array}{ccc} M_1 & & \\ & M_2 & \\ & & M_3 \end{array} \right] \left\{ \begin{array}{c} \ddot{u}_1 \\ \ddot{u}_2 \\ \ddot{u}_3 \end{array} \right\} = -[M]\{\ddot{u}\}. \tag{2.2.70}$$

For dynamic equilibrium,

$$\{F(t)\} + \{F_I(t)\} = 0. \tag{2.2.71}$$

Combining Eqs. (2.2.65) – (2.2.71) yields

$$[M]\{\ddot{u}\} + [C]\{\dot{u}\} + [K]\{u\} = \{\mathbf{F}_A(t)\}. \tag{2.2.72}$$

For a relatively simple system, such as the one used in this example, this direct method is easy to apply and almost intuitively obvious. However, for more complicated systems the direct method ceases to be convenient. Therefore, one has to use more systematic methods for formulating dynamic equations of equilibrium.

Two of the most useful methods of this type are based upon Hamilton's principle and Lagrange's equations. In the following sections some background material required for the proper understanding of Hamilton's principle and Lagrange's equations will be presented. This background material consists of

(1) review of elementary concepts of elasticity and definition of strain energy;
(2) brief review of the calculus of variations;
(3) principle of virtual work static case for an elastic body; and
(4) principle of minimum potential energy.

## 2.3    Review of Elasticity and Energy Concepts

Recall from elementary courses in structural mechanics that when dealing with an infinitesimal rectangular parallelepiped with faces parallel to the coordinate planes of $x_1, x_2,$ and $x_3$ one encounters nine force-intensity components acting on the six faces of the element; these are nine components of the stress tensor

$$\sigma_{ij} = \begin{bmatrix} \sigma_{11} & \sigma_{12} & \sigma_{13} \\ \sigma_{21} & \sigma_{22} & \sigma_{23} \\ \sigma_{31} & \sigma_{32} & \sigma_{33} \end{bmatrix}, \tag{2.3.1}$$

where the first index represents the normal of the area element and the second subscript gives the direction of the force intensity itself. From elementary considerations it can be shown that the stress tensor is symmetric, that is,

$$\sigma_{ij} = \sigma_{ji} \quad \text{[results from moment equilibrium]}. \tag{2.3.2}$$

Isolating such a three-dimensional element (infinitesimal) of the continuum and considering its equilibrium in the $1, 2, 3$ – directions, respectively, results in the equilibrium equations of the theory of elasticity:

$$\sigma_{ij,j} + f_i = 0 \quad i = 1, 2, 3, \tag{2.3.3}$$

where the components of $f_i$ represent a body force. In expanded form, Eq. (2.3.3) is written as

$$\frac{\partial \sigma_{11}}{\partial x_1} + \frac{\partial \sigma_{12}}{\partial x_2} + \frac{\partial \sigma_{13}}{\partial x_3} + f_1 = 0 \tag{2.3.4}$$

$$\frac{\partial \sigma_{21}}{\partial x_1} + \frac{\partial \sigma_{22}}{\partial x_2} + \frac{\partial \sigma_{23}}{\partial x_3} + f_2 = 0$$

$$\frac{\partial \sigma_{31}}{\partial x_1} + \frac{\partial \sigma_{32}}{\partial x_2} + \frac{\partial \sigma_{33}}{\partial x_3} + f_3 = 0.$$

In tensor notation, repeated indices imply summation.

In addition to the internal element of the continuum one can also isolate an element, which represents a boundary-type element, shown in Figure 2.23, where the sides $BPC, APC,$ and $ABP$ are internal, and the side $ABC$ coincides approximately with the boundary of the elastic continuum on which a surface traction vector $\overset{\rightarrow(n)}{T}$ is acting. The superscript $n$ indicates the outward pointing unit vector of the boundary surface $ABC$.

Consideration of the equilibrium of a boundary-type infinitesimal element yields the boundary conditions for the problem.

$$T_i = \sigma_{ij} n_j \quad (i = 1, 2, 3). \tag{2.3.5}$$

This equation relates the tractions on the boundary to the stresses directly next to the boundary. The $n_j$s are the direction cosines of the unit normal of the boundary on which the traction is required.

**Figure 2.23** Boundary element illustrating the relations between surface tractions and internal stress components

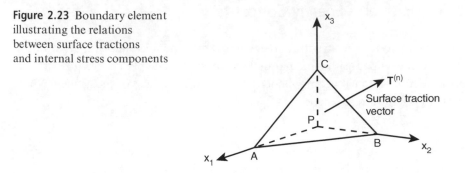

In addition to the stresses, one has to deal also with quantities which describe the deformation of the elastic continuum. As shown in elementary courses in structural mechanics the deformation of the continuum (structure) with respect to the original (**undeformed**) coordinate system can be represented by the components of the strain tensor $\epsilon_{ij}$ (Lagrangian strain tensor) given by

$$\epsilon_{ij} = \frac{1}{2}\left(u_{i,j} + u_{j,i} + u_{k,i}u_{k,j}\right). \quad (i,j,k = 1,2,3) \tag{2.3.6}$$

In the last expression, commas indicate derivatives with respect to the spatial variable. It is useful to write two components so as to illustrate the meaning of this expression. Again note that repeated indices imply summation.

$$\epsilon_{11} = \frac{1}{2}\left(u_{1,1} + u_{1,1} + u_{1,1}u_{1,1} + u_{2,1}u_{2,1} + u_{3,1}u_{3,1}\right) \tag{2.3.7}$$

$$= u_{1,1} + \frac{1}{2}\left(u_{1,1}u_{1,1} + u_{2,1}u_{2,1} + u_{3,1}u_{3,1}\right).$$

Similarly,

$$\epsilon_{12} = \frac{1}{2}\left(u_{1,2} + u_{2,1} + u_{1,1}u_{1,2} + u_{2,1}u_{2,2} + u_{3,1}u_{3,2}\right) \tag{2.3.8}$$

$$= \frac{1}{2}(u_{1,2} + u_{2,1}) + \frac{1}{2}\left(u_{1,1}u_{1,2} + u_{2,1}u_{2,2} + u_{3,1}u_{3,2}\right);$$

by inspection, it is clear that the strain components are also symmetric, thus the strain tensor

$$\epsilon_{ij} = \begin{bmatrix} \epsilon_{11} & \epsilon_{12} & \epsilon_{13} \\ \epsilon_{21} & \epsilon_{22} & \epsilon_{23} \\ \epsilon_{31} & \epsilon_{32} & \epsilon_{33} \end{bmatrix} \tag{2.3.9}$$

is also a symmetric tensor.

At this point, it should be noted that the $\epsilon_{ii}$ components of the strain represent the axial or normal strains, while the mixed components $\epsilon_{ij}$, where $i \neq j$ represent the shear strains, represent the change in angle between two originally perpendicular lines in the undeformed structure.

When dealing with small strains and small displacements it is common practice to neglect the nonlinear terms (nonlinear in terms of the components of the displacement vector), thus Eqs. (2.3.7) and (2.3.8) reduce to

$$\epsilon_{ij} = \frac{1}{2}\left(u_{i,j} + u_{j,i}\right) \quad (j, i = 1, 2, 3) \,. \tag{2.3.10}$$

To complete the formulation of the simplest (i.e. linear) elasticity problem, constitutive laws relating strains to stresses are also required. For the case of a linear elastic isotropic solid, the generalized version of Hooke's law is used. This can be written as

$$\epsilon_{11} = \frac{1}{E}\left[\sigma_{11} - \nu\left(\sigma_{33} + \sigma_{22}\right)\right]$$

$$\epsilon_{22} = \frac{1}{E}\left[\sigma_{22} - \nu\left(\sigma_{11} + \sigma_{33}\right)\right] \tag{2.3.11}$$

$$\epsilon_{33} = \frac{1}{E}\left[\sigma_{33} - \nu\left(\sigma_{11} + \sigma_{22}\right)\right]$$

$$\epsilon_{12} = \frac{1}{2G}\sigma_{12}; \ \epsilon_{13} = \frac{1}{2G}\sigma_{13}; \ \text{and} \ \epsilon_{23} = \frac{1}{2G}\sigma_{23} \tag{2.3.12}$$

or in the general tensor form

$$\epsilon_{ij} = \frac{1}{E}\left[(1 + \nu)\sigma_{ij} - \nu\delta_{ij}\sigma_{kk}\right] \quad (i, j, k = 1, 2, 3), \tag{2.3.13}$$

where $\nu$ is Poisson's ratio and $\delta_{ij}$ is the Kroenecker delta

$$\delta_{ij} = \begin{cases} 0 & i \neq j \\ 1 & i = j. \end{cases} \tag{2.3.14}$$

For some applications, it is convenient to express the stress components in terms of the strain components, then

$$\sigma_{ij} = 2\mu\epsilon_{ij} + \lambda\delta_{ij}\epsilon_{kk} \quad (i, j, k = 1, 2, 3) \tag{2.3.15}$$

and

$$\mu = \frac{E}{2(1 + \nu)} = G; \ \lambda = \frac{E\nu}{(1 + \nu)(1 - 2\nu)}. \tag{2.3.16}$$

$G$ is the shear modulus and $\lambda$ is Lame's constant.

## Strain Energy

In this brief review, some basic concepts associated with a structure undergoing elastic deformation are discussed.

For simplicity it will be assumed that the elastic continuum is initially unstressed. We shall also assume that the rate of loading is *slow*, thus there is no heat dissipation associated with the process of deformation. Furthermore it will be assumed that a state of equilibrium is maintained during the loading process. In this case the change in kinetic energy is zero, and the work of external forces $W_E$ is equal to the change in internal energy, due to the deformation $U$, that is,

$$W_E = U \tag{2.3.17}$$

Equation (2.3.17) represents the law of conservation of energy, that is, the mechanical work done by the applied loads is equal to the change in internal energy. This stored energy is called *strain energy*. Consider an element of $dV$ of the structure that is subject to a single stress component $\sigma_{11}$. In the process of deformation the normal stress $\sigma_{11}$ and strain increase from zero (initially unstressed) to their final values $\sigma_{11}$ and $\epsilon_{11}$. During an increment of strain $\epsilon_{11}$ the work done is (where $U_0$ is called the strain energy density) as illustrated in Figure 2.24.

**Figure 2.24**
Strain energy when a single component of axial strain is acting

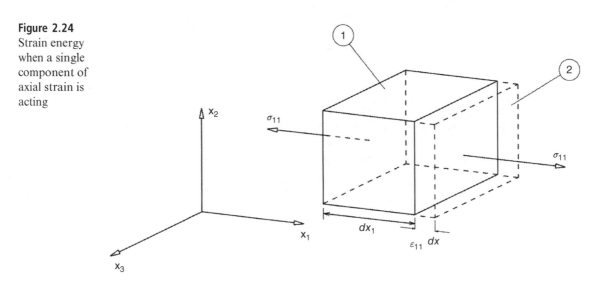

$$\text{work} = \underbrace{\sigma_{11}dx_2dx_3}_{\text{force}} \cdot \underbrace{d\epsilon_{11}dx_1}_{\text{displacement}} = dU_0dV \qquad (2.3.18)$$

$$U_0dV = \int_0^{\epsilon_{11}} \sigma_{11}d\epsilon_{11}dx_1dx_2dx_3 = \int_0^{\epsilon_{11}} \sigma_{11}d\epsilon_{11}dV$$

and the strain energy $U$ is given by

$$U = \int_V \left( \int_0^{\epsilon_{11}} \sigma_{11}d\epsilon_{11} \right) dV = \int_V U_0dV. \qquad (2.3.19)$$

One can also go through a more accurate analysis by considering first the first three axial components of the stress and strain.

Considering faces 1 and 2 as shown, where the displacement of face 1 in the $x_1$ direction is given by $u_1$ and the displacement of face 2 in the $x_1$ direction is given by $u_1 + \frac{\partial u_1}{\partial x_1}dx_1$; thus the increment in mechanical work done by the stresses on the element during a deformation is given by

$$-\sigma_{11} du_1 dx_2 dx_3 + \left(\sigma_{11} + \frac{\partial \sigma_{11}}{\partial x_1} dx_1\right) d\left(u_1 + \frac{\partial u_1}{\partial x} dx_1\right) dx_2 dx_3$$

$$+ f_1 dx_1 dy_1 dz_1 d\left(u_1 + \kappa \frac{\partial u_1}{\partial x_1} dx_1\right) \cong \qquad (2.3.20)$$

$$\left\{\sigma_{11} d\left(\frac{\partial u_1}{\partial x_1}\right) + \left(\frac{\partial \sigma_{11}}{\partial x_1} + f_1\right) du_1\right\} dx_1 dx_2 dx_3,$$

where higher-order terms (H.O.T.) in Eq. (2.3.20) were neglected and $0 < \kappa < 1.0$ is a constant associated body forces and their variation. The last term in Eq. (2.3.20) is zero by virtue of the equations of equilibrium. Thus,

$$\sigma_{11} d\left(\frac{\partial v_1}{\partial x_1}\right) dx_1 dx_2 dx_3 = \sigma_{11} d\epsilon_{11} dV. \qquad (2.3.21)$$

Thus for the normal stresses on an element, the increment of mechanical work is given by

$$(dU_0)_{\text{normal}} dV = (\sigma_{11} d\epsilon_{11} + \sigma_{22} d\epsilon_{22} + \sigma_{33} d\epsilon_{33}) dV. \qquad (2.3.22)$$

Equation (2.3.22) is a direct application of the principle of superposition for a linear system.

Next consider the case of pure shear strain as shown in Figure 2.25:

$$\gamma_{12} = 2\epsilon_{12} = \frac{\partial u_1}{\partial x_2} + \frac{\partial u_2}{\partial x_1}, \qquad (2.3.23)$$

where $\gamma_{ij}$ gives the change in the right angle of vanishingly small line segments originally in the **i** and **j** directions at a point.

**Figure 2.25**
Geometry of problem for calculating strain energy during shear deformation only

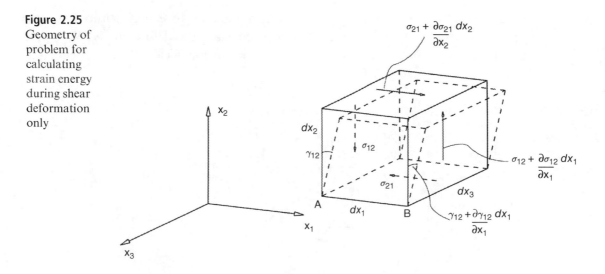

For the problem shown, the mechanical increment of work can be given as follows:

$$\left[\left(\sigma_{21} + \frac{\partial \sigma_{21}}{\partial x_2}dx_2\right)dx_1dx_3\right]dx_2\left[\gamma_{12} + \beta\left(\frac{\partial \gamma_{12}}{\partial x_1}\right)dx_1\right]$$

$$+f_1dx_1dx_2dx_3d\left(\gamma_{12} + \eta\,\frac{\partial \gamma_{12}}{\partial x_1}dx_1\right)(\kappa\,dx_2) \cong \qquad (2.3.24)$$

$$= \sigma_{12}d\gamma_{12}dx_1dx_2dx_3 = 2\sigma_{12}d\epsilon_{12}dV,$$

where again $\beta, \eta, \kappa$ are between 0 and 1 and H.O.T. in Eq. (2.3.24) are neglected.

Using pure shear on all faces, one has

$$(dU_0)_{shear}\,dV = 2\left(\sigma_{12}d\epsilon_{12} + \sigma_{13}d\epsilon_{13} + \sigma_{23}d\epsilon_{23}\right)dV. \qquad (2.3.25)$$

Finally, combining Eqs. (2.3.22) and (2.3.24), one has

$$U_0 = \int_0^{\epsilon_{11}} \sigma_{11}d\epsilon_{11} + \int_0^{\epsilon_{22}} \sigma_{22}d\epsilon_{22} + \int_0^{\epsilon_{33}} \sigma_{33}d\epsilon_{33} \qquad (2.3.26)$$

$$+ 2\left[\int_0^{\epsilon_{12}} \sigma_{12}d\epsilon_{12} + \int_0^{\epsilon_{13}} \sigma_{13}d\epsilon_{13} + \int_0^{\epsilon_{23}} \sigma_{23}d\epsilon_{23}\right] = \int_0^{\epsilon_{ij}} \sigma_{ij}d\epsilon_{ij}$$

and

$$U = \int_V U_0 dV. \qquad (2.3.27)$$

Furthermore it is important to note that in order to have a **strain energy density function** $U_o$ must be a point function, that is, the integral must be independent of the path which means that $\sigma_{ij}d\epsilon_{ij}$ is a perfect differential for this case also

$$\frac{\partial U_0}{\partial \epsilon_{ij}} = \sigma_{ij}. \qquad (2.3.28)$$

This can be shown from thermodynamic considerations. Strain energy density functions are **positive-definite** functions of the strains associated with small deformations. With these basic relations established, it is now possible to represent graphically the meaning of Eq. (2.3.19), as shown in Figure 2.26.

**Figure 2.26** Illustration of strain energy and complementary strain energy for the case of a single axial strain component

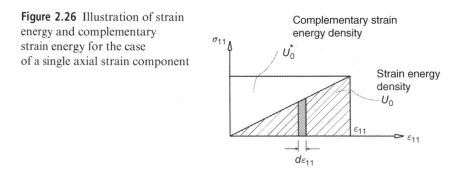

Using Hooke's law, the strain energy density function can be expressed in terms of the strain or the stress components by combining Eqs. (2.3.26) and (2.3.15).

$$U_0 = \int_0^{\epsilon_{ij}} \sigma_{ij} d\epsilon_{ij} = \int_0^{\epsilon_{ij}} \left(2\mu\epsilon_{ij} + \lambda\delta_{ij}\epsilon_{kk}\right) d\epsilon_{ij}$$
$$= \left(\mu\epsilon_{ij}\epsilon_{ij} + \frac{\lambda}{2}\epsilon_{kk}^2\right) \tag{2.3.29}$$

or when expressed in terms of stress components

$$U_0 = \int_V \left(\frac{1}{4}\mu\sigma_{ij}\sigma_{ij} - \frac{\lambda}{4\mu\left(2\mu + 3\lambda\right)}\sigma_{kk}^2\right) dV. \tag{2.3.30}$$

## Examples of Strain Energy for Simple Structural Members

The elastic strain energy is given by

$$U = \int_V \int_0^{\epsilon_{ij}} \sigma_{ij} d\epsilon_{ij} dV. \tag{2.3.31}$$

The strain energy expressions for three simple structural elements are given below.

### Axial Deformation of a Bar

For this case, only one component of stress and strain exist, thus

$$\sigma_{11} = \frac{P}{A} = E\epsilon_{11} = E\frac{du_1}{dx} \tag{2.3.32}$$

as shown in Figure 2.27.

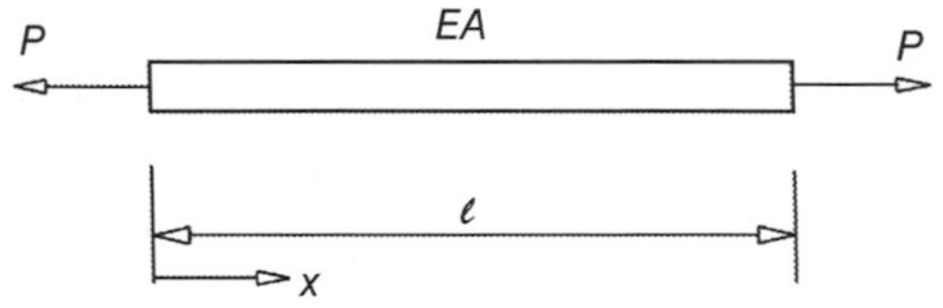

**Figure 2.27** Geometry for calculating strain energy for a axially loaded bar

From Eq. (2.3.31),

$$U = \int_0^l \int_0^{\epsilon_{11}} \sigma_{11} d\epsilon_{11} A dx = \int_0^l E\frac{\epsilon_{11}^2}{2} A dx = \int_0^l \frac{\sigma_{11}\epsilon_{11}}{2} A dx$$
$$= \int_0^l \frac{\sigma_{11}^2}{2E} A dx. \tag{2.3.33}$$

Combining Eqs. (2.3.32) and (2.3.33),

$$U = \frac{1}{2}\int_0^l A\frac{P}{A}\frac{P}{AE} dx = \frac{1}{2}\frac{P^2 L}{AE}. \tag{2.3.34}$$

### Flexural Member (Beam Bending)

Consider next the bending of a beam in absence of axial loads, about one of its principal axes, $x_3$- or $z$-axis, as shown in Figure 2.28. Clearly this implies symmetry of the cross section about the $x_3$- or $z$-axis. Bending of the beam is due to a distributed load $p_y(x)$ as well as concentrated loads $P_i$ acting in the $x-y$ or $x_1-x_2$ plane. The bending moment and shear force acting at any arbitrary cross section $x$ (or $x_1$) are denoted by $M = M(x_1)$ and $V = V(x_1)$ respectively, as shown in Figure 2.28. From beam theory based on the Euler–Bernoulli assumptions – that cross sections remain plane and perpendicular to the deformed neutral axis of the beam

$$\sigma_{11} = -\frac{Mx_2}{I} \qquad \sigma_{12} = \frac{VQ}{Ib} \tag{2.3.35}$$

and all other stress components $\sigma_{ij}$ are assumed to be zero, that is, $\sigma_{ij} = 0$, where $I = \int_A x_2^2 dA$ and $Q$ is the first moment of the area A* about the $x_3$ (or $z$) axis. Equation (2.3.31) in terms of the stress components becomes

$$U = \int_V \left( \frac{1}{2E}\sigma_{11}^2 + \frac{1}{2G}\sigma_{12}^2 \right) dV$$
$$= \int_V \left( \frac{1}{2E} \frac{M^2 x_2^2}{I^2} + \frac{1}{2G} \frac{V^2 Q^2}{I^2 b^2} \right) dV. \tag{2.3.36}$$

Since $M$ and $V$ are functions of $x_1$ only, while $Q$ is a function of $x_2$ only, Eq. (2.3.36) may be simplified to

$$U = \int_0^l \frac{M^2 dx_1}{2EI} + \int_0^l \frac{V^2 dx_1}{2GI^2} \int_A \frac{Q^2}{b^2} dA. \tag{2.3.37}$$

**Figure 2.28**
Geometry for calculating strain energy for a beam in bending

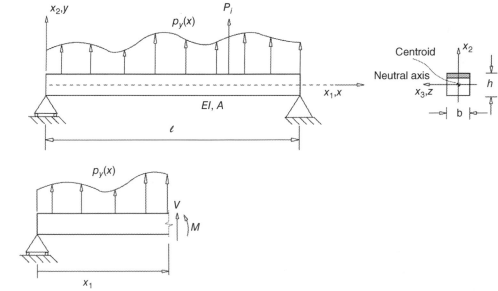

The first term in Eq. (2.3.37) represents strain energy due to bending, while the second term represents strain energy due to shear. For typical beam geometries (slender beams) and loading conditions the strain energy due to shear is negligible when compared to the strain energy due to bending. Therefore in most structural dynamics applications of Eq. (2.3.37) the strain energy due to shear is neglected. A convenient, alternate expression for the strain energy associated with bending can be obtained by using the moment curvature expression of beam theory

$$\frac{M}{EI} = \frac{d^2v}{dx_1^2},$$

(2.3.38)

where $v$ is the displacement of the point on the neutral axis in the $x_2$ direction. Combining Eqs. (2.3.37) and (2.3.38) yields

$$U = \int_0^l \frac{EI}{2}\left(\frac{d^2v}{dx_1^2}\right)^2 dx_1.$$

(2.3.39)

## Torsional Member

Consider next the strain energy associated with a twisted circular bar, and using a polar coordinate system, $(\varphi, r)$ the only nonzero stress and strain components are $\sigma_{\varphi x}$ and $\gamma_{\varphi x}$ as shown in figure 2.29.

**Figure 2.29** Geometry for calculating the strain energy in torsion of a bar with circular cross section

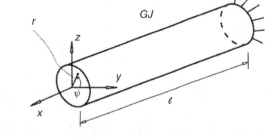

From simple considerations it can be shown that

$$\gamma_{\varphi x} = r\theta = r\frac{d\varphi}{dx}$$

(2.3.40)

and all other strain components are zero, and $\theta$ is the twisting angle per unit length.

$$\sigma_{\varphi x} = G\gamma_{\varphi x}.$$

(2.3.41)

The strain energy is given by

$$U = \frac{1}{2}\int_V \sigma_{\varphi x}\gamma_{\varphi x}dV = \int_0^l \frac{G}{2}\int_A \gamma_{\varphi x}^2 dxdA$$
$$= \int_0^l \int_A \frac{G}{2}r^2\left(\frac{d\varphi}{dx}\right)^2 dxdA = \frac{GJ}{2}\int_0^l \left(\frac{d\varphi}{dx}\right)^2 dx,$$

(2.3.42)

where $J = \int_A r^2 dA$ = polar area moment of inertia of the cross section. The torsional moment can also be written as

$$M_T = GJ\frac{d\varphi}{dx}.$$

(2.3.43)

Combining Eqs. (2.3.42) and (2.3.43) yields the alternate expression

$$U = \frac{1}{2GJ} \int_0^l M_T^2 dx.$$

(2.3.44)

## 2.4    Brief Review of the Calculus of Variations

In structural mechanics, structural stability, and structural dynamics extensive use is made of the principle of virtual work, principle of minimum potential energy, and Hamilton's principle. These are essentially variational principles and therefore it is both useful and important to develop some familiarity with basic concepts of the calculus of variations.

The basic problem of the calculus of variations is to find, among a set of admissible functions (these are functions which satisfy certain boundary conditions), a function that leads to the *stationary condition* of a certain functional (i.e. a function of functions). Consider the integral $I(u)$ for which the stationary value is sought

$$I(u) = \int_a^b \underbrace{F(u, u_x, x)\, dx}_{\text{functional}},$$

(2.4.1)

where $u(x)$ is the *argument function*. Furthermore $u(x)$ must satisfy the conditions

$$u(x = a) = u_a \quad \text{and} \quad u(x = b) = u_b,$$

(2.4.2)

which are prescribed end conditions. Note that $F(u, u_x, x)$ is given and $u(x)$ is sought. For a given $u(x)$, $I(u)$ yields a numerical value. The problem is to determine the particular function which minimizes $I(u)$.

Consider the geometry of the problem shown in Figure 2.30 and let us assume that $u(x)$ is the actual unknown minimizing curve. Consider the behavior of $I(u)$ when $u(x)$ is replaced by *a slightly* different curve $\bar{u}(x)$, where $\bar{u}(x)$ is called a comparison function defined by the following relation:

**Figure 2.30** Illustration of the argument and comparison functions

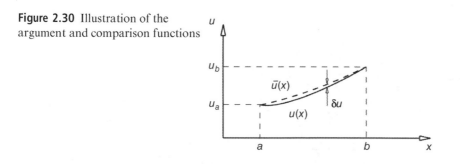

$$\bar{u}(x) = u(x) + \epsilon\eta(x), \tag{2.4.3}$$

where $\epsilon$ *is a small parameter* and $\eta(x)$ is an arbitrary function, which must satisfy the conditions

$$\eta(a) = \eta(b) = 0. \tag{2.4.4}$$

The quantity $\delta u$ is the *variation in $u(x)$* and is defined by

$$\delta u = \bar{u}(x) - u(x) = \epsilon\eta(x). \tag{2.4.5}$$

Denote the difference in slope of $u(x)$ and $\bar{u}(x)$ by $\delta u_x$ and call it *variation in slope*, thus,

$$\delta\left(\frac{du}{dx}\right) = \frac{d}{dx}(\bar{u}(x)) - \frac{d}{dx}(u(x)). \tag{2.4.6}$$

Differentiation of Eq. (2.4.5) yields

$$\frac{d}{dx}(\delta u) = \epsilon\eta_x(x) = \bar{u}_x(x) - u_x(x). \tag{2.4.7}$$

Comparing Eqs. (2.4.6) and (2.4.7),

$$\frac{d}{dx}(\delta u) = \delta\left(\frac{du}{dx}\right). \tag{2.4.8}$$

Hence the process of *variation and differentiation is permutable.*

Consider the behavior of $F(u, u_x, x)$ in the neighborhood of $u(x)$. For a fixed value of $x$, $F$ is a function of $u_x, u$, thus

$$\Delta F = F(\bar{u}, \bar{u}_x, x) - F(u, u_x, x) = F(u + \delta u, u_x + \delta u_x, x) -$$
$$- F(u, u_x, x). \tag{2.4.9}$$

Expanding $F(u + \delta u, u_x + \delta u_x, x)$ into Taylor Series

$$F(u + \delta u, u_x + \delta u_x, x) = F(u, u_x, x) + \frac{\partial F}{\partial u}\delta u + \frac{\partial F}{\partial u_x}\delta u_x +$$
$$+ \frac{1}{2!}\left(\frac{\partial^2 F}{\partial u^2}\delta u^2 + 2\frac{\partial^2 F}{\partial u\partial u_x}\delta u\delta u_x + \frac{\partial^2 F}{\partial u_x^2}\delta u_x^2\right) + \cdots \tag{2.4.10}$$

or combining Eqs. (2.4.9) and (2.4.10)

$$\Delta F = \delta F + \frac{1}{2!}\delta^2 F + \cdots, \tag{2.4.11}$$

where

$$\delta F \equiv \text{first variation of } F = \frac{\partial F}{\partial u}\delta u + \frac{\partial F}{\partial u_x}\delta u_x \tag{2.4.12}$$

$$\delta^2 F = \delta(\delta F) = \frac{\partial^2 F}{\partial u^2}\delta u^2 + 2\frac{\partial^2 F}{\partial u\partial u_x}\delta u\delta u_x + \frac{\partial^2 F}{\partial u_x^2}\delta u_x^2 \tag{2.4.13}$$

$$\equiv \text{ second variation of } F$$

therefore,

$$\Delta I = I(\bar{u}, \bar{u}_x, x) - I(u, u_x, x)$$
$$= \int_a^b F(\bar{u}, \bar{u}_x, x) - \int_a^b F(u, u_x, x)dx$$
$$= \int_a^b \Delta F dx. \tag{2.4.14}$$

Combining Eqs. (2.4.13) and (2.4.14) yields

$$\Delta I = \int_a^b (\delta F + \frac{1}{2!}\delta^2 F + \cdots)dx. \tag{2.4.15}$$

The first and second variations of the integral are defined respectively as

$$\delta I = \int_a^b \delta F dx \tag{2.4.16}$$

$$\delta^2 I = \int_a^b \delta^2 F dx. \tag{2.4.17}$$

From Eqs. (2.4.15) to (2.4.17),

$$\Delta I = \delta I + \frac{1}{2!}\delta^2 I + \cdots \tag{2.4.18}$$

Based on our original hypothesis, $u(x)$ is the function which *minimizes* $I(u)$. It follows therefore that $\Delta I \geq 0$. Note that as $\epsilon \to 0$ when $\bar{u}(x) \to u(x)$ and $\Delta I \to 0$.

Combining Eqs. (2.4.7), (2.4.5), (2.4.16), and (2.4.17), one has

$$\delta I = \int_a^b \left(\frac{\partial F}{\partial u}\delta u + \frac{\partial F}{\partial u_x}\delta u_x\right) dx$$
$$= \epsilon \int_a^b \left(\frac{\partial F}{\partial u}\eta + \frac{\partial F}{\partial u_x}\eta_x\right) dx. \tag{2.4.19}$$

Similarly,

$$\delta^2 I = \int_a^b \left(\frac{\partial^2 F}{\partial u^2}\delta u^2 + 2\frac{\partial^2 F}{\partial u_x \partial u}\delta u \delta u_x + \frac{\partial^2 F}{\partial u_x^2}\delta u_x^2\right) dx$$
$$= \epsilon^2 \int_a^b \left(\frac{\partial^2 F}{\partial u^2}\eta^2 + 2\frac{\partial^2 F}{\partial u_x \partial u}\eta\eta_x + \frac{\partial^2 F}{\partial u_x^2}\delta\eta_x^2\right) dx. \tag{2.4.20}$$

Thus $\delta I$ is proportional to $\epsilon$ and $\delta^2 I$ is proportional to $\epsilon^2$. Thus, second- and higher-order variations in the series expansion for $\Delta I$ are negligible compared to $\delta I$ when $\epsilon$ is sufficiently small. Accordingly the necessary condition for $I$ to have a minimum is that $\delta I$ vanish identically. This is a necessary condition. The determination regarding the minimum can be made from $\delta^2 I$, but when only the stationary condition is sought this is unnecessary.

## 2.4.1 Euler–Lagrange Equations

From the condition for the stationary value $\delta I = 0$

$$\delta I = \int_a^b \left( \frac{\partial F}{\partial u}\delta u + \frac{\partial F}{\partial u_x}\delta u_x \right) dx = 0. \tag{2.4.21}$$

Note that $\delta u_x = \frac{d}{dx}(\delta u)$, thus the second term in Eq. (2.4.21) can be integrated by parts

$$\int_a^b \frac{\partial F}{\partial u_x}\delta u_x dx = \left[ \frac{\partial F}{\partial u_x}\delta u \right]_a^b - \int_a^b \frac{d}{dx}\left( \frac{\partial F}{\partial u_x} \right)\delta u\, dx \tag{2.4.22}$$

and

$$\delta I = \int_a^b \left[ \frac{\partial F}{\partial u} - \frac{d}{dx}\left( \frac{\partial F}{\partial u_x} \right) \right]\delta u\, dx + \left[ \frac{\partial F}{\partial u_x}\delta u \right]_a^b = 0. \tag{2.4.23}$$

Since the variation $\delta u = \bar{u}(x) - u(x) = 0$ at $x = a, b$, the last term in Eq. (2.4.23) vanishes. Thus

$$\delta I = \int_a^b \left[ \frac{\partial F}{\partial u} - \frac{d}{dx}\left( \frac{\partial F}{\partial u_x} \right) \right]\delta u\, dx = 0. \tag{2.4.24}$$

Furthermore, $\delta u$ is arbitrary between $a < x < b$, therefore the term in the square bracket in Eq. (2.4.24) must vanish, thus

$$\frac{\partial F}{\partial u} - \frac{d}{dx}\left( \frac{\partial F}{\partial u_x} \right) = 0. \tag{2.4.25}$$

This differential equation is the Euler–Lagrange equation. It represents a necessary *but not sufficient condition*, which $u(x)$ must satisfy if it is to yield a stationary value for $I(u)$. It is important to realize that the Euler–Lagrange equation can be generalized to the case when $F$ contains higher-order derivatives of $u$, for example, when

$$I = \int_a^b F(u, u_x, u_{xx}, x)dx \tag{2.4.26}$$

$$\delta F = \frac{\partial F}{\partial u}\delta u + \frac{\partial F}{\partial u_x}\delta u_x + \frac{\partial F}{\partial u_{xx}}\delta u_{xx} \tag{2.4.27}$$

so that

$$\delta I = 0 = \int_a^b \left( \frac{\partial F}{\partial u}\delta u + \frac{\partial F}{\partial u_x}\delta u_x + \frac{\partial F}{\partial u_{xx}}\delta u_{xx} \right) dx = 0 \tag{2.4.28}$$

and integration by parts yields.

$$\delta I = \int_a^b \left[ \frac{\partial F}{\partial u} - \frac{d}{dx}\left( \frac{\partial F}{\partial u_x} \right) + \frac{d^2}{dx^2}\left( \frac{\partial F}{\partial u_{xx}} \right) \right] dx\delta u$$
$$+ \left[ \left\{ \frac{\partial F}{\partial u_x} - \frac{d}{dx}\left( \frac{\partial F}{\partial u_{xx}} \right) \right\}\delta u \right]_a^b + \left[ \frac{\partial F}{\partial u_{xx}}\delta u_x \right]_a^b = 0. \tag{2.4.29}$$

If $u$ satisfies the conditions that $u$ and $u_x$ are given at $x = a, b$ then $\delta u = \delta u_x = 0$ at $x = a, b$ and since $\delta u$ is arbitrary Eq. (2.4.29) yields

$$\frac{\partial F}{\partial u} - \frac{d}{dx}\left(\frac{\partial F}{\partial u_x}\right) + \frac{d^2}{dx^2}\left(\frac{\partial F}{\partial u_{xx}}\right) = 0. \tag{2.4.30}$$

The above ideas can be easily extended to situations involving several dependent and/or several independent variables.

## Natural Boundary Conditions

Until now it was postulated that the argument function $u(x)$ had prescribed values at the end points $x = a, b$. If the boundary values are left unspecified, then the variational approach leads to a set of boundary conditions which must be satisfied if the functional is to possess an extremum. These boundary conditions are called *natural boundary conditions*. Consider again Eq. (2.4.1)

$$I(u) = \int_a^b F(u, u_x, x)dx$$

but without imposing the boundary condition which were previously specified by Eq. (2.4.2). The necessary condition for $I$ to attain a stationary value is still given by Eq. (2.4.23), that is,

$$\delta I = \int_a^b \left[\frac{\partial F}{\partial u} - \frac{d}{dx}\left(\frac{\partial F}{\partial u_x}\right)\right]\delta u\, dx + \left[\frac{\partial F}{\partial u_x}\delta u\right]_a^b = 0. \tag{2.4.31}$$

Since $\delta u$ is arbitrary over the region $a < x < b$, the Euler–Lagrange equation, Eq. (2.4.25), must still be satisfied. However, Eq. (2.4.23) requires now that either

$$\delta u = 0 \text{ or } \frac{\partial F}{\partial u_x} = 0. \tag{2.4.32}$$

Hence, if $u$ is not prescribed at the boundary (in which case $\delta u \neq 0$), then a necessary condition for an extremum is $\partial F/\partial u_x = 0$. The first type of boundary condition is called a *rigid boundary condition* (geometric or kinematic boundary condition), whereas the latter is called a *natural boundary condition*.

Similarly for Eqs. (2.4.29) and (2.4.30), $u(x)$ must satisfy the following conditions at $x = a$ and $x = b$; either

$$\delta u = 0 \text{ or } \frac{\partial F}{\partial u_x} - \frac{d}{dx}\left(\frac{\partial F}{\partial u_{xx}}\right) = 0 \tag{2.4.33}$$

and also either

$$\delta u_x = 0 \text{ or } \frac{\partial F}{\partial u_{xx}} = 0. \tag{2.4.34}$$

## 2.5    Principle of Virtual Work for an Elastic Body

Consider a structure, restrained against rigid body rotation and subject to a system of generalized forces $Q_i$ (which can be both forces or moments), described in Figure 2.31. The structure has a volume $V$ and is enclosed by a surface $S$. The stress at an arbitrary point $P$, caused by the loads $Q_i$, is specified by the stress tensor $\sigma_{ij}$. Since the body is in equilibrium the stress tensor components satisfy the equations of equilibrium, that is,

$$\sigma_{ij,j} + f_i = 0. \qquad (2.5.1)$$

**Figure 2.31** A constrained, loaded structure, undergoing a virtual deformation

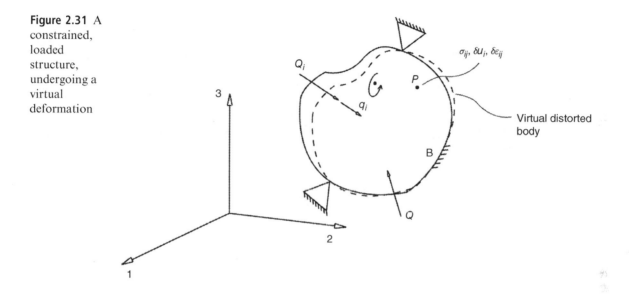

Consider the virtually distorted structure shown by the dashed line. The virtual distorted body must be consistent with the *kinematically admissible displacements*. The virtual deformations are assumed to have occurred after the real displacements have taken place. At this stage, one can assume arbitrary material behavior (conservative) and also assume that the structure is subject to arbitrary surface and body loads.

$\delta u_i$ = virtual displacements are very small displacements (infinitesimal) which are consistent with the geometric (rigid or kinematic) boundary conditions of the structure (i.e. kinematically admissible displacements), and arbitrary

$\delta q_i$ = virtual displacement in the $Q_i$ direction, that is, variation in a generalized coordinate associated with the load $Q_i$.

It is important to emphasize that during a virtual displacement the applied loads remain unchanged.

The virtual strains corresponding to the virtual displacements $\delta u_i$ are given by

$$\delta \epsilon_{ij} = \frac{1}{2}(\delta u_{i,j} + \delta u_{j,i}). \tag{2.5.2}$$

Note that the operator $\delta$ is the variational operator which has been defined and used in the previous section on the calculus of variations. As shown previously for algebraic purposes the $\delta$ operator is similar to the $d$ operator of differential calculus; furthermore recall that the operators $d$ and $\delta$ are permutable.

The external virtual work is given by

$$\underbrace{\delta W_E}_{\text{external virtual work}} = \int_{S_1} \underbrace{T_i \; \delta u_i \; dS}_{\text{work of surface tractions}} + \int_V \underbrace{f_i \; \delta u_i \; dV}_{\text{work of body forces}} . \tag{2.5.3}$$

$S_1$ is the part of the boundary on which surface tractions are specified. Note also that during a virtual displacement, body forces and surface tractions remain unchanged.

From the equilibrium boundary conditions, Eq. (2.3.5),

$$T_i = \sigma_{ij} n_j, \tag{2.5.4}$$

where $i, j = 1, 2, 3$ and $n_j = n_j(\ell, m, n)$ are the direction cosines of the normal to the surface of the structure with the $x_1, x_2, x_3$ coordinate.

Using Gauss' divergence theorem

$$\int_V \boldsymbol{\nabla} . \mathbf{A} dV = \int_S \mathbf{n} . \mathbf{A} dS, \tag{2.5.5}$$

where

$$\mathbf{A} = A_1 \mathbf{i} + A_2 \mathbf{j} + A_3 \mathbf{k}, \tag{2.5.6}$$

where $\mathbf{i}, \mathbf{j}, \mathbf{k}$ are unit vectors in the $x_1, x_2$, and $x_3$ directions, respectively, and

$$\left. \begin{aligned} \boldsymbol{\nabla} &= \frac{\partial}{\partial x_1}\mathbf{i} + \frac{\partial}{\partial x_2}\mathbf{j} + \frac{\partial}{\partial x_3}\mathbf{k} \\ \mathbf{n} &= n_1 \mathbf{i} + n_2 \mathbf{j} + n_3 \mathbf{k} \end{aligned} \right\}. \tag{2.5.7}$$

Equation (2.5.5) is thus equivalent to

$$\int_V \left( \frac{\partial A_1}{\partial x_1} + \frac{\partial A_2}{\partial x_2} + \frac{\partial A_3}{\partial x_3} \right) dV = \oint_s (n_1 A_1 + n_2 A_2 + n_3 A_3) dS. \tag{2.5.8}$$

Applying Eq. (2.5.5) on the first term of Eq. (2.5.3) yields

$$\oint_S T_i \delta u_i dS = \int_S \sigma_{ij} n_j \delta u_i dS = \int_V (\sigma_{ij} \delta u_i)_{,j} dV$$
$$= \int_V [\sigma_{ij,j} \delta u_i + \sigma_{ij} \delta u_{i,j}] dV. \tag{2.5.9}$$

Combining Eqs. (2.5.3) and (2.5.9)

$$\delta W_E = \int_V \left[ (\sigma_{ij,j} + f_i) \, \delta u_i + \sigma_{ij} \delta u_{i,j} \right] dV. \tag{2.5.10}$$

Recall that the structure is maintained in a state of equilibrium, thus due to Eq. (2.3.3) the first term in Eq. (2.5.10) is zero, thus

$$\delta W_E = \int_V \sigma_{ij}\delta u_{i,j} dV. \tag{2.5.11}$$

Using Eq. (2.5.2)

$$2\delta\epsilon_{ij} = \delta u_{i,j} + \delta u_{j,i}$$

and using the fact that $\sigma_{ij} = \sigma_{ji}$

$$2\sigma_{ij}\delta\epsilon_{ij} = \sigma_{ij}\delta u_{i,j} + \sigma_{ji}\delta u_{j,i} =$$
$$= \sigma_{1j}\delta u_{1,j} + \sigma_{2j}\delta u_{2,j} + \sigma_{3j}\delta u_{3,j} +$$
$$+\sigma_{1i}\delta u_{1,i} + \sigma_{2i}\delta u_{2,i} + \sigma_{3i}\delta u_{3,i} \tag{2.5.12}$$

Since both $i$ and $j$ represent dummy indices in Eq. (2.5.12), both parts of the expression, that is, first three terms and the last three terms, are equal to each other. Thus

$$\sigma_{ij}\delta u_{i,j} = \sigma_{ji}\delta u_{j,i} = \sigma_{ij}\delta\epsilon_{ij}. \tag{2.5.13}$$

Using Eqs. (2.5.13) and (2.5.11)

$$\delta W_E = \int_V \sigma_{ij}\delta\epsilon_{ij} dV = \delta U, \tag{2.5.14}$$

where $\delta U$ is the variation in strain energy.

Thus we have shown that the virtual work done by the external forces is equal to the internal strain energy variation due to virtual distortion of the elastic structure.

Now recall

$$\delta W_E = \text{external } V.W = \int_{S_1} T_i\delta u_i dS + \int_V f_i\delta u_i dV.$$

$\delta U = \int_V \sigma_{ij}\delta\epsilon_{ij} dV = $ internal virtual work of the *actual stresses* during a virtual distortion.

## Principle of Virtual Work

If a structure is in equilibrium and remains in equilibrium while subject to a virtual distortion the external virtual work $\delta W_E$ done by the external forces acting on the structure is equal to the internal virtual work $\delta U$ done by the internal stresses. The previous sentence represents a formal statement of the principle of virtual work (PVW).

Note that the opposite is also true, that is, if $\delta W_E = \delta U$, it implies that the structure is in a state of equilibrium, this conclusion can be also written as

$$\delta(W_E - U) = 0. \tag{2.5.15}$$

When the structure is subject to a finite number of discrete generalized forces and $f_i = 0$

$$\delta W_E = \int_S T_i\delta u_i dS = \sum_{i=1}^{n} Q_i\delta q_i \tag{2.5.16}$$

in this work expression, there is no $\left(\frac{1}{2}\right)$ since the forces $Q_i$ remain unchanged during a virtual distortion.

## 2.6    Principle of Minimum Potential Energy (PMPE)

Consider again the same structure which has been used in Section 2.5. The deformed configuration is described by the displacement field $u_i$ at each point of the structure. Next introduce a virtual displacement field $\delta u_i$ consistent with all the constraints imposed upon the body. Thus the total displacement field is given by

$$\bar{u}_i = u_i + \delta u_i.$$

According to the section on the calculus of variations, $\delta u_i$ must vanish at points where constraints (i.e. supports) exist. The internal virtual work from Eq. (2.5.14) is

$$\delta U = \int_V \sigma_{ij}\delta\epsilon_{ij}dV. \tag{2.6.2}$$

It can be shown that for an *elastic body* the internal virtual work can be interpreted as the first variation of the strain energy resulting from the variation of the displacement field. It is important to note that in Section 2.5 the virtual displacement field $\delta u_i$ was a priori not related to the stress field $\sigma_{ij}$; when applying the principle of virtual work, one can link these fields through the constitutive law. It is also important to emphasize that since no constitutive relation was used in the principle of virtual work the principle is general. Due to the use of the constitutive relations the principle of minimum potential energy (PMPE) is more limited. With this assumption one can write

$$U = \int_V (\mu\epsilon_{ij}\epsilon_{ij} + \frac{\lambda}{2}\epsilon_{kk}^2)dV = \int_V U_0 dV \tag{2.6.3}$$

$$\delta U = \delta \int_V (\mu\epsilon_{ij}\epsilon_{ij} + \frac{\lambda}{2}\epsilon_{kk}^2)dV$$

$$= \int_V (2\mu\epsilon_{ij}\delta\epsilon_{ij} + \frac{\lambda}{2}2\epsilon_{kk}\delta\epsilon_{kk})dV$$

$$= \int_V (2\mu\epsilon_{ij} + \lambda\delta_{ij}\epsilon_{kk})\delta\epsilon_{ij}dV$$

$$= \int_V \sigma_{ij}\delta\epsilon_{ij}dV = \int_V f_i\delta u_i dV + \int_s T_i\delta u_i dS, \tag{2.6.4}$$

where we have used Eq. (2.3.15) in Eq. (2.6.3).

Next define the potential energy $V_E$ of the applied loads as a function of the displacement field $u_i$ and the applied loads as follows:

$$V_E = -\int_V f_i u_i dV - \int_{S_1} T_i u_i dS = -W_E. \tag{2.6.5}$$

Conservative loads can be obtained from a potential function, when forces are independent of the deformation of the structure

$$\left. \begin{array}{ll} T_i = -\frac{\partial G}{\partial u_i} & \text{and} \quad G = -T_i u_i \\ f_i = -\frac{\partial g}{\partial u_i} & \text{and} \quad g = -f_i u_i \end{array} \right\}$$
(2.6.6)

then

$$\delta W_E = -\delta V_E = \int_S T_i \delta u_i dS + \int_V f_i \delta u_i dV$$
(2.6.7)

Thus for a structure that possesses strain energy $U$ and an external potential $V_E$ the PVW may be written as

$$\delta(U + V_E) = \delta\Pi = 0,$$
(2.6.8)

where $\Pi = U + V_E =$ is the total potential energy of the structure.

Eq. (2.6.8) is the *principle of stationary potential energy*. The formal statement of the principle is as follows:

A kinematically admissible displacement field, being related through some constitutive law to a stress field satisfying equilibrium requirements in a body acted upon by statically compatible external loads, must yield an extremum of the total potential energy with respect to all other kinematically admissible displacement fields. If one considers the total potential as a functional in the sense of the calculus of variations, the Euler–Lagrange equations corresponding to the variational statement are the equilibrium relations and the boundary conditions are given by Eq. (2.5.1).

For a stable equilibrium configuration, it can be shown that this stationary value is a local minimum by considering the second variation.

## 2.7    Hamilton's Principle

Hamilton's principle is perhaps the most fundamental principle in structural dynamics. It serves as a powerful tool for deriving partial differential equations of motion and corresponding natural boundary conditions for complicated, continuous parameter, structural dynamic systems. Hamilton's principle can be viewed as an extension of the principle of potential energy to the dynamic case, or a general dynamic principle of virtual work. It should be noted that Hamilton's principle was formulated by W. R. Hamilton, mathematician and Andrews Professor of Astronomy, Trinity College, Dublin, in 1834–1835. However, despite its age, some more recent literature on structural dynamics still contains a number of papers which debate the precise meaning of this principle and its extension to nonconservative systems.

Consider the structure treated in Section 2.5 and described in Figure 2.31. Assume that the system is subjected to *dynamic loading conditions*. Using D'Alembert's principle the inertia loads due to the dynamic loading conditions can be handled by representing them as body forces. Denote by $\rho$ the mass density of the material and by $u_i$ the components of the displacement vector. The body forces corresponding to the inertia loads can be represented as $-\rho\frac{\partial^2 u_i}{\partial t^2}$ and the equations of equilibrium for this case are

$$\sigma_{ij,j} + f_i = \rho\frac{\partial^2 u_i}{\partial t^2}. \tag{2.7.1}$$

Consider next the virtual work $\delta W_E$ done by the applied forces $T_i$ and $f_i$ during a virtual displacement $\delta u_i$. The deformation and displacement are now time dependent, that is,

$$u = u(x_j, t).$$

In this case, $\delta u_i$ will be further restricted by demanding that $\delta u_i$ be zero at some initial time $t_1$ and a final time $t_2$, thus

$$\delta u_i(x_j, t_1) = \delta u_i(x_j, t_2) = 0. \tag{2.7.2}$$

Note that Eq. (2.7.2) represents a further restriction on the kinematically admissible virtual distortion $\delta u_i$. The physical meaning of Eq. (2.7.2) is that we assume that the displacement field is known at time instants $t_1$ and $t_2$ and the problem is to find the motion of the body during the interval $t_1 < t < t_2$.

According to Eq. (2.5.3) the external virtual work is given by,

$$\delta W_E = \int_S T_i\delta u_i dS + \int_V f_i\delta u_i dV \tag{2.7.3}$$

during a virtual distortion $\delta u_i$. Again from the equilibrium boundary conditions $T_i = \sigma_{ij}n_j$ and thus

$$\int_S T_i\delta u_i dS = \int_S \sigma_{ij}n_j\delta u_i ds = \int_V (\sigma_{ij}\delta u_i)_{,j}\, dV =$$

$$= \int_V [\sigma_{ij,j}\delta u_i + \sigma_{ij}\delta u_{i,j}]dV. \tag{2.7.4}$$

Combining Eqs. (2.7.3) and (2.7.4)

$$\delta W_E = \int_V \left[(\sigma_{ij,j} + f_i)\,\delta u_i + \sigma_{ij}\delta u_{i,j}\right]dV. \tag{2.7.5}$$

Furthermore recall $\sigma_{ij} = \sigma_{ji}$ and

$$\delta\epsilon_{ij} = \frac{1}{2}(\delta u_{i,j} + \delta u_{j,i})$$

$$2\sigma_{ij}\delta\epsilon_{ij} = \sigma_{ij}(\delta u_{i,j} + \delta u_{j,i}) = \sigma_{ij}\delta u_{i,j} + \sigma_{ji}\delta u_{j,i}$$

thus

$$\int_V \sigma_{ij}\delta u_{i,j}dV = \int_V \sigma_{ij}\delta\epsilon_{ij}dV = \delta U. \tag{2.7.6}$$

Combining Eqs. (2.7.5), (2.7.6), and (2.7.1),

$$\delta W_E = \int_V [\sigma_{ij}\delta\epsilon_{ij} + \rho\frac{\partial^2 u_i}{\partial t^2}\delta u_i]dV$$

$$= \delta U + \int_V \rho\frac{\partial^2 u_i}{\partial t^2}\delta u_i dV. \tag{2.7.7}$$

To obtain Hamilton's principle integrate Eq. (2.7.7) with respect to time from $t_1$ to $t_2$; this yields

$$\int_{t_1}^{t_2} (\delta W_E - \delta U)dt = \int_{t_1}^{t_2} \int_V \rho\frac{\partial^2 u_i}{\partial t^2}\delta u_i dV dt. \tag{2.7.8}$$

Interchanging the order of integrations in the last term of Eq. (2.7.8),

$$\int_{t_1}^{t_2} (\delta W_E - \delta U)dt = \underbrace{\int_V \rho\frac{\partial u_i}{\partial t}\delta u_i dV\Big|_{t_1}^{t_2}} - \int_V \int_{t_1}^{t_2} \rho\frac{\partial u_i}{\partial t}\frac{\partial \delta u_i}{\partial t}dV dt$$

this is zero due to Eq. 2.7.2

$$= -\int_V \int_{t_1}^{t_2} \rho\frac{\partial u_i}{\partial t}\delta\left(\frac{\partial u_i}{\partial t}\right)dV dt$$

$$= -\int_{t_1}^{t_2} \delta \underbrace{\int_V \frac{\rho}{2}\left(\frac{\partial u_i}{\partial t}\right)\left(\frac{\partial u_i}{\partial t}\right)dV}\ dt$$

kinetic energy T

thus

$$\int_{t_1}^{t_2} (\delta W_E - \delta U + \delta T)dt = 0, \tag{2.7.9}$$

where the kinetic energy

$$T = \int_V \frac{\rho}{2}\left(\frac{\partial u_i}{\partial t}\right)^2 dV.$$

Equation (2.7.9) is the general statement of *Hamilton's principle*. For an elastic structure which process a strain energy $U$ and is subject to external forces which are derivable from a scalar potential (conservative forces).
$T_i = -\frac{\partial G}{\partial u_i}$ and $f_i = -\frac{\partial g}{\partial u_i}$ then Eq. (2.7.9) can be rewritten as

$$\delta \int_{t_1}^{t_2} (U - T + V_E)\, dt = 0, \tag{2.7.10}$$

where

$$\delta V_E = -\delta W_E = -\int_S T_i\delta u_i dS - \int_V f_i\delta u_i dV \tag{2.7.11}$$

is the virtual change in *potential of external forces.*

If the external forces $T_i$ and $f_i$ are independent of the elastic displacements $u_i$, which is generally the case except in aeroelasticity

$$\delta \int_{t_1}^{t_2} (U - T + V_E) \, dt = 0. \tag{2.7.12}$$

For the case of discrete applied loads that can exist in addition to the distributed loads or instead of the distributed loads, the potential of external loads that are independent of the displacements can be written as

$$V_E = -\sum_{i=1}^{n} Q_i q_i, \tag{2.7.13}$$

where $Q_i$ are discrete generalized forces and $q_i$ are corresponding displacements.

Hamilton's principle: The motion of an elastic structure during the time interval $t_1 < t < t_2$ is such that the time integral of the *total dynamic potential* $U - T + V_E$ is an extremum.

When $T = 0$, Hamilton's principle reduces to the principle of minimum potential energy.

## 2.8    Lagrange's Equations

A direct derivation of Lagrange's equations can be obtained from Hamilton's principle. Consider an elastic system, which can be described by a discrete set of generalized coordinates shown in Figure 2.32. Recall Hamilton's principle

$$\int_{t_1}^{t_2} (\delta W_E - \delta U + \delta T) dt = 0 \tag{2.8.1}$$

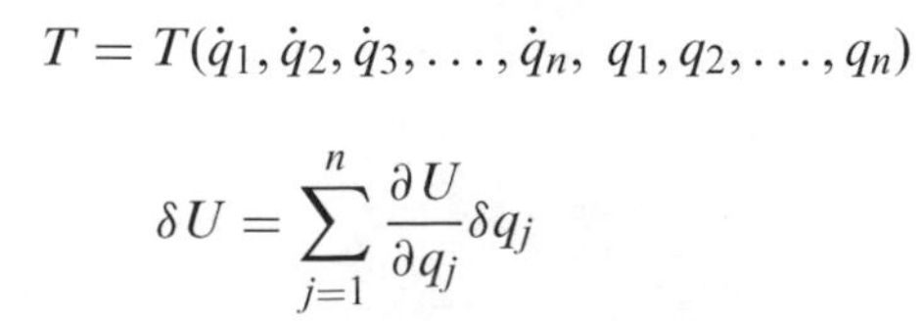

**Figure 2.32** Combined flexible beam and mass system for deriving Lagrange's equations

then

$$U = U(q_1, q_2, \ldots, q_n)$$
$$T = T(\dot{q}_1, \dot{q}_2, \dot{q}_3, \ldots, \dot{q}_n, \ q_1, q_2, \ldots, q_n)$$

$$\delta U = \sum_{j=1}^{n} \frac{\partial U}{\partial q_j} \delta q_j \tag{2.8.2}$$

$$\delta T = \sum_{j=1}^{n} \left( \frac{\partial T}{\partial \dot{q}_j} \delta \dot{q}_j + \frac{\partial T}{\partial q_j} \delta q_j \right) \tag{2.8.3}$$

$$\delta W_E = \sum_{j=1}^{n} Q_j \delta q_j. \tag{2.8.4}$$

Combining Eqs. (2.8.1) through (2.8.4)

$$\int_{t_1}^{t_2} \sum_{j=1}^{n} \left\{ \frac{\partial T}{\partial \dot{q}_j} \delta \dot{q}_j + \frac{\partial T}{\partial q_j} \delta q_j - \frac{\partial U}{\partial q_j} \delta q_j + Q_j \delta q_j \right\} dt = 0. \tag{2.8.5}$$

Consider the first term in Eq. (2.8.5) and integrate it by parts

$$\int_{t_1}^{t_2} \sum_{j=1}^{n} \left( \frac{\partial T}{\partial \dot{q}_j} \delta \dot{q}_j dt \right) = \sum_{j=1}^{n} \frac{\partial T}{\partial \dot{q}_j} \delta q_j \bigg|_{t_1}^{t_2} - \int_{t_1}^{t_2} \sum_{j=1}^{n} \left\{ \frac{d}{dt} \left( \frac{\partial T}{\partial \dot{q}_j} \right) \delta q_j \right\} dt. \tag{2.8.6}$$

Recall also that one of the fundamental relations used in deriving Hamilton's principle was

$$\delta q_j(t_1) = \delta q_j(t_2) = 0, \tag{2.8.7}$$

Therefore, the first term in Eq. (2.8.6) is zero; from Eqs. (2.8.5) and (2.8.6),

$$\int_{t_1}^{t_2} \sum_{j=1}^{n} \left\{ \left[ -\frac{d}{dt} \left( \frac{\partial T}{\partial \dot{q}_j} \right) + \frac{\partial T}{\partial q_j} - \frac{\partial U}{\partial q_j} + Q_j \right] \delta q_j \right\} dt = 0. \tag{2.8.8}$$

Since generalized coordinates are independent, this implies

$$\frac{d}{dt} \left( \frac{\partial T}{\partial \dot{q}_j} \right) - \frac{\partial T}{\partial q_j} + \frac{\partial U}{\partial q_j} = Q_j, \qquad j = 1, 2, \ldots, n. \tag{2.8.9}$$

Equation (2.8.9) is Lagrange's equations of motion in generalized coordinates. The generalized load or force $Q_j$ can further be decomposed into several parts

$$Q_j = Q_{Aj} + Q_{Dj},$$

where $Q_{Aj}$ are the applied generalized forces, externally reacted, and $Q_{Dj}$ are damping forces.

Recall now that for a structure governed by linear stress strain relation, the Principle of Minimum Potential Energy applies, and from it one obtains *Castigliano's first theorem*

$$Q_{Ej} \text{ internal elastic force } = -\frac{\partial U}{\partial q_j}$$

and Eq. (2.8.9) can be rewritten as

$$\frac{d}{dt} \left( \frac{\partial T}{\partial \dot{q}_j} \right) - \frac{\partial T}{\partial q_j} = Q_{Ej} + Q_{Aj} + Q_{Dj} = Q_j. \tag{2.8.10}$$

If the generalized loads are derivable from an external potential such that

$$Q_j = -\frac{\partial V}{\partial q_j} \tag{2.8.11}$$

Equation (2.8.10) becomes

$$\frac{d}{dt} \left( \frac{\partial T}{\partial \dot{q}_j} \right) - \frac{\partial T}{\partial q_j} + \frac{\partial V}{\partial q_j} = 0. \tag{2.8.12}$$

Equation (2.8.12) is sometimes called Lagrange's equation for a *conservative* system. Define $L$ = Lagrangian, where $L = T - V$; furthermore, $V = V(q_1, q_2, \ldots, q_n)$. Thus, Eq. (2.8.12) can be rewritten as

$$\frac{d}{dt}\left(\frac{\partial L}{\partial \dot{q}_j}\right) - \frac{\partial L}{\partial q_j} = 0. \tag{2.8.13}$$

It is important to realize that Lagrange's equations are analogous to the Euler equations of a variational statement for a dynamical system. Thus, Eq. (2.8.13) can be considered to be similar to Eq. (2.4.5) found in Section 2.3. The statement corresponding to

$$I(u) = \int_a^b F(u, u_x, x)dx$$

is

$$I(q_i) = \int_{t_1}^{t2} (W_E - U + T)dt$$

and the condition for the stationary value of $I(u)$ is Hamilton's principle (Eq. (2.7.12)).

Finally, it is important to emphasize that Lagrange's equations are not restricted to the conservative case and they can accommodate damping. In structural dynamics it is common practice to use two types of damping. The first type of damping is *structural damping*

$$Q_D = i\gamma Q_E = -i\gamma \frac{\partial U}{\partial q_j}.$$

For this case, Lagrange's equations can be written as

$$\frac{d}{dt}\left(\frac{\partial T}{\partial \dot{q}_j}\right) - \frac{\partial T}{\partial q_j} + (1 + i\gamma)\frac{\partial U}{\partial q_j} = Q_{Aj}, \quad j = 1, 2, \ldots, n. \tag{2.8.14}$$

The second type of damping widely used in structural dynamics is the *viscous type of damping* that can be conveniently represented using the *Rayleigh dissipation* function $R$.

$$R = \frac{1}{2}\int_V c_{ij}\dot{q}_i\dot{q}_j dV = \frac{1}{2}\sum_i\sum_j c_{ij}\dot{q}_i\dot{q}_j$$

and

$$\frac{\partial R}{\partial \dot{q}_j} = \sum_k c_{jk}\dot{q}_k, \qquad Q_D = -\frac{\partial R}{\partial \dot{q}_j}.$$

Lagrange's equation with this type of damping can be written as

$$\frac{\partial}{\partial t}\left(\frac{\partial T}{\partial \dot{q}_j}\right) - \frac{\partial T}{\partial q_j} + \frac{\partial U}{\partial q_j} + \frac{\partial R}{\partial \dot{q}_j} = Q_{Aj}. \tag{2.8.15}$$

## 2.9    Application of Hamilton's Principle and Lagrange's Equations

### Example 1

For the application of Hamilton's principle to a simple system consider a uniform simply supported beam without any external loading, shown in Figure 2.33

**Figure 2.33** Simply supported uniform beam

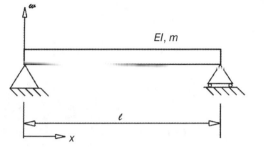

Kinetic energy given by

$$T = \frac{1}{2} \int_0^\ell m\dot{w}^2 dx. \tag{2.9.1}$$

Strain energy given by

$$U = \frac{1}{2} \int_0^\ell EIw_{,xx}^2 dx. \tag{2.9.2}$$

Hamilton's principle

$$\int_{t_1}^{t_2} (\delta W_E - \delta U + \delta T)dt = 0.$$

This is a conservative system with no applied loads, thus $\delta W_E = 0$

$$\int_{t_1}^{t_2} \delta(T - U)dt = 0 \tag{2.9.3}$$

$$\int_{t_1}^{t_2} \delta T dt = \int_{t_1}^{t_2} \int_0^\ell m\dot{w}\delta\left(\frac{\partial w}{\partial t}\right) dx dt = \int_0^\ell \left\{ m\dot{w}\delta w \Big|_{t_1}^{t_2} - \right.$$

$$\left. - \int_{t_1}^{t_2} m\ddot{w}\delta w dt \right\} dx.$$

In Hamilton's principle, the variation is such that

$$\delta w|_{t_1} = \delta w|_{t_2} \equiv 0$$

thus

$$\int_{t_1}^{t_2} \delta T dt = - \int_0^\ell \int_{t_1}^{t_2} m\ddot{w}\delta w dt dx. \tag{2.9.4}$$

Similar integrations by parts are carried out on the strain energy terms of Hamilton's principle; the objective is always to have the spanwise integrals multiplied by arbitrary variation of the displacement quantity, which in this case is $w$; thus,

$$\int_{t_1}^{t_2} \delta U dt = \int_{t_1}^{t_2} \int_0^\ell EI w_{,xx} \frac{\partial}{\partial x} \left( \delta \frac{\partial w}{\partial x} \right) dx\, dt. \tag{2.9.5}$$

Performing two consecutive integrations by parts on Eq. (2.9.5), one has

$$\int_{t_1}^{t_2} \delta U dt = \int_{t_1}^{t_2} \left\{ EI w_{,xx} \delta \left( \frac{\partial w}{\partial x} \right) \Big|_{x=0}^{x=\ell} \right.$$

$$\left. - \int_0^\ell EI w_{,xxx} \delta \left( \frac{\partial w}{\partial x} \right) dx \right\} dt$$

$$\int_{t_1}^{t_2} \delta U dt = \int_{t_1}^{t_2} \left\{ EI w_{,xx} \delta \left( \frac{\partial w}{\partial x} \right) \Big|_{x=0}^{x=\ell} \right.$$

$$\left. - EI w_{,xxx} \delta w \Big|_0^\ell + \int_0^\ell EI \frac{\partial^4 w}{\partial x^4} dx \delta w \right\} dt. \tag{2.9.6}$$

The first term on the right-hand side of Eq. (2.9.6) represents the virtual work of the bending moment at the ends of a simply supported beam. The second term represents the virtual work of the shearing force at the ends. The kinematic boundary conditions or geometric boundary conditions for this problem are

$$w(0) = w(\ell) = 0.$$

Since the displacements at the ends are given, it implies that

$$\delta w(0) = \delta w(\ell) = 0$$

and thus

$$EI w_{,xxx} \delta w \Big|_{x=0} = EI w_{,xxx} \delta w \Big|_{x=\ell} = 0. \tag{2.9.7}$$

Combining Eqs. (2.9.3), (2.9.4), (2.9.6), and (2.9.7) yields

$$\int_{t_1}^{t_2} (\delta T - \delta U) dt = - \int_{t_1}^{t_2} \int_0^\ell \left[ m\ddot{w} + EI \frac{\partial^4 w}{\partial x^4} - \right] \delta w\, dx\, dt$$

$$+ \int_{t_1}^{t_2} \left\{ -EI w_{,xx} \delta(w_{,x}) \Big|_0^\ell \right\} dt = 0. \tag{2.9.8}$$

Recall that $\delta w$ is arbitrary for any value of $x$, except at the boundaries where $\delta w(0) = \delta w(\ell) = 0$. Thus, in order to satisfy Eq. (2.9.8), the following relations must exist

$$m\ddot{w} + EI \frac{\partial^4 w}{\partial x^4} = 0 \tag{2.9.9}$$

for $0 < x < \ell$ and

$$EI w_{,xx} = 0 \text{ at } x = 0, \ell. \tag{2.9.10}$$

Equation (2.9.9) represents the partial differential equation of equilibrium for the system, while Eq. (2.9.10) represents the *natural boundary conditions* (NBC) for the system at $x = 0, \ell$.

## Example 2

To examine the effect of axial force on a vibrating beam, consider Figure 2.34 that illustrates the deformed configuration of the beam under the action of an axial load $P$, which always remains vertical during deformation (i.e. dead load). Note that a dead load is a conservative load. The differential equation of equilibrium for the transverse vibrations of this beam is required together with the appropriate boundary conditions. Again the starting point is Hamilton's principle, Eq. (2.7.9),

$$\int_{t_1}^{t_2} (\delta W_E - \delta U + \delta T) dt = 0.$$

**Figure 2.34** Axially loaded beam, with uniform mass and stiffness

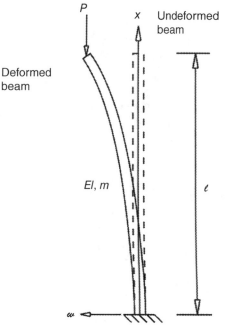

The treatment of strain energy and kinetic energy is similar to the previous example, Eqs. (2.9.6) and (2.9.4),

$$\int_{t_1}^{t_2} \delta T dt = -\int_0^\ell \int_{t_1}^{t_2} m \ddot{w} \delta w \, dt \, dx$$

$$\int_{t_1}^{t_2} \delta U dt = \int_{t_1}^{t_2} \left\{ EI w_{,xx} \delta \left( \frac{\partial w}{\partial x} \right) \bigg|_0^\ell - EI w_{,xxx} \delta w \bigg|_0^\ell + \right.$$

$$\left. + \int_0^\ell EI \frac{\partial^4 w}{\partial x^4} dx \delta w \right\} dt.$$

The main difference between this and the previous example is the presence of the axial load which introduces a term due to external work $W_E$. In order to evaluate this term, one has to distinguish between the deformed and undeformed configurations of this structural system. In its *undeformed state* the beam was coincident with the $x$-axis. Furthermore it should be noted that distinction between the deformed and undeformed beam configurations implies the need for using a nonlinear strain displacement relation that has been given previously in Eq. (2.3.6)

$$\epsilon_{ij} = \frac{1}{2}(u_{i,j} + u_{j,i} + u_{k,i}u_{k,j})$$

in this particular case $u_1 = u$ ; $u_2 = w$ ; $u_3 = 0$; thus

$$\epsilon_{xx} = u_{,x} + \frac{1}{2}\left(w_{,x}^2 + u_{,x}^2\right). \tag{2.9.11}$$

Next introduce the assumption that bending of the beam occurs in an *inextensional manner*, which physically implies that the length of the elastic axis in the deformed state and undeformed state are equal, that is,

$$\epsilon_{xx} = 0 = u_{,x} + \frac{1}{2}w_{,x}^2 dx \tag{2.9.12}$$

thus the axial displacement at the tip of the beam

$$u(x = \ell) = -\frac{1}{2}\int_0^\ell w_{,x}^2 dx. \tag{2.9.13}$$

Definition of the work of the external load $W_E$ is as follows. The external work is equal to the load (which is constant) multiplied by the displacement in the direction of the load. If the load and displacement are in the *same* direction $W_E$ is positive. If the load and the displacement are in *opposite* directions $W_E$ is negative, thus

$$W_E = +Pu(x = \ell) = \frac{P}{2}\int_0^\ell w_{,x}^2 dx \tag{2.9.14}$$

$$\int_{t_1}^{t_2} \delta W_E dt = \int_{t_1}^{t_2}\int_0^\ell Pw_{,x}\delta w_{,x}dx\, dt$$

$$= \int_{t_1}^{t_2}\left[ Pw_{,x}\delta w\Big|_0^\ell - \int_0^\ell Pw_{,xx}\delta w dx\right] dt, \tag{2.9.15}$$

where in Eq. (2.9.15), one integration by parts with respect to $x$ has been carried out. Combining Eqs. (2.7.9), (2.9.4), (2.9.6), and (2.9.15) yields

$$\int_{t_1}^{t_2}\left\{ Pw_{,x}\delta w\Big|_0^\ell - \int_0^\ell Pw_{,xx}\delta w dx - EIw_{,xx}\delta w_{,x}\Big|_0^\ell + \right.$$

$$\left. EIw_{,xxx}\delta w\Big|_0^\ell - \int_0^\ell EI\frac{\partial^4 w}{\partial x^4}dx\delta w - \int_0^\ell m\ddot{w}\delta w dx\right\} dt = 0$$

or

$$\int_{t_1}^{t_2} \left\{ (Pw_{,x} + EIw_{,xxx})\delta w \Big|_0^{\ell} - EIw_{,xx}\delta w_{,x} \Big|_0^{\ell} \right.$$

$$\left. - \int_0^{\ell} (Pw_{,xx} + EI\frac{\partial^4 w}{\partial x^4} + m\ddot{w})\delta w dx \right\} dt = 0. \qquad (2.9.16)$$

The geometric or kinematic boundary condition for a cantilevered beam

$$w(0) = w_{,x}(0) = 0, \qquad (2.9.17)$$

which implies $\delta w(0) = \delta w_{,x}(0) = 0$

Thus Eq. (2.9.16) yields the natural boundary conditions for this problem given by

$$(Pw_{,x} + EIw_{,xxx})\delta w(\ell) = 0$$

$$EIw_{,xx}\delta w_{,x}(\ell) = 0.$$

Since $\delta w(\ell)$ and $\delta w_{,x}(\ell)$ are arbitrary the last two equations imply

$$Pw_{,x} + EIw_{,xxx} = 0 \quad \text{and} \quad x = \ell \qquad (2.9.18)$$

$$EIw_{,xx} = 0 \quad \text{at} \quad x = \ell. \qquad (2.9.19)$$

Equations (2.9.18) and (2.9.19) are the natural boundary conditions for this problem and they represent, respectively, shear force equilibrium at the tip of the beam and the condition for zero moment at the end of the beam.

The differential equation of equilibrium or the equation of motion for this structural dynamic system is obtained from the last integral of Eq. (2.9.16). Since $\delta w$ is an arbitrary variation at any station $x$, one has

$$m\ddot{w} + EI\frac{\partial^4 w}{\partial x^4} + P\frac{\partial^2 w}{\partial x^2} = 0. \qquad (2.9.20)$$

## Example 3

The formulation of the equations of motion for a three degree-of-freedom system is shown in Figure 2.35 using Lagrange's equations. For convenience, replace

$$\{ u \} = \begin{Bmatrix} u_1 \\ u_2 \\ u_3 \end{Bmatrix} \quad \text{by} \quad \{ q \} = \begin{Bmatrix} q_1 \\ q_2 \\ q_3 \end{Bmatrix}.$$

Using the matrices obtained in Section 2.1, Eq. (2.2.69),

$$U = \frac{1}{2}\{ q \}^T EI \begin{bmatrix} k_{11} & k_{12} & k_{13} \\ k_{21} & k_{22} & k_{23} \\ k_{31} & k_{32} & k_{33} \end{bmatrix} \{ q \} = \frac{1}{2}\{ q \}^T [ k ]\{ q \}. \qquad (2.9.21)$$

**Figure 2.35** Three degree-of-freedom mass, damper, elastic beam system

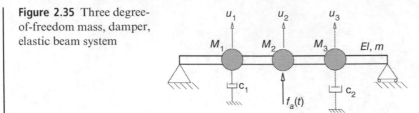

From Eq. (2.2.68), using Rayleigh's dissipation function

$$R = \frac{1}{2} \{ \dot{q} \}^T \begin{bmatrix} C_1 & 0 & 0 \\ 0 & 0 & 0 \\ 0 & 0 & C_2 \end{bmatrix} \{ \dot{q} \} = \frac{1}{2} \{ \dot{q} \}^T [ C ] \{ \dot{q} \} \tag{2.9.22}$$

and from Eqs. (2.2.67) and (2.2.70),

$$\{ \mathbf{F}_A(t) \} = \left\{ \begin{array}{c} 0 \\ f_A(t) \\ 0 \end{array} \right\} \quad ; \quad T = \frac{1}{2} \{ \dot{q} \}^T [ M ] \{ \dot{q} \}, \tag{2.9.23}$$

where

$$[M] = \begin{bmatrix} M_1 & 0 & 0 \\ 0 & M_2 & 0 \\ 0 & 0 & M_3 \end{bmatrix}. \tag{2.9.24}$$

Substituting these equations into Lagrange's equations of motion

$$\frac{d}{dt} \left( \frac{\partial T}{\partial \dot{q}_i} \right) - \frac{\partial T}{\partial q_i} + \frac{\partial U}{\partial q_i} + \frac{\partial R}{\partial \dot{q}_i} = F_{Ai}, \qquad \text{where } i = 1, 2, 3 \tag{2.9.25}$$

yields

$$[M]\{\ddot{q}\} + [C]\{\dot{q}\} + [K]\{q\} = \{\mathbf{F}_A(t)\}. \tag{2.9.26}$$

## 2.10  Extension of Hamilton's Principle to Nonconservative Systems

Based on the discussion of Hamilton's principle it was mentioned that it is limited to conservative systems. However, it can be shown that it is relatively simple to extend it to nonconservative systems by using the principle of virtual work. When applying the principle in presence of nonconservative loads, the principle has to be modified by adding the virtual work of the nonconservative loads. With this modification the principle can be written as

$$\delta \int_{t_1}^{t_2} (T - \Pi_p)dt + \int_{t_1}^{t_2} \sum_{n=1}^{N} \mathbf{F}_n^{n.c.} \cdot \delta r_n^{n.c.} dt = 0, \tag{2.10.1}$$

where $F_n^{n.c.}$ is the nonconservative load vector, and $\delta r_n^{n.c.}$ is the virtual displacement vector associated with the nonconservative load, where $T$ is the kinetic energy, $U + V_E = \Pi_p$ total potential and $V_E = -W_E$, $V_E$ is potential and virtual work of the conservative loads. Note the $n$ in the last equation implies the inclusion of a number of nonconservative loads for the example shown next $n = 1$. The application of this extension of Hamilton's principle is best illustrated by a simple example, described next.

Consider the formulation of the equation of motion and boundary condition for a cantilevered beam loaded by a follower force, as well as a typical dead load due to gravity, shown in Figure 2.36.

**Figure 2.36** Axially loaded beam with a follower force

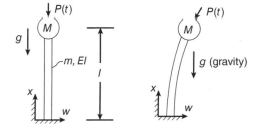

To simplify the problem it will be assumed that the slope $\frac{\partial w}{\partial x}$ of the deformed beam is small so that

$$\sin\left(\frac{\partial w}{\partial x}\right) \cong \frac{\partial w}{\partial x} \quad ; \text{and} \quad \cos\left(\frac{\partial w}{\partial x}\right) \cong 1.0.$$

It is also assumed that gravity acts only on the large mass $M$ while its effect on the distributed mass per unit length $m$ is negligible. The beam has uniform stiffness $EI$ and length $l$ and we are interested in the vibrations of this structural dynamic system in the $w$-direction. Furthermore the rotary inertia of the mass $M$ can also be neglected.

Applying Hamilton's principle

$$T = \frac{1}{2}\int_0^l m\dot{w}^2 dx + \frac{1}{2}M\dot{w}(l) \tag{2.10.2}$$

$$U = \frac{1}{2}\int_0^l EIw_{,xx}^2 dx \tag{2.10.3}$$

$$W_E = \frac{P_x + Mg}{2}\int_0^l w_{,x}^2 dx = \frac{P + Mg}{2}\int_0^l w_{,x}^2 dx. \tag{2.10.4}$$

For small slopes, $P_x = P\cos\left[\frac{\partial w}{\partial x}(l)\right] \cong P$.

The problem involves a follower force, representing nonconservative loading, requiring a modified version of Hamilton's principle represented by Eq. (2.10.1), which for the current example can be rewritten as

$$\delta\int_{t_1}^{t_2}[T - (U + V_E)]\,dt + \int_{t_1}^{t_2} -P\frac{\partial w}{\partial x}(l)\delta w(l)dt = 0. \tag{2.10.5}$$

The minus sign in the second term of Eq. (2.10.5) is due to the fact that the lateral load and the displacement at the tip of the beam are in opposite directions. Carrying out the usual manipulations, and using the geometric BC's $w(0) = w_{,x}(0) = 0$,

$$
\int_{t_1}^{t_2} \delta T dt = \int_{t_1}^{t_2} \left[ \int_0^l m\dot{w}\delta\dot{w}dx + M\delta\dot{w}(l)\dot{w}(l) \right] dt
$$

$$
= \int_{t_1}^{t_2} \left[ \int_0^l -m\ddot{w}\delta w dx - M\ddot{w}\delta w(l) \right] dt \qquad (2.10.6)
$$

$$
\int_{t_1}^{t_2} \delta U dt = \int_{t_1}^{t_2} \int_0^l EIw_{,xx}\delta w_{,xx}dtdx
$$

$$
= \int_{t_1}^{t_2} \left[ EIw_{,xx}\delta w_{,x}(l) - EIw_{,xxx}\delta w(l) + \int_0^l EI\frac{\partial^4 w}{\partial x^4}\delta w dx \right] dt \qquad (2.10.7)
$$

$$
\int_{t_1}^{t_2} \delta V_E dt = -\int_{t_1}^{t_2} \delta W_E dt = -\int_{t_1}^{t_2} \left[ (P + Mg)\int_0^l w_{,x}\delta w_{,x}dx \right] dt
$$

$$
= \int_{t_1}^{t_2} \left[ (P + Mg)w_{,x}(l)\delta w(l) - (P + Mg)\int_0^l w_{,xx}\delta w dx \right] dt. \qquad (2.10.8)
$$

Using Eqs. (2.10.7) and (2.10.8) one obtains

$$
\int_{t_1}^{t_2} \left\{ -M\ddot{w}\delta w(l) - \int_0^e \left[ m\ddot{w} + EI\frac{\partial^4 w}{\partial x^4} + (P + Mg)w_{,xx} \right]\delta w \right.
$$

$$
- EIw_{,xx}\delta w_{,x}(l) + EIw_{,xxx}\delta w(l)
$$

$$
\left. + (P + Mg)w_{,x}\delta w_{,x}(l) - Pw_{,x}(l)\delta w(l) \right\} dt = 0. \qquad (2.10.9)
$$

From the last equation, one obtains the equation of motion and the natural boundary conditions

$$
m\ddot{w} + EI\frac{\partial^4 w}{\partial x^4} + (P + Mg)\frac{\partial^2 w}{\partial x^2} = 0 \qquad (2.10.10)
$$

$$
- M\ddot{w}(l) + EIw_{,xxx}(l) + (P + Mg)w_{,x}(l) - Pw_{,x}(l) = 0,
$$

which yields

$$
EIw_{,xxx} + Mgw_{,x} - M\ddot{w} = 0 \qquad (2.10.11)
$$

$$
EI\frac{\partial^2 w}{\partial x^2} = 0. \qquad (2.10.12)
$$

The nonconservative nature of the problem manifests itself in the boundary condition given by Eq. (2.10.11). The natural moment boundary condition is given by Eq. (2.10.12).

## BIBLIOGRAPHY

Dym, C. L. and Shames, I. H. (1973). *Solid Mechanics: A Variational Approach*. McGraw-Hill Book Co.

Geradin, M. and Rixen, D. (2015). *Mechanical Vibrations Theory and Application to Structural Dynamics*. John Wiley & Sons, 3rd edition.

Gould, P. L. (1994). *Introduction to Linear Elasticity*. Springer, 2nd edition.

Greenwood, D. T. (1988). *Principles of Dynamics*. Prentice Hall, 2nd edition.

Hamilton, W. R. (1834). On a general method in dynamics, by which the study of motions of all free systems of attracting or repelling points is reduced to the search and differentiation of one central relation, or characteristic function. *Philosophical Transaction of the Royal Society, London*, 124: 247–308.

Hamilton, W. R. (1835). Second essay on a general method in dynamics. *Philosophical Transactions of the Royal Society, London*, 125: 95–144.

Hurty, W. and Rubinstein, M. (1964). *Dynamics of Structures*. Prentice Hall.

Kausel, E. (2017). *Advanced Structural Dynamics*. Cambridge University Press.

Langhaar, H. L. (1962). *Energy Methods in Applied Mechanics*. John Wiley & Sons.

Meirovitch, L. (1967). *Analytical Methods in Vibrations*. The Macmillan Co.

Shames, I. H. and Dym, C. (1985). *Energy and Finite Element Methods in Structural Mechanics*. McGraw-Hill Book Co., Hemisphere Publishing co-edition.

Tauchert, T. R. (1974). *Energy Principles in Structural Mechanics*. McGraw-Hill Book Co.

Washizu, K. (1982). *Variational Methods in Elasticity and Plasticity*. Pergamon Press, 3rd edition.

## PROBLEMS

1. If the square matrix $[A]$ is symmetric show that

$$\{a\}^T[A]\{b\} = \{b\}^T[A]\{a\}.$$

2. The beam in the figure is inextensible. It is restrained at its ends by two elastic springs. Then mass per unit length of the beam is $m$. Express the displacement of the beam by

$$u(x) = \phi_1(x)q_1$$
$$w(x) = \phi_2(x)q_2 + \phi_3(x)q_3 + \phi_4(x)q_4$$
$$= \sum_{1=2}^{4} \phi_i(x)q_i,$$

where

$$\phi_1(x) = 1$$
$$\phi_2(x) = \left(\frac{x}{l}\right)$$
$$\phi_3(x) = \left(\frac{x}{l}\right)^2$$
$$\phi_4(x) = \left(\frac{x}{l}\right)^3.$$

The kinetic energy of the beam in motion is given by

$$T = \frac{1}{2} \int_0^1 m \left( \dot{w}^2 + \dot{u}^2 \right) dx.$$

Starting with this expression, derive the generalized mass matrix of the beam in the $q$ coordinate system. In the resulting matrix

$$m_{j1} = m_{1j} = 0 \quad \text{for} \quad j = 2, 3, 4.$$

Discuss why this is to be expected in the present problem.

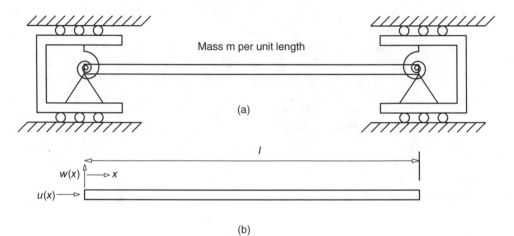

Mass m per unit length

(a)

(b)

3.  A mass $m$ is connected to two springs of stiffness $k_1$ and $k_2$ which are in turn attached to supports that are undergoing harmonic motions of the form

$$u_{g1}(t) = a_1 \sin p_1 t$$
$$u_{g2}(t) = a_2 \sin p_2 t,$$

where $a_1$ and $a_2$ are independent amplitudes.
(a)  Write the equation of motion of this system in terms of the absolute displacement $u(t)$ for given motions of the supports.
(b)  Write down the particular solution for the absolute displacement $u_p(t)$ due to motions of the support.

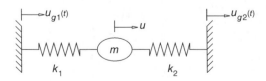

(c) If these two sinusoidal responses comprising the particular solution in (b) were combined into a single response, what are (1) the beat frequency, (2) the time dependent amplitude, and (3) the phase angle $\psi$ of this beat motion in terms of the parameters.

$$m = 1.0 \text{ kip sec}^2/\text{ft}$$

$$k_1 = 0.40 \text{ kip/ft}$$

$$k_2 = 0.60 \text{ kip/ft}$$

$$a_1 = 0.5 \text{ ft}$$

$$a_2 = 0.375 \text{ ft}$$

$$p_1 = 0.4472 \text{ rad/sec}$$

$$p_2 = 0.500 \text{ rad/sec}$$

4.  (a) Consider the system shown below. Choose coordinates $x$ for the displacement of point c and $\theta$-clockwise for the rotation of the uniform bar shown in the figure below. Derive the equations of motion for the free vibrations of this system, assuming that the mass of the bar is uniform with mass $m$ per unit length and length $\ell$.

    (b) Determine the natural frequencies and mode shapes, assuming that the beam (bar) is rigid.

    (c) How would you modify the answer to (a) given that the beam is flexible and has a bending stiffness EI?

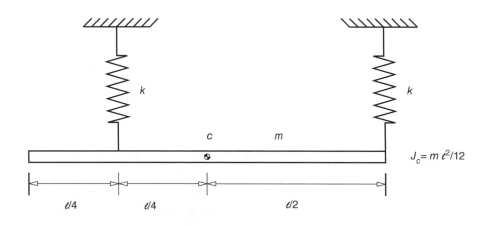

5.  An airfoil to be tested in a wind tunnel is supported by a linear spring $k$ and a torsional spring $K$, as shown in the figure below. If the center of gravity of the section is a distance $e$ ahead of the point of support, determine the differential equations of motion of the

system. The moment of inertia of the airfoil about the elastic axis, which is the point where the springs are attached, $J_0$ is the rotational inertia of the airfoil and $m$ is the mass of the airfoil.

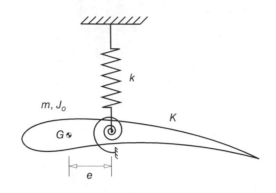

6.  Using Lagrange's equations, formulate the equations of motion for the mass-spring system shown. The uniform disk has a mass $m$ per unit volume and rolls without slip in the plane of the paper. Mass $m_2$ translates without friction in the horizontal direction and is acted upon by force $F_2(t)$.

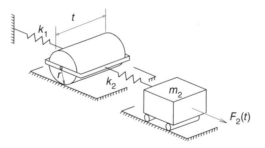

7.  Consider a uniform airplane wing of semispan $l$ cantilevered, as shown. The wing carries a tip-tank of mass $m$ and mass radius of gyration $\rho$ taken relative to a lateral axis through its center of mass. The center of mass of the tank lies behind the elastic axis of the wing at a distance $e$. The stiffness moduli of the wing are constant and are $EI$ and $GJ$ in flexure and torsion, respectively. Considering only vertical bending and twisting of the wing and neglecting its mass, formulate the equations of motion using Lagrange's equations.

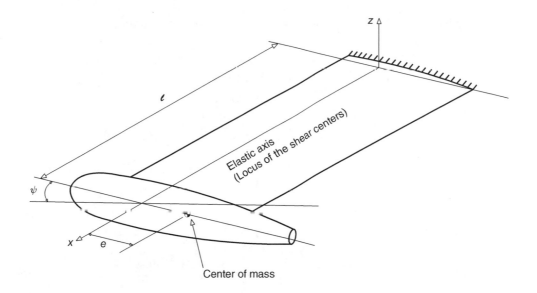

8. Using Lagrange's equations, derive the equations of motion for the following system.

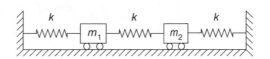

9. Consider a beam on an elastic foundation shown below, with one end clamped. The geometric boundary conditions are

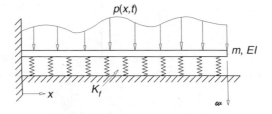

$$w\,(x = 0) = w'\,(x = 0) = 0$$

$K_f$ = elastic foundation spring constant/per unit length

$p\,(x, t)$ = load/unit length

$m$ = mass/unit length;

$EI$ = stiffness

Using Hamilton's principle, derive the differential equations of motion for the beam and the appropriate boundary conditions.

How would you modify the answer to the previous problem by the addition of a simple support and an axial load $P$ as indicated below?

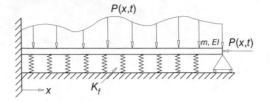

10. Consider the structural system shown below (consider the mass $M$ to be a concentrated mass).

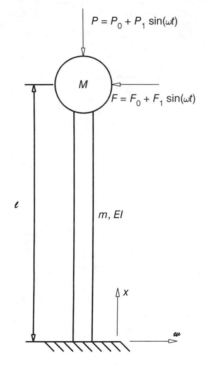

(a) Write the equations of motion and the boundary condition describing both forced and the free vibrations of this system using Hamilton's principle.

(b) Assume that the lateral displacements of this system can be expressed by two distributed coordinates

$$w(x,t) = \sum_{i}^{2} = \phi_i(x)q_i(t).$$

Using these, formulate the equations of motion for the free and forced vibration of the system using Lagrange's equations. Set up the equations in a general form, and then specialize them by assuming

$$\phi_1 = \left(\frac{x}{l}\right)^2 \left[3 - \left(\frac{x}{l}\right)\right]\frac{1}{2}; \quad \phi_2 = \left(\frac{x}{l}\right)^3 \left[3 - \left(\frac{x}{l}\right)\right]\frac{1}{2}.$$

(c) What fundamental difference do you notice by applying the forcing

$$F = F_o + F_1 \sin \omega t$$

as opposed to

$$P = P_o + P_1 \sin \omega t.$$

(d) What would you expect to occur if the force $P$ or the mass $M$ is increased significantly?

11. A spherical water tank is mounted eccentrically on a tower having bending and torsional rigidity given by $EI(x)$, $GJ(x)$ and mass distribution $m(x)$ and polar mass moment of inertia $I_p(x)$ per unit length as shown in the figure. The eccentricity given by $e$ is in a plane parallel to the $z-y$ plane.

The mass of the water tank is $M$ and its rotational inertias are $I_z$ and

$$I_y = I_x = I_z + e^2 M.$$

Note that the CG of the water tank does not coincide with the elastic axis of the tower.

Furthermore, while the effect of the gravitational force on the mass $M$ is important, the effect of gravitation on $m(x)$ can be neglected.

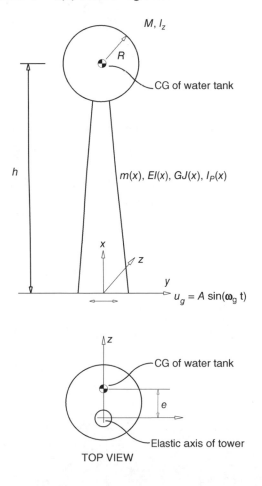

The system is excited by ground motion such as an earthquake in the $y$ direction

$$U_g = A \sin \omega_g t,$$

where $\omega_g$ is the frequency of the ground motion (rad/sec).

(a) Using any method suitable (such as Hamilton's principle, for example), derive the partial differential equations of motion and boundary conditions for this structure.

(b) Using one appropriate mode shape (approximate, assumed) for each of the governing degrees of freedom of this problem, set up the forced vibration problem, in symbolic form, such that the spatial dependence of the problem is eliminated.

(c) How would you modify the answers to (a) and (b) if $e = 0$?

12. Consider the structural dynamic system shown below which consists of a cantilevered beam with uniform mass per unit length and stiffness properties $m$ and $EI$, respectively. A mass $M$ is located at the middle of the beam, and at the right end a spring with spring stiffness $K$ is attached.

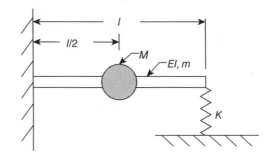

(a) Using Hamilton's principle, derive the equations of motion and boundary conditions for this structural dynamic system.

(b) Write down the orthogonality conditions with respect to mass and stiffness for this system.

(c) Using Rayleigh's quotient estimate the fundamental frequency of this system. Justify the selection of the shape function.

# 3  Natural Modes of Vibration: Continuous Systems

## 3.1  Introduction

Free vibrations of an ideal elastic structure with no damping can be considered as a superposition of *natural modes* (also called "normal modes," "principal modes," or "eigenmodes"). With proper initial conditions, it is possible to cause the structure to vibrate in just one of its natural modes. Each point on the structure then undergoes simple harmonic motion about its static equilibrium position, and all points are in phase with each other. Thus, a natural mode is a *standing wave*.

In the absence of damping, free vibrations – once started – would continue forever. This is of course not observed in actual structures; however, many structures have such small amounts of damping that "free vibration" can persist for a considerable time. The term *free vibrations* here refers to vibrations in the absence of external time-dependent forces (but possibly including internal damping forces). Mathematically, free vibration of a structural system is determined by solving the *homogeneous* part of the equation of motion, with appropriate boundary conditions and initial conditions. In the simplest case, the effect of damping is to cause a decay of the modal amplitudes *without* altering the mode shape or introducing phase differences between the various points on the structure. It is then reasonable to speak of "damped natural modes," since the amplitude of an eigenmode of a linear system is arbitrary.

As indicated, "classical" damped normal modes (i.e. uncoupled standing waves) exist only if the damping is of a special type, so that the damping operators are diagonalizable by a congruence transformation involving the eigenvectors of the free vibration problem. For example, the so-called "proportional" or "Rayleigh" damping is of this type. If the damping matrices or damping operators are not diagonalizable by the eigenvectors (normal modes) of the *undamped* system, then normal modes do not exist in the sense discussed earlier. Some typical examples of undamped modes for discretized continuous system are shown in Figure 3.1.

### 3.1.1  Basic Assumptions

Here and in what follows, the following assumptions are introduced:

(1) linear $\sigma_{ij}, \epsilon_{ij}$ relations (Hooke's law)
(2) linear displacement–strain relations:

$$\epsilon_{ij} = \frac{1}{2}\Big(u_{i,j} + u_{j,i}\Big).$$

**Figure 3.1**
Mode shapes
of a discrete
spring-mass
system (left)
and mode
shapes of a
continuous
system (right)

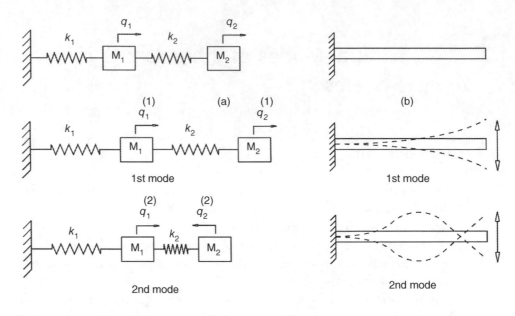

These two assumptions result in *linear* equations of motion for the structure, and allow the use of the *principle of superposition* in solving vibration problems. In the case of free vibrations, for example, the superposition principle implies that the sum of any two free vibration solutions is also a permissible free vibration solution. If this process is repeated, one obtains the classical principle of the *coexistence of small motions*, first stated in a clear manner by Daniel Bernoulli (see Rayleigh 1894, 105).

Many engineering structures can be analyzed with the above assumptions; however, it would be misleading to give the impression that a linear analysis is adequate for all cases of engineering relevance. Often, one encounters structures where the first assumption (Hooke's law) holds, but the second assumption (linear strain–displacement relations) does not. These problems require the use of *nonlinear* strain–displacement relations in order to get realistic equations, because the displacements $u_i$ cannot be assumed sufficiently small to neglect the nonlinear terms in the strain–displacement relations:

$$\epsilon_{ij} = \frac{1}{2}\left(u_{i,j} + u_{j,i} + u_{i,k}u_{j,k}\right). \tag{3.1.1}$$

Examples of such structures include long, very flexible beams, helicopter blades, wings of high altitude long endurance aircraft (HALE), cables with significant "sag," and other problems that are *geometrically nonlinear*. In most problems, the equations of motion can be linearized about the static (or mean) equilibrium position, resulting in *linear* equations for the dynamic perturbation equations:

$$u_i\left(x, y, z, t\right)_{\text{tot}} = u_{0_i}\left(x, y, z\right) + u_i\left(x, y, z, t\right). \tag{3.1.2}$$

Here, the $u_{0_i}$s represent the static equilibrium displacements and the $u_i$s the oscillations about this equilibrium. Generally, the so-called "variational equations" for the perturbation displacements $u_i$ will depend on the equilibrium solutions $u_{0_i}$, and partial differential equations with variable coefficients are obtained. The equations are nevertheless linear, in contrast to the equilibrium equations for the static displacement field $u_{0_i}$, which would be nonlinear.

### 3.1.2  Common Methods of Analysis

A number of different methods have been developed for obtaining the free vibration solutions, including the normal modes, for engineering structures. The methods can be classified according to the approach taken in formulating and solving the homogeneous boundary value problem:

(1) The Differential Equation Method;
(2) The Integral Equation Method;
(3) The Rayleigh–Ritz Method;
(4) Galerkin's Method;
(5) Finite Element Methods;
(6) Finite Difference Methods.

The first two are "exact" within the original assumptions, whereas the last four are approximate methods.

In this book, the differential equation method will be favored over the integral equation method, because the power of the latter cannot be fully appreciated without the use of the theory of integral equations.

## 3.2  Free Vibrations of Axial Members

It is convenient to start by considering the *exact* solution of a simple problem: the natural vibrations of an axial member, or a bar as shown in Figure 3.2.

**Figure 3.2** Geometry of an axially loaded bar

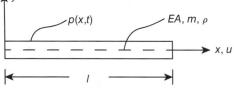

In this figure, $p(x, t)$ is a distributed load per unit length, $E$ is Young's modulus, $A$ is the cross-sectional area of the bar, $l$ is the length of the bar, $m$ is the mass per unit length, and $\rho$ is the material density of the bar. The equation of motion is obtained from Hamilton's principle described in Chapter 2

$$\int_{t_1}^{t_2} (\delta W_E - \delta U + \delta T)\, dt = 0 \tag{3.2.1}$$

$$U = \frac{1}{2} \int_0^l EA \left( \frac{\partial u}{\partial x} \right)^2 dx \tag{3.2.2}$$

$$T = \frac{1}{2} \int_0^l m \left( \frac{\partial u}{\partial t} \right)^2 dx \tag{3.2.3}$$

$$W_E = \int_0^l p(x,t)u dx, \tag{3.2.4}$$

where $m = \rho A$. Taking the variations of Eqs. (3.2.2–3.2.4)

$$\int_{t_1}^{t_2} \delta U dt = \int_{t_1}^{t_2} \left\{ \int_0^l EA u_{,x} \delta u_{,x} dx \right\} dt$$

$$= \int_{t_1}^{t_2} \left\{ EA u_{,x} \delta u \Big|_0^l - \int_0^l EA u_{,xx} \delta u dx \right\} dt \tag{3.2.5}$$

$$\int_{t_1}^{t_2} \delta T dt = \int_{t_1}^{t_2} \int_0^l m \ddot{u} \delta u dx dt$$

$$= m \dot{u} \delta u \Big|_{t_1}^{t_2} - \int_0^l \int_{t_1}^{t_2} m \ddot{u} \delta u dx = - \int_{t_1}^{t_2} \int_0^l m \ddot{u} \delta u dx dt. \tag{3.2.6}$$

Since in Hamilton's principle $\delta u(t_1) = \delta u(t_2) = 0$,

$$\int_{t_1}^{t_2} \delta W_E dt = \int_{t_1}^{t_2} \int_0^l p(x,t) \delta u dx dt. \tag{3.2.7}$$

The geometric boundary conditions for this case, where the left end of the bar is restrained, is $u(0) = 0$. Substituting Eqs. (3.2.5) through (3.2.7) into Hamilton's principle, Eq. (3.2.1) yields

$$\int_{t_1}^{t_2} \left\{ -EA u_{,x} \delta u(l) + \int_0^l \left[ EA u_{,xx} - m\ddot{u} + p(x,t) \right] \delta u dx \right\} dt = 0. \tag{3.2.8}$$

Since the variation $\delta u$ is arbitrary, Eq. (3.2.8) yields the differential equation of motion and the natural boundary conditions (NBC)

$$EA \frac{\partial^2 u}{\partial x^2} - m \frac{\partial^2 u}{\partial t^2} + p(x,t) = 0, \tag{3.2.9}$$

the NBC at $x = l$ is

$$EA \frac{\partial u}{\partial x} = 0, \tag{3.2.10}$$

and the geometric BC is $u(0) = 0$.

For the case of free vibrations $p(x,t) = 0$, and Eq. (3.2.9) reduces to

$$EA \frac{\partial^2 u}{\partial x^2} = \rho A \frac{\partial^2 u}{\partial t^2}. \tag{3.2.11}$$

Equation (3.2.11) is the famous **wave equation** which governs the speed of wave propagation in a solid, and it can be rewritten as

$$\frac{\partial^2 u}{\partial x^2} = \frac{\rho}{E}\frac{\partial^2 u}{\partial t^2} = \frac{1}{c^2}\frac{\partial^2 u}{\partial t^2}, \tag{3.2.12}$$

where $c^2 = \frac{E}{\rho}$ and $c = \sqrt{\frac{E}{\rho}}$ is the speed of sound in the solid.

Solution of Eq. (3.2.11) is known as D'Alembert's solution

$$u(x,t) = f_1(x+ct) + f_2(x-ct), \tag{3.2.13}$$

where $f_1$ and $f_2$ are arbitrary functions, determined from initial and boundary conditions. D'Alembert's solution is most useful in wave propagation analyses, which is not the objective of this section. Instead, the main objective here is to obtain harmonic solutions in time, representing natural modes, which are **standing waves** and not travelling waves. The solution is obtained by separation of variables. Let $u(x,t)$ be

$$u(x,t) = \bar{u}(x)T(t) \tag{3.2.14}$$

$$u_{,xx} = \bar{u}_{,xx}T(t) \text{ and } \frac{\partial u}{\partial t^2} = \bar{u}(x)\ddot{T}. \tag{3.2.15}$$

Substituting relations Eq. (3.2.15) into Eq. (3.2.11) yields

$$\frac{\bar{u}_{,xx}}{\bar{u}(x)} = \frac{1}{c^2}\frac{\ddot{T}(t)}{T(t)} = \text{const.} = -k^2 = -\frac{\omega^2}{c^2}. \tag{3.2.16}$$

Since each side of Eq. (3.2.16) depends on a different independent variable, thus

$$\ddot{T} + \omega^2 T = 0 \tag{3.2.17}$$

$$\text{and} \quad \bar{u}_{,xx} + k\bar{u} = 0. \tag{3.2.18}$$

The solution of Eq. (3.2.17) is

$$T(t) = A\sin\omega t + B\cos\omega t \tag{3.2.19}$$

$$\text{and} \quad u(x,t) = \bar{u}(x)(A\sin\omega t + B\cos\omega t), \tag{3.2.20}$$

where $A$ and $B$ are determined from the initial conditions and $\bar{u}(x)$ is obtained from Eq. (3.2.18).

$$\frac{d^2\bar{u}}{dx^2} + \frac{\omega^2}{c^2}\bar{u} = 0. \tag{3.2.21}$$

Its solution is

$$\bar{u}(x) = c_1\sin\frac{\omega}{c}x + c_2\cos\frac{\omega}{c}x, \tag{3.2.22}$$

where $c_1$ and $c_2$ are determined from the boundary conditions.

## 3.2.1 Boundary Conditions

Pinned end or free, shown in Figure 3.3.

**Figure 3.3** Illustration of pinned boundary condition

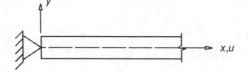

$$u(0, t) = 0 \quad \text{for all} \quad t \quad \text{which implies} \quad \bar{u}(0, t) = 0. \tag{3.2.23}$$

Free end, shown in Figure 3.4.

**Figure 3.4** Illustration of free end boundary condition

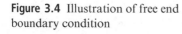

$$P = \sigma_{xx} A = 0 = EA\epsilon_{xx} = EAu_{,x}$$

$$u_{,x} = \frac{\partial u}{\partial x} = \frac{\partial \bar{u}}{\partial x} \cdot T(t) \quad \text{which implies} \quad \bar{u}'(0) = 0. \tag{3.2.24}$$

Equation (3.2.11) together with the appropriate boundary conditions constitutes an eigenvalue problem; the eigenvalues determine the natural frequencies, and the eigenfunctions represent the natural mode shapes.

## Example: Fixed-free Member, Shown in Figure 3.5

**Figure 3.5** Bar with fixed free ends

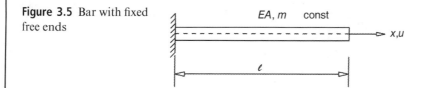

Boundary conditions are

$$\bar{u}(0) = \bar{u}'(\ell) = 0 \tag{3.2.25}$$

and from (3.2.22):  $\bar{u}(0) = C_2 \cos\left(\frac{\omega}{c} \cdot 0\right) = 0 \implies C_2 = 0$

$$\bar{u}'(\ell) = \frac{\omega}{c} \cdot C_1 \cos\frac{\omega}{c}\ell = 0.$$

For nontrivial solutions, $C_1 \neq 0$, thus

$$\cos\frac{\omega}{c}\ell = 0 \implies \frac{\omega}{c}\ell = (2n - 1) \cdot \frac{\pi}{2} \ (= k\ell) \tag{3.2.26}$$

and   $\omega_n = (2n-1)\frac{\pi}{2\ell}\sqrt{\frac{E}{\rho}}$ ; $n = 1, 2, \ldots$

are the natural frequencies of axial vibrations of the bar.

The displacement field corresponding to the $n$th natural mode is then

$$u_n(x, t) = \sin\left[(2n-1)\frac{\pi x}{2\ell}\right]\{A_n \sin \omega_n t + B_n \cos \omega_n t\}$$
$$= \phi_n(x) \cdot T_n(t), \qquad (3.2.27)$$

where $\phi_n(x) = \bar{u}_n(x)$ are called the *mode shapes*, and these are shown in Figure 3.6.

**Figure 3.6** Mode shapes of a bar with fixed free-end conditions

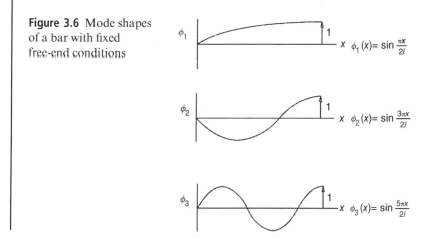

$\phi_1(x) = \sin \frac{\pi x}{2l}$

$\phi_2(x) = \sin \frac{3\pi x}{2l}$

$\phi_3(x) = \sin \frac{5\pi x}{2l}$

## 3.3 The Eigenvalue Problem: For Beam Vibrations

### 3.3.1 Exact Solution of the Eigenvalue Problem for Beams

Recall that the equation of motion for free vibrations of a slender beam using Hamilton's principle was obtained earlier and is given by

$$\frac{\partial^2}{\partial x^2}\left[EI(x)\frac{\partial^2 w(x,t)}{\partial x^2}\right] = -m(x)\frac{\partial^2 w(x,t)}{\partial t^2}. \qquad (3.3.1)$$

Assume solution to Eq. (3.3.1) to be separable in time and space

$$w(x, t) = W(x)f(t). \qquad (3.3.2)$$

From Eqs. (3.3.1) and (3.3.2),

$$\underbrace{\frac{1}{m(x)W(x)}\frac{d^2}{dx^2}\left[EI\frac{d^2 W(x)}{dx^2}\right]}_{x\text{-dependent}} = \underbrace{-\frac{1}{f(t)}\frac{d^2 f(t)}{dt^2}}_{t\text{-dependent}} \qquad (3.3.3)$$

$x$ and $t$ are independent variables; thus, Eq. (3.3.3) has a solution when both sides are constant, choosing the constant to be $\omega^2$. This value of the constant is selected to obtain simple harmonic motion in time, which is a basic requirement for free vibrations.

$$\frac{d^2}{dx^2}\left[EI(x)\frac{d^2W}{dx^2}\right] - \omega^2 m(x)W(x) = 0 \tag{3.3.4}$$

$$\frac{d^2f}{dt^2} + \omega^2 f(t) = 0. \tag{3.3.5}$$

Note that Eq. (3.3.4) is a fourth-order homogeneous differential equation which must have four boundary conditions.

The problem of determining the value of $\omega^2$ for which the homogeneous differential equation (3.3.4) has a nontrivial solution $W(x)$, satisfying the boundary conditions, is called the eigenvalue problem. The values of $\omega^2$s are the eigenvalues and $W(x)$ are eigenfunctions.

Equation (3.3.4) is a homogeneous differential equation; thus the solution can only be determined as multiplied by an arbitrary constant. Exact solution of Eq. (3.3.4) with arbitrary mass and stiffness variation cannot be obtained.

To further illustrate this problem consider a slender *cantilevered beam* with uniform mass and stiffness $m(x) = m$; $EI(x) = EI$ from Eq. (3.3.4).

$$\frac{d^2}{dx^2}\left(EI\frac{d^2W}{dx^2}\right) - \omega^2 mW = 0$$

or

$$\frac{d^4W}{dx^4} - \frac{\omega^2 m}{EI}W = 0$$

where

$$\frac{d^4W}{dx^4} - \beta^4 W = 0, \tag{3.3.6}$$

and $\beta^4 = \omega^2 m/EI$.

A solution of the form $W = e^{rx}$ satisfies Eq. (3.3.6); thus, the general solution has the form of

$$W(x) = A_1 e^{r_1 x} + A_2 e^{r_2 x} + A_3 e^{r_3 x} + A_4 e^{r_4 x} \tag{3.3.7}$$

and corresponds to the solution of the characteristic equation

$$r^4 - \beta^4 = (r^2 + \beta^2)(r^2 - \beta^2) = 0,$$

which yields

$$W = A_1 e^{\beta x} + A_2 e^{-\beta x} + A_3 e^{i\beta x} + A_4 e^{-i\beta x}. \tag{3.3.8}$$

Equation (3.3.8) is frequently rewritten in a more convenient form as

$$W(x) = c_1 \cosh \beta x + c_2 \sinh \beta x + c_3 \cos \beta x + c_4 \sin \beta x. \tag{3.3.9}$$

The boundary conditions for the cantilevered beam are

$$W(0) = \frac{dW(0)}{dx} = 0$$

$$\frac{d^2 W}{dx^2}(x = \ell) = \frac{d^3 W}{dx^3}(x = \ell) = 0. \tag{3.3.10}$$

Taking the various derivatives of Eq. (3.3.9) as needed for the implementation of the boundary conditions, one has

$$\frac{dW}{dx} = \beta \left[ c_1 \sinh \beta x + c_2 \cosh \beta x - c_3 \sin \beta x + c_4 \cos \beta x \right]$$

$$\frac{d^2 W}{dx^2} = \beta^2 \left[ c_1 \cosh \beta x + c_2 \sinh \beta x - c_3 \cos \beta x - c_4 \sin \beta x \right] \tag{3.3.11}$$

$$\frac{d^3 W}{dx^4} = \beta^3 \left[ c_1 \sinh \beta x + c_2 \cosh \beta x + c_3 \sin \beta x - c_4 \cos \beta x \right]$$

satisfying the boundary conditions, one has

$$\begin{bmatrix} 1 & 0 & 1 & 0 \\ 0 & 1 & 0 & 1 \\ \cosh \beta \ell & \sinh \beta \ell & -\cos \beta \ell & -\sin \beta \ell \\ \sinh \beta \ell & \cosh \beta \ell & \sin \beta \ell & -\cos \beta \ell \end{bmatrix} \begin{Bmatrix} c_1 \\ c_2 \\ c_3 \\ c_4 \end{Bmatrix} = 0, \tag{3.3.12a}$$

which can be written more compactly as

$$[B_c]\{c\} = 0, \tag{3.3.12b}$$

where the definitions of $[B_c]$ and $\{c\}$ are evident from comparing Eqs. (3.3.12a) and (3.3.12b). The homogeneous linear system represented by Eq. (3.3.12b) has a nontrivial solution only for

$$\det \left| B_c \right| = 0. \tag{3.3.13}$$

Expanding the determinant yields

$$\cos^2 \beta \ell + \sin^2 \beta \ell + \sin \beta \ell \sinh \beta \ell + \cosh \beta \ell \cos \beta \ell + \cosh \beta \ell \cos \beta \ell$$
$$- \sinh \beta \ell \sin \beta \ell + \cosh^2 \beta \ell - \sinh^2 \beta \ell = 0$$

or

$$\cosh \beta \ell \cos \beta \ell + 1 = 0, \tag{3.3.14a}$$

which can also be written as

$$\cosh \beta \ell = -\frac{1}{\cos \beta \ell} = -\sec \beta \ell. \tag{3.3.14b}$$

Solutions to a transcendental equation such as Eq. (3.3.14a) can be obtained easily by numerical means or graphical means (Hurty and Rubinstein 1964, 200). Typical results for the first three frequencies are (Young and Felgar 1949, Chang and Craig 1969): $\beta_1 \ell = 1.8751$; $\beta_2 \ell = 4.6941$; $\beta_3 \ell = 7.8548$; and

$$\omega_n = \frac{(\beta_n \ell)^2}{\ell^2} \sqrt{\frac{EI}{m}}$$

$$\omega_1 = \frac{3.516}{\ell^2} \sqrt{\frac{EI}{m}}$$

$$\omega_2 = \frac{22.03}{\ell^2} \sqrt{\frac{EI}{m}}$$

$$\omega_3 = \frac{61.70}{\ell^2} \sqrt{\frac{EI}{m}}.$$

The mode shapes are obtained by using three of the four equations represented by Eq. (3.3.12a), to express the constants $c$ in terms of the remaining fourth constant which remains arbitrary. From the first two equations in Eq. (3.3.12a), we have

$$c_3 = -c_1$$

$$c_4 = -c_2 \qquad (3.3.15)$$

and from the third line of the equation in Eq. (3.3.12a)

$$c_1 \cosh \beta_n \ell + c_2 \sinh \beta_n \ell - c_3 \cos \beta_n \ell - c_4 \sin \beta_n \ell = 0,$$

which combined with Eq. (3.3.15) yields

$$c_1 \cosh \beta_n l + c_2 \sinh \beta_n l + c_1 \cos \beta_n l + c_2 \sin \beta_n l = 0$$

$$c_2 = -\frac{c_1 (\cosh \beta_n \ell + \cos \beta_n \ell)}{(\sinh \beta_n \ell + \sin \beta_n \ell)}. \qquad (3.3.16)$$

Combining Eqs. (3.3.15) and (3.3.16) with Eq. (3.3.9) yields the mode shape

$$W_n(x) = c_1 \left[ \left( \frac{\cosh \beta_n \ell + \cos \beta_n \ell}{\sinh \beta_n \ell + \sin \beta_n \ell} \right) (\sin \beta_n x - \sinh \beta_n x) + (\cosh \beta_n x - \cos \beta_n x) \right],$$
$$(3.3.17)$$

where it should be understood that $c_1$ is an arbitrary coefficient which could be also replaced by a quantity $c = c_1$.

The complete solution for free vibration in any mode will yield

$$f_n(t) = A \sin \omega_n t + B \cos \omega_n t, \qquad (3.3.18)$$

where $A$ and $B$ are determined from the initial conditions on displacement and velocity at time $t = t_0$, and the complete motion will be given by

$$w(x, t) = W_n(x) f_n(t).$$

The first three modes of a typical uniform cantilevered beam are illustrated in Figure 3.7 (Young and Felgar 1949, Chang and Craig 1969).

**Figure 3.7** First three mode shapes of a uniform cantilevered beam

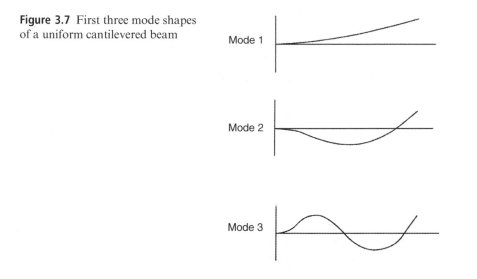

Mode 1

Mode 2

Mode 3

## 3.3.2    Orthogonality Conditions for a Vibrating Beam

Recall Eq. (3.3.4) of Section 3.3.1

$$[EI(x)W,_{xx}(x)],_{xx} - \omega^2 m(x)W(x) = 0$$

and consider two distinct eigenvalues and the associated eigenfunctions $W_n(x)$ and $W_m(x)$, respectively, which satisfy the last equation:

$$\left[EI(x)W_{m,xx}\right],_{xx} - m(x)\omega_m^2 W_m = 0 \qquad (3.3.19\text{a})$$

$$\left[EI(x)W_{n,xx}\right],_{xx} - m(x)\omega_n^2 W_n = 0. \qquad (3.3.19\text{b})$$

Multiplying Eq. (3.3.19a) by $W_n$ and Eq. (3.3.19b) by $W_m$ and integrating over the span of the beam yields

$$\int_0^\ell \left[EI(x)W_{m,xx}\right],_{xx} W_n dx = \omega_m^2 \int_0^\ell m(x)W_m W_n dx$$

$$\int_0^\ell \left[EI(x)W_{n,xx}\right],_{xx} W_m dx = \omega_n^2 \int_0^\ell m(x)W_n W_m dx. \qquad (3.3.20)$$

Subtracting the right sides of these equations from each other and performing two consecutive integration by parts on the left side gives the following expression:

$$(\omega_m^2 - \omega_n^2) \int_0^\ell m(x) W_m(x) W_n(x) dx$$

$$= \int_0^\ell \left\{ \left[ EI(x) W_{m,xx} \right]_{,xx} W_n - \left[ EI(x) W_{n,xx} \right]_{,xx} W_m \right\} dx$$

$$= \left[ EI(x) W_{m,xx} \right]_{,x} W_n \Big|_0^\ell - \int_0^\ell \left[ EI(x) W_{m,xx} \right]_{,x} W_{n,x} dx$$

$$- \left[ EI(x) W_{n,xx} \right]_{,x} W_m \Big|_0^\ell + \int_0^\ell \left[ EI(x) W_{n,xx} \right]_{,x} W_{m,x} dx$$

$$= \left[ EI(x) W_{m,xx} \right]_{,x} W_n \Big|_0^\ell - \left[ EI(x) W_{m,xx} \right] W_{n,x} \Big|_0^\ell + \int_0^\ell EI(x) W_{m,xx} W_{n,xx} dx$$

$$- \left[ EI(x) W_{n,xx} \right]_{,x} W_m \Big|_0^\ell + \left[ EI(x) W_{n,xx} \right] W_{m,x} \Big|_0^\ell$$

$$- \int_0^\ell EI(x) W_{m,xx} W_{n,xx} dx. \tag{3.3.21}$$

The right-hand side of Eq. (3.3.21) vanishes, if at each end of the beam one prescribes at least one of the following pairs of boundary conditions:

(a) $W = 0$ and $W_{,x} = 0$, that is, cantilevered end (Figure 3.8)

**Figure 3.8** Cantilevered support at left end

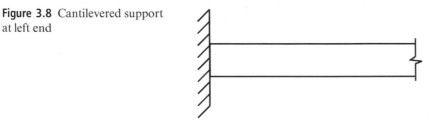

(b) $W = 0$ and $EI(x) W_{,xx} = 0$, that is, simple support (Figure 3.9)

**Figure 3.9** Simple support at left end

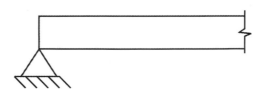

(c) $W_{,x} = 0$ and $\left[ EI(x) W_{,xx} \right]_{,x} = 0$, which corresponds to a sliding support (Figure 3.10).

**Figure 3.10** Sliding support at left end

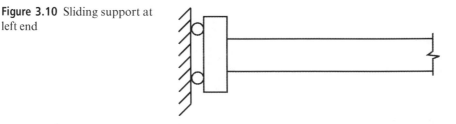

(d) $EI(x) W_{,xx} = 0$ and $[EI(x) W_{,xx}]_{,x} = 0$, which corresponds to a free end (Figure 3.11).

**Figure 3.11** Free end on left side

The four pairs of boundary conditions described earlier are sometimes denoted as classical boundary conditions for beam. When these boundary conditions are satisfied and $\omega_m^2 \neq \omega_n^2$, Eq. (3.3.21) reduces to

$$\left(\omega_m^2 - \omega_n^2\right) \int_0^\ell W_m(x) W_n(x) m(x) dx = 0$$

or

$$\int_0^\ell m(x) W_m(x) W_n(x) dx = 0 \quad \text{for } m \neq n. \tag{3.3.22}$$

Equation (3.3.22) represents the orthogonality property for eigenfunctions of a vibrating beam, with respect to the mass distribution $m(x)$. Next returning to Eqs. (3.3.20) and using Eq. (3.3.22), we obtain

$$\int_0^\ell \left[EI(x) W_{m,xx}\right]_{,xx} W_n dx = 0 \quad \text{for } \omega_n \neq \omega_m. \tag{3.3.23}$$

Integrating Eq. (3.3.23) by parts twice and using the boundary condition pairs (a) through (d) given above, one has

$$\int_0^\ell EI(x) W_{m,xx} W_{n,xx} dx = 0. \tag{3.3.24}$$

Equation (3.3.24) is another orthogonality condition for the eigenfunctions of a vibrating beam with respect to the stiffness distribution.

## 3.4 General Formulation of the Eigenvalue Problem

### 3.4.1 Introduction

The preceding example clearly illustrates that the variables representing the motion of a continuous system (which is also frequently denoted as a distributed parameter system) depend explicitly on space and time and thus the motion of the vibrating structure is described by partial differential equations. Furthermore, for the case of free vibrations, the solution is obtained by solving a boundary value problem. In the preceding example, we have obtained an exact, closed-form solution for the free vibrations of a uniform beam. Such closed-form solutions can only be obtained for a limited number of simple structural systems, which are usually characterized by uniform mass and stiffness distributions. Many special cases of vibrating systems that belong to this category are documented in Gorman (1975) and Blevins (1979).

As mentioned, the number of actual cases when exact solutions can be obtained is limited and therefore in most practical cases encountered in structural dynamics approximate solutions are sought. However, the exact solutions are extremely valuable because they serve as a basis for evaluating the accuracy of approximate solutions. Exact solutions also provide us with the basic shape functions which are needed for constructing approximate solutions, as will be shown later in this section.

The beam problem treated in Section 3.3.1 is the only one example of a wide class of problems lending themselves to the same general formulation. The treatment of continuous systems based upon such a general formulation is useful for identifying the fundamental common aspects of the structural dynamics eigenvalue problem.

## 3.4.2    General Mathematical Form of the Eigenvalue Problem

The eigenvalue problem can be formulated in general form by using a linear homogeneous differential expression that applies to one or two-dimensional problems (extension to three dimensions is obvious). Consider the expression

$$L[w] = A_{00}(x,y) + A_{1x}(x,y)\frac{\partial w}{\partial x} + A_{1y}(x,y)\frac{\partial w}{\partial y}$$
$$+ A_{2x}(x,y)\frac{\partial^2 w}{\partial x^2} + A_{2y}(x,y)\frac{\partial^2 w}{\partial y^2} + A_{2xy}(x,y)\frac{\partial^2 w}{\partial x \partial y} +$$
$$+ A_{2px}(x,y)\frac{\partial^{2p} w}{\partial x^{2p}}, \tag{3.4.1}$$

where $A_{00}(x,y), A_{1x}(x,y), A_{1y}(x,y), \ldots, A_{2px}, \ldots$, etc., are known functions of the spatial variables $x$ and $y$. If the expression $L[w]$ is a linear homogeneous combination of the function $w$ and its derivatives up to order $2p$, then the differential expression is said to be of order $2p$. Implicit in Eq. (3.4.1) is also the definition of a linear differential operator of the form

$$L[\ ] = A_{00}(x,y) + A_{1x}(x,y)\frac{\partial(\ )}{\partial x} + A_{1y}(x,y)\frac{\partial(\ )}{\partial y}$$
$$+ A_{2x}(x,y)\frac{\partial^2(\ )}{\partial x^2} + A_{2y}(x,y)\frac{\partial^2(\ )}{\partial y^2} + A_{2xy}(x,y)\frac{\partial^2(\ )}{\partial x \partial y}$$
$$+ \cdots A_{2px}(x,y)\frac{\partial^{2p}(\ )}{\partial x^{2p}} + \cdots, \tag{3.4.2}$$

where the linearity of the operator implies the property

$$L[C_1 w_1 + C_2 w_2] = C_1 L[w_1] + C_2 L[w_2]. \tag{3.4.3}$$

When the problem depends only on one space variable say $x$, the expression simplifies to

$$L[w] = A_0(x) + A_1(x)\frac{dw}{dx} + A_2(x)\frac{d^2 w}{dx^2} + \cdots A_{2p}(x)\frac{d^{2p} w}{dx^{2p}}. \tag{3.4.4}$$

One can also define an operator $M[\ ]$, which is similar to $L[\ ]$, except that its highest derivative is of order $2q$, where $q < p$. These two operators can be combined to form a differential equation

$$L[w] = \lambda M[w] \tag{3.4.5}$$

in which $\lambda$ is a parameter.

In addition to the differential equation that is valid for the domain $D$ (for one-dimensional structures $0 < x < \ell$), the function $w$ must also satisfy boundary conditions at ends $x = 0$ and $x = \ell$. These boundary conditions can be symbolically represented by

$$
\begin{aligned}
B_i[w] &= 0 & i &= 1, 2, \ldots, k, & \text{satisfied at } x = 0, \ell & & n = 1, 2, \ldots, p - k \\
B_j[w] &= \lambda C_j[w] & j &= 1, 2, \ldots, n & \text{and } n = p - k, \ldots, p & & x = 0, \ell,
\end{aligned} \tag{3.4.6}
$$

where $B_i[\ ]$, $B_j[\ ]$, and $C_j[\ ]$ are also linear homogeneous differential operators.

Note that the boundary conditions represented by Eq. (3.4.6) consist of boundary conditions that depend on the parameter $\lambda$ and those which do not depend on $\lambda$. The highest derivative in the boundary conditions, which determines the order of $B_i[\ ]$ or $B_j[\ ]$ and $C_j[\ ]$, is $2p - 1$ and $2q - 1$, respectively. The boundary conditions at the ends of the domain, $x = 0, \ell$, can be different and can have different orders. Thus, $B_i[\ ]$ at $x = 0$ can be different from $B_i[\ ]$ at $x = \ell$.

Solution of the eigenvalue problem implies determination of the eigenvalues $\lambda$ for which the eigenfunctions $w$ satisfy the differential Eq. (3.4.5) given over the open domain $D$, $(0 < x < \ell)$, together with the boundary conditions, Eq. (3.4.6), at every point of the boundary $S$ enclosing the domain $D$.

In many structural dynamics applications, the form of the $M[\ ]$ operator is simply a function $m(x)$, representing mass per unit length, corresponding to an arbitrary mass distribution. Furthermore, the boundary conditions are frequently independent of the eigenvalue $\lambda$. For these cases, a simpler representation of the eigenvalue problem is convenient.

$$L[w] = \lambda m(x) w \quad 0 < x < \ell \quad B_i[w] = 0 \quad i = 1, 2, \ldots, p; \quad x = 0, \ell. \tag{3.4.7}$$

In the general treatment of eigenvalue problems, one encounters three types of functions that are useful for establishing general features of the solutions. These functions also have an important role in the systematic generation of approximate solutions to the eigenvalue problem. These three types of functions are described as follows.

(a) Admissible functions

These are arbitrary functions that are $p$ times differentiable over the domain $D$ and satisfy the geometric boundary conditions on the boundary $S$.

(b) Comparison functions

These are arbitrary functions, which are $2p$ times differentiable over the domain $D$ and satisfy both the geometric and NBC on the boundary $S$.

(c) Eigenfunctions

These represent exact solutions of the eigenvalue problem, governed by Eqs. (3.4.5) and (3.4.6).

The eigenvalue problem representing the structural dynamics free vibration problem, with no damping, has two important properties: (1) it is self-adjoint, and (2) it is positive definite. These properties imply that the eigenvalues $\lambda$ are real and positive.

The eigenvalue problem is self-adjoint, if for any two comparison functions $u$ and $v$, the statements given below are satisfied.

$$\int_0^\ell uL[v]dx + \sum_{j=1}^n uB_j[v]\Big|_0^\ell = \int_0^\ell vL[u]dx + \sum_{j=1}^n vB_j[u]\Big|_0^\ell \qquad (3.4.8a)$$

$$\int_0^\ell uM[v]dx + \sum_{j=1}^n uC_j[v]\Big|_0^\ell = \int_0^\ell vM[u]dx + \sum_{j=1}^n vC_j[u]\Big|_0^\ell. \qquad (3.4.8b)$$

When the boundary conditions are independent of the eigenvalue $\lambda$, the condition for self-adjointness reduces to

$$(u, L[v]) = (v, L[u])$$
$$(u, M[v]) = (v, M[u]). \qquad (3.4.9)$$

The notation used in Eq. (3.4.9) employs the definition of the inner product of two functions. The inner product of two functions $f_1(x), f_2(x)$ defined over the domain $0 < x < \ell$ is defined as

$$(f_1, f_2) = \int_0^\ell f_1(x)f_2(x)dx. \qquad (3.4.10)$$

Self-adjointness in continuous systems corresponds to the symmetry of the stiffness and mass matrices encountered in the discrete representation of structural dynamic systems. Self-adjointness of structural dynamic systems is shown by performing integration by parts with respect to the spatial variables as illustrated in the examples presented after this section.

If for any comparison function $u$, the expressions

$$\int_0^\ell uL[u]dx + \sum_{j=1}^n uB_j[u]\Big|_0^\ell \geq 0$$
$$\int_0^\ell uM[u]dx + \sum_{j=1}^n uC_j[u]\Big|_0^\ell \geq 0 \qquad (3.4.11)$$

vanish only for $u$ identically zero and are positive otherwise, then both $L[\ ]$ and $M[\ ]$ are positive definite. Again for boundary conditions independent of $\lambda$, Eqs. (3.4.11) reduce to

$$(u, L[u]) \geq 0$$
$$(u, M[u]) \geq 0. \qquad (3.4.12)$$

A positive definite, self-adjoint, eigenvalue problem yields an infinite sequence of positive real eigenvalues $\lambda_r$ and corresponding eigenfunctions $w_r$.

## Generalized Orthogonality

A generalized orthogonality condition can be obtained for the eigenvalue problem defined by Eqs. (3.4.5) and (3.4.6). Let $\lambda_r$ and $\lambda_s$ be two distinct eigenvalues and let $w_r$ and $w_s$ be the corresponding eigenfunctions of the self-adjoint eigenvalue problem, which implies

$$L[w_r] = \lambda_r M[w_r] \tag{3.4.13a}$$

$$L[w_s] = \lambda_s M[w_s]. \tag{3.4.13b}$$

Multiply Eq. (3.4.13a) by $w_r$ and (3.4.13b) by $w_s$, subtract the second expression from the first and integrate over the domain $D(0 < x < \ell)$, to obtain

$$\int_0^\ell \left( w_s L[w_r] - w_r L[w_s] \right) dx - \int_0^\ell \left( \lambda_r w_s M[w_r] - \lambda_s w_r M[w_s] \right) dx \tag{3.4.14}$$

Since the problem is self-adjoint, one can use Eq. (3.4.8a)

$$\int_0^\ell \left( w_s L[w_r] - w_r L[w_s] \right) dx = \left( \sum_{j=1}^n w_r B_j[w_s] - \sum_{j=1}^n w_s B_j[w_r] \right)\Big|_0^\ell \tag{3.4.15}$$

and from Eq. (3.4.8b)

$$\int_0^\ell w_s M[w_r] dx = \int_0^\ell w_r M[w_s] dx + \sum_{j=1}^n \left( w_r C_j[w_s] - w_s C_j[w_r] \right)\Big|_0^\ell. \tag{3.4.16}$$

Substituting Eqs. (3.4.15) and (3.4.16) into Eq. (3.4.14) yields

$$\sum_{j=1}^n \left( w_r B_j[w_s] - w_s B_j[w_r] \right)\Big|_0^\ell$$
$$= \int_0^\ell \{\lambda_r w_r M[w_s] - \lambda_s w_r M[w_s]\}\, dx$$
$$+ \lambda_r \sum_{j=1}^n \left( w_r C_j[w_s] - w_s C_j[w_r] \right)\Big|_0^\ell. \tag{3.4.17}$$

Using Eq. (3.4.6) in the left-hand side of Eq. (3.4.17) yields

$$\sum_{j=1}^n \left( w_r \lambda_s C_j[w_s] - w_s \lambda_r C_j[w_r] \right)\Big|_0^\ell$$
$$= \left( \lambda_r - \lambda_s \right) \int_0^\ell w_r M[w_s] dx + \lambda_r \sum_{j=1}^n \left( w_r C_j[w_s] - w_s C_j[w_r] \right)\Big|_0^\ell$$
$$\left( \lambda_r - \lambda_s \right) \left\{ \int_0^\ell w_r M[w_s] dx + \sum_{j=1}^n \left( w_r C_j[w_s] \right)\Big|_0^\ell \right\} = 0. \tag{3.4.18}$$

For $\lambda_r \neq \lambda_s$, we obtain a general orthogonality relation

$$\int_0^\ell w_r M[w_s] dx + \sum_{j=1}^n \left( w_r C_j[w_s] \right)\Big|_0^\ell = 0, \tag{3.4.19}$$

when the boundary conditions do not depend on the eigenvalue $\lambda$ the orthogonality condition reduces to

$$\left(w_r, M[w_s]\right) = 0, \tag{3.4.20}$$

which also implies

$$\left(w_r, L[w_s]\right) = 0. \tag{3.4.21}$$

Finally, it should be noted that for multiple eigenvalues, the corresponding eigenfunctions are not orthogonal to each other, although they are orthogonal to the remaining eigenfunctions of the set.

In structural dynamics, the eigenfunctions $w_r$ are frequently normalized with respect to $M[\ ]$, when Eq. (3.4.20) applies, thus,

$$\int_0^\ell w_r M[w_r] dx = 1 \quad r = 1, 2 \tag{3.4.22}$$

and

$$\int_0^\ell w_r M[w_s] dx = \delta_{rs}, \tag{3.4.23}$$

where $\delta_{rs}$ is the Kroenecker delta. When $M[\ ]$ is not a differential operator, that is, $m(x) = M[\ ]$,

$$\int_0^\ell m(x) w_r w_s dx = \delta_{rs} \tag{3.4.24}$$

in this case, the functions $\sqrt{m}\, w_r$ and $\sqrt{m}\, w_s$ are orthogonal in the ordinary sense.

**Examples**

To illustrate the application of the material presented in the previous sections, a number of examples are provided:

## Example 1

Consider a beam having some variable mass and stiffness distribution $m(x)$ and $EI(x)$, respectively, which rests on an elastic foundation having a spring stiffness $\beta$ per unit length, shown in Figure 3.12.

**Figure 3.12** Simply supported nonuniform beam on a spring foundation

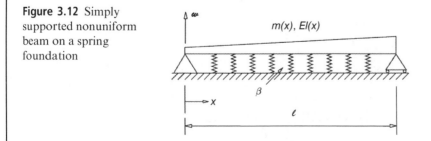

The equations of motion and boundary conditions for this system can be obtained from Hamilton's principle. Recall

$$U = \frac{1}{2}\int_0^\ell EI(x)\left(\frac{d^2w}{dx^2}\right)^2 dx + \frac{1}{2}\int_0^\ell \beta w^2 dx \qquad (3.4.1.\text{Ex}1)$$

$$T = \frac{1}{2}\int_0^\ell m(x)\dot{w}^2 dx. \qquad (3.4.2.\text{Ex}1)$$

Substituting these expressions into

$$\int_{t_1}^{t_2}\left(\delta T - \delta U\right)dt = 0 \qquad (3.4.3.\text{Ex}1)$$

gives the differential equation of equilibrium and boundary conditions:

$$m(x)\frac{\partial^2 w}{\partial t^2} + \frac{\partial^2}{\partial x^2}\left(EI\frac{\partial^2 w}{\partial x^2}\right) + \beta w = 0, \qquad (3.4.4.\text{Ex}1)$$

and the boundary conditions are

$$w = 0 \quad \text{at} \quad x = 0, \ell. \qquad (3.4.5.\text{Ex}1a)$$

$$EI(x)\frac{d^2 w}{dx^2} = 0 \quad \text{at} \quad x = 0, \ell. \qquad (3.4.5.\text{Ex}1b)$$

Since the free vibration solution has the form of

$$w(x,t) = W(x)e^{i\omega t}. \qquad (3.4.6.\text{Ex}1)$$

Substitution of this expression into Eq. (3.4.3.Ex1) yields

$$\frac{d^2}{dx^2}\left[EI(x)\frac{d^2 W}{dx^2}\right] + \beta W = m(x)\omega^2 W(x). \qquad (3.4.7.\text{Ex}1)$$

Equation (3.4.7.Ex1) is used to identify the operators $M[\ ]$ and $L[W]$ in typical eigenvalue problems, when $\lambda = \omega^2$, thus

$$M[\ ] = m(x) \qquad (3.4.8.\text{Ex}1)$$

$$L[\ ] = \frac{d^2}{dx^2}EI(x)\frac{d^2}{dx^2} + \beta \qquad (3.4.9.\text{Ex}1)$$

$$B_1[\ ] = 1 \quad \text{at} \quad x = 0, \ell$$
$$B_2[\ ] = EI(x)\frac{d^2}{dx^2} \quad \text{at} \quad x = 0, \ell \qquad (3.4.10.\text{Ex}1)$$

and since in beam problems $L[\ ]$ is always a fourth-order operator, $p = 2$.

Consider the self-adjointness of this eigenvalue problem. The $M[\,]$ operator is self-adjoint by inspection. Select two comparison functions $W_r$ and $W_s$. Using Eq. (3.4.9.Ex1) and integrating by parts twice

$$
\int_0^\ell W_r L[W_s]dx = \int_0^\ell W_r\left\{\frac{d^2}{dx^2}\left[EI(x)\frac{d^2W_s}{dx^2}\right] + \beta W_s\right\}dx
$$
$$
= W_r\frac{d}{dx}\left[EI(x)\frac{d^2W_s}{dx^2}\right]\Big|_0^\ell - \frac{dW_r}{dx}EI\frac{d^2W_s}{dx^2}\Big|_0^\ell
$$
$$
+ \int_0^\ell EI(x)\frac{d^2W_r}{dx^2}\frac{d^2W_s}{dx^2}dx + \int_0^\ell \beta W_r W_s dx. \quad (3.4.11.\text{Ex1})
$$

Due to the boundary conditions, Eqs. (3.4.5.Ex1a) and (3.4.5.Ex1b), which are satisfied by all comparison functions, the boundary terms in Eq. (3.4.11.Ex1) vanish and thus

$$
\int_0^\ell W_r L[W_s]dx = \int_0^\ell EI(x)\frac{d^2W_r}{dx^2}\frac{d^2W_s}{dx^2}dx + \int_0^\ell \beta W_r W_s dx. \quad (3.4.12.\text{Ex1})
$$

Thus, the $L[\,]$ operator is self-adjoint and the $M[\,]$ is also self-adjoint.

Positive definiteness of the $M[\,]$ and $L[\,]$ is shown by taking $W_r = W_s$ in Eq. (3.4.12.Ex1)

$$
\int_0^\ell W_r L[W_r]dx = \int_0^\ell EI(x)\left(\frac{d^2W_r}{dx^2}\right)^2 dx + \int_0^\ell \beta W_r^2 dx \geq 0
$$
$$
\int_0^\ell m(x)W_r^2(x)dx \geq 0. \quad (3.4.13.\text{Ex1})
$$

Therefore, the eigenvalue problem is self-adjoint and positive definite.

## Example 2

Consider next the small vibrations of a string in tension, shown in Figure 3.13, where tension and mass per unit length, denoted $T(x)$ and $\rho(x)$ respectively, can vary as a function of the variable $x$.

**Figure 3.13** Vibrating string under tension

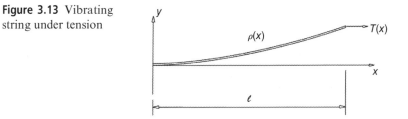

Recall from the section on Hamilton's principle that the equations of motion for an axially loaded cantilever beam were

$$
m\ddot{w} + EI\frac{\partial^4 w}{\partial x^4} + P\frac{\partial^2 w}{\partial x^2} = 0 \quad (3.4.1.\text{Ex2})
$$

$$w = \frac{\partial w}{\partial x} = 0 \quad \text{at} \quad x = 0$$

$$EI\frac{\partial^2 w}{\partial x^2} = 0; \quad P\frac{\partial w}{\partial x} + EI\frac{\partial^3 w}{\partial x^3} = 0 \quad \text{at} \quad x = \ell. \tag{3.4.2.Ex2}$$

There are two differences between the string problem and the beam problem which was discussed previously. For a string compression is replaced by tension

$$P = P(x) = -T(x)$$

and the stiffness $EI = 0$; also due to change in notation $\rho(x) = m$ and $w = y$, therefore Eq. (3.4.1.Ex2) is replaced by

$$\rho\ddot{y} - \frac{\partial}{\partial x}\left(T\frac{\partial y}{\partial x}\right) - 0, \tag{3.4.3.Ex2}$$

and the boundary conditions are replaced by

$$y = 0 \quad \text{at} \quad x = 0$$

$$T(x)\frac{dy}{dx} = 0 \quad \text{at} \quad x = \ell. \tag{3.4.4.Ex2}$$

Again by substituting $y(x, t) = Y(x)e^{i\omega t}$ into Eq. (3.4.3.Ex2), one has

$$-\frac{d}{dx}\left[T(x)\frac{dY}{dx}\right] = \omega^2 \rho(x) Y(x), \tag{3.4.5.Ex2}$$

and the boundary conditions are

$$Y(0) = 0$$

$$T(x)\frac{dY}{dx} = 0 \quad x = \ell. \tag{3.4.6.Ex2}$$

This particular eigenvalue problem is also called the Sturm–Liouville problem. The operators for this problem can be identified as

$$L[\ ] = -\frac{d}{dx}\left(T(x)\frac{d}{dx}\right)$$

$$M[\ ] = \rho(x); \quad \lambda = \omega^2 \tag{3.4.7.Ex2}$$

for this case $p = 1$ and the boundary conditions are

$$B_1[\ ] = 1 \quad \text{at} \quad x = 0$$

$$B_1[\ ] = T(x)\frac{d}{dx} \quad \text{at} \quad x = \ell.$$

Let $Y_r(x)$ and $Y_s(x)$ be two arbitrary comparison functions for a vibrating string problem. Consider the self-adjointness and positive definiteness of this eigenvalue problem. The operator $M[\ ]$ is self-adjoint and positive definite by inspection. Using integration by parts on $L[\ ]$, one has

$$\int_0^\ell Y_r L[Y_s] dx = -\int_0^\ell Y_r \frac{d}{dx}\left(T\frac{dY_s}{dx}\right) dx = -Y_r\left(T\frac{dY_s}{dx}\right)\Big|_0^\ell$$

$$+\int_0^\ell \frac{dY_r}{dx} T(x)\frac{dY_s}{dx} dx = \int_0^\ell T(x)\frac{dY_r}{dx}\frac{dY_s}{dx} dx \qquad (3.4.8.\text{Ex2})$$

and therefore $\left(Y_r, L[Y_s]\right) = \left(Y_s, L[Y_r]\right)$.

Positive definiteness is obtained by setting $r = s$ in Eq. (3.4.8.Ex2)

$$\int_0^\ell T(x)\left(\frac{dY_r}{dx}\right)^2 dx \geq 0 \qquad (3.4.9.\text{Ex2})$$

for the case of tension in the string $T(x) > 0$.

The string problem is one of the simpler problems for which closed-form solutions can be obtained. Assume a constant mass distribution $\rho = \text{const}$, constant tension $T = T(x)$, and fixed boundary conditions at both ends. For this case, the eigenvalue problem reduces to

$$\frac{d^2 Y}{dx^2} + \beta^2 Y = 0; \quad \text{and} \quad \beta^2 = \omega^2\frac{\rho}{T} \qquad (3.4.10.\text{Ex2})$$

with boundary conditions

$$Y(0) = Y(\ell) = 0. \qquad (3.4.11.\text{Ex2})$$

Solution of Eq. (3.4.10.Ex2) is

$$Y(x) = A_1 \sin \beta x + A_2 \cos \beta x \qquad (3.4.12.\text{Ex2})$$

from $Y(0) = 0$, one obtains $A_2 = 0$, and from the second boundary condition $Y(\ell) = 0$, the equation for the frequency is obtained

$$\sin \beta\ell = 0. \qquad (3.4.13.\text{Ex2})$$

Equation (3.4.13.Ex2) has an infinite number of solutions

$$\beta_n = \frac{n\pi}{\ell}; \quad n = 1, 2, \ldots, \qquad (3.4.14.\text{Ex2})$$

which yields the natural frequencies of the system

$$\omega_n = n\pi\sqrt{\frac{T}{\rho\ell^2}} \quad n = 1, 2, \ldots \qquad (3.4.15.\text{Ex2})$$

The corresponding eigenfunctions are

$$Y_n(x) = A_n \sin\left(\frac{n\pi x}{\ell}\right) \qquad (3.4.16.\text{Ex2})$$

and since the eigenfunctions are trigonometric functions they have the mathematical property of orthogonality. The eigenfunctions can be normalized by writing

$$\int_0^\ell \rho Y_n^2(x)dx = 1 \quad \text{for} \quad n = 1, 2, \ldots$$

$$\int_0^\ell \rho A_n^2 \sin^2\left(\frac{n\pi x}{\ell}\right)dx = \rho A_n^2 \frac{\ell}{2} = 1$$

$$A_n = \sqrt{\frac{2}{\rho\ell}}. \tag{3.4.17.Ex2}$$

The orthonormal set of eigenfunctions can be written as

$$Y_n(x) = \sqrt{\frac{2}{\rho\ell}} \sin\left(\frac{n\pi x}{\ell}\right) \quad n = 1, 2, \ldots \tag{3.4.18.Ex2}$$

The first three modes of a string fixed at both ends are shown in Figure 3.14.

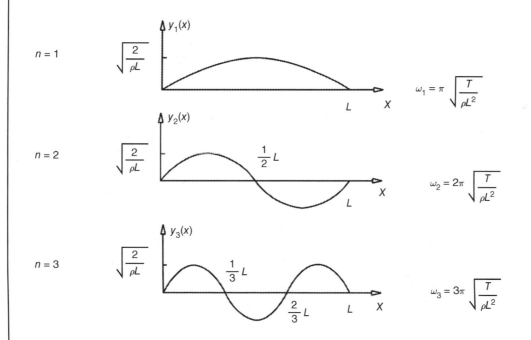

**Figure 3.14** First three free vibration modes of a string in tension

## Alternate Derivation of the String Equation of Motion

The string has no stiffness; thus $U \equiv 0$. Its cross-sectional area is $A$ and the material density is $\rho$. The tension is $T$ and the kinetic energy is $T_k$

$$W_E = -\frac{1}{2}\int_0^\ell T v_{,x}^2 \, dx \tag{3.4.19.Ex2}$$

$$T_k = \frac{1}{2} \int_0^\ell \rho A \dot{v}^2 dx \qquad (3.4.20.\text{Ex2})$$

Hamilton's principle

$$\int_{t_1}^{t_2} (\delta T_k + \delta W_E)\, dt = 0 \qquad (3.4.21.\text{Ex2})$$

$$\int_{t_1}^{t_2} \delta W_E dt = -\int_{t_1}^{t_2} \int_0^\ell T(x) v_{,x}\, \delta v_{,x}\, dx dt$$

$$= \int_{t_1}^{t_2} \left\{ -T(x)v_{,x}\, \delta v\big|_0^\ell + \int_0^\ell [T(x)v_{,x}]_{,x}\, \delta v dx \right\} dt$$

$$= \int_{t_1}^{t_2} \left\{ -T(x)v_{,x}\, \delta v(\ell) + \int_0^\ell [T(x)v_{,x}]_{,x}\, \delta v dx \right\} dt \qquad (3.4.22.\text{Ex2})$$

$$\int_{t_1}^{t_2} \delta T_k dt = \int_{t_1}^{t_2} \left[ \int_0^\ell \rho A \dot{v}\delta\dot{v} dx \right] dt.$$

Recall that $\delta v(t_1) = \delta v(t_2) = 0$, and integrate by parts

$$\int_{t_1}^{t_2} \delta T_k dt = -\int_0^\ell \int_{t_1}^{t_2} \rho A \ddot{v}\delta v\, dt dx. \qquad (3.4.23.\text{Ex2})$$

Combining Eqs (3.4.22.Ex2) and (3.4.23.Ex2) yields

$$\int_{t_1}^{t_2} \left\{ \int_0^\ell \left[ \frac{\partial}{\partial x}(T(x)v_{,x}) - m\ddot{v} \right] dx - T(x)v_{,x}\, \delta v(\ell) \right\} dt = 0,$$

which yields the differential equation of motion

$$\frac{\partial}{\partial x}[T(x)v_{,x}] = m\ddot{v} \qquad (3.4.24.\text{Ex2})$$

and the boundary conditions are $v(0) = 0$ at $x = 0$, and $T(\ell)v_{,x}(\ell) = 0$.

## Example 3

In this example, a more complicated system is treated to emphasize a case in which nonclassical boundary conditions appear. Consider a beam with mass $m$ (per unit length) and a concentrated mass $M$ at its end, which is spring supported by a spring having stiffness $k$ is subjected to an axial load $P$. This structural dynamic system, shown in Figure 3.15, can be used to model a steel pier of a highway bridge. It is necessary to know its natural frequencies and mode shapes to determine its dynamic behavior.

**Figure 3.15** Beam example illustrating nonclassical boundary conditions

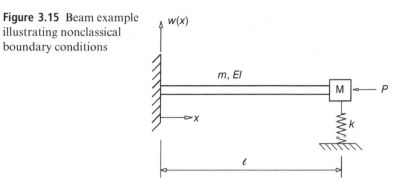

The equations of motion and boundary conditions for this case can be obtained by applying Hamilton's principle.

$$T = \frac{1}{2}\int_0^\ell m\dot{w}^2 dx + \frac{1}{2}M\dot{w}(\ell)^2 \tag{3.4.1.Ex3}$$

$$U = \frac{1}{2}\int_0^\ell EIw_{,xx}^2 dx + \frac{1}{2}kw(\ell)^2 \tag{3.4.2.Ex3}$$

$$W_E = \frac{P}{2}\int_0^\ell w_{,x}^2 dx \tag{3.4.3.Ex3}$$

$$\int_{t_1}^{t_2}\delta w_E dt = \int_{t_1}^{t_2}\left[ Pw_{,x}\delta w\Big|_0^\ell - P\int_0^\ell w_{,xx}\delta w dx \right] dt \tag{3.4.4.Ex3}$$

$$\int_{t_1}^{t_2}\delta T dt = -\int_0^\ell\int_{t_1}^{t_2} m\ddot{w}\delta w dt dx - \int_{t_1}^{t_2} M\ddot{w}(\ell)\delta w(\ell) dt \tag{3.4.5.Ex3}$$

$$\int_{t_1}^{t_2}\delta U dt = \int_{t_1}^{t_2}\left[ EIw_{,xx}\delta w_{,x}\Big|_0^\ell - EIw_{,xxx}\delta w\Big|_0^\ell + \int_0^\ell EI\frac{\partial^4 w}{\partial x^4} dx\delta w \right] dt$$
$$+ \int_{t_1}^{t_2} kw(\ell)\delta w(\ell) dt. \tag{3.4.6.Ex3}$$

The geometric boundary conditions are given by

$$w(0) = w_{,x}(0) = 0 \quad \text{thus} \quad \delta w(0) = \delta w_{,x}(0) = 0. \tag{3.4.7.Ex3}$$

Substituting into Hamilton's principle yields

$$\int_{t_1}^{t_2} \left[ \delta W_E - \delta U + \delta T \right] dt = 0$$

$$\int_{t_1}^{t_2} \left\{ \int_0^\ell \left( -Pw_{,xx}\delta w - m\ddot{w}\delta w - EI\frac{\partial^4 w}{\partial x^4}\delta w \right) dx \right.$$
$$+ Pw_{,x}\delta w(\ell) - M\ddot{w}(\ell)\delta w(\ell) - EIw_{,xx}(\ell)\delta w_{,x}(\ell)$$
$$\left. + EIw_{,xxx}(\ell)\delta w(\ell) - kw(\ell)\delta w(\ell) \right\} dt = 0; \qquad (3.4.8.\text{Ex}3)$$

thus, the differential equation is

$$EI\,\frac{\partial^4 w}{\partial x^4} + P\frac{\partial^2 w}{\partial x^2} + m\ddot{w} = 0, \qquad (3.4.9.\text{Ex}3)$$

and the NBC are

$$EIw_{,xx}(\ell) = 0 \quad \text{or} \quad w_{,xx}(\ell) = 0 \qquad (3.4.10.\text{Ex}3)$$

$$Pw_{,x}(\ell) + EIw_{,xxx}(\ell) - M\ddot{w}(\ell) - kw(\ell) = 0. \qquad (3.4.11.\text{Ex}3)$$

Assume the free vibration solution having the form of

$$w(x,t) = W(x)e^{i\omega t}. \qquad (3.4.12.\text{Ex}3)$$

Substituting Eq. (3.4.12.Ex3) into Eq. (3.4.9.Ex3) yields

$$EI\frac{d^4 W}{dx^4} + P\frac{d^2 W}{dx^2} - m\omega^2 W = 0, \qquad (3.4.13.\text{Ex}3)$$

and substituting Eq. (3.4.12.Ex3) into Eq. (3.4.11.Ex3) yields

$$PW_{,x}(\ell) + EIW_{,xxx}(\ell) + M\omega^2 W - kW(\ell) = 0. \qquad (3.4.14.\text{Ex}3)$$

Next identify the $L[\ ]$, $M[\ ]$, $B[\ ]$, and $C[\ ]$ operators for this problem:

$$M = [\ ] = m; \quad \lambda = \omega^2$$

$$L[\ ] = EI\frac{d^4}{dx^4} + P\frac{d^2}{dx^2} \qquad (3.4.15.\text{Ex}3)$$

$$B_1[\ ] = 0; \quad x = 0$$

$$B_2[\ ] = \frac{d}{dx}; \quad x = 0$$

$$B_1[\ ] = \frac{d^2}{dx^2}; \quad x = \ell \qquad (3.4.16.\text{Ex}3)$$

$$B_2[\ ] = -\left[ P\frac{d}{dx} + EI\frac{d^3}{dx^2} - k \right] \quad \text{at} \quad x = \ell$$

$$C_2[\ ] = M \quad \text{at} \quad x = \ell. \qquad (3.4.17.\text{Ex}3)$$

Next, examine self-adjointness for the system by selecting two comparison functions $u$ and $v$, and implementing Eqs. (3.4.8a) and (3.4.8b) of Section 3.4.8. Consider first Eq. (3.4.8a)

$$\int_0^\ell uL[v]dx + uB_2[v]\Big|_\ell = \int_0^\ell vL[u]dx + vB_2[u]\Big|_\ell. \tag{3.4.18.Ex3}$$

Taking the left-hand side of this expression and integrating by parts,

$$\int_0^\ell u\left[EI\frac{d^4v}{dx^4} + P\frac{d^2v}{dx^2}\right]dx - uP\frac{dv}{dx}\Big|_\ell - uEI\frac{d^3v}{dx^3}\Big|_\ell + ukv\Big|_\ell$$

$$= uEI\frac{d^3v}{dx^3}\Big|_0^\ell - \frac{du}{dx}EI\frac{d^2v}{dx^2}\Big|_\ell + \int_0^\ell EI\frac{d^2u}{dx^2}\frac{d^2v}{dx^2}dx$$

$$+uP\frac{dv}{dx}\Big|_0^\ell - \int_0^\ell P\frac{du}{dx}\frac{dv}{dx}dx - uP\frac{dv}{dx}\Big|_\ell - uEI\frac{d^3v}{dx^3}\Big|_\ell$$

$$+ukv\Big|_\ell = uEI\frac{d^3v}{dx^3}\Big|_\ell + \int_0^\ell EI\frac{d^2u}{dx^2}\frac{d^2v}{dx^2}dx + uP\frac{dv}{dx}\Big|_\ell$$

$$- \int_0^\ell P\frac{du}{dx}\frac{dv}{dx}dx - uP\frac{dv}{dx}\Big|_\ell - uEI\frac{d^3v}{dx^3}\Big|_\ell + kuv\Big|_\ell$$

$$= \int_0^\ell EI\frac{d^2u}{dx^2}\frac{d^2v}{dx^2}dx - \int_0^\ell P\frac{du}{dx}\frac{dv}{dx}dx + kuv\Big|_\ell. \tag{3.4.19.Ex3}$$

Equation (3.4.19.Ex3) is obviously self-adjoint. For the mass operator, conside Eq. (3.4.8b):

$$\int_0^\ell uM[v]dx + uC_2[v]\Big|_\ell = \int_0^\ell umvdx + uMv\Big|_\ell. \tag{3.4.20.Ex3}$$

Equation (3.4.20.Ex3) is obviously self-adjoint.

To establish positive definiteness, consider Eq. (3.4.19.Ex3) with $u = v$; then

$$\int_0^\ell EI\left(\frac{d^2u}{dx^2}\right)^2 dx - \int_0^\ell P\left(\frac{du}{dx}\right)^2 dx + ku^2(\ell) \geq 0. \tag{3.4.21.Ex3}$$

Equation (3.4.21.Ex3) is positive only when

$$\int_0^\ell EI\left(\frac{d^2u}{dx^2}\right)^2 dx + ku^2(\ell) > \int_0^\ell P\left(\frac{du}{dx}\right)^2 dx, \tag{3.4.22.Ex3}$$

which implies that for positive definiteness the axial load should be less than the buckling load for this problem.

Next, consider the orthogonality conditions that exist for this system. Using the general orthogonality relation, Eq. (3.4.19), which was proved in Section 3.3 earlier and introducing the specific terms for the mass operator for our example, we obtain

$$\int_0^\ell mw_rw_sdx + Mw_r(\ell)w_s(\ell) = 0, \quad \text{for} \quad r = s. \tag{3.4.23.Ex3}$$

It is interesting to note that this orthogonality relation, which can be written down immediately by using the general orthogonality condition, Eq. (3.4.19), can also be obtained without recourse to the general expression. To prove this statement, consider two eigenfunctions $W_r$ and $W_s$ that are based on Eq. (3.4.13.Ex3).

$$EI\frac{d^4 W_r}{dx^4} + P\frac{d^2 W_r}{dx^2} = m\omega_r^2 W_r \qquad (3.4.24.\text{Ex3})$$

$$EI\frac{d^4 W_s}{dx^4} + P\frac{d^2 W_s}{dx^2} = m\omega_s^2 W_s. \qquad (3.4.25.\text{Ex3})$$

Multiplying by $W_s$ and $W_r$ respectively and integrating, one has

$$\int_0^\ell \left(EI\frac{d^4 W_r}{dx^4} + P\frac{d^2 W_r}{dx^2}\right) W_s dx = \int_0^\ell m\omega_r^2 W_r W_s dx \qquad (3.4.26.\text{Ex3})$$

$$\int_0^\ell \left(EI\frac{d^4 W_s}{dx^4} + P\frac{d^2 W_s}{dx^2}\right) W_r dx = \int_0^\ell m\omega_s^2 W_r W_s dx. \qquad (3.4.27.\text{Ex3})$$

Subtract (3.4.27.Ex3) from (3.4.26.Ex3) and add $M(\omega_r^2 - \omega_s^2)W_r(\ell)W_s(\ell)$ to both sides

$$\left(\omega_r^2 - \omega_s^2\right)\left\{\int_0^\ell mW_r W_s dx + MW_r(\ell)W_s(\ell)\right\}$$
$$= \int_0^\ell \left(EI\frac{d^4 W_r}{dx^4} + P\frac{d^2 W_r}{dx^2}\right) W_s dx - \int_0^\ell \left(EI\frac{d^4 W_s}{dx^4} + P\frac{d^2 W_s}{dx^2}\right) W_r dx$$
$$+ M\left(\omega_r^2 - \omega_s^2\right) W_r(\ell)W_s(\ell). \qquad (3.4.28.\text{Ex3})$$

Integrating by parts on the right-hand side (RHS) of Eq. (3.4.28.Ex3) and using the boundary conditions

$$\text{RHS} = \left(EI\frac{d^3 W_r}{dx^3} + P\frac{dW_r}{dx}\right) W_s(\ell) - EI\frac{d^2 W_r}{dx^2}\frac{dW_s}{dx}\Big|_0^\ell$$
$$+ \int_0^\ell \left(EI\frac{d^2 W_r}{dx^2}\frac{d^2 W_s}{dx^2} - P\frac{dW_r}{dx}\frac{dW_s}{dx}\right)dx$$
$$- \left(EI\frac{d^3 W_s}{dx^3} + P\frac{dW_s}{dx}\right) W_r(\ell) + EI\frac{d^2 W_s}{dx^2}\frac{dW_r}{dx}\Big|_0^\ell$$
$$+ \int_0^\ell \left(- EI\frac{d^2 W_s}{dx^2}\frac{d^2 W_r}{dx^2} + P\frac{dW_s}{dx}\frac{dW_r}{dx}\right)dx$$
$$+ MW_r(\ell)W_s(\ell)\left(\omega_r^2 - \omega_s^2\right) \qquad (3.4.29.\text{Ex3})$$

from the boundary conditions at $x = \ell$

$$\left.\begin{array}{rcl} P\frac{dW_r}{dx} + EI\frac{d^3 W_r}{dx^3} + M\omega_r^2 W_r(\ell) - kW_r\ell &=& 0 \\ P\frac{dW_s}{dx} + EI\frac{d^3 W_s}{dx^3} + M\omega_s^2 W_s(\ell) - kW_s\ell &=& 0 \end{array}\right\}. \qquad (3.4.30.\text{Ex3})$$

Combining Eqs. (3.4.29.Ex3) and (3.4.30.Ex3)

$$\text{RHS} = \left[-M\omega_r^2 W_r(\ell) + kW_r(\ell)\right] W_s(\ell) - \left[-M\omega_s^2 W_s(\ell) + kW_s(\ell)\right] W_r(\ell)$$
$$+ MW_r(\ell)W_s(\ell)\left(\omega_r^2 - \omega_s^2\right) W_r(\ell)W_s(\ell) \tag{3.4.31.Ex3}$$
$$+ kW_r(\ell)W_s(\ell) - kW_r(\ell)W_s(\ell) + M\left(\omega_r^2 - \omega_s^2\right) W_r(\ell)W_s(\ell) = 0.$$

Thus, the orthogonality condition for $\omega_r^s \neq \omega_s^2$ with respect to mass is

$$\int_0^\ell mW_r W_s dx + MW_r(\ell)W_s(\ell) = 0. \tag{3.4.32.Ex3}$$

Multiplying Eq. (3.4.32.Ex3) by $\omega_r^2$

$$\omega_r^2 \int_0^\ell mW_r W_s dx + M\omega_r^2 W_r(\ell)W_s(\ell) = 0 \tag{3.4.33.Ex3}$$

from Eq. (3.4.24.Ex3) integrating by parts

$$\left(EI\frac{d^3 W_r}{dx^3} + P\frac{dw_r}{dx}\right) W_s\bigg|_0^\ell - \int_0^\ell \left(EI\frac{d^3 W_r}{dx^3} + P\frac{dW_r}{dx}\right)\frac{dW_s}{dx} dx$$
$$= \left(EI\frac{d^3 W_r}{dx^3} + P\frac{dW_r}{dx}\right)\bigg|_{x=\ell} W_s(\ell) - EI\frac{d^2 W_r}{dx^2}\frac{dW_s}{dx}\bigg|_0^\ell \tag{3.4.34.Ex3}$$
$$+ \int_0^\ell \left(EI\frac{d^2 W_r}{dx^2}\frac{d^2 W_s}{dx^2} - P\frac{dW_r}{dx}\frac{dW_s}{dx}\right) dx = \int_0^\ell mW_r W_s \omega_r^2 dx.$$

Using boundary conditions at $x = \ell$, Eq. (3.4.30.Ex3)

$$-M\omega_r^2 W_r(\ell)W_s(\ell) + kW_r(\ell)W_s(\ell)$$
$$+ \int_0^\ell \left(EI\frac{d^2 W_r}{dx^2}\frac{d^2 W_s}{dx^2} - P\frac{dW_r}{dx}\frac{dW_s}{dx}\right) dx$$
$$= \omega_r^2 \int_0^\ell mW_r W_s dx \qquad \text{or}$$
$$kW_r(\ell)W_s(\ell) + \int_0^\ell \left(EI\frac{d^2 W_r}{dx^2}\frac{d^2 W_s}{dx^2} - P\frac{dW_r}{dx}\frac{dW_s}{dx}\right) dx$$
$$= \omega_r^2\left[MW_r(\ell)W_s(\ell) + \int_0^\ell mW_r W_s dx\right]$$

$$\tag{3.4.35.Ex3}$$

for $\omega_r \neq \omega_s$, the last term in Eq. (3.4.35.Ex3) is zero due to Eq. (3.4.32.Ex3). Thus, orthogonality with respect to stiffness is given by

$$\int_0^\ell EI\frac{d^2 W_r}{dx^2}\frac{d^2 W_s}{dx^2} dx - P\int_0^\ell \frac{dW_r}{dx}\frac{dW_s}{dx} dx + kW_r(\ell)W_s(\ell) = 0. \tag{3.4.36.Ex3}$$

Since we have spent a considerable amount of time on this example, it is worth considering the form of an exact solution to the structural dynamic eigenvalue problem for this case. For convenience, introduce a nondimensional variable into Eqs. (3.4.13.Ex3), $\zeta = x/\ell$,

$$\frac{d^4 W}{d\zeta^4} + \frac{P\ell^2}{EI}\frac{d^2 W}{d\zeta^2} - \frac{m\omega^2\ell^4}{EI}W = 0. \tag{3.4.37.Ex3}$$

Define

$$\beta = \frac{P\ell^2}{EI} = \frac{\pi^2}{4}\mu; \quad \mu = \frac{P}{P_{cr}}; P_{cr} = \frac{\pi^2 EI}{4\ell^2}$$

also $\lambda^4 = \frac{m\ell^4\omega^2}{EI}$; then

$$\frac{d^4 W}{d\zeta^4} + \beta\frac{d^2 W}{d\zeta^2} - \lambda^4 W = 0. \tag{3.4.38.Ex3}$$

Solutions to Eq. (3.4.38.Ex3) have the form $W = ce^{r\zeta}$. Substitution yields

$$r^4 + \beta r^2 - \lambda^4 = 0, \tag{3.4.39.Ex3}$$

and the four roots of Eq. (3.4.39.Ex3) are

$$r_{1,2} = \pm\sqrt{\frac{1}{2}\left(-\beta + \sqrt{\beta^2 + 4\lambda^4}\right)} = \pm\delta$$

$$r_{3,4} = \pm\sqrt{\frac{1}{2}\left(-\beta - \sqrt{\beta^2 + 4\lambda^4}\right)} = \pm\gamma. \tag{3.4.40.Ex3}$$

The general solution for $W(\zeta)$ has the form

$$W(\zeta) = A\cos h\delta\zeta + B\sin h\delta\zeta + C\cos\gamma\zeta + D\sin\gamma\zeta. \tag{3.4.41.Ex3}$$

Substituting Eq. (3.4.41.Ex3) into the boundary conditions

$$W(0) = \frac{dW}{d\zeta} = 0 \quad \text{at} \quad x = 0. \tag{3.4.42.Ex3}$$

At $x = \ell$,

$$\frac{d^2 W}{d\zeta^2} = 0; \quad \frac{EI}{\ell^3}\frac{d^3 W}{d\zeta^3} + \frac{P}{\ell}\frac{dW}{d\zeta} + \left(M\omega^2 - k\right)W = 0 \tag{3.4.43.Ex3}$$

yields a $4 \times 4$ matrix equation for A, B, C, and D; and the expansion of the determinant yields the following frequency equation:

$$\delta\left\{\delta^4 + \gamma^4 + 2\delta^2\gamma^2\cos h\delta\cos\gamma + \delta\gamma(\gamma^2 - \delta^2)\sin h\delta\sin\gamma\right\}$$

$$+(\lambda^4\epsilon - \kappa)\left\{(\delta^2 + \gamma^2)\sin h\delta\cos\gamma - (\delta/\gamma)(\delta^2 + \gamma^2)\cos h\delta\sin\gamma\right\}$$

$$-\beta\delta\left\{(\delta^2 - \gamma^2) - (\delta^2 - \gamma^2)\cos h\delta\cos\gamma - 2\delta\gamma\sin h\delta\sin\gamma\right\} = 0, \tag{3.4.44.Ex3}$$

where

$$\epsilon = \frac{M}{m\ell} \quad \text{and} \quad \kappa = \frac{k\ell^3}{EI}, \quad \beta = \frac{\pi^2\mu}{4}$$

## 3.5    Approximate Methods

### 3.5.1    Rayleigh's Quotient

It was shown that the eigenvalue problem for a continuous structure can be written as

$$L[w] = \lambda M[w], \tag{3.5.1}$$

where $L$ and $M$ are linear differential homogeneous operators of order $2p$ and $2q$ ($p > q$), respectively, that have been discussed in Section 3.4.2.

Any eigenvalue $\lambda_r$ and the corresponding eigenfunction $w_r$ will satisfy Eq. (3.5.1).

$$L[w_r] = \lambda_r M[w_r] \tag{3.5.2}$$

multiplying both sides by $w_r$ and integrating over the domain $D$

$$\lambda_r = \frac{\int_D w_r L[w_r] dD}{\int_D w_r M[w_r] dD}. \tag{3.5.3}$$

With boundary conditions independent of $\omega^2$, the problem is positive definite and self-adjoint.

If the eigenfunctions in Eq. (3.5.3) are replaced by arbitrary comparison functions say $u$, one obtains Rayleigh's quotient

$$\omega^2 = R(u) = \frac{\int_D u L[u] dD}{\int_D u M[u] dD}. \tag{3.5.4}$$

Rayleigh's quotient has a stationary value in the neighborhood of the eigenfunction, which is an important and useful property. The arbitrary comparison function $u$ can be expanded in terms of the eigenfunctions (this is known as the expansion theorem).

$$u = \sum_{i=1}^{\infty} c_i w_i \tag{3.5.5}$$

and assume $w_i$ is normalized with respect to mass, this implies

$$\int_D w_i M[w_i] dD = 1 \quad \text{for } i = 1, 2, \ldots \tag{3.5.6}$$

then

$$\int_D w_i L[w_i] dD = \lambda_i. \tag{3.5.7}$$

Substituting into Eq. (3.5.4) yields

$$R(u) = \frac{\int_D \sum_{i=1}^{\infty} c_i w_i L[\sum_{j=1}^{\infty} c_j w_j] dD}{\int_D \sum_{i=1}^{\infty} c_i w_i M[\sum_{j=1}^{\infty} c_j w_j] dD} \tag{3.5.8}$$

$$= \frac{\sum_{i=1}^{\infty} \sum_{j=1}^{\infty} c_i c_j \int_D w_i L[w_j] dD}{\sum_{i=1}^{\infty} \sum_{j=1}^{\infty} c_i c_j \int_D w_i M[w_j] dD}. \tag{3.5.9}$$

If $u$ differs only by a small amount from the eigenfunction $w_r$, then in the expansion represented by Eq. (3.5.5) the coefficient $c_r$ is much larger than $c_i$ (for $i \neq r$) and

$$\left| \frac{c_i}{c_r} \right| = \epsilon_i \ll 1 \quad \text{for } i \neq r, \tag{3.5.10}$$

where $\epsilon_i$ is a small number, thus from Eqs. (3.5.6)–(3.5.9)

$$R(u) = \sum_{i=1}^{\infty} \frac{c_i^2 \lambda_i}{\sum_{i=1}^{\infty} c_i^2}. \tag{3.5.11}$$

Equation (3.5.11) is a result of orthogonality of the eigenfunctions, using Eqs. (3.5.10) and (3.5.11)

$$R(u) = \frac{\lambda_r + \sum_{\substack{i=1 \\ i \neq r}}^{\infty} \epsilon_i^2 \lambda_i}{1 + \sum_{\substack{i=1 \\ i \neq r}}^{\infty} \epsilon_i^2} = \lambda_r[1 + o(\epsilon_i^2) + o(\epsilon_i^4) + \ldots]. \tag{3.5.12}$$

Equation (3.5.12) indicates that if the comparison function differs from the eigenfunction $w_r$ by a small amount [quantity of $0(\epsilon_i)$], Rayleigh's quotient $R(u)$ differs from the eigenvalue $\lambda_r$ by a quantity of $0(\epsilon_i^2)$. Hence, Rayleigh's quotient has a *stationary value* in the neighborhood of an eigenfunction.

This stationary value is actually a minimum in the neighborhood of the lowest, or fundamental mode [eigenfunction] $r = 1$. To show this return to Eq. (3.5.12)

$$R(u) = \frac{\lambda_1 + \sum_{i=2}^{\infty} \epsilon_i^2 \lambda_i}{1 + \sum_{i=2}^{\infty} \epsilon_i^2} \cong \lambda_1 + \sum_{i=2}^{\infty} \epsilon_i^2 \lambda_i - \lambda_1 \sum_{i=2}^{\infty} \epsilon_i^2 + 0(\epsilon_i^4) \tag{3.5.13}$$

$$R(u) = \lambda_1 + \sum_{i=2}^{\infty} (\lambda_i - \lambda_1) \epsilon_i^2. \tag{3.5.14}$$

But in general $\lambda_i > \lambda_1 (i = 2, 3, \ldots, \infty)$ so that $\lambda_i - \lambda_1 > 0$ and

$$R(u) \geq \lambda_1 \tag{3.5.15}$$

that is, Rayleigh's quotient is never lower than the lowest eigenvalue (i.e. minimum for $\lambda_1$). However if $r = n$, from Eq. (3.5.12) then

$$R(u) \cong \lambda_n - \sum_{i=1}^{\infty} (\lambda_n - \lambda_i)\epsilon_2^2 + o(\epsilon_i^4). \tag{3.5.16}$$

In general $\lambda_n > \lambda_i$ $(i = 1, 2, \ldots, \infty)$ $[n \to \infty]$, that is, Rayleigh's quotient is never higher than the highest eigenvalue. Hence, Rayleigh's quotient provides an upper bound for $\lambda_1$ and a lower bound for $\lambda_n$.

A physical interpretation of Eq. (3.5.3) is that it represents a ratio between the strain and kinetic energy density in a structural dynamic system undergoing free vibrations. This is usually denoted by the name Rayleigh's principle, which states:

The frequency of vibration of a conservative system, vibrating in simple harmonic motion, about an equilibrium position has a stationary value in the neighborhood of a natural mode.

$$\omega^2 = R(u) = \frac{V_{\max}}{T^*}, \tag{3.5.17}$$

where $V_{\max} = $ max value of potential or strain energy of the vibrating system, and $T^*$ kinetic energy density of the system, $T^* = T/\omega^2$.

### 3.5.2 The Rayleigh–Ritz Method

It is a very effective approximate method for solving partial differential equations. The method consists of taking $u_i$-comparison functions, that is, function which are $2p$-differentiable and satisfy all the boundary conditions, and representing the eigenfunction by

$$w_n = \sum_{i=1}^{n} a_i u_i. \tag{3.5.18}$$

This approximation is based upon the minimization of Rayleigh's quotient; in this process, the $a_i$'s are unknowns, and the $u_i$'s are called the generating set. The continuous system is being approximated because $a_{n+1} = a_{n+2} = \cdots = 0$. Thus, the continuous system that has an infinite number of degrees of freedom (DOFs) is approximated by a finite number of DOFs. Substituting Eq. (3.5.18) into Rayleigh's quotient yields

$$R(w_n) = \frac{\int_D w_n L[w_n] dD}{\int_D w_n M[w_n] dD} = \frac{N(w_n)}{D(w_n)}. \tag{3.5.19}$$

The conditions for the minimum are

$$\frac{\partial R(w_n)}{\partial a_j} = \frac{D(w_n)\frac{\partial N(w_n)}{\partial a_j} - N(w_n)\frac{\partial D(w_n)}{\partial a_j}}{D^2(w_n)} = 0 \quad ; j = 1, 2, \ldots, n. \tag{3.5.20}$$

$R(w_n)$ has a stationary value in the vicinity of $w_n$; this estimated value of Rayleigh's quotient can be denoted by

$$\min \quad R(w_n) = {}^n\Lambda. \tag{3.5.21}$$

With this relation, Eq. (3.5.20) can be rewritten as

$$\frac{\partial N(w_n)}{\partial a_j} - {}^n\Lambda \frac{\partial D(w_n)}{\partial a_j} = 0 \quad j = 1, 2, 3, \ldots, n. \tag{3.5.22}$$

Now introduce the notation

$$k_{ij} = \int_D u_i L[u_j] dD$$
$$m_{ij} = \int_D u_i M[u_j] dD \qquad i, j = 1, 2, \ldots, n \tag{3.5.23}$$

if the system is self-adjoint

$$k_{ij} = k_{ji}; \quad m_{ij} = m_{ji}. \tag{3.5.24}$$

Furthermore, the operators $L$ and $M$ are linear such that

$$N = \int_D \sum_{i=1}^n a_i u_i L \left[ \sum_j a_j u_j \right] dD$$
$$= \sum_i \sum_j a_i a_j \int_D u_i L\left[u_j\right] dD$$
$$= \sum_i \sum_j k_{ij} a_i a_j. \tag{3.5.25}$$

Similarly,

$$D = \int_D \sum_i a_i u_i M \left[ \sum_j a_j u_j \right] dD = \sum_i \sum_j m_{ij} a_i a_j. \tag{3.5.26}$$

Thus

$$\frac{\partial N}{\partial a_r} = \sum_{i=1}^n \sum_{j=1}^n \left( k_{ij} \frac{\partial a_i}{\partial a_r} a_j + k_{ij} a_i \frac{\partial a_j}{\partial a_r} \right)$$
$$= \sum_i^n \sum_j^n \left( k_{ij} \delta_{ir} a_j + k_{ij} a_i \delta_{jr} \right)$$
$$= \sum_{j=1}^n \left( k_{rj} a_j + k_{ir} a_i \right) = 2 \sum_{j=1}^n k_{rj} a_j. \tag{3.5.27}$$

The last expression in Eq. (3.5.27) is a result of the symmetry of $k_{ij}$, which represents elements of the stiffness matrix. Similarly,

$$\frac{\partial D}{\partial a_r} = 2 \sum_{j=1}^{n} m_{rj} a_j \quad r = 1, 2, \ldots, n, \tag{3.5.28}$$

where $m_{ij}$ represent the elements of a mass matrix. Combining Eqs. (3.5.22), (3.5.27), and (3.5.28), we have

$$\sum_{j=1}^{n} \left( k_{rj} - {}^{n}\Lambda m_{rj} \right) a_j = 0 \quad r = 1, 2, \ldots, n. \tag{3.5.29}$$

This homogeneous algebraic system for $a_j$ can be written in eigenvalue form

$$[k]\{a\} = {}^{n}\Lambda [m]\{a\} \tag{3.5.30}$$

that represents the approximate eigenvalue problem obtained from applying the Rayleigh–Ritz method.

From the solution of Eq. (3.5.30), $n$ eigenvalues ${}^{n}\Lambda$, $r = 1, 2, \ldots, n$ are obtained together with the corresponding eigenvectors $\{a^{(r)}\}$, these are upper bounds for the true eigenvalues $\lambda_r$.

$$^{n}\Lambda_r \geq \lambda_r. \tag{3.5.31}$$

The corresponding eigenfunctions are given by

$$w_n^{(r)} = \sum_{i=1}^{n} a_i^{(r)} u_i. \tag{3.5.32}$$

It can be shown that the Rayleigh–Ritz method converges for self-adjoint problems, that is,

$$\lim_{n \to \infty} {}^{n}\Lambda_r \to \lambda_r \quad \text{and} \quad \lim_{n \to \infty} w_n^{(r)} = w^{(r)}, \tag{3.5.33}$$

which means that Eq. (3.5.18) represents a minimizing sequence.

### Choice of Functions for the Rayleigh–Ritz Method

Recall that comparison functions must be $2p$ times differentiable in the domain $D$ and must also satisfy all the boundary conditions. These requirements can sometimes be difficult to satisfy. We have also indicated that Rayleigh's quotient, for the structural dynamics eigenvalue problems, represents the ratio between strain energy and kinetic energy density per $\omega^2$, thus using Eq. (3.5.34) the requirement on the functions that can be used to

$$R(u) = \frac{V_{\max}}{T_{\max}^*} \tag{3.5.34}$$

construct the minimizing sequence can be relaxed and instead of using comparison functions it is possible to use admissible functions (which have to be only $p$ times differentiable in D, and need to satisfy only the geometric boundary conditions). This is due to the fact that strain energy of the structural system only contains derivatives of order $p$.

## Example

The choice of functions for the Rayleigh–Ritz method is best understood by using an example. Consider a beam with nonuniform mass, $m(x)$ and stiffness distribution $EI(x)$, shown in Figure 3.16. Comparison functions for this case are

**Figure 3.16** Simply supported nonuniform beam

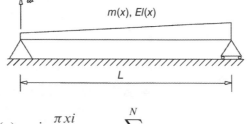

$$u_i(x) = \sin\frac{\pi x i}{L}; \quad w_n = \sum_{n=1}^{N} a_i u_i, \qquad (3.5.35\text{Example})$$

which satisfy all the boundary conditions

$$w(0) = w(L) \quad \text{and} \quad w''EI(x=0) = w''EI(x=L) = 0 \qquad (3.5.36\text{Example})$$

$$R(w_n) = \frac{\int_0^L w_n EI\frac{d^4 w_n}{dx^4}dx}{\int_0^L w_n mw_n dx} = \frac{V_{max}}{T^*} = \frac{\int_0^L EI\left(\frac{d^2 w_n}{dx^2}\right)^2 dx}{\int_0^L mw_n^2 dx}. \qquad (3.5.37\text{Example})$$

Note that the simpler expression with lower derivatives represented by the last term in Eq. (3.5.37Example) can be obtained from integration by parts. Obviously for the last term, a function which is $p$ differentiable can be used, thus polynomials of the type

$$u_1 = x(L-x)a_1; \quad u_2 = x(L-x)^2 \quad \text{etc., can be used.} \qquad (3.5.38\text{Example})$$

## 3.5.3 Galerkin's Method

Galerkin's method is a general method for solving equations in partial differential or ordinary differential form. Galerkin's method has a number of features which makes it quite useful not only for the solution of the structural dynamics eigenvalue problem but also in more complicated structural dynamics problems such as aeroelastic problems in their linear and nonlinear form.

The purpose of the present section is to illustrate the application of Galerkin's method to the structural dynamics eigenvalue problem. First it is important to note that Galerkin's method assumes the solution of the eigenvalue problem in the form of a series of $n$ comparison functions satisfying all the boundary conditions and being $2p$ times differentiable

$$w_n = \sum_{j=1}^{n} a_j u_j. \qquad (3.5.1)$$

In general, this series solution will not satisfy, exactly, the eigenvalue problem

$$L[w] = \lambda M[w], \tag{3.5.2}$$

where $L$ and $M$ are self-adjoint linear homogeneous differential operators of order $2p$ and $2q$ respectively as defined before. The function $w$ is subject to boundary conditions which, in general, do not depend on the eigenvalue $\lambda$. The coefficients $a_j$ of Eq. (3.5.1) are to be determined while $u_j$ are assumed comparison functions. When Eq. (3.5.1) is substituted into Eq. (3.5.2) an error function $\epsilon$ results, as denoted by

$$\epsilon = L[w_n] - {}^n\Lambda M[w_n], \tag{3.5.3}$$

where ${}^n\Lambda$ represents an estimate of the eigenvalue $\lambda$ based on $n$-comparison functions. Galerkin's method consists of the requirement that the error $\epsilon$ be orthogonal to the assumed comparison functions over the domain of existence of the problem. Sometimes an equivalent statement is made requiring that the weighted error integrated over the domain should be zero, that is

$$\int_D \epsilon u_r dD = 0 \quad \text{for} \quad r = 1, 2, 3, \ldots, n. \tag{3.5.4}$$

In applying Eq. (3.5.4), the $L$ and $M$ operators can be considered separately

$$\int_D u_r L[w_n] dD = \sum_{j=1}^{n} a_j \int_D u_r L[u_j] dD = \sum_{j=1}^{n} k_{rj} a_j \quad r = 1, 2, 3, \ldots, n, \tag{3.5.5}$$

where the coefficients $k_{rj}$ are symmetric

$$k_{rj} = k_{jr} = \int_D u_r L[u_j] dD \tag{3.5.6}$$

because $L$ is self-adjoint. Similarly,

$$\int_D u_r M[w_n] dD = \sum_{j=1}^{n} a_j \int_D u_r M[u_j] dD = \sum_{j=1}^{n} m_{rj} a_j \quad r = 1, 2, 3, \ldots, n, \tag{3.5.7}$$

where the coefficients $m_{rj}$ are also symmetric

$$m_{rj} = m_{jr} = \int_D u_r M[u_j] dD \tag{3.5.8}$$

because $M$ is self-adjoint. Again, $k_{rj}$ and $m_{rj}$ are elements of the stiffness and mass matrices, respectively.

Combining Eqs. (3.5.2) through (3.5.8), one has

$$\sum_{j=1}^{n} \left( k_{rj} - {}^n\Lambda m_{rj} \right) a_j = 0 \quad r = 1, 2, 3 \ldots, n, \tag{3.5.9}$$

which yields the regular eigenvalue problem in matrix form

$$[k]\{a\} = {}^n\Lambda [m]\{a\}. \tag{3.5.10}$$

Equations (3.5.9) and (3.5.10) are called Galerkin's equations and represent the eigen-value problem for an $n$ degree-of-freedom system. These equations are similar to those one would obtain by applying the Rayleigh–Ritz method to the same problem, except that the Rayleigh–Ritz method can be based upon admissible functions while Galerkin's method requires the use of comparison functions. When both methods employ comparison functions for the minimizing sequence the resulting eigenvalue problems represented by Eq. (3.5.10) will be identical.

With the comparison functions selected the eigenvalue problem represented by Eqs. (3.5.10) can be formulated and solved using the methods described previously. The solution for the eigenvalues $^n\Lambda_r$ and the eigenvector $\{a^{(r)}\}$ results in the following expression for the eigenfunctions

$$w_n^{(r)} = \sum_{j=1}^{n} a_j^{(r)} u_j. \tag{3.5.11}$$

Furthermore, it should be noted that the convergence of Galerkin's method for linear self-adjoint boundary value problems can be proved (Leipholz 1977). For non-selfadjoint, nonconservative problems, such as aeroelastic stability problems proofs of convergence are more difficult and there is no documented evidence indicating that correct application of the method has resulted in solutions which fail to converge (Leipholz 1977, 128–144).

## 3.6    Vibrations of Timoshenko Beams

### 3.6.1    Background Theory of Timoshenko Beams

Consider first a static simple theory for beams which takes into account the effect of shear. This theory is applicable in particular to short stubby beams, undergoing deformation in the plane of the page. However, it should be mentioned that it can be extended to plate-type structures. When applied to plate structures it is known as Mindlin plate theory.

Consider the shear deformation alone, described in Figure 3.17, and introduce the following assumptions:

**Figure 3.17** Geometry illustrating the shear deformation of a Timoshenko beam

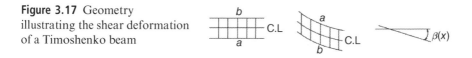

(a) A typical line element, such as $a-b$ normal to the centerline in the undeformed state, moves only in a vertical direction, and also remains vertical during shear deformation.

(b) Line elements tangent to the centerline undergo a rotation $\beta(x)$, where $\beta(x)$ gives the shear angle $\gamma_{xz}$ at a point along the centerline. With these assumptions, the total slope $\frac{dw}{dx}$ of the centerline due to shear deformation and bending deformation respectively can be given by

$$\frac{dw}{dx} = \beta(x) + \psi(x), \tag{3.6.1}$$

where $\psi(x)$ is the rotation of line elements along the centerline due to bending only.

(c) Assume shear strain to be the same at all points over a given cross section of the beam. That is the angle $\beta(x)$ used for rotation of the elements along the centerline is considered to measure the shear angle at all points in the cross section of the beam. This assumption is equivalent to assuming constant shear over the face of the cross section, which is physically impossible and will be corrected later.

With these assumptions, the displacement field for the beam, shown in Figure 3.18 can be written as

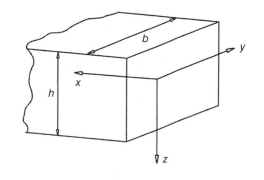

**Figure 3.18** Geometry and coordinate system for a Timoshenko beam

$$u_x = u_1(x, y, z) = -z\psi(x) = -z\left[\frac{dw}{dx} - \beta(x)\right] \tag{3.6.2}$$

$$u_y = u_2(x, y, z) = 0 \tag{3.6.3}$$

$$u_z = u_3(x, y, z) = w(x) \tag{3.6.4}$$

$$\left.\begin{aligned}\epsilon_{xx} &= -z\frac{d\psi}{dx}\\ \epsilon_{xz} &= \tfrac{1}{2}\beta(x)\end{aligned}\right\}. \tag{3.6.5}$$

Next, assume that Poisson's ratio $\nu = 0$, this assumption is similar to the one made in Euler–Bernoulli beam theory. The axial stress and the moment $M$ are

$$\sigma_{xx} = E\epsilon_{xx} \tag{3.6.6}$$

$$M = \int_{\frac{-h}{2}}^{\frac{h}{2}} b\sigma_{xx}z\,dz = -\int_{\frac{-h}{2}}^{\frac{h}{2}} Ez^2\frac{d\psi}{dx}b\,dz = -EI\frac{d\psi}{dx} \tag{3.6.7}$$

and the shear force is

$$V_s = \int_{\frac{-h}{2}}^{\frac{h}{2}} \sigma_{xz}b\,dz = \sigma_{xz}\int_{\frac{-h}{2}}^{\frac{h}{2}} b\,dz = \sigma_{xz}A = GA\beta. \tag{3.6.8}$$

Recall also

$$\epsilon_{xz} = \frac{1}{2}\left(\frac{\partial u_x}{\partial z} + \frac{\partial u_z}{\partial x}\right) = \frac{1}{2}\left[-\frac{dw}{dx} + \beta(x) + \frac{dw}{dx}\right] = \frac{1}{2}\beta(x)$$

$$\frac{\beta}{2} = \epsilon_{xz} = \frac{1}{2G}\sigma_{xz} \tag{3.6.9}$$

thus

$$\sigma_{xz} = \frac{V}{A} = G\beta(x). \tag{3.6.10}$$

Actually, a more accurate statement for $\sigma_{xz}$ would be

$$\sigma_{xz} = G\beta(x,z). \tag{3.6.11}$$

Such a procedure would unnecessarily complicate the problem, and destroy its one-dimensional nature. The situation can be improved somewhat by introducing an additional factor $k$, which is a cross-sectional shear correction factor, as follows

$$\sigma_{xz} = kG\beta(x) \tag{3.6.12}$$

$$V(x) = kG\beta(x)A. \tag{3.6.13}$$

There is a considerable amount of literature, on calculating $k$ for various cross sections (Dym and Shames 1973, 187–194).

The total potential for the assumed deformation field of the Timoshenko beam is given by

$$\pi_p = \frac{1}{2}\int_0^L \int_{\frac{-h}{2}}^{\frac{h}{2}} \sigma_{xx}\epsilon_{xx}dxbdz + \int_0^L \int_{\frac{-h}{2}}^{\frac{h}{2}} \sigma_{xz}\epsilon_{xz}dxbdz - \int_0^L p(x)wdx. \tag{3.6.14}$$

Using

$$\sigma_{xx} = E\epsilon_{xx} = -Ez\frac{d\psi}{dx} \tag{3.6.15}$$

$$\sigma_{xz} = kG\beta(x)$$

$$\pi_p = \frac{1}{2}\int_0^L EI\left(\frac{d\psi}{dx}\right)^2 dx + \frac{GA}{2}\int_0^L k\beta^2 dx - \int_0^L pwdx$$

$$= \frac{1}{2}\int_0^L EI\left(\frac{d\psi}{dx}\right)^2 dx + \frac{kGA}{2}\int_0^L \left(\frac{dw}{dx} - \psi\right)^2 dx - \int_0^L pwdx$$

or

$$\pi_p = \int_0^L \left[\frac{EI}{2}\left(\frac{d\psi}{dx}\right)^2 + \frac{kGA}{2}\left(\frac{dw}{dx} - \psi\right)^2 - pw\right]dx. \tag{3.6.16}$$

Thus, the functional depends on two functions $\psi$ and $w$. Performing the appropriate variations one obtains, the Euler–Lagrange equations, or the equations of equilibrium for the system.

$$\delta \pi_p = \int_0^L EI \frac{d\psi}{dx} \frac{d\delta\psi}{dx} dx + kAG \int_0^L \left[ \left( \frac{dw}{dx} - \psi \right) \frac{d\delta w}{dx} - \left( \frac{dw}{dx} - \psi \right) \delta\psi \right] dx$$
$$- \int_0^L p \delta w \, dx = 0 \tag{3.6.17}$$

Integrating by parts

$$EI \frac{d\psi}{dx} \delta\psi \Big|_0^L - \int_0^L \frac{d}{dx} \left( EI \frac{d\psi}{dx} \right) \delta\psi \, dx + kAG \left( \frac{dw}{dx} - \psi \right) \delta w \Big|_0^L$$
$$- kAG \int_0^L \frac{d}{dx} \left( \frac{dw}{dx} - \psi \right) \delta w \, dx - \int_0^L kAG \left( \frac{dw}{dx} - \psi \right) \delta\psi \, dx$$
$$- \int_0^L p \delta w \, dx = 0, \tag{3.6.18}$$

which can be rewritten as

$$- \int_0^L \left[ kAG \frac{d}{dx} \left( \frac{dw}{dx} - \psi \right) + p \right] \delta w \, dx$$
$$- \int_0^L \left[ \frac{d}{dx} \left( EI \frac{d\psi}{dx} \right) + kAG \left( \frac{dw}{dx} - \psi \right) \right] dx \delta\psi$$
$$+ EI \frac{d\psi}{dx} \delta\psi \Big|_0^L + kAG \left( \frac{dw}{dx} - \psi \right) \delta w \Big|_0^L = 0. \tag{3.6.19}$$

The equations of equilibrium are

$$\frac{d}{dx} \left( EI \frac{d\psi}{dx} \right) + kAG \left( \frac{dw}{dx} - \psi \right) = 0 \tag{3.6.20}$$

$$kAG \frac{d}{dx} \left( \frac{dw}{dx} - \psi \right) + p = 0 \tag{3.6.21}$$

for constant $EI$ and $kAG$ these reduce to

$$EI \frac{d^2\psi}{dx^2} + kAG \left( \frac{dw}{dx} - \psi \right) = 0 \tag{3.6.22}$$

$$kAG \left( \frac{d\psi}{dx} - \frac{d^2w}{dx^2} \right) = p. \tag{3.6.23}$$

The boundary conditions at $x = 0, L$ are

(1) either $\psi$ is specified or $EI \frac{d\psi}{dx} = 0$;

(2) either $w$ is specified or $kAG \left( \frac{dw}{dx} - \psi \right) = 0$.

From a physical point of view, these statements are equivalent to

(1) either $\psi$ is specified or $M = 0$ at $x = 0, L$;
(2) either $w$ is specified or $V = 0$ at $x = 0, L$.

Finally, it is important to note that, the differential equations for $w$ and $\psi$ can be decoupled. For example, differentiating Eq. (3.6.22)

$$EI\frac{d^3\psi}{dx^3} + kAG\left(\frac{d^2w}{dx^2} - \frac{d\psi}{dx}\right) = 0. \tag{3.6.24}$$

Substituting into (3.6.23) yields

$$EI\frac{d^3\psi}{dx^3} = p. \tag{3.6.25}$$

Next, differentiate (3.6.23) twice and solve for $\psi'''(x)$

$$kGA\left(\frac{d^3\psi}{dx^3} - \frac{d^4w}{dx^4}\right) = \frac{d^2p}{dx^2}$$

$$\frac{d^3\psi}{dx^3} = \frac{1}{kGA}\frac{d^2p}{dx^2} + \frac{d^4w}{dx^4}. \tag{3.6.26}$$

Combine Eqs. (3.6.25) and (3.6.26)

$$EI\frac{d^4w}{dx^4} = p - \frac{EI}{kGA}\frac{d^2p}{dx^2}. \tag{3.6.27}$$

Equations (3.6.25) and (3.6.27) represent decoupled equations for $\psi$ and w.

Next, consider the dynamic situation. To use Hamilton's principle, the kinetic energy, $T_K$, is needed in addition to the potential $\pi_p$.

Using Eqs. (3.6.2)–(3.6.4), the kinetic energy is given by

$$T_K = \frac{1}{2}\int_0^L \int_{\frac{-h}{2}}^{\frac{h}{2}} \rho \dot{u}_i^2 dxb\, dz$$

$$T_K = \frac{1}{2}\int_0^L \int_{\frac{-h}{2}}^{\frac{h}{2}} b\rho\left[z^2\left(\frac{\partial\psi}{\partial t}\right)^2 + \left(\frac{\partial w}{\partial t}\right)^2\right]dxdz$$

$$= \frac{1}{2}\int_0^L \left[\rho I\left(\frac{\partial\psi}{\partial t}\right)^2 + \rho A\left(\frac{\partial w}{\partial t}\right)^2\right]dx, \tag{3.6.28}$$

where $I$ = area moment of inertia and $A$ = area of beam cross section. Obviously, Eq. (3.6.28) contains the kinetic energy due to the rotation associated with shear. Combining the expression for $\pi_p$ in Eq. (3.6.16) with Eq. (3.6.28), and substituting the result in Hamilton's principle yields the dynamic equations of equilibrium

$$\delta\int_{t_1}^{t_2}\left(T_K - \pi_p\right)dt = \delta\int_{t_1}^{t_2} L dt = 0 \tag{3.6.29}$$

$$\delta \int_{t_1}^{t_2} L dt = \delta \int_{t_1}^{t_2} \int_0^L \left\{ \frac{1}{2} \left[ \rho I \left( \frac{\partial \psi}{\partial t} \right)^2 + \rho A \left( \frac{\partial w}{\partial t} \right)^2 \right] \right.$$

$$\left. - \frac{EI}{2} \left( \frac{\partial \psi}{\partial x} \right)^2 - \frac{kGA}{2} \left( \frac{\partial w}{\partial x} - \psi \right)^2 + pw \right\} dxdt = 0. \tag{3.6.30}$$

The treatment of the terms in this expression, which are associated with $\pi_p$, has been considered earlier. The variation of the terms associated with the kinetic energy is

$$\delta \int_{t_1}^{t_2} T_K dt = \int_{t_1}^{t_2} \int_0^L \left\{ \rho I \frac{\partial \psi}{\partial t} \frac{\partial \delta \psi}{\partial t} + \rho A \frac{\partial w}{\partial t} \frac{\partial \delta w}{\partial t} \right\} dxdt. \tag{3.6.31}$$

Integrating by parts with respect to $t$

$$\int_0^L \left[ \rho I \frac{\partial \psi}{\partial t} \delta \psi \Big|_{t_1}^{t_2} + \rho A \frac{\partial w}{\partial t} \delta w \Big|_{t_1}^{t_2} \right] dx$$

$$- \int_{t_1}^{t_2} \int_0^L \left[ \rho I \frac{\partial^2 \psi}{\partial t^2} \delta \psi + \rho A \frac{\partial^2 w}{\partial t^2} \delta w \right] dtdx = \delta \int_{t_1}^{t_2} T_K dt. \tag{3.6.32}$$

According to Hamilton's principle,

$$\delta w \Big|_{t_1}^{t_2} = \delta \psi \Big|_{t_1}^{t_2} = 0.$$

Thus, Eq. (3.6.32) when combined with Eq. (3.6.19) yields

$$\left. \begin{array}{c} -\rho A \frac{\partial^2 w}{\partial t^2} + \frac{\partial}{\partial x} \left[ kGA \left( \frac{\partial w}{\partial x} - \psi \right) \right] + p = 0 \\ -\rho I \frac{\partial^2 \psi}{\partial t^2} + \frac{\partial}{\partial x} \left( EI \frac{\partial \psi}{\partial x} \right) + kAG \left( \frac{\partial w}{\partial x} - \psi \right) = 0 \end{array} \right\} \tag{3.6.33}$$

for constant values of $EI, kAG$

$$\rho A \ddot{w} - kAG \left( \frac{\partial^2 w}{\partial x^2} - \frac{\partial \psi}{\partial x} \right) - p(x, t) = 0$$

$$\rho I \ddot{\psi} - EI \frac{\partial^2 \psi}{\partial x^2} - kAG \left( \frac{\partial w}{\partial x} - \psi \right) = 0. \tag{3.6.34}$$

Boundary conditions for this case have been already given. Recall also that

$$M = -EI \frac{\partial \psi}{\partial x} \quad \text{and} \quad V_s = kGA\beta(x) = kGA \left( \frac{\partial w}{\partial x} - \psi \right). \tag{3.6.35}$$

Using these expressions for the shear and moment, the equations of motion Eq. (3.6.33) can be also expressed in terms of these quantities, thus

$$\left. \begin{array}{c} \frac{\partial V_s}{\partial x} = \rho A \ddot{w} - p(x, t) \\ V_s - \frac{\partial M}{\partial x} = \rho I \ddot{\psi} \end{array} \right\} \tag{3.6.36}$$

for this case the boundary conditions at $x = 0, L$ are

either $V_s = 0$ or $w$ is specified
either $M = 0$ or $\psi$ is specified.

**Some Additional Material on the Vibrations of Timoshenko Beams**

Consider Eqs. (3.6.34), for the case of a beam with uniform mass and stiffness properties

$$\rho A \ddot{w} - kAG(w_{,xx} - \psi_{,x}) - p(x,t) = 0 \tag{3.6.37a}$$

$$\rho I \ddot{\psi} - EI \psi_{,xx} - kAG(w_{,x} - \psi) = 0. \tag{3.6.37b}$$

From Eq. (3.6.37a)

$$\psi_{,x} = -\rho \frac{\ddot{w}}{kG} + w_{,xx} + \frac{p}{kGA}. \tag{3.6.38}$$

Differentiating Eq. (3.6.37b)

$$\rho I \ddot{\psi}_{,x} - EI \psi_{,xxx} - kAG(w_{,xx} - \psi_{,x}) = 0. \tag{3.6.39}$$

Differentiating Eq. (3.6.38) twice

$$\psi_{,xxx} = -\frac{\rho}{kG} \ddot{w}_{,xx} + w_{,xxxx} + \frac{p_{,xx}}{kGA}. \tag{3.6.40}$$

Differentiating Eq. (3.6.38) twice with respect to time, one has

$$\ddot{\psi}_{,x} = -\rho \frac{\ddddot{w}}{kG} + \ddot{w}_{,xx} + \frac{\ddot{p}}{kGA}. \tag{3.6.41}$$

Next substitute Eqs. (3.6.38), (3.6.40), and (3.6.41) into Eq. (3.6.39)

$$\rho I \left[ -\rho \frac{\ddddot{w}}{kG} + \ddot{w}_{,xx} + \frac{\ddot{p}}{kGA} \right] - EI \left[ -\frac{\rho}{kG} \ddot{w}_{,xx} + w_{,xxxx} + \frac{p_{,xx}}{kGA} \right]$$
$$- kGA \left\{ w_{,xx} + \rho \frac{\ddot{w}}{kG} - w_{,xx} - \frac{p}{kGA} \right\} = 0, \tag{3.6.42}$$

which can be simplified to yield

$$EI \frac{\partial^4 w}{\partial x^4} + \rho A \frac{\partial^2 w}{\partial t^2} - \rho I \left( 1 + \frac{E}{kG} \right) \frac{\partial^4 w}{\partial x^2 \partial t^2} + \frac{\rho^2 I}{kG} \frac{\partial^4 w}{\partial t^4}$$
$$= p(x,t) + \frac{\rho I}{kGA} \frac{\partial^2 p}{\partial t^2} - \frac{EI}{kGA} \frac{\partial^2 p}{\partial x^2}. \tag{3.6.43}$$

The last equation illustrates decoupling between $w$ and $\psi$. When $kGA \to \infty$, this reduces to Euler–Bernoulli beam theory.

Consider the free vibration problem $p(x,t) = 0$

$$EI \frac{\partial^4 w}{\partial x^4} + \rho A \frac{\partial^2 w}{\partial t^2} - \rho I \left( 1 + \frac{E}{kG} \right) \frac{\partial^4 w}{\partial x^2 \partial t^2} + \frac{\rho^2 I}{kG} \frac{\partial^4 w}{\partial t^4} = 0. \tag{3.6.44}$$

Similarly to previous free vibration problems, one can demonstrate separation of variables

$$w(x,t) = W(x)f(t) \tag{3.6.45}$$

Substitute into Eq. (3.6.44)

$$EIW''''f + \rho A W \ddot{f} - \rho I \left(1 + \frac{E}{kG}\right) W'' \ddot{f} + \frac{\rho^2 I}{kG} W \ddddot{f} = 0, \tag{3.6.46}$$

where $(\ )' = \frac{\partial}{\partial x}$ and $(\dot{\ }) = \frac{\partial}{\partial t}$, and divide Eq. 3.6.46 by $W\ddot{f}$

$$EI \left(\frac{W''''}{W}\right)\left(\frac{f}{\ddot{f}}\right) + \rho A - \rho I \left(1 + \frac{E}{kG}\right)\left(\frac{W''}{W}\right) + \frac{\rho^2 I}{kG}\left(\frac{\ddddot{f}}{\ddot{f}}\right) = 0. \tag{3.6.47}$$

Differentiate Eq. (3.6.47) with respect to $x$:

$$EI \left(\frac{W'''''W - W''''W'}{W^2}\right)\left(\frac{f}{\ddot{f}}\right) - \rho I \left(1 + \frac{F}{kG}\right)\left(\frac{W'''W - W''W'}{W^2}\right) = 0. \tag{3.6.48}$$

Rearrange Eq. (3.6.48) to get

$$\frac{EI}{I\rho\left(1 + \frac{E}{kG}\right)}\left(\frac{W'''''W - W''''W'}{W'''W - W''W'}\right) = \frac{\ddot{f}}{f} = -\omega^2, \tag{3.6.49}$$

which yields

$$\ddot{f} + \omega^2 f = 0, \tag{3.6.50}$$

which implies simple harmonic motion, that is,

$$f(t) = e^{i\omega t}. \tag{3.6.51}$$

Substitute Eq. (3.6.51) into Eq. (3.6.44) to get

$$EI\frac{\partial^4 W}{\partial x^4} - \rho A\omega^2 W(x) + \omega^2\rho I\left(1 + \frac{E}{kG}\right)\frac{\partial^2 W}{\partial x^2} + \omega^4\frac{\rho^2 I}{kG}W(x) = 0. \tag{3.6.52}$$

Following the approach used earlier, one can assume a solution in the form of

$$W(x) = \bar{A}e^{rx}, \tag{3.6.53}$$

which yields a characteristic equation

$$r^4 EI - \omega^2\rho A + r^2\omega^2\rho I\left(1 + \frac{E}{kG}\right) + \omega^4\frac{\rho^2 I}{kG} = 0. \tag{3.6.54}$$

The general solution corresponding to this equation is

$$W(x) = \bar{A}e^{r_1 x} + \bar{B}e^{r_2 x} + \bar{C}e^{r_3 x} + \bar{D}e^{r_4 x}, \tag{3.6.55}$$

which can also be rewritten as

$$W(x) = A\sin\lambda x + B\cos\lambda x + C\sinh\lambda x + D\cosh\lambda x. \tag{3.6.56}$$

At this stage, one has to select some boundary conditions. Assume that the Timoshenko beam is simply supported at both ends, which implies

$$\text{either } \psi \text{ is specified or } EI\psi,_x = 0 \text{ at } x = 0, \ell$$

$$\text{or } w \text{ is specified or } kGA(w,_x - \psi) = 0 \tag{3.6.57}$$

for a simply supported beam $w(0) = w(\ell) = 0$, and according to the moment boundary conditions

$$EI\psi,_x(0) = EI\psi,_x(\ell) = 0 \tag{3.6.58}$$

from Eq. (3.6.38)

$$\psi,_x = -\rho\frac{\ddot{w}}{kG} + w,_{xx} + \frac{p(x,t)}{kGA} \tag{3.6.59}$$

for free vibrations $p(x,t) = 0$ and $\ddot{w} = -\omega^2 f(t)W(x)$. Thus the boundary conditions at $x = 0, \ell$ can be written using Eq. (3.6.38)

$$EI\psi,_x(0) = \left[\rho\omega^2\frac{f(t)W(0)}{kG} + f(t)W,_{xx}(0)\right]EI = 0 \tag{3.6.60}$$

and a similar equation at $x = l$ which implies

$$\left.\begin{array}{l} EI\left[\frac{\rho\omega^2 W(0)}{kG} + W,_{xx}(0)\right] = 0 \\[2mm] EI\left[\frac{\rho\omega^2 W(\ell)}{kG} + W,_{xx}(\ell)\right] = 0 \end{array}\right\}. \tag{3.6.61}$$

The simply supported boundary conditions,

$$w(x = 0) = w(x = \ell) = 0 \tag{3.6.62}$$

also imply

$$f(t)W(0) = f(t)W(\ell) = 0$$
$$W(0) = W(\ell) = 0. \tag{3.6.63}$$

Using Eqs. (3.6.63) and (3.6.61), it is clear that the boundary conditions (3.6.61) reduce to

$$W,_{xx}(0) = W,_{xx}(\ell) = 0. \tag{3.6.64}$$

For these boundary conditions $B = C = D = 0$ in Eq. (3.6.56) and $A \sin \lambda\ell = 0$ which yields $\lambda = \frac{n\pi}{\ell}$ for $n = 1, 2, \ldots$

Therefore, the mode shapes are given by

$$W_n = A_n \sin\frac{n\pi x}{\ell}. \tag{3.6.65}$$

Next, substitute Eq. (3.6.65) into Eq. (3.6.54)

$$EI\left(\frac{n\pi}{\ell}\right)^4 - \omega_n^2\rho A - \omega_n^2\rho I\left(1 + \frac{E}{kG}\right)\left(\frac{n\pi}{\ell}\right)^2 + \omega_n^4\frac{\rho^2 I}{kG} = 0. \tag{3.6.66}$$

Dividing by $EI/\ell^4$

$$- \frac{\rho^2 \ell^4}{EkG} \omega_n^4 + \omega_n^2 \left[ \frac{\rho A \ell^4}{EI} + \frac{\rho \ell^2}{E} \left(1 + \frac{E}{kG}\right) n^2 \pi^2 \right] - n^4 \pi^4 = 0. \qquad (3.6.67)$$

Introduce the following definition

$$\left. \begin{array}{lll} \Omega_n^2 & = & \frac{\rho I}{EA} \omega_n^2 = \text{dimensionless frequency squared} \\[2mm] r^2 & = & \frac{I}{A} = \text{radius of gyration squared} \end{array} \right\} \qquad (3.6.68)$$

into Eq. (3.6.67) to get

$$- \left(\frac{EA}{\rho I}\right)^2 \Omega_n^4 \frac{\rho^2 \ell^4}{EkG} + \frac{EA}{\rho I} \Omega_n^2 \left[ \frac{\rho A \ell^4}{EI} + \frac{\rho \ell^2}{E} \left(1 + \frac{E}{kG}\right) n^2 \pi^2 \right] - (n\pi)^4 = 0 \qquad (3.6.69)$$

or

$$- \Omega_n^4 \frac{E}{kG} \left(\frac{\ell}{r}\right)^4 + \Omega_n^2 \left[ \left(\frac{\ell}{r}\right)^4 + \left(\frac{\ell}{r}\right)^2 \left(1 + \frac{E}{kG}\right) n^2 \pi^2 \right] - (n\pi)^4 = 0 \qquad (3.6.70)$$

If in Eq. (3.6.70) the portions

$$- \frac{E}{kG} \left(\frac{\ell}{r}\right)^4 \quad \text{as well as} \quad \left(\frac{\ell}{r}\right)^2 \left(1 + \frac{E}{kG}\right) \qquad (3.6.71)$$

are set equal to zero, one has

$$\left(\frac{\ell}{r}\right)^4 \Omega_n^2 - (n\pi)^4 = 0 \qquad (3.6.72)$$

$\Omega_{nEB}^2 = \frac{n^4 \pi^4 r^4}{\ell^4}$ which is equivalent to $\omega_n^2 = \frac{(n\pi)^4 EI}{\ell^4 \rho A}$ which is the natural frequency of a simply supported Euler–Bernoulli beam. Among the terms which have been neglected those containing the term $k$ pertain to the transverse shear effect. Thus if one wants to include only rotary inertia effects, one would set equal to zero only $\frac{-E}{kG} \left(\frac{\ell}{r}\right)^4$ and $\left(\frac{\ell}{r}\right)^2 \frac{E}{kG}$ and then from Eq. (3.6.70) one has

$$\left[ \left(\frac{\ell}{r}\right)^4 + \left(\frac{\ell}{r}\right)^2 (n\pi)^2 \right] \Omega_n^2 - (n\pi)^4 = 0. \qquad (3.6.73)$$

Thus

$$\Omega_{n\,\text{rot inertia}}^2 = \frac{(n\pi)^4}{\left(\frac{\ell}{r}\right)^4 \left[1 + (n\pi)^2 \left(\frac{r}{\ell}\right)^2\right]} = \frac{\Omega_{nEB}^2}{\left[1 + (n\pi)^2 \left(\frac{r}{\ell}\right)^2\right]}, \qquad (3.6.74)$$

where $\Omega_{nEB}^2$ represents the dimensionless frequency of an Euler–Bernoulli beam.

Figure 3.19 depicts the effect of rotatory inertia for various values of $n$ shown as a function of the slenderness ratio $\ell/r$. For small $n$ and for large slenderness ratios $\ell/r$ there is only a limited difference between, Euler–Bernoulli and this "partial" Timoshenko beam theory which includes only the effect of rotary inertia. However for short stubby beams and higher frequencies this effect can be significant.

**Figure 3.19**
Effect of
rotary inertia
alone on
modal
frequencies of
Timoshenko
beams

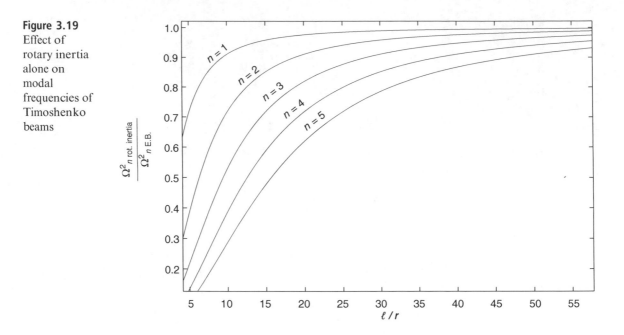

Next, consider the effects of shear only, this time we delete the terms $\left(\frac{\ell}{r}\right)^2 (n\pi)^2$ and $-\left(\frac{E}{kG}\right)\left(\frac{\ell}{r}\right)^4 \Omega_n^4$ since the latter contains both shear and rotatory inertia as can be seen by inspecting the expression for $T_k$ that indicates the source of this term. With this approximation the frequency equation becomes

$$\left[\left(\frac{\ell}{r}\right)^4 + \frac{E}{kG}\left(\frac{\ell}{r}\right)^2 (n\pi)^2\right]\left(\Omega_n^2\right)_{\text{shear}} - (n\pi)^4 = 0 \tag{3.6.75}$$

or

$$\left(\Omega_n^2\right)_{\text{shear}} = \frac{(n\pi)^4}{\left(\frac{\ell}{r}\right)^4 \left[1 + \frac{E}{kG}\left(\frac{r}{\ell}\right)^2 (n\pi)^2\right]} = \frac{\Omega_{nEB}^2}{\left[1 + \frac{E}{kG}\left(\frac{r}{\ell}\right)^2 (n\pi)^2\right]}. \tag{3.6.76}$$

A plot of $\left(\Omega_n^2\right)_{\text{shear}} / \Omega_{nEB}^2$ versus $\ell/r$ for $E/kG = 3.00$ and various values of $n$ is given in Figure 3.20. Again for large values of $n$ and small values of $\ell/r$ one observes definite shear effects which must be accounted for.

The complete Eq. (3.6.70) is considered next. One can write the solution in the form

$$\Omega_n^2 = Z \qquad \Omega_n^4 = Z^2$$

$$\frac{E}{kG}\left(\frac{\ell}{r}\right)^4 Z^2 - Z\left[\left(\frac{\ell}{r}\right)^4 + \left(\frac{\ell}{r}\right)^2 \left(1 + \frac{E}{kG}\right) n^2\pi^2\right] + (n\pi)^4 = 0$$

**Figure 3.20**
Effect of shear
alone on modal
frequencies of
Timoshenko
beams

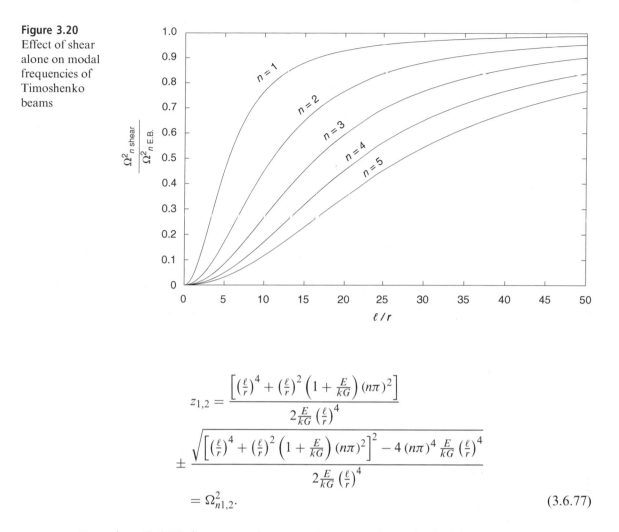

$$z_{1,2} = \frac{\left[ \left(\frac{\ell}{r}\right)^4 + \left(\frac{\ell}{r}\right)^2 \left(1 + \frac{E}{kG}\right)(n\pi)^2 \right]}{2\frac{E}{kG}\left(\frac{\ell}{r}\right)^4}$$

$$\pm \frac{\sqrt{\left[ \left(\frac{\ell}{r}\right)^4 + \left(\frac{\ell}{r}\right)^2 \left(1 + \frac{E}{kG}\right)(n\pi)^2 \right]^2 - 4(n\pi)^4 \frac{E}{kG}\left(\frac{\ell}{r}\right)^4}}{2\frac{E}{kG}\left(\frac{\ell}{r}\right)^4}$$

$$= \Omega_{n1,2}^2. \tag{3.6.77}$$

Equation (3.6.77) is too cumbersome for convenient physical interpretation and an approximation is introduced for illustrative purposes

$$\Omega^2 = \Omega_{EB}^2 + \delta^2 \tag{3.6.78}$$

and in subsequent calculations higher order terms in $\delta^2$ will be neglected. This approach will limit the validity of the results to those cases when the combination of rotatory inertia and shear still yield frequencies which are relatively close to Euler–Bernoulli beam theory. Substituting Eq. (3.6.78) into Eq. (3.6.70) and neglecting terms of order $\delta^4$ or higher yields

$$-\left(\frac{E}{kG}\right)\left(\frac{\ell}{r}\right)^4 \left(\Omega_{EB}^4 + 2\delta^2 \Omega_{EB}^2\right) + \left[ \left(\frac{\ell}{r}\right)^4 + \left(\frac{\ell}{r}\right)^2 \left(1 + \frac{E}{kG}\right)(n\pi)^2 \right]$$

$$\left(\Omega_{EB}^2 + \delta^2\right) - (n\pi)^4 = 0. \tag{3.6.79}$$

Solving for $\delta^2$ from the last equation

$$\delta^2 = \frac{\left(\frac{E}{kG}\right)\left(\frac{\ell}{r}\right)^4 \Omega_{EB}^4 - \left[\left(\frac{\ell}{r}\right)^4 + \left(\frac{\ell}{r}\right)^2 \left(1 + \frac{E}{kG}\right)(n\pi)^2\right]\Omega_{EB}^2 + (n\pi)^4}{-2\frac{E}{kG}\left(\frac{\ell}{r}\right)^4 \Omega_{EB}^2 + \left[\left(\frac{\ell}{r}\right)^4 + \left(\frac{\ell}{r}\right)^2 \left(1 + \frac{E}{kG}\right)(n\pi)^2\right]} \qquad (3.6.80)$$

Combining Eqs. (3.6.78) and (3.6.80)

$$\Omega^2 = \Omega_{EB}^2 + \delta^2 = \Omega_{EB}^2 \left[1 + \delta^2/\Omega_{EB}^2\right]$$

$$= \Omega_{EB}^2 \left\{1 + \frac{\left(\frac{E}{kG}\right)\left(\frac{\ell}{r}\right)^4 \Omega_{EB}^2 - \left[\left(\frac{\ell}{r}\right)^4 + \left(\frac{\ell}{r}\right)^2 \left(1 + \frac{E}{kG}\right)(n\pi)^2\right] + \frac{(n\pi)^4}{\Omega_{EB}^2}}{-2\frac{E}{kG}\left(\frac{\ell}{r}\right)^4 \Omega_{EB}^2 + \left[\left(\frac{\ell}{r}\right)^4 + \left(\frac{\ell}{r}\right)^2 \left(1 + \frac{E}{kG}\right)(n\pi)^2\right]}\right\}$$

$$= \Omega_{EB}^2 \left\{\frac{-2\frac{E}{kG}\left(\frac{\ell}{r}\right)^4 \Omega_{EB}^2 + \left[\left(\frac{\ell}{r}\right)^4 + \left(\frac{\ell}{r}\right)^2 \left(1 + \frac{E}{kG}\right)(n\pi)^2\right] + \frac{E}{kG}\left(\frac{\ell}{r}\right)^4 \Omega_{EB}^2}{-2\frac{E}{kG}\left(\frac{\ell}{r}\right)^4 \Omega_{EB}^2 + \left[\left(\frac{\ell}{r}\right)^4 + \left(\frac{\ell}{r}\right)^2 \left(1 + \frac{E}{kG}\right)(n\pi)^2\right]}\right.$$

$$\left. - \frac{\left[\left(\frac{\ell}{r}\right)^4 + \left(\frac{\ell}{r}\right)^2 \left(1 + \frac{E}{kG}\right)(n\pi)^2\right] + \frac{(n\pi)^4}{\Omega_{EB}^2}}{-2\frac{E}{kG}\left(\frac{\ell}{r}\right)^4 \Omega_{EB}^2 + \left[\left(\frac{\ell}{r}\right)^4 + \left(\frac{\ell}{r}\right)^2 \left(1 + \frac{E}{kG}\right)(n\pi)^2\right]}\right\}$$

$$= \Omega_{EB}^2 \left\{\frac{-\frac{E}{kG}\left(\frac{l}{r}\right)^4 \Omega_{EB}^2 + \frac{(n\pi)^4}{\Omega_{EB}^2}}{-\frac{2E}{kG}\left(\frac{l}{r}\right)^4 \Omega_{EB}^2 + \left[\left(\frac{\ell}{r}\right)^4 + \left(\frac{\ell}{r}\right)^2 \left(1 + \frac{E}{kG}\right)(n\pi)^2\right]}\right\} \qquad (3.6.81)$$

Recall $\Omega_{nEB}^2 = n^4\pi^4\left(\frac{r}{\ell}\right)^4$ and substituting this relation into Eq. (3.6.81) finally yields

$$\Omega^2 = \Omega_{EB}^2 \left\{\frac{-\frac{E}{kG}n^4\pi^4 + \left(\frac{\ell}{r}\right)^4}{-\frac{2E}{kG}n^4\pi^4 + \left[\left(\frac{\ell}{r}\right)^4 + \left(\frac{\ell}{r}\right)^2 \left(1 + \frac{E}{kG}\right)(n\pi)^2\right]}\right\} \qquad (3.6.82)$$

multiply numerator and denominator by $(r/\ell)^4$

$$\Omega^2 = \Omega_{EB}^2 \left\{\frac{1 - \frac{E}{kG}n^4\pi^4\left(\frac{r}{\ell}\right)^4}{1 + \left(\frac{r}{\ell}\right)^2 \left(1 + \frac{E}{kG}\right)(n\pi)^2 - \frac{2E}{kG}(n\pi)^4\left(\frac{r}{\ell}\right)^4}\right\} \qquad (3.6.83)$$

for $\frac{r}{\ell} < 1$, that is, relatively slender beams where this approximation applies one may drop the $\left(\frac{r}{\ell}\right)^4$ terms in Eq. (3.6.83) to yield

$$\Omega_{AP}^2 = \Omega_{EB}^2 \left\{\frac{1}{1 + \left(\frac{r}{\ell}\right)^2 \left(1 + \frac{E}{kG}\right)(n\pi)^2}\right\}$$

$$\cong \Omega_{EB}^2 \left[1 - \left(1 + \frac{E}{kG}\right)\left(\frac{r}{\ell}\right)^2 (n\pi)^2\right]. \qquad (3.6.84)$$

A typical plot of Eq. (3.6.84) based on $E/kG = 3.00$ is given in Figure 3.21.

**Figure 3.21**
Combined
effect of rotary
inertia and
shear on the
approximate
modal
frequencies of
Timoshenko
beams

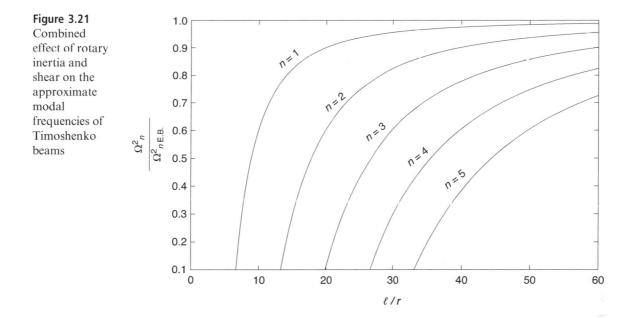

## 3.7 Vibrations of Plates

Up to this point only one-dimensional structural dynamic systems were considered. However, plate and shell structures are used in many aerospace and mechanical applications. In this section the vibrations of flat plates are considered because they contain many features associated with two-dimensional structures. Consideration of curved plates introduces complications which require the treatment of shell structures that are beyond the scope of this book.

### 3.7.1 Classical Kirchhoff Plate Theory

**Kinematics of Deformation**

Consider a simple model for the deformations of a plate that resembles the Bernoulli–Euler beam theory for one-dimensional structures. The consequence of such simplifications is that one needs to consider only the deformation of the midplane of the plate so as to obtain information representing the complete structure. However, these simplifications imply "hidden" constraints imposed on the displacements of the structure, which artificially increase the stiffness.

The geometry of a typical, two-dimensional, plate structure is shown in Figure 3.22. The $x$–$y$ plane is assumed to coincide with the midplane of the undeformed plate.

The top of the plate carries the applied load $q(x, y)$, acting normal to the midsurface and represents a load per unit area. At the edge of the plate one can have a shear force distribution $Q$ and an edge moment distribution $M$.

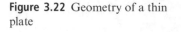

**Figure 3.22** Geometry of a thin plate

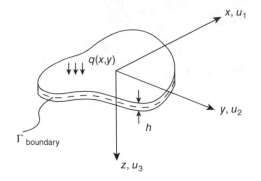

The basic assumptions of classical plate theory are listed below.

(1) The plate is assumed to be thin, that is, small, where $a, b$ represent typical plate dimensions in the middle surface.

$$\frac{h}{(a, b)} << 1.$$

This assumption implies that the vertical displacement of any point in the plate is identical to the vertical displacement of the midsurface.

$$u_3(x_1, x_2, x_3) = w(x, y). \tag{3.7.1}$$

(2) For the displacements parallel to the midsurface two contributions are considered.
   (a) First – stretching actions due to loads *at the edge of the plate*, these loads act parallel to the midsurface. Denote these displacement components by $u_{1s}, u_{2s}$.
   Next it is assumed that displacements due to stretching at a point $x_1, x_2, x_3$ have the same magnitude at any point above or below the midsurface.

$$[u_1(x_1, x_2, x_3)]_s = u(x, y)_s \tag{3.7.2a}$$

$$[u_2(x_1, x_2, x_3)]_s = v(x, y)_s. \tag{3.7.2b}$$

Thus lines normal to the midsurface of the plate translate horizontally due to stretching.
   (b) A second contribution is due to *bending*. It is assumed that lines normal to the midsurface of the plate in the undeformed state remain normal to the midsurface after deformation.

$$[u_1(x_1, x_2, x_3)]_b = -z\frac{\partial w(x, y)}{\partial x} \tag{3.7.3a}$$

$$[u_2(x_1, x_2, x_3)]_b = -z\frac{\partial w(x, y)}{\partial y}, \tag{3.7.3b}$$

where the subscript $b$ implies displacement due to bending.

The total displacement field is a combination of bending and stretching.

$$u_1 = u_{1s}(x, y) - z\frac{\partial w(x, y)}{\partial x} \tag{3.7.4a}$$

$$u_2 = u_{2s}(x, y) - z\frac{\partial w(x, y)}{\partial y} \tag{3.7.4b}$$

$$u_3 = w(x, y). \tag{3.7.4c}$$

Note that the displacement field $u_1, u_2, u_3$ is completely described in terms of the deformation of the midsurface ($u_s, v_s, w$).

Using the displacements defined in Eqs. (3.7.4a)–(3.7.4c) allows one to obtain the strain in a straightforward manner.

$$\varepsilon_{xx} = \frac{\partial u_s}{\partial x} - z\frac{\partial^2 w}{\partial x^2} \tag{3.7.5a}$$

$$\varepsilon_{yy} = \frac{\partial v_s}{\partial y} - z\frac{\partial^2 w}{\partial y^2} \tag{3.7.5b}$$

$$\varepsilon_{xy} = \frac{1}{2}\left(\frac{\partial u_s}{\partial y} + \frac{\partial v_s}{\partial x}\right) - z\frac{\partial^2 w}{\partial x \partial y} = \frac{1}{2}\left(\frac{\partial u_1}{\partial x_2} + \frac{\partial u_2}{\partial x_1}\right). \tag{3.7.5c}$$

All other strains are assumed to be zero. Note this assumption introduces an inconsistency. The transverse shear stresses $\sigma_{xy}$, $\sigma_{yz}$ will be zero due to the assumed displacement and strain field. That these quantities cannot be always zero is evident from simple equilibrium considerations. However this difficulty is also encountered in Euler–Bernoulli beam theory and we will "live with it" for the present.

## 3.7.2 Stress Resultant Intensity Functions

In the study of plates, it is useful to introduce shear and moment *distributions* per unit length. A plate element is shown in Figure 3.23, all stresses acting on the faces of the element are shown, they vary as a function of $z$. First note the presence of $\sigma_{xz}$ and $\sigma_{yz}$ which is not consistent with the assumed strain displacement field.

**Figure 3.23** Plate element for definition of shear and moment intensity functions

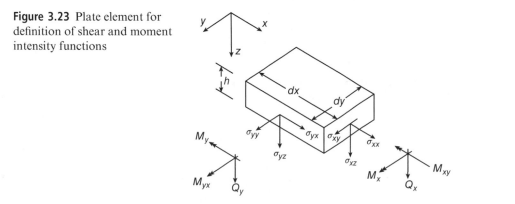

Next, define shear force intensities per unit length by the following expressions:

$$Q_x = \int_{-\frac{h}{2}}^{\frac{h}{2}} \sigma_{xz} \, dz \qquad (3.7.6a)$$

$$Q_y = \int_{-\frac{h}{2}}^{\frac{h}{2}} \sigma_{yz} \, dz. \qquad (3.7.6b)$$

Similarly, bending moment intensities per unit length are defined as

$$M_x = \int_{-\frac{h}{2}}^{\frac{h}{2}} \sigma_{xx} z \, dz \qquad (3.7.7a)$$

$$M_y = \int_{-\frac{h}{2}}^{\frac{h}{2}} \sigma_{yy} z \, dz. \qquad (3.7.7b)$$

Next, two *twisting moment intensities* per unit length are introduced by

$$M_{xy} = \int_{-\frac{h}{2}}^{\frac{h}{2}} \sigma_{xy} z \, dz \qquad (3.7.8a)$$

$$M_{yx} = \int_{-\frac{h}{2}}^{\frac{h}{2}} \sigma_{yx} z \, dz. \qquad (3.7.8b)$$

Obviously $M_{xy} = M_{yx}$ because $\sigma_{xy} = \sigma_{yx}$.

### 3.7.3    Derivation of Plate Equilibrium Equations Using a Variational Approach

In order to be consistent with the derivation of the equations of equilibrium obtained for various structural configurations considered earlier in this book, the equations of equilibrium for a plate-type structure are derived using the principle of the minimum potential energy (PMPE). Recall the strain energy $U$ of a plate, for linear elastic isotropic behavior is given by the following integral:

$$U = \frac{1}{2} \iint_R \int_{-\frac{h}{2}}^{\frac{h}{2}} \sigma_{ij} \varepsilon_{ij} \, dx \, dy \, dz. \qquad (3.7.9)$$

Using linear stress–strain relations obtained from Hooke's law

$$\begin{aligned}
\sigma_{xx} &= \frac{E}{1 - v^2} (\varepsilon_{xx} + v\varepsilon_{yy}) \\
\sigma_{yy} &= \frac{E}{1 - v^2} (\varepsilon_{yy} + v\varepsilon_{xx}) \\
\sigma_{xy} &= 2G\varepsilon_{xy}.
\end{aligned} \qquad (3.7.10)$$

Combining Eqs. (3.7.9) and (3.7.10), the strain energy is obtained

$$U = \frac{1}{2} \iint_R \int_{-\frac{h}{2}}^{\frac{h}{2}} (\sigma_{xx}\varepsilon_{xx} + 2\sigma_{xy}\varepsilon_{xy} + \sigma_{yy}\varepsilon_{yy}) \, dxdydz$$

$$= \frac{1}{2} \iint_R \int_{-\frac{h}{2}}^{\frac{h}{2}} \left[ \frac{E}{1-\nu^2}(\varepsilon_{xx}^2 + \varepsilon_{yy}^2 + 2\nu\varepsilon_{xx}\varepsilon_{yy}) + 4G\varepsilon_{xy}^2 \right] dxdydz, \qquad (3.7.11)$$

where the shear modulus $G$ is given by

$$G = \frac{E}{2(1+\nu)} = \frac{E(1-\nu)}{2(1-\nu^2)}. \qquad (3.7.12)$$

Combining the last two equations yields

$$U = \frac{E}{2(1-\nu^2)} \iint_R \int_{-\frac{h}{2}}^{\frac{h}{2}} \left[ \varepsilon_{xx}^2 + \varepsilon_{yy}^2 + 2\nu\varepsilon_{xx}\varepsilon_{yy} + 2(1-\nu)\varepsilon_{xy}^2 \right] dxdydz. \qquad (3.7.13)$$

The potential energy of the external loads is given by

$$V = - \iint_R q(x,y) \, w(x,y) \, dxdy. \qquad (3.7.14)$$

Substituting the strains given by Eqs. (3.7.5a)–(3.7.5c) into Eq. (3.7.13) and combining with Eq. (3.7.14) yields the total potential

$$\Pi_p = \frac{E}{2(1-\nu^2)} \iint_R \int_{-\frac{h}{2}}^{\frac{h}{2}} \left\{ \left( \frac{\partial u_s}{\partial x} - z\frac{\partial^2 w}{\partial x^2} \right)^2 + \left( \frac{\partial v_s}{\partial y} - z\frac{\partial^2 w}{\partial y^2} \right)^2 \right.$$

$$+ 2\nu \left( \frac{\partial u_s}{\partial x} - z\frac{\partial^2 w}{\partial x^2} \right) \left( \frac{\partial v_s}{\partial y} - z\frac{\partial^2 w}{\partial y^2} \right)$$

$$\left. + 2(1-\nu) \left[ \frac{1}{2}\left( \frac{\partial u_s}{\partial y} + \frac{\partial v_s}{\partial x} \right) - z\frac{\partial^2 w}{\partial x \partial y} \right]^2 \right\} dxdydz$$

$$- \iint_R qw \, dxdydz. \qquad (3.7.15)$$

Expanding the expressions in the brackets and integrating over the thickness of the plate

$$\int_{-\frac{h}{2}}^{\frac{h}{2}} \left\{ 1, z, z^2 \right\} dz = \left\{ h, 0, \frac{h^3}{12} \right\} \qquad (3.7.16)$$

one obtains

$$\Pi_p = \frac{C}{2} \iint_R \left[ \left(\frac{\partial u_s}{\partial x}\right)^2 + \left(\frac{\partial v_s}{\partial y}\right)^2 + 2v\frac{\partial u_s}{\partial x}\frac{\partial v_s}{\partial y} + \frac{1-v}{2}\left(\frac{\partial u_s}{\partial x} + \frac{\partial v_s}{\partial y}\right)^2 \right] dxdy$$

$$+ \frac{D}{2} \iint_R \left[ \left(\frac{\partial^2 w}{\partial x^2}\right)^2 + \left(\frac{\partial^2 w}{\partial y^2}\right)^2 + 2v\left(\frac{\partial^2 w}{\partial x^2}\right)\left(\frac{\partial^2 w}{\partial y^2}\right) \right.$$

$$\left. + 2(1-v)\left(\frac{\partial^2 w}{\partial x\partial y}\right)^2 \right] dxdy - \iint_R qw\,dxdy, \tag{3.7.17}$$

where

$$D = \frac{Eh^3}{12(1-v^2)} = \text{bending rigidity of the plate}$$

and

$$C = \frac{Eh}{1-v^2} = \text{extensional stiffness of the plate}$$

the total potential functional has three independent variables $u_s$, $v_s$, and $w$.

The equations of equilibrium are obtained by taking $\delta\Pi_p = 0$. It can be shown when performing this variation the equations for $u_s$ and $v_s$ become decoupled from the equation for $w$, which is similar to what one observes in Euler–Bernoulli beam theory. The equations for $u_s$ and $v_s$ can be shown to be

$$\frac{\partial^2 u_s}{\partial x^2} + \frac{1-v}{2}\frac{\partial^2 u_s}{\partial y^2} + \frac{1+v}{2}\frac{\partial^2 v_s}{\partial x\partial y} = 0 \tag{3.7.18}$$

$$\frac{\partial^2 v_s}{\partial y^2} + \frac{1-v}{2}\frac{\partial^2 v_s}{\partial x^2} + \frac{1+v}{2}\frac{\partial^2 u_s}{\partial x\partial y} = 0. \tag{3.7.19}$$

Next, consider the variation of $\Pi_{p1}$ after deleting the inplane displacement components for $u_s$ and $v_s$.

$$\Pi_{p1} = \frac{D}{2} \iint_R \left\{ \left(\nabla^2 w\right)^2 + 2(1-v)\left[ \left(\frac{\partial^2 w}{\partial x\partial y}\right)^2 - \left(\frac{\partial^2 w}{\partial x^2}\right)\left(\frac{\partial^2 w}{\partial y^2}\right) \right] \right\} dxdy$$

$$- \iint_R qw\,dxdy, \tag{3.7.20}$$

where

$$\nabla^2 = \frac{\partial^2}{\partial x^2} + \frac{\partial^2}{\partial y^2}$$

and Eq. (3.7.20) contains two added terms

$$2\frac{\partial^2 w}{\partial x^2}\frac{\partial^2 w}{\partial y^2} - 2\frac{\partial^2 w}{\partial x^2}\frac{\partial^2 w}{\partial y^2},$$

which cancel.

$$\delta\Pi_{p1} = 0 = \frac{D}{2}\iint_R \left\{ 2\left(\nabla^2 w\right)\left[\frac{\partial^2(\delta w)}{\partial x^2} + \frac{\partial^2(\delta w)}{\partial y^2}\right]\right.$$

$$+ (1-v)\left[2\frac{\partial^2 w}{\partial x\partial y}\frac{\partial^2(\delta w)}{\partial x\partial y} + 2\frac{\partial^2 w}{\partial y\partial x}\frac{\partial^2(\delta w)}{\partial y\partial x} - 2\left(\frac{\partial^2 w}{\partial x^2}\right)\frac{\partial^2\delta w}{\partial y^2}\right.$$

$$\left.\left.-2\left(\frac{\partial^2 w}{\partial y^2}\right)\left(\frac{\partial^2\delta w}{\partial x^2}\right)\right]\right\}\,dxdy - \iint_R q\,\delta w\,dxdy, \tag{3.7.21}$$

where in Eq. (3.7.21)

$$2\left(\frac{\partial^2 w}{\partial x\partial y}\right)^2 = \left(\frac{\partial^2 w}{\partial x\partial y}\right)^2 + \left(\frac{\partial^2 w}{\partial y\partial x}\right)^2$$

has been split into two parts as indicated above.

Next Green's theorem is applied, recall from Chapter 2.

$$\iint_S \left(\frac{\partial\phi}{\partial x} + \frac{\partial\psi}{\partial y}\right)dA = \oint_\Gamma \left(\phi a_{vx} + \psi a_{vy}\right)dl \tag{3.7.22}$$

$a_{vx} = v_1$ = etc., are the direction cosines of the outward pointing normal to the boundary. Integration by parts of Eq. (3.7.20) yields

$$\iint_S u\frac{\partial w}{\partial x_i}\,dA = \int_\Gamma (uw)q_{vx_i}dl - \iint_S w\frac{\partial u}{\partial x_i}dA. \tag{3.7.23}$$

Using Green's theorem and Eq. (3.7.21), one has

$$\iint_R \left(D\nabla^4 w - q\right)dwdxdy$$

$$+ D\oint_\Gamma \left(\frac{\partial^2 w}{\partial x^2} + v\frac{\partial^2 w}{\partial y^2}\right)\frac{\partial\delta w}{\partial x}dy - D\oint_\Gamma \left(\frac{\partial^2 w}{\partial y^2} + v\frac{\partial^2 w}{\partial x^2}\right)\frac{\partial\delta w}{\partial y}dx$$

$$+ D\oint_\Gamma (1-v)\frac{\partial^2 w}{\partial x\partial y}\frac{\partial\delta w}{\partial y}dy - D\oint_\Gamma \left(\frac{\partial^3 w}{\partial x^3} + v\frac{\partial^3 w}{\partial x\partial y^2}\right)\delta wdy$$

$$+ D\oint_\Gamma (1-v)\frac{\partial^3 w}{\partial x^2\partial y}\delta wdy - D\oint_\Gamma (1-v)\frac{\partial^3 w}{\partial x\partial y^2}\delta wdy \quad = 0. \tag{3.7.24}$$

Using the definitions of the shear and moment intensities defined in Eqs. (3.7.6a)–(3.7.8b), when these equations are expressed in terms of displacements, one has

$$
M_x = \int_{-\frac{h}{2}}^{\frac{h}{2}} \sigma_{xx} z \, dz = \int_{-\frac{h}{2}}^{\frac{h}{2}} \frac{E}{1 - v^2} \left( \varepsilon_{xx} + v \varepsilon_{yy} \right) dz
$$

$$
= \int_{-\frac{h}{2}}^{\frac{h}{2}} \frac{E}{1 - v^2} \left[ -z \frac{\partial^2 w}{\partial x^2} - v z \frac{\partial^2 w}{\partial y^2} \right] dz = -D \left( \frac{\partial^2 w}{\partial x^2} + v \frac{\partial^2 w}{\partial y^2} \right) \tag{3.7.25}
$$

$$
M_y = \int_{-\frac{h}{2}}^{\frac{h}{2}} \sigma_{yy} z \, dz = \int_{-\frac{h}{2}}^{\frac{h}{2}} \frac{E}{1 - v^2} \left( \varepsilon_{yy} + v \varepsilon_{xx} \right) z \, dz
$$

$$
= \int_{-\frac{h}{2}}^{\frac{h}{2}} \frac{E}{1 - v^2} \left( -z \frac{\partial^2 w}{\partial y^2} - v z \frac{\partial^2 w}{\partial x^2} \right) dz = -D \left( \frac{\partial^2 w}{\partial y^2} + v \frac{\partial^2 w}{\partial x^2} \right) \tag{3.7.26}
$$

$$
M_{xy} = \int_{-\frac{h}{2}}^{\frac{h}{2}} 2 G \varepsilon_{xy} z \, dz = \frac{1 - v}{1 - v^2} \int_{-\frac{h}{2}}^{\frac{h}{2}} -z^2 \frac{\partial^2 w}{\partial x \partial y} dz = -(1 - v) D \frac{\partial^2 w}{\partial x \partial y} \tag{3.7.27}
$$

using these equations, Eq. (3.7.24) is rewritten as

$$
\iint_R \left( D \nabla^4 w - q \right) dw \, dx \, dy - \oint_\Gamma M_x \frac{\partial \delta w}{\partial x} dy + \oint_\Gamma M_y \frac{\partial \delta w}{\partial y} dx
$$

$$
- \oint_\Gamma M_{xy} \frac{\partial \delta w}{\delta y} dy + \oint_\Gamma M_{xy} \frac{\partial \delta w}{\partial x} dx
$$

$$
- \oint_\Gamma \left( \frac{\partial M_y}{\partial y} + \frac{\partial M_{xy}}{\partial x} \right) \delta w \, dx + \oint_\Gamma \left( \frac{\partial M_x}{\partial x} + \frac{\partial M_{xy}}{\partial y} \right) \delta w \, dy \quad = 0. \tag{3.7.28}
$$

From simple equilibrium considerations, it can be shown that the last two terms in Eq. (3.7.28) represent the boundary work of shear stress resultants $Q_x$ and $Q_y$, that is,

$$
Q_y = \frac{\partial M_y}{\partial y} + \frac{\partial M_{xy}}{\partial x} \tag{3.7.29}
$$

$$
Q_x = \frac{\partial M_x}{\partial x} + \frac{\partial M_{xy}}{\partial y} \tag{3.7.30}
$$

thus Eq. (3.7.28) finally becomes

$$
\iint_R \left( D \nabla^4 w - q \right) dw \, dx \, dy - \oint_\Gamma M_x \frac{\partial \delta w}{\partial x} dy + \oint_\Gamma M_y \frac{\partial \delta w}{\partial y} dx
$$

$$
- \oint_\Gamma M_{xy} \frac{\partial \delta w}{\delta y} dy + \oint_\Gamma M_{xy} \frac{\partial \delta w}{\partial x} dx - \oint_\Gamma Q_y \delta w \, dx + \oint_\Gamma Q_x \delta w \, dy \quad = 0. \tag{3.7.31}
$$

Further simplification of this functional is possible by considering a portion of the boundary $\Gamma$ shown in Figure 3.24. Consider $s$ and $n$ to be a rectangular coordinate system at a point on the boundary of the plate, that is tangential and normal to the curved boundary

$$
\cos \varphi = a_{nx} = \frac{dn}{dx} \tag{3.7.32}
$$

$$
\sin \varphi = a_{ny} = -\frac{ds}{dx}. \tag{3.7.33}
$$

**Figure 3.24** Relations between the $x, y$ system and a coordinate system tangential and normal to the curved boundary of the plate

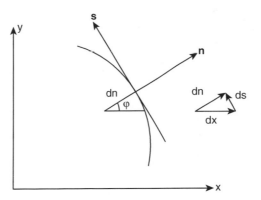

The minus in Eq. (3.7.33) implies that moving along the positive direction for $s$ on $\Gamma$ results in a motion in the $-x$ direction, similarly the geometry in Figure 3.24 can be used, from which

$$\cos \varphi = a_{nx} = \frac{ds}{dy} \tag{3.7.34}$$

$$\sin \varphi = a_{ny} = -\frac{dn}{dy}. \tag{3.7.35}$$

Using these expressions together with a considerable amount of algebraic manipulation (Dym and Shames 1973, 294–297; Figure 3.25), one can show that Eq. (3.7.31) reduces to

$$\delta \Pi_{p1} = \iint_R \left( D\nabla^4 w - q \right) \delta w \, dx \, dy - \oint_\Gamma M_n \delta \left( \frac{\partial w}{\partial n} \right) ds$$

$$+ \oint_\Gamma \left( Q_n + \frac{\partial M_{ns}}{\partial s} \right) \delta w \, ds = 0 \tag{3.7.36}$$

**Figure 3.25** Relations between the $x, y$ system and a coordinate system tangential and normal to the curved boundary of the plate

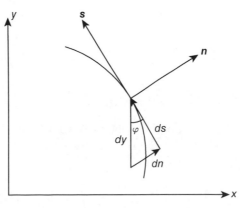

The equations of equilibrium and the boundary conditions for a plate with a general curved boundary are obtained from Eq. (3.7.36). The equation of equilibrium is given by

$$\nabla^4 w = \frac{\partial^4 w}{\partial x^4} + 2\frac{\partial^4 w}{\partial x^2 \partial y^2} + \frac{\partial^4 w}{\partial y^4} = \frac{q(x,y)}{D} \tag{3.7.37}$$

and on the boundary $\Gamma$

| | | | | |
|---|---|---|---|---|
| either | $M_n = 0$ | or | $\dfrac{\partial w}{\partial n}$ is prescribed, and | (3.7.38a) |
| either | $Q_n + \dfrac{\partial M_{ns}}{\partial s} = 0$ | or | $w$ is prescribed. | (3.7.38b) |

For the case of a rectangular plate shown in Figure 3.26, the appropriate boundary conditions are given on $x = 0, a$ and $y = 0, b$.

**Figure 3.26** Geometry of a rectangular plate

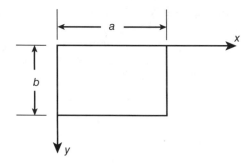

Along $x = 0$ and $x = a$

| | | | | |
|---|---|---|---|---|
| either | $M_x = 0$ | or | $\dfrac{\partial w}{\partial x}$ is prescribed, and | |
| either | $Q_x + \dfrac{\partial M_{xy}}{\partial y} = 0$ | or | $w$ is prescribed. | (3.7.38c) |

Along $y = 0$ and $y = b$,

| | | | | |
|---|---|---|---|---|
| either | $M_y = 0$ | or | $\dfrac{\partial w}{\partial y}$ is prescribed, and | |
| either | $Q_y + \dfrac{\partial M_{xy}}{\partial x} = 0$ | or | $w$ is prescribed. | (3.7.38d) |

### 3.7.4 Dynamic Equations of Equilibrium

To complete the formulation of the plate vibration problem for structural dynamic applications the kinetic energy has to be added to the total potential. The kinetic energy of a plate, neglecting rotary inertia, is given by

$$T = \frac{1}{2}\iint_R h\rho \dot{w}^2 dx dy. \tag{3.7.39}$$

Applying Hamilton's principle, one obtains

$$\delta \int_{t_1}^{t_2} \left( T - \Pi_p \right) dt = 0 \tag{3.7.40}$$

$$\delta \int_{t_1}^{t_2} T dt = \int_{t_1}^{t_2} \iint_R \frac{1}{2} \rho \, \dot{w} \, 2 \, \delta \dot{w} \, dxdy$$

$$= \iint_R \rho \, \dot{w} \, \delta w |_{t_1}^{t_2} \, dxdy - \int_{t_1}^{t_2} \iint_R \rho \, \ddot{w} \, \delta w \, dtdxdy$$

$$\text{and } \delta w(t_1) = \delta w(t_2) = 0. \tag{3.7.41}$$

Combining the relevant equations, one obtains

$$, D\nabla^4 w + m\ddot{w} = q(x, y, t) \tag{3.7.42}$$

where $m = \rho h$ is the mass of the plate per unit area, and the boundary conditions are given by Eqs. (3.7.38a) and (3.7.38b) as well as two additional similar relations. Note that Eq. (3.7.42) represents the plate vibration problem in the absence of axial loads.

### 3.7.5 Solution of Typical Plate Vibration Problems

Consider a thin simply supported plate with dimensions $a, b$, and thickness $h$. The differential equation for the free vibration problem is given by Eq. (3.7.42), which can be rewritten as

$$\beta^2 \nabla^4 w + \ddot{w} = 0, \tag{3.7.43}$$

$$\text{where } \beta^2 = \left( \frac{D}{m} \right) \text{ and } \nabla^4 = \frac{\partial^4}{\partial x^4} + 2\frac{\partial^4}{\partial x^2 \partial y^2} + \frac{\partial^4}{\partial y^4}.$$

Using separation of variables, that has also been used for the beam vibration problem, one has

$$w = W(x, y)T(t). \tag{3.7.44}$$

Substitution yields

$$\beta^2 T(t)\nabla^4 W(x, y) + W(x, y)\ddot{T}(t) = 0$$

or

$$\beta^2 \frac{\nabla^4 W(x, y)}{W(x, y)} = -\frac{\ddot{T}(t)}{T(t)}. \tag{3.7.45}$$

Setting each side equal to a constant $\omega^2$ yields

$$\ddot{T} + \omega^2 T = 0 \tag{3.7.46}$$

$$\beta^2 \nabla^4 W - \omega^2 W = 0. \tag{3.7.47}$$

Solution of Eq. (3.7.46) is

$$T = A \cos \omega t + B \sin \omega t. \qquad (3.7.48)$$

This problem resembles the simply supported beam vibration problem, and the solution to the problem is easy to identify

$$W(x, y) = X(x)Y(y) = \sin \frac{n\pi x}{a} \sin \frac{m\pi y}{b}. \qquad (3.7.49)$$

The expression given by Eq. (3.7.49) satisfies the boundary conditions

$$W = M_x = 0 \qquad \text{at } x = 0, a \qquad (3.7.50a)$$
$$W = M_y = 0 \qquad \text{at } y = 0, b. \qquad (3.7.50b)$$

Since this is a free vibration problem, substitution of Eq. (3.7.49) into Eq. (3.7.43) yields

$$\left[ \left( \frac{n\pi}{a} \right)^4 + 2 \left( \frac{n\pi}{a} \right)^2 \left( \frac{m\pi}{b} \right)^2 + \left( \frac{m\pi}{b} \right)^4 - \frac{w_{nm}^2}{\beta^2} \right] \sin \frac{n\pi x}{a} \sin \frac{m\pi x}{b} = 0$$

from which one obtains the well-known frequency equation

$$\omega_{nm}^2 = \beta^2 \left[ \left( \frac{n\pi}{a} \right)^4 + 2 \left( \frac{n\pi}{a} \right)^2 \left( \frac{m\pi}{b} \right)^2 + \left( \frac{m\pi}{b} \right)^4 \right], \qquad (3.7.51)$$

which can also be written as

$$\omega_{nm}^2 = \beta^2 \pi^4 \left[ \left( \frac{n}{a} \right)^2 + \left( \frac{m}{b} \right)^2 \right]^2. \qquad (3.7.52)$$

For the $n, m$th mode the mode shape is given by

$$W_{nm}(x, y) = \sin \frac{n\pi x}{a} \sin \frac{m\pi y}{b} \qquad (3.7.53)$$

and the solution to the free vibration problem is given by

$$w = \sum_{m=1}^{\infty} \sum_{n=1}^{\infty} \left( \sin \frac{n\pi x}{a} \sin \frac{m\pi y}{b} \right) \left( A_{mn} \sin \omega_{nm} t + B_{nm} \cos \omega_{nm} t \right). \qquad (3.7.54)$$

Equation (3.7.54) is the general solution to the problem and it must satisfy the initial conditions. The constants $A_{nm}$ and $B_{nm}$ are determined so as to satisfy the initial conditions, given by

$$w(x, y, 0) = \phi(x, y)$$
$$\frac{\partial w}{\partial t}(x, y, 0) = \psi(x, y). \qquad (3.7.55)$$

From Eqs. (3.7.54) and (3.7.55)

$$\sum_{n=1}^{\infty} \sum_{m=1}^{\infty} \sin \frac{n\pi x}{a} \sin \frac{m\pi y}{b} B_{nm} = \phi(x, y) \tag{3.7.56a}$$

$$\sum_{n=1}^{\infty} \sum_{m=1}^{\infty} \sin \frac{n\pi x}{a} \sin \frac{m\pi y}{b} \omega_{nm} A_{nm} = \psi(x, y). \tag{3.7.56b}$$

Using the Kronecker delta $\delta_{rs} = 0$ for $r \neq s$, and $\delta_{rs} = 1$ for $r = s$

$$\begin{aligned} \int_0^a \sin \frac{r\pi x}{a} \sin \frac{s\pi x}{a} dx &= \frac{a}{2} \delta_{rs} \\ \int_0^b \sin \frac{p\pi y}{b} \sin \frac{q\pi y}{b} dx &= \frac{b}{2} \delta_{pq} \end{aligned} \tag{3.7.57}$$

one finds that

$$\iint_R (W_{rs}) (W_{pq}) \, dxdy = \frac{ab}{4} \delta_{rs} \delta_{pq}. \tag{3.7.58}$$

Multiplying Eq. (56a) by $\sin \frac{p\pi x}{a} \sin \frac{q\pi y}{b}$ and integrating over the area of the $R$, one has

$$\iint_R \left[ \sum_{m=1}^{\infty} \sum_{n=1}^{\infty} \sin \frac{n\pi x}{a} \sin \frac{m\pi y}{b} \sin \frac{p\pi x}{a} \sin \frac{q\pi y}{b} B_{nm} \right] dxdy$$

$$= \iint_R \phi(x, y) \sin \frac{p\pi x}{a} \sin \frac{g\pi y}{b} dxdy = B_{pq} \frac{ab}{4}$$

thus

$$B_{pq} = \frac{4}{ab} \iint_R \phi(x, y) \sin \frac{p\pi x}{a} \sin \frac{q\pi y}{b} dxdy \tag{3.7.59}$$

and similarly

$$A_{pq} = \frac{4}{ab\omega_{pq}} \iint_R \psi(x, y) \sin \frac{p\pi x}{a} \sin \frac{q\pi y}{b} dxdy. \tag{3.7.60}$$

## 3.7.6  General Rayleigh–Ritz Procedure for Plates

Exact solutions to plate vibration problems involving more complicated boundary conditions or shapes or mass and stiffness distributions are difficult to obtain. Many solutions to plate problems can be found in Leissa (1969). For many cases, one has to use approximate methods such as the Rayleigh–Ritz, Galerkin, or finite element method (FEM) in order to obtain solutions to the free or forced vibration problem of plate-type structures. As indicated earlier in this book, both methods are similar; however, while the Rayleigh–Ritz method is based upon Rayleigh's quotient, Galerkin's method is an approximate method for obtaining solutions to partial differential equations. Another widely used method for obtaining accurate approximate solutions is the FEM discussed in Chapter 4.

Consider a typical application of the Rayleigh–Ritz method for constructing approximate solutions to rectangular plate vibration problems.

The displacement of the plate is approximated by

$$W_n(x, y) = \sum_{i=1}^{n} A_i \phi_i(x, y),$$  (3.7.61)

where $A_i$s are undetermined coefficients and the accuracy of the solution depends on the number of terms $n$ used in the approximation. Recall that Rayleigh's quotient can be written in two possible ways, which are identical. One may use the energy formulation, where the approximate eigenvalue (or frequency squared) $^n\Lambda$ is given by Eq. (3.7.62)

$$^n\Lambda = \frac{U_{\max}}{T^*_{\max}},$$  (3.7.62)

where $U_{\max}$ is the maximum value of strain energy and $T^*$ is the maximum value of kinetic energy density.

An alternative approach is based on the operator notation, see Eq. (3.5.19). Recall that Rayleigh quotient in operator notation is written as

$$\lambda_r = \frac{\int_D w_r L[w_r] \, dD}{\int_D w_r M[w_r] \, dD}.$$  (3.7.63)

For plate problems, assuming that

$$w(x, y, t) = W(x, y) \cos \omega t,$$  (3.7.64)

where the $L[\,]$ operator is identified from Eq. (3.7.47) and $M[\,] = m = \rho h$

$$L[\,] = D\nabla^4 = D\left(\frac{\partial^4}{\partial x^4} + 2\frac{\partial^4}{\partial x^2 \partial y^2} + \frac{\partial^4}{\partial y^4}\right)$$  (3.7.65)

and $M[\,] = \rho h$; thus,

$$\lambda_r = \frac{\iint_R W D\nabla^4 W \, dx dy}{\iint_R \rho h W^2 \, dx dy}.$$  (3.7.66)

Combining Eqs. (3.7.62), (3.7.43), and (3.7.20), an alternate expression for Rayleigh's quotient is obtained

$$^n\Lambda = \frac{D \iint_R \left\{(\nabla^2 W)^2 + 2(1-v)\left[\left(\frac{\partial^2 W}{\partial x \partial y}\right)^2 - \left(\frac{\partial^2 W}{\partial x^2}\right)\left(\frac{\partial^2 W}{\partial y^2}\right)\right]\right\} dx dy}{\iint_R \rho h W^2 \, dx dy}.$$  (3.7.67)

The statements made earlier in the book regarding the choice of functions that can be used for obtaining approximate free vibration frequencies and mode shapes also apply to this case. When Eq. (3.7.63) is used in order to construct approximate expressions for the stiffness and mass matrices comparison functions have to be used. These have to satisfy all the boundary conditions of the problem, both natural and geometric. When expression (3.7.62) is used to obtain the solution of the approximate free vibration problem, the requirements on the assumed functions $\phi_i(x, y)$ can be relaxed, and admissible functions, satisfying only the geometric boundary conditions, can be employed.

Consider a solution of a typical plate vibration problem based on the application of Eq. (3.7.67), with the displacement function given by Eq. (3.7.61)

$$
^n\Lambda = \frac{D \iint_R \left\{ \left( \sum_{i=1}^n A_i \nabla^2 \phi_i \right)^2 + 2(1 - v)B \right\} dxdy}{\iint_R m \left( \sum_{i=1}^n A_i \phi_i \right)^2 dxdy}, \tag{3.7.68}
$$

where

$$
B = \left[ \left( \sum_{i=1}^n A_i \frac{\partial^2 \phi_i}{\partial x \partial y} \right)^2 - \left( \sum_{i=1}^n A_i \frac{\partial^2 \phi_i}{\partial x^2} \right) \left( \sum_{i=1}^n A_i \frac{\partial^2 \phi_i}{\partial y^2} \right) \right]
$$

or using a more compact notation

$$
^n\Lambda = R(W_n) = \frac{N(W_n)}{D(W_n)}. \tag{3.7.69}
$$

The solution of the approximate eigenvalue problem is obtained from the requirement that $R(W_n)$ should be a minimum in the vicinity of the natural frequencies of the system, which is expressed by

$$
\frac{\partial R(W_n)}{\partial A_j} = \frac{D(W_n)\dfrac{\partial N(W_n)}{\partial A_j} - N(W_n)\dfrac{\partial D(W_n)}{\partial A_j}}{[D(W_n)]^2} = 0, \tag{3.7.70}
$$

which is equivalent to

$$
\frac{\partial N(W_n)}{\partial A_j} - {}^n\Lambda \frac{\partial D(W_n)}{\partial A_j} = 0, \quad \text{for } j = 1, 2, \ldots, n. \tag{3.7.71}
$$

Define elements of the approximate mass matrix by

$$
m_{ij} = \iint_R h\rho\phi_i\phi_j dxdy \tag{3.7.72}
$$

then $D(W_n) = \sum_{i=1}^n \sum_{j=1}^n m_{ij} A_i A_j.$ \tag{3.7.73}

Similarly, define approximate elements of the stiffness matrix by

$$
\begin{aligned}
k_{ij} = D \iint_R \Bigg[ & \frac{\partial^2 \phi_i}{\partial x^2} \frac{\partial^2 \phi_j}{\partial x^2} + \frac{\partial^2 \phi_i}{\partial y^2} \frac{\partial^2 \phi_j}{\partial y^2} \\
& + 2v \frac{\partial^2 \phi_i}{\partial x^2} \frac{\partial^2 \phi_j}{\partial y^2} + 2(1 - v) \frac{\partial^2 \phi_i}{\partial x \partial y} \frac{\partial^2 \phi_j}{\partial x \partial y} \Bigg] dxdy
\end{aligned}
\tag{3.7.74}
$$

thus

$$
N(W_n) = \sum_{i=1}^{n} \sum_{i=1}^{n} k_{ij} A_i A_y.
\tag{3.7.75}
$$

Substituting Eqs. (3.7.73) and (3.7.75) into Eq. (3.7.71) yields

$$
k_{ij} A_j - {}^n\Lambda m_{ij} A_j = 0 \quad \text{for } j = 1, 2, ..., n.
\tag{3.7.76}
$$

Equation (3.7.76) written in matrix form yields the approximate eigenvalue problem

$$
[k]\{A\} = {}^n\Lambda [m]\{A\}
\tag{3.7.77}
$$

from which the first $n$ approximate eigenvalues or natural frequencies of the plate vibration problem can be obtained. For a particular eigenvalue ${}^n\Lambda_r$, the corresponding mode shape is given by

$$
W_n^{(r)} = \sum_{i=1}^{n} A_i^{(r)} \phi_i(x, y),
\tag{3.7.78}
$$

where $A_i^{(r)}$ are the elements of the eigenvector corresponding to ${}^n\Lambda_r$.

## 3.8    The von Kármán Theory of Flat Plates

In the classical linear theory of flat plates (Kirchhoff theory), the quadratic terms in the strain tensor are neglected. This approximation fails to account for an important source of nonlinearity which is frequently present in plate behavior. The in-plane stretching of the plate, denoted "membrane effect," produces tensions in the deflected plate producing force components that oppose the applied lateral loads. This effect becomes important when the deflections at the middle plane of the thin plate are of the same order of magnitude as the thickness $h$ of the plate. This situation is remedied by von Kármán's plate theory that is an approximate theory that accounts for the presence of the $\left(\frac{\partial w}{\partial x}\right)^2$ and $\left(\frac{\partial w}{\partial y}\right)^2$ terms in the strain tensor. Recall that the nonlinear strain displacement relations can be written as

$$\epsilon_x = \frac{\partial u}{\partial x} + \frac{1}{2}\left[\left(\frac{\partial u}{\partial x}\right)^2 + \left(\frac{\partial v}{\partial x}\right)^2 + \left(\frac{\partial w}{\partial x}\right)^2\right] \tag{3.8.1a}$$

$$\epsilon_y = \frac{\partial v}{\partial y} + \frac{1}{2}\left[\left(\frac{\partial u}{\partial y}\right)^2 + \left(\frac{\partial v}{\partial y}\right)^2 + \left(\frac{\partial w}{\partial y}\right)^2\right] \tag{3.8.1b}$$

$$\epsilon_z = \frac{\partial w}{\partial z} + \frac{1}{2}\left[\left(\frac{\partial u}{\partial z}\right)^2 + \left(\frac{\partial v}{\partial z}\right)^2 + \left(\frac{\partial w}{\partial z}\right)^2\right] \tag{3.8.1c}$$

$$\epsilon_{zy} = \left(\frac{\partial w}{\partial y} + \frac{\partial v}{\partial z}\right) + \frac{\partial u}{\partial y}\frac{\partial u}{\partial z} + \frac{\partial v}{\partial y}\frac{\partial v}{\partial z} + \frac{\partial w}{\partial y}\frac{\partial w}{\partial z} \tag{3.8.1d}$$

$$\epsilon_{xz} = \left(\frac{\partial u}{\partial z} + \frac{\partial w}{\partial x}\right) + \frac{\partial u}{\partial z}\frac{\partial u}{\partial x} + \frac{\partial v}{\partial z}\frac{\partial v}{\partial x} + \frac{\partial w}{\partial z}\frac{\partial w}{\partial x} \tag{3.8.1e}$$

$$\epsilon_{yx} = \left(\frac{\partial v}{\partial x} + \frac{\partial u}{\partial y}\right) + \frac{\partial u}{\partial x}\frac{\partial u}{\partial y} + \frac{\partial v}{\partial x}\frac{\partial v}{\partial y} + \frac{\partial w}{\partial x}\frac{\partial w}{\partial y}. \tag{3.8.1f}$$

Concisely stated, the von Kármán plate theory is based on the retention of the quadratic terms representing the slopes of the plate while the rest of the quadratic terms in Eqs. (3.8.1a)–(3.8.1f) are neglected. This assumption is reasonable, when one recognizes that in most cases the axial strain $u_{,x}$ is small compared to the slope of the plate when undergoing moderate deflections. Where moderate deflections implies $w/h \approx O(1)$, where $O()$ denotes order of magnitude. Therefore, in Eq. (3.8.1a), $u_{,x}^2$ and $v_{,x}^2$ are negligible compared to $w_{,x}^2$ since $u_{,x}$ and $v_{,x}$ in most engineering applications are $O(10^{-2}), O(10^{-3})$. Therefore, they are negligible compared to slopes. Similar arguments apply to $v_{,y}$ and $u_{,y}$. The same argument also applies to $\frac{\partial v}{\partial y}\frac{\partial v}{\partial z}, \frac{\partial u}{\partial z}\frac{\partial u}{\partial x}, \frac{\partial w}{\partial y}\frac{\partial w}{\partial z}, \frac{\partial w}{\partial z}\frac{\partial w}{\partial x}, \frac{\partial u}{\partial x}\frac{\partial u}{\partial y}$, and $\frac{\partial v}{\partial x}\frac{\partial v}{\partial y}$ terms in Eqs. (3.8.1d) through (3.8.1f). The other quadratic terms in $u$ and $v$ are dropped because they have the same magnitude as the squares of the axial strain components and the terms which have been discarded. The theory is denoted as a moderate deflection theory, because it is based on the assumption that the slopes of the middle surface of the plate are moderate, that is, $w/h \approx O(1)$ as opposed to a large deflection theory where the displacements can exceed the thickness of the plate.

For convenience, denote the displacement of an arbitrary point $x, y, z$ in the plate by $u, v$, and $w$, while the displacement of a point on the middle plane of the plate is denoted by $u_0, v_0$, and $w_0$. Similarly, $\epsilon_x, \epsilon_y$, and $\epsilon_{xy}$ denote the strain components at any point $x, y$, and $z$ while $\epsilon_{x_0}, \epsilon_{y_0}$, and $\epsilon_{xy_0}$ denote the corresponding strain components at the point $(x, y, 0)$. After neglecting all quadratic terms in Eqs. (3.8.1) except $\left(\frac{\partial w}{\partial x}\right)^2, \left(\frac{\partial w}{\partial y}\right)^2$, and $\left(\frac{\partial w}{\partial x}\right)\left(\frac{\partial w}{\partial y}\right)$, Eqs. (3.8.1a) through (3.8.1f) simplify to

$$\epsilon_x = \frac{\partial u}{\partial x} + \frac{1}{2}\left(\frac{\partial w}{\partial x}\right)^2 \qquad\qquad \epsilon_{yz} = \frac{\partial w}{\partial y} + \frac{\partial v}{\partial z}$$

$$\epsilon_y = \frac{\partial v}{\partial y} + \frac{1}{2}\left(\frac{\partial w}{\partial y}\right)^2 \qquad\qquad \epsilon_{zx} = \frac{\partial u}{\partial z} + \frac{\partial w}{\partial x} \tag{3.8.2}$$

$$\epsilon_z = \frac{\partial w}{\partial z} \qquad\qquad\qquad\qquad \epsilon_{xy} = \frac{\partial v}{\partial x} + \frac{\partial u}{\partial y} + \frac{\partial w}{\partial x}\frac{\partial w}{\partial y}.$$

Additional simplification of Eq. (3.8.2) is needed. The variation of $u$ and $v$ through the thickness $z$ can be determined by introducing the assumption $\epsilon_z = 0$, which implies that the $w$ displacement of the plate is represented by the deflection at the middle plane. This assumption implies that change in $w$ as a function of $z$ has a small and negligible effect on $u$ and $v$. Thus, $\epsilon_z = 0$ implies $w = w_0$. If there are no external shearing forces applied at the faces of the plate, and the material is isotropic $\epsilon_{yz}$ and $\epsilon_{zx}$ vanishes at the faces, since $\sigma_{yz} = G\epsilon_{yz}$ and $\sigma_{zx} = G\epsilon_{zx}$. Assuming that the plate is thin implies that $\epsilon_{yz}$ and $\epsilon_{zx}$ are small everywhere, and therefore one can assume $\epsilon_{yz} = \epsilon_{zx} = 0$. The relations $\epsilon_{yz} = \epsilon_{zx} = \epsilon_z = 0$ imply that lines normal to the middle plane, before deformation, remain straight and normal during the deformation. This assumption is analogous to Bernoulli's assumption for beams and it is known as the Kirchhoff approximation in the theory of plates. Introducing these assumptions into Eq. (3.8.2) yields the displacement field

$$
\begin{aligned}
u &= u_0 - z\frac{\partial w}{\partial x} \\
v &= v_0 - z\frac{\partial w}{\partial y} \\
w &= w_0
\end{aligned}
\tag{3.8.3}
$$

and the corresponding strain–displacement relations are

$$
\begin{aligned}
\epsilon_x &= \epsilon_{x_0} - z\frac{\partial^2 w}{\partial x^2} \\
\epsilon_y &= \epsilon_{y_0} - z\frac{\partial^2 w}{\partial y^2} \\
\epsilon_{xy} &= \epsilon_{xy_0} - 2z\frac{\partial^2 w}{\partial x \partial y},
\end{aligned}
\tag{3.8.4}
$$

where

$$
\begin{aligned}
\epsilon_{x_0} &= \frac{\partial u_0}{\partial x} + \frac{1}{2}\left(\frac{\partial w}{\partial x}\right)^2 \\
\epsilon_{y_0} &= \frac{\partial v_0}{\partial y} + \frac{1}{2}\left(\frac{\partial w}{\partial y}\right)^2 \\
\epsilon_{xy_0} &= \left(\frac{\partial u_0}{\partial y} + \frac{\partial v_0}{\partial x}\right) + \frac{\partial w}{\partial x}\frac{\partial w}{\partial y}.
\end{aligned}
\tag{3.8.5}
$$

The shearing stresses $\sigma_{xz}$ and $\sigma_{yz}$ have been already neglected. In the stress–strain relations $\sigma_z$ is also neglected, since it is approximately of the same magnitude as the external pressures applied to the faces of the plate. The assumption made implies that for a thin plate acted on by loads in the plane of symmetry can be considered to be in a state of plane stress. Furthermore, assuming isotropic plate properties combined with Hooke's law

$$
\epsilon_x = \frac{1}{E}\left(\sigma_x - \nu\sigma_y\right)
$$

$$\epsilon_y = \frac{1}{E}\left(\sigma_y - \nu\sigma_x\right) \tag{3.8.6}$$

$$\epsilon_{xy} = \frac{1}{G}\sigma_{xy}$$

or an alternative form of these equations

$$\sigma_x = \frac{E}{1-\nu^2}\left(\epsilon_x + \nu\epsilon_y\right)$$

$$\sigma_y = \frac{E}{1-\nu^2}\left(\epsilon_y + \nu\epsilon_x\right) \tag{3.8.7}$$

$$\sigma_{xy} = G\epsilon_{xy}$$

and expressing the strain energy density in terms of the stress components

$$U_0 = \frac{1}{2E}\left[\sigma_x^2 + \sigma_y^2 - 2\nu\left(\sigma_x\sigma_y\right) + 2(1+\nu)\sigma_{xy}^2\right] \tag{3.8.8}$$

alternatively Eq. (3.8.8) can be expressed in terms of strain components

$$U_0 = \frac{E}{2(1-\nu^2)}\left[\epsilon_x^2 + \epsilon_y^2 + 2\nu\epsilon_x\epsilon_y + \frac{1}{2}(1-\nu)\epsilon_{xy}^2\right]. \tag{3.8.9}$$

The strain energy is given by

$$U = \iint_R \left(\int_{-\frac{h}{2}}^{\frac{h}{2}} U_0 dz\right) dxdy. \tag{3.8.10}$$

When Eqs. (3.8.4) and (3.8.9) are substituted into Eq. (3.8.10) the strain energy can be separated into two contributions $U = U_m + U_b$, where $U_m$ = membrane strain energy which is associated with stretching and is linear in h, and $U_b$ = bending strain energy which is cubic in $h$. The results of the integrations with respect to $z$ yields (with $D = \dfrac{Eh^3}{12(1-\nu^2)}$)

$$U_m = \frac{Eh}{2(1-\nu^2)}\iint\left[\epsilon_{x_0}^2 + \epsilon_{y_0}^2 + 2\nu\epsilon_{x_0}\epsilon_{y_0} + \frac{1}{2}(1-\nu)\epsilon_{xy_0}^2\right] dxdy \tag{3.8.11}$$

$$U_b = \frac{D}{2}\iint_R\left[\left(\frac{\partial^2 w}{\partial x^2}\right)^2 + \left(\frac{\partial^2 w}{\partial y^2}\right)^2 + 2\nu\frac{\partial^2 w}{\partial x^2}\frac{\partial^2 w}{\partial y^2}\right.$$

$$\left. + 2(1-\nu)\frac{\partial^2 w}{\partial x \partial y}\right] dxdy. \tag{3.8.12}$$

Equation (3.8.11), which represents the membrane energy, can be manipulated further by considering a rectangular element of the plate shown in Figure 3.27 with dimensions $dx$ and $dy$, subjected to forces $N_x dy$, $N_y dx$, $N_{xy} dy$, and $N_{yx} dx$

The tractions $N_x$, $N_y$, $N_{xy}$, and $N_{yx}$ are the stress resultants given by

$$N_x = \int_{-\frac{h}{2}}^{\frac{h}{2}} \sigma_x dz \;\;;\;\; N_y = \int_{-\frac{h}{2}}^{\frac{h}{2}} \sigma_y dz \;\;;\;\; N_{xy} = N_{yx} = \int_{-\frac{h}{2}}^{\frac{h}{2}} \sigma_{xy} dz. \tag{3.8.13}$$

**Figure 3.27** Rectangular
plate element

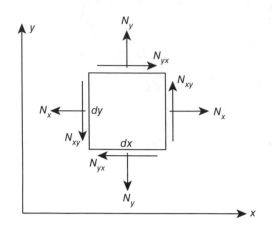

Substituting Eqs. (3.8.4) and (3.8.7) into Eq. (3.8.13) yields

$$N_x = \frac{Eh}{1 - v^2}(\epsilon_{x_0} + v\epsilon_{y_0})$$

$$N_y = \frac{Eh}{1 - v^2}(\epsilon_{y_0} + v\epsilon_{x_0}) \qquad (3.8.14)$$

$$N_{xy} = N_{yx} = Gh\epsilon_{xy_0}.$$

Combining Eq. (3.8.14) with Eq. (3.8.11) yields

$$U_m = \frac{1}{2Eh} \iint \left[ N_x^2 + N_y^2 - 2vN_xN_y + 2(1 + v)N_{xy}^2 \right] dxdy$$

$$= \frac{1}{2} \iint \left[ N_x\epsilon_{x_0} + N_y\epsilon_{y_0} + N_{xy}\epsilon_{xy_0} \right] dxdy. \qquad (3.8.15)$$

Recognizing that $U = U_m + U_b$, the variation of the strain energy can be written as

$$\delta U = \iint \left[ N_x\delta\epsilon_{x_0} + N_y\delta\epsilon_{y_0} + N_{xy}\delta\epsilon_{yx_0} \right] dxdy$$

$$+ D \iint \left[ \frac{\partial^2 w}{\partial x^2}\delta\left(\frac{\partial^2 w}{\partial x^2}\right) + \frac{\partial^2 w}{\partial y^2}\delta\left(\frac{\partial^2 w}{\partial y^2}\right) + v\frac{\partial^2 w}{\partial x^2}\delta\frac{\partial^2 w}{\partial y^2} \right.$$

$$\left. + v\frac{\partial^2 w}{\partial y^2}\delta\frac{\partial^2 w}{\partial x^2} + 2(1 - v)\frac{\partial^2 w}{\partial x\partial y}\delta\frac{\partial^2 w}{\partial x\partial y} \right] dxdy. \quad (3.8.16)$$

Furthermore from Eq. (3.8.5)

$$\delta\epsilon_{x_0} = \delta\left(\frac{\partial u}{\partial x}\right) + \frac{\partial w}{\partial x}\delta\left(\frac{\partial w}{\partial x}\right)$$

$$\delta\epsilon_{y_0} = \delta\left(\frac{\partial v}{\partial y}\right) + \frac{\partial w}{\partial y}\delta\left(\frac{\partial w}{\partial y}\right) \qquad (3.8.17)$$

$$\delta\epsilon_{xy_0} = \delta\left(\frac{\partial v}{\partial x}\right) + \delta\left(\frac{\partial u}{\partial y}\right) + \frac{\partial w}{\partial x}\delta\left(\frac{\partial w}{\partial y}\right) + \frac{\partial w}{\partial y}\delta\left(\frac{\partial w}{\partial x}\right).$$

Using Eqs. (3.8.16) and (3.8.17) and integrating several terms by parts and combining the appropriate terms one finally obtains

$$
\begin{aligned}
\delta U = - \iint & \left\{ \left( \frac{\partial N_x}{\partial x} + \frac{\partial N_{xy}}{\partial y} \right) \delta u + \left( \frac{\partial N_y}{\partial y} + \frac{\partial N_{xy}}{\partial x} \right) \delta v \right. \\
& \left. + \left[ \frac{\partial}{\partial x} \left( N_x \frac{\partial w}{\partial y} \right) + \frac{\partial}{\partial y} \left( N_y \frac{\partial w}{\partial y} \right) + \frac{\partial}{\partial y} \left( N_{xy} \frac{\partial w}{\partial x} \right) + \frac{\partial}{\partial x} \left( N_{xy} \frac{\partial w}{\partial y} \right) \right] \delta w \right\} dxdy \\
& + \int \left\{ N_x \delta u + N_{xy} \delta v + \left[ N_x \frac{\partial w}{\partial x} + N_{xy} \frac{\partial w}{\partial y} \right] \delta w \right\} \Big|_0^a dy \\
& + \int \left\{ N_y \delta v + N_{xy} \delta u + \left[ N_y \frac{\partial w}{\partial y} + N_{xy} \frac{\partial w}{\partial x} \right] \delta w \right\} \Big|_0^b dx \\
& + D \iint \nabla^4 w \delta w dxdy + D \int \left\{ \left( \frac{\partial^2 w}{\partial x^2} + v \frac{\partial^2 w}{\partial y^2} \right) \delta \left( \frac{\partial w}{\partial x} \right) \right. \\
& + \left. \left[ -\frac{\partial^3 w}{\partial x^3} - (2 - v)\frac{\partial^3 w}{\partial x \partial y^2} \right] \delta w \right\} \Big|_0^a dy + D \int \left\{ \left( \frac{\partial^2 w}{\partial y^2} + v \frac{\partial^2 w}{\partial x^2} \right) \delta \left( \frac{\partial w}{\partial y} \right) \right. \\
& + \left. \left[ -\frac{\partial^3 w}{\partial y^3} - (2 - v)\frac{\partial^3 w}{\partial y \partial x^2} \right] \delta w \right\} \Big|_0^b dx + 2D(1 - v)\frac{\partial^2 w}{\partial x \partial y} \delta w \Big|_0^a \Big|_0^b .
\end{aligned}
\tag{3.8.18}
$$

The kinetic energy is given by Eq. (3.8.19), where $m$ is the mass per unit area of the plate

$$
T = \frac{1}{2} \iint m \left( \frac{\partial w}{\partial t} \right)^2 dxdy
\tag{3.8.19}
$$

and the work of the external loads is positive since the load and deflection are in the same direction.

$$
W_E = \iint q(x, y)w dxdy.
\tag{3.8.20}
$$

The variations of $T$ and $W_E$ are

$$
\delta T = \iint m \frac{\partial w}{\partial t} \delta \left( \frac{\partial w}{\partial t} \right) dxdy
\tag{3.8.21}
$$

$$
\delta W_E = \iint q \delta w dxdy.
\tag{3.8.22}
$$

Substituting Eqs. (3.8.18), (3.8.20), and (3.8.21) into Hamilton's principle

$$
\int_{t_1}^{t_2} (\delta T - \delta U - \delta W_E)dt = 0
\tag{3.8.23}
$$

and integrating $\delta T$ by parts with respect to time yields the differential equations and boundary conditions for this problem. For convenience the various contributions to the variation are identified below.

$$\delta u: \quad \frac{\partial N_x}{\partial x} + \frac{\partial N_{xy}}{\partial y} = 0 \tag{3.8.24}$$

$$\delta v: \quad \frac{\partial N_y}{\partial y} + \frac{\partial N_{xy}}{\partial x} = 0 \tag{3.8.25}$$

related boundary conditions:

| | | | | | |
|---|---|---|---|---|---|
| on | $x = 0, a$ | either | $N_x = 0$ | or | $u$ is specified |
| on | $y = 0, b$ | either | $N_{xy} = 0$ | or | $u$ is specified |
| on | $x = 0, a$ | either | $N_{xy} = 0$ | or | $v$ is specified |
| on | $y = 0, b$ | either | $N_y = 0$ | or | $v$ is specified |

$$\delta w: D\nabla^4 w - N_x \frac{\partial^2 w}{\partial x^2} - N_y \frac{\partial^2 w}{\partial y^2} - 2N_{xy}\frac{\partial^2 w}{\partial x \partial y} + m\frac{\partial^2 w}{\partial t^2} - q(x,y) = 0, \tag{3.8.26}$$

where Eqs. (3.8.24) and (3.8.25) have been used to simplify Eq. (3.8.26)

on $x = 0, a$

$$\text{either } D\left(\frac{\partial^2 w}{\partial x^2} + v\frac{\partial^2 w}{\partial y^2}\right) = 0 \text{ or } \frac{\partial w}{\partial x} \text{ is specified}$$

$$\text{and } -D\left[\frac{\partial^3 w}{\partial x^3} + (2-v)\frac{\partial^3 w}{\partial x \partial y^2}\right] + N_x\frac{\partial w}{\partial x} + N_{xy}\frac{\partial w}{\partial y} = 0 \text{ or } w \text{ is specified}$$

on $y = 0, b$

$$\text{either } D\left(\frac{\partial^2 w}{\partial y^2} + v\frac{\partial^2 w}{\partial x^2}\right) = 0 \text{ or } \frac{\partial w}{\partial y} \text{ is specified}$$

$$\text{and } -D\left[\frac{\partial^3 w}{\partial y^3} + (2-v)\frac{\partial^3 w}{\partial y \partial x^2}\right] + N_y\frac{\partial w}{\partial y} + N_{xy}\frac{\partial w}{\partial x} = 0 \text{ or } w \text{ is specified}$$

on $x = 0, a$ and $y = 0, b$

$$2D(1-v)\frac{\partial^2 w}{\partial x \partial y} = 0 \text{ or } w \text{ are specified.}$$

Note that this analysis is valid for a plate having constant thickness and a rectangular planform. More complicated shapes can be treated in a similar manner.

As indicated earlier, thin plates, under these loading conditions, can be considered to be under a state of generalized plane stress (Dym and Shames 1973, 50) and therefore one can introduce the Airy stress function $F(x, y)$, which allows one to write

$$N_x = \frac{\partial^2 F}{\partial y^2} \quad ; \quad N_y = \frac{\partial^2 F}{\partial x^2} \quad ; \quad N_{xy} = -\frac{\partial^2 F}{\partial x \partial y}. \tag{3.8.27}$$

Using the Airy stress function, Eq. (3.8.25) is rewritten as

$$D\nabla^4 w = q(x,y) - m\frac{\partial^2 w}{\partial t^2} + \frac{\partial^2 F}{\partial y^2}\frac{\partial^2 w}{\partial x^2} + \frac{\partial^2 F}{\partial x^2}\frac{\partial^2 w}{\partial y^2} - 2\frac{\partial^2 F}{\partial x \partial y}\frac{\partial^2 w}{\partial x \partial y}. \tag{3.8.28}$$

For the complete formulation of the von Kármán plate equations a second relation between $F(x,y)$ and $w$ is required. The additional relation is obtained by using the compatibility equations. The compatibility equations for *linear strain* at the middle plane of the plate are

$$\frac{\partial^2 \epsilon_{y0}}{\partial x^2} + \frac{\partial^2 \epsilon_{x0}}{\partial y^2} - \frac{\partial^2 \epsilon_{xy0}}{\partial x \partial y} = 0. \tag{3.8.29}$$

Using Eq. (3.8.5) these relations can be rewritten as

$$\frac{\partial \epsilon_{x0}}{\partial y} = \frac{\partial^2 u_0}{\partial x \partial y} + \left(\frac{\partial w}{\partial x}\right)\left(\frac{\partial^2 w}{\partial x \partial y}\right)$$

$$\frac{\partial^2 \epsilon_{x0}}{\partial y^2} = \frac{\partial^3 u_0}{\partial x \partial y^2} + \left(\frac{\partial^2 w}{\partial x \partial y}\right)^2 + \frac{\partial w}{\partial x}\frac{\partial^3 w}{\partial x \partial y^2} \tag{3.8.30a}$$

$$\frac{\partial \epsilon_{y0}}{\partial x} = \frac{\partial^2 v_0}{\partial x \partial y} + \left(\frac{\partial w}{\partial y}\right)\left(\frac{\partial^2 w}{\partial x \partial y}\right)$$

$$\frac{\partial^2 \epsilon_{y0}}{\partial x^2} = \frac{\partial^3 v_0}{\partial x^2 \partial y} + \left(\frac{\partial^2 w}{\partial x \partial y}\right)^2 + \frac{\partial w}{\partial y}\frac{\partial^3 w}{\partial x^2 \partial y} \tag{3.8.30b}$$

$$\frac{\partial \epsilon_{xy0}}{\partial x} = \left(\frac{\partial^2 u_0}{\partial x \partial y} + \frac{\partial^2 v_0}{\partial x^2}\right) + \frac{\partial^2 w}{\partial x^2}\frac{\partial w}{\partial y} + \frac{\partial w}{\partial x}\frac{\partial^2 w}{\partial x \partial y}$$

$$\frac{\partial^2 \epsilon_{xy0}}{\partial x \partial y} = \left(\frac{\partial^3 u_0}{\partial x \partial y^2} + \frac{\partial^3 v_0}{\partial x^2 \partial y}\right) + \frac{\partial^3 w}{\partial y \partial x^2}\frac{\partial w}{\partial y} + \frac{\partial^2 w}{\partial x^2}\frac{\partial^2 w}{\partial y^2}$$

$$+ \left(\frac{\partial^2 w}{\partial x \partial y}\right)^2 + \frac{\partial w}{\partial x}\frac{\partial^3 w}{\partial x \partial y^2}. \tag{3.8.30c}$$

Substituting Eqs. (3.8.30a), (3.8.30b), and (3.8.30c) into Eq. (3.8.29) yields

$$\left(\frac{\partial^2 w}{\partial x \partial y}\right)^2 - \frac{\partial^2 w}{\partial x^2}\frac{\partial^2 w}{\partial y^2} = 0. \tag{3.8.31}$$

Solving Eq. (3.8.14) for $h\epsilon_{x0}$, $h\epsilon_{x0}$ yields

$$h\epsilon_{x0} = \frac{1}{E}(N_x - \nu N_y)$$

$$h\epsilon_{y0} = \frac{1}{E}(N_y - \nu N_x) \tag{3.8.32}$$

$$h\epsilon_{xy0} = \frac{N_{xy}}{G}.$$

Using Eqs. (3.8.27) and (3.8.32) yields

$$h\epsilon_{x_0} = \frac{1}{E}\left(\frac{\partial^2 F}{\partial y^2} - \nu\frac{\partial^2 F}{\partial x^2}\right)$$

$$h\epsilon_{y_0} = \frac{1}{E}\left(\frac{\partial^2 F}{\partial x^2} - \nu\frac{\partial^2 F}{\partial y^2}\right) \tag{3.8.33}$$

$$h\epsilon_{xy_0} = -\frac{1}{G}\frac{\partial^2 F}{\partial x\partial y}.$$

Equations (3.8.29) and (3.8.31) imply

$$\frac{\partial^2\epsilon_{y_0}}{\partial x^2} + \frac{\partial^2\epsilon_{x_0}}{\partial y^2} - \frac{\partial^2\epsilon_{xy_0}}{\partial x\partial y} = \left(\frac{\partial^2 w}{\partial x\partial y}\right)^2 - \frac{\partial^2 w}{\partial x^2}\frac{\partial^2 w}{\partial y^2}. \tag{3.8.34}$$

Substituting Eq. (3.8.33) into Eq. (3.8.34) yields (recall $G = \frac{E}{2(1+\nu)}$)

$$\frac{1}{Eh}\left(\frac{\partial^4 F}{\partial x^4} - \nu\frac{\partial^4 F}{\partial x^2\partial y^2}\right) + \frac{1}{Eh}\left(\frac{\partial^4 F}{\partial y^4} - \nu\frac{\partial^4 F}{\partial x^2\partial y^2}\right) + \frac{2(1+\nu)}{hE}\frac{\partial^4 F}{\partial x^2\partial y^2}$$

$$= \left(\frac{\partial^2 w}{\partial x\partial y}\right)^2 - \frac{\partial^2 w}{\partial x^2}\frac{\partial^2 w}{\partial y^2} + \frac{\partial^4 F}{\partial x^4} + \frac{\partial^4 F}{\partial y^4} + 2\frac{\partial^4 F}{\partial x^2\partial y^2}$$

$$+ Eh\left[\frac{\partial^2 w}{\partial x^2}\frac{\partial^2 w}{\partial y^2} - \left(\frac{\partial^2 w}{\partial x\partial y}\right)^2\right] = 0$$

or in a more compact manner

$$\nabla^4 F + Eh\left[\frac{\partial^2 w}{\partial x^2}\frac{\partial^2 w}{\partial y^2} - \left(\frac{\partial^2 w}{\partial x\partial y}\right)^2\right] = 0. \tag{3.8.35}$$

Equations (3.8.28) and (3.8.35) are the fundamental relations for the von Kármán theory of plates.

When the axial loads are given, denoted by constant values

$$N_x = \overline{N_x} \ ; \ N_y = \overline{N_y} \ ; \ N_{xy} = \overline{N_{xy}} \tag{3.8.36}$$

then Eqs. (3.8.24) and (3.8.25) are satisfied identically and $\nabla^4 F = 0$. Thus Eq. (3.8.35) is not needed. For this case the equation of motion of the axially loaded plate is given by Eq. (3.8.26), with the appropriate substitutions implied by Eq. (3.8.36).

## BIBLIOGRAPHY

Blevins, R. D. (1979). *Formulas for Natural Frequency and Mode Shape*. Van Norstrand Reinhold.
Bolotin, V. V. (1963). *Nonconservative Problems of the Theory of Elastic Stability*. Pergamon Press.
Chang, T. C. and Craig, R. R. (1969). Normal modes of uniform beams. *Proceedings of ASCE*, 95 (EM4): 1025–1031.

Dowell, E. (1975). *Aeroelasticity of Plates and Shells.* Noordhoff International Publishing.

Dym, C. L. and Shames, I. H. (1973). *Solid Mechanics: A Variational Approach.* McGraw-Hill Book Co.

Gorman, D. J. (1975). *Free Vibration Analysis of Beams and Shafts.* John Wiley & Sons.

Hurty, W. and Rubinstein, M. (1964). *Dynamics of Structures.* Prentice Hall.

Leipholz, L. (1977). *Direct Variational Methods and Eigenvalue Problems in Engineering.* Noordhoff International Publishing Co.

Leissa, A. W. (1969). Vibration of plates. Technical report, NASA SP-160, Ohio State University.

Meirovitch, L. (1967). *Analytical Methods in Vibrations.* The Macmillan Co.

Rayleigh, J. W. S. B. (1894). *The Theory of Sound.* Macmillan.

Young, D. and Felgar, R. P. (1949). Tables of characteristic functions representing normal modes of vibrations of a beam. Eng, res. series no. 44, bureau of eng. research, The University of Texas, Austin.

## PROBLEMS

1.  Consider the torsional vibrations of a circular shaft, shown below having uniform torsional stiffness $GJ$ and polar mass moment of inertial $I$ per unit length. BCs at $x = 0; \theta = 0$.

    (a) Derive the differential equations of motion and boundary condition for this system. Identify L[ ], M[ ], B[ ] and show self-adjointness and positive definiteness.

    (b) Derive the exact eigenvalues and eigenfunctions for this case.

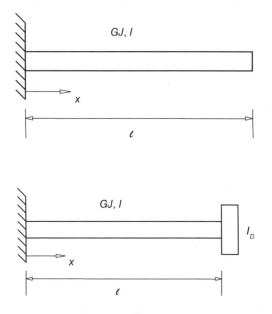

    (c) For the same problem as in (a) except having a round disk at $x = l$ with polar mass moment inertial $I_D$ repeat (a) and (b) of above. Hint: solution to this part of the problem requires the solution of a transcendental equation, which can be done using MATLAB or Mathematica.

2.  Using the differential equation method, derive the frequency equation and show how the mode shapes can be computed for the torsional free vibration of the shaft shown in the figure.

    The shaft is fixed at one end and is connected rigidly to two beams at its other end (forming a cross) so that the end of the shaft and the beams undergo equal rotation at the junction $x = 0$. Neglect the mass of the two beams. The mass density of the shaft is $\rho$.

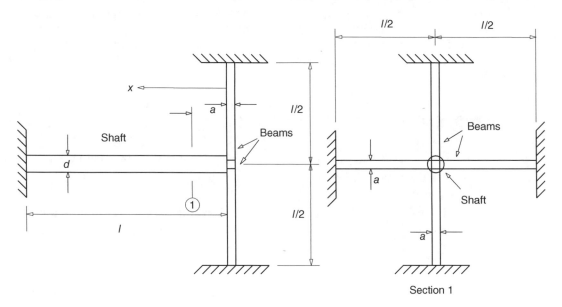

Section 1

3.  Consider the beam problem shown above, where the beam is on an elastic foundation having spring constant $\beta$ per unit length and linearly varying mass and stiffness distributions.

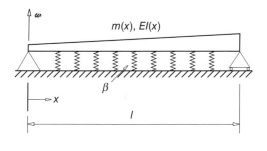

$$m(x) = m_0 \left(1 + x/l\right)$$
$$EI(x) = EI_0 \left(1 + 2\frac{x}{l}\right)$$

Don't worry about how such a beam could exist in practice!

where $m_0$ is the mass per unit length at $x = 0$ and $EI_0$ is the bending stiffness at $x = 0$

Assume $l = 100''$    $E = 10^7$ $lb/in^2$    $\beta = 10^4$ $lb/in^2$

$m_0 = 1.12$ $\frac{lbs^2}{in^2}$   A = cross-sectional area is a square

$$A = 10'' \times 10''$$

(a) Use the Rayleigh–Ritz method based upon *two shape* functions ($n = 2$)

$$w(x) = C_1\phi_1(x) + C_2\phi_2(x) = \sum_{i-1}^{n} C_i\phi_i(x),$$

where

$$\phi_i = \sin\frac{\pi i x}{l}$$

to evaluate the first two approximate mode shapes and frequencies.

(b) Repeat the calculation for $n = 3$, that is,

$$w(x) = \sum_{i=1}^{3} C_i\phi_i(x).$$

Hint: solution of the eigenvalue problem for this case can be accomplished using MATLAB or Mathematica

(c) Repeat part (a) using Galerkin's method and compare your results with the Rayleigh–Ritz method.

4. Consider the vibrations of a Timoshenko beam shown below, which has mass distribution $m$ per unit length, stiffness $EI$ in bending, and shear stiffness $kAG$. The length of the beam is $L$, and it is loaded by an axial load $P$.

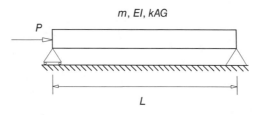

$m$, $EI$, $kAG$

$P$

$L$

- Derive the equations of motion using Hamilton's principle selecting $w$ and $\psi$ as the DOFs.
- Estimate the fundamental natural frequency using Rayleigh's quotient; can you use this approach?
- Assume that the axial load $P \neq 0$ and show that separation of variables can be used in this case too, like in the case of an Euler–Bernoulli beam discussed in your class notes.

5.  A spherical mass having diameter $D$ and density $\rho$ is rolling, without slipping, with uniform velocity $V$ on a Timoshenko beam that is simply supported at its end $x = 0$, and clamped at its end $x = L$. The load introduced on the beam by this rolling mass is weight (*mass* $* g$) applied at a different location at every instant of time. The properties of the Timoshenko beam are bending stiffness $EI$, shear modulus $G$, shear correction constant $k$, cross-sectional area $A$ and mass per unit length $m$. The geometry of this problem is shown in the figure below.

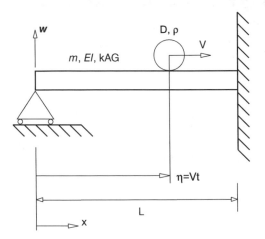

Using Hamilton's principle, derive the partial differential equations and boundary conditions governing the behavior of this structural dynamic system.

Note: In the figure below, the position of the center of mass of the rolling sphere, relative to the left support, is given by $\eta = Vt$.

Hint: Mathematically, singularities can be represented using the Dirac delta function, which has the following properties:

$$\int_0^L \delta(x - \eta)dx = 1$$
$$\int_0^L g(x)\delta(x - \eta)dx = g(\eta).$$

6.  The structural dynamic system shown below represents the axial free vibrations of a rod with a tip mass and a spring at $x = L$.

    The mass per unit length of the bar is $m$, its stiffness is $EA$, and the mass $M$ has the same order of magnitude as $mL$. A spring with stiffness $K$ is located between the mass $M$ and the wall.

    (a) Use Hamilton's principle to derive the equation of motion and boundary conditions for this structural dynamic system. **Note** that in this problem you can neglect the effect of gravity.

(b) Identify the $L[\ ]$ and $M[\ ]$ operators for this problem. Is the problem self-adjoint and positive definite? Explain why.

(c) Write down the orthogonality conditions for this system.

(d) Estimate the fundamental frequency using a suitably selected shape function. Justify your selection of the function.

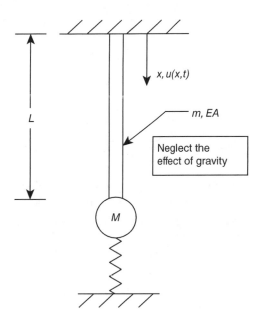

7.  Consider the vibrations of a rectangular simply supported plate shown below. The plate has dimensions $a$ and $b$ as shown in the figure. It has uniform thickness $t$ and the density of the material is $\rho$. At the center of the plate a spherical mass is attached, with properties $M = 4\rho tab$ and a diameter $d = 2r = 2\left(\frac{3tab}{\pi}\right)^{\frac{1}{3}}$ and a mass moment of inertia $I_p$. The edges of the plate $x = 0$ and $x = a$ are loaded by a uniform axial load resultant (force per unit length) given by $N_x$ where

$$N_x = \gamma N_{xcr} = \gamma D \left(\frac{a}{\pi}\right)^2 \left[\frac{\pi^2}{a^2} + \frac{\pi^2}{b^2}\right]^2$$

$$D = \frac{Et^3}{12\left(1 - \nu^2\right)}; \quad \gamma = \frac{N_x}{N_{xcr}},$$

where $E$ is Young's modulus, $\nu$ is Poisson's ratio, $D$ is the stiffness of the plate, and $N_{xcr}$ is the lowest buckling load, under the compressive loads shown.

The Kirchhoff plate theory equations for this problem are given by

The total potential $U + V$ is given by

$$\Pi_p = U + V$$

$$= \iint_A \frac{D}{2}\left[w_{,xx}^2 + w_{,yy}^2 + 2vw_{,xx}w_{,yy} + 2(1 - v)w_{,xy}^2\right]dxdy$$

$$- \frac{N_x}{2}\iint_A w_{,x}^2 dxdy$$

and $(\ )_{,x}$ and $(\ )_{,y}$ are derivatives with respect to $x$ and $y$. The kinetic energy density $T^*$ is given by

$$T^* = \frac{1}{2}\iint_A \rho t w^2 dxdy + \frac{1}{2}Mw^2\left(x = \frac{a}{2}, y = \frac{b}{2}\right)$$

$$+ \frac{1}{2}I_p\left[w_{,x}^2\left(x = \frac{a}{2}, y = \frac{b}{2}\right) + w_{,y}^2\left(x = \frac{a}{2}, y = \frac{b}{2}\right)\right]$$

and $A$ is the area of the plate over which the integrations have to be carried out.

The boundary conditions for this simple supported plate are the same as for a simply supported Euler–Bernoulli beam, that is, zero deflection and zero moment along the edges. Thus the boundary conditions are given by

- Along $x = 0$ and $x = a$ the bending moment intensity (i.e. moment per unit length) $M_x = 0$, where $M_x = -D(w_{,xx}^2 + vw_{,yy}^2)$
- Along $y = 0$ and $y = b$ the bending moment intensity $M_y = 0$, where $M_y = -D(w_{,yy}^2 + vw_{,xx}^2)$

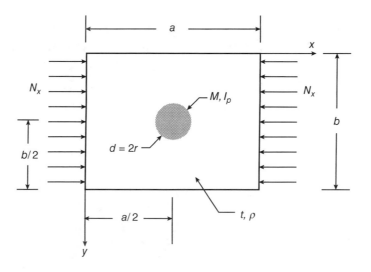

(a) Using Rayleigh's quotient $\omega^2 \cong \frac{\Pi_{\max}}{T_{\max}^*}$ estimate the fundamental frequency for this structural dynamic system, using a suitably selected shape function, that is, assume $w(x, y) = a_{11}\sin\left(\frac{\pi x}{a}\right)\sin\left(\frac{\pi y}{b}\right)$. Show that this shape function satisfies the boundary conditions and use it to estimate the fundamental frequency.

(b) Using the expression obtained in (a), calculate the fundamental frequency for $\gamma = 0$; $\gamma = 0.5$; and $\gamma = 1.0$.

Hint: You may need the following integrals

$$\int_0^a \int_0^b \sin^2\left(\frac{\pi x}{a}\right) \sin^2\left(\frac{\pi y}{b}\right) dxdy = \frac{ab}{4}$$

$$\int_0^a \int_0^b \cos^2\left(\frac{\pi x}{a}\right) \cos^2\left(\frac{\pi y}{b}\right) dxdy = \frac{ab}{4}$$

# 4     Finite Element Method and Model Reduction

In the context of structural mechanics and structural dynamics, the evolution of the finite element (FE) method was closely linked to the evolution of the computers and their application to the analysis large complicated structures used in aerospace and civil engineering. A landmark paper was published in 1956, at the dawn of the first computer revolution in engineering (Turner et al. 1956). It is relevant to note that the paper presented at the 22nd Annual Meeting of the Institute of Aeronautical Sciences (predecessor of AIAA) in New York, January 25–29, 1954, in the Aeroelasticity Session. It represented a collaboration between Industry (Boeing) and the academia (UC Berkeley and the University of Washington, Seattle). The paper clearly states that one of its objectives is to provide a methodology for structural analysis that can take advantage of the potential of digital computers.

Since then, an extensive literature on the subject has evolved consisting of thousands of papers and hundreds of books. This literature is extremely rich and detailed. Only a limited number of the excellent available references is provided at the end of this chapter to enable further study of the topic. The material provided in this chapter is essentially an attempt to introduce the topic with an emphasis on structural dynamics.

It is also important to mention that during the last 70 years since the inception of the method several powerful and comprehensive commercially available computer codes have been developed that allow one to perform a comprehensive static and dynamic analysis of complicated structural systems having millions of degrees of freedom. These codes are capable of both linear and nonlinear analyses of structural systems, undergoing moderate as well as large deformation, built from a variety of materials. The codes are capable of modeling modern composite structures used extensively in the aerospace and automobile industry. The more popular commercial computer codes are NASTRAN, ABAQUS, ANSYS, and CATIA to name a few. There are several other codes developed in Europe and other places with which the authors are not familiar.

Finally, it should be noted the FE method that started as a computational structural mechanics tool has evolved during the last 50 years into a general purpose computational tool used in other disciplines such as computational fluid mechanics (CFD) and heat transfer. In many cases it replaced the finite difference methods used previously, due to its superior capability to represent complicated geometric configurations. For more advanced treatment of the topic the reader is encouraged to examine the references provided at the end of this chapter.

The basic concepts underlying FE analysis methods can be summarized as follows:

(1) model the continuum (solid or fluid) as a finite collection of discrete elements;

(2) make simplifying assumptions about the displacement field (or the stress field) within the elements and across element boundaries; and

(3) minimize the error introduced by assumption (2) using a suitable method.

If step (2) approximates displacements, the method is denoted by the name displacement or stiffness method; if step (2) approximates stresses, the method is referred to as the force or flexibility method; there is a third variant called the hybrid method where in step (2) displacements are approximated on the boundary and stresses inside the domain of the element.

In the modern context, step (3) in the FE formulation is often accomplished using a variational principle, for example, Hamilton's principle, precisely as is done in the Rayleigh–Ritz method. In some cases, it is convenient or necessary to work directly with the governing differential equations and use the method of weighted residuals to get the "best" approximation. This produces a Galerkin-type FE model. Finite element formulations are therefore classified into three different classes, according to the approach used:

(1) *Direct method*: based on basic structural principles and Newton's Laws;
(2) *Variational method(s)*: based on minimizing or finding the stationary values of a functional; and
(3) *Method of weighted residuals*: based directly on the governing differential equation(s).

Examples of FE modeling are shown in Figures 4.1 through 4.3.
   (1) Truss (Figure 4.1);

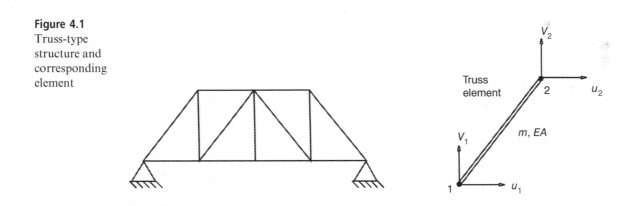

**Figure 4.1**
Truss-type
structure and
corresponding
element

Note: Elements are interconnected at nodes only.
(2) Thin plate (Figure 4.2); and
(3) Solid body (Figure 4.3).

The formulation must address inter-element compatibility, that is, no gaps; no kinks during the assembly.

**Figure 4.2**
Thin plate

**Figure 4.3**
Solid body

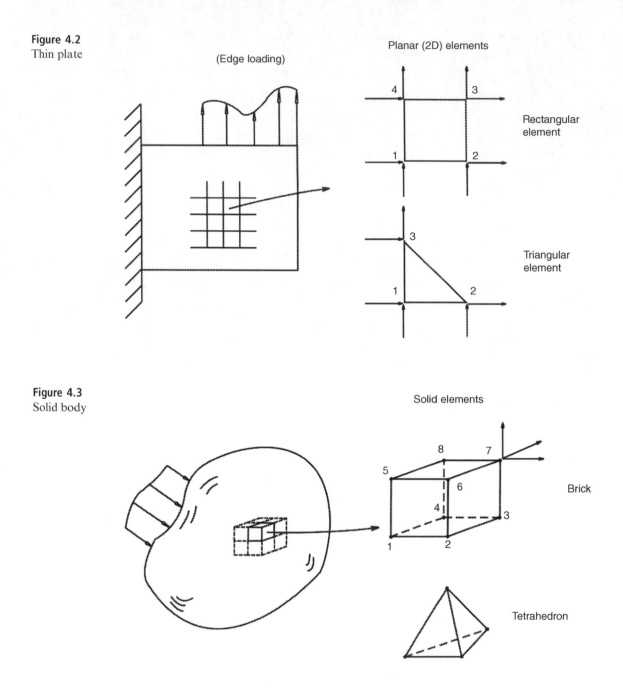

Finite element formulations of structural dynamic systems result in models consisting of large mass and stiffness matrices from which many free vibration modes can be computed. For a variety of applications the dynamic behavior of the system can be captured in a satisfactory manner by using a limited number of modes. Therefore approaches capable of producing reduced order models in a systematic manner are connected to the FE approach.

## 4.1    Finite Element Formulation for Axial Member

Before examining energy-based formulation of stiffness and mass matrices, it is useful to derive the stiffness and mass matrices for the simplest element as represented by an axial member.

Note that while the static problem of a bar loaded by end forces has only two degrees of freedom, the dynamic problem for the same bar has an infinite number of degrees of freedom. The geometry of the bar is illustrated in Figure 4.4, where $E$ is Young's modulus, $A$ is the cross-sectional area, and $\gamma$ is the mass per unit length.

**Figure 4.4** Geometry of a bar in axial extension

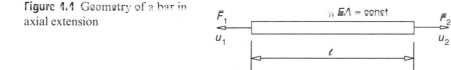

Consider a FE (displacement) formulation for the bar. Assume a displacement field approximated by a linear polynomial in $x$.

$$u(x) = a_1 + a_2 x = \lfloor 1 \quad x \rfloor \left\{ \begin{array}{c} a_1 \\ a_2 \end{array} \right\}. \tag{4.1.1}$$

This is a suitable approximation because when the static equilibrium equation of a bar $EAu_{,xx} = 0$ is integrated, it yields $u = a_1 + a_2 x$ where the generalized coordinates $a_1, a_2$ can be evaluated in terms of end displacements, or nodal displacements:

$$u(0) = u_1 = a_1 \quad u(\ell) = u_2 = a_1 + a_2\ell \Rightarrow a_2 = (u_2 - u_1)/\ell. \tag{4.1.2}$$

In matrix form,

$$\left\{ \begin{array}{c} a_1 \\ a_2 \end{array} \right\} = \left[ \begin{array}{cc} 1 & 0 \\ -1/\ell & 1/\ell \end{array} \right] \left\{ \begin{array}{c} u_1 \\ u_2 \end{array} \right\}. \tag{4.1.3}$$

Combining Eqs. (4.1.1) through (4.1.3) yields

$$u(x) = \lfloor 1 \quad x \rfloor \left[ \begin{array}{cc} 1 & 0 \\ -1/\ell & 1/\ell \end{array} \right] \left\{ \begin{array}{c} u_1 \\ u_2 \end{array} \right\} = \lfloor N_1(x) \quad N_2(x) \rfloor \left\{ \begin{array}{c} u_1 \\ u_2 \end{array} \right\}, \tag{4.1.4}$$

where the shape functions are $N_1(x) = 1 - \frac{x}{\ell}$ and $N_2(x) = x/\ell$ are denoted as interpolation functions

Strain:

$$\epsilon_{xx} = u_{,x} = \lfloor -\frac{1}{\ell} \quad \frac{1}{\ell} \rfloor \left\{ \begin{array}{c} u_1 \\ u_2 \end{array} \right\} \tag{4.1.5}$$

Stress:

$$\sigma_{xx} = E\epsilon_{xx} = \frac{E}{\ell} = \lfloor -1 \quad 1 \rfloor \left\{ \begin{array}{c} u_1 \\ u_2 \end{array} \right\} = \frac{E}{\ell}(-u_1 + u_2) \tag{4.1.6}$$

Joint (nodal) forces:

$$\left\{ \begin{array}{c} F_1 \\ F_2 \end{array} \right\} = A\sigma_{xx} \left\{ \begin{array}{c} -1 \\ 1 \end{array} \right\} = \frac{AE}{\ell} \left[ \begin{array}{cc} 1 & -1 \\ -1 & 1 \end{array} \right] \left\{ \begin{array}{c} u_1 \\ u_2 \end{array} \right\} \tag{4.1.7}$$

or

$$\{F\} = [K]\{u\}, \tag{4.1.8}$$

where

$$[K] = \frac{AE}{\ell} \left[ \begin{array}{cc} 1 & -1 \\ -1 & 1 \end{array} \right] = \begin{array}{l} \text{element stiffness} \\ \text{matrix (singular).} \end{array} \tag{4.1.9}$$

Note that $[K]$ in Eq. (4.1.9) is singular since rigid-body motion is not prevented. This completes the static analysis. For the dynamic case, a mass matrix is required.

## 4.1.1 Lumped Mass Matrix

Mass of element is distributed as point masses at nodes (where the displacement DOFs are defined) as shown in Figure 4.5.

**Figure 4.5** Lumped mass representation of the axial element

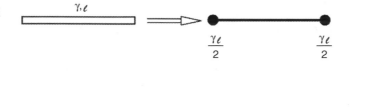

$$[M]_{\text{lumped}} = \frac{\gamma\ell}{2} \left[ \begin{array}{cc} 1 & 0 \\ 0 & 1 \end{array} \right]. \tag{4.1.10}$$

## 4.1.2 Consistent Mass Matrix

Recall that the mass matrix [m] relates inertial forces to accelerations; from D'Alembert's principle

$$\left\{ F^I \right\} = -[m]\{\ddot{u}\} \quad \text{(D'Alembert's principle).} \tag{4.1.11}$$

In the one-dimensional element under consideration, the inertial force appears as a distributed load:

$$p^I(x,t) = -\gamma\,\ddot{u}(x,t). \tag{4.1.12}$$

To find the equivalent nodal forces, apply the principle of virtual work. Consider a virtual displacement $\delta u_1$ to $u_1$; then, from Eq. (4.1.12), the corresponding (consistent) internal virtual displacement field is

$$\delta u_1(x) = \left(1 - \frac{x}{\ell}\right)\delta u_1 = N_1(x)\delta u_1. \tag{4.1.13}$$

To find the equivalent nodal force $F_1$ due to a distributed load $p(x)$, equate their virtual work due to $\delta u_1$:

$$\delta W = F_1 \cdot \delta u_1 = \int_0^\ell p(x)\delta u_1(x)dx = \delta u_1 \cdot \int_0^\ell p(x)N_1(x)dx. \qquad (4.1.14)$$

Thus

$$F_1 = \int_0^\ell p(x)N_1(x)dx \qquad (4.1.15)$$

is the equivalent consistent nodal force at node (1) due to $p(x)$. Recall from the principle of virtual work that if $\delta W = F \cdot \delta u = 0; \delta u$ is arbitrary, then $\mathbf{F} = \mathbf{0}$ (equilibrium).

Similarly, to find $F_2$ apply $\delta u_2$ to $u_2$, with corresponding displacement field from Eq. (4.1.4):

$$\delta u_2(x) = \frac{x}{\ell}\delta u_2 = N_2(x)\delta u_2 \qquad (4.1.16)$$

then

$$F_2 \cdot \delta u_2 = \delta u_2 \int_0^\ell p(x)N_2(x)dx$$

$$F_2 = \int_0^\ell p(x)N_2(x)dx. \qquad (4.1.17)$$

The mass matrix can now be evaluated from Eqs. (4.1.15) and (4.1.17)

$$p^I(x,t) = -\gamma \ddot{u}(x,t) = -\gamma N_1(x)\ddot{u}_1 - \gamma N_2(x)\ddot{u}_2,$$

where

$$N_1(x) = 1 - x/\ell; \quad N_2(x) = x/\ell$$

using Eqs. (4.1.15) and (4.1.17),

$$F_1^I = -\ddot{u}_1 \cdot \int_0^\ell \gamma N_1^2(x)dx - \ddot{u}_2 \int_0^\ell \gamma N_1(x)N_2(x)dx \qquad (4.1.18a)$$

and

$$F_2^I = -\ddot{u}_1 \int_0^\ell \gamma N_2(x)N_1(x)dx - \ddot{u}_2 \int_0^\ell \gamma N_2^2(x)dx. \qquad (4.1.18b)$$

In matrix form,

$$\left\{ \begin{array}{c} F_1^I \\ F_2^I \end{array} \right\} = - \left[ \begin{array}{cc} m_{11} & m_{12} \\ m_{21} & m_{22} \end{array} \right] \left\{ \begin{array}{c} \ddot{u}_1 \\ \ddot{u}_2 \end{array} \right\},$$

where

$$m_{ij} = \int_0^\ell \gamma N_i(x)N_j(x)dx, \qquad (i,j = 1,2) \qquad (4.1.19)$$

are the elements of the *consistent mass matrix*.

### 4.1.3  Truss Element

Evaluating appropriate terms, using Eq. (4.1.19)

$$m_{11} = \gamma \int_0^\ell \left(1 - \frac{x}{\ell}\right)^2 dx = \gamma \left(x - \frac{x^2}{\ell} + \frac{x^3}{3\ell^2}\right)\Big|_0^\ell$$

$$= \frac{\gamma\ell}{3}$$

$$m_{22} = \gamma \int_0^\ell \left(\frac{x}{\ell}\right)^2 dx = \frac{\gamma\ell}{3}$$

$$m_{12} = m_{21} = \gamma \int_0^\ell \frac{x}{\ell}\left(1 - \frac{x}{\ell}\right) dx = \frac{\gamma\ell}{6}. \qquad (4.1.20)$$

Thus, the consistent mass matrix is given by

$$[m] = \frac{\gamma\ell}{6}\begin{bmatrix} 2 & 1 \\ 1 & 2 \end{bmatrix}. \qquad (4.1.21)$$

Finally it should be mentioned that the relations for the consistent nodal forces and the mass matrix, Eqs. (4.1.15), (4.1.17), and (4.1.19), are general and apply to other elements as well. In practice, the consistent mass matrix is most easily derived from the kinetic energy, as shown in the next section.

$$m_{ij} = \frac{\partial^2 T}{\partial \dot{q}_i \partial \dot{q}_j} = \int_0^\ell \gamma N_i(x) N_j(x) dx. \qquad (4.1.22)$$

## Example: Free Vibrations of a Bar (Axial Member)

Recall that the stiffness and consistent mass matrices for an axial (truss) member obtained earlier are

$$[k] = \frac{AE}{\ell}\begin{bmatrix} 1 & -1 \\ -1 & 1 \end{bmatrix} \qquad [m] = \frac{\gamma\ell}{6}\begin{bmatrix} 2 & 1 \\ 1 & 2 \end{bmatrix}. \qquad (4.1.1.Ex1)$$

Consider the element shown in Figure 4.6 in which the entire fixed-free member is represented by one element. Since $u_1 = 0$, there is only one DOF, $u_2$, and the governing differential equation is

**Figure 4.6** Free vibration of a bar using one element

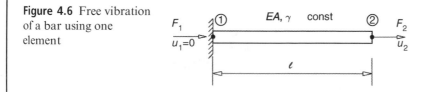

$$m_{22}\ddot{u}_2 + k_{22}u_2 = 0, \qquad (4.1.2.\text{Ex}1)$$

where $m_{22} = \gamma\ell/3;\ k_{22} = AE/\ell$.

For natural vibrations, $\ddot{u}_2 = -\omega^2 u_2$, thus

$$\left(-\omega^2 m_{22} + k_{22}\right)\cdot u_2 = 0 \quad \Rightarrow \omega = \sqrt{\frac{k_{22}}{m_{22}}} = \sqrt{\frac{3AE}{\gamma\ell^2}} \qquad (4.1.3.\text{Ex}1)$$

or

$$\omega = \frac{\sqrt{3}}{\ell}\cdot\sqrt{\frac{E}{\rho}} \simeq \frac{1.73}{\ell}\sqrt{\frac{E}{\rho}}. \qquad (4.1.4.\text{Ex}1)$$

If lumped mass matrix is used,

$$m_{22} = \frac{\gamma\ell}{2}; \quad \omega = \frac{\sqrt{2}}{\ell}\sqrt{\frac{E}{\rho}} \simeq \frac{1.41}{\ell}\sqrt{\frac{E}{\rho}}. \qquad (4.1.5.\text{Ex}1)$$

Note that the exact solution is $\omega = \frac{\pi}{2\ell}\sqrt{\frac{E}{\rho}} \simeq \frac{1.57}{\ell}\sqrt{\frac{E}{\rho}}$.

Next consider the same problem as before and model the bar with two truss elements shown in Figure 4.7. In this case $u_1 = 0$, and the two degrees of freedom in the problem are

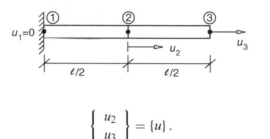

**Figure 4.7** Free vibrations of a bar represented by two elements

$$\left\{\begin{matrix} u_2 \\ u_3 \end{matrix}\right\} = \{u\}.$$

Each element of length $\ell/2$ has stiffness and mass matrices

$$[k] = \frac{2AE}{\ell}\begin{bmatrix} 1 & -1 \\ -1 & 1 \end{bmatrix} \qquad (4.1.6.\text{Ex}1)$$

$$[m] = \frac{\gamma\ell}{12}\begin{bmatrix} 2 & 1 \\ 1 & 2 \end{bmatrix}. \atop \text{(consistent)} \qquad (4.1.7.\text{Ex}1)$$

One must address the assembly problem, that is, how do we connect the elements "mathematically"?

Since the structure is represented by two elements, in this simple example it is easy to convince oneself by writing the equilibrium equation for node 2, that the mass and stiffness elements for the $i$th force and $j$th DOF can be obtained by adding the contribution from each element. Thus the global mass and stiffness matrices are, by this "direct method,"

$$[M] = \frac{\gamma \ell}{12} \begin{bmatrix} 4 & 1 \\ 1 & 2 \end{bmatrix}; \quad [K] = \frac{2AE}{\ell} \begin{bmatrix} 2 & -1 \\ -1 & 1 \end{bmatrix} \qquad (4.1.8.\text{Ex}1)$$

and the governing differential equation becomes

$$[M] \left\{ \begin{array}{c} \ddot{u}_2 \\ \ddot{u}_3 \end{array} \right\} + [K] \left\{ \begin{array}{c} u_2 \\ u_3 \end{array} \right\} = 0. \qquad (4.1.9.\text{Ex}1)$$

Set $\{\ddot{u}\} = -\omega^2 \{u\}$, and let $\lambda = \frac{\omega^2 \rho \ell^2}{24EA}$

$$\left( -\omega^2 [M] + [K] \right) \{u\} = 0$$

or

$$[A]\{u\} = 0, \quad \text{where} \quad [A] = \begin{bmatrix} (2 - 4\lambda) & -(1 + \lambda) \\ -(1 + \lambda) & (1 - 2\lambda) \end{bmatrix}. \qquad (4.1.10.\text{Ex}1)$$

This is an algebraic eigenvalue problem, and nontrivial solutions exist only if

$$\det[A] = 0 = 2(1 - 2\lambda)^2 - (1 + \lambda)^2 \qquad (4.1.11.\text{Ex}1)$$

from which $\sqrt{2} \cdot (1 - 2\lambda) = \pm(1 + \lambda)$

$$\lambda_1 = \frac{\sqrt{2} - 1}{2\sqrt{2} + 1} \cong 0.1082; \quad \lambda_2 = \frac{\sqrt{2} + 1}{2\sqrt{2} - 1} \cong 1.320$$

$$\omega_1 = \sqrt{\frac{24E\lambda_1}{\rho \ell^2}} \cong \frac{1.61}{\ell} \sqrt{\frac{E}{\rho}} \quad \left( \text{exact} = \frac{\pi}{2\ell} \sqrt{\frac{E}{\rho}} \right)$$

$$\omega_2 = \frac{5.63}{\ell} \sqrt{\frac{E}{\rho}} \quad \left( \text{exact} = \frac{3\pi}{2\ell} \sqrt{\frac{E}{\rho}} \cong \frac{4.71}{\ell} \sqrt{\frac{E}{\rho}} \right). \qquad (4.1.12.\text{Ex}1)$$

If one normalizes the mode shapes (eigenvectors of (4.1.10.Ex1)) by setting $u_3 = 1$, then

$$\{u\} = \left\{ \begin{array}{c} u_2 \\ u_3 \end{array} \right\} = \left\{ \begin{array}{c} \frac{1+\lambda}{2(1-2\lambda)} \\ 1 \end{array} \right\} \Rightarrow \{u\}_1 = \left\{ \begin{array}{c} \frac{\sqrt{2}}{2} \\ 1 \end{array} \right\} \qquad (4.1.13.\text{Ex}1)$$

$$\{u\}_2 = \left\{ \begin{array}{c} -\frac{\sqrt{2}}{2} \\ 1 \end{array} \right\}.$$

Note that for this example, the nodal displacements $u_2$ and $u_3$ are the exact values for the corresponding exact mode shapes.

## 4.2    Beam Element

Engineering beam theory is based on the Euler–Bernoulli assumption, which implies that shear deformations and shear strains are neglected (Figure 4.8). Let $EI, \gamma$ be constant stiffness and mass distribution along the element. Consider bending only (axial loads can be incorporated via the axial element considered earlier in this chapter). Assume a displacement field of the form

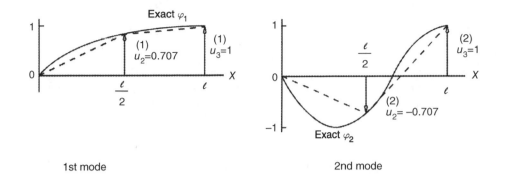

Exact $\varphi_1$

(1) $u_2=0.707$  (1) $u_3=1$

$\frac{\ell}{2}$  $\ell$

(2) $u_3=1$

$\frac{\ell}{2}$

(2) $u_2=-0.707$

Exact $\varphi_2$

1st mode                                2nd mode

– – – – – Displacement (interpolation) field used in finite element model

**Figure 4.8** Comparison of the approximate shape functions and the exact mode shapes

$$w(x,t) = a_1 + a_2 x + a_3 x^2 + a_4 x^3 = \{\phi\}^T \{a\}, \tag{4.2.1}$$

which is exact for a statically loaded beam with end forces only. That is, the assumed displacement field satisfies the equilibrium equation from statics,

$$EIw_{,xxxx} = 0. \tag{4.2.2}$$

In the dynamic case, the "constants" $a_i$ are functions of time and have the role of generalized coordinates. They can be related to the nodal (end) displacements as follows:

$$w(0,t) = q_1 = a_1$$
$$w'(0,t) = q_2 = a_2$$
$$w(l,t) = q_3 = a_1 + a_2 l + a_3 l^2 + a_4 l^3$$
$$w'(l,t) = q_4 = a_2 + 2a_3 l + 3a_4 l^2. \tag{4.2.3}$$

Small displacements and small slopes have been assumed (Figure 4.9). In matrix form, Eq. (4.2.3) can be expressed as

$$\{q\} = \begin{Bmatrix} q_1 \\ q_2 \\ q_3 \\ q_4 \end{Bmatrix} = \begin{bmatrix} 1 & 0 & 0 & 0 \\ 0 & 1 & 0 & 0 \\ 1 & l & l^2 & l^3 \\ 0 & 1 & 2l & 3l^2 \end{bmatrix} \begin{Bmatrix} a_1 \\ a_2 \\ a_3 \\ a_4 \end{Bmatrix} = [G]\{a\}. \tag{4.2.4}$$

**Figure 4.9** Geometry for a typical beam-type finite element

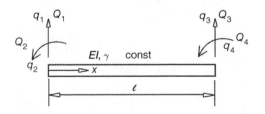

$q_1, Q_1$                    $q_3, Q_3$

$Q_2$                                $Q_4$ $q_4$

$q_2$    $EI, \gamma$   const    $x$

$\ell$

Solve for the $a_i$s:

$$\{a\} = [G]^{-1}\{q\}, \tag{4.2.5}$$

where

$$[G]^{-1} = \begin{bmatrix} 1 & 0 & 0 & 0 \\ 0 & 1 & 0 & 0 \\ -3/l^2 & -2/l & 3/l^2 & -1/1 \\ 2/l^3 & 1/l^2 & -2/l^3 & 1/l^2 \end{bmatrix}. \tag{4.2.6}$$

Back-substituting Eqs. (4.2.5) and (4.2.6) into Eq. (4.2.1) yields $w(x,t)$ in terms of the nodal displacements:

$$w(x,t) = \{\phi\}^T [G]^{-1}\{q\} = \{N(x)\}^T \{q(t)\}, \tag{4.2.7}$$

where

$$\{\phi\}^T = \left\lfloor 1\ x\ x^2\ x^3 \right\rfloor \tag{4.2.8}$$

and

$$N_1(x) = 1 - 3(x/l)^2 + 2(x/l)^3$$
$$N_2(x) = x - 2x^2/l + x^3/l^2$$
$$N_3(x) = 3(x/l)^2 - 2(x/l)^3$$
$$N_4(x) = -x^2/l + x^3/l^2 \tag{4.2.9}$$

are *shape* or *interpolation* functions associated with the coordinates $q_i$, which yields a detailed form of Eq. (4.2.7)

$$w(x,t) = N_1(x)q_1(t) + N_2(x)q_2(t) + N_3(x)q_3(t) + N_4(x)q_4(t). \tag{4.2.10}$$

The $N_i(x)$ expressions are the so-called Hermite (cubic) polynomials and they are shown in Figure 4.10. They are cubic functions that are defined in terms of the values of the function and its first derivative (slope) at the two end points of the interval $[0, l]$.

**Figure 4.10** Geometric interpretation of the interpolation functions

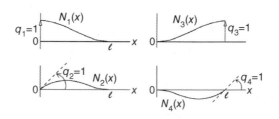

*Strain Energy*: The strain energy in the beam is

$$U = \frac{1}{2} \int_0^l EIw_{,xx}^2 \, dx \tag{4.2.11}$$

and

$$w_{,xx} = \{\phi''\}^T [G]^{-1} \{q\} = \{N''\}^T \{q\} = \{q\}^T \{N''\} \tag{4.2.12}$$

$$w_{,xx}^2 = \{q\}^T \{N''\} \{N''\}^T \{q\} = \{q\}^T \left([G]^{-1}\right)^T \{\phi''\} \{\phi''\}^T [G]^{-1} \{q\} . \tag{4.2.13}$$

The strain energy can now be expressed in a form that allows direct identification of the stiffness matrix:

$$U = \frac{1}{2} \{q\}^T \left(\int_0^l EI \{N''\} \{N''\}^T dx\right) \{q\} = \frac{1}{2} \{q\}^T [k] \{q\} . \tag{4.2.14}$$

The stiffness matrix for the beam element is therefore

$$[k] = \int_0^l EI \{N''(x)\} \{N''(x)\}^T dx \tag{4.2.15}$$

or, when expressed in terms of the vector $\phi$,

$$[k] = \left([G]^{-1}\right)^T \left(\int_0^l EI \{\phi''\} \{\phi''\}^T dx\right) [G]^{-1}. \tag{4.2.16}$$

The elements of $[k]$ are readily evaluated, using Eq. (4.2.15) ,

$$k_{ij} = \frac{\partial^2 U}{\partial q_i \partial q_j} = \int_0^l EI N_i''(x) N_j''(x) dx. \tag{4.2.17}$$

Substituting from Eq. (4.2.9) and carrying out the integrations yield the $4 \times 4$ stiffness matrix for a Bernoulli–Euler beam element:

$$[k] = \frac{EI}{l^3} \begin{bmatrix} 12 & 6l & -12 & 6l \\ 6l & 4l^2 & -6l & 2l^2 \\ -12 & -6l & 12 & -6l \\ 6l & 2l^2 & -6l & 4l^2 \end{bmatrix}. \tag{4.2.18}$$

*Kinetic Energy*: For a slender beam, the kinetic energy contributed by the angular velocity of the cross section can be neglected and

$$T = \frac{1}{2} \int_0^l \gamma \dot{w}^2 dx. \tag{4.2.19}$$

By expressing the velocity term as

$$\dot{w}^2 = \{\dot{q}\}^T \{N(x)\} \{N(x)\}^T \{\dot{q}\} \tag{4.2.20}$$

one can write

$$\begin{aligned} T &= \frac{1}{2} \{\dot{q}\}^T \left(\int_0^l \gamma \{N\} \{N\}^T dx\right) \{\dot{q}\} \\ &= \frac{1}{2} \{\dot{q}\}^T [m] \{\dot{q}\} , \end{aligned} \tag{4.2.21}$$

where an inertial reference frame has been assumed. By direct comparison of the last two expressions, the mass matrix for the beam is identified and given by

$$[m] = \int_0^l \gamma \{N(x)\} \{N(x)\}^T dx \qquad (4.2.22)$$

with elements

$$m_{ij} = \frac{\partial^2 T}{\partial \dot{q}_i \partial \dot{q}_j} = \int_0^l \gamma N_i(x) N_j(x) dx. \qquad (4.2.23)$$

Of course, $[m]$ can also be expressed in terms of the column vector $\{\phi\}$:

$$[m] = \left([G]^{-1}\right)^T \left(\int_0^l \gamma \{\phi\} \{\phi\}^T dx\right) [G]^{-1}. \qquad (4.2.24)$$

Evaluating the integrals in Eq. (4.2.23) yields the mass matrix for the slender Bernoulli–Euler beam element:

$$[m] = \frac{\gamma l}{420} \begin{bmatrix} 156 & 22l & 54 & -13l \\ 22l & 4l^2 & 13l & -3l^2 \\ 54 & 13l & 156 & -22l \\ -13l & -3l^2 & -22l & 4l^2 \end{bmatrix}. \qquad (4.2.25)$$

This matrix is the so-called *consistent mass matrix*, because the velocity field used in calculating the kinetic energy is consistent with the assumed displacement field.

A lumped mass matrix can also be constructed, by lumping half of the beam mass at each end (at the $q_1$ and $q_3$ coordinates) and associating the moment of inertia from half of the beam with each of the $q_2$ and $q_4$ coordinates. The resulting lumped mass matrix becomes

$$[m] = \frac{\gamma l}{24} \begin{bmatrix} 12 & 0 & 0 & 0 \\ 0 & l^2 & 0 & 0 \\ 0 & 0 & 12 & 0 \\ 0 & 0 & 0 & l^2 \end{bmatrix}. \qquad (4.2.26)$$

The consistent mass matrix appears to be preferable in applications, since it is based on somewhat more "rational" arguments. Thus one would expect better results in numerical calculations of frequencies and mode shapes, for example, if one used the consistent rather than the lumped mass matrix. However, this is not always the case. Lumped mass matrices typically overestimate the inertias associated with the individual degrees of freedom, that is, the diagonal terms in $[m]$. This tends to cancel the excess stiffness inherent in consistent mass matrices, with a net result that the use of lumped mass matrices often gives calculated frequencies closer to the exact values, for a typical mesh size. Use of the consistent mass matrix, on the other hand, always yields frequencies that are above the corresponding exact natural frequencies. Furthermore, the convergence to the exact frequencies is monotonic and from above, permitting error bounds to be established. Because convergence from above cannot be guaranteed if lumped mass matrices are used (or if inter element compatibility is violated), use of consistent mass matrices is preferred in most FE calculations. However, there is by no means a general consensus on this issue, and the lumped mass matrix does have a strong following in certain circles.

### 4.2.1 Comment

It is important to mention that in the calculation of the elements of the stiffness and mass matrices for the beam or similar simple one-dimensional structural components calculation can be performed in closed form. This approach is not practical anymore for more complicated structural components such as Timoshenko beam elements or two-dimensional elements like plates. For such cases a numerical method like Gaussian quadrature is preferred. Actually from a practical point of view it is more convenient to use a numerical method like Gaussian quadrature to evaluate the integrals, since algebraic errors can be avoided. **Gaussian quadrature** is an effective integration procedure in which both the positions of the sampling points and the weights are optimized. The typical form of the Gaussian integration process is

$$\int_a^b F(r)dr = \alpha_1 F(r_1) + \alpha_2 F(r_2) + \cdots + \alpha_n F(r_n) + R_n,$$

where both the weight $\alpha_1, \alpha_2, \ldots \alpha_n$ and the sampling points are variables. Using a Gaussian quadrature based on $n$ points means that the required integral $\int_a^b F(r)dr$ is approximated by integrating a polynomial of order $2n - 1$ instead of $F(r)$. A detailed description of Gaussian quadrature is provided in Appendix C.

### 4.2.2 Consistent Load Vector

In actual applications, some of the loads may be distributed over one or more beam elements. Recall that in the FE approach, all forces and displacements must eventually be referred to the nodes of the FE model. The principle of virtual work is used to identify the consistent nodal forces associated with a distributed force $p_z(x, t)$, as illustrated in Figure 4.11.

$$\delta W_E = \int_0^l p_z(x, t) \delta w(x, t) dx \equiv \{Q\}^T \{\delta q\} . \qquad (4.2.27)$$

But

$$\delta w = \{N\}^T \{\delta q\} \qquad (4.2.28)$$

**Figure 4.11**
Geometry for deriving the consistent loading vector

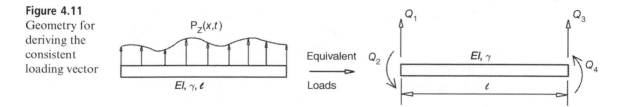

hence

$$\delta W_E = \int_0^l p_z(x,t) \{N\}^T \{\delta q\}\, dx \tag{4.2.29}$$

from which it follows that the consistent load vector is given by

$$\{Q\} = \int_0^l p_z(x,t) \{N(x)\}\, dx \tag{4.2.30}$$

with components

$$Q_i(t) = \int_0^l p_z(x,t) N_i(x) dx \tag{4.2.31}$$

with $i = 1, 2, 3, 4$.

## 4.3    Assembly of Elements

Two methods of performing element assembly are presented:

(1) the direct method and
(2) the congruence transformation method.

Additional useful information on these methods of assembly can found in Gallagher (1975, 53–72).

In the direct method, we consider equilibrium of elastic or inertial forces at each node by summing the contributions from all elements joining at the node. This method is often used in simple calculations, and is best illustrated through examples. The congruent transformation method offers a systematic, but somewhat less efficient method of synthesizing the global mass and stiffness matrices. It is useful in problems where the geometry or connectivity of the structure makes the direct method cumbersome, for example, where the elements do not join at 90-degree angles. It has the additional advantage of being relatively easy to program.

### 4.3.1    Direct Method

The direct method, also sometimes referred to as the *direct stiffness method*, is best illustrated through an example. Consider the beam-spring-mass system illustrated in Figure 4.12, modeled with four beam type FE.

By considering force and moment equilibrium at each node, it is easy to show that the stiffness (or mass) terms simply add algebraically at each node. Note that global coordinates $q_i$ have been chosen such that the assembly process is simplified; indeed, the assembled mass and stiffness matrices can be synthesized by simply overlaying (superimposing) the individual element matrices, as indicated in Figure 4.13.

Note: Cross-hatched regions indicate where elements of the sub-matrices add algebraically at a node.

**Figure 4.12** System illustrating the direct stiffness method of assembly

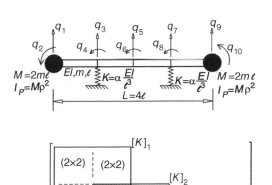

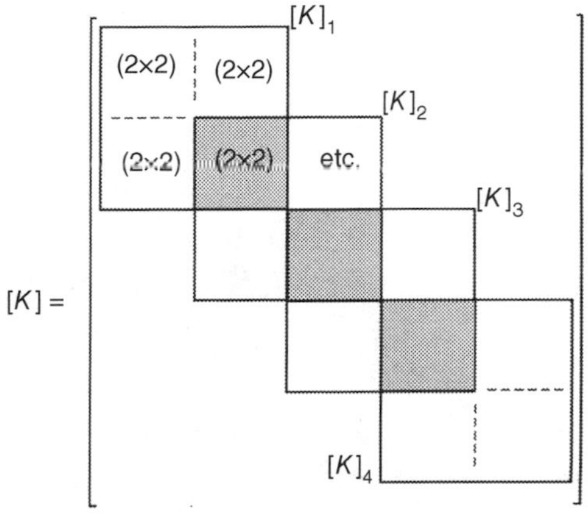

**Figure 4.13** Illustration of the direct stiffness method of element assembly

Thus, the assembled or *global* stiffness and mass matrices for the structure become

$$[K] = \frac{EI}{l^3}\begin{bmatrix} \begin{array}{cc} 12 & 6\ell \\ 6\ell & 4\ell^2 \\ -12 & -6\ell \\ 6\ell & 2\ell^2 \end{array} & \begin{array}{cc} -12 & 6\ell \\ -6\ell & 2\ell^2 \\ (24+\alpha) & 0 \\ 0 & 8\ell^2 \\ -12 & -6\ell \\ 6\ell & 2\ell^2 \end{array} & \begin{array}{cc} [0] \\ -12 & 6\ell \\ -6\ell & 2\ell^2 \\ 24 & 0 \\ 0 & 8\ell^2 \\ -12 & -6\ell \\ 6\ell & 2\ell^2 \end{array} & \begin{array}{cc} [0] \\ [0] \\ -12 & 6\ell \\ -6\ell & 2\ell^2 \\ (24+\alpha) & 0 \\ 0 & 8\ell^2 \\ -12 & -6\ell \\ 12 & -6\ell \\ -6\ell & 4\ell^2 \end{array} \end{bmatrix}$$ (4.3.1)

$$[M] = \frac{ml}{420}\begin{bmatrix} \begin{array}{cc} (156+210) & 22\ell \\ 22\ell & (4\ell^2+210\rho^2) \\ 54 & 13\ell \\ -13\ell & -3\ell^2 \end{array} & \begin{array}{cc} 54 & -13\ell \\ 13\ell & -3\ell^2 \\ 312 & 0 \\ 0 & 8\ell^2 \\ 54 & 13\ell \\ -13\ell & -3\ell^2 \end{array} & \begin{array}{cc} [0] \\ 54 & -13\ell \\ 13\ell & -3\ell^2 \\ 312 & 0 \\ 0 & 8\ell^2 \\ 54 & 13\ell \\ -13\ell & -3\ell^2 \end{array} & \begin{array}{cc} [0] \\ [0] \\ 54 & -13\ell \\ 13\ell & -3\ell^2 \\ 312 & 0 \\ 0 & 8\ell^2 \\ 54 & 13\ell \\ (156+210) & 22\ell \\ 22\ell & (4\ell^2+210\rho^2) \end{array} \end{bmatrix}$$ (4.3.2)

In situations where it is not possible to arrange the coordinate numbering so that a direct superposition of the individual element matrices will yield the global matrices, one can use an alternate approach. First express the total strain energy and the total kinetic energy in terms of the global coordinates, by adding the respective energies in the individual elements:

$$U = \sum_k U_k(q_i) = \frac{1}{2}\sum_k \{q\}_k^T [k]_k \{q\}_k$$ (4.3.3)

$$T = \sum_k T_k (\dot{q}_i) = \frac{1}{2} \sum_k \{\dot{q}\}_k^T [m]_k \{\dot{q}\}_k , \qquad (4.3.4)$$

where the subscript $k$ refers to the $k$th element, and the $k$th element coordinates are expressed in terms of the global coordinates $q_i$. It is assumed that no transformations are needed, or the element matrices would first have to be transformed to the new coordinates. The global mass and stiffness matrices are synthesized by calculating

$$K_{ij} = \frac{\partial^2 U}{\partial q_i \partial q_j} \qquad (4.3.5)$$

$$M_{ij} = \frac{\partial^2 T}{\partial \dot{q}_i \partial \dot{q}_j}, \qquad (4.3.6)$$

where an inertial reference frame has been assumed. Finally, the consistent load vector is found by calculating the total potential $V_E$ and setting

$$Q_i = -\frac{\partial V_E}{\partial q_i} \qquad (4.3.7)$$

or, in the case of nonpotential forces, by identifying the $Q_i$s from the virtual work

$$\delta W_E = \sum_i^n Q_i \delta q_i. \qquad (4.3.8)$$

### 4.3.2  Congruence Transformation Method

It is often convenient to formulate the mass and stiffness matrices for a given element in a *local* coordinate system, and later transform back to the global coordinate system (Figure 4.14). This is accomplished through a transformation of the form

$$\{\bar{x}\} = [T]\{x\}, \qquad (4.3.9)$$

where

$$\begin{aligned} \{\bar{x}\}^T &= \{\bar{x} \quad \bar{y} \quad \bar{z}\} \qquad \text{local} \\ \{x\}^T &= \{x \quad y \quad z\} \quad \text{global} \end{aligned} \qquad (4.3.10)$$

are the local and the global coordinate systems, respectively. Note that, in general, one would have a different local coordinate system for each element in the assembly.

**Figure 4.14** Illustration showing elements with local coordinate systems that are different from the global coordinates

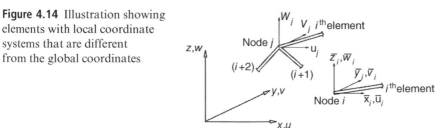

Similarly, at any node of the structure, say the $j$th, one can write, for the $i$th element:

$$\left\{ \begin{array}{c} \bar{u} \\ \bar{v} \\ \bar{w} \end{array} \right\}_i = [T]_i\,\{q\}_j \tag{4.3.11}$$

$$\{\bar{F}\}_i \;=\; [T]_i\,\{F\}_j \tag{4.3.12}$$

## Example: Rotation in 2-space

When the two coordinate systems are oriented at an angle $\theta$ relative to each other as shown in Figure 4.15

$$\left\{ \begin{array}{c} \bar{x} \\ \bar{y} \end{array} \right\} = \left[ \begin{array}{cc} \cos\theta & \sin\theta \\ -\sin\theta & \cos\theta \end{array} \right] \left\{ \begin{array}{c} x \\ y \end{array} \right\}. \tag{4.3.13}$$

For this case the transformation matrix is orthogonal, $[T]^{-1} = [T]^T$. Let $[\bar{m}]_i, [\bar{k}]_i, \{\bar{q}_i\}\ldots$ denote the mass, stiffness, and generalized coordinate matrices of the $i$th element, respectively. Construct the "unassembled" mass, stiffness, ... matrices:

**Figure 4.15** Coordinate systems rotated by an angle $\theta$ with respect to each other

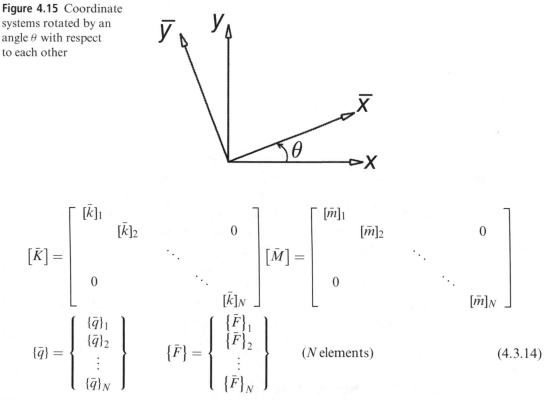

$$[\bar{K}] = \left[ \begin{array}{ccccc} [\bar{k}]_1 & & & & \\ & [\bar{k}]_2 & & 0 & \\ & & \ddots & & \\ & 0 & & \ddots & \\ & & & & [\bar{k}]_N \end{array} \right] \quad [\bar{M}] = \left[ \begin{array}{ccccc} [\bar{m}]_1 & & & & \\ & [\bar{m}]_2 & & 0 & \\ & & \ddots & & \\ & 0 & & \ddots & \\ & & & & [\bar{m}]_N \end{array} \right]$$

$$\{\bar{q}\} = \left\{ \begin{array}{c} \{\bar{q}\}_1 \\ \{\bar{q}\}_2 \\ \vdots \\ \{\bar{q}\}_N \end{array} \right\} \quad \{\bar{F}\} = \left\{ \begin{array}{c} \{\bar{F}\}_1 \\ \{\bar{F}\}_2 \\ \vdots \\ \{\bar{F}\}_N \end{array} \right\} \quad (N \text{ elements}) \tag{4.3.14}$$

When connecting the elements to form an assembly, there exists a relation of the form

$$\{\bar{q}\} = [\beta]\,\{q\} \tag{4.3.15}$$

between the local coordinates $\{\bar{q}\}$ and the global coordinates $\{q\}$. The matrix $[\beta]$ is often referred to as the *compatibility* or *connectivity* matrix. Note that $[\beta]$ may include constraints, rotations, etc., and is not in general a square matrix.

Because the strain energy $U$, the kinetic energy $T$, and the virtual work $\delta W_E$ are invariants, one can write

$$U = \sum_i^N U_i = \sum_i^N \frac{1}{2} \{\bar{q}\}^T [\bar{k}] \{\bar{q}\} = \frac{1}{2} \{q\}^T [\beta]^T [\bar{k}][\beta] \{q\} \tag{4.3.16}$$

$$T = \sum_i^N T_i = \sum_i^N \frac{1}{2} \{\dot{\bar{q}}\}^T [\bar{m}] \{\dot{\bar{q}}\} = \frac{1}{2} \{\dot{q}\}^T [\beta]^T [\bar{m}][\beta] \{\dot{q}\} \tag{4.3.17}$$

$$\delta W_E = \{\delta \bar{q}\}^T \{\bar{F}\} = \{\delta q\}^T [\beta]^T \{\bar{F}\} . \tag{4.3.18}$$

From these expressions, one can immediately conclude that the global mass matrix $[M]$, stiffness matrix $[K]$, and consistent load vector $\{Q\}$ are

$$[M] = [\beta]^T [\bar{m}][\beta] \tag{4.3.19}$$

$$[K] = [\beta]^T [\bar{k}][\beta] \tag{4.3.20}$$

$$\{Q\} = [\beta]^T \{\bar{F}\} . \tag{4.3.21}$$

Note that the mass and stiffness matrices are obtained through a congruence transformation. Clearly, the individual element matrices could also have first been transformed to global coordinates by a transformation of the form (4.3.9); however, it is often just as convenient to include *all* transformations in the connectivity matrix $[\beta]$.

The resultant global mass and stiffness matrices are typically banded matrices as shown in Figure 4.16. The bandwidth depends on the ordering of the nodes (coordinates). Large-scale FE programs usually have bandwidth minimization routines that reorder the coordinates in an optimum manner. This is important in large problems, because special algorithms have been developed for inverting and processing so-called *sparse* matrices.

Although the congruent transformation method is easy to understand and program, it is less efficient that the "direct" method of assembly, especially for large problems. The reason for this is that the unassembled matrices are sparse; consequently, a large number of wasteful multiplications of zero by zero occurs. Of course, a clever programmer may be able to avoid some of these by keeping track of the nodal connectivity of the structure, as shown in the next example.

## Example Illustrating the Congruent Method of Assembly

Two massless beam elements are joined to form a frame, as shown in Figure 4.17. A mass $M$ with mass moment of inertia $J$ is attached at a point $B$.

**Figure 4.16** Typical banded matrices encountered in the global assembly process

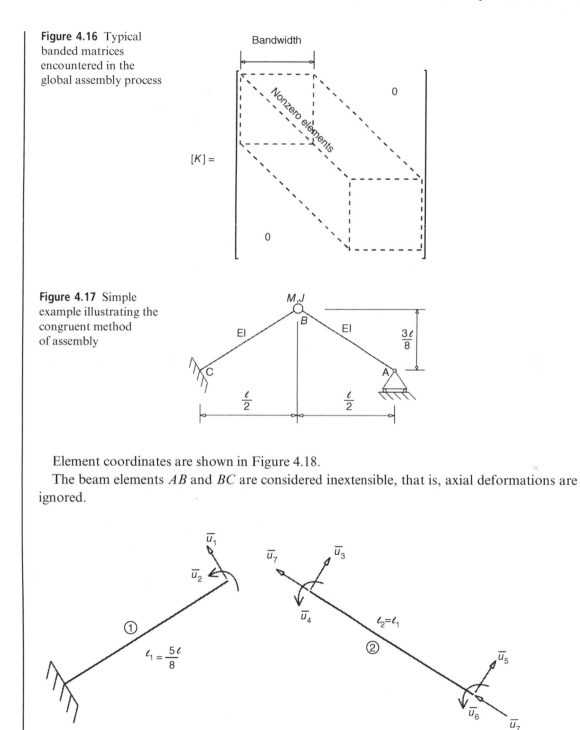

**Figure 4.17** Simple example illustrating the congruent method of assembly

Element coordinates are shown in Figure 4.18.

The beam elements $AB$ and $BC$ are considered inextensible, that is, axial deformations are ignored.

**Figure 4.18** Element coordinates for the example shown in Figure 4.17

For element 1, one has

$$\{\bar{u}\} = \begin{Bmatrix} \bar{u}_1 \\ \bar{u}_2 \end{Bmatrix}; \quad [k]_1 = \frac{EI}{\ell_1^3} \begin{bmatrix} 12 & -6\ell_1 \\ -6\ell_1 & 4\ell_1^2 \end{bmatrix}.$$

The global coordinates are shown in Figure 4.19.

**Figure 4.19** Global coordinates for the example shown in Figure 4.17

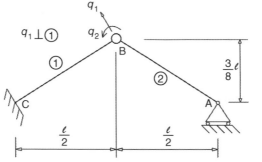

For element 2, one has

$$\{\bar{u}\}_2 = \begin{Bmatrix} \bar{u}_3 \\ \bar{u}_4 \\ \bar{u}_5 \\ \bar{u}_6 \\ \bar{u}_7 \end{Bmatrix}; \quad [k]_2 = \frac{EI}{\ell_2^3} \begin{bmatrix} 12 & 6\ell_2 & -12 & 6\ell_2 & 0 \\ & 4\ell_2^2 & -6\ell_2 & 2\ell_2^2 & 0 \\ & & 12 & -6\ell_2 & 0 \\ & \text{Symmetric} & & 4\ell_2^2 & 0 \\ & & & & 0 \end{bmatrix}$$

Note that $\bar{u}_7$ DOF has no stiffness associated with it, but must be introduced to permit axial rigid-body motion of element 2.

From the geometry shown in Figure 4.20, the following relations between various coordinates can be identified.

**Figure 4.20** Final assembly illustrating the relations between the various coordinates

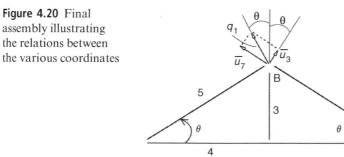

$$\bar{u}_3 = q_1 \cos 2\theta = q_1(\cos^2\theta - \sin^2\theta)$$

$$= q_1\left[\left(\frac{4}{5}\right)^2 - \left(\frac{3}{5}\right)^2\right] = \frac{7}{25}q_1$$

$$\bar{u}_7 = q_1 \sin 2\theta = q_1 \cdot 2\sin\theta\cos\theta = q_1 \cdot 2 \cdot \left(\frac{3}{5}\right)\cdot\left(\frac{4}{5}\right)$$

$$= \left(\frac{24}{25}\right)q_1$$

$$\bar{u}_1 = q_1; \qquad \bar{u}_2 = q_2; \qquad \bar{u}_4 = q_2$$

$\bar{u}_5$ and $\bar{u}_6$ can be determined by using the boundary conditions at support A.

$\frac{4}{5}\bar{u}_5 + \left(\frac{3}{5}\right)\bar{u}_7 = 0$ (roller stays on ground)

$\bar{u}_5 = -\frac{3}{4}\bar{u}_7 = -\frac{3}{4}\cdot\frac{24}{25}q_1 = -\frac{18}{25}q_1$

and

$$\bar{F}_6 = 0 = 6\ell_2\bar{u}_3 + 2\ell_2^2\bar{u}_4 - 6\ell_2\bar{u}_5 + 4\ell_2^2\bar{u}_6$$

$$\bar{u}_6 = -\frac{3}{2}\cdot\frac{7}{25}\left(\frac{q_1}{\ell_2}\right) - \frac{1}{2}q_2 + \frac{3}{2}\left(\frac{-18}{25}\right)\left(\frac{q_1}{\ell_2}\right) = -\frac{3}{2}\left(\frac{q_1}{\ell_2}\right) - \frac{1}{2}q_2.$$

Hence the compatibility relation becomes

$$\begin{Bmatrix} \bar{u}_1 \\ \bar{u}_2 \\ \bar{u}_3 \\ \bar{u}_4 \\ \bar{u}_5 \\ \bar{u}_6 \\ \bar{u}_7 \end{Bmatrix} = \begin{bmatrix} 1 & 0 \\ 0 & 1 \\ 7/25 & 0 \\ 0 & 1 \\ -18/25 & 0 \\ -3/(2\ell_2) & -1/2 \\ 24/25 & 0 \end{bmatrix} \begin{Bmatrix} q_1 \\ q_2 \end{Bmatrix}$$

note that $\ell_1 = \ell_2 = \frac{5}{8}\ell$ and $\{\bar{u}\} = [\beta]\{q\}$

System stiffness matrix: $[K] = [\beta]^T[\bar{k}][\beta]$,

where the unassembled stiffness matrix is given by $[\bar{k}] = \begin{bmatrix} [\bar{k}]_1 & \\ & [\bar{k}]_2 \end{bmatrix}$

$$[\bar{k}][\beta] = \frac{IE}{\ell_1^3}\begin{bmatrix} 12 & -6\ell_1 & & & & \\ -6\ell_1 & 4\ell_1^2 & & & & \\ & & 12 & 6\ell_1 & -12 & 6\ell_1 & 0 \\ & & 6\ell_1 & 4\ell_1^2 & -6\ell_1 & 2\ell_1^2 & 0 \\ & & -12 & -6\ell_1 & 12 & -6\ell_1 & 0 \\ & & 6\ell_1 & 2\ell_1^2 & -6\ell_1 & 4\ell_1^2 & 0 \\ & & 0 & 0 & 0 & 0 & 0 \end{bmatrix}\begin{bmatrix} 1 & 0 \\ 0 & 1 \\ 7/25 & 0 \\ 0 & 1 \\ -18/25 & 0 \\ -3/2\ell_1 & -1/2 \\ 24/25 & 0 \end{bmatrix}$$

$$
= \frac{EI}{\ell_1^3}
\begin{pmatrix}
12 & -6\ell_1 \\
-6\ell_1 & 4\ell_1^2 \\
\left(\frac{84}{25} + \frac{216}{25} - 9\right) & (6-3)\,\ell_1 \\
\left(\frac{42}{25} + \frac{108}{25} - 3\right)\ell_1 & (4-1)\,\ell_1^2 \\
-\left(\frac{84}{25} + \frac{216}{25} - 9\right) & (-6+3)\,\ell_1 \\
\left(\frac{42}{25} + \frac{108}{25} - 6\right)\ell_1 & (2-2)\,\ell_1^2 \\
0 & 0
\end{pmatrix}
= \frac{EI}{\ell_1^3}
\begin{pmatrix}
12 & -6\ell_1 \\
-6\ell_1 & -4\ell_1^2 \\
3 & 3\ell_1 \\
3\ell_1 & 3\ell_1^2 \\
-3 & -3\ell_1 \\
0 & 0 \\
0 & 0
\end{pmatrix}
$$

and the final assembled stiffness matrix is given by

$$
[\beta]^T [\bar{k}][\beta] = [K]
$$

$$
[K] =
\begin{bmatrix}
1 & 0 & \frac{7}{25} & 0 & -\frac{18}{25} & -\frac{3}{2\ell_1} & \frac{24}{25} \\
0 & 1 & 0 & 1 & 0 & -\frac{1}{2} & 0
\end{bmatrix}
\begin{bmatrix}
12 & -6\ell_1 \\
-6\ell_1 & 4\ell_1^2 \\
3 & 3\ell_1 \\
3\ell_1 & 3\ell_1^2 \\
-3 & -3\ell_1 \\
0 & 0 \\
0 & 0
\end{bmatrix}
\cdot \frac{EI}{\ell_1^3}
$$

$$
= \frac{EI}{\ell_1^3}
\begin{bmatrix}
\left(12 + \frac{21+54}{25}\right) & \left(-6 + \frac{21+54}{25}\right) \\
(-6 + 3 + 0) & \left(4\ell_1^2 + 3\ell_1^2\right)
\end{bmatrix}
$$

$$
= \frac{EI}{\ell_1^3}
\begin{bmatrix}
15 & -3\ell_1 \\
-3\ell_1 & 7\ell_1^2
\end{bmatrix}.
$$

The mass matrix is given by

$$
[M] =
\begin{bmatrix}
M & 0 \\
0 & J
\end{bmatrix}
= M
\begin{bmatrix}
1 & 0 \\
0 & \rho^2
\end{bmatrix},
$$

where $\rho$ is the radius of gyration.
Equations of motion:

$$
[M]\{\ddot{q}\} + [K]\{q\} = 0
$$

it implies that for free vibration one has simple harmonic motion, that is, $\{\ddot{q}\} = -\omega^2\{q\}$, and the eigenvalue problem is given by $\left(-\omega^2[M] + [K]\right)\{q\} = 0$.

Let $\lambda = \frac{\omega^2 M \ell_1^3}{EI}$, then the characteristic equation becomes

$$\begin{vmatrix} (15-\lambda) & -3\ell_1 \\ -3\ell_1 & (7\ell_1^2 - \rho^2\lambda) \end{vmatrix} = 0$$

$$(15-\lambda)\left(7\ell_1^2 - \rho^2\lambda\right) - 9\ell_1^2 = 105\ell_1^2 - 7\ell_1^2\lambda - 15\rho^2\lambda + \rho^2\lambda^2 - 9\ell_1^2 = 0$$

$$\lambda^2 - \left[15 + 7\left(\ell_1/\rho\right)^2\right]\lambda + 96\left(\ell_1/\rho\right)^2 = 0$$

$$\lambda_{1,2} = \frac{1}{2}\left[15 + 7\left(\frac{\ell_1}{\rho}\right)^2\right] \pm \frac{1}{2}\sqrt{\left[15 + 7\left(\frac{\ell_1}{\rho}\right)^2\right]^2 - 384\left(\frac{\ell_1}{\rho}\right)^2}$$

$$\omega_{1,2} = \sqrt{\frac{EI}{M\ell_1^3}\lambda_{1,2}}.$$

For a point mass, $\rho = 0$; $\lambda_1 = \frac{96}{7}$ and $\omega_1 = \sqrt{\frac{96EI}{7M\ell_1^3}} = \sqrt{\frac{49152EI}{875M\ell^3}}$ or $\omega_1 \approx 7.4949\sqrt{\frac{EI}{M\ell^3}}$.

## 4.4 Comments on the Choice of Trial Functions

This is only an elementary discussion of the requirement for element interpolation functions. A more detailed and accurate discussion can be found in modern texts on FE listed at the end of this chapter. Comments of the choice of trial functions in the FE method can be made from two points of view.

Using mathematical considerations, where the FE method is considered an approximation tool used to discretize the partial differential equation. In this case the partial differential equation has to be identified first (e.g. elliptic, parabolic or hyperbolic) and mathematical conditions are imposed on the trial functions to achieve the desired rate of convergence.

For structural dynamic problems of self-adjoint type, advantage can be taken of the variational principles which are employed in formulating the element properties such as the principle of minimum potential energy and Hamilton's principle in order to obtain guidelines on the choice of interpolation functions (or fields) used to describe element behavior.

(1) Basic requirements of shape functions also known as interpolation functions are provided below.

(a) Continuity:
   (i) Within element: continuous up through highest derivative occurring in the functional used in variational principle.
   (ii) Across element boundaries: continuous up through one order less than highest derivative appearing in functional.
(b) Rigid-Body Modes: all appropriate rigid-body modes must be represented. That is, when nodal DOFs are given values corresponding to a state of rigid-body motion, zero strain energy (zero nodal forces) should result

(c) Constant Strain State: the assumed functions should be capable of representing constant values of all stresses and strains.

### 4.4.1 Comments

- One or more of these requirements have often been violated in formulating elements, sometimes deliberately and sometimes inadvertently.
- Condition (a)(i) is easy to satisfy, while condition (a)(ii) can be difficult. Many successful elements have been formulated by relaxing the inter element continuity requirement (a)(ii) on some derivative, thus making the element more flexible. For such cases, convergence is generally no longer monotonic.
- Condition (b) is obvious, but can be hard to satisfy in complicated elements in curvilinear coordinates (shells), and is sometimes violated "slightly" to simplify formulation. Serious problems occur when condition (c) is violated, often resulting in convergence toward an incorrect solution. This can occur because, as the grid is refined, the strain state of an individual element should approach a constant value but cannot since such a constant value is not represented in the shape function(s).

## 4.5    Finite Element for Timoshenko Beam Free Vibrations

### 4.5.1 Introduction

A considerable number of FE models capable of simulating the behavior of Timoshenko beams exist, however several of these models have experienced numerical difficulties, known as "locking" (or excessive shear stiffness) when the element is approaching the thin beam regime (Tessler and Dong 1981). While such numerical problems can be alleviated by reduced (selective) integration (Hughes 1987b), the approach provided in Tessler and Dong (1981) presents a more appropriate approach based on physical considerations.

### 4.5.2 Interdependent Variable Interpolation

In Timoshenko beam theory the transverse displacement $w$, the bending slope $\psi$, and the shear angle $\beta$ are related

$$\beta = w_{,x} - \psi. \tag{4.5.1}$$

Therefore, if $w$ and $\psi$ are selected as the independent variables, the FE formulation for the Timoshenko beam should be based on the same order of interpolation for $w_{,x}$ and $\psi$. This implies that the interpolation polynomial for $w$ should be one order higher than the interpolation polynomial for $\psi$.

**Constrained Elements**

Requiring a continuous shear constraint allows one to reduce the degrees of freedom of a virgin element, without lowering the order of the element. The elements obtained in this manner are denoted "constrained elements." The interpolation field for the virgin element is expressed by

$$w = \sum_{n=0}^{p+1} a_n x^n \; ; \; \psi = \sum_{n=0}^{p} b_n x^n. \tag{4.5.2}$$

Clearly Eq. (4.5.2) implies that the interpolating polynomial for the shear angle is one order lower than for the transverse displacement

$$\beta = w_{,x} - \psi = (a_1 - b_0) + (2a_2 - b_1)x + \ldots [(p+1)a_{p+1} - b_p]x^p \tag{4.5.3}$$

by imposing the constraint represented by Eq. (4.5.2)

$$b_p = (p+1)a_{p+1}\, , \; b_{p-1} = pa_p, \ldots \text{etc.} \tag{4.5.4}$$

the order of the shear angle variation can be lowered sequentially. Among the various elements discussed in Tessler and Dong (1981) consider the T2CL6 element – which denotes an element, which is constrained, with $p = 2$, linear $\beta$ and six nodal degrees of freedom as shown in Figure 4.21.

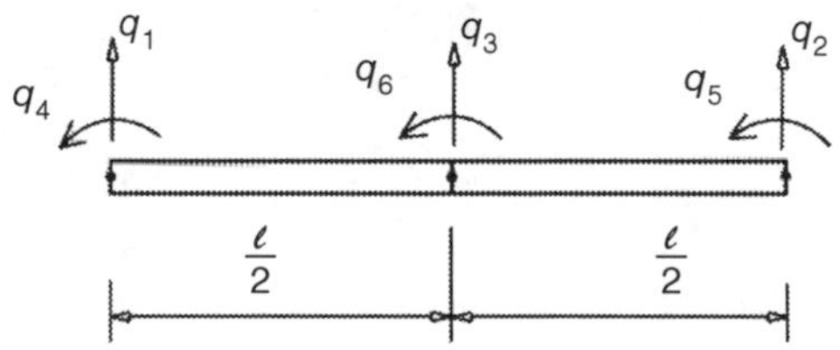

**Figure 4.21** The geometry and degrees of freedom of the T2CL6 element

$$q_1 = w_1 \;\; ; q_4 = \psi_1$$
$$q_2 = w_2 \;\; ; q_5 = \psi_2$$
$$q_3 = w_3 \;\; ; q_6 = \psi_3$$

Since $p = 2$ assume that

$$w = \sum_{n=0}^{3} a_n x^n = a_0 + a_1 x + a_2 x^2 + a_3 x^3 = \{\phi_w\}^T \{a\} \tag{4.5.5}$$

$$\psi = \sum_{n=0}^{2} b_n x^n = b_0 + b_1 x + b_2 x^2 = \{\phi_\psi\}^T \{b\} , \tag{4.5.6}$$

where

$$\begin{aligned}
\{\phi_w\}^T &= \lfloor 1 \; x \; x^2 \; x^3 \rfloor \\
\{\phi_\psi\}^T &= \lfloor 1 \; x \; x^2 \rfloor \\
\{a\}^T &= \lfloor a_0 \; a_1 \; a_2 \; a_3 \rfloor \\
\{b\}^T &= \lfloor b_0 \; b_1 \; b_2 \rfloor.
\end{aligned}$$

The virgin element has seven degrees of freedom, while the constrained element T2CL6 has six DOFs, therefore one continuous shear constraint should be applied

$$b_p = (p+1)a_{p+1} \qquad (4.5.7)$$

for $p = 2$ it reduces to

$$b_2 = 3a_3. \qquad (4.5.8)$$

The procedure used to obtain the constrained displacement field constrained displacement field with the highest order term in the polynomial for shear angle variation removed (Tessler and Dong 1981) involves the derivation of the shape functions $N_w$ and $N_\psi$ for the virgin element. This, in turn, requires the evaluation of $w$ at the additional nodal degree of freedom $w_4$, although this additional $w$ degree of freedom is eliminated subsequently by enforcing the shear constraint. However, it is more convenient to apply the continuous shear constraint to the displacement field expressed by Eqs. (4.5.5) and (4.5.6) and then derive the constrained shape functions directly.

Thus, substitute Eq. (4.5.8) into Eq. (4.5.6)

$$\psi = b_0 + b_1 x + 3a_3 x^2 \qquad (4.5.9)$$

and evaluate $w$ and $\psi$ at each nodal DOF

$$\left.\begin{array}{l} w_1 = w(0) = a_0 \\ w_2 = w(l) = a_0 + a_1\ell = a_0 + a_1\ell + a_2\ell^2 + a_3\ell^3 \\ w_3 = w\left(\frac{\ell}{2}\right) = a_0 + a_1\left(\frac{\ell}{2}\right) + a_2\left(\frac{\ell}{2}\right)^2 + a_3\left(\frac{\ell}{2}\right)^3 \\ \psi_1 = \psi(0) = b_0 \\ \psi_2 = \psi(\ell) = b_0 + b_1\ell + 3a_3\ell^2 \\ \psi_3 = \psi\left(\frac{\ell}{2}\right) = b_0 + b_1\left(\frac{\ell}{2}\right) + 3a_3\left(\frac{\ell}{2}\right)^2 \end{array}\right\}. \qquad (4.5.10)$$

The last equation can be written in matrix form, with all generalized coordinates written in the same dimensions

$$\begin{Bmatrix} w_1 \\ w_2 \\ w_3 \\ \ell\psi_1 \\ \ell\psi_2 \\ \ell\psi_3 \end{Bmatrix} = \begin{bmatrix} 1 & 0 & 0 & 0 & 0 & 0 \\ 1 & \ell & \ell^2 & \ell^3 & 0 & 0 \\ 1 & \frac{\ell}{2} & \frac{\ell^2}{4} & \frac{\ell^3}{8} & 0 & 0 \\ 0 & 0 & 0 & 0 & \ell & 0 \\ 0 & 0 & 0 & 3\ell^3 & \ell & \ell^2 \\ 0 & 0 & 0 & \frac{3\ell^3}{4} & \ell & \frac{\ell^2}{2} \end{bmatrix} \begin{Bmatrix} a_0 \\ a_1 \\ a_2 \\ a_3 \\ b_0 \\ b_1 \end{Bmatrix}. \qquad (4.5.11)$$

Carrying out the appropriate matrix inversion one obtains

$$
\begin{Bmatrix} a_0 \\ a_1 \\ a_2 \\ a_3 \\ b_0 \\ b_1 \end{Bmatrix} = \begin{bmatrix} 1 & 0 & 0 & 0 & 0 & 0 \\ -3/\ell & -1/\ell & 4/\ell & 1/3\ell & 1/3\ell & -2/3\ell \\ 2/\ell^2 & 2/\ell^2 & -4/\ell^2 & -1/\ell^2 & -1/\ell^2 & 2/\ell^2 \\ 0 & 0 & 0 & 2/3\ell^3 & 2/3\ell^3 & -4/3\ell^3 \\ 0 & 0 & 0 & 1/\ell & 0 & 0 \\ 0 & 0 & 0 & -3/\ell^2 & -1/\ell & 4/\ell^2 \end{bmatrix} \begin{Bmatrix} w_1 \\ w_2 \\ w_3 \\ \ell\psi_1 \\ \ell\psi_2 \\ \ell\psi_3 \end{Bmatrix}. \tag{4.5.12}
$$

Thus the $\{a\}$ and $\{b\}$ matrices can be identified from Eq. (4.5.12):

$$
\{a\} = \begin{Bmatrix} a_0 \\ a_1 \\ a_2 \\ a_3 \end{Bmatrix} \begin{bmatrix} 1 & 0 & 0 & 0 & 0 & 0 \\ -3/\ell & -1/\ell & 4/\ell & 1/3\ell & 1/3\ell & -2/3\ell \\ 2/\ell^2 & 2/\ell^2 & -4/\ell^2 & -1/\ell^2 & -1/\ell^2 & 2/\ell^2 \\ 0 & 0 & 0 & 2/3\ell^3 & 2/3\ell^3 & -4/3\ell^3 \end{bmatrix} \tag{4.5.13}
$$

$$
\{b\} = \begin{Bmatrix} b_0 \\ b_1 \\ b_2 \end{Bmatrix} = \begin{Bmatrix} b_0 \\ b_1 \\ 3a_3 \end{Bmatrix} = \begin{bmatrix} 0 & 0 & 0 & 1/\ell & 0 & 0 \\ 0 & 0 & 0 & -3/\ell^2 & -1/\ell^2 & 4/\ell^2 \\ 0 & 0 & 0 & 2/\ell^3 & 2/\ell^3 & -4/\ell^3 \end{bmatrix} \begin{Bmatrix} w_1 \\ w_2 \\ w_3 \\ \ell\psi_1 \\ \ell\psi_2 \\ \ell\psi_3 \end{Bmatrix}. \tag{4.5.14}
$$

Substitute Eq. (4.5.13) into Eq. (4.5.5)

$$
w = \{\phi_w\}^T \{a\} =
$$

$$
= \lfloor 1 \ x \ x^2 \ x^3 \rfloor \begin{bmatrix} 1 & 0 & 0 & 0 & 0 & 0 \\ -3/\ell & -1/\ell & 4/\ell & 1/3\ell & 1/3\ell & -2/3\ell \\ 2/\ell^2 & 2/\ell^2 & -4/\ell^2 & -1/\ell^2 & -1/\ell^2 & 2/\ell^2 \\ 0 & 0 & 0 & 2/3\ell^3 & 2/3\ell^3 & -4/3\ell^3 \end{bmatrix} \begin{Bmatrix} w_1 \\ w_2 \\ w_3 \\ \ell\psi_1 \\ \ell\psi_2 \\ \ell\psi_3 \end{Bmatrix} \tag{4.5.15}
$$

$$
= \{N_w(x)\}^T \{q(t)\},
$$

where

$$
\begin{aligned}
\{N_w(x)\}^T &= \lfloor N_{w_1} \ N_{w_2} \ N_{w_3} \ N_{w_4} \ N_{w_5} \ N_{w_6} \rfloor \\
\{q(t)\}^T &= \lfloor q_1 \ q_2 \ q_3 \ \ell q_4 \ \ell q_5 \ \ell q_6 \rfloor \\
&= \lfloor w_1 \ w_2 \ w_3 \ \ell\psi_1 \ \ell\psi_2 \ \ell\psi_3 \rfloor \\
N_{w_1}(x) &= 1 - 3\left(\tfrac{x}{\ell}\right) + 2\left(\tfrac{x}{\ell}\right)^2 \\
N_{w_2}(x) &= -\left(\tfrac{x}{\ell}\right) + 2\left(\tfrac{x}{\ell}\right)^2 \\
N_{w_3}(x) &= 4\left(\tfrac{x}{\ell}\right) - 4\left(\tfrac{x}{\ell}\right)^2 \\
N_{w_4}(x) &= \tfrac{1}{3}\left(\tfrac{x}{\ell}\right) - \left(\tfrac{x}{\ell}\right)^2 + \tfrac{2}{3}\left(\tfrac{x}{\ell}\right)^3 \\
N_{w_5}(x) &= \tfrac{1}{3}\left(\tfrac{x}{\ell}\right) - \left(\tfrac{x}{\ell}\right)^2 + \tfrac{2}{3}\left(\tfrac{x}{\ell}\right)^3 \\
N_{w_6}(x) &= -\tfrac{2}{3}\left(\tfrac{x}{\ell}\right) + 2\left(\tfrac{x}{\ell}\right)^2 - \tfrac{4}{3}\left(\tfrac{x}{\ell}\right)^3 .
\end{aligned} \tag{4.5.16}
$$

Substitute Eq. (4.5.14) into Eq. (4.5.6)

$$
\psi = \{\phi_\psi\}^T \{b\} =
$$

$$
= \lfloor 1 \ x \ x^2 \rfloor
\begin{bmatrix}
0 & 0 & 0 & 1/\ell & 0 & 0 \\
0 & 0 & 0 & -3/\ell^2 & -1/\ell^2 & 4/\ell^2 \\
0 & 0 & 0 & 2/\ell^3 & 2/\ell^3 & -4/\ell^3
\end{bmatrix}
\begin{Bmatrix}
w_1 \\ w_2 \\ w_3 \\ \ell\psi_1 \\ \ell\psi_2 \\ \ell\psi_3
\end{Bmatrix}
=
$$

$$
= \{N_\psi(x)\}^T \{q(t)\}, \tag{4.5.17}
$$

where

$$
\begin{aligned}
\{N_\psi(x)\}^T &= \lfloor N_{\psi_1} N_{\psi_2} N_{\psi_3} N_{\psi_4} N_{\psi_5} N_{\psi_6} \rfloor \\
N_{\psi_1}(x) &= N_{\psi_2}(x) = N_{\psi_3}(x) = 0 \\
N_{\psi_4}(x) &= \tfrac{1}{\ell}\left[ 1 - 3\left(\tfrac{x}{\ell}\right) + 2\left(\tfrac{x}{\ell}\right)^2 \right] \\
N_{\psi_5}(x) &= \tfrac{1}{\ell}\left[ -\left(\tfrac{x}{\ell}\right) + 2\left(\tfrac{x}{\ell}\right)^2 \right] \\
N_{\psi_6}(x) &= \tfrac{1}{\ell}\left[ 4\left(\tfrac{x}{\ell}\right) - 4\left(\tfrac{x}{\ell}\right)^2 \right].
\end{aligned} \tag{4.5.18}
$$

The strain energy is given by

$$
\begin{aligned}
U &= \int_0^\ell \left[ \frac{EI}{2}\psi_{,x}^2 + \frac{kGA}{2}(w_{,x} - \psi)^2 \right] dx = \\
&= \tfrac{1}{2}\int_0^\ell \left\{ EI\psi_{,x}^2 + kGA\left[ w_{,x}^2 - 2w_{,x}\psi + \psi^2 \right] \right\} dx = \\
&= \tfrac{1}{2}\int_0^\ell \Big( EI\{q\}^T \{N'_\psi\}^T \{N'_\psi\}^T \{q\} + kGA\{q\}^T \{N'_w\}\{N'_w\}\{q\} \\
&\quad -2kGA\{q\}^T \{N'_w\}\{N_\psi\}^T \{q\} + kGA\{q\}^T \{N_\psi\}\{N_\psi^T\}\{q\} \Big) dx
\end{aligned} \tag{4.5.19}
$$

$$
U = \frac{1}{2}\{q\}^T [k]\{q\}. \tag{4.5.20}
$$

Comparing Eqs. (4.5.19) and (4.5.20) yields the element stiffness matrix

$$[k] = \int_0^\ell (EI\{N_\psi'\}\{N_\psi'\}^T + kGA[\{N_w'\}\{N_w'\}^T - \{N_w'\}\{N_\psi\}^T + \{N_\psi\}\{N_\psi\}^T])dx. \quad (4.5.21)$$

The kinetic energy is given by

$$
\begin{aligned}
T_K &= \tfrac{1}{2}\int_0^\ell (\rho I\dot\psi^2 + \rho A\dot w^2)dx = \\
&= \tfrac{1}{2}\int_0^\ell [\rho I\{\dot q\}^T\{N_\psi\}\{N_\psi\}^T\{\dot q\} + \rho A\{\dot q\}^T\{N_w\}\{N_w\}^T\{\dot q\}]dx = \\
&= \tfrac{1}{2}\{\dot q\}^T[m]\{\dot q\}.
\end{aligned}
\quad (4.5.22)
$$

Thus the mass matrix is given by

$$[m] = \int_0^\ell \rho I\{N_\psi\}\{N_\psi\}^T dx + \int_0^\ell \rho A\{N_w\}\{N_w\}^T dx.$$

The shape functions $\{N_w\}$ and $\{N_\psi\}$ in Eqs. (4.5.21) and (4.5.22) were defined in Eqs. (4.5.16) and (4.5.18), respectively.

### The Approaches Used in Tessler and Dong (1981) and Kapur (1966)

(1) In Tessler and Dong (1981) the total transverse displacement $w$ and the bending rotation $\psi$ were used as independent variables, while Kapur (1966) used the transverse displacement due to bending $w_b$ and the transverse shear displacement ($w_s$) as the independent variables.

(2) In Tessler and Dong (1981) $w$ and $\psi$ are used as the nodal degrees of freedom, which require only $C_0$ continuity for full inter-element continuity. In Kapur (1966) $w_b, w_{b,x}, w_s, w_{s,x}$ are used as the nodal degrees of freedom, which requires $C_1$ continuity for full inter-element compatibility. For problems involving changes in beam cross-sectional properties, there is a discontinuity of shear strain (corresponding to $w_{s,x}$), therefore the Kapur (1966) formulation is inappropriate because it enforces continuity in $w_{s,x}$. However, Tessler and Dong (1981) is appropriate for such cases.

(3) The formulation in Kapur (1966) leads to an uncoupled element stiffness matrix

$$[k] = \begin{bmatrix} [k_B] & [0] \\ [0] & [k_S] \end{bmatrix} \quad (4.5.23)$$

as was shown. However, Tessler and Dong (1981) will produce a fully populated stiffness matrix as evident from Eq. (4.5.21).

The mass matrices are fully coupled for both Tessler and Dong (1981) and Kapur (1966).

(4) From the comparison above it is clear that Tessler and Dong (1981) has the advantage of not having to impose higher continuity requirements for inter-element boundaries. Also the constrained elements developed in Tessler and Dong (1981) perform well in the range of small thickness/length ratios.

## 4.6    Treatment of Axially Loaded Beam Elements

Axial loads modify the effective stiffness of structural elements, therefore from a structural dynamic viewpoint the natural frequencies are affected (Figure 4.22). Axial loads can be loads applied in a direct manner. Another type of axial loading is associated with centrifugal effects present in rotating structures. These loads can be also treated as axial loads, however, they are different because they vary with the distance from the axis of rotation and speed of rotation. Axial loads can be incorporated in FE models by using the so-called "geometric stiffness matrix." An important aspect of axially loaded structures is due to the fact that the axial load has to be incorporated by using the second-order-strain – displacement relation, namely

$$\epsilon_{iJ} = \frac{1}{2}(u_{i,J} + u_{J,i} + u_{k,i}u_{k,J})$$

or

$$\epsilon_{xx} = \frac{\partial u}{\partial x} + \frac{1}{2}\left[u_{,x}^2 + v_{,x}^2 + w_{,x}^2\right],$$

where $u$, $v$, and $w$ are respectively the displacements in the $x$-, $y$-, and $z$-directions and $\epsilon_{xx}$ is the axial strain.

**Figure 4.22** Illustration of the geometry for an axially loaded finite element

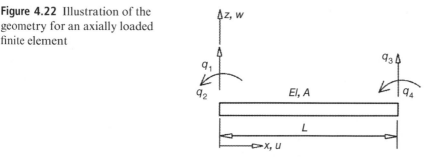

When using the last relation for axially loaded beam-type structural elements where only $u$ and $w$ displacements exist, the $u_{,x}^2$ term is usually much smaller than the square of the slope of the beam and therefore it can be neglected, thus

$$\epsilon_{xx} = \frac{\partial u}{\partial x} + \frac{1}{2}w_{,x}^2. \tag{4.6.1}$$

For beam theory one can introduce the Euler–Bernoulli assumption

$$\frac{\partial u}{\partial x} = \frac{\partial \bar{u}}{\partial x} - zw_{,xx}, \tag{4.6.2}$$

where $\bar{u}$ is the axial displacement on the elastic axis of the beam and $w$ is the displacement of a point on the elastic axis in the $z$-direction. Combining Eqs. (4.6.1) and (4.6.2) yields

$$\epsilon_{xx} = \frac{d\bar{u}}{dx} - zw_{,xx} + \frac{1}{2}w_{,x}^2 \tag{4.6.3}$$

The strain energy of the beam for this case is given by

$$U^e = \frac{1}{2} \int_v E\epsilon_{xx}^2 dV \qquad (4.6.4)$$

and $dV = dAdx$, where $A$ is the cross-sectional area

Combining Eqs. (4.6.3) and (4.6.4)

$$U^e = \frac{1}{2} \int_L \int_A \left[ \left(\frac{d\bar{u}}{dx}\right)^2 + z^2 \left(\frac{\partial^2 w}{dx^2}\right)^2 + \frac{1}{4}\left(\frac{dw}{dx}\right)^4 - 2z\frac{du}{dx}\frac{d^2 w}{dx^2} \right.$$
$$\left. - z\left(\frac{d^2 w}{dx^2}\right)\left(\frac{dw}{dx}\right)^2 + \left(\frac{d\bar{u}}{dx}\right)\left(\frac{dw}{dx}\right)^2 \right] EdAdx \qquad (4.6.5)$$

for $z$-measured from the centroid

$$\int_A dA = A \quad ; \quad \int_A zdA = 0 \quad ; \quad \int_A z^2 dA = I \qquad (4.6.6)$$

due to these relations the fourth and fifth term in Eq. (4.6.5) are eliminated and the equation reduces to Eq. (4.6.7)

$$U^e = \frac{1}{2}\int_L \left[ A\left(\frac{d\bar{u}}{dx}\right)^2 + I\left(\frac{d^2 w}{dx^2}\right)^2 + A\left(\frac{d\bar{u}}{dx}\right)\left(\frac{dw}{dx}\right)^2 + \frac{A}{4}\left(\frac{dw}{dx}\right)^4 \right] Edx. \qquad (4.6.7)$$

Next it is important to realize that the formulation for the axial load, developed in this section, is a **linear one**, in terms of the displacement $w$. This implies the one deals with relatively small slopes, that is, the term $\frac{A}{4}\left(\frac{dw}{dx}\right)^4$ is negligible when compared to the other terms in Eq. (4.6.7). Furthermore we **assume** that the axial load is independent of the bending deformation and related only to the axial deformation by the relation

$$P = EA\frac{d\bar{u}}{dx}, \qquad (4.6.8)$$

where for tension value $P$ is taken as positive. Equations (4.6.7) and (4.6.8) can be combined to give

$$U^e = \frac{1}{2}\int_0^L \left[ \underbrace{A\left(\frac{d\bar{u}}{dx}\right)^2}_{\text{Axial strain energy}} + \underbrace{EI\left(\frac{d^2 w}{dx^2}\right)^2 + P\left(\frac{dw}{dx}\right)^2}_{\text{Flexural strain energy}} \right] dx \qquad (4.6.9)$$

thus it is important to note that the axial and flexural strain energies are uncoupled.

$$U^e = U_a^e + U_b^e$$
$$U_a^e = \frac{1}{2}\int_0^L EA\left(\frac{d\bar{u}}{dx}\right)^2 dx$$
$$U_b^e = \frac{1}{2}\int_0^L \left[ EI\left(\frac{d^2 w}{dx^2}\right)^2 + P\left(\frac{dw}{dx}\right)^2 \right] dx \qquad (4.6.10)$$

the kinetic energy is given by

$$T = \frac{1}{2} \int_0^L m\dot{w}^2 dx. \tag{4.6.11}$$

Applying Hamilton's principle and using Eqs. (4.6.10) and (4.6.11) yields

$$m\frac{\partial^2 w}{\partial t^2} + EI\frac{\partial^4 w}{\partial x^4} - P\frac{d^2 w}{dx^2} = 0. \tag{4.6.12}$$

The mass matrix can be constructed in a manner identical to the beam vibrations discussed earlier. The only difference is in the stiffness matrix based on Eq. (4.6.10). It should be noted that the contribution to the strain energy $U_a^e$ is unimportant, unless one wants to deal with the *axial* vibrations of the beam.

Recall from the derivation of the beam element

$$w = \{\phi\}^T\{\alpha\} = \alpha_1 + \alpha_2 x + \alpha_3 x^2 + \alpha_4 x^3 \tag{4.6.13}$$

$$\{\phi\}^T = \lfloor 1 \quad x \quad x^2 \quad x^3 \rfloor. \tag{4.6.14}$$

Furthermore the nodal degrees of freedom $\{q\}$ have been related to the coefficients $\{\alpha\}$ by

$$\{\alpha\} = [B]\{q\}, \tag{4.6.15}$$

where

$$\{q\} = \begin{Bmatrix} w(x=0) \\ \frac{dw}{dx}(x=0) \\ w(x=L) \\ \frac{dw}{dx}(x=L) \end{Bmatrix} = \begin{bmatrix} 1 & 0 & 0 & 0 \\ 0 & 1 & 0 & 0 \\ 1 & L & L^2 & L^3 \\ 0 & 1 & 2L & 3L^2 \end{bmatrix} \begin{Bmatrix} \alpha \end{Bmatrix} = [A]\{\alpha\} \tag{4.6.16}$$

$$\{\alpha\} = [A]^{-1}\{q\} = [B]\{q\}$$

$$[B] = [A]^{-1} = \begin{bmatrix} 1 & 0 & 0 & 0 \\ 0 & 1 & 0 & 0 \\ -3/L^2 & -2/L & 3/L^2 & -1/L \\ 2/L^3 & 1/L^2 & -2/L^3 & 1/L^2 \end{bmatrix} \tag{4.6.17}$$

$$w = \{\phi\}^T[B]\{q\}. \tag{4.6.18}$$

Thus

$$\frac{d^2 w}{dx^2} = \{\phi''\}^T[B]\{q\} \tag{4.6.19}$$

$$\frac{dw}{dx} = \{\phi'\}^T[B]\{q\} \tag{4.6.20}$$

Combining equations (4.6.11), (4.6.19), and (4.6.20)

$$U_b^e = \frac{1}{2}\{q\}^T [k]\{q\} = \frac{1}{2}\{q\}^T \int_0^L EI[B]^T \{\phi''\}\{\phi''\}^T [B]dx\{q\}$$

$$+ \frac{1}{2}\{q\}^T \int_0^L P[B]^T \{\phi'\}\{\phi'\}^T [B]dx\{q\} =$$

$$= \frac{1}{2}\{q\}^T [k_b]\{q\} + \frac{1}{2}\{q\}^T [k_g]\{q\}, \qquad (4.6.21)$$

where

$$[k_b] = \int_0^L EI[B]^T \{\phi''\}\{\phi''\}^T [B]dx \qquad (4.6.22)$$

$$[k_g] = \int_0^L P[B]^T \{\phi'\}\{\phi'\}^T [B]dx, \qquad (4.6.23)$$

where $[k_b]$ is the conventional element flexural stiffness matrix. The new matrix $[k_g]$ introduces considerations related to axial loading of the structure, under free vibrations. It is usually denoted by name **incremental stiffness** matrix because it represents an addition to the conventional stiffness matrix. Sometimes it is also called the geometric stiffness matrix. Furthermore it should be noted that when one deals with compressive forces $P$ is negative.

Using Eqs. (4.6.23) and (4.6.20) the various elements of the geometric stiffness matrix can be calculated and are given below.

$$[k_G] = \frac{P}{L}\begin{bmatrix} \frac{6}{5} & & & \\ \frac{L}{10} & \frac{2}{15}L^2 & Symmetric & \\ -\frac{6}{5} & -\frac{L}{10} & \frac{6}{5} & \\ \frac{L}{10} & -\frac{L^2}{30} & \frac{-L}{10} & \frac{2}{15}L^2 \end{bmatrix}. \qquad (4.6.24)$$

When assembling the element properties into master stiffness matrices, one can write symbolically, with **J** representing a symbolic matrix performing the assembly of the elements

$$[K] = \mathbf{J}^T \lceil \mathbf{k}_1 \quad \mathbf{k}_2 \quad \dots \mathbf{k}_N \rfloor \mathbf{J}$$

$$= [K_b] + [K_g] \qquad (4.6.25)$$

$$[M] = \mathbf{J}^T \lceil \mathbf{m}_1 \quad \mathbf{m}_2 \quad \dots \mathbf{m}_N \rfloor \mathbf{J} \qquad (4.6.26)$$

and the structural dynamics eigenvalue problem for an axially loaded structure can be written as (4.6.27), where $\mathbf{q}^*$ represents the DOF vector with geometric boundary conditions enforced.

$$[M]\{\ddot{q}^*\} + \left([K_b] + [K_g]\right)\{q^*\} = 0. \qquad (4.6.27)$$

Now recall that $[K_g]$ depends on the axial load $P$, which can be factored out

$$[K_g] = P[\bar{K}_g] \qquad (4.6.28)$$

and Eq. (4.6.27) becomes

$$-\omega^2[M]\{q^*\} + \left([K_b] + P[\bar{K}_g]\right)\{q^*\} = 0. \qquad (4.6.29)$$

From Eq. (4.6.29) a number of physically important conclusion can be immediately obtained:

(1) When $P$ is a tension force, Eq. (4.6.29) predicts an incremental increase in stiffness, increasing the natural frequency $\omega$.
(2) When $P$ is a compression force, an incremental reduction in stiffness occurs, yielding lower natural frequency.
(3) When $P = P_{cr}$ ($P_{cr} =$ is the critical buckling load)

$$\det\left([K_b] = P[\bar{K}_g]\right) = 0$$

or $P_{cr}$ is the solution to the following eigenvalue problem

$$[K_b]\{q\} = P[\bar{K}_g]\{q\}.$$

Thus when $P \rightarrow P_{cr}$, $\omega \rightarrow 0$, or when the axially compressed system approaches the buckling stability boundary, the natural frequency of free vibrations for the system approaches zero.

Before leaving this topic, it should be noted that the geometric stiffness matrix can be easily extended to two-dimensional structures, such as plates, enabling one to deal with axially loaded plates, in a manner that resembles the current discussion.

Finally, the geometric stiffness matrix is a useful tool when dealing with rotating structures. The effects of rotation as represented by the centrifugal forces are axial loads, which vary along the span of the structural element and can be included using the geometric stiffness matrix.

## 4.7    Finite Element for Plate Vibrations

The main objective of this section is to illustrate the application of the FE method for the simple structural elements considered in this book. Note that the plate element described next is a natural extension of the beam element described earlier and it is not the most modern or efficient plate element available. For more advanced treatment of the topic the reader is encouraged to examine the references provided at the end of this chapter.

### 4.7.1    Rectangular Plate Elements

An interesting aspect of the FE method is that the development of plate elements is more complicated than the bar and beam elements considered earlier in this chapter. The uncoupled bending problem, where inplane stretching effects are neglected, is considered. The expression for the strain energy for this problem has been derived in the previous sections and is given by

$$U_b = \frac{D}{2} \iint_R \left\{ (\nabla^2 w)^2 + 2(1-v) \left[ \left( \frac{\partial^2 w}{\partial x \partial y} \right)^2 - \left( \frac{\partial^2 w}{\partial x^2} \right) \left( \frac{\partial^2 w}{\partial y^2} \right) \right] \right\} dxdy$$

$$= \frac{D}{2} \iint_R \left[ w_{,xx}^2 + w_{,yy}^2 + 2v w_{,xx} w_{,yy} + 2(1-v) w_{,xy}^2 \right] dxdy. \tag{4.7.1}$$

The strain energy written in matrix form is more suitable for manipulation when constructing a FE model for the plate vibration problem.

A stress vector can be defined, using the moment resultant intensity functions that have been obtained previously Eq. (3.7.25) through Eq. (3.7.27), thus

$$\{\sigma\} = \left\{ \begin{array}{c} M_{xx} \\ M_{yy} \\ M_{xy} \end{array} \right\} = -D \left\{ \begin{array}{c} \dfrac{\partial^2 w}{\partial x^2} + v \dfrac{\partial^2 w}{\partial y^2} \\[2mm] \dfrac{\partial^2 w}{\partial y^2} + v \dfrac{\partial^2 w}{\partial x^2} \\[2mm] (1-v) \dfrac{\partial^2 w}{\partial x \partial y} \end{array} \right\}. \tag{4.7.2}$$

The strain components can also be replaced by more suitable quantities. The fundamental assumption of Kirchhoff plate bending theory is that lines initially normal to the midplane remain such during deformation of the plate. The rates of change of the angular displacements of these normals are the bending ($\kappa_x, \kappa_y$) and the twisting ($\kappa_{xy}$) curvatures, and these are assumed to be adequately approximated by the following second derivatives of the transverse displacement, $w$:

$$\kappa_{xx} = -\frac{\partial^2 w}{\partial x^2} \quad ; \quad \kappa_{yy} = -\frac{\partial^2 w}{\partial y^2} \quad ; \quad \kappa_{xy} = -2\frac{\partial^2 w}{\partial x \partial y}. \tag{4.7.3}$$

These curvatures are the basic measure of deformation in thin plate flexure, hence the vector of curvatures $\{\kappa\}^T = \lfloor \kappa_{xx}, \kappa_{yy}, \kappa_{xy} \rfloor$ replaces the strain vector $\{\varepsilon\}^T = \lfloor \varepsilon_{xx}, \varepsilon_{yy}, \varepsilon_{xy} \rfloor$ of plane stress analysis.

Thus one can construct a further analogy to plane elasticity by introducing the *constitutive relationship* of thin plane flexure:

$$\{M\} = [E_f]\{\kappa\}. \tag{4.7.4}$$

Where for an isotropic, thin plate

$$[E_f] = D \begin{bmatrix} 1 & v & 0 \\ v & 1 & 0 \\ 0 & 0 & \frac{(1-v)}{2} \end{bmatrix}. \tag{4.7.5}$$

With these relations, using Eqs. (2.2.1) and (2.2.3) the strain energy of the plate can be written as

$$U_B = \frac{D}{2} \iint_R \left[ \kappa_{xx}^2 + \kappa_{yy}^2 + 2v\kappa_{xx}\kappa_{yy} + (1-v)\frac{\kappa_{xy}^2}{2} \right] dxdy \tag{4.7.6}$$

or

$$U = \frac{1}{2} \iint_R \{\kappa\}^T [E_f] \{\kappa\} dx dy. \tag{4.7.7}$$

A rectangular plate element, that resembles the beam-bending element considered earlier, was developed in Schmit et al. (1965) and Bogner et al. (1965). The element is suitable for both static and dynamic applications. A concise description of this element is given next. The element has 16 degrees of freedom or 4 nodal degrees of freedom at each node as shown below. It is rectangular with dimensions $a$ in the $x$-direction and $b$ in the $y$-direction.

The nodal degrees of freedom are given by

$$w \;\; ; \;\; \theta_x = \frac{\partial w}{\partial x} \;\; ; \;\; \theta_y = \frac{\partial w}{\partial y} \;\; ; \;\; \theta_{xy} = \frac{\partial^2 w}{\partial x \partial y}. \tag{4.7.8}$$

These represent displacement and slopes in bending and twist.

### 4.7.2  Interpolation Relations for This Element

In order to satisfy the geometric compatibility requirements it is convenient to pick element modes such that the undetermined coefficients are related to the geometrically important quantities along the edges. For a rectangular plate, one assumes

$$w(x, y) \cong \tilde{w}(x, y) = \sum_{i=0}^{N} \sum_{j=0}^{M} a_{ij} x^i y^j. \tag{4.7.9}$$

Using these relations, one can construct the algebraic relations

$$\tilde{w}(x_p, y_q) = \sum_{i=0}^{N} \sum_{j=0}^{M} a_{ij} x_p^i y_q^j \tag{4.7.10}$$

$$\frac{\partial \tilde{w}(x_p, y_q)}{\partial x} = \sum_{i=1}^{N} \sum_{j=0}^{M} i a_{ij} x_p^{i-1} y_q^j \tag{4.7.11}$$

$$\frac{\partial \tilde{w}(x_p, y_q)}{\partial y} = \sum_{i=0}^{N} \sum_{j=1}^{M} j a_{ij} x_p^i y_q^{j-1} \tag{4.7.12}$$

$$\frac{\partial^2 \tilde{w}(x_p, y_q)}{\partial x \partial y} = \sum_{i=1}^{N} \sum_{j=1}^{M} i j a_{ij} x_p^{i-1} y_q^{j-1} \tag{4.7.13}$$

$$\vdots$$

etc.,

where $p, q = 1, 2$ are the nodal values at the corners shown in Figure 4.23. This procedure is followed until $N \cdot M$ equations are obtained through which the generalized coordinates or the nodal degrees of freedom are related to the coefficients $a_{ij}$ of the interpolating polynomial. These equations may sometimes, **but not always**, be solved to express $a_{ij}$ in terms of $\tilde{w}(x_p, y_q)$, $\frac{\partial \tilde{w}}{\partial x}(x_p, y_q)$ ... etc.

**Figure 4.23** Geometry of a rectangular plate element including nodal numbering

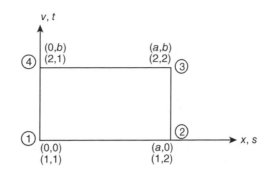

Even when this can be done, the problem of inter-element compatibility is still unresolved because in general, setting $\tilde{w}(x_p, y_q)$, $\frac{\partial \tilde{w}}{\partial x}(x_p, y_q)$, and $\frac{\partial \tilde{w}}{\partial y}(x_p, y_q)$ equal between adjacent elements at the corner does not ensure displacement compatibility along the entire edge; this is primarily due to the fact that the nodes do not in general possess the property that the displacements and the slopes along the edge depend entirely upon the values at the corner associated with that edge.

A remedy to the difficulties mentioned is to use modes, or interpolating polynomials, which are products of one-dimensional **Hermite interpolation formulas**. The appropriate interpolation for these degrees of freedom can be written as

$$\tilde{w}(x, y) = \sum_{i=1}^{2} \sum_{j=1}^{2} \left[ H_{0i}^{(1)}(x) H_{0j}^{(1)}(y) w_{ij} + H_{1i}^{(1)}(x) H_{0j}^{(1)}(y) w_{xij} \right.$$
$$\left. + H_{0i}^{(1)}(x) H_{ij}^{(1)}(y) w_{yij} + H_{1i}^{(1)}(x) H_{ij}^{(1)}(y) w_{xyij} \right],$$
(4.7.14)

where the general notation has the following meaning:

$s$ = running coordinate, being used for interpolation, shown in Figure 4.23;

$N$ = order of the derivative being interpolated by the polynomial,

in this case (1) indicates that the function and its first

derivative are matched; *and*

$i, j$ = refer to the nodes of the element shown in Figure 4.23.

The following polynomials required for $x$ are given below

$$H_{01}^{(1)}(x) = \frac{1}{a^3}(2x^3 - 3ax^2 + a^3)$$

$$H_{02}^{(1)}(x) = -\frac{1}{a^3}(2x^3 - 3ax^2)$$

$$H_{11}^{(1)}(x) = \frac{1}{a^2}(x^3 - 2ax^2 + a^2 x)$$
(4.7.15)

$$H_{12}^{(1)}(x) = \frac{1}{a^2}(x^3 - ax^2)$$

and similarly $H_{01}^{(1)}(y)$, $H_{02}^{(1)}(y)$, $H_{11}^{(1)}(y)$, $H_{12}^{(1)}(y)$ are defined by replacing $x$ with $y$ in Eqs. (4.7.15) and $a$ by $b$.

This assumed displacement mode is capable of representing exactly any displacement of the form

$$\tilde{w}(x,y) = \sum_{r=0}^{3}\sum_{s=0}^{3} \alpha_{rs}x^r y^s, \tag{4.7.16}$$

which means that the interpolation is complete up to terms of the third power.

In terms of Hermite polynomials, Eq. (4.7.16) can be written as

$$\tilde{w}(x,y) = \sum_{i=1}^{2}\sum_{j=1}^{2} \Big[ H_{0i}^{(1)}(x)H_{0j}^{(1)}(y)\alpha_{ij} + H_{1i}^{(1)}(x)H_{0j}^{(1)}(y)\beta_{ij}$$
$$+ H_{0i}^{(1)}(x)H_{1j}^{(1)}(y)\gamma_{ij} + H_{1i}^{(1)}(x)H_{1j}^{(1)}(y)\eta_{ij} \Big], \tag{4.7.17}$$

where $i,j$ refer to the node of the element. Evaluating $w(x,y)$ and its derivatives at the four nodes of the element and using the properties of the Hermite polynomials

$$H_{0i}^{(1)}(x_k) = \delta_{ik} \qquad H_{1i}^{(1)}(x_k) = 0$$
$$\frac{dH_{0i}^{(1)}(x_k)}{dx} = 0 \quad ; \quad \frac{dH_{1i}^{(1)}(x_k)}{dx} = \delta_{ik} \tag{4.7.18}$$

it is found that the 16 undetermined parameters $\alpha_{ij}$, $\beta_{ij}$, $\gamma_{ij}$, and $\eta_{ij}$ are exactly the 16 nodal quantities $w_{ij}$, $w_{xij}$, $w_{yij}$, and $w_{xyij}$, respectively; thus Eq. (4.7.14) is obtained.

Expanding Eq. (4.7.14) partially, one has

$$\tilde{w}(x,y) = \sum_{i=1}^{2} \Big[ H_{0i}^{(1)}(x)H_{01}^{(1)}(y)w_{i1} + H_{0i}^{(1)}(x)H_{02}^{(1)}(y)w_{i2}$$
$$+ H_{1i}^{(1)}(x)H_{01}^{(1)}(y)w_{xi1} + H_{1i}^{(1)}(x)H_{02}^{(1)}(y)w_{xi2}$$
$$+ H_{0i}^{(1)}(x)H_{11}^{(1)}(y)w_{yi1} + H_{0i}^{(1)}(x)H_{12}^{(1)}(y)w_{yi2}$$
$$+ H_{1i}^{(1)}(x)H_{11}^{(1)}(y)w_{xyi1} + H_{1i}^{(1)}(x)H_{12}^{(1)}(y)w_{xyi2} \Big]$$
$$= \{N\}^T_{(1\times16)} \{q\}_{(16\times1)}, \tag{4.7.19}$$

where

$$\{q\}^T = \lfloor w_{11}, w_{21}, w_{12}, w_{22}, w_{x11}, w_{x21}, w_{x12}, w_{x22},$$
$$w_{y11}, w_{y21}, w_{y12}, w_{y22}, w_{xy11}, w_{xy21}, w_{xy12}, w_{xy22} \rfloor \tag{4.7.20}$$

and the elements of $\{N\}$ are given by combinations of the appropriate Hermite polynomials.

It is important to realize that this element includes the important uniform twisting effect $\frac{\partial^2}{\partial x \partial y}$ in addition to the constant strain state of plates.

## 4.8 Element Mass and Stiffness Properties

Using Eqs. 4.7.3 and 4.7.20

$$
\{\kappa\} = \left\{ \begin{array}{c} \kappa_{xx} \\[2ex] \kappa_{yy} \\[2ex] \kappa_{xy} \end{array} \right\} = \left\{ \begin{array}{c} -\dfrac{\partial^2 w}{\partial x^2} \\[2ex] -\dfrac{\partial^2 w}{\partial y^2} \\[2ex] -2\dfrac{\partial^2 w}{\partial x \partial y} \end{array} \right\} = \underset{3\times 16}{\left\{ \begin{array}{c} -\{N_{xx}\}^T \\[2ex] -\{N_{yy}\}^T \\[2ex] -2\{N_{xy}\}^T \end{array} \right\}} \underset{16\times 1}{\{q\}} = \underset{3\times 16}{[B]} \{q\} \tag{4.8.1}
$$

$$
\{\kappa\}^T = \{q\}^T [B]^T. \tag{4.8.2a}
$$

Combining Eqs. (4.7.5), (4.7.7), (4.8.1), and (4.8.3)

$$
U = \frac{1}{2}\{q\}^T [k]\{q\} = \frac{1}{2} \iint_{R_n} \{q\}^T [B]^T [E_f][B]\{q\} dx dy \tag{4.8.3}
$$

from which (where $R_n$ denotes area of the element)

$$
\underset{16\times 16}{[k]} = \iint_{R_n} \underset{16\times 3}{[B]^T} \underset{3\times 3}{[E_f]} \underset{3\times 16}{[B]}\, dx dy. \tag{4.8.4}
$$

In a similar manner, it is possible to obtain consistent mass matrix for this element

$$
T = \frac{1}{2} \iint_{R_n} \rho h \dot{w}^2 dx dy
$$
$$
w = \{N\}^T \{q\}
$$
$$
\dot{w} = \{N\}^T \{\dot{q}\}
$$
$$
T = \frac{1}{2} \iint_{R_n} \rho h \{\dot{q}\}^T \{N\}\{N\}^T \{\dot{q}\} dx dy
$$
$$
[m] = \iint_{R_n} \rho h \{N\}\{N\}^T dx dy. \tag{4.8.5}
$$

Additional information on the stiffness element, where the numerical values can be obtained by Gaussian quadrature (Appendix C) can be found in Bogner et al. (1965) and in Shames and Dym (1985, 606–612).

For a **simply supported plate** using only a quarterplate based on 4 or 16 elements, the following results are obtained (Bogner et al. 1965) as shown in Figure 4.24.

**Figure 4.24** Simply supported quarter plate modeled by four elements

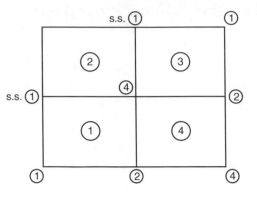

| No. of elements | No. of DOFs | $\omega_{11}$ | $\omega_{12}, \omega_{21}$ | $\omega_{22}$ | $\omega_{31}, \omega_{13}$ |
|---|---|---|---|---|---|
| 4 | 16 | 1037 | 2777 | 4399 | 6222 |
| 16 | 64 | 1035 | 2593 | 4144 | 5251 |
| Exact | $\infty$ | 1035 | 2587 | 4138 | 5173 |

Again due to the consistent nature of this FE model, convergence from above is observed. Details on the mass and stiffness elements can be found in Appendix A of Bogner et al. (1965). Note that while the polynomials lend themselves to closed-form integration, when evaluating $[k]$ and $[m]$ from Eqs. (4.8.4) or (4.8.5). Actually it is more convenient to use Gaussian quadrature to evaluate the integrals numerically. **Gaussian quadrature** is an integration procedure in which both the positions of the sampling points and the weights are optimized. The typical form of the Gaussian integration process is

$$\int_a^b F(r)dr = \alpha_1 F(r_1) + \alpha_2 F(r_2) + \cdots + \alpha_n F(r_n) + R_n, \qquad (4.8.6)$$

where both the weight $\alpha_1, \alpha_2, \ldots, \alpha_n$, and the sampling points are variables. Using a Gaussian quadrature based on $n$ points means that the required integral $\int_a^b F(r)dr$ is approximated by integrating a polynomial of order $2n - 1$ instead of $F(r)$. A detailed description of Gaussian quadrature is provided in Appendix C.

## 4.9    Model Reduction

Finite element models of structures are constructed by combining component models, and the resulting global models can have very large numbers of degrees of freedom. Solving such large models repetitively for different load conditions can be challenging and is inefficient, since accurate response predictions can be obtained from models having far fewer degrees of freedom. Furthermore, structures are often subject to just a few inputs, and structural dynamicists are concerned with the response at a few key locations over a limited frequency range. Thus, several methods have been developed to reduce the size of structural dynamics models.

**Model reduction** approaches seek to reduce the number of degrees of freedom in a model while preserving accurate dynamic input–output relationships in a desired frequency range. Classical normal mode models provide an example of an approach to model reduction that might begin with a detailed global model of an entire structure. Other approaches tailored specifically to model reduction are **component mode synthesis** and **domain decomposition**.

Component mode synthesis (CMS) (or *Dynamic Substructuring*) aims to decrease model complexity at the substructure level using a relatively small number of "modes" (not limited to normal modes) to represent its response. Subsequently, a global structural dynamic model is obtained by coupling the substructure models. Such a decomposition is pursued hierarchically, on multiple levels. For instance, the response of a substructure can be represented by a small number of modes themselves determined from a FE model. Alternatively, substructure models can be obtained experimentally.

Domain Decomposition addresses the analysis of a complex structure by dividing it into components of comparable *computational* size. Response analysis is then pursued through the parallel solution of multiple local models. Thus, multiple smaller problems are employed, while simultaneously considering additional interface constraints between them.

An objective of structural dynamics analysis is to obtain accurate response results at minimal computational cost. By minimizing the computational cost of each individual analysis, additional load cases can be considered, leading to a better overall design. This section and the next section introduce concepts of model reduction and CMS used in response analysis.

Consider an undamped system described by Eq. (4.9.1)

$$[M]\{\ddot{q}\} + [K]\{q\} = \{Q(t)\}.$$  (4.9.1)

The generalized forces exciting the structure, $\{Q\}$, can often be represented in terms of a few inputs, $\{u\}$. An input (*actuator*) equation distributes these inputs to the forces as follows:

$$\{Q(t)\} = [b]\{u(t)\},$$  (4.9.2)

where $[b]$ is the *input matrix* and $\{u\}$ is the *input* (or *control*) *vector*.

The dynamic response is usually sought at a limited number of locations. An output (*sensor*) equation collects these outputs, $\{y\}$, from the generalized coordinates, $\{q\}$, as shown in the following relation:

$$\{y(t)\} = [c]\{q(t)\},$$  (4.9.3)

where $[c]$ is the *output matrix*.

The dynamic model described by Eqs. (4.9.1)–(4.9.3) is often realized in the frequency (Laplace) domain, especially for representing input–output transfer functions used in the design of vibration control systems (Preumont, 2011). This model uses the concept of **dynamic stiffness**, $Z(s)$:

$$[Z(s)]\{q(s)\} = \left[[M]s^2 + [K]\right]\{q(s)\} = \{Q(s)\},$$  (4.9.4)

which, in combination with the input and output equations (Eqs. (4.9.2) and (4.9.3)), yields the input–output transfer function:

$$\{y(s)\} = [c] [Z(s)]^{-1} [b] \{u(s)\}. \tag{4.9.5}$$

### 4.9.1    Reduced Subspaces

Model reduction methods seek response solutions by representing a large set of degrees of freedom, $\{q\}$, as a linear combination of a smaller, *reduced* set, $\{q_R\}$:

$$\{q\} \simeq [T] \{q_R\}, \tag{4.9.6}$$

where $\{q\}$ is $N \times 1$, $\{q_R\}$ is $n \times 1$, and $n \ll N$, typically. The range of solutions for $\{q\}$ lies in the range of the reduction matrix $[T]$. Clearly, some error is introduced by this reduction.

The number of individual equations of motion in Eq. (4.9.1) can be reduced to a smaller set using the reduced basis of Eq. (4.9.6) as follows:

$$[T]^T [M] [T] \{\ddot{q}_R\} + [T]^T [K] [T] \{q_R\} = [T]^T \{Q(t)\}$$
$$[M_{RR}] \{\ddot{q}_R\} + [K_{RR}] \{q_R\} = \{Q_R(t)\}. \tag{4.9.7}$$

Note that each term has been pre-multiplied by $[T]^T$; this reduces the number of equations and preserves symmetry of the reduced mass and stiffness matrices, $[M_{RR}]$ and $[K_{RR}]$. The corresponding reduced input and output equations are

$$\{Q_R(t)\} = [T]^T [b] \{u(t)\} \tag{4.9.8}$$
$$\{y(t)\} \simeq [c] [T] \{q_R\}. \tag{4.9.9}$$

The reduced-order model 4.9.7 is considered to be *accurate* if it adequately preserves the input–output behavior of the original model over a desired frequency range. In terms of the transfer function, Eq. (4.9.5), that is,

$$\{y(s)\} = [c] [Z(s)]^{-1} [b] \{u(s)\} \approx [c] [T] \left[ [T]^T [Z(s)] [T] \right]^{-1} [T]^T [b] \{u(s)\}$$
$$\{y(s)\} = [c] [Z(s)]^{-1} [b] \{u(s)\} \approx [c] [Z_R(s)]^{-1} [b] \{u(s)\} \tag{4.9.10}$$
$$\text{for} \quad i \omega_{low} \leq s = i\omega \leq i \omega_{high},$$

and where $[Z_R(s)]$ is the reduced dynamic stiffness. Various approaches to model reduction differ from one another largely by the generation of appropriate subspaces.

### 4.9.2    Truncated Normal Mode Model

The most common reduced-order model is constructed from a subset of normal modes, as described in Section 7.1. As described in Chapter 5, normal modes are found from the solution of an undamped eigenvalue problem. The structural dynamic response $\{q(t)\}$ may be represented as the superposition of responses of *all* $N$ individual modes, $q_j^m(t)$:

$$\{q\} = \sum_{j=1}^{N} \{\phi_j\} q_j^m = [\Phi]\{q^m\}, \qquad (4.9.11)$$

where $\{\phi_j\}$ is the $j^{\text{th}}$ modal vector and $q_j^m$ is the corresponding modal coordinate. The matrix $[\Phi]$ ($N \times N$) contains the modal vectors, and the vector of modal coordinates $\{q^m\}$ is the same size as $\{q\}(N \times 1)$. (For a valuable perspective on discretization, see (Hughes, 1987a).) Assuming the modal vectors are normalized to yield unit modal mass, the modal equations of motion are

$$[M]\{\ddot{q}\} + [K]\{q\} = \{Q(t)\}$$
$$[\Phi]^T[M][\Phi]\{\ddot{q}^m\} + [\Phi]^T[K][\Phi]\{q^m\} = [\Phi]^T\{Q(t)\}$$
$$\{\ddot{q}^m\} + \lceil\Omega^2\rfloor\{q^m\} = \{Q^m(t)\}, \qquad (4.9.12)$$

where $\lceil\Omega^2\rfloor$ is a diagonal matrix of squared natural frequencies and $\{Q^m\}$ is a vector of modal forces. The input and output equations become

$$\{Q^m(t)\} = [\Phi]^T[b]\{u(t)\} \qquad (4.9.13)$$
$$\{y(t)\} = [c][\Phi]\{q^m\}. \qquad (4.9.14)$$

And the input–output transfer function is given by

$$\{y(s)\} = [c][\Phi]\left[s^2 + \lceil\Omega^2\rfloor\right]^{-1}[\Phi]^T[b]\{u(s)\} \qquad (4.9.15)$$
$$= \sum_{j=1}^{N} \frac{[c]\{\phi_j\}\{\phi_j\}^T[b]}{s^2 + \omega_j^2}\{u(s)\}.$$

Typically, a modal model is **truncated** to use fewer modes, with the number of modes retained, $M \ll N$.

$$\{q\} \simeq \sum_{j=1}^{M \ll N} \{\phi_j\} q_j^m = [\Phi_R]\{q_R^m\}, \qquad (4.9.16)$$

where the modes are indexed in order of increasing frequency. The matrix $[\Phi_R]$ ($N \times M$) contains fewer modal vectors than $[\Phi]$, while the reduced vector of modal coordinates $\{q_R^m\}$ ($M \times 1$) is smaller than $\{q\}$. Such an approach is reasonable because the dynamic response of interest is typically dominated by the responses of a few low-frequency modes. A rule of thumb is to retain modes having natural frequencies up to about 1.5 times the highest frequency of interest, $\omega_{\text{max}}$.

The input–output transfer function for the truncated model is

$$\{y(s)\} \approx \sum_{j=1}^{M \ll N} \frac{[c]\{\phi_j\}\{\phi_j\}^T[b]}{s^2 + \omega_j^2} \{u(s)\}$$

$$= [c][\Phi_R]\left[s^2 + \lceil\Omega_R^2\rfloor\right]^{-1}[\Phi_R]^T[b]\{u(s)\}. \tag{4.9.17}$$

There are more advanced approaches to **mode selection** than frequency-based truncation (Skelton et al., 1982). Such approaches include additional considerations such as whether a given mode can be excited from the input or observed from the output, and how damping affects its contributions to the physical response.

### 4.9.3  Static Correction to the Normal Mode Model

In a truncated modal model, some modes are retained and some are discarded. Since the natural frequencies of the discarded modes are typically much higher than the highest response frequency of interest, their contributions to dynamic response are usually insignificant. However, the static responses of these discarded modes can sometimes be significant, affecting, for instance, the zeros of transfer functions (Lee and Tsuha, 1994). The complete input–output transfer function of a modal model is given by a version of Eq. (4.9.15):

$$\{y(s)\} = \sum_{j=1}^{M} \frac{[c]\{\phi_j\}\{\phi_j\}^T[b]}{s^2 + \omega_j^2} \{u(s)\} + \sum_{j=M+1}^{N} \frac{[c]\{\phi_j\}\{\phi_j\}^T[b]}{s^2 + \omega_j^2} \{u(s)\}. \tag{4.9.18}$$

The second sum in Eq. (4.9.18) is the residual error of modal truncation. Recognizing that the modal frequencies, $\omega_j$, are much higher than the excitation frequencies of interest, for $M + 1 \leq j \leq N$, this sum can be approximated by its static (zero-frequency) response, called the *static correction*.

$$\{y(s)\} \simeq \sum_{j=1}^{M} \frac{[c]\{\phi_j\}\{\phi_j\}^T[b]}{s^2 + \omega_j^2} \{u(s)\} + \sum_{j=M+1}^{N} \frac{[c]\{\phi_j\}\{\phi_j\}^T[b]}{\omega_j^2} \{u(s)\}$$

$$= [c][\Phi_R]\left[s^2 + \lceil\Omega_R^2\rfloor\right]^{-1}[\Phi_R]^T[b]\{u(s)\} + [c][\Phi_D]\lceil\Omega_D^2\rfloor^{-1}[\Phi_D]^T[b]\{u(s)\}. \tag{4.9.19}$$

Here, $[\Phi_D]$ collects the truncated or discarded mode vectors and $\lceil\Omega_D^2\rfloor$ the corresponding squared natural frequencies. Note that the second sum in Eq. (4.9.19) relates the input to the output with no dynamics.

For large FE models, determining this static correction should not require having to calculate all the modes of the full model, which could be computationally prohibitive. Consider the static version of Eq. (4.9.1):

$$[K]\{q\} = \{Q\} = [b]\{u\}. \tag{4.9.20}$$

For a structure having no rigid-body modes (all $\omega_j > 0$), the *attachment modes* represent static deflections due to unit loads at each input $\{u\}$. The attachment modes, columns of the matrix $[T_A]$, can be determined from

$$[K][T_A] = [b]$$

$$[T_A] = [K]^{-1}[b] = \sum_{j=1}^{N} \frac{\{\phi_j\}\{\phi_j\}^T[b]}{\omega_j^2}. \tag{4.9.21}$$

The *residual attachment modes* remove the contributions to $[T_A]$ from the $M$ retained modes:

$$[T_{AR}] = [K]^{-1}[b] - \sum_{j=1}^{M \ll N} \frac{\{\phi_j\}\{\phi_j\}^T[b]}{\omega_j^2}$$

$$[T_{AR}] = [K]^{-1}[b] - [\Phi_R]\lceil\Omega_R^2\rfloor^{-1}[\Phi_R]^T[b]. \tag{4.9.22}$$

With this result, Eq. (4.9.19) becomes

$$\{y(s)\} \simeq [c][\Phi_R]\left[s^2 + \lceil\Omega_R^2\rfloor\right]^{-1}[\Phi_R]^T[b]\{u(s)\} + [c][T_{AR}]\{u(s)\}, \tag{4.9.23}$$

where the second term is the **static correction**.

### 4.9.4 Static Condensation

This approach to model reduction was first developed for statics problems, then applied to dynamics (Guyan, 1965). The global equations of motion, Eq. (4.9.1) and generalized coordinates $\{q\}$ may be *partitioned* with respect to those which are forced (usually exterior), $\{q_F\}$, and those which are not (usually interior), $\{q_C\}$:

$$[M]\{\ddot{q}\} + [K]\{q\} = \{Q(t)\}$$

$$\begin{bmatrix} [M_{FF}] & [M_{FC}] \\ [M_{CF}] & [M_{CC}] \end{bmatrix}\begin{Bmatrix} \{\ddot{q}_F\} \\ \{\ddot{q}_C\} \end{Bmatrix} + \begin{bmatrix} [K_{FF}] & [K_{FC}] \\ [K_{CF}] & [K_{CC}] \end{bmatrix}\begin{Bmatrix} \{q_F\} \\ \{q_C\} \end{Bmatrix} = \begin{Bmatrix} \{Q_F\} \\ \{0\} \end{Bmatrix}. \tag{4.9.24}$$

Now, assume that the *condensed* coordinates, $\{q_C\}$, are linearly related to the *forced* coordinates, $\{q_F\}$, and use the second row of Eq. (4.9.24), ignoring the acceleration terms, to find

$$\{q_C\} = -[K_{CC}]^{-1}[K_{CF}]\{q_F\}. \tag{4.9.25}$$

Note that Eq. (4.9.25) can be recast as a transformation to a reduced set of coordinates:

$$\{q\} = \begin{Bmatrix} \{q_F\} \\ \{q_C\} \end{Bmatrix} = \begin{bmatrix} [I] \\ -[K_{CC}]^{-1}[K_{CF}] \end{bmatrix}\{q_F\}$$

$$= [T_G]\{q_F\}. \tag{4.9.26}$$

This **static condensation** is also known as *Guyan reduction*. Using this result, Eq. (4.9.24) becomes

$$\left[[M_{FF}] - [M_{FC}][K_{CC}]^{-1}[K_{CF}]\right]\{\ddot{q}_F\} + \left[[K_{FF}] - [K_{FC}][K_{CC}]^{-1}[K_{CF}]\right]\{q_F\} = \{Q_F\} \tag{4.9.27}$$

Because this model reduction approach was initially developed for statics problems, it does not assure validity for dynamic response over a particular frequency range.

### 4.9.5  Craig–Bampton Method

In order to better control the frequency range over which a statically-reduced model is accurate, Hurty (1965) and Craig and Bampton (1968) introduced *fixed interface modes* to supplement the static reduction basis. These fixed interface modes are normal modes of the structure under the conditions $\{q_F\} = \{0\}$ and capture some dynamics of the initially condensed coordinates. This method of analysis is also suitable for complex structures that can be divided into interconnected components.

With $\{q_F\} = \{0\}$ (like a boundary condition), Eq. (4.9.24) simplifies to

$$[M_{CC}]\{\ddot{q}_C\} + [K_{CC}]\{q_C\} = \{0\}. \tag{4.9.28}$$

The solution to the eigenvalue problem corresponding to Eq. (4.9.28) yields a set of modal vectors for the fixed interface modes that can be collected as columns of $[\Phi_C]$, usually in order of increasing modal frequency. The number of fixed interface modes retained in $[\Phi_C]$ depends on the response frequency range of interest.

The Craig–Bampton model reduction basis then has the form:

$$\{q\} = \left\{ \begin{array}{c} \{q_F\} \\ \{q_C\} \end{array} \right\} = \left[ \begin{array}{cc} [I] & [0] \\ -[K_{CC}]^{-1}[K_{CF}] & [\Phi_C] \end{array} \right] \left\{ \begin{array}{c} \{q_F\} \\ \{q_C^m\} \end{array} \right\}$$

$$= [T_{CB}] \left\{ \begin{array}{c} \{q_F\} \\ \{q_C^m\} \end{array} \right\}, \tag{4.9.29}$$

where the coordinates $\{q_C^m\}$ correspond to the fixed interface modes in $[\Phi_C]$.

Using this result, Eq. (4.9.24) becomes

$$[T_{CB}]^T[M][T_{CB}] \left\{ \begin{array}{c} \{\ddot{q}_F\} \\ \{\ddot{q}_C^m\} \end{array} \right\} + [T_{CB}]^T[K][T_{CB}] \left\{ \begin{array}{c} \{q_F\} \\ \{q_C^m\} \end{array} \right\} = [T_{CB}]^T \left\{ \begin{array}{c} \{Q_F\} \\ \{0\} \end{array} \right\}, \tag{4.9.30}$$

Alternative reduction bases, *for example, loaded interface modes* rather than fixed interface modes, may be developed that are consistent with the main goal of the Craig–Bampton approach, namely the more accurate prediction of structural dynamic response (Craig, 2000). These methods launched a family of modeling approaches known as Component Mode Synthesis.

## 4.10  Component Mode Synthesis

**Substructures** are parts of a global structural model that, joined together, represent the entire structure. Substructures interact with adjacent substructures via **interfaces**. **Component mode synthesis** is a process for constructing a global dynamic model from multiple substructure models, sometimes also called *superelements* (de Klerk et al., 2008). In general, superelements are reduced models, so that the response at all coordinates is described by a linear combination of various "modes" (or *basis vectors*) and associated generalized coordinates. Some approaches to CMS consider separate physical interface structures that connect individual substructures to each other; this results in a model coupled by the dynamic stiffnesses of the interfaces (Farhat and Geradin, 1994).

When substructures are directly coupled, two conditions must always be satisfied at interfaces, regardless of the coupling method used:

(1) *Compatibility.* Displacements of points on an interface shared by substructures are the same.
(2) *Equilibrium.* Internal structural forces on opposite sides of an interface, at points on the interface shared by substructures, are self-equilibrated (equal and opposite).

### 4.10.1 Model Assembly

The process of assembling a global FE model from individual element models illustrates the general considerations in CMS. Equilibrium of individual elements is represented by

$$[M_i]\{\ddot{q}_i\} + [K_i]\{q_i\} = \{Q\} = \left\{Q_i^E\right\} + \left\{Q_i^I\right\} \qquad (4.10.1)$$

for each element $i$, where $\{Q_i^E\}$ are external forces acting on element $i$, and $\{Q_i^I\}$ are internal interface forces from adjacent elements. All of the element equations of motion can be collected as

$$[M]\{\ddot{q}\} + [K]\{q\} = \left\{Q^E\right\} + \left\{Q^I\right\}, \qquad (4.10.2)$$

where $\{q\}$ here is the collection of all element *local* coordinates.

For illustration, consider a structure modeled using three rod finite elements, as shown in Figure 4.25.

**Figure 4.25** Structure modeled using three rod elements

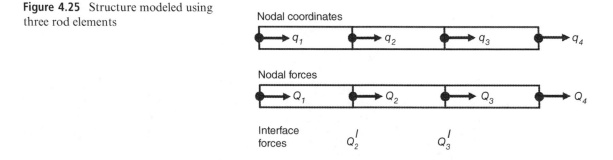

Nodal coordinates

Nodal forces

Interface forces

This FE model has four global coordinates that represent the axial displacements at nodes, along with four corresponding nodal forces associated with external loads. The interface (internal) forces that act at node 2 (between elements 1 and 2) and at node 3 (between elements 2 and 3) are labeled $Q_2^I$ and $Q_3^I$, respectively.

Figure 4.26 shows the local coordinates associated with individual elements. The local forces, not shown, are labeled consistently. Note that the local nodal forces represent both distributed and discrete external forces that act on the structure, as well as internal interface forces that arise as the result of interactions with adjacent elements. The equation of motion specialized from Eq. (4.10.1) for each these elements "$i$" is

**Figure 4.26**
Individual rod
elements

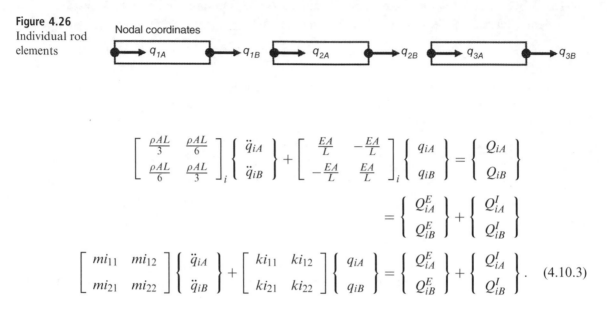

$$\left[ \begin{array}{cc} \frac{\rho A L}{3} & \frac{\rho A L}{6} \\ \frac{\rho A L}{6} & \frac{\rho A L}{3} \end{array} \right]_i \left\{ \begin{array}{c} \ddot{q}_{iA} \\ \ddot{q}_{iB} \end{array} \right\} + \left[ \begin{array}{cc} \frac{EA}{L} & -\frac{EA}{L} \\ -\frac{EA}{L} & \frac{EA}{L} \end{array} \right]_i \left\{ \begin{array}{c} q_{iA} \\ q_{iB} \end{array} \right\} = \left\{ \begin{array}{c} Q_{iA} \\ Q_{iB} \end{array} \right\}$$

$$= \left\{ \begin{array}{c} Q_{iA}^E \\ Q_{iB}^E \end{array} \right\} + \left\{ \begin{array}{c} Q_{iA}^I \\ Q_{iB}^I \end{array} \right\}$$

$$\left[ \begin{array}{cc} mi_{11} & mi_{12} \\ mi_{21} & mi_{22} \end{array} \right] \left\{ \begin{array}{c} \ddot{q}_{iA} \\ \ddot{q}_{iB} \end{array} \right\} + \left[ \begin{array}{cc} ki_{11} & ki_{12} \\ ki_{21} & ki_{22} \end{array} \right] \left\{ \begin{array}{c} q_{iA} \\ q_{iB} \end{array} \right\} = \left\{ \begin{array}{c} Q_{iA}^E \\ Q_{iB}^E \end{array} \right\} + \left\{ \begin{array}{c} Q_{iA}^I \\ Q_{iB}^I \end{array} \right\}. \quad (4.10.3)$$

Collected together, as in Eq. (4.10.2), these are

$$\left[ \begin{array}{cccccc} m1_{11} & m1_{12} & 0 & 0 & 0 & 0 \\ m1_{21} & m1_{22} & 0 & 0 & 0 & 0 \\ 0 & 0 & m2_{11} & m2_{12} & 0 & 0 \\ 0 & 0 & m2_{21} & m2_{22} & 0 & 0 \\ 0 & 0 & 0 & 0 & m3_{11} & m3_{12} \\ 0 & 0 & 0 & 0 & m3_{21} & m3_{22} \end{array} \right] \left\{ \begin{array}{c} \ddot{q}_{1A} \\ \ddot{q}_{1B} \\ \ddot{q}_{2A} \\ \ddot{q}_{2B} \\ \ddot{q}_{3A} \\ \ddot{q}_{3B} \end{array} \right\}$$

$$+ \left[ \begin{array}{cccccc} k1_{11} & k1_{12} & 0 & 0 & 0 & 0 \\ k1_{21} & k1_{22} & 0 & 0 & 0 & 0 \\ 0 & 0 & k2_{11} & k2_{12} & 0 & 0 \\ 0 & 0 & k2_{21} & k2_{22} & 0 & 0 \\ 0 & 0 & 0 & 0 & k3_{11} & k3_{12} \\ 0 & 0 & 0 & 0 & k3_{21} & k3_{22} \end{array} \right] \left\{ \begin{array}{c} q_{1A} \\ q_{1B} \\ q_{2A} \\ q_{2B} \\ q_{3A} \\ q_{3B} \end{array} \right\} = \left\{ \begin{array}{c} Q_{1A}^E \\ Q_{1B}^E \\ Q_{2A}^E \\ Q_{2B}^E \\ Q_{3A}^E \\ Q_{3B}^E \end{array} \right\} + \left\{ \begin{array}{c} Q_{1A}^I \\ Q_{1B}^I \\ Q_{2A}^I \\ Q_{2B}^I \\ Q_{3A}^I \\ Q_{3B}^I \end{array} \right\}$$

$$(4.10.4)$$

Compatibility of interface displacements is represented as

$$[B]\{q\} = \{0\}. \quad (4.10.5)$$

If displacements of connecting elements at common interfaces have the same direction, the **compatibility matrix** $[B]$ is a signed Boolean matrix–but this is not so in general. For the three-rod-element example, where $\{q_{1B}\} = \{q_{2A}\}$ and $\{q_{2B}\} = \{q_{3A}\}$, Eq. (4.10.5) is

$$\begin{bmatrix} 0 & +1 & -1 & 0 & 0 & 0 \\ 0 & 0 & 0 & +1 & -1 & 0 \end{bmatrix} \begin{Bmatrix} q_{1A} \\ q_{1B} \\ q_{2A} \\ q_{2B} \\ q_{3A} \\ q_{3B} \end{Bmatrix} = \begin{Bmatrix} 0 \\ 0 \end{Bmatrix}. \qquad (4.10.6)$$

A unique set of global generalized coordinates, $\{q_G\}$, can be defined with relation to the collection of element coordinates, $\{q\}$, using a **global transformation matrix** $[L]$. Again, if local coordinates are aligned with the global system, $[L]$ may be Boolean.

$$\{q\} = [L]\{q_G\}. \qquad (4.10.7)$$

For the three-rod-element example, Eq. (4.10.7) is

$$\begin{Bmatrix} q_{1A} \\ q_{1B} \\ q_{2A} \\ q_{2B} \\ q_{3A} \\ q_{3B} \end{Bmatrix} = \begin{bmatrix} 1 & 0 & 0 & 0 \\ 0 & 1 & 0 & 0 \\ 0 & 1 & 0 & 0 \\ 0 & 0 & 1 & 0 \\ 0 & 0 & 1 & 0 \\ 0 & 0 & 0 & 1 \end{bmatrix} \begin{Bmatrix} q_1 \\ q_2 \\ q_3 \\ q_4 \end{Bmatrix}. \qquad (4.10.8)$$

Equilibrium at element interfaces is obtained by summing internal forces at all unique global nodes:

$$[L]^T \{Q^I\} = \{0\}. \qquad (4.10.9)$$

For the three-rod-element example, Eq. (4.10.9) is

$$\begin{bmatrix} 1 & 0 & 0 & 0 & 0 & 0 \\ 0 & 1 & 1 & 0 & 0 & 0 \\ 0 & 0 & 0 & 1 & 1 & 0 \\ 0 & 0 & 0 & 0 & 0 & 1 \end{bmatrix} \begin{Bmatrix} Q^I_{1A} \\ Q^I_{1B} \\ Q^I_{2A} \\ Q^I_{2B} \\ Q^I_{3A} \\ Q^I_{3B} \end{Bmatrix} = \begin{Bmatrix} 0 \\ 0 \\ 0 \\ 0 \end{Bmatrix}. \qquad (4.10.10)$$

So, interface equilibrium requires that $Q_{1A} = 0$, $Q_{1B} = -Q_{2A}$, $Q_{2B} = -Q_{3A}$, and $Q_{3B} = 0$, as expected.

In light of Eqs. (4.10.5) and (4.10.9), which must hold for any set of global coordinates,

$$[B][L]\{q_G\} = \{0\} \quad \forall \quad \{q_G\}$$
$$[B][L] = [0] \qquad (4.10.11)$$

a result which is readily verified for the three-rod-element example.

A complete system model is described by the collection of equations for element equilibrium, Eq. (4.10.2), interface displacement compatibility, Eq. (4.10.5), and interface equilibrium, Eq. (4.10.9). This applies equally well to substructures as to individual finite elements.

## 4.10.2 Primal Assembly

**Primal assembly** of the global system equations of motion proceeds by substituting Eq. (4.10.7) into Eq. (4.10.2) to express the equations in terms of global coordinates:

$$[M][L]\{\ddot{q}_G\} + [K][L]\{q_G\} = \left\{Q^E\right\} + \left\{Q^I\right\}. \tag{4.10.12}$$

Note that global compatibility, Eq. (4.10.5) is automatically satisfied. To eliminate the interface forces per Eq. (4.10.9), pre-multiplication by $[L]^T$ yields the *primal* global equations of motion.

$$[L]^T[M][L]\{\ddot{q}_G\} + [L]^T[K][L]\{q_G\} = [L]^T\left\{Q^E\right\}$$
$$[M_G]\{\ddot{q}_G\} + [K_G]\{q_G\} = \{Q_G\}. \tag{4.10.13}$$

Note that the global mass and stiffness matrices, $[M_G]$ and $[K_G]$, are square and symmetric. This is the familiar result of global matrix assembly using the FE method.

For the three-rod-element example, the assembled global mass matrix is given by

$$[M_G] = \begin{bmatrix} m1_{11} & m1_{12} & 0 & 0 \\ m1_{21} & m1_{22}+m2_{11} & m2_{12} & 0 \\ 0 & m2_{21} & m2_{22}+m3_{11} & m3_{12} \\ 0 & 0 & m3_{21} & m3_{22} \end{bmatrix}. \tag{4.10.14}$$

## 4.10.3 Dual Assembly

An alternative approach to assembly of the system model *explicitly retains the interface forces* rather than eliminating them. This is useful for some solution approaches and when those forces are of intrinsic interest. Self-equilibrating interface forces $\{\lambda\}$ come in equal and opposite pairs so that the interface force vector $\{Q^I\}$ can be represented as

$$\left\{Q^I\right\} = -[B]^T\{\lambda\}. \tag{4.10.15}$$

Then, in light of Eq. (4.10.11), Eq. (4.10.9) for interface equilibrium is automatically satisfied, as $[L]^T[B]^T = [[B][L]]^T = [0]$. With Eq. (4.10.5) for interface compatibility, this yields the **dual assembly** form of the governing equations:

$$[M]\{\ddot{q}\} + [K]\{q\} + [B]^T\{\lambda\} = \left\{Q^E\right\}$$
$$[B]\{q\} = \{0\} \tag{4.10.16}$$

or, in block form:

$$\begin{bmatrix} [M] & \\ & \end{bmatrix} \begin{Bmatrix} \{\ddot{q}(t)\} \\ \end{Bmatrix} + \begin{bmatrix} [K] & [B]^T \\ [B] & [0] \end{bmatrix} \begin{Bmatrix} \{q(t)\} \\ \{\lambda(t)\} \end{Bmatrix} = \begin{Bmatrix} \{Q^E(t)\} \\ \{0\} \end{Bmatrix}. \qquad (4.10.17)$$

Note that $\{q\}$ is the collection of all local coordinates, as in Eqs. (4.10.2) and (4.10.14) – not a set of global coordinates, $\{q_G\}$ used in Eq. (4.10.13). A solution process determines responses for $\{q(t)\}$ and $\{\lambda(t)\}$ concurrently.

## 4.10.4 Dual Coupling in Reduced Spaces

One of the powerful features of CMS is the ability to model each substructure separately prior to coupling the substructure models in a global model. This ability can be particularly useful when only a part of the system changes from case to case, so that most of the global model need only be developed once. A substructure might initially be modeled using a large FE model, but its order would typically be reduced prior to coupling in the global model.

The response of substructure "$s$" is modeled using a reduced set of basis vectors ("modes"), represented as columns of a reduction matrix $[R^{(s)}]$.

$$\left\{ q^{(s)} \right\} \simeq \left[ R^{(s)} \right] \left\{ q_R^{(s)} \right\}. \qquad (4.10.18)$$

Depending on the situation, basis vectors determined under different assumed boundary and loading conditions might be used (e.g. eigenvectors or static deflections determined with free-free or fixed-interface boundaries).

The *full* equation of motion for substructure $s$ is

$$\left[ M^{(s)} \right] \left\{ \ddot{q}^{(s)} \right\} + \left[ K^{(s)} \right] \left\{ q^{(s)} \right\} = \left\{ Q^{E(s)} \right\} + \left\{ Q^{I(s)} \right\}. \qquad (4.10.19)$$

Approximating this in a reduced space defined by Eq. (4.10.18) introduces an error represented as a residual force, $\{Q_{\text{res}}^{(s)}\}$:

$$\left[ M^{(s)} \right] \left[ R^{(s)} \right] \left\{ \ddot{q}_R^{(s)} \right\} + \left[ K^{(s)} \right] \left[ R^{(s)} \right] \left\{ q_R^{(s)} \right\} = \left\{ Q^{E(s)} \right\} + \left\{ Q^{I(s)} \right\} + \left\{ Q_{\text{res}}^{(s)} \right\}. \qquad (4.10.20)$$

Premultiplying this by a term orthogonal to the residual force yields a reduced equation of motion:

$$\left[ R^{(s)} \right]^T \left[ M^{(s)} \right] \left[ R^{(s)} \right] \left\{ \ddot{q}_R^{(s)} \right\} + \left[ R^{(s)} \right]^T \left[ K^{(s)} \right] \left[ R^{(s)} \right] \left\{ q_R^{(s)} \right\}$$

$$= \left[ R^{(s)} \right]^T \left\{ Q^{E(s)} \right\} + \left[ R^{(s)} \right]^T \left\{ Q^{I(s)} \right\}$$

$$\left[ M_R^{(s)} \right] \left\{ \ddot{q}_R^{(s)} \right\} + \left[ K_R^{(s)} \right] \left\{ q_R^{(s)} \right\} = \left\{ Q_R^{E(s)} \right\} + \left\{ Q_R^{I(s)} \right\}. \qquad (4.10.21)$$

The compatibility condition in the reduced space is

$$\left[ B^{(s)} \right] \left[ R^{(s)} \right] \left\{ q_R^{(s)} \right\} = \left[ B_R^{(s)} \right] \left\{ q_R^{(s)} \right\} = \{0\} \tag{4.10.22}$$

and the **reduced substructure model** can be written in **dual assembly form** as

$$\left[ \begin{matrix} \left[ M_R^{(s)} \right] & \end{matrix} \right] \left\{ \begin{matrix} \left\{ \ddot{q}_R^{(s)} \right\} \end{matrix} \right\} + \left[ \begin{matrix} K_R^{(s)} & \left[ B_R^{(s)} \right]^T \\ B_R^{(s)} & [0] \end{matrix} \right] \left\{ \begin{matrix} \left\{ q_R^{(s)} \right\} \\ \left\{ \lambda_R^{(s)} \right\} \end{matrix} \right\} = \left\{ \begin{matrix} \left\{ Q_R^{E(s)} \right\} \\ \{0\} \end{matrix} \right\}. \tag{4.10.23}$$

The reduction space $[R^{(s)}]$ for each *substructure* should be built from vectors that can properly represent the response of the *assembled system* in the required frequency range.

## 4.10.5 Interface Compatibility between Substructures

Enforcing displacement compatibility of reduced substructure models can be challenging. If two substructure FE models are developed to share common nodes at an interface, then reduced models of each are developed as described in Section 4.10.4, but compatibility can be described in terms of the reduced coordinates for each substructure. Then, a global model can be assembled in either **primal form** (using global reduced coordinates and eliminating interface forces) or **dual form** (using substructure reduced coordinates and retaining interface forces).

Let $\{q_G^{(r)}\}$ and $\{q_G^{(s)}\}$ be global coordinates for substructures "$r$" and "$s$." Following Eq. (4.10.5), interface compatibility can be expressed as

$$\left[ B_G^{(rs)} \right] \left\{ \begin{matrix} \left\{ q_G^{(r)} \right\} \\ \left\{ q_G^{(s)} \right\} \end{matrix} \right\} = \left[ \begin{matrix} \left[ B_{G(r)}^{(rs)} \right] & \left[ B_{G(s)}^{(rs)} \right] \end{matrix} \right] \left\{ \begin{matrix} \left\{ q_G^{(r)} \right\} \\ \left\{ q_G^{(s)} \right\} \end{matrix} \right\} = \{0\}. \tag{4.10.24}$$

Note that this interface compatibility condition describes perfectly matched displacements on the interface. Using the substructure reduction spaces, $\{q_G^{(i)}\} = [R_G^{(i)}]\{q_{GR}^{(i)}\}$, this compatibility condition can be expressed in terms of the reduced coordinates:

$$\left[ \begin{matrix} \left[ B_{G(r)}^{(rs)} \right] & \left[ B_{G(s)}^{(rs)} \right] \end{matrix} \right] \left\{ \begin{matrix} \left[ R_G^{(r)} \right] \left\{ q_{GR}^{(r)} \right\} \\ \left[ R_G^{(s)} \right] \left\{ q_{GR}^{(s)} \right\} \end{matrix} \right\} =$$

$$\left[ \begin{matrix} \left[ B_{G(r)}^{(rs)} \right] \left[ R_G^{(r)} \right] & \left[ B_{G(s)}^{(rs)} \right] \left[ R_G^{(s)} \right] \end{matrix} \right] \left\{ \begin{matrix} \left\{ q_{GR}^{(r)} \right\} \\ \left\{ q_{GR}^{(s)} \right\} \end{matrix} \right\} =$$

$$\left[ \begin{matrix} \left[ B_{GR(r)}^{(rs)} \right] & \left[ B_{GR(s)}^{(rs)} \right] \end{matrix} \right] \left\{ \begin{matrix} \left\{ q_{GR}^{(r)} \right\} \\ \left\{ q_{GR}^{(s)} \right\} \end{matrix} \right\} = \{0\}. \tag{4.10.25}$$

### Substructures in Primal Assembly Form

With this result for compatibility of reduced substructure models, the governing equations for two coupled reduced substructures $r$ and $s$ individually obtained in *primal* assembly form (as in Eq. (4.10.13)) are

$$
\begin{bmatrix} \left[M_{GR}^{(r)}\right] & 0 \\ 0 & \left[M_{GR}^{(s)}\right] \end{bmatrix} \left\{ \begin{array}{c} \left\{\ddot{q}_{GR}^{(r)}\right\} \\ \left\{\ddot{q}_{GR}^{(s)}\right\} \end{array} \right\} +
$$

$$
\begin{bmatrix} \left[K_{GR}^{(r)}\right] & 0 & \left[B_{GR(r)}^{(rs)}\right]^T \\ 0 & \left[K_{GR}^{(s)}\right] & \left[B_{GR(s)}^{(rs)}\right]^T \\ \left[B_{GR(r)}^{(rs)}\right] & \left[B_{GR(s)}^{(rs)}\right] & 0 \end{bmatrix} \left\{ \begin{array}{c} \left\{q_{GR}^{(r)}\right\} \\ \left\{q_{GR}^{(s)}\right\} \\ \left\{\lambda_{GR}^{(rs)}\right\} \end{array} \right\} = \left\{ \begin{array}{c} \left\{Q_{GR}^{(r)}\right\} \\ \left\{Q_{GR}^{(s)}\right\} \\ \{0\} \end{array} \right\}. \tag{4.10.26}
$$

Since the interface motion of each substructure is now represented by a relatively small number of "modes" – modes that are different for each substructure – this could result in **locking**, or an unintentionally stiff model (Balmes, 1996). This can be addressed by **enriching** the reduction subspaces with additional basis vectors – such as displacements due to static forces applied at interface nodes, or by weakening the interface compatibility requirement and enforcing it approximately.

## Example: Clamped-pinned Beam

Consider the beam structure shown in Figure 4.27. It is fixed at the left end and simply supported at the right end. Component mode synthesis will be used to estimate the first few natural frequencies of transverse vibration of this structure.

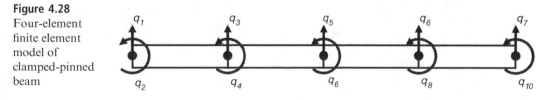

**Figure 4.27** Clamped-pinned beam

To provide a baseline for comparison, a FE model is first developed for the overall structure. To keep matrix sizes reasonable for display on the page, four elements are used, as shown in Figure 4.28. For simplicity, the total length of the beam is $L_{\text{tot}} = 1$, the flexural rigidity is $EA = 1$, and the mass per unit length is $\rho A = 1$.

**Figure 4.28**
Four-element
finite element
model of
clamped-pinned
beam

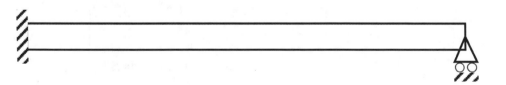

Eigenvalue analysis of this model provides estimates for the first three natural frequencies of vibration as

$$\underbrace{\omega_{ref}}_{4 \text{ elements}} \simeq \begin{bmatrix} 15.428 & 50.277 & 106.60 \end{bmatrix}.$$

Note that the precise numerical values of the first three natural frequencies of this beam structure are $\omega = [15.418, 49.965, 104.25]$, so the four-element FE estimates are high by $0.06\%, 0.62\%$, and $0.97\%$, respectively.

Now, the overall structure is modeled using two substructures, defined by dividing the overall beam structure in half. As shown in Figure 4.29, *Substructure 1* is the left side of the structure, with fixed-free boundary conditions, and *Substructure 2* is the right side, with free-pinned boundary conditions. Again, to keep matrix sizes reasonable for display, two finite elements are used to model each substructure.

**Figure 4.29** Clamped-pinned beam modeled using two substructures

A normal mode (eigenvalue) analysis is performed for each of the substructures to yield a set of natural frequencies and corresponding eigenvectors (discrete mode shapes). The first substructure has four normal modes and corresponding mass-normalized eigenvectors:

$$\underbrace{\omega_{G1}}_{\text{Substructure 1; 2 elements}} \simeq \begin{bmatrix} 14.071 & 88.886 & 300.63 & 872.55 \end{bmatrix}$$

$$[\Phi_{G1}] = \begin{bmatrix} 0.9612 & 2.0617 & 0.3231 & 1.3507 \\ 6.5856 & -2.4818 & -48.587 & 55.525 \\ 2.8312 & -2.8563 & 3.1767 & 5.3345 \\ 7.7945 & -27.503 & 61.274 & 206.23 \end{bmatrix}.$$

The second substructure has five normal modes and corresponding eigenvectors:

$$\underbrace{\omega_{G2}}_{\text{Substructure 2; 2 elements}} \simeq \begin{bmatrix} 0 & 62.057 & 223.29 & 542.24 & 996.92 \end{bmatrix}$$

$$[\Phi_{G2}] = \begin{bmatrix} 2.4495 & 2.8603 & -2.8916 & -3.2133 & -4.7798 \\ -4.8990 & -22.5393 & 45.803 & 93.923 & 196.21 \\ 1.2247 & -1.6788 & 0.6857 & -1.5959 & -0.6156 \\ -4.8990 & -7.6502 & -33.577 & -30.351 & 80.252 \\ -4.8990 & 15.438 & 37.701 & -82.862 & 62.937 \end{bmatrix}.$$

Displacement compatibility requires that the displacement and slope at the substructure connection point be the same on both sides. For this example, Eq. (4.10.24) is

$$\left[\, [B_{G1}]\quad [B_{G2}]\,\right]\left\{\begin{array}{c} \{q_{G1}\} \\ \{q_{G2}\} \end{array}\right\} = \{0\}$$

$$\left[\begin{bmatrix} 0 & 0 & 1 & 0 \\ 0 & 0 & 0 & 1 \end{bmatrix}\quad \begin{bmatrix} -1 & 0 & 0 & 0 & 0 \\ 0 & -1 & 0 & 0 & 0 \end{bmatrix}\right]\left\{\begin{array}{c} \{q_{G1}\} \\ \{q_{G2}\} \end{array}\right\} = \left\{\begin{array}{c} 0 \\ 0 \end{array}\right\}$$

To enforce compatibility of displacement and slope, each substructure must be represented using at least two modes. Using this minimum of four modes, minus the two compatibility constraints, would yield estimates for two natural frequencies of the joined structure. To obtain estimates of three natural frequencies, at least five substructure modes must be used.

For this example, retain the two lowest frequency modes for Substructure 1 and the three lowest frequency modes for Substructure 2. These modes are the first columns of $[\Phi_{G1}]$ and $[\Phi_{G2}]$ above, the matrices $[R_{G1}]$ and $[R_{G2}]$ are subsets of the corresponding $[\Phi]$, and $\{q_{GR1}\}$ and $\{q_{GR2}\}$ are the corresponding modal coefficients.

Following Eq. (4.10.25), the compatibility conditions in the reduced subspace are

$$\left[\, [B_{G1}]\quad [B_{G2}]\,\right]\left\{\begin{array}{c} [R_{G1}]\{q_{GR1}\} \\ [R_{G2}]\{q_{GR2}\} \end{array}\right\} =$$

$$\left[\, [B_{GR1}]\quad [B_{GR2}]\,\right]\left\{\begin{array}{c} \{q_{GR1}\} \\ \{q_{GR2}\} \end{array}\right\} =$$

$$\left[\begin{bmatrix} 2.8312 & -2.8563 \\ 7.7945 & -27.503 \end{bmatrix}\quad \begin{bmatrix} -2.4495 & -2.8603 & 2.8916 \\ 4.8990 & 22.539 & -45.803 \end{bmatrix}\right]\left\{\begin{array}{c} \{q_{GR1}\} \\ \{q_{GR2}\} \end{array}\right\} = \left\{\begin{array}{c} 0 \\ 0 \end{array}\right\}.$$

The substructure (component) mass and stiffness matrices are readily determined as $[R_G]^T[M_G][R_G]$ and $[R_G]^T[K_G][R_G]$ and, since normal modes are used as the reduced space, are diagonal. The coupled equations of motion can be written in the form shown in Eq. (4.10.26) and a corresponding eigenvalue problem is addressed. This yields the following estimates for the first three natural frequencies of the coupled structure:

$$\underbrace{\omega_{\text{coupled}}}_{\text{S1: 2 modes | S2: 3 modes}} \simeq \left[\, 17.766\quad 52.945\quad 139.31\,\right].$$

These are higher than those of the reference full four-element model, $\omega_{ref}$, by 15.2%, 5.3%, and 30.68%, respectively.

The **modal assurance criterion** (MAC) provides a way to compare mass-normalized vectors. The triple product $\{\phi_i\}^T[M]\{\phi_j\}$ would equal 1 if the two vectors $\{\phi_i\}$ and $\{\phi_j\}$ are the same and 0 if they are orthogonal with respect to the mass matrix. The closer the value is to 1, the more similar are the two vectors. In this example, comparing the mass orthogonality of the three corresponding eigenvectors to those of the full model yields

$$[\phi_{\text{CMS}}]^T[M][\phi_{\text{ref}}] = \begin{bmatrix} 0.9963 & -0.0586 & 0.0583 \\ 0.0496 & 0.9893 & 0.1223 \\ -0.0442 & -0.0794 & 0.8786 \end{bmatrix}$$

and shows that the third eigenvector is poorly represented, consistent with the frequency results.

Including all available modes–four for Substructure 1 and five for Substructure 2–recovers the reference values exactly. Including all available modes but one for each substructure leads to the following estimates for the first three natural frequencies of the coupled structure:

$$\underbrace{\omega_{\text{coupled}}}_{\text{S1: 3 modes | S2: 4 modes}} \simeq \begin{bmatrix} 16.592 & 51.319 & 119.93 \end{bmatrix},$$

which are higher than those of the reference model by 7.5%, 2.1%, and 12.5%.

The reduced subspaces can also be enriched using "modes" other than nominal normal modes. These could include normal modes obtained under a variety of boundary conditions as well as static deflections obtained under a variety of boundary conditions and forcing. For example, consider the two substructures under the static load conditions shown in Figure 4.30. For these two cases, the resulting mass-normalized static displacement vectors are

Substructure 1    Substructure 2

**Figure 4.30** Substructure static load cases to generate additional "modes"

$$[\Phi_{S1}] = \begin{bmatrix} 0 \\ 0 \\ -4.4721 \\ -35.777 \end{bmatrix} \qquad [\Phi_{S2}] = \begin{bmatrix} 0 \\ 20.494 \\ 1.9213 \\ -2.5617 \\ -10.247 \end{bmatrix}.$$

These modes are added as columns of $[R_{G1}]$ and $[R_{G2}]$ respectively, and the analysis proceeds as before. Using two normal modes and one static mode for Substructure 1 and three normal modes and one static mode for Substructure 2 yields the following improved estimates for the first three natural frequencies of the coupled structure:

$$\underbrace{\omega_{\text{coupled}}}_{\text{S1: 2 normal, 1 static | S2: 3 normal, 1 static}} \simeq \begin{bmatrix} 15.595 & 50.415 & 107.95 \end{bmatrix}.$$

These are higher than those of the reference model by 1.08%, 0.27%, 1.27%, respectively – a very good improvement, and better than that obtained by including additional normal modes. The corresponding MAC of the eigenvectors with those of the full model is

$$[\phi_{\text{CMS}}]^T[M][\phi_{\text{ref}}] = \begin{bmatrix} 1.0000 & -0.0037 & 0.0035 \\ 0.0036 & 0.9999 & 0.0070 \\ -0.0034 & -0.0067 & 0.9992 \end{bmatrix},$$

which shows good improvement in the quality of the eigenvectors as well as the natural frequencies.

## Primal Assembly Form for Joined Substructures Using Global Coordinates

While it is not necessary, if a set of global coordinates for the two coupled reduced substructures is defined such that

$$\left\{ \begin{Bmatrix} q_{GR}^{(r)} \\ q_{GR}^{(s)} \end{Bmatrix} \right\} = \left[ L_{GR}^{(rs)} \right] \left\{ q_{GR}^{(rs)} \right\} = \begin{bmatrix} L_{GR(r)}^{(rs)} \\ L_{GR(s)}^{(rs)} \end{bmatrix} \left\{ q_{GR}^{(rs)} \right\} \tag{4.10.27}$$

then, in light of Eq. (4.10.11), the interface forces can be eliminated from Eq. (4.10.26) to yield

$$\begin{bmatrix} L_{GR(r)}^{(rs)} \\ L_{GR(s)}^{(rs)} \end{bmatrix}^T \begin{bmatrix} M_{GR}^{(r)} & 0 \\ 0 & M_{GR}^{(s)} \end{bmatrix} \begin{bmatrix} L_{GR(r)}^{(rs)} \\ L_{GR(s)}^{(rs)} \end{bmatrix} \left\{ \ddot{q}_{GR}^{(rs)} \right\} +$$

$$\begin{bmatrix} L_{GR(r)}^{(rs)} \\ L_{GR(s)}^{(rs)} \end{bmatrix}^T \begin{bmatrix} K_{GR}^{(r)} & 0 \\ 0 & K_{GR}^{(s)} \end{bmatrix} \begin{bmatrix} L_{GR(r)}^{(rs)} \\ L_{GR(s)}^{(rs)} \end{bmatrix} \left\{ q_{GR}^{(rs)} \right\} =$$

$$\begin{bmatrix} L_{GR(r)}^{(rs)} \\ L_{GR(s)}^{(rs)} \end{bmatrix}^T \left\{ \begin{Bmatrix} Q_{GR}^{(r)} \\ Q_{GR}^{(s)} \end{Bmatrix} \right\} - \begin{bmatrix} L_{GR(r)}^{(rs)} \\ L_{GR(s)}^{(rs)} \end{bmatrix}^T \begin{bmatrix} B_{GR(r)}^{(rs)T} \\ B_{GR(s)}^{(rs)T} \end{bmatrix} \left\{ \lambda_{GR}^{(rs)} \right\} =$$

$$\left[ \left[ L_{GR(r)}^{(rs)} \right]^T \left[ M_{GR}^{(r)} \right] \left[ L_{GR(r)}^{(rs)} \right] + \left[ L_{GR(s)}^{(rs)} \right]^T \left[ M_{GR}^{(s)} \right] \left[ L_{GR(s)}^{(rs)} \right] \right] \left\{ \ddot{q}_{GR}^{(rs)} \right\} +$$

$$\left[ \left[ L_{GR(r)}^{(rs)} \right]^T \left[ K_{GR}^{(r)} \right] \left[ L_{GR(r)}^{(rs)} \right] + \left[ L_{GR(s)}^{(rs)} \right]^T \left[ K_{GR}^{(s)} \right] \left[ L_{GR(s)}^{(rs)} \right] \right] \left\{ q_{GR}^{(rs)} \right\} =$$

$$\left\{ \left[ L_{GR(r)}^{(rs)} \right]^T \left\{ Q_{GR}^{(r)} \right\} + \left[ L_{GR(s)}^{(rs)} \right]^T \left\{ Q_{GR}^{(s)} \right\} \right\}$$

$$\left[ M_{GR}^{(rs)} \right] \left\{ \ddot{q}_{GR}^{(rs)} \right\} + \left[ K_{GR}^{(rs)} \right] \left\{ q_{GR}^{(rs)} \right\} = \left\{ Q_{GR}^{(rs)} \right\}. \tag{4.10.28}$$

Note that an independent set of global coordinates can be selected from $\{q_{GR}^{(r)}\}$ and $\{q_{GR}^{(s)}\}$ while treating a number equal to the number of interface constraints $\{\lambda_{GR}^{(rs)}\}$ as *dependent*.

## Substructures in Dual Assembly Form

If the substructure models are obtained individually in *dual* assembly form (as in Eq. (4.10.23)) and then reduced, compatibility can be enforced in a similar manner to yield the coupled reduced global model as

$$
\begin{bmatrix} \left[ M_R^{(r)} \right] & 0 \\ 0 & \left[ M_R^{(s)} \right] \\ & & \end{bmatrix} \left\{ \begin{array}{c} \left\{ \ddot{q}_R^{(r)} \right\} \\ \left\{ \ddot{q}_R^{(s)} \right\} \end{array} \right\} +
$$

$$
\begin{bmatrix} \left[ K_R^{(r)} \right] & 0 & \left[ B_R^{(r)} \right]^T & 0 & \left[ B_{R(r)}^{(rs)} \right]^T \\ 0 & \left[ K_R^{(s)} \right] & 0 & \left[ B_R^{(s)} \right]^T & \left[ B_{R(s)}^{(rs)} \right]^T \\ \left[ B_R^{(r)} \right] & 0 & 0 & 0 & 0 \\ 0 & \left[ B_R^{(s)} \right] & 0 & 0 & 0 \\ \left[ B_{R(r)}^{(rs)} \right] & \left[ B_{R(s)}^{(rs)} \right] & 0 & 0 & 0 \end{bmatrix} \left\{ \begin{array}{c} \left\{ q_R^{(r)} \right\} \\ \left\{ q_R^{(s)} \right\} \\ \left\{ \lambda_R^{(r)} \right\} \\ \left\{ \lambda_R^{(s)} \right\} \\ \left\{ \lambda_R^{(rs)} \right\} \end{array} \right\} =
$$

$$
\left\{ \begin{array}{c} \left\{ Q_R^{E(r)} \right\} \\ \left\{ Q_R^{E(s)} \right\} \\ \{0\} \\ \{0\} \\ \{0\} \end{array} \right\}, \qquad (4.10.29)
$$

where interface compatibility is expressed in terms of the local substructure coordinates $\{q^{(r)}\}$ and $\{q^{(s)}\}$ as

$$
\left[ B^{(rs)} \right] \left\{ \begin{array}{c} \{q^{(r)}\} \\ \{q^{(s)}\} \end{array} \right\} = \left[ \left[ B_{(r)}^{(rs)} \right] \quad \left[ B_{(s)}^{(rs)} \right] \right] \left\{ \begin{array}{c} \{q^{(r)}\} \\ \{q^{(s)}\} \end{array} \right\} = \{0\} \qquad (4.10.30)
$$

and in terms of the reduced coordinates (Eq. (4.10.18)) as

$$
\left[ \left[ B_{(r)}^{(rs)} \right] \quad \left[ B_{(s)}^{(rs)} \right] \right] \left\{ \begin{array}{c} \left[ R^{(r)} \right] \left\{ q_R^{(r)} \right\} \\ \left[ R^{(s)} \right] \left\{ q_R^{(s)} \right\} \end{array} \right\} =
$$

$$
\left[ \left[ B_{R(r)}^{(rs)} \right] \quad \left[ B_{R(s)}^{(rs)} \right] \right] \left\{ \begin{array}{c} \left\{ q_R^{(r)} \right\} \\ \left\{ q_R^{(s)} \right\} \end{array} \right\} = \{0\}. \qquad (4.10.31)
$$

# BIBLIOGRAPHY

Balmes, E. (1996). Use of generalized interface degrees of freedom in component mode synthesis. *Proceedings, International Modal Analysis Conference (IMAC)*, pages 1–7.

Bathe, K. J. (1966). *Finite Element Procedures*. Prentice Hall.

Bathe, K. J. and Wilson, E. L. (1976). *Numerical Methods in Finite Element Analysis*. Prentice Hall.

Bogner, F. K., Fox, R. L., and Schmit, L. A. (1965). The generation of inter-element-compatible stiffness and mass matrices by use of interpolation formulas, pages 397–443. AFFDL-TR-066-80, 1965, *Proceedings of the First Conference on Matrix Methods in Structural Engineering*.

Craig, R., Jr. and Bampton, M. (1968). Coupling of substructures for dynamic analyses. *AIAA Journal*, 6(7): 1313–1319.

Craig, R. R., Jr. (2000). Coupling of substructures for dynamic analyses – an overview. In *41st AIAA Structures, Structural Dynamics, and Materials Conference*. AIAA-2000-1573.

de Klerk, D., Rixen, D. J., and Voormeeren, S. N. (2008). General framework for dynamic substructuring: History, review, and classification of techniques. *AIAA Journal*, 46(5): 1169–1181.

Doyle, J. F. (2015). *Nonlinear Structural Dynamics Using FE Methods*. Cambridge University Press.

Farhat, C. and Geradin, M. (1994). On a component mode synthesis method and its application to incompatible substructures. *Comput. Struct. (UK)*, 51(5): 459–473.

Gallagher, R. H. (1975). *Finite Element Analysis Fundamentals*. Prentice Hall.

Geradin, M. and Rixen, D. (2015). *Mechanical Vibrations Theory and Application to Structural Dynamics*. John Wiley & Sons, 3rd edition.

Guyan, R. J. (1965). Reduction of stiffness and mass matrices. *AIAA Journal*, 3(2): 380–381.

Hughes, P. (1987a). Space structure vibration modes: How many exist? Which ones are important? *IEEE Control Systems Magazine*, 7(1): 22–28.

Hughes, T. (1987b). *The Finite Element Method – Linear Static and Dynamic Finite Element Analysis*. Prentice Hall.

Hurty, W. (1965). Dynamic analysis of structural systems using component modes. AIAA Journal, 3(4): 678–685.

Kapur, K. K. (1966). Vibrations of Timoshenko beams, using finite element approach. *Journal of Acoustical Society of America*, 40(5): 1058–1063.

Lee, A. and Tsuha, W. (1994). Model reduction methodology for articulated, multiflexible body structures. *Journal of Guidance, Control, and Dynamics*, 17(1): 69–75.

Preumont, A. (2011). *Vibration Control of Active Structures*. Springer Dordrecht, 3rd edition.

Schmit, L. A., Bogner, F. K., and Fox, R. L. (1965). Finite deflection structural analysis using plate and shell discrete elements. *AIAA Journal*, 6(5): 781–791.

Shames, I. H. and Dym, C. (1985). *Energy and Finite Element Methods in Structural Mechanics*. McGraw-Hill Book Co., Hemisphere Publishing co-edition.

Skelton, R. E., Hughes, P. C., and Hablani, H. B. (1982). Order reduction for models of space structures using modal cost analysis. *Journal of Guidance, Control, and Dynamics*, 5(4): 351–357.

Tessler, A. and Dong, S. B. (1981). On a hierarchy of conforming Timoshenko beam elements. *Computers and Structures*, 14(3–4): 335–344.

Turner, M. J., Clough, R. W., Martin, H. C., and Topp, L. J. (1956). Stiffness and deflection analysis of complex structures. *Journal of Aeronautical Sciences*, 23(9): 805–823, 854.

Zienkiewicz, O. C. and Taylor, R. L. (1989). *The Finite Element Method, volume 1: Basic Formulation and Linear Problems*. McGraw-Hill Book Co., 4th edition.

## PROBLEMS

1. A FE model for the representation of torsional dynamics is needed, for a **coupled bending-torsion** problem in structural dynamics. The bending degree of freedom is

represented by a conventional beam-bending element that was derived earlier in this chapter. To represent the torsional degree of freedom two candidate elements need to be considered.

**Candidate A**

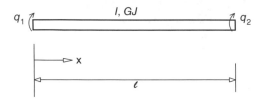

Is a torsional element, having two DOFs, consisting of rotation $q_1$ and $q_2$ at the ends, $GJ$ is the torsional stiffness, and $I$ is the mass polar moment of inertia per unit length. Select the appropriate interpolation function, and derive the consistent mass and stiffness matrix required for the representation of the torsional degree of freedom.

**Candidate B**

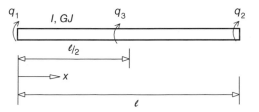

Is a torsional element having three nodal degrees of freedom consisting of $q_1$ and $q_2$, as for **Candidate A** together with an additional node representing the rotation $q_3$ at the midpoint. Select the appropriate interpolation function, and derive the consistent mass and stiffness matrix required for the representation of the torsional degree of freedom. Please be explicit regarding the treatment of the internal degree of freedom.
- Which of the two elements would you select for the treatment of the coupled bending-torsion problem? Justify your answer carefully.
- Calculate the first few torsional modes and frequencies for the torsional vibrations of the fixed end bar shown below, assuming that it consists of three FE. Solve this problem using both candidates A and B, and compare your results with the exact solution that can be found in books on structural dynamics.

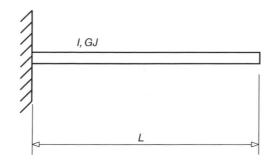

2. Using the FE method, derive the stiffness and consistent mass matrix of a beam on an elastic foundation shown in the figure below. The beam has a square cross section.

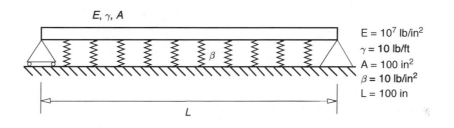

$$E = 10^7 \text{ lb/in}^2$$
$$\gamma = 10 \text{ lb/ft}$$
$$A = 100 \text{ in}^2$$
$$\beta = 10 \text{ lb/in}^2$$
$$L = 100 \text{ in}$$

To derive element properties use geometry given below.

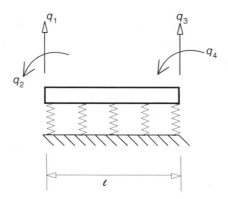

Set up eigenvalue problem in general form when beam is modeled using two of these elements.

(a) Calculate the fundamental mode shape using Raleigh's quotient in discrete form.

(b) Solve the eigenvalue problem using two and four elements and compare your results.

3. Consider the structure shown below, composed of a beam with a tip mass $M$, stiffness $EI$, and distributed mass $m$ per unit length. The beam has a linear spring in bending $k_B$ and a torsional spring $k_T$ at $x = L$.

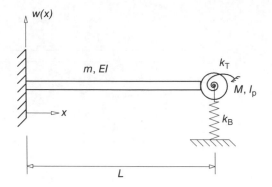

(a) Assume that the motion of the structure can be represented by a distributed coordinate

$$w(x) = \phi(x)q(t),$$

where

$$\phi(x) = \left(\frac{1}{2}\right)\left(\frac{x}{l}\right)^2\left[3 - \left(\frac{x}{l}\right)\right].$$

Using Raleigh's quotient estimate the fundamental frequency of this structural dynamic system.

(b) Modeling the same structure as composed of two beam-type elements, calculate the mode shapes and frequencies of the system. Assume the following numerical values in your calculations

$$E = 10^7 \ \text{lb/in}^2; \quad I = \frac{1250}{12} \ \text{in}^4; \quad l = 100'';$$

$$I_p = \frac{10^3(25)}{2} \ \text{lb.sec}^2.\text{in}; \quad m = 10 \ \text{lb.sec}^2/\text{in}^2;$$

$$k_T = 10^6 \ \text{lb.in/rad}; \quad k_B = 10^5 \ \text{lb/in}; \quad M = 10^3 \ \text{lb.sec}^2/\text{in}$$

(c) Compare the results from (a) and (b).

4. Consider the structural dynamic system shown in the figure. It consists of a combination of several springs, masses, and two beam segments that are described below:
   - Two uniform beam segments of length $l_1$ and $l_2$ having stiffness $EI_1$ and $EI_2$, together with mass per unit length of $m_1$ and $m_2$, respectively. The lower beam segment is cantilevered to the base.
   - Two fairly large masses $M_1$ and $M_2$ are attached to the tip of the beam and at an intermediate position located at $x = l_2$. These masses have polar moments of inertia $I_{p1}$ and $I_{p2}$ respectively about an axis perpendicular to the plane of the paper. *These*

*masses are quite heavy and the effect of gravity on these masses should be incorporated when formulating the structural dynamics problem. On the other hand, the effect of gravity on the distributed masses $m_1$ and $m_2$ can be neglected.*

- Three springs: Two linear springs having spring constants $K_1$ and $K_2$, respectively are located at $x = l_2$ and $x = l_1 + l_2$, respectively as shown in the figure; and a torsional spring with spring constant $K_{T1}$ is located at the top of the beam $x = l_1 + l_2$.
- A time-dependent load $F(t)$ is acting as shown at $x = l_2$.

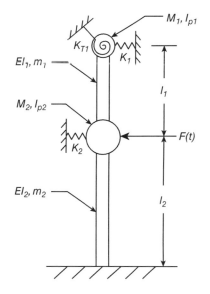

(a) Using the FE method, formulate the matrix equations of motion for this structural dynamic system as it vibrates in the lateral direction under the action of the excitation $F(t)$.

(b) Using the energy from of Rayleigh's quotient, that is,

$$\omega_1^2 \cong \frac{V_{\max}}{T^*_{\max}}$$

find an approximation to the fundamental frequency using an appropriately selected approximation to the fundamental mode shape. Justify the selection of this mode shape.

(c) What is the *minimal* number of DOFs required to model this system, assuming that you model the system using a single beam-type FE for each beam?

(d) Assume that the system has the following physical properties:

$l_1 = 4.0$ ft; cross-sectional area is 1 inch $\times$ 1 inch; $l_2 = 6.0$ ft; cross-sectional area 2 inch $\times$ 2 inch.

The beams and the spherical masses are made of aluminum, the material properties of aluminum are $E = 1.0 \times 10^7$ psi, and density $\rho = 0.104$ lb/in$^3$.

$M_1$ is a spherical mass with radius of 10 inch, so given the density the mass and rotational inertia $I_{p1}$ can be calculated; the spherical mass $M_2$ has a radius of 15 inch, so given the mass and the rotational inertia $I_{p2}$ can be calculated; the spring constants $K_1 = K_2 = 10^5$ lb/in; and the torsional spring $K_T = 10^6$ lb-inch/rad. Calculate the frequencies of this structural dynamic system:

- First assume that each beam is modeled using one element.
- Next assume that each beam is modeled by two elements.

Compare the results obtained for the frequencies from part (b) and (d) of the problem.

# 5 Natural Modes of Vibration: Discrete Systems

## 5.1 General Discussion

Natural modes are properties of structural dynamic systems undergoing free vibrations, with low damping or in absence of damping. In this section undamped modes, that is, the case when the damping matrix $\mathbf{C} = 0$, are considered. A natural mode is a deformed configuration of the structure in which the motion of each point is **harmonic**, and in which the oscillations occur at a specific natural frequency associated with that mode.

Recall that a continuous elastic structure has an infinite number of degrees of freedom. When a continuous structure is modeled by a finite number of degrees of freedom, it implies that the model is artificially stiffer than the original structure and the frequencies predicted for such a model will be somewhat higher than the exact frequencies.

Consider the undamped free vibration problem for an $n$-degree-of-freedom system for which the equation of motion is represented by

$$\mathbf{M}\ddot{\mathbf{q}} + \mathbf{K}\mathbf{q} = 0, \tag{5.1.1}$$

where the $\mathbf{M}$ and $\mathbf{K}$ matrices for the system have been obtained either from a finite element approach, or a global approximation technique such as Rayleigh–Ritz or Galerkin methods. For simple harmonic motion

$$\mathbf{q}(t) = \mathbf{q}_0 e^{i\omega t} \tag{5.1.2}$$

and

$$\ddot{\mathbf{q}}(t) = \mathbf{q}_0 e^{i\omega t}(-\omega^2) = -\omega^2 \mathbf{q}(t). \tag{5.1.3}$$

Combining Eqs. (5.1.1) and (5.1.3) yields

$$\mathbf{K}\mathbf{q} = \omega^2 \mathbf{M}\mathbf{q}. \tag{5.1.4}$$

Equation (5.1.4) represents a variant of the **algebraic eigenvalue problem** known as the **structural dynamics eigenvalue** problem. In texts on mathematics Eq. (5.1.4) is denoted as the **generalized eigenvalue problem**, or in short the **generalized eigenproblem**. When $\mathbf{M}$ is an identity matrix, Equation (5.1.4) simplifies to

$$\mathbf{K}\mathbf{q} = \omega^2 \mathbf{q} \tag{5.1.5}$$

and the generalized eigenvalue problem becomes the **standard eigenvalue problem**.

## 5.2    Structural Dynamics Eigenvalue Problem

The structural dynamics eigenvalue problem, for cases when the stiffness matrix is non-singular, implying that the matrix contains no rigid-body modes, can be rewritten as

$$\frac{1}{\omega^2}\mathbf{q} = \mathbf{K}^{-1}\mathbf{Mq} = \mathbf{AMq} = \mathbf{Dq}, \tag{5.2.1}$$

where $\mathbf{D} = \mathbf{K}^{-1}\mathbf{M}$ is called the **dynamic matrix** and $\mathbf{A}$ in Eq. (5.2.1) is the flexibility matrix. The last equation can be also written as

$$(\mathbf{D} - \lambda\mathbf{I})\,\mathbf{q} = 0, \tag{5.2.2}$$

where $\lambda = 1/\omega^2$. Equation (5.2.1) or (5.2.2) represents the **standard eigenvalue problem**, for which many numerical methods of solution have been developed by numerical analysts. Note that while $\mathbf{K}$ and $\mathbf{M}$ are symmetric, $\mathbf{D}$ in general is nonsymmetric. From linear algebra, the eigenvalues are obtained from the characteristic determinant of the system

$$\det(\mathbf{D} - \lambda\mathbf{I}) \equiv \Delta = 0, \tag{5.2.3}$$

which can be expanded to yield the characteristic polynomial, a polynomial of degree $n$ in $\lambda$

$$\lambda^n + c_1\lambda^{n-1} + c_2\lambda^{n-2} + \dots c_n = 0, \tag{5.2.4}$$

which can be also written as

$$(\lambda - \lambda_1)(\lambda - \lambda_2)(\lambda - \lambda_3)\dots(\lambda - \lambda_n) = 0. \tag{5.2.5}$$

In the treatment of the continuous problem, it was shown that the structural dynamic eigenvalue problem is self-adjoint and positive definite. Therefore, the $\mathbf{M}$ and $\mathbf{K}$ matrices are symmetric $\mathbf{M} = \mathbf{M}^T$ and $\mathbf{K} = \mathbf{K}^T$, and positive definite, which implies

$$\begin{aligned} \mathbf{x}^T\mathbf{Mx} > 0 \\ \mathbf{x}^T\mathbf{Kx} > 0 \end{aligned} \quad \text{for } \mathbf{x} \neq 0. \tag{5.2.6}$$

For each eigenvalue $\lambda_j$, which corresponds to a frequency $1/\omega_j^2$, one can return to the homogeneous problem, Eq. (5.2.1), to calculate the **mode shape**, or eigenvector, which represents the modal pattern of vibration corresponding to a particular eigenvalue. Since $n$ homogeneous equations are **linearly dependent**, for each $\lambda_i$, one may assign an arbitrary value to one of the elements in $\mathbf{q}_j$, say the first element, and express the remaining elements as ratios of this assigned value. In other words, only $(n-1)$ equations are linearly independent.

It should be noted that the stiffness matrix can be semi-positive definite, which implies that it contains rigid-body modes. The rigid-body modes are special eigenvectors associated with a zero frequency. When rigid-body modes are present, the stiffness matrix $\mathbf{K}$ cannot be inverted and the formulation Eq. (5.2.1) is no longer valid.

## 5.3    Orthogonality Condition: Free Vibration Modes

Based on the orthogonality properties of the structural dynamic eigenvalue problem, discussed for continuous systems, it follows that orthogonality conditions, with respect to the mass and stiffness matrices, can be also shown to exist for the discrete case.

Each eigenvalue and eigenvector satisfies the free vibration problem, thus, for the $r$th mode

$$\omega_r^2 \mathbf{M} \mathbf{u}_r = \mathbf{K} \mathbf{u}_r. \tag{5.3.1}$$

Pre-multiply the last equation by another eigenvector $\mathbf{u}_s$, then

$$\omega_r^2 \mathbf{u}_s^T \mathbf{M} \mathbf{u}_r = \mathbf{u}_s^T \mathbf{K} \mathbf{u}_r. \tag{5.3.2}$$

Since $\mathbf{M}$ and $\mathbf{K}$ are symmetric, one can transpose Eq. (5.3.2)

$$\omega_r^2 \mathbf{u}_r^T \mathbf{M} \mathbf{u}_s = \mathbf{u}_r^T \mathbf{K} \mathbf{u}_s. \tag{5.3.3}$$

Rewriting the eigenvalue problem for the $s$th mode

$$\omega_s^2 \mathbf{M} \mathbf{u}_s = \mathbf{K} \mathbf{u}_s \tag{5.3.4}$$

and pre-multiplying by $\mathbf{u}_r^T$

$$\omega_s^2 \mathbf{u}_r^T \mathbf{M} \mathbf{u}_s = \mathbf{u}_r^T \mathbf{K} \mathbf{u}_s. \tag{5.3.5}$$

Subtracting Eq. (5.3.5) from Eq. (5.3.3) yields

$$(\omega_r^2 - \omega_s^2) \mathbf{u}_r^T \mathbf{M} \mathbf{u}_s = 0 \tag{5.3.6}$$

for $\omega_r^2 \neq \omega_s^2$

$$\mathbf{u}_r^T \mathbf{M} \mathbf{u}_s = 0. \tag{5.3.7}$$

Similarly, from Eqs. (5.3.2) and (5.3.7)

$$\mathbf{u}_r^T \mathbf{K} \mathbf{u}_s = 0. \tag{5.3.8}$$

Thus orthogonality of the natural modes for an undamped system with respect to the mass and stiffness matrix has been demonstrated.

When $r = s$, $\omega_r^2 - \omega_s^2 = 0$ and

$$M_r = \mathbf{u}_r^T \mathbf{M} \mathbf{u}_r \tag{5.3.9}$$

$$K_r = \mathbf{u}_r^T \mathbf{K} \mathbf{u}_r, \tag{5.3.10}$$

where $M_r$ and $K_r$ are the generalized mass and stiffness associated with the $r$th mode, and

$$K_r = \omega_r^2 M_r. \tag{5.3.11}$$

The orthogonality relations are not direct, in a strict mathematical sense, but represent orthogonality with respect to the **weighting matrices M** and **K**.

Since the eigenvector $\mathbf{u}_r$ contains an arbitrary constant $\alpha_r$, the eigenvector can be always adjusted so that it is normalized or orthonormal. To this end, each element of $\mathbf{u}_r$ is divided by $\sqrt{M_r}$, then $\bar{\mathbf{u}}_r = \dfrac{1}{\sqrt{M_r}}\, \mathbf{u}_r$ and

$$\bar{\mathbf{u}}_r^T \mathbf{M} \bar{\mathbf{u}}_r = 1. \tag{5.3.12}$$

An eigenvector that satisfies Eq. (5.3.12) is **mass normalized**.

## 5.4    Geometric Interpretation of Eigenvalue Problem

Consider the eigenvalue problem

$$\mathbf{A}\mathbf{x} = \lambda \mathbf{x}, \tag{5.4.1}$$

where $\mathbf{A}$ is an $n \times n$ symmetric matrix consisting of real elements and $\mathbf{x}$ is an $n$-dimensional vector. Note that this is a special type of eigenvalue problem. The geometric aspect of the eigenvalue problem becomes evident after multiplying Eq. (5.4.1) by $\mathbf{x}^T$

$$\mathbf{x}^T \mathbf{A}\mathbf{x} = \lambda \mathbf{x}^T \mathbf{x}. \tag{5.4.2}$$

The vector $\mathbf{x}$. can be normalized by requiring

$$\mathbf{x}^T \mathbf{x} = \sum_{i=1}^{n} x_i^2 \equiv r^2 = 1 \tag{5.4.3}$$

such that

$$\mathbf{x}^T \mathbf{A}\mathbf{x} = \lambda, \tag{5.4.4}$$

which is a quadratic form of the coordinates $x_1, x_2, \ldots, x_n$. For $n = 2$, Eq. (5.4.4) yields

$$\frac{1}{\lambda} \lfloor x_1 \quad x_2 \rfloor \begin{bmatrix} a_{11} & a_{12} \\ a_{21} & a_{22} \end{bmatrix} \begin{Bmatrix} x_1 \\ x_2 \end{Bmatrix} = \frac{1}{\lambda}\left[ a_{11}x_1^2 + (a_{12} + a_{21})\, x_1 x_2 + a_{22} x_2^2 \right]$$

$$= \frac{1}{\lambda}\left[ a_{11}x_1^2 + 2a_{12}x_1 x_2 + a_{22}x_2^2 \right] = 1. \tag{5.4.5}$$

For the coefficients $a_{ij}$ associated with small oscillations, Eq. (5.4.5) represents the equation for an ellipse as shown in Figure 5.1. The coordinates $x_1, x_2$ do not coincide with the major and minor axes of the ellipse $\xi_1$ and $\xi_2$ as shown. Denoting by $\theta$ the angle between these directions a coordinate transformation can be introduced

$$\mathbf{x} = \begin{Bmatrix} x_1 \\ x_2 \end{Bmatrix} = \begin{bmatrix} \cos\theta & -\sin\theta \\ \sin\theta & \cos\theta \end{bmatrix} \begin{Bmatrix} \xi_1 \\ \xi_2 \end{Bmatrix} = \mathbf{R}\boldsymbol{\xi}. \tag{5.4.6}$$

Note that $\mathbf{R}$ is a rotation matrix which has the property that

$$\mathbf{R}^T \mathbf{R} = \mathbf{R}\mathbf{R}^T = \mathbf{I} \tag{5.4.7}$$

**Figure 5.1** Geometric interpretation of a $2 \times 2$ symmetric eigenvalue problem

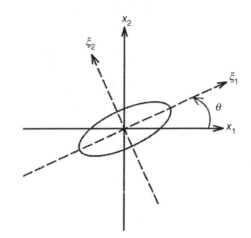

or $\mathbf{R}^T = \mathbf{R}^{-1}$, that is, the transformation (5.4.6) is **orthonormal**. Substituting Eq. (5.4.6) into Eq. (5.4.4) yields

$$\frac{1}{\lambda}\boldsymbol{\xi}^T\mathbf{R}^T\mathbf{A}\mathbf{R}\boldsymbol{\xi} = \frac{1}{\lambda}\boldsymbol{\xi}^T\mathbf{B}\boldsymbol{\xi} = 1, \tag{5.4.8}$$

where

$$\mathbf{B} = \mathbf{R}^T\mathbf{A}\mathbf{R} = \begin{bmatrix} \cos\theta & \sin\theta \\ -\sin\theta & \cos\theta \end{bmatrix} \begin{bmatrix} a_{11} & a_{12} \\ a_{21} & a_{22} \end{bmatrix} \begin{bmatrix} \cos\theta & -\sin\theta \\ \sin\theta & \cos\theta \end{bmatrix} \tag{5.4.9}$$

and

$$\begin{aligned} b_{11} &= a_{11}\cos^2\theta + 2a_{12}\sin\theta\cos\theta + a_{22}\sin^2\theta \\ b_{12} &= b_{21} = -(a_{11} - a_{22})\sin\theta\cos\theta + a_{12}(\cos^2\theta - \sin^2\theta) \\ b_{22} &= a_{11}\sin^2\theta - 2a_{12}\sin\theta\cos\theta + a_{22}\cos^2\theta. \end{aligned} \tag{5.4.10}$$

The transformed matrix is reduced to diagonal form when $b_{12} = b_{21} = 0$. This condition is valid when $\theta$ is obtained from Eq. (5.4.11)

$$-(a_{11} - a_{22})\sin\theta\cos\theta + a_{12}(\cos^2\theta - \sin^2\theta) = 0$$

$$\text{or} \quad -(a_{11} - a_{22})\frac{1}{2}\sin 2\theta + a_{12}\cos 2\theta = 0$$

$$\text{or} \quad \tan 2\theta = \frac{2a_{12}}{a_{11} - a_{22}}. \tag{5.4.11}$$

Two values of $\theta$ satisfying Eq. (5.4.11) differ by $\pi/2$ corresponding to $\xi_1$ and $\xi_2$. When $\theta$ satisfies Eq. (5.4.11), Eq. (5.4.8) reduces to

$$\frac{1}{\lambda} \lfloor \xi_1 \quad \xi_2 \rfloor \begin{bmatrix} b_{11} & 0 \\ 0 & b_{22} \end{bmatrix} \begin{Bmatrix} \xi_1 \\ \xi_2 \end{Bmatrix} = 1$$

$$\text{or} \qquad \boxed{\frac{b_{11}}{\lambda} \xi_1^2 + \frac{b_{22}}{\lambda} \xi_2^2 = 1} \qquad (5.4.12)$$

Equation (5.4.12) is the equation of the ellipse in canonical form, so that the principal axes coincide with the directions $\xi_1$ and $\xi_2$. Combining Eqs. (5.4.1) and (5.4.6)

$$\mathbf{A R \xi} = \lambda \mathbf{R \xi}$$

$$\text{or} \qquad \mathbf{R}^T \mathbf{A R \xi} = \lambda \mathbf{R}^T \mathbf{R \xi}$$

$$\mathbf{B \xi} = \lambda \xi, \qquad (5.4.13)$$

which represents an uncoupled system of equations. Its solution is

$$\begin{aligned} \lambda_1 = b_{11}: & \quad \boldsymbol{\xi}_1 = [1, 0]^T \\ \lambda_2 = b_{22}: & \quad \boldsymbol{\xi}_2 = [0, 1]^T \end{aligned} \qquad (5.4.14)$$

$$\begin{aligned} \mathbf{x}_1 = \mathbf{R} \boldsymbol{\xi}_1 = [\cos\theta, \ \sin\theta]^T \\ \mathbf{x}_2 = \mathbf{R} \boldsymbol{\xi}_2 = [-\sin\theta, \ \cos\theta]^T \end{aligned}, \qquad (5.4.15)$$

where $\lambda_1$ and $\lambda_2$ are the eigenvalues of the problem and $\mathbf{x}_1$ and $\mathbf{x}_2$ are the eigenvectors, or normal modes.

This simple example indicates that the solution of the eigenvalue problem consists of reducing the quadratic expression to its canonical form. Equation (5.4.12) can also be rewritten using Eq. (5.4.14) as eigenvalues are obviously inversely proportional to the squares of the semimajor and semiminor axes, respectively.

$$\lambda_1 \xi_1^2 + \lambda_2 \xi_2^2 = \lambda. \qquad (5.4.16)$$

For $n = 3$, Eq. (5.1.1) represents a quadratic surface with the center at the origin. This time one has to find the principal axes of an ellipsoid. For this case a transformation is required such that

$$\mathbf{x} = \mathbf{U \xi} \qquad (5.4.17)$$

and

$$\mathbf{U}^T \mathbf{A U} = \mathbf{\Lambda}, \qquad$$

where $\mathbf{\Lambda}$ is a diagonal matrix containing eigenvalues $\lambda_1, \lambda_2, \ldots, \lambda_3$. The equation of the ellipsoid in its canonical form becomes

$$\lambda_1 \xi_1^2 + \lambda_2 \xi_2^2 + \lambda_3 \xi_3^2 = \lambda. \qquad (5.4.18)$$

## 5.5    Example of Solving the Eigenvalue Problem

A concrete example is presented to illustrate the analytical solution of a $3 \times 3$ structural dynamics eigenvalue problem. This eigenproblem will be revisited and solved numerically when numerical algorithms are introduced in the later sections of this chapter.

Consider a simplified three-story building shown in Figure 5.2. The stiffness of the girders is sufficiently high so that they can be considered to be rigid relative to the stiffness of the columns. The girder column rotation are rigid (i.e. no joint rotation). The stiffness matrix for this problem can be obtained by assembling three beam-type elements such as derived in Eq. (4.2.18). Recall that the stiffness matrix is

$$\mathbf{k} = \frac{EI}{\ell^3} \begin{bmatrix} 12 & 6\ell & -12 & 6\ell \\ & 4\ell^2 & -6\ell & 2\ell^2 \\ & & 12 & -6\ell \\ \text{Symm.} & & & 4\ell^2 \end{bmatrix} \tag{E5.1}$$

**Figure 5.2** A simplified three-story building

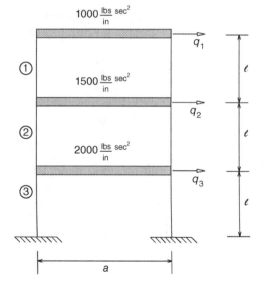

Restraining the appropriate degrees of freedom associated with the rotational DOFs, and using the direct stiffness method of assembly, yields

$$\mathbf{k}_1 = \frac{EI_1}{\ell^3} \begin{bmatrix} 12 & -12 \\ -12 & 12 \end{bmatrix} = \frac{12EI_1}{\ell^3} \begin{bmatrix} 1 & -1 \\ -1 & 1 \end{bmatrix}$$

$$\mathbf{k}_2 = \frac{EI_2}{\ell^3} \begin{bmatrix} 12 & -12 \\ -12 & 12 \end{bmatrix} = \frac{12EI_2}{\ell^3} \begin{bmatrix} 1 & -1 \\ -1 & 1 \end{bmatrix}$$

$$\mathbf{k}_3 = \frac{12EI_3}{\ell^3}.$$

Assume that $I_1 = I_0$, $I_2 = 2I_0$, $I_3 = 3I_0$, then the assembled stiffness matrix can be written as

$$\mathbf{K} = \frac{12EI_0}{\ell^3}\begin{bmatrix} 1 & -1 & 0 \\ -1 & (1+2) & -2 \\ 0 & -2 & (2+3) \end{bmatrix} = \frac{12EI_0}{\ell^3}\begin{bmatrix} 1 & -1 & 0 \\ -1 & 3 & -2 \\ 0 & -2 & 5 \end{bmatrix} \quad (E5.2)$$

assuming numerical values for

$$12\frac{EI_0}{\ell^3} = 600\frac{\text{k lbs}}{\text{in}}.$$

The mass matrix can be obtained by simply lumping the mass of the girder associated with each generalized degree of freedom

$$\mathbf{M} = \begin{bmatrix} 1 & 0 & 0 \\ 0 & 1.5 & 0 \\ 0 & 0 & 2.0 \end{bmatrix}\frac{\text{k lbs.sec}^2}{\text{in}} \quad (E5.3)$$

The free vibration problem for this structure is given by

$$\begin{bmatrix} 1 & -1 & 0 \\ -1 & 3 & -2 \\ 0 & -2 & 5 \end{bmatrix}\begin{Bmatrix} q_1 \\ q_2 \\ q_3 \end{Bmatrix} = \frac{\omega^2}{600}\begin{bmatrix} 1 & 0 & 0 \\ 0 & 1.5 & 0 \\ 0 & 0 & 2.0 \end{bmatrix}\begin{Bmatrix} q_1 \\ q_2 \\ q_3 \end{Bmatrix}. \quad (E5.4)$$

Let $\lambda = \omega^2/600$, the characteristic equation is then

$$\begin{vmatrix} 1-\lambda & -1 & 0 \\ -1 & 3-1.5\lambda & -2 \\ 0 & -2 & 5-2.0\lambda \end{vmatrix} = \lambda^3 \quad 5.5\lambda^2 + 7.5\lambda - 2 = 0. \quad (E5.5)$$

The roots of this cubic equation are listed in Table 5.1.

Mode shapes are obtained in the following manner. First, set $q_1 = 1$ and solve for remaining displacements

$$\begin{bmatrix} K_{11}-\lambda M_{11} & | & K_{12} & K_{13} \\ K_{21} & | & K_{22}-\lambda M_{22} & K_{23} \\ K_{31} & | & K_{32} & K_{33}-\lambda M_{33} \end{bmatrix}\begin{Bmatrix} q_1=1 \\ q_2 \\ q_3 \end{Bmatrix} = 0 \quad (E5.6)$$

which simplifies to

$$\begin{bmatrix} K_{22}-\lambda M_{22} & K_{23} \\ K_{32} & K_{33}-\lambda M_{33} \end{bmatrix}\begin{Bmatrix} q_2 \\ q_3 \end{Bmatrix} = -\begin{Bmatrix} K_{21} \\ K_{31} \end{Bmatrix}. \quad (E5.7)$$

## Table 5.1 Eigenvalues of the three-story building problem

| Mode | $\lambda$ | $\omega$, rad/s | $T$, s |
|---|---|---|---|
| 1 | 0.35 | 14.5 | 0.43 |
| 2 | 1.61 | 31.1 | 0.20 |
| 3 | 3.54 | 46.1 | 0.14 |

Setting $\lambda = \lambda_1 = 0.35$ in Eq. (E5.7) yields the first mode shape

$$\begin{bmatrix} 3 - (1.5)(0.35) & -2 \\ -2 & 5 - (2)(0.35) \end{bmatrix} \begin{Bmatrix} q_2 \\ q_3 \end{Bmatrix} = \begin{Bmatrix} 1 \\ 0 \end{Bmatrix} \Rightarrow \begin{Bmatrix} q_2 \\ q_3 \end{Bmatrix} = \begin{Bmatrix} 0.647 \\ 0.301 \end{Bmatrix}. \qquad (E5.8)$$

Similarly the second mode shape is given by

$$\begin{bmatrix} 3 - (1.5)(1.61) & -2 \\ -2 & 5 - (2)(1.61) \end{bmatrix} \begin{Bmatrix} q_2 \\ q_3 \end{Bmatrix} = \begin{Bmatrix} 1 \\ 0 \end{Bmatrix} \Rightarrow \begin{Bmatrix} q_2 \\ q_3 \end{Bmatrix} = \begin{Bmatrix} -0.602 \\ -0.676 \end{Bmatrix} \qquad (E5.9)$$

and the third mode

$$\begin{bmatrix} 3 - (1.5)(3.54) & -2 \\ -2 & 5 - (2)(3.54) \end{bmatrix} \begin{Bmatrix} q_2 \\ q_3 \end{Bmatrix} = \begin{Bmatrix} 1 \\ 0 \end{Bmatrix} \Rightarrow \begin{Bmatrix} q_2 \\ q_3 \end{Bmatrix} = \begin{Bmatrix} -2.58 \\ 2.49 \end{Bmatrix}. \qquad (E5.10)$$

To summarize, the eigenvectors, normalized w.r.t. the largest element of the vector, are

$$\begin{bmatrix} 1.000 & 1.000 & -0.388 \\ 0.647 & -0.602 & 1.000 \\ 0.301 & -0.676 & -0.965 \end{bmatrix}. \qquad (E5.11)$$

If the normalization w.r.t. the mass matrix using Eq. (5.3.12) is employed, the eigenvectors are

$$\begin{bmatrix} 0.743 & 0.636 & -0.210 \\ 0.482 & -0.386 & 0.535 \\ 0.224 & -0.432 & -0.513 \end{bmatrix}. \qquad (E5.12)$$

The mode shapes are shown in Figure 5.3.

**Figure 5.3**
Mode shapes of the three-story building

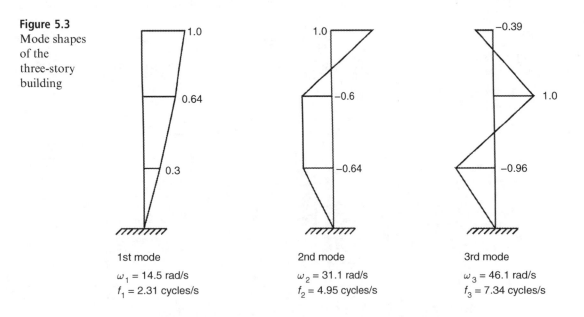

1st mode
$\omega_1 = 14.5$ rad/s
$f_1 = 2.31$ cycles/s

2nd mode
$\omega_2 = 31.1$ rad/s
$f_2 = 4.95$ cycles/s

3rd mode
$\omega_3 = 46.1$ rad/s
$f_3 = 7.34$ cycles/s

## 5.6    Numerical Methods for Solving the Eigenvalue Problem

Modal analysis is a fundamental problem in structural dynamics. Determination of the modes and frequencies for modal analysis implies the need to compute several eigenvectors and eigenvalues, that is eigenpairs, of the structural dynamic eigenvalue problem.

The development of an effective eigenvalue solver is challenging when compared to the development of a linear solver, that is, a solver for a system of linear equations. This is not only due to the fact that an eigenvalue solver itself typically requires a linear solver, but also due to the iterative nature of eigenvalue algorithms (Trefethen and Bau III, 1997; Bai et al., 2000). As shown in Eq. (5.2.3), finding the eigenvalues of an $n \times n$ matrix is equivalent to finding the roots of a characteristic polynomial of degree $n$. However, since early nineteenth century, it is well known that the roots of polynomials of degree 5 or higher cannot be found in a finite number of steps employing elementary mathematical operations. Therefore, it is impossible to find the eigenvalues of a matrix having a size $n \geq 5$ in a finite number of steps.

Therefore, **an eigenvalue algorithm must be iterative**, and its goal is to produce a sequence of numbers and vectors that converge to the required eigenvalues and eigenvectors rapidly.

Several efficient and robust eigenvalue algorithms have been developed and implemented in modern finite element codes. In Figure 5.4, these algorithms are roughly divided into three categories, according to the degrees of freedom of the eigenproblem and the number of required eigenpairs. The three categories and their characteristics are illustrated in Figure 5.4 and explained with additional details below (Meirovitch, 1997; Craig Jr and Kurdila, 2006; Géradin and Rixen, 2014),

(1) Vector iteration methods, represented by the power iteration and **inverse iteration** algorithms. These typically require only a matrix-vector multiplication and are applicable

**Figure 5.4** Three categories of eigenvalue algorithms

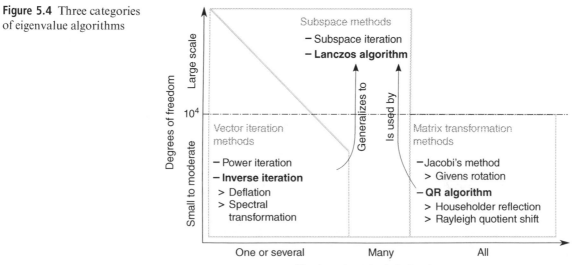

to eigenproblems of any size. However, such methods are only suitable for extracting one or a limited number of eigenpairs.

(2) Matrix transformation methods, represented by the Jacobi's method and the **QR algorithm**. These generate all the eigenpairs but require the knowledge of the values of all the elements of a matrix. Such methods are suitable for small to moderate eigenproblems where the number of DOFs is below $10^4$.

(3) Subspace methods, represented by the subspace iteration method and the **Lanczos algorithm**. These are clever combinations of vector iteration and matrix transformation methods. Such methods enable the extraction of many eigenpairs of large-scale eigenproblems where the number of DOFs is greater than $10^4$.

This section describes the three representative eigenvalue algorithms in each category. Combined with the presentation of the algorithms, several important mathematical techniques for eigenvalue algorithms, such as deflation, spectral transformation, Givens rotation, and Householder reflection, are also introduced. However, the readers are not advised to develop their own eigenvalue solvers based on this section. Developing an efficient and robust eigenvalue solver requires long-term dedicated effort and critical feedback from a wide user base. Such eigenvalue solvers have been made available via various commercial and open-source software. The interested readers are referred to Géradin and Rixen (2014) for a comprehensive discussion of algorithms for generalized eigenvalue problems in the context of structural dynamics, as well as Trefethen and Bau III (1997) and Bai et al. (2000) for a general mathematical and algorithmic treatment of algorithms for generalized eigenvalue problems.

### 5.6.1 Properties of Generalized Eigenvalue Problems

**Solution Structure**

In this section, the generalized eigenvalue problem, which is described by Eq. (5.1.4), is considered,

$$\mathbf{K}\mathbf{u} = \omega^2 \mathbf{M}\mathbf{u}, \tag{5.6.1}$$

where $\mathbf{M}$ and $\mathbf{K}$ are assumed to be positive definite. Therefore, the eigenvalues $\omega^2$ are positive real numbers

$$0 < \omega_1^2 \leq \omega_2^2 \leq \cdots \leq \omega_n^2 \tag{5.6.2}$$

with the corresponding eigenvectors denoted as $\mathbf{U} = [\mathbf{u}_1, \ldots, \mathbf{u}_n]$. Note that multiple eigenvectors may correspond to the same nonzero eigenvalue, which is typically due to the symmetry in the structural configuration.

Due to orthogonality conditions, Eqs. (5.3.7)–(5.3.10), the $\mathbf{K}$ and $\mathbf{M}$ matrices can be diagonalized as

$$\mathbf{U}^T \mathbf{K} \mathbf{U} = \mathbf{\Lambda}, \quad \mathbf{U}^T \mathbf{M} \mathbf{U} = \mathbf{I}, \tag{5.6.3}$$

where $\mathbf{\Lambda}$ is a $n \times n$ diagonal matrix containing the eigenvalues.

### Rayleigh's Quotient (Discrete Case)

An important concept associated with a generalized eigenvalue problem is the Rayleigh's quotient, as the discrete analogy of the Rayleigh's quotient introduced in Section 3.5.1. It is a convenient mathematical device in many eigensolvers for estimating eigenvalues given approximate eigenvectors.

For any eigenvalue $\omega_k^2$ and the corresponding eigenvector $\mathbf{u}_k$ one has

$$\mathbf{K}\mathbf{u}_k = \omega_k^2 \mathbf{M}\mathbf{u}_k. \tag{5.6.4}$$

Pre-multiply by $\mathbf{u}_k^T$ and rewrite Eq. (5.6.4) as

$$\omega_k^2 = \frac{\mathbf{u}_k^T \mathbf{K}\mathbf{u}_k}{\mathbf{u}_k^T \mathbf{M}\mathbf{u}_k} \equiv \frac{\text{Strain Energy}}{\text{Kinetic Energy/Frequency}^2}. \tag{5.6.5}$$

Equation (5.6.5) implies that the eigenvalue $\omega_k^2$ represents the ratio of two quadratic forms, in this case the strain energy and the kinetic energy, associated with a particular mode $\mathbf{u}_k$. The denominator in Eq. (5.6.5) is sometimes called **kinetic energy density**.

Assume an **arbitrary** vector $\mathbf{u}$ and form a corresponding expression,

$$\omega^2 = R(\mathbf{u}) = \frac{\mathbf{u}^T \mathbf{K}\mathbf{u}}{\mathbf{u}^T \mathbf{M}\mathbf{u}} \tag{5.6.6}$$

and the expression $R(\mathbf{u})$ is called **Rayleigh's quotient**. When $\mathbf{u}$ coincides with one of the eigenvectors, $\omega^2$ reduces to the associated eigenvalue,

$$\omega^2 = R(\mathbf{u}_k) = \omega_k^2. \tag{5.6.7}$$

An important property of Rayleigh's quotient is that it has **a stationary value** when the arbitrary vector $\mathbf{u}$ is in the neighborhood of an eigenvector $\mathbf{u}_k$. This stationarity property has been shown for the continuous systems in Section 3.5.1. For discrete systems, the stationarity of Rayleigh's quotient is shown using the Taylor series expansion, as described below.

First, the Taylor series expansion of $R(\mathbf{u})$ at $\mathbf{u} = \mathbf{u}_k$ is

$$R(\mathbf{u}) = R(\mathbf{u}_k) + (\mathbf{u} - \mathbf{u}_k)^T \nabla R(\mathbf{u}_k) + \frac{1}{2}(\mathbf{u} - \mathbf{u}_k)^T \nabla^2 R(\mathbf{u}_k)(\mathbf{u} - \mathbf{u}_k)$$
$$+ O(\|\mathbf{u} - \mathbf{u}_k\|^3), \tag{5.6.8}$$

where $\nabla R$ and $\nabla^2 R$ are the gradient vector and Hessian matrix of $R$, respectively. The $O$ notation means order of magnitude and the norm $\|\mathbf{x}\| = \sqrt{\mathbf{x}^T \mathbf{x}}$. Additional discussion of the gradient and Hessian of a multivariate function is provided in Appendix A.5.

The gradient $R(\mathbf{u}_k)$ in Eq. (5.6.8) is in fact zero. To show this, one may utilize the identity for a symmetric matrix $\mathbf{A}$, Eq. (A.5.9)

$$\nabla(\mathbf{u}^T \mathbf{A}\mathbf{u}) = 2\mathbf{A}\mathbf{u} \tag{5.6.9}$$

and compute the gradient of $R(\mathbf{u})$

$$
\begin{aligned}
\nabla R(\mathbf{u}) &= \frac{(\mathbf{u}^T\mathbf{Mu})\nabla(\mathbf{u}^T\mathbf{Ku}) - \nabla(\mathbf{u}^T\mathbf{Mu})(\mathbf{u}^T\mathbf{Ku})}{(\mathbf{u}^T\mathbf{Mu})^2} \\
&= \frac{(\mathbf{u}^T\mathbf{Mu})(2\mathbf{Ku}) - (2\mathbf{Mu})(\mathbf{u}^T\mathbf{Ku})}{(\mathbf{u}^T\mathbf{Mu})^2} \\
&= \frac{(2\mathbf{Ku}) - (2\mathbf{Mu})\frac{\mathbf{u}^T\mathbf{Ku}}{\mathbf{u}^T\mathbf{Mu}}}{\mathbf{u}^T\mathbf{Mu}} \\
&= 2\frac{\mathbf{Ku} - R(\mathbf{u})\mathbf{Mu}}{\mathbf{u}^T\mathbf{Mu}}.
\end{aligned}
\tag{5.6.10}
$$

Next, using Eqs. (5.6.4) and (5.6.7), when $\mathbf{u} = \mathbf{u}_k$, Eq. (5.6.10) simplifies as

$$
\nabla R(\mathbf{u}_k) = 2\frac{\mathbf{Ku}_k - R(\mathbf{u}_k)\mathbf{Mu}_k}{\mathbf{u}_k^T\mathbf{Mu}_k} = 2\frac{\mathbf{Ku}_k - \omega_k^2\mathbf{Mu}_k}{\mathbf{u}_k^T\mathbf{Mu}_k} = 0.
\tag{5.6.11}
$$

Hence, Eq. (5.6.8) simplifies as

$$
\begin{aligned}
R(\mathbf{u}) &= R(\mathbf{u}_k) + \frac{1}{2}(\mathbf{u} - \mathbf{u}_k)^T\nabla^2 R(\mathbf{u}_k)(\mathbf{u} - \mathbf{u}_k) + O(\|\mathbf{u} - \mathbf{u}_k\|^3) \\
&= \omega_k^2 + O(\|\mathbf{u} - \mathbf{u}_k\|^2).
\end{aligned}
\tag{5.6.12}
$$

Equation (5.6.12) indicates that if the arbitrary vector $\mathbf{u}$ differs from the eigenvector $\mathbf{u}_k$ by a small quantity of order $O(\|\mathbf{u} - \mathbf{u}_k\|)$, $R(\mathbf{u})$ differs from $\omega_k^2$ by a quantity of $O(\|\mathbf{u} - \mathbf{u}_k\|^2)$. Hence Rayleigh's quotient has a stationary value in the neighborhood of an eigenvector. Furthermore, this property also renders Rayleigh's quotient to be an effective means for obtaining a **quadratically accurate estimate** of an eigenvalue given an approximate eigenvector.

Finally, Rayleigh's quotient of the discrete system has a **Min–Max** property, which is similar to the continuous case as shown in Eqs. (3.5.15) and (3.5.16). The Min–Max property indicates that the lower and upper bounds of Rayleigh's quotient are $\omega_1^2$ and $\omega_n^2$, respectively, that is,

$$
\omega_1^2 \le R(\mathbf{u}) \le \omega_n^2.
\tag{5.6.13}
$$

Furthermore, if $\mathbf{u}$ consists only of vectors that are orthogonal to the first $(s-1)$ eigenvectors, that is, $\mathbf{u}^T\mathbf{Mu}_i = 0$ for $i = 1, 2, \cdots, (s-1)$, then

$$
\omega_s^2 \le R(\mathbf{u}) \le \omega_n^2.
\tag{5.6.14}
$$

### Conversion to Standard Eigenproblem Form

Sometimes it is more convenient to transform the generalized eigenproblem to standard eigenproblem form, since there are many established and efficient eigensolvers for the latter problem. A convenient transformation is accomplished using the dynamic matrix $\mathbf{D} = \mathbf{K}^{-1}\mathbf{M}$, as was done in Eq. (5.2.1)

$$\mathbf{Du} = \lambda\mathbf{u}, \quad \lambda = \frac{1}{\omega^2}. \tag{5.6.15}$$

The new eigenproblem Eq. (5.6.15) has the same eigenvectors as the generalized eigenproblem, Eq. (5.6.1), but has a different set of eigenvalues

$$\lambda_1 \geq \lambda_2 \geq \cdots \geq \lambda_n \geq 0. \tag{5.6.16}$$

Note that in the new eigenproblem, the order of the eigenvalues is reversed: the smallest eigenvalue $\omega_1^2$ in the generalized eigenproblem becomes the largest eigenvalue $\lambda_1$.

Typically the structural modes associated with the lowest frequencies, that is, the **smallest** few eigenvalues, are of major interest. Therefore, the transformation using the dynamic matrix, Eq. (5.6.15), is particularly useful in modal analysis. This is because, the standard eigensolvers by default only extract the **largest** few eigenvalues and their eigenvectors. However, utilizing the dynamic matrix, such standard eigensolvers can be applied to solve the new eigenproblem Eq. (5.6.15) without any modification to obtain the **largest** eigenvalues, $\lambda_1, \lambda_2, \ldots$, which correspond to the **smallest** eigenvalues, $\omega_1^2, \omega_2^2, \ldots$, of the generalized eigenproblem, Eq. (5.6.1).

Despite the desirable property of the transformation Eq. (5.6.15), the dynamic matrix $\mathbf{D}$ destroys the symmetric structure of the original formulation. The nonsymmetric property of $\mathbf{D}$ can slow down the convergence rate of some eigensolvers, particularly those that require the knowledge of every element of $\mathbf{D}$. An improved transformation from the generalized eigenproblem to a standard eigenproblem can be accomplished using Cholesky decomposition (see Appendix A.3.7) of the mass matrix $\mathbf{M}$,

$$\mathbf{M} = \mathbf{LL}^T, \tag{5.6.17}$$

where $\mathbf{L}$ is a lower triangular matrix. Such decomposition is always possible since $\mathbf{M}$ is a positive-definite matrix. Using Eq. (5.6.17), the formulation Eq. (5.6.15) becomes

$$\mathbf{K}^{-1}\mathbf{LL}^T\mathbf{u} = \lambda\mathbf{u}. \tag{5.6.18}$$

Pre-multiplying Eq. (5.6.18) with $\mathbf{L}^T$, one obtains a new symmetric eigenproblem,

$$\mathbf{S\tilde{u}} = \lambda\tilde{\mathbf{u}}, \tag{5.6.19}$$

where $\tilde{\mathbf{u}} = \mathbf{L}^T\mathbf{u}$ and $\mathbf{S} = \mathbf{L}^T\mathbf{K}^{-1}\mathbf{L}$. Since $\mathbf{K}$ is symmetric, $\mathbf{K}^{-1}$ and hence $\mathbf{S}$ are symmetric. Comparing Eq. (5.6.19) with the nonsymmetric formulation Eq. (5.6.15), $\mathbf{S}$ has the same eigenvalues as the dynamic matrix $\mathbf{D}$. An eigenvector $\tilde{\mathbf{u}}_i$ of the eigenproblem in Eq. (5.6.19) is converted to an eigenvector of the original generalized eigenproblem by solving the following equation

$$\mathbf{L}^T\mathbf{u}_i = \tilde{\mathbf{u}}_i, \tag{5.6.20}$$

which is an upper triangular linear system that can be solved efficiently using back substitution (see Appendix A.3.8).

### 5.6.2  Vector Iteration Methods

#### Basic Power Iteration

The power iteration is the simplest algorithm for eigenproblems, since it only requires repeated matrix-vector multiplications. In this section, the basic power iteration method is used to solve a standard eigenproblem $\mathbf{Au} = \lambda\mathbf{u}$. Starting with an arbitrary initial guess $\mathbf{z}^{(0)}$, the eigenvector associated with the largest eigenvalue is found by forming the successive iterates $\mathbf{z}^{(k)}$ by

$$\mathbf{z}^{(k)} = \mathbf{Az}^{(k-1)}. \tag{5.6.21}$$

To show how power iteration works, consider the case where $\mathbf{A}$ has $n$ real eigenvalues $|\lambda_1| > |\lambda_2| \geq \cdots \geq |\lambda_n|$ and corresponding eigenvectors $\mathbf{u}_1, \mathbf{u}_2, \ldots, \mathbf{u}_n$. The eigenvectors form an orthonormal basis of the $n$-dimensional vector space, and therefore the initial guess $\mathbf{z}^{(0)}$ can be written as a linear combination of the eigenvectors,

$$\mathbf{z}^{(0)} = c_1\mathbf{u}_1 + c_2\mathbf{u}_2 + \cdots + c_n\mathbf{u}_n, \tag{5.6.22}$$

where $c_i = \mathbf{u}_i^T\mathbf{z}^{(0)}$ and it is assumed that $c_1 \neq 0$. Applying the iteration Eq. (5.6.21) $k$ times,

$$\begin{aligned}
\mathbf{z}^{(k)} = \mathbf{Az}^{(k-1)} &= \mathbf{A}^2\mathbf{z}^{(k-2)} = \cdots = \mathbf{A}^k\mathbf{z}^{(0)} \\
&= c_1\mathbf{A}^k\mathbf{u}_1 + c_2\mathbf{A}^k\mathbf{u}_2 + \cdots + c_n\mathbf{A}^k\mathbf{u}_n \\
&= c_1\lambda_1^k\mathbf{u}_1 + c_2\lambda_2^k\mathbf{u}_2 + \cdots + c_n\lambda_n^k\mathbf{u}_n \\
&= \lambda_1^k\left[c_1\mathbf{u}_1 + c_2(\lambda_2/\lambda_1)^k\mathbf{u}_2 + \cdots + c_n(\lambda_n/\lambda_1)^k\mathbf{u}_n\right].
\end{aligned} \tag{5.6.23}$$

Since $\lambda_1$ is the magnitude of the largest eigenvalue, the ratios $|\lambda_2/\lambda_1|, \ldots, |\lambda_n/\lambda_1|$ are all smaller than 1. As the iteration progresses, the factors $(\lambda_2/\lambda_1)^k, \ldots, (\lambda_n/\lambda_1)^k$ approach zero, and

$$\lim_{k\to\infty} \mathbf{z}^{(k)} = c_1\lambda_1^k\mathbf{u}_1. \tag{5.6.24}$$

Therefore, in basic power iteration, the iterate $\mathbf{z}^{(k)}$ converges to the eigenvector associated with the largest eigenvalue of $\mathbf{A}$. Furthermore, Eq. (5.6.23) indicates that the relative difference between the iterate $\mathbf{z}^{(k)}$ and the eigenvector $\mathbf{u}_1$ approaches zero at a rate of $|\lambda_2/\lambda_1|$,

$$\frac{\|\mathbf{z}^{(k)} - c_1\lambda_1^k\mathbf{u}_1\|}{\|c_1\lambda_1^k\mathbf{u}_1\|} = \frac{\|\mathbf{z}^{(k)} - c_1\lambda_1^k\mathbf{u}_1\|}{c_1\lambda_1^k} = O\left(\left|\frac{\lambda_2}{\lambda_1}\right|^k\right). \tag{5.6.25}$$

The implication of Eq. (5.6.25) is that, when the largest eigenvalue of $\mathbf{A}$ is significantly larger than the second largest eigenvalue, the iterations converge quickly to the largest eigenvector, regardless of the size of the matrix $\mathbf{A}$.

Next, utilizing Rayleigh's quotient, one can estimate $\lambda_1$ using $\mathbf{z}^{(k)}$,

$$\lambda_1 \approx R(\mathbf{z}^{(k)}) = \frac{\mathbf{z}^{(k)T}\mathbf{Az}^{(k)}}{\mathbf{z}^{(k)T}\mathbf{z}^{(k)}}. \tag{5.6.26}$$

Based on Eqs. (5.6.12) and (5.6.25), $R(\mathbf{z}^{(k)})$ approaches $\lambda_1$ at a rate of $|\lambda_2/\lambda_1|^2$,

$$\left\| R(\mathbf{z}^{(k)}) - \lambda_1 \right\| = O\left( \left| \frac{\lambda_2}{\lambda_1} \right|^{2k} \right). \tag{5.6.27}$$

Finally, from Eq. (5.6.23), it is clear that the magnitude of the iterate $\mathbf{z}^{(k)}$ is proportional to $\lambda_1^k$. As the iteration proceeds, the factor $\lambda_1^k$ may become exceedingly large or small and cause numerical inaccuracies in a computer. Therefore, for the practical implementation of the basic power iteration, it is necessary to replace the iteration Eq. (5.6.21) with a two-step iteration,

(1) Matrix-vector multiplication: $\tilde{\mathbf{z}} = \mathbf{A}\mathbf{z}^{(k-1)}$
(2) Vector normalization: $\mathbf{z}^{(k)} = \tilde{\mathbf{z}}/\left\| \tilde{\mathbf{z}} \right\|$,

where $\left\| \tilde{\mathbf{z}} \right\| = \sqrt{\tilde{\mathbf{z}}^T \tilde{\mathbf{z}}}$ is the vector norm to normalize the iterate $\mathbf{z}^{(k)}$. The normalization ensures that each iterate $\mathbf{z}^{(k)}$ is of unit length, and when the iteration converges, the eigenvector $\mathbf{u}_1$ is also of unit length.

While the basic power iteration is convenient to implement, there are several issues that limit its effectiveness. First, it can only produce the largest eigenpair. When the largest eigenvalue corresponds to multiple eigenvectors, the power iteration can only find a linear combination of those eigenvectors. In fact, when the starting vector contains a linear combination of such multiple eigenvectors, then this combination remains unchanged during the iterations. Upon convergence, this linear combination of eigenvectors becomes the eigenvector found by basic power iteration. Second, the convergence rate of power iteration is slow, particularly when the largest two eigenvalues are close to each other in magnitude. Next, several techniques to address these issues are described.

### Inverse Iteration with Deflation

To solve a generalized eigenproblem using power iteration, one needs to first convert the generalized eigenproblem to a standard one. The generalized eigenproblem is typically converted to a nonsymmetric formulation using the dynamic matrix, Eq. (5.6.15), instead of the symmetric formulation, Eq. (5.6.19). Due to the inclusion of the matrix inversion $\mathbf{K}^{-1}$ in the dynamic matrix, the power iteration applied to the eigenproblem $\mathbf{D}\mathbf{u} = \lambda\mathbf{u}$ is referred to as **inverse iteration**.

The choice of the nonsymmetric formulation over the symmetric one is due to the fact that the power iteration requires only matrix-vector multiplications, and the multiplication by the dynamic matrix $\mathbf{D}$ can be accomplished conveniently without explicitly computing the inverse of $\mathbf{K}$. In fact, the product of $\mathbf{D}$ with an arbitrary vector $\mathbf{u}$,

$$\mathbf{x} = \mathbf{D}\mathbf{u} = \mathbf{K}^{-1}\mathbf{M}\mathbf{u} \tag{5.6.28}$$

is the solution to the following system of linear equations,

$$\mathbf{K}\mathbf{x} = \mathbf{M}\mathbf{u}. \tag{5.6.29}$$

If one treats the right-hand side $\mathbf{Mu}$ as a load vector, Eq. (5.6.29) is effectively a problem of static structural deformation. Using the Cholesky decomposition $\mathbf{K} = \mathbf{L}_K \mathbf{L}_K^T$, the **static problem** Eq. (5.6.29) can be solved efficiently using forward and back substitution, as illustrated in Appendix A.3.8.

To implement the inverse iteration, the basic power iteration is modified as follows:

(1) Matrix-vector multiplication: To compute $\tilde{\mathbf{z}} = \mathbf{D}\mathbf{z}^{(k-1)} = \mathbf{K}^{-1}\mathbf{M}\mathbf{z}^{(k-1)}$
  (a) Multiplication with mass matrix: $\mathbf{y} = \mathbf{M}\mathbf{z}^{(k-1)}$
  (b) Solve a static problem: $\mathbf{K}\tilde{\mathbf{z}} = \mathbf{y}$
(2) Vector normalization: $\mathbf{z}^{(k)} = \tilde{\mathbf{z}} / \|\tilde{\mathbf{z}}\|_{\mathbf{M}}$,

where the vector $\mathbf{z}^{(k)}$ is normalized using the $\mathbf{M}$-norm $\|\tilde{\mathbf{z}}\|_{\mathbf{M}} = \sqrt{\tilde{\mathbf{z}}^T \mathbf{M} \tilde{\mathbf{z}}}$. The normalization ensures that every iterate is orthonormal with respect to the mass matrix $\mathbf{M}$, as required by the orthonormal condition (Eq. (5.3.12)).

Inverse iteration produces the eigenvector $\mathbf{u}_1$ associated with the largest eigenvalue $\lambda_1$ of $\mathbf{D}$, or equivalently the lowest mode of the structural dynamic system with a fundamental frequency $\omega_1^2$. Next, to obtain the second structural mode $\mathbf{u}_2$, one can restart the inverse iteration with a judiciously chosen initial guess $\mathbf{z}_2^{(0)}$. If the initial guess contains a component of the first eigenvector $\mathbf{u}_1$, then the inverse iteration converges to $\mathbf{u}_1$ again. Therefore, one needs to **sweep out** the mode $\mathbf{u}_1$ from the initial guess, which means making $\mathbf{z}_2^{(0)}$ orthogonal to $\mathbf{u}_1$, that is, $\mathbf{u}_1^T \mathbf{M} \mathbf{z}_2^{(0)} = 0$ (Hurty and Rubinstein, 1964). Sweeping out $\mathbf{u}_1$ ensures that all the subsequent iterates $\mathbf{z}_2^{(k)}$ are orthogonal to $\mathbf{u}_1$ and hence converges to the second eigenvector $\mathbf{u}_2$.

The sweeping procedure is achieved using a mathematical device called the **projection matrix**. A projection matrix $\mathbf{P}$ associated with $\mathbf{u}_1$ is defined as

$$\mathbf{P}(\mathbf{u}_1) = \mathbf{I} - \mathbf{u}_1 \mathbf{u}_1^T \mathbf{M}, \tag{5.6.30}$$

where $\mathbf{u}_1$ is assumed to be normalized, that is, $\mathbf{u}_1^T \mathbf{M} \mathbf{u}_1 = 1$. The most important property of the projection matrix is that, for any vector $\tilde{\mathbf{z}}$, the product $\mathbf{P}(\mathbf{u}_1)\tilde{\mathbf{z}}$ is always orthogonal to $\mathbf{u}_1$. This property can be verified by checking the orthogonality,

$$\mathbf{u}_1^T \mathbf{M} [\mathbf{P}(\mathbf{u}_1)\tilde{\mathbf{z}}] = \mathbf{u}_1^T \mathbf{M} \left( \tilde{\mathbf{z}} - \mathbf{u}_1 \mathbf{u}_1^T \mathbf{M} \tilde{\mathbf{z}} \right)$$
$$= \mathbf{u}_1^T \mathbf{M} \tilde{\mathbf{z}} - \underbrace{\mathbf{u}_1^T \mathbf{M} \mathbf{u}_1}_{=1} (\mathbf{u}_1^T \mathbf{M} \tilde{\mathbf{z}}) = 0. \tag{5.6.31}$$

Therefore, to ensure the orthogonality with respect to $\mathbf{u}_1$, the initial guess $\mathbf{z}_2^{(0)}$ is constructed as

$$\mathbf{z}_2^{(0)} = \mathbf{P}(\mathbf{u}_1)\tilde{\mathbf{z}} = \tilde{\mathbf{z}} - (\mathbf{u}_1^T \mathbf{M} \tilde{\mathbf{z}})\mathbf{u}_1. \tag{5.6.32}$$

In actual computations, $\mathbf{u}_1$ is known only approximately, and the round-off errors could compromise the orthogonality of the iterates with respect to $\mathbf{u}_1$. Therefore, one needs to apply the sweeping procedure to $\mathbf{z}_2^{(k)}$ in every iteration to ensure the orthogonality condition

$\mathbf{u}_1^T \mathbf{M} \mathbf{z}_2^{(k)} = 0$. Starting from an initial guess $\mathbf{z}^{(0)}$, the steps of inverse iteration are modified as follows,

(1) Sweeping procedure to eliminate $\mathbf{u}_1$: $\mathbf{z}^* = \mathbf{P}(\mathbf{u}_1)\mathbf{z}^{(k-1)}$
(2) Matrix-vector multiplication: $\tilde{\mathbf{z}} = \mathbf{D}\mathbf{z}^*$
(3) Vector normalization: $\mathbf{z}^{(k)} = \tilde{\mathbf{z}}/\|\tilde{\mathbf{z}}\|_{\mathbf{M}}$.

The sweeping procedure is closely related to the concept of **deflation**, as explained in the following section. The steps of sweeping procedure and matrix-vector multiplication are combined as

$$\begin{aligned}
\tilde{\mathbf{z}} = \mathbf{D}\mathbf{z}^* = \mathbf{D}\mathbf{P}(\mathbf{u}_1)\mathbf{z}^{(k-1)} &= \mathbf{D}\left[\mathbf{I} - \mathbf{u}_1\mathbf{u}_1^T\mathbf{M}\right]\mathbf{z}^{(k-1)} \\
&= \left(\mathbf{D} - \mathbf{D}\mathbf{u}_1\mathbf{u}_1^T\mathbf{M}\right)\mathbf{z}^{(k-1)} = \left(\mathbf{D} - \lambda_1\mathbf{u}_1\mathbf{u}_1^T\mathbf{M}\right)\mathbf{z}^{(k-1)} \\
&\equiv \mathbf{D}_1\mathbf{z}^{(k-1)},
\end{aligned} \tag{5.6.33}$$

where $\mathbf{D}_1$ is called a **deflated** matrix and defined as

$$\mathbf{D}_1 = \mathbf{D}\mathbf{P}(\mathbf{u}_1) = \mathbf{D} - \lambda_1\mathbf{u}_1\mathbf{u}_1^T\mathbf{M} \tag{5.6.34}$$

and the procedure of multiplying $\mathbf{D}$ with a projection matrix associated with its eigenvector, $\mathbf{P}(\mathbf{u}_1)$, is called **deflation**.

Deflating $\mathbf{D}$ using $\mathbf{P}(\mathbf{u}_1)$ makes its largest eigenvalue zero, while keeping all the other eigenpairs the same. In fact,

$$\mathbf{D}_1\mathbf{u}_1 = \mathbf{D}\mathbf{u}_1 - \lambda_1\mathbf{u}_1\underbrace{\mathbf{u}_1^T\mathbf{M}\mathbf{u}_1}_{=1} = \lambda_1\mathbf{u}_1 - \lambda_1\mathbf{u}_1 = 0\mathbf{u}_1 \tag{5.6.35}$$

$$\mathbf{D}_1\mathbf{u}_i = \mathbf{D}\mathbf{u}_i - \lambda_1\mathbf{u}_1\underbrace{\mathbf{u}_1^T\mathbf{M}\mathbf{u}_i}_{=0} = \lambda_i\mathbf{u}_i, \quad (i > 1). \tag{5.6.36}$$

Therefore, after deflation, the **second** largest eigenvalue of $\mathbf{D}$ becomes the largest eigenvalue of $\mathbf{D}_1$.

The concept of deflation leads to an alternative understanding of the sweeping procedure. Equation (5.6.33) shows that the inverse iteration of $\mathbf{D}$ with the projection matrix $\mathbf{P}(\mathbf{u}_1)$ is equivalent to the inverse iteration of the deflated matrix $\mathbf{D}_1$. The inverse iteration of $\mathbf{D}_1$ is expected to produce the largest eigenpair of $\mathbf{D}_1$, that is, the second largest eigenpair of $\mathbf{D}$, and thus fulfills the quest to find the second largest structural mode.

Finally, the sweeping procedure can be generalized to multiple modes. A projection matrix associated with a set of $m$ modes, $\mathbf{U}_m = [\mathbf{u}_1, \mathbf{u}_2, \ldots, \mathbf{u}_m]$, $\mathbf{U}_m^T\mathbf{M}\mathbf{U}_m = \mathbf{I}$, is defined as

$$\mathbf{P}(\mathbf{U}_m) = \mathbf{I} - \mathbf{U}_m\mathbf{U}_m^T\mathbf{M}. \tag{5.6.37}$$

The matrix-vector product $\mathbf{P}(\mathbf{U}_m)\tilde{\mathbf{z}}$ results in a vector that is orthogonal to all the $m$ modes. In fact,

$$\mathbf{U}_m^T\mathbf{M}[\mathbf{P}(\mathbf{U}_m)\tilde{\mathbf{z}}] = \mathbf{U}_m^T\mathbf{M}\tilde{\mathbf{z}} - \underbrace{\mathbf{U}_m^T\mathbf{M}\mathbf{U}_m}_{=\mathbf{I}}(\mathbf{U}_m^T\mathbf{M}\tilde{\mathbf{z}}) = 0. \tag{5.6.38}$$

When the first $m$ modes are known, the $(m+1)$th mode can be found using inverse iteration with deflation, where the step of sweeping procedure is

$$\mathbf{z}_{m+1}^{(0)} = \mathbf{P}(\mathbf{U}_m)\tilde{\mathbf{z}}. \tag{5.6.39}$$

Again, the inverse iteration of $\mathbf{D}$ with the projection matrix $\mathbf{P}(\mathbf{U}_m)$ is equivalent to inverse iteration of a deflated matrix,

$$\mathbf{D}_m = \mathbf{DP}(\mathbf{U}_m) = \mathbf{D} - \sum_{i=1}^{m} \lambda_i \mathbf{u}_i \mathbf{u}_i^T \mathbf{M}, \tag{5.6.40}$$

which causes the first $m$ eigenvalues of $\mathbf{D}$ to be zero.

To summarize, the deflation technique, or the sweeping procedure, is a remedy for the fact that power iteration only produces the largest eigenpair. Note that care must be exercised because deflation requires careful accuracy control. The eigenpairs are computed in a sequential manner, therefore the numerical errors of the eigenpairs can accumulate during the deflation and compromise the accuracy of the eigenpairs computed in the subsequent iterations.

## Inverse Iteration with Spectral Shift

As has been shown in Eq. (5.6.25), the convergence of the power iteration relies on the spectrum of an eigenproblem, that is, the distribution of all of its eigenvalues over the real axis, or more specifically the ratio of the largest two eigenvalues $\lambda_2/\lambda_1$. One can apply a technique called **spectral transformation** to transform the original eigenproblem to a new equivalent problem, whose spectrum is more favorable for power iteration (Chapter 3 in Bai et al. (2000)). For example, if the ratio $\lambda_2/\lambda_1$ in the transformed problem is 10 times smaller than the eigenvalue ratio in the original problem, then the power iteration for the new problem should converge 10 times faster. Subsequently, one can transform the eigensolution of the new problem back to the original eigenproblem to find the eigenvalue and eigenvector of interest.

In this section the spectral transformation using a **shift-and-invert** operator $(\mathbf{K} - \mu\mathbf{M})^{-1}$, where $\mu$ is called a shift, is considered. To illustrate the effect of this operator, one can start from the original generalized eigenproblem and subtract $\mu\mathbf{Mu}$ from both sides,

$$\mathbf{Ku} - \mu\mathbf{Mu} = \omega^2\mathbf{Mu} - \mu\mathbf{Mu}$$
$$(\mathbf{K} - \mu\mathbf{M})\mathbf{u} = (\omega^2 - \mu)\mathbf{Mu}$$
$$(\mathbf{K} - \mu\mathbf{M})^{-1}\mathbf{Mu} = (\omega^2 - \mu)^{-1}\mathbf{u}$$
$$\mathbf{D}_\mu\mathbf{u} = \lambda_\mu\mathbf{u}, \tag{5.6.41}$$

where

$$\mathbf{D}_\mu = (\mathbf{K} - \mu\mathbf{M})^{-1}\mathbf{M}, \quad \lambda_\mu = (\omega^2 - \mu)^{-1}. \tag{5.6.42}$$

Therefore, the shift-and-invert operator converts the generalized eigenproblem to a standard one having the same eigenvectors but different eigenvalues $\lambda_\mu = (\omega^2 - \mu)^{-1}$. Note that the eigenproblem with dynamic matrix $\mathbf{D}\mathbf{u} = \lambda\mathbf{u}$ is a special case of the shift-and-invert problem with a shift of $\mu = 0$. Following from the terminology of inverse iteration, the power iteration applied to Eq. (5.6.41) with a constant shift $\mu$ is referred to as **inverse iteration with spectral shift**.

Now if the shift $\mu$ is sufficiently close to the $m$th eigenvalue $\omega_m^2$, then $(\omega_m^2 - \mu)^{-1}$ is the eigenvalue of the largest magnitude in the spectrum of $\mathbf{D}_\mu = (\mathbf{K} - \mu\mathbf{M})^{-1}\mathbf{M}$. Furthermore, the new eigenvalue is well separated from the rest of the eigenvalues and hence the convergence of power iteration applied to the new eigenproblem with $\mathbf{D}_\mu$ is fast. Based on the convergence rate of the power iteration, Eq. (5.6.25), the convergence rate of inverse iteration with spectral shift is

$$\lim_{k \to \infty} \left\| \mathbf{z}^{(k)} - \mathbf{u}_m \right\| = O\left( \left| \frac{\omega_m^2 - \mu}{\omega_n^2 - \mu} \right|^k \right) = 0, \tag{5.6.43}$$

where $\omega_m^2$ is the closest eigenvalue of $\mathbf{D}_\mu$ to $\mu$ and $\omega_n^2$ is the second closest.

Rayleigh's quotient can be used to estimate the eigenvalue in the process of iteration. However, a more efficient use of Rayleigh's quotient is to use it as the new shift in the next iteration. When the shift is sufficiently close to the target eigenvalue, in each iteration the quotient would in principle move closer to this eigenvalue. The combination of Rayleigh's quotient and the shift-and-invert operator leads to a technique called **Rayleigh quotient iteration**. Each iteration of this algorithm is as follows:

(1) Setting the new shift: $\mu^{(k)} = R(\mathbf{z}^{(k-1)})$
(2) Matrix-vector multiplication: To compute $\tilde{\mathbf{z}} = \mathbf{D}_\mu \mathbf{z}^{(k-1)}$
    (a) Multiplication with mass matrix: $\mathbf{y} = \mathbf{M}\mathbf{z}^{(k-1)}$
    (b) Solve a static problem: $(\mathbf{K} - \mu^{(k)}\mathbf{M})\tilde{\mathbf{z}} = \mathbf{y}$
(3) Vector normalization: $\mathbf{z}^{(k)} = \tilde{\mathbf{z}} / \left\| \tilde{\mathbf{z}} \right\|_\mathbf{M}$.

Rayleigh quotient iteration achieves a remarkable cubic convergence rate for both the eigenvector and the eigenvalue; however, the price for the accelerated convergence is the requirement to solve the static problem with a new matrix $(\mathbf{K} - \mu^{(k)}\mathbf{M})$ in each iteration. Furthermore, note that when the initial shift is not sufficiently close to the target eigenvalue, Rayleigh quotient iteration may skip this eigenvalue and converge to a different eigenvalue.

The complete algorithm of inverse iteration with deflation and spectral shift is summarized in Algorithm 5.1 on page 265. The algorithm is presented in the form of **pseudocode**. The notation used in the pseudocode is explained in Appendix D.

### Simultaneous Iteration

Before considering a numerical example for the vector iteration methods, it is worth mentioning another technique for obtaining multiple eigenpairs using inverse iteration, called

---

**Algorithm 5.1** Inverse iteration with deflation and spectral shift

---

**Require:** Matrices $\mathbf{K}$ and $\mathbf{M}$, number of requested eigenpairs $m$, maximum number of iteration $N_{\max}$, tolerance $\epsilon$.

1: Initialize an empty matrix for storing the eigenvectors $\mathbf{U} = []$, and an empty list for storing the eigenvalues $\mathbf{w} = []$.

2: **for** $i = 1, \ldots, m$ **do**                          $\triangleright$ Compute the $m$ eigenpairs one by one

3:     Initialize the shift: If $i = 1$, $\mu_i^{(1)} = 0$; otherwise $\mu_i^{(1)} = \lambda_{i-1}^{(j)}$

4:     Factorize the matrix: $(\mathbf{K} - \mu_i^{(1)}\mathbf{M}) = \mathbf{L}_\mu \mathbf{L}_\mu^T$

5:     Pick a random vector $\mathbf{z}$

6:     **if** $i > 1$ **then**

7:         Sweep out the existing modes from $\mathbf{z}$. $\mathbf{z} \leftarrow \mathbf{z} - \mathbf{U}\mathbf{U}^T\mathbf{M}\mathbf{z}$

8:     **end if**

9:     Normalize $\mathbf{z}$ to set the initial guess: $\mathbf{z}_i^{(0)} = \mathbf{z}/\|\mathbf{z}\|_\mathbf{M}$

10:     **for** $j = 1, \ldots, N_{\max}$ **do**                   $\triangleright$ Inverse iteration for the $i$th eigenpair

11:         Matrix-vector multiplication: $\mathbf{y} = \mathbf{M}\mathbf{z}_i^{(j-1)}$

12:         Solve $(\mathbf{K} - \mu_i^{(j)}\mathbf{M})\tilde{\mathbf{z}} = \mathbf{y}$ using the factor $\mathbf{L}_\mu$

13:         Normalize $\tilde{\mathbf{z}}$ to get the new iterate: $\mathbf{z}_i^{(j)} = \tilde{\mathbf{z}}/\|\tilde{\mathbf{z}}\|_\mathbf{M}$

14:         Compute the new estimate of eigenvalue: $\lambda_i^{(j)} = R(\mathbf{z}_i^{(j)})$

15:         **if** $\left|\lambda_i^{(j)} - \lambda_i^{(j-1)}\right| < \epsilon \left|\lambda_i^{(j)}\right|$ **then**

16:             Break

17:         **end if**

18:         **if** using Rayleigh quotient iteration **then**

19:             Update the shift $\mu_i^{(j+1)} = \lambda_i^{(j)}$

20:             Factorize the matrix $(\mathbf{K} - \mu_i^{(j+1)}\mathbf{M}) = \mathbf{L}_\mu \mathbf{L}_\mu^T$

21:         **else**

22:             Keep using the same shift $\mu_i^{(j+1)} = \mu_i^{(j)}$

23:         **end if**

24:     **end for**

25:     Store the eigenvector $\mathbf{U} \leftarrow [\mathbf{U}; \mathbf{z}_i^{(j)}]$

26:     Store the eigenvalue $\mathbf{w} \leftarrow [\mathbf{w}, 1/\lambda_i^{(j)}]$

27: **end for**

28: **return** The requested $m$ eigenvectors as columns of matrix $\mathbf{U}$ and $m$ eigenvalues $\mathbf{w}$

---

the **simultaneous iteration** method, where the deflation technique is not employed. The simultaneous iteration method finds all the eigenpairs of interest simultaneously, avoiding the sequential procedures of inverse iteration with deflation that may cause the accumulation of numerical errors.

In simultaneous iteration, the dynamic matrix $\mathbf{D}$ is multiplied by an $n \times m$ matrix $\mathbf{Z}$, $m > 1$, as if applying the inverse iteration simultaneously to $m$ initial guesses. Besides normalizing each column of the $n \times m$ matrix, one also needs to ensure that the $m$ column vectors are orthogonal to each other, so that the $i$th column vector converges to the $i$th eigenvector. The normalization and orthogonalization are achieved using QR decomposition, a numerically stable implementation of the Gram–Schmidt process. More details on the Gram–Schmidt process and QR decomposition are provided in Appendices A.1.6 and A.3.7, respectively.

Simultaneous iteration starts with an $n \times m$ orthonormal matrix $\mathbf{Z}^{(0)}$, that is, a matrix that satisfies $\mathbf{Z}^{(0)T}\mathbf{Z}^{(0)} = \mathbf{I}_m$, and subsequently the following iterations are carried out:

(1) Matrix-matrix multiplication: $\tilde{\mathbf{Z}} = \mathbf{D}\mathbf{Z}^{(k-1)}$
(2) QR decomposition: $\mathbf{Z}^{(k)}\mathbf{R}^{(k)} = \tilde{\mathbf{Z}}$,

where $\mathbf{Z}^{(k)} \in \mathbb{R}^{n \times m}$ is an orthonormal matrix and $\mathbf{R}^{(k)} \in \mathbb{R}^{m \times m}$ is an upper triangular matrix. Generally, simultaneous iteration produces a sequence of matrices $\{\mathbf{Z}^{(k)}\}$ that converges to the first $m$ eigenvectors of the eigenproblem. Subsequently, the $i$th eigenvalue is computed using Rayleigh's quotient,

$$\lambda_i = R(\mathbf{z}_i), \tag{5.6.44}$$

where $\mathbf{z}_i$ is the $i$th column of the iterate $\mathbf{Z}^{(k)}$.

The simultaneous iteration method not only provides a useful algorithm for solving generalized eigenproblems, but also forms the basis for two important eigenvalue algorithms, the QR algorithm and the subspace iteration method, which are described later in Sections 5.6.3 and 5.6.4, respectively.

## Example Illustrating the Inverse Iteration with Spectral Shift

In this section, the algorithm of inverse iteration with spectral shift, Algorithm 5.1, is used to solve the generalized eigenvalue problem described in Section 5.5. The matrices used in this example are identical to those used in Section 5.5,

$$\mathbf{K} = 600 \begin{bmatrix} 1 & -1 & 0 \\ -1 & 3 & -2 \\ 0 & -2 & 5 \end{bmatrix}, \quad \mathbf{M} = \begin{bmatrix} 1 & 0 & 0 \\ 0 & 1.5 & 0 \\ 0 & 0 & 2.0 \end{bmatrix}. \tag{E5.13}$$

The Rayleigh quotient iteration technique is employed, and the shift is updated using the latest estimate of eigenvalue in each iteration.

The iteration procedure starts by picking an initial guess for the first eigenvector. This is accomplished by selecting an arbitrary initial vector, $\mathbf{z} = [1, 1, 1]^T$ for this $3 \times 3$ problem, and normalizing it w.r.t. the mass matrix,

$$\|\mathbf{z}\|_{\mathbf{M}} = \sqrt{\mathbf{z}^T\mathbf{M}\mathbf{z}} = \sqrt{\begin{bmatrix} 1 & 1 & 1 \end{bmatrix} \begin{bmatrix} 1 & 0 & 0 \\ 0 & 1.5 & 0 \\ 0 & 0 & 2.0 \end{bmatrix} \begin{bmatrix} 1 \\ 1 \\ 1 \end{bmatrix}}$$

$$= \sqrt{1.0 + 1.5 + 2.0} = 2.1213203 \tag{E5.14}$$

$$\mathbf{z}_1^{(0)} = \frac{\mathbf{z}}{\|\mathbf{z}\|_{\mathbf{M}}} = \frac{1}{2.1213203} \begin{bmatrix} 1 \\ 1 \\ 1 \end{bmatrix} = \begin{bmatrix} 4.7140452e{-}01 \\ 4.7140452e{-}01 \\ 4.7140452e{-}01 \end{bmatrix}, \tag{E5.15}$$

where the subscript 1 of $\mathbf{z}_1^{(0)}$ indicates it is the iterate for the first eigenvector, while the superscript (0) indicates the number of iteration, 0 for initial guess.

With the initial vector selected, the first iteration is carried out using the three steps of inverse iteration with shift:

(1) Setting the new shift: For the initial shift,

$$\mu_1^{(1)} = 0. \tag{E5.16}$$

(2) Matrix-vector multiplication: To compute $\tilde{\mathbf{z}} = (\mathbf{K} - \mu_1^{(1)}\mathbf{M})^{-1}\mathbf{M}\mathbf{z}_1^{(0)}$
  (a) Multiplication by the mass matrix:

$$\mathbf{y} = \mathbf{M}\mathbf{z}_1^{(0)} = \begin{bmatrix} 1 & 0 & 0 \\ 0 & 1.5 & 0 \\ 0 & 0 & 2.0 \end{bmatrix} \begin{bmatrix} 4.7140452e{-}01 \\ 4.7140452e{-}01 \\ 4.7140452e{-}01 \end{bmatrix}$$

$$= [4.7140452e{-}01, 7.0710678e{-}01, 9.4280904e{-}01]^T. \tag{E5.17}$$

  (b) Solve a static problem:

$$\tilde{\mathbf{z}} = (\mathbf{K} - \mu_1^{(1)}\mathbf{M})^{-1}\mathbf{y}$$

$$= \left( 600 \begin{bmatrix} 1 & -1 & 0 \\ -1 & 3 & -2 \\ 0 & -2 & 5 \end{bmatrix} - 0 \begin{bmatrix} 1 & 0 & 0 \\ 0 & 1.5 & 0 \\ 0 & 0 & 2.0 \end{bmatrix} \right)^{-1} \begin{bmatrix} 4.7140452e{-}01 \\ 7.0710678e{-}01 \\ 9.4280904e{-}01 \end{bmatrix}$$

$$= [3.4476753e{-}03, 2.3047995e{-}03, 1.1047799e{-}03]^T, \tag{E5.18}$$

  where the matrix inversion is accomplished using the Cholesky decomposition, described in Appendix A.3.8.

(3) Vector normalization to find the next iterate:

$$\mathbf{z}_1^{(1)} = \frac{\tilde{\mathbf{z}}}{\|\tilde{\mathbf{z}}\|_{\mathbf{M}}}$$

$$= [6.8572548e{-}01, 5.0286535e{-}01, 2.7429019e{-}01]^T. \tag{E5.19}$$

After the first iteration, one can estimate the eigenvalue from the first iterate $\mathbf{z}_1^{(1)}$ using Rayleigh's quotient,

$$\lambda_1^{(1)} = \frac{\mathbf{z}_1^{(1)T}\mathbf{K}\mathbf{z}_1^{(1)}}{\mathbf{z}_1^{(1)T}\mathbf{M}\mathbf{z}_1^{(1)}} = 2.1818182e{+}02. \tag{E5.20}$$

Subsequently, the second iteration is carried out again using the steps shown in Eqs. (E5.16)–(E5.19):

(1) Setting the new shift using Rayleigh's quotient, that is, the previous estimate of eigenvalue

$$\mu_1^{(2)} = \lambda_1^{(1)} = 2.1818182e{+}02. \tag{E5.21}$$

(2) Matrix-vector multiplication: To compute $\tilde{\mathbf{z}} = (\mathbf{K} - \mu_1^{(2)}\mathbf{M})^{-1}\mathbf{M}\mathbf{z}_1^{(1)}$

(a) Multiplication by the mass matrix:
$$\mathbf{y} = \mathbf{M}\mathbf{z}_1^{(1)}$$
$$= [\, 6.8572548e\text{--}01, \, 7.5429802e\text{--}01, \, 5.4858038e\text{--}01]^T. \tag{E5.22}$$

(b) Solve a static problem:
$$\tilde{\mathbf{z}} = (\mathbf{K} - \mu_1^{(2)}\mathbf{M})^{-1}\mathbf{y}$$
$$= [-1.0131294e\text{--}01, \, -6.5614745e\text{--}02, \, -3.0499300e\text{--}02]^T. \tag{E5.23}$$

(3) Vector normalization to find the next iterate:
$$\mathbf{z}_1^{(2)} = \frac{\tilde{\mathbf{z}}}{\|\tilde{\mathbf{z}}\|_{\mathbf{M}}}$$
$$= [-7.4320885e\text{--}01, \, -4.8133496e\text{--}01, \, -2.2373598e\text{--}01]^T. \tag{E5.24}$$

After the second iteration, one can estimate the eigenvalue using Rayleigh's quotient again,

$$\lambda_1^{(2)} = \frac{\mathbf{z}_1^{(2)T}\mathbf{K}\mathbf{z}_1^{(2)}}{\mathbf{z}_1^{(2)T}\mathbf{M}\mathbf{z}_1^{(2)}} = 2.1087946e\text{+}02. \tag{E5.25}$$

The rest of the iterations are carried out by repeating the steps provided in Eqs. (E5.21)–(E5.25). The results of all the iterations are listed in Table 5.2. The convergence of inverse iteration accelerated by Rayleigh quotient iteration is remarkable. Within three iterations, the first eigenvalue has been found with a relative error less than $10^{-8}$. The first eigenpair is

$$\mathbf{u}_1 = [\, 7.4265357e\text{--}01, \, 4.8163703e\text{--}01, \, 2.2416995e\text{--}01]^T$$
$$\lambda_1 = 2.1087884e\text{+}02, \text{ or } \omega_1 = 1.4521668e\text{+}01 \text{ rad/s}. \tag{E5.26}$$

After obtaining the first eigenpair, one can proceed to compute the second eigenpair using the deflation technique. To prepare an initial guess, pick an arbitrary vector $\mathbf{z} = [1, 1, 1]^T$ and sweep out the first mode shape $\mathbf{u}_1$,

$$\mathbf{z}^* = \mathbf{z} - \mathbf{u}_1\mathbf{u}_1^T\mathbf{M}\mathbf{z}$$

$$= \begin{bmatrix} 1 \\ 1 \\ 1 \end{bmatrix} - \begin{bmatrix} 7.4265357e\text{--}01 \\ 4.8163703e\text{--}01 \\ 2.2416995e\text{--}01 \end{bmatrix} \begin{bmatrix} 7.4265357e\text{--}01 \\ 4.8163703e\text{--}01 \\ 2.2416995e\text{--}01 \end{bmatrix}^T \begin{bmatrix} 1 & 0 & 0 \\ 0 & 1.5 & 0 \\ 0 & 0 & 2.0 \end{bmatrix} \begin{bmatrix} 1 \\ 1 \\ 1 \end{bmatrix}$$

$$= [-4.2102973e\text{--}01, \, 7.8412094e\text{--}02, \, 5.7106224e\text{--}01]^T. \tag{E5.27}$$

**Table 5.2 Convergence history of the inverse iteration for the first eigenpair**

| Iteration, $i$ | Mode shape, $\mathbf{z}_1^{(i)T}$ | Eigenvalue, $\lambda_1^{(i)}$ |
|---|---|---|
| 1 | [ 6.8572548e–01, 5.0286535e–01, 2.7429019e–01] | 2.1818182e+02 |
| 2 | [–7.4320885e–01, –4.8133496e–01, –2.2373598e–01] | 2.1087946e+02 |
| 3 | [ 7.4265357e–01, 4.8163703e–01, 2.2416995e–01] | 2.1087884e+02 |
| 4 | [ 7.4265357e–01, 4.8163703e–01, 2.2416995e–01] | 2.1087884e+02 |

Next, normalize $\mathbf{z}^*$ to obtain the initial guess,

$$\mathbf{z}_2^{(0)} = \mathbf{z}^* / \|\mathbf{z}^*\|_{\mathbf{M}}$$
$$= [-4.5973346e{-}01,\, 8.5620231e{-}02,\, 6.2355790e{-}01]^T. \qquad (E5.28)$$

Subsequently, the first iteration for the second eigenvector is carried out using the following four steps:

(1) Setting the new shift: For the initial shift, use the previous eigenvalue
$$\mu_2^{(1)} = \lambda_1 = 2.1087884e{+}02. \qquad (E5.29)$$

(2) Matrix-vector multiplication: To compute $\tilde{\mathbf{z}} = (\mathbf{K} - \mu_2^{(1)}\mathbf{M})^{-1}\mathbf{M}\mathbf{z}_2^{(0)}$
  (a) Multiplication by the mass matrix:
$$\mathbf{y} = \mathbf{M}\mathbf{z}_2^{(0)}$$
$$= [-4.5973346e{-}01,\, 1.2843035e{-}01,\, 1.2471158e00]^T. \qquad (E5.30)$$

  (b) Solve a static problem:
$$\tilde{\mathbf{z}} = (\mathbf{K} - \mu_2^{(1)}\mathbf{M})^{-1}\mathbf{y}$$
$$= [\, 6.4775430e{-}02,\, 4.2972938e{-}02,\, 2.0411235e{-}02]^T. \qquad (E5.31)$$

(3) Sweep out the first mode $\mathbf{u}_1$ from the matrix-vector product $\tilde{\mathbf{z}}$,
$$\mathbf{z}^* = \tilde{\mathbf{z}} - \mathbf{u}_1\mathbf{u}_1^T\mathbf{M}\tilde{\mathbf{z}}$$
$$= [-8.0304348e{-}04,\, 4.4298557e{-}04,\, 6.1637549e{-}04]^T \qquad (E5.32)$$

(4) Vector normalization to find the next iterate:
$$\mathbf{z}_2^{(1)} = \frac{\mathbf{z}^*}{\|\mathbf{z}^*\|_{\mathbf{M}}}$$
$$= [-5.7827256e{-}01,\, 2.6628341e{-}01,\, 5.2879134e{-}01]^T. \qquad (E5.33)$$

After this iteration, estimate the eigenvalue associated with $\mathbf{z}_2^{(1)}$ using Rayleigh's quotient,

$$\lambda_2^{(1)} = \frac{\mathbf{z}_2^{(1)T}\mathbf{K}\mathbf{z}_2^{(1)}}{\mathbf{z}_2^{(1)T}\mathbf{M}\mathbf{z}_2^{(1)}} = 1.0139739e{+}03. \qquad (E5.34)$$

The eigenvalue estimate $\lambda_2^{(1)}$ becomes the new shift $\mu_2^{(2)}$ for the next iteration.

The subsequent iterations are performed by repeating the steps Eqs. (E5.29)–(E5.34). The results of all the iterations are listed in Table 5.3. Again the second eigenpair converges quickly within three iterations and is found to be

$$\mathbf{u}_2 = [\, 6.3577474e{-}01,\, -3.8566038e{-}01,\, -4.3167673e{-}01]^T$$
$$\lambda_2 = 9.6395946e{+}02,\ \text{or}\ \omega_2 = 3.1047697e{+}01\ \text{rad/s}. \qquad (E5.35)$$

The third eigenpair is found using the same iteration procedure. However, in this particular example problem, there are only three eigenvectors, and hence the third eigenpair is

**Table 5.3 Convergence history of the inverse iteration for the second eigenpair**

| Iteration, $i$ | Mode shape, $\mathbf{z}_2^{(i)T}$ | Eigenvalue, $\lambda_2^{(i)}$ |
|---|---|---|
| 1 | [−5.7827256e−01, 2.6628341e−01, 5.2879134e−01] | 1.0139739e+03 |
| 2 | [ 6.3775452e−01, −3.9074894e−01, −4.2675641e−01] | 9.6406533e+02 |
| 3 | [−6.3577455e−01, 3.8565991e−01, 4.3167717e−01] | 9.6395946e+02 |
| 4 | [ 6.3577474e−01, −3.8566038e−01, −4.3167673e−01] | 9.6395946e+02 |

easy to compute. Sweeping out the first two eigenvectors $\mathbf{U} = [\mathbf{u}_1, \mathbf{u}_2]$ from an initial vector directly results in the third eigenvector

$$\mathbf{z}^* = \mathbf{z} - \mathbf{U}\mathbf{U}^T\mathbf{M}\mathbf{z}$$
$$= [\, 9.1448752e{-}02, -2.3245689e{-}01, 2.2310089e{-}01]^T \quad (E5.36)$$
$$\mathbf{z}_3 = \mathbf{z}^* / \|\mathbf{z}^*\|_{\mathbf{M}}$$
$$= [\, 2.1037148e{-}01, -5.3475088e{-}01, 5.1322806e{-}01]^T. \quad (E5.37)$$

The eigenvalue is found using Rayleigh's quotient,

$$\lambda_3 = \frac{\mathbf{z}_3^T\mathbf{K}\mathbf{z}_3}{\mathbf{z}_3^T\mathbf{M}\mathbf{z}_3} = 2.1251617e{+}03. \quad (E5.38)$$

The third eigenpair is therefore,

$$\mathbf{u}_3 = [\, 2.1037148e{-}01, -5.3475088e{-}01, 5.1322806e{-}01]^T$$
$$\lambda_3 = 2.1251617e{+}03, \text{ or } \omega_3 = 4.6099476e{+}01 \text{ rad/s}. \quad (E5.39)$$

The three eigenpairs, Eqs. (E5.26), (E5.35), and (E5.39), are identical to the values found in the analytical example presented in Section 5.5, and these verify the correctness of the inverse iteration algorithm.

Finally, the convergence history of the inverse iteration without spectral shift is compared to that of the Rayleigh quotient iteration in Figure 5.5. Without using the shift during the inverse iteration, the number of iterations required for convergence is significantly increased, which underlines the importance of using spectral shift in inverse iteration. For eigenproblems of larger size, the acceleration in the computer time achieved using spectral shift becomes more evident, even though with spectral shift one needs to perform Cholesky decomposition at each iteration (Chapter 27 of Trefethen and Bau III (1997)).

To summarize, in this example the generalized eigenproblem is solved using the inverse iteration method with spectral shift. The method centers around two operations. One is the repeated multiplication of a matrix, constructed using the stiffness and mass matrices, to a vector iterate, which leads to an eigenvector. The other is the sweep-out or deflation procedure that eliminates the known eigenvectors from the vector iterate, so as to find the remaining unknown eigenvectors. In the end, it also illustrates the computational acceleration achieved by the spectral shift technique, when compared to the inverse iteration without any shift.

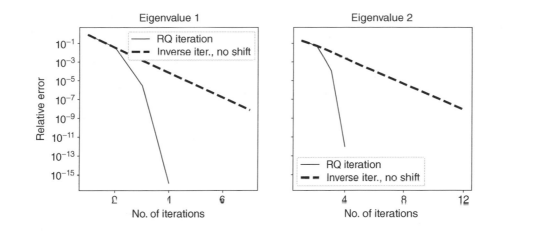

**Figure 5.5** Convergence history of inverse iteration without spectral shift and Rayleigh quotient iteration

### 5.6.3 Matrix Transformation Methods

This section describes matrix transformation methods, including the Jacobi's method and the QR algorithm. These methods are based on similarity transformation and generate all the eigenpairs simultaneously. The matrix transformation methods are computationally superior to vector iteration methods when a significant number of the eigenpairs is required for moderate-size problems.

The matrix transformation methods are applicable to both symmetric and nonsymmetric eigenproblems, but these methods are computationally more efficient for the symmetric problems. Therefore, throughout this section, the symmetric form of the eigenproblem $\mathbf{S}\mathbf{u} = \lambda\mathbf{u}$ is presented, instead of the nonsymmetric formulation using dynamic matrix $\mathbf{D}$. The matrix transformation methods for nonsymmetric eigenproblems are briefly discussed in Section 5.6.5.

#### Similarity Transformation and Jacobi's Method

The similarity transformation of a matrix $\mathbf{S}$ is defined as $\tilde{\mathbf{S}} = \mathbf{Q}^{-1}\mathbf{S}\mathbf{Q}$, where $\mathbf{Q}$ is called a transformation matrix. To maintain the symmetry during the transformation, the matrix $\mathbf{Q}$ has to be orthonormal, that is, $\mathbf{Q}^{-1} = \mathbf{Q}^T$, and therefore $\tilde{\mathbf{S}} = \mathbf{Q}^T\mathbf{S}\mathbf{Q}$. More details on similarity transformation are provided in Appendix A.4.

The transformed matrix $\tilde{\mathbf{S}}$ has the same eigenvalues as the original matrix $\mathbf{S}$. In fact, pre-multiplying $\mathbf{Q}^T$ the equation $\mathbf{S}\mathbf{u} = \lambda\mathbf{u}$ results in

$$\mathbf{Q}^T\mathbf{S}\mathbf{u} = \lambda\mathbf{Q}^T\mathbf{u}$$
$$\mathbf{Q}^T\mathbf{S}\underbrace{\mathbf{Q}\mathbf{Q}^T}_{=\mathbf{I}}\mathbf{u} = \lambda\mathbf{Q}^T\mathbf{u}$$
$$(\mathbf{Q}^T\mathbf{S}\mathbf{Q})(\mathbf{Q}^T\mathbf{u}) = \lambda\mathbf{Q}^T\mathbf{u}$$
$$\tilde{\mathbf{S}}\tilde{\mathbf{u}} = \lambda\tilde{\mathbf{u}}. \tag{5.6.45}$$

Therefore, $\mathbf{S}$ and $\tilde{\mathbf{S}}$ have the same eigenvalues, and the eigenvectors of $\mathbf{S}$ and $\tilde{\mathbf{S}}$ are related by the transformation $\mathbf{u} = \mathbf{Q}\tilde{\mathbf{u}}$.

The underlying idea of the matrix transformation methods is to apply a succession of similarity transformations to the initial matrix $\mathbf{S}^{(0)} = \mathbf{S}$, with the transformation matrices $\{\mathbf{Q}^{(i)}\}_{i=1}^{k}$ chosen such that the transformed matrix gradually approaches the diagonal form. The succession of transformations is written explicitly as

$$
\begin{aligned}
\mathbf{S}^{(k)} &= \mathbf{Q}^{(k)T}\mathbf{S}^{(k-1)}\mathbf{Q}^{(k)} \\
&= \mathbf{Q}^{(k)T}\mathbf{Q}^{(k-1)T}\mathbf{S}^{(k-2)}\mathbf{Q}^{(k-1)}\mathbf{Q}^{(k)} \\
&= \underbrace{\mathbf{Q}^{(k)T}\mathbf{Q}^{(k-1)T}\cdots\mathbf{Q}^{(1)T}}_{\equiv \mathbf{Q}^T}\,\mathbf{S}^{(0)}\,\underbrace{\mathbf{Q}^{(1)}\cdots\mathbf{Q}^{(k-1)}\mathbf{Q}^{(k)}}_{\equiv \mathbf{Q}} \\
&= \mathbf{Q}^T\mathbf{S}\mathbf{Q}.
\end{aligned}
\tag{5.6.46}
$$

Once $\mathbf{S}^{(k)}$ becomes sufficiently close to a diagonal matrix, that is, having off-diagonal elements close to zero, one can identify the eigenvalues of $\mathbf{S}$ from the diagonal elements of $\mathbf{S}^{(k)}$, and the eigenvectors from the columns of transformation matrix product $\mathbf{Q} = \mathbf{Q}^{(1)}\cdots\mathbf{Q}^{(k-1)}\mathbf{Q}^{(k)}$.

A classical example of the matrix transformation method is Jacobi's method, which is in fact one of the oldest methods for symmetric eigenproblems. The method is developed by Carl Gustav Jacob Jacobi in 1846, but gains popularity only after the advent of computers (Forsythe and Henrici, 1960). The Jacobi's method relies on the application of Givens rotation matrix $\mathbf{G}$ as the transformation matrix,

$$
\mathbf{G}(i,j,\theta) = \begin{array}{c} \\ \\ i\text{th row} \\ \\ j\text{th row} \\ \\ \\ \end{array}
\begin{bmatrix}
1 & \cdots & 0 & \cdots & 0 & \cdots & 0 \\
\vdots & \ddots & \vdots & & \vdots & & \vdots \\
0 & \cdots & \cos(\theta) & \cdots & -\sin(\theta) & \cdots & 0 \\
\vdots & & \vdots & \ddots & \vdots & & \vdots \\
0 & \cdots & \sin(\theta) & \cdots & \cos(\theta) & \cdots & 0 \\
\vdots & & \vdots & & \vdots & \ddots & \vdots \\
0 & \cdots & 0 & \cdots & 0 & \cdots & 1
\end{bmatrix},
\tag{5.6.47}
$$

where the cos and sin terms appear at the intersections the $i$th and $j$th rows and columns. The Givens rotation matrix is orthonormal and thus easy to invert

$$
\mathbf{G}(i,j,\theta)^{-1} = \mathbf{G}(i,j,-\theta) = \mathbf{G}(i,j,\theta)^T.
\tag{5.6.48}
$$

In the $k$th iteration of Jacobi's method, the rotation matrix is constructed such that the **pivot**, the largest off-diagonal element of $\mathbf{S}^{(k-1)}$, becomes zero in $\mathbf{S}^{(k)}$. Assuming the pivot

$S_{ij}^{(k-1)}$ is on the $i$th row and $j$th column, it can be shown that $S_{ij}^{(k)} = 0$ if the parameter $\theta^{(k)}$ is chosen such that

$$\tan(2\theta^{(k)}) = \frac{2S_{ij}^{(k-1)}}{S_{jj}^{(k-1)} - S_{ii}^{(k-1)}}, \tag{5.6.49}$$

where $\theta^{(k)} = \pi/4$ when $S_{jj}^{(k-1)} = S_{ii}^{(k-1)}$.

Thus, the $k$th iteration of Jacobi's method consists of the following steps:

(1) Identify the pivot $S_{ij}^{(k-1)}$ from $\mathbf{S}^{k-1}$.
(2) Compute $\theta^{(k)}$ using Eq. (5.6.49).
(3) Let $\mathbf{Q}^{(k)} = \mathbf{G}(i,j,\theta^{(k)})$.
(4) Compute $\mathbf{S}^{(k)} = \mathbf{Q}^{(k)T}\mathbf{S}^{(k-1)}\mathbf{Q}^{(k)}$,

Note that when computing $\mathbf{S}^{(k)}$, one only needs to modify the $i$th and $j$th rows and columns of $\mathbf{S}^{(k-1)}$.

Mathematically speaking, an infinite number of rotations are required to ensure that the matrix $\mathbf{S}$ diagonal. However, in practice only a finite number of multiplications are necessary to achieve the diagonalization with machine precision. To test for convergence, one can check if the maximum off-diagonal element is smaller than the prescribed tolerance.

Alternatively, a more refined error estimate of the eigenvalues can be obtained using the Gershgorin circle theorem. For a real symmetric matrix $\mathbf{A}$, the theorem states that the sum of absolute values of the off-diagonal elements, $R_i$, can serve as an error estimate of the eigenvalue $\lambda_i$. Apply the theorem to the $k$th iteration $\mathbf{S}^{(k)}$ of Jacobi's method,

$$\left| \lambda - S_{ii}^{(k)} \right| \le R_i = \sum_{j \neq i} \left| S_{ij}^{(k)} \right|. \tag{5.6.50}$$

As the iterations proceed, $R_i$ should approach zero. When $R_i$ is smaller than the prescribed tolerance, one can assume that convergence has been achieved.

## Tridiagonalization and Householder's Transformation

While Jacobi's method requires only the multiplication of Givens matrix and is easy to implement, it has an undesirable property since the transformation matrices do not maintain the zero structure during iteration. Every time a new pair of off-diagonal elements are zeroed out, the previously zeroed elements may become nonzero again. Direct transformation to a diagonal matrix may not be the most efficient approach to solve a symmetric eigenproblem. Instead, a two-step procedure is a better alternative:

$$\begin{bmatrix} \times & \times & \times & \times \\ \times & \times & \times & \times \\ \times & \times & \times & \times \\ \times & \times & \times & \times \end{bmatrix} \xrightarrow{\text{Tridiagonalize}} \begin{bmatrix} \times & \times & & \\ \times & \times & \times & \\ & \times & \times & \times \\ & & \times & \times \end{bmatrix} \xrightarrow{\text{Diagonalize}} \begin{bmatrix} \times & & & \\ & \times & & \\ & & \times & \\ & & & \times \end{bmatrix}$$

The procedure starts by transforming a real symmetric matrix into a tridiagonal matrix; the tridiagonalization can be done exactly in a **finite** number of similarity transformations. The second step is to solve the tridiagonal eigenproblem, for which several efficient eigenvalue algorithms exist.

There are two typical approaches to tridiagonalizing a **dense** real symmetric matrix. One approach consists of employing the Givens rotation (Givens, 1953). Unlike the strategy in Jacobi's method, the Givens rotation is applied to zero out only the non-tridiagonal elements. Givens rotation needs to be performed exactly $(n-1)(n-2)/2$ times to tridiagonalize an $n \times n$ matrix, because it zeros out one element at a time. A more efficient method for tridiagonalization, however, is the Householder's transformation (Householder, 1958). It zeros out at once all the desired entries in one column, and the tridiagonalization of an $n \times n$ matrix is achieved using only $(n-2)$ transformations. Therefore, in the following, the Householder's transformation and its application to tridiagonalization are presented.

Householder's transformation is done using the Householder reflection matrix, which is defined as

$$\mathbf{H}(\mathbf{u}) = \mathbf{I} - 2\mathbf{u}\mathbf{u}^T, \quad \mathbf{u}^T\mathbf{u} = 1. \tag{5.6.51}$$

The reflection matrix is symmetric,

$$\mathbf{H}(\mathbf{u})^T = (\mathbf{I} - 2\mathbf{u}\mathbf{u}^T)^T = \mathbf{I} - 2\mathbf{u}\mathbf{u}^T = \mathbf{H}(\mathbf{u}) \tag{5.6.52}$$

as well as orthonormal,

$$\begin{aligned}
\mathbf{H}(\mathbf{u})^2 &= \mathbf{H}(\mathbf{u})^T\mathbf{H}(\mathbf{u}) = \mathbf{H}(\mathbf{u})\mathbf{H}(\mathbf{u})^T \\
&= (\mathbf{I} - 2\mathbf{u}\mathbf{u}^T)(\mathbf{I} - 2\mathbf{u}\mathbf{u}^T) \\
&= \mathbf{I} - 2\mathbf{u}\mathbf{u}^T - 2\mathbf{u}\mathbf{u}^T + 4\mathbf{u}\underbrace{\mathbf{u}^T\mathbf{u}}_{=1}\mathbf{u}^T = \mathbf{I},
\end{aligned}$$

which indicates

$$\mathbf{H}(\mathbf{u})^T = \mathbf{H}(\mathbf{u})^{-1}. \tag{5.6.53}$$

Therefore, the reflection matrix is suitable for the similarity transformation of a symmetric matrix.

In Figure 5.6, a geometrical example is provided to illustrate the effect of the matrix $\mathbf{H}(\mathbf{u})$. Figure 5.6 shows an arbitrary vector $\mathbf{q} = \alpha\mathbf{u} + \beta\mathbf{w}$, where $\alpha$ and $\beta$ are two constants and $\mathbf{u}$ and $\mathbf{w}$ are orthonormal vectors. Pre-multiplying $\mathbf{H}(\mathbf{u})$ with $\mathbf{q}$, one obtains a new vector $\tilde{\mathbf{q}}$

**Figure 5.6** Illustrative example of Householder reflection

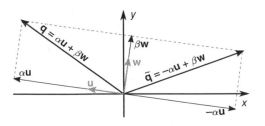

$$\tilde{\mathbf{q}} = \mathbf{H}(\mathbf{u})\mathbf{q} = (\mathbf{I} - 2\mathbf{u}\mathbf{u}^T)(\alpha\mathbf{u} + \beta\mathbf{w})$$
$$= \alpha(\mathbf{u} - 2\mathbf{u}\underbrace{\mathbf{u}^T\mathbf{u}}_{=1}) + \beta(\mathbf{w} - 2\mathbf{u}\underbrace{\mathbf{u}^T\mathbf{w}}_{=0})$$
$$= -\alpha\mathbf{u} + \beta\mathbf{w}. \tag{5.6.54}$$

When compared to $\mathbf{q}$, the component of $\tilde{\mathbf{q}}$ in the $\mathbf{w}$ direction remains the same, but its component in the $\mathbf{u}$ direction becomes negative. In other words, the matrix $\mathbf{H}(\mathbf{u})$ reflects a vector $\mathbf{q}$ along the direction of $\mathbf{u}$, and obtains its "mirror" $\tilde{\mathbf{q}}$. Therefore, $\mathbf{H}(\mathbf{u})$ is referred to as reflection. Also, note that after the reflection, the length of the vector remains the same.

Conversely, if one is given two vectors, $\mathbf{q}$ and $\tilde{\mathbf{q}}$, of the same length, and the transformation of $\mathbf{q}$ to $\tilde{\mathbf{q}}$ using a reflection matrix $\mathbf{H}(\mathbf{u})$ is required, then the vector $\mathbf{u}$ should be chosen as

$$\mathbf{u} = \pm\frac{\mathbf{q} - \tilde{\mathbf{q}}}{\|\mathbf{q} - \tilde{\mathbf{q}}\|}. \tag{5.6.55}$$

Next, assume that a sequence of $(k-1)$ Householder's transformations have been applied to a symmetric matrix $\mathbf{S}^{(0)} = \mathbf{S}$,

$$\mathbf{S}^{(i)} = \mathbf{H}^{(i)}\mathbf{S}^{(i-1)}\mathbf{H}^{(i)}, \quad \text{for } i = 1, \cdots, k-1 \tag{5.6.56}$$

such that the $(k-1)$th iterate becomes

$$\mathbf{S}^{(k-1)} = \begin{bmatrix} \times & \times & & & & & & & \\ \times & \times & \ddots & & & & & & \\ & \ddots & \ddots & & \times & & & & \\ & & \times & S_{k,k}^{(k-1)} & S_{k,k+1}^{(k-1)} & S_{k,k+2}^{(k-1)} & \cdots & S_{k,n}^{(k-1)} \\ & & & S_{k+1,k}^{(k-1)} & & & & \\ & & & S_{k+2,k}^{(k-1)} & & \text{Fully populated block of} & & \\ & & & \vdots & & (n-k+1) \times (n-k+1) & & \\ & & & S_{n,k}^{(k-1)} & & & & \end{bmatrix}, \tag{5.6.57}$$

where the upper left block of $(k-1) \times (k-1)$ has been tridiagonalized and the lower right block of $(n-k+1) \times (n-k+1)$ remains fully populated. Next, using Eq. (5.6.55), a reflection matrix $\mathbf{H}^{(k)}$ is constructed to reflect the $k$th column (or equivalently the $k$th row)

$$\mathbf{q} = [\underbrace{0, \ldots, 0}_{k-2 \text{ zeros}}, S_{k-1,k}^{(k-1)}, S_{k,k}^{(k-1)}, \underbrace{S_{k+1,k}^{(k-1)}}_{\substack{\text{First} \\ \text{sub-diagonal}}}, S_{k+2,k}^{(k-1)}, \ldots, S_{n,k}^{(k-1)}]^T \tag{5.6.58}$$

to a vector $\tilde{\mathbf{q}}$ containing mostly zeros

$$\tilde{\mathbf{q}} = [\underbrace{0, \ldots, 0, S_{k-1,k}^{(k-1)}, S_{k,k}^{(k-1)}}_{\text{Unchanged}}, \underbrace{S_{k+1,k}^{(k)}}_{\text{Modified}}, \underbrace{0, \ldots, 0}_{n-k-1 \text{ new zeros}}]^T, \tag{5.6.59}$$

where

$$S_{k+1,k}^{(k)} = -\text{sign}(S_{k+1,k}^{(k-1)}) \left( \sum_{i=k+1}^{n} \left[ S_{i,k}^{(k-1)} \right]^2 \right)^{1/2}. \tag{5.6.60}$$

This reflection matrix $\mathbf{H}^{(k)}$ is employed for the $k$th Householder's transformation, which zeros out $(n - k - 1)$ nonzero elements each on the $k$th column and row, and produces a matrix having a larger tridiagonal block of $k \times k$,

$$\mathbf{S}^{(k)} = \left[ \begin{array}{ccccc|cccc}
\times & \times & & & & & & & \\
\times & \times & \ddots & & & & & & \\
& \ddots & \ddots & & \times & & & & \\
& & \times & S_{k,k}^{(k-1)} & S_{k,k+1}^{(k)} & 0 & \cdots & & 0 \\
\hline
& & & S_{k+1,k}^{(k)} & & & & & \\
& & & 0 & & \multicolumn{4}{c}{\text{Fully populated block of}} \\
& & & \vdots & & \multicolumn{4}{c}{(n-k) \times (n-k)} \\
& & & 0 & & & & &
\end{array} \right].$$

Next, reflection matrices $\mathbf{H}^{(k+1)}$, $\mathbf{H}^{(k+2)}$, ..., $\mathbf{H}^{(n-2)}$ are constructed in a similar manner to bring the initial matrix $\mathbf{S}$ to a tridiagonal form $\mathbf{S}^{(n-2)}$. The product of all $(n - 2)$ reflection matrices

$$\mathbf{Q} = \mathbf{H}^{(1)} \ldots \mathbf{H}^{(n-2)} \tag{5.6.61}$$

is a similarity transformation matrix that tridiagonalizes $\mathbf{S}$,

$$\mathbf{S}^{(n-2)} = \mathbf{H}^{(n-2)} \ldots \mathbf{H}^{(1)} \mathbf{S}^{(0)} \mathbf{H}^{(1)} \ldots \mathbf{H}^{(n-2)} \equiv \mathbf{Q}^T \mathbf{S} \mathbf{Q}. \tag{5.6.62}$$

### Basic QR Algorithm

Having transformed a symmetric matrix into tridiagonal form, one can proceed to solve a symmetric tridiagonal eigenproblem. This section presents the well-known **QR algorithm** (Rutishauser, 1958; Francis, 1961, 1962). The QR algorithm is applicable to both symmetric and nonsymmetric eigenproblems, and it is by no means limited to tridiagonal problems. It not only contains concepts and ideas that are the basis of eigenvalue algorithms for large-scale problems, but also is of great practical importance in many standard eigensolvers, particularly in commercial finite element codes. There are also several algorithms designed specially for symmetric tridiagonal eigenproblems. The interested readers are referred to Géradin and Rixen (2014) for a discussion on these algorithms.

The basic QR algorithm consists of a two-step iteration. Considering the symmetric case in Eq. (5.6.19), $\mathbf{Su} = \lambda\mathbf{u}$, and setting $\mathbf{S}^{(0)} = \mathbf{S}$,

(1) QR decomposition:

$$\mathbf{Q}^{(k)}\mathbf{R}^{(k)} = \mathbf{S}^{(k-1)} \tag{5.6.63}$$

(2) Matrix reconstruction:

$$\mathbf{S}^{(k)} = \mathbf{R}^{(k)}\mathbf{Q}^{(k)}. \tag{5.6.64}$$

Upon convergence, the iterate $\mathbf{S}^{(k)}$ becomes a diagonal matrix containing the eigenvalues of $\mathbf{S}$, while the eigenvectors are the columns of the matrix product

$$\mathbf{P}_k = \mathbf{Q}^{(1)}\dots\mathbf{Q}^{(k)}. \tag{5.6.65}$$

Note that $\mathbf{P}_k$ is an orthonormal matrix because the $\mathbf{Q}^{(k)}$ matrices are orthonormal.

There are two key ideas pertaining to how the basic QR algorithm works. First, the matrix product $\mathbf{P}_k$ is a similarity transformation matrix that transforms the original matrix $\mathbf{S}$ to the iterate $\mathbf{S}^{(k)}$. In fact, since $\mathbf{Q}^{(k)}$ is orthonormal, the QR decomposition step, Eq. (5.6.63), can be written as,

$$\mathbf{R}^{(k)} = \mathbf{Q}^{(k)T}\mathbf{S}^{(k-1)}. \tag{5.6.66}$$

Subsequently, the matrix reconstruction step, Eq. (5.6.64), can be written as

$$\begin{aligned} \mathbf{S}^{(k)} &= \mathbf{R}^{(k)}\mathbf{Q}^{(k)} \\ &= \mathbf{Q}^{(k)T}\mathbf{S}^{(k-1)}\mathbf{Q}^{(k)} \\ &= \mathbf{Q}^{(k)T}\dots\mathbf{Q}^{(1)T}\mathbf{S}^{(0)}\mathbf{Q}^{(1)}\dots\mathbf{Q}^{(k)} \\ &= \mathbf{P}_k^T\mathbf{S}\mathbf{P}_k, \end{aligned} \tag{5.6.67}$$

which represents a similarity transformation from $\mathbf{S}$ to $\mathbf{S}^{(k)}$.

The second key idea of the basic QR algorithm is that the iterate $\mathbf{S}^{(k)}$ converges to an upper triangular matrix. In fact, using Eq. (5.6.67), the QR decomposition step, Eq. (5.6.63), can be written as

$$\mathbf{Q}^{(k)}\mathbf{R}^{(k)} = \mathbf{S}^{(k-1)} = \mathbf{P}_{k-1}^T\mathbf{S}\mathbf{P}_{k-1}. \tag{5.6.68}$$

Pre-multiply Eq. (5.6.68) by $\mathbf{P}_{k-1}$, and note that $\mathbf{P}_k = \mathbf{P}_{k-1}\mathbf{Q}^{(k)}$,

$$\mathbf{P}_{k-1}\mathbf{Q}^{(k)}\mathbf{R}^{(k)} = \underbrace{\mathbf{P}_{k-1}\mathbf{P}_{k-1}^T}_{=\mathbf{I}}\mathbf{S}\mathbf{P}_{k-1}$$

$$\mathbf{P}_k\mathbf{R}^{(k)} = \mathbf{S}\mathbf{P}_{k-1}. \tag{5.6.69}$$

Next, pre-multiply Eq. (5.6.69) by $\mathbf{P}_k^T$,

$$\mathbf{R}^{(k)} = \mathbf{P}_k^T\mathbf{S}\mathbf{P}_{k-1}. \tag{5.6.70}$$

If the matrix sequence $\{\mathbf{P}_k\}$ converges to a matrix $\mathbf{P}$, that is, $\mathbf{P}_k = \mathbf{P}_{k-1} = \mathbf{P}$ as $k \to \infty$, one can find the limits of the matrix sequences $\{\mathbf{S}^{(k)}\}$ and $\{\mathbf{R}^{(k)}\}$ using Eqs. (5.6.67) and (5.6.70), respectively,

$$\lim_{k \to \infty} \mathbf{S}^{(k)} = \lim_{k \to \infty} \mathbf{P}_k^T \mathbf{S} \mathbf{P}_k = \mathbf{P}^T \mathbf{S} \mathbf{P} \tag{5.6.71}$$

$$\lim_{k \to \infty} \mathbf{R}^{(k)} = \lim_{k \to \infty} \mathbf{P}_k^T \mathbf{S} \mathbf{P}_{k-1} = \mathbf{P}^T \mathbf{S} \mathbf{P}. \tag{5.6.72}$$

In other words, the iterate $\mathbf{S}^{(k)}$ converges to an upper triangular matrix $\mathbf{R}$,

$$\lim_{k \to \infty} \mathbf{S}^{(k)} = \mathbf{P}^T \mathbf{S} \mathbf{P} = \lim_{k \to \infty} \mathbf{R}^{(k)} \equiv \mathbf{R} \tag{5.6.73}$$

or equivalently,

$$\mathbf{S} = \mathbf{P} \mathbf{R} \mathbf{P}^T. \tag{5.6.74}$$

Note that Eq. (5.6.74) represents a similarity transformation and therefore the matrices $\mathbf{S}$ and $\mathbf{R}$ possess the same eigenvalues. Furthermore, since $\mathbf{S}$ is a symmetric matrix, $\mathbf{R}$ is symmetric and hence diagonal. This means that the diagonal elements of $\mathbf{R}$ and the column vectors of $\mathbf{P}$ are the eigenvalues and eigenvectors of $\mathbf{S}$, respectively, and that Eq. (5.6.74) represents the eigenvalue decomposition of the matrix $\mathbf{S}$.

To summarize, the basic QR algorithm successively constructs an orthonormal matrix $\mathbf{P}$ that diagonalizes the symmetric matrix $\mathbf{S}$.

An alternative perspective to understanding the basic QR algorithm is via the simultaneous iteration method, which was described in Section 5.6.2. The basic QR algorithm is equivalent to the simultaneous iteration method with $n$ initial guesses $\mathbf{Z}^{(0)} = \mathbf{I}$. Combining the steps of matrix-matrix multiplication and QR decomposition in simultaneous iteration, its $(k + 1)$th iteration is written as,

$$\mathbf{Z}^{(k+1)} \mathbf{R}^{(k+1)} = \mathbf{S} \mathbf{Z}^{(k)}$$
$$\mathbf{R}^{(k+1)} = \mathbf{Z}^{(k+1)T} \mathbf{S} \mathbf{Z}^{(k)}. \tag{5.6.75}$$

Equation (5.6.75) is identical to Eq. (5.6.70) obtained from the QR algorithm, with $\mathbf{Z}^{(k)} = \mathbf{P}_k$. This shows that the iterates $\mathbf{Z}^{(k)}$ in simultaneous iteration converge to the same similarity transformation matrix found by the basic QR algorithm.

### Practical QR Algorithm

The basic QR algorithm has been improved in several aspects in order to become a practical eigenvalue algorithm. First, the most computational intensive step in the algorithm is the QR decomposition, whose operation count is proportional to $n^3$, denoted by $O(n^3)$, for a fully populated $n \times n$ matrix. It is therefore recommended to tridiagonalize a symmetric matrix $\mathbf{S}$ before applying the QR algorithm, since the QR decomposition of a tridiagonal matrix has a lower operation count of $O(n)$.

Next, a standard technique to accelerate the convergence of the QR algorithm is to use the spectral shift in Rayleigh quotient iteration. This is not surprising, since the QR algorithm

---

**Algorithm 5.2** The QR algorithm with tridiagonalization and shift for standard symmetric eigenproblem

---

**Require:** Symmetric matrix $\mathbf{S}$ of size $n \times n$, maximum number of iteration $N_{max}$, tolerance $\epsilon$.

1: Initialize an identity matrix for storing the eigenvectors $\mathbf{Q} = \mathbf{I}$, and an empty list for storing the eigenvalues $\mathbf{w} = []$.
2: Set $\mathbf{T}^{(0)} = \mathbf{S}$ ▷ Householder tridiagonalization
3: **for** $i = 1, \ldots, n-2$ **do**
4:     Create a partial column vector $\mathbf{u}^{(i)} = [T_{i+1,i}^{(i-1)}, \ldots, T_{n,i}^{(i-1)}]^T$
5:     Define reflection matrix $\mathbf{H}^{(i)} = \mathbf{I} - 2\mathbf{u}^{(i)}\mathbf{u}^{(i)T}$
6:     Update the last $(n-i+1)$ block of $\mathbf{T}^{(i)}$: $\mathbf{T}^{(i)} = \mathbf{H}^{(i)}\mathbf{T}^{(i-1)}\mathbf{H}^{(i)}$
7:     Update the last $(n-i+1)$ columns of $\mathbf{Q}$. $\mathbf{Q} \leftarrow \mathbf{Q}\mathbf{H}^{(i)}$
8: **end for**
9: Set $\mathbf{S}_1^{(0)} = \mathbf{T}^{(n-2)}$ ▷ QR algorithm for tridiagonal matrix
10: **for** $i = n, \ldots, 2$ **do**
11:     Let $k = n - i + 1$
12:     **for** $j = 1, \ldots, N_{max}$ **do**
13:         Set the shift $\mu^{(j)} = S_{ii}^{(j-1)}$
14:         Tridiagonal QR decomposition: $\mathbf{Q}_k^{(j)}\mathbf{R}_k^{(j)} = \mathbf{S}_k^{(j-1)} - \mu^{(j)}\mathbf{I}$
15:         Matrix reconstruction: $\mathbf{S}_k^{(j)} = \mathbf{R}_k^{(j)}\mathbf{Q}_k^{(j)} + \mu^{(j)}\mathbf{I}$.
16:         Update the first $i$ columns of $\mathbf{Q}$: $\mathbf{Q} \leftarrow \mathbf{Q}\mathbf{Q}_k^{(j)}$
17:         **if** $2\left|S_{i,i-1}^{(j)}\right| < \epsilon\left(\left|S_{ii}^{(j)}\right| + \left|S_{i-1,i-1}^{(j)}\right|\right)$ **then**
18:             Store the eigenvalue $\mathbf{w} \leftarrow [\mathbf{w}, S_{ii}^{(j)}]$
19:             Break
20:         **end if**
21:     **end for**
22:     Let $\mathbf{S}_{k+1}^{(0)}$ be $\mathbf{S}_k^{(j)}$ with the last row and column removed
23: **end for**
24: **return** The $n$ eigenvectors as columns of matrix $\mathbf{Q}$ and $n$ eigenvalues $\mathbf{w}$

---

is equivalent to simultaneous iteration, and any vector iteration method in principle can be accelerated using spectral shift.

In the **QR algorithm with shift**, one iteration is performed as

(1) Choose a shift $\mu^{(k)}$
(2) QR decomposition: $\mathbf{Q}^{(k)}\mathbf{R}^{(k)} = \mathbf{S}^{(k-1)} - \mu^{(k)}\mathbf{I}$
(3) Matrix reconstruction: $\mathbf{S}^{(k)} = \mathbf{R}^{(k)}\mathbf{Q}^{(k)} + \mu^{(k)}\mathbf{I}$.

The shift can be simply chosen as the last element in $\mathbf{S}^{(k-1)}$ (i.e. $\mu^{(k)} = S_{nn}^{(k-1)}$); it can be shown that such a shift is equivalent to the Rayleigh's quotient associated with the estimate of the eigenvector in the current iteration (Trefethen and Bau III, 1997).

The QR algorithm with tridiagonalization and shift for **standard symmetric eigenproblem** is summarized in Algorithm 5.2, where the shift using Rayleigh's quotient is assumed.

---

**Algorithm 5.3** Solve generalized eigenproblem using the QR algorithm

---

**Require:** Matrices $\mathbf{K}$ and $\mathbf{M}$ of size $n \times n$.

1: Factorize $\mathbf{K} = \mathbf{L}_K \mathbf{L}_K^T$ and $\mathbf{M} = \mathbf{L}_M \mathbf{L}_M^T$
2: Compute factor $\mathbf{L} = \mathbf{L}_K^{-1} \mathbf{L}_M$
3: Compute the symmetric matrix $\mathbf{S} = \mathbf{L}^T \mathbf{L}$
4: Solve the standard symmetric eigenproblem associated with $\mathbf{S}$, and find the eigenvectors $\mathbf{U}$ and eigenvalues $\mathbf{w}$ using the QR algorithm, Algorithm 5.2.
5: Compute the eigenvectors for generalized eigenproblem $\mathbf{U} \leftarrow \mathbf{L}^{-T} \mathbf{U}$
6: Compute the eigenvalues for generalized eigenproblem $\mathbf{w} \leftarrow 1/\mathbf{w}$
7: **return** $\mathbf{U}, \mathbf{w}$

---

Subsequently, based on Algorithm 5.2, the QR algorithm for a **generalized eigenproblem** is summarized in Algorithm 5.3.

## An Example Illustrating the QR Algorithm

In this section the QR algorithm described by Algorithm 5.3 is used to solve the generalized eigenvalue problem in Section 5.5 with the matrices,

$$\mathbf{K} = 600 \begin{bmatrix} 1 & -1 & 0 \\ -1 & 3 & -2 \\ 0 & -2 & 5 \end{bmatrix}, \quad \mathbf{M} = \begin{bmatrix} 1 & 0 & 0 \\ 0 & 1.5 & 0 \\ 0 & 0 & 2.0 \end{bmatrix}.$$

Note that the QR algorithm is a powerful numerical tool that can be applied to solve eigenproblems of much larger size. The $3 \times 3$ problem is chosen here as a simple illustration of the QR algorithm.

The solution is accomplished in four stages.

**Stage 1: Conversion to Standard Eigenproblem** The generalized eigenproblem is converted to a standard symmetric eigenproblem. The mass matrix is factorized using Cholesky decomposition $\mathbf{M} = \mathbf{L}_M \mathbf{L}_M^T$,

$$\mathbf{L}_M = \begin{bmatrix} 1.0000000e+00 & 0.0000000e+00 & 0.0000000e+00 \\ 0.0000000e+00 & 1.2247449e+00 & 0.0000000e+00 \\ 0.0000000e+00 & 0.0000000e+00 & 1.4142136e+00 \end{bmatrix}. \quad (E5.40)$$

The symmetric matrix is computed as

$$\mathbf{S} = \mathbf{L}_M^T \mathbf{K}^{-1} \mathbf{L}_M$$

$$= \begin{bmatrix} 3.0555556e\text{--}03 & 1.7010345e\text{--}03 & 7.8567420e\text{--}04 \\ 1.7010345e\text{--}03 & 2.0833333e\text{--}03 & 9.6225045e\text{--}04 \\ 7.8567420e\text{--}04 & 9.6225045e\text{--}04 & 1.1111111e\text{--}03 \end{bmatrix}. \quad (E5.41)$$

**Stage 2: Tridiagonalization** Since $\mathbf{S}$ is a fully populated matrix, it needs to be tridiagonalized before applying the QR algorithm. For a $3 \times 3$ matrix like $\mathbf{S}$, only one Householder reflection is needed for tridiagonalization, because only the bottom left and upper right elements, $S_{31}$

and $S_{13}$, need to be zeroed out. Let $\mathbf{q}$ be the first column of $\mathbf{S}$,

$$\mathbf{q} = [S_{11}, S_{21}, S_{31}]^T$$
$$= [\, 3.0555556e\text{--}03,\ 1.7010345e\text{--}03,\ 7.8567420e\text{--}04]^T \tag{E5.42}$$

and using Eqs. (5.6.59) and (5.6.60), the first column of $\mathbf{S}$ should transform to

$$\tilde{\mathbf{q}} = [S_{11}, -\sqrt{S_{21}^2 + S_{31}^2}, 0]^T = [\, 3.0555556e\text{--}03, -1.8737136e\text{--}03, 0]^T. \tag{E5.43}$$

Using Eqs. (5.6.51) and (5.6.55), the reflection vector and matrix are constructed as, respectively,

$$\mathbf{u} = (\mathbf{q} - \tilde{\mathbf{q}})/ \,\|\mathbf{q} - \tilde{\mathbf{q}}\| = [\, 0,\ 9.7668861e\text{--}01,\ 2.1466101e\text{--}01]^T \tag{E5.44}$$

and

$$\mathbf{H} = \mathbf{I} - 2\mathbf{u}\mathbf{u}^T$$
$$= \begin{bmatrix} 1.0000000e+00 & 0.0000000e+00 & 0.0000000e+00 \\ 0.0000000e+00 & -9.0784130e\text{--}01 & -4.1931393e\text{--}01 \\ 0.0000000e+00 & -4.1931393e\text{--}01 & 9.0784130e\text{--}01 \end{bmatrix}. \tag{E5.45}$$

The symmetric matrix $\mathbf{S}$ is tridiagonalized as

$$\mathbf{T} = \mathbf{HSH}$$
$$= \begin{bmatrix} \mathbf{3.0555556e\text{--}03} & \mathit{-1.8737136e\text{--}03} & \mathit{0.0000000e+00} \\ \mathit{-1.8737136e\text{--}03} & 2.6449939e\text{--}03 & -2.5378034e\text{--}04 \\ \mathit{0.0000000e+00} & -2.5378034e\text{--}04 & 5.4945055e\text{--}04 \end{bmatrix}. \tag{E5.46}$$

Note that when comparing the tridiagonal matrix Eq. (E5.46) to the original symmetric matrix $\mathbf{S}$ in Eq. (E5.41), the first element of $\mathbf{T}$, highlighted in boldface, is unchanged; the rest of the first row and column of $\mathbf{T}$, shown in italic, are determined by the vector $\tilde{\mathbf{q}}$ in Eq. (E5.43); only the lower right $2 \times 2$ block of $\mathbf{T}$ needs to be recomputed.

**Stage 3: QR Algorithm** After the tridiagonalization, the QR algorithm is applied to solve the symmetric eigenproblem $\mathbf{Tx} = \lambda\mathbf{x}$. The first iterate is initialized as

$$\mathbf{S}_1^{(0)} = \mathbf{T} = \begin{bmatrix} 3.0555556e\text{--}03 & -1.8737136e\text{--}03 & 0.0000000e+00 \\ -1.8737136e\text{--}03 & 2.6449939e\text{--}03 & -2.5378034e\text{--}04 \\ 0.0000000e+00 & -2.5378034e\text{--}04 & 5.4945055e\text{--}04 \end{bmatrix} \tag{E5.47}$$

and the transformation matrix is initialized as $\mathbf{P}_0 = \mathbf{I}$. A numerical tolerance of $10^{-8}$ is assumed.

The first iteration of QR algorithm is performed in the following three steps:

(1) Choose a shift using Rayleigh's quotient:

$$\mu^{(1)} = S_{33}^{(0)} = 5.4945055e\text{--}04. \tag{E5.48}$$

(2) Apply QR decomposition to a shifted matrix:

$$\mathbf{Q}_1^{(1)}\mathbf{R}_1^{(1)} = \mathbf{S}_1^{(0)} - \mu^{(1)}\mathbf{I} \tag{E5.49}$$

$$\mathbf{Q}_1^{(1)} = \begin{bmatrix} -8.0089881e{-}01 & -5.4479507e{-}01 & 2.4851443e{-}01 \\ 5.9879971e{-}01 & -7.2866723e{-}01 & 3.3238979e{-}01 \\ 0.0000000e{+}00 & 4.1502095e{-}01 & 9.0981185e{-}01 \end{bmatrix} \tag{E5.50}$$

$$\mathbf{R}_1^{(1)} = \begin{bmatrix} -3.1291157e{-}03 & 2.7554657e{-}03 & -1.5196359e{-}04 \\ 0.0000000e{+}00 & -6.1148801e{-}04 & 1.8492142e{-}04 \\ 0.0000000e{+}00 & 0.0000000e{+}00 & -8.4353993e{-}05 \end{bmatrix}. \tag{E5.51}$$

(3) Reconstruct the tridiagonal matrix for the next iteration:

$$\mathbf{S}_1^{(1)} = \mathbf{R}_1^{(1)}\mathbf{Q}_1^{(1)} + \mu^{(1)}\mathbf{I}$$

$$= \begin{bmatrix} 4.7055276e{-}03 & -3.6615884e{-}04 & 0.0000000e{+}00 \\ -3.6615884e{-}04 & 1.0717681e{-}03 & \mathbf{-3.5008675e{-}05} \\ 0.0000000e{+}00 & \mathbf{-3.5008675e{-}05} & 4.7270429e{-}04 \end{bmatrix}. \tag{E5.52}$$

Note that the tridiagonal pattern is preserved in the reconstructed matrix $\mathbf{S}_1^{(1)}$, and when compared to $\mathbf{S}_1^{(0)}$ in Eq. (E5.47), the magnitude of the last off-diagonal element of $\mathbf{S}_1^{(1)}$, highlighted in boldface, is reduced. At the end of this iteration, the transformation matrix is

$$\mathbf{P}_1 = \mathbf{P}_0\mathbf{Q}_1^{(1)}$$

$$= \begin{bmatrix} -8.0089881e{-}01 & -5.4479507e{-}01 & 2.4851443e{-}01 \\ 5.9879971e{-}01 & -7.2866723e{-}01 & 3.3238979e{-}01 \\ 0.0000000e{+}00 & 4.1502095e{-}01 & 9.0981185e{-}01 \end{bmatrix}. \tag{E5.53}$$

The above procedure, Eqs. (E5.48)–(E5.53), is repeated until the last off-diagonal element of $\mathbf{S}_1^{(k)}$ becomes sufficiently small. For this particular problem, two more iterations are required,

$$\mathbf{S}_1^{(2)} = \begin{bmatrix} 4.7414163e{-}03 & -4.8810596e{-}05 & 0.0000000e{+}00 \\ -4.8810596e{-}05 & 1.0380312e{-}03 & -1.3326648e{-}07 \\ 0.0000000e{+}00 & -1.3326648e{-}07 & 4.7055246e{-}04 \end{bmatrix} \tag{E5.54}$$

$$\mathbf{S}_1^{(3)} = \begin{bmatrix} 4.7420482e{-}03 & -6.4783461e{-}06 & 0.0000000e{+}00 \\ -6.4783461e{-}06 & 1.0373994e{-}03 & -7.3646011e{-}15 \\ 0.0000000e{+}00 & -7.3645518e{-}15 & 4.7055243e{-}04 \end{bmatrix}. \tag{E5.55}$$

The iterations associated with $\mathbf{S}_1^{(2)}$ and $\mathbf{S}_1^{(3)}$ generate two new orthonormal matrices $\mathbf{Q}_1^{(2)}$ and $\mathbf{Q}_1^{(3)}$, respectively, and the transformation matrix is now

$$\mathbf{P}_3 = \mathbf{P}_1\mathbf{Q}_1^{(2)}\mathbf{Q}_1^{(3)}$$

$$= \begin{bmatrix} -7.4376421e{-}01 & -6.3447509e{-}01 & 2.1037148e{-}01 \\ 6.6725383e{-}01 & -6.8595768e{-}01 & 2.9023163e{-}01 \\ -3.9838807e{-}02 & 3.5623508e{-}01 & 9.3354670e{-}01 \end{bmatrix} \tag{E5.56}$$

The last diagonal element of $\mathbf{S}_1^{(3)}$ becomes the first eigenvalue, which corresponds to the largest eigenvalue of the original generalized eigenproblem,

$$\lambda_3 = 4.7055243\text{e--}04, \quad \omega_3 = 1/\sqrt{\lambda_3} = 4.6099476\text{e+}01 \text{ rad/s}. \tag{E5.57}$$

Given the numerical error tolerance, the matrix $\mathbf{S}_1^{(3)}$ has been partially diagonalized, and it is no longer necessary to keep the last row and column of $\mathbf{S}_1^{(3)}$ in the QR iteration. Therefore, the QR algorithm is subsequently applied to the upper left $2 \times 2$ block of $\mathbf{S}_1^{(3)}$, denoted

$$\mathbf{S}_2^{(0)} = \begin{bmatrix} 4.7420482\text{e--}03 & -6.4783461\text{e--}06 \\ -6.4783461\text{e--}06 & 1.0373994\text{e--}03 \end{bmatrix}. \tag{E5.58}$$

The procedure, Eqs. (E5.48)–(E5.53), is subsequently applied to $\mathbf{S}_2^{(0)}$. The first iteration is the following.

(1) Choose a shift using Rayleigh's quotient:

$$\mu^{(1)} = S_{22}^{(0)} = 1.0373994\text{e--}03. \tag{E5.59}$$

(2) Apply QR decomposition to a shifted matrix:

$$\mathbf{Q}_2^{(1)} \mathbf{R}_2^{(1)} = \mathbf{S}_2^{(0)} - \mu^{(1)} \mathbf{I} \tag{E5.60}$$

$$\mathbf{Q}_2^{(1)} = \begin{bmatrix} -9.9999847\text{e--}01 & 1.7487045\text{e--}03 \\ 1.7487045\text{e--}03 & 9.9999847\text{e--}01 \end{bmatrix} \tag{E5.61}$$

$$\mathbf{R}_2^{(1)} = \begin{bmatrix} -3.7046545\text{e--}03 & 6.4783362\text{e--}06 \\ 0.0000000\text{e+}00 & -1.1328713\text{e--}08 \end{bmatrix}. \tag{E5.62}$$

(3) Reconstruct the tridiagonal matrix for the next iteration:

$$\begin{aligned} \mathbf{S}_2^{(1)} &= \mathbf{R}_2^{(1)} \mathbf{Q}_2^{(1)} + \mu^{(1)} \mathbf{I} \\ &= \begin{bmatrix} 4.7420595\text{e--}03 & -1.9810572\text{e--}11 \\ -1.9810572\text{e--}11 & 1.0373880\text{e--}03 \end{bmatrix}. \end{aligned} \tag{E5.63}$$

At the end of this iteration, the transformation matrix $\mathbf{Q}$ needs to be updated. Since only the upper left $2 \times 2$ block of $\mathbf{S}_2^{(0)}$ is modified, one only needs to update the first two columns of $\mathbf{P}_3$,

$$\begin{aligned} \mathbf{P}_4 &= \mathbf{P}_3 \begin{bmatrix} \mathbf{Q}_2^{(1)} & \mathbf{0} \\ \mathbf{0}^T & 1 \end{bmatrix} \\ &= \begin{bmatrix} 7.4265356\text{e--}01 & -6.3577474\text{e--}01 & 2.1037148\text{e--}01 \\ -6.6845234\text{e--}01 & -6.8478980\text{e--}01 & 2.9023163\text{e--}01 \\ 4.0461695\text{e--}02 & 3.5616487\text{e--}01 & 9.3354670\text{e--}01 \end{bmatrix}, \end{aligned} \tag{E5.64}$$

where $\mathbf{0}^T = [0, 0]$.

For the current $3 \times 3$ problem, only one iteration has reduced the off-diagonal element to be close to zero, and the symmetric matrix is diagonalized. The remaining two eigenvalues are found from Eq. (E5.63) to be

$$\lambda_2 = 1.0373880e{-}03, \quad \omega_2 = 1/\sqrt{\lambda_2} = 3.1047696e{+}01 \text{ rad/s} \tag{E5.65}$$

$$\lambda_1 = 4.7420595e{-}03, \quad \omega_1 = 1/\sqrt{\lambda_1} = 1.4521668e{+}01 \text{ rad/s.} \tag{E5.66}$$

**Stage 4: Recover the Generalized Eigensolution** The eigenvalues of the original generalized eigenproblem are the same as those for the standard eigenproblem, given in Eqs. (E5.57), (E5.65), and (E5.66). The eigenvectors of the original generalized eigenproblem are recovered using the transformation matrices $\mathbf{L}_M$, $\mathbf{H}$, $\mathbf{P}_4$ from Eqs. (E5.40), (E5.45), and (E5.64), respectively, via two successive transformations,

$$\mathbf{U} = \mathbf{L}_M^{-T}\mathbf{H}\mathbf{P}_4$$

$$= \begin{bmatrix} 7.4265356e{-}01 & -6.3577474e{-}01 & 2.1037148e{-}01 \\ 4.8163704e{-}01 & 3.8566038e{-}01 & -5.3475088e{-}01 \\ 2.2416995e{-}01 & 4.3167672e{-}01 & 5.1322806e{-}01 \end{bmatrix}. \tag{E5.67}$$

The eigenvalues in Eqs. (E5.57), (E5.65), (E5.66) and the three eigenvectors as the columns of Eq. (E5.67) compare well with the analytical solutions found in the example in Section 5.5.

To summarize, the QR algorithm is applied to solve a generalized eigenproblem via a series of matrix transformations. The goal of transformation is to convert the original problem to an eigenproblem with a simpler form that is easier to solve. First, it converted the original problem to a standard eigenproblem with a symmetric matrix. Then, the symmetric matrix is tridiagonalized using Householder transformation. Subsequently, the eigenproblem with the tridiagonal matrix is solved by the QR algorithm with shift. Finally, the solution of the standard tridiagonal eigenproblem is transformed to yield the solution of the original generalized eigenproblem.

## 5.6.4  Subspace Methods

Several useful eigenvalue algorithms have been discussed in this chapter, including the inverse iteration method and the QR algorithm. The inverse iteration method is designed for finding one particular eigenpair, with the convergence accelerated using the spectral transformation technique. However, the computational efficiency of this method is reduced when the objective is to compute a set of eigenpairs using deflation. This is because the iteration process starts from scratch for every new eigenpair and ignores the results from the previous iterations. However, these neglected results contain potentially useful information that can be used to accelerate the convergence.

The QR algorithm efficiently computes all the eigenpairs of an eigenproblem. However, the application of QR algorithm requires the complete knowledge of the matrices $\mathbf{K}$ and $\mathbf{M}$, which may be impractical for large-scale structural analysis. Even if the matrices are known, the tridiagonalization and QR decomposition can become computationally impractical for such large matrices. Furthermore, in many applications not all eigenpairs are of interest, and hence some of the computational cost in the QR algorithm produces limited benefits.

The subspace methods represent a class of eigenvalue algorithms that overcome these limitations. This class of algorithms builds on the concept of subspace, that is, a $m$-dimensional vector space in a $n$-dimensional Euclidean space $\mathbb{R}^n$, where $m < n$ (see Appendices A.1.2–A.1.4 for Euclidean and vector spaces). In other words, a subspace $\mathcal{S}$ is the **set** of vectors **u** that satisfy

$$\mathbf{u} = \mathbf{Z}\mathbf{w} = w_1\mathbf{z}_1 + w_2\mathbf{z}_2 + \cdots + w_m\mathbf{z}_m, \tag{5.6.76}$$

where $\mathbf{Z} = [\mathbf{z}_1, \mathbf{z}_2, \dots, \mathbf{z}_m] \in \mathbb{R}^{n \times m}$ consists of $m$ linearly independent column vectors as the basis vectors of the subspace $\mathcal{S}$, and $\mathbf{w} = [w_1, w_2, \dots, w_m] \in \mathbb{R}^m$ is a vector of coordinates. The subspace $\mathcal{S}$ is said to be spanned by $\mathbf{Z}$. A vector $\mathbf{u}^*$ belongs to the subspace $\mathcal{S}$ if there exists a coordinate vector $\mathbf{w}^*$ such that $\mathbf{u}^* = \mathbf{Z}\mathbf{w}^*$.

Unlike the inverse iteration and QR algorithms that compute sequences of vectors that eventually converge to the eigenvectors, the subspace methods construct a sequence of subspaces that eventually converge to a subspace containing the eigenvectors. As a result, the subspace methods generate the eigenpairs of interest simultaneously in an efficient manner. Furthermore, with the introduction of subspaces, the original large-scale eigenproblem is converted to a series of small eigenproblems that can be solved efficiently and accurately using the matrix transformation methods, such as the QR algorithm.

This section introduces two typical subspace methods, namely the subspace iteration method (Dong et al., 1972; Bathe and Wilson, 1972) and the Lanczos algorithm (Lanczos, 1950; Arnoldi, 1951). Both methods are widely used in commercial finite element codes.

## Subspace Iteration Method

**Motivation** For an $n$-dimensional eigenproblem, the goal of subspace iteration is to find an $m$-dimensional subspace, spanned by a $n \times m$ matrix $\tilde{\mathbf{Z}}$, that contains $m$ eigenvectors of interest. That means, for the $i$th eigenvector $\mathbf{u}_i$, one can find a coordinate vector $\mathbf{w}_i \in \mathbb{R}^m$ such that

$$\mathbf{u}_i = \tilde{\mathbf{Z}}\mathbf{w}_i \tag{5.6.77}$$

and the $m$ eigenvectors that reside in this subspace can be represented as

$$\mathbf{U} = [\mathbf{u}_1, \dots, \mathbf{u}_m] = [\tilde{\mathbf{Z}}\mathbf{w}_1, \tilde{\mathbf{Z}}\mathbf{w}_2, \dots, \tilde{\mathbf{Z}}\mathbf{w}_m] \equiv \tilde{\mathbf{Z}}\mathbf{W}, \tag{5.6.78}$$

where $\mathbf{W} = [\mathbf{w}_1, \mathbf{w}_2, \cdots, \mathbf{w}_m] \in \mathbb{R}^{m \times m}$ consists of $m$ coordinate vectors. In other words, each eigenvector is a linear combination of the column vectors of $\tilde{\mathbf{Z}}$.

The $m$ eigenvectors satisfy the generalized eigenproblem,

$$\mathbf{K}\mathbf{U} = \mathbf{M}\mathbf{U}\boldsymbol{\Lambda}_m, \tag{5.6.79}$$

where $\boldsymbol{\Lambda}_m$ is a diagonal matrix containing the $m$ eigenvalues corresponding to the $m$ eigenvectors. Combining Eqs. (5.6.78) and (5.6.79), one obtains

$$\mathbf{K}\tilde{\mathbf{Z}}\mathbf{W} = \mathbf{M}\tilde{\mathbf{Z}}\mathbf{W}\boldsymbol{\Lambda}. \tag{5.6.80}$$

Pre-multiplying Eq. (5.6.80) by $\tilde{\mathbf{Z}}^T$, one obtains a reduced eigenproblem

$$\mathbf{K}_r \mathbf{W} = \mathbf{M}_r \mathbf{W} \boldsymbol{\Lambda}, \tag{5.6.81}$$

where the reduced mass and stiffness matrices, $\mathbf{M}_r$ and $\mathbf{K}_r$, are respectively

$$\mathbf{M}_r = \tilde{\mathbf{Z}}^T \mathbf{M} \tilde{\mathbf{Z}}, \quad \mathbf{K}_r = \tilde{\mathbf{Z}}^T \mathbf{K} \tilde{\mathbf{Z}}. \tag{5.6.82}$$

The derivation presented in Eqs. (5.6.78)–(5.6.82) shows that, with a properly chosen subspace $\tilde{\mathbf{Z}}$, the solution of the reduced eigenproblem Eq. (5.6.81) yields $m$ eigenvectors $\mathbf{U} = \tilde{\mathbf{Z}} \mathbf{W}$ of the generalized eigenproblem as well as their $m$-associated eigenvalues.

**Formulation of the Subspace Iteration Method**    In practice, it is unlikely that one can guess the subspace $\tilde{\mathbf{Z}}$ that contains all the $m$ eigenvectors of interest. In the subspace iteration method, such subspace is found in an iterative manner. One starts with an initial guess of the subspace $\mathbf{Z}^{(0)}$, and solves a series of multiple reduced eigenproblems to update the subspace $\mathbf{Z}^{(k)}$; such procedure is repeated until the subspace $\mathbf{Z}^{(k)}$ converges to $\tilde{\mathbf{Z}}$.

To obtain a good initial guess, the columns of $\mathbf{Z}^{(0)}$ can be chosen to approximate the fundamental modal behavior, as found from the solution of

$$\mathbf{KT} = \mathbf{P}, \tag{5.6.83}$$

where $\mathbf{P} \in \mathbb{R}^{n \times m}$ is a rectangular matrix of linearly independent static load patterns. Ideally, these patterns should approximate the distribution of the inertial loads of the first few modes, so as to yield the approximate modal patterns. A direct solution of Eq. (5.6.83) is feasible, even for large systems, because $\mathbf{K}$ is usually sparse.

Subsequently, considering the standard eigenproblem with dynamic matrix $\mathbf{D}$, the $k$th iteration of the subspace iteration method is performed as

(1) Matrix-matrix multiplication: $\tilde{\mathbf{Z}} = \mathbf{D} \mathbf{Z}^{(k-1)}$.
(2) Construct reduced mass and stiffness matrices $\mathbf{M}_r$ and $\mathbf{K}_r$ from $\tilde{\mathbf{Z}}$ using Eq. (5.6.82).
(3) Solve the reduced eigenproblem $\mathbf{K}_r \mathbf{W}^{(k)} = \mathbf{M}_r \mathbf{W}^{(k)} \boldsymbol{\Lambda}^{(k)}$ using, for example, QR algorithm.
(4) Update subspace $\mathbf{Z}^{(k)} = \tilde{\mathbf{Z}} \mathbf{W}^{(k)}$.

Note that the subspace iteration method resembles the simultaneous iteration method discussed in Section 5.6.2. In fact, steps 2–4 in subspace iteration are a replacement of the QR decomposition step in simultaneous iteration for an improved estimate of the eigenvectors. Both methods iteratively generate a matrix $\mathbf{Z}^{(k)}$ with $m$ columns to approximate $m$ eigenvectors of interest. The difference is that, in simultaneous iteration, the QR decomposition is employed to orthogonalize the new iterate $\tilde{\mathbf{Z}}$, such that the $i$th column of $\tilde{\mathbf{Z}}$ is designated to approximate the $i$th eigenvector $\mathbf{u}_i$ only. While in subspace iteration, the $m$ eigenvectors are approximated using $m$ linear combinations of **all** the columns of the iterate $\tilde{\mathbf{Z}}$.

Next, the convergence characteristics of the subspace iteration method are discussed briefly. Assuming $m$ eigenpairs are sought, the convergence rate of the $i$th eigenvector is

$\omega_i^2/\omega_{m+1}^2$ in subspace iteration. Typically, $\omega_{m+1}^2 \gg \omega_{i+1}^2 > \omega_i^2$. Therefore, the convergence rate of $\omega_i^2/\omega_{m+1}^2$ is a significant improvement over the convergence rate of $\omega_i^2/\omega_{i+1}^2$ in inverse iteration or simultaneous iteration.

It is possible to accelerate the convergence of subspace iteration using a technique called **buffer vectors** that employs an augmented subspace of dimension $p$, $p > m$, to find $m$ eigenpairs. Using the buffer vectors, the convergence rate of the $i$th eigenvector improves from $\omega_i^2/\omega_{m+1}^2$ to $\omega_i^2/\omega_{p+1}^2$, where $\omega_{p+1}^2 > \omega_{m+1}^2$. Note that the parameter $p$ needs to be chosen so as to reduce the required number of iterations, and without causing a significant increase in the computational cost per iteration. A typical choice of $p$ is $p = \min\{2m, m+8\}$ (Bathe, 2006, 2013).

The complete algorithm for subspace iteration is summarized in Algorithm 5.4, where the reduced eigenproblem is solved using Algorithm 5.3.

## Lanczos Algorithm

The Lanczos algorithm has advantages when compared to the subspace iteration method for large sparse eigenproblems frequently encountered in practice. The advantage is due to the fact that the Lanczos algorithm does not require $\mathbf{K}$ and $\mathbf{M}$ explicitly, instead it only uses the matrix-vector products of these matrices. Both the subspace iteration method and the Lanczos algorithm iteratively generate a sequence of subspaces while solving a series of reduced eigenproblems. The difference lies in how the subspace is constructed and updated during the iterations.

---

**Algorithm 5.4** Subspace iteration method

---

**Require:** Matrices $\mathbf{K}$ and $\mathbf{M}$, number of requested eigenpairs $m$, maximum number of iteration $N_{\max}$, tolerance $\epsilon$.

1: Determine the size of buffer vectors: $p = \min\{2m, m+8\}$
2: Pick an orthonormal matrix $\mathbf{Z}^{(0)} \in \mathbb{R}^{n \times p}$ as the initial subspace
3: Factorize $\mathbf{K} = \mathbf{L}_K \mathbf{L}_K^T$
4: **for** $i = 1, \ldots, N_{\max}$ **do**
5:     Matrix-matrix multiplication: $\mathbf{Y} = \mathbf{M}\mathbf{Z}^{(i-1)}$
6:     Solve $\mathbf{K}\tilde{\mathbf{Z}} = \mathbf{Y}$ using the factor $\mathbf{L}_K$
7:     Construct the matrices $\mathbf{M}_r$ and $\mathbf{K}_r$ from $\tilde{\mathbf{Z}}$ using Eq. (5.6.82)
8:     Solve the reduced eigenproblem $\mathbf{K}_r \mathbf{W}^{(i)} = \mathbf{M}_r \mathbf{W}^{(i)} \mathbf{\Lambda}^{(i)}$
9:     Update the subspace $\mathbf{Z}^{(i)} = \tilde{\mathbf{Z}} \mathbf{W}^{(i)}$
10:     **if** $\max_j \left| (\Lambda_{jj}^{(i)} - \Lambda_{jj}^{(i-1)})/\Lambda_{jj}^{(i)} \right| < \epsilon$ **then**
11:         Break
12:     **end if**
13: **end for**
14: Set $\mathbf{w} = [\Lambda_{11}^{(i)}, \ldots, \Lambda_{mm}^{(i)}]$
15: Set $\mathbf{U}$ as the first $m$ column vectors of $\mathbf{Z}^{(i)}$
16: **return** The requested $m$ eigenvectors $\mathbf{U}$ and the $m$ eigenvalues $\mathbf{w}$

---

**Motivation**    The Lanczos algorithm in some sense is a more advanced version of the inverse iteration method that is discussed in Section 5.6.2. It is therefore beneficial to review the inverse iteration method to motivate the development of the Lanczos algorithm.

In the inverse iteration method, one starts from an arbitrary initial guess vector $\mathbf{z}^{(0)}$, and iteratively generates the matrix-vector products of $\mathbf{z}^{(0)}$ and the powers of the dynamic matrix $\mathbf{D} = \mathbf{K}^{-1}\mathbf{M}$, leading to a sequence of vectors,

$$\{\mathbf{z}^{(0)}, \mathbf{D}\mathbf{z}^{(0)}, \mathbf{D}^2\mathbf{z}^{(0)}, \ldots, \mathbf{D}^m\mathbf{z}^{(0)}, \ldots\}. \tag{5.6.84}$$

It was shown in Section 5.6.2 that the iterates converge to the eigenvector associated with the largest eigenvalue of $\mathbf{D}$, that is, the lowest frequency of the structural dynamic eigenproblem. Typically, such convergence is achieved within a few iterations, even for an eigenproblem having millions of degrees of freedom.

Taking a closer look at the inverse iteration and recalling Eq. (5.6.23), the $m$th iterate may be decomposed into a linear combination of eigenvectors as

$$\mathbf{D}^m\mathbf{z}^{(0)} = \lambda_1^m \left[ c_1\mathbf{u}_1 + c_2(\lambda_2/\lambda_1)^m\mathbf{u}_2 + c_3(\lambda_3/\lambda_1)^m\mathbf{u}_3 + \cdots \right.$$
$$\left. + c_{n-1}(\lambda_{n-1}/\lambda_1)^m\mathbf{u}_{n-1} + c_n(\lambda_n/\lambda_1)^m\mathbf{u}_n \right] \tag{5.6.85}$$

where $\mathbf{u}_i$ is the eigenvector associated with the $i$th eigenvalue $\lambda_i$ of $\mathbf{D}$, $c_i = \mathbf{u}_i^T\mathbf{z}^{(0)}$, and

$$\lambda_1 \geq \lambda_2 \geq \cdots \geq \lambda_n > 0. \tag{5.6.86}$$

In a typical large-scale structural dynamics problem, the distribution of the eigenvalues of the dynamic matrix is clustered near the origin point, as illustrated in Figure 5.7. As a result, most of the ratios $(\lambda_i/\lambda_1)$ in Eq. (5.6.85) are close to zero. Therefore, one may assume that after a few inverse iterations, and before the convergence, only the first, $k$ factors $(\lambda_i/\lambda_1)^m$, $i \leq k$, remain nonzero, and the $m$th iterate $\mathbf{D}^m\mathbf{z}^{(0)}$ consists of only the first $k$ eigenvectors,

$$\mathbf{D}^m\mathbf{z}^{(0)} \approx \lambda_1^m \left[ c_1\mathbf{u}_1 + c_2(\lambda_2/\lambda_1)^m\mathbf{u}_2 + \cdots + c_k(\lambda_k/\lambda_1)^m\mathbf{u}_k \right]. \tag{5.6.87}$$

The implication of Eq. (5.6.87) is significant. It shows that the vectors from inverse iteration Eq. (5.6.84)

$$\tilde{\mathbf{Z}} = [\mathbf{z}^{(0)}, \mathbf{D}\mathbf{z}^{(0)}, \mathbf{D}^2\mathbf{z}^{(0)}, \ldots, \mathbf{D}^m\mathbf{z}^{(0)}] \tag{5.6.88}$$

span a subspace that contains the first $k$ eigenvectors. This subspace $\tilde{\mathbf{Z}}$ is **exactly** what a subspace method looks for to solve a generalized eigenproblem. In other words, one may use the matrix $\tilde{\mathbf{Z}}$ to construct a reduced eigenproblem, as was done in Eq. (5.6.81) in the subspace iteration method, to produce the first $k$ eigenpairs of the generalized eigenproblem.

**Figure 5.7** Distribution of eigenvalues of a $n$-dimensional structural dynamics eigenproblem on the real axis

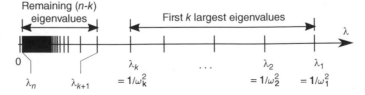

Due to their mathematical importance, the special vector sequence Eq. (5.6.84) and the subspace spanned by this sequence are referred to as the **Krylov sequence** and the **Krylov subspace**, respectively (Krylov, 1931). The Krylov subspace plays an important role in computational mathematics, such as the solution of systems of linear equations and optimization problems. This section focuses on the application of the Krylov subspace to the solution of symmetric generalized eigenproblems, assuming the lowest frequencies are of major interest.

While the idea of using the Krylov sequence as a vector basis is mathematically sound, it is not practical in the numerical sense. As the size of the Krylov sequence increases, the last few vectors in the sequence will be dominated by the eigenvector associated with the lowest frequency, and thus become almost linearly dependent. This linear dependency is inevitable as a result of inverse iteration and renders the Krylov sequence inappropriate as a vector basis for the Krylov subspace.

What Lanczos algorithm does essentially is the iterative construction of orthonormal basis vectors for the Krylov subspace. Using such a vector basis, the reduced eigenproblems may be constructed and solved in a numerically accurate and stable manner. The solutions of the reduced eigenproblems are subsequently converted to the eigenpairs of the generalized eigenproblem.

**Derivation of the Lanczos Algorithm**    To construct an orthonormal basis for the Krylov subspace $\{\mathbf{z}^{(i)}\}_{i=0}^{k+1}$, the Lanczos algorithm combines the inverse iteration with orthogonalization of the successive iterates using the following relation

$$\gamma_{k+1}\mathbf{z}^{(k+1)} = \mathbf{D}\mathbf{z}^{(k)} - \alpha_k\mathbf{z}^{(k)} - \beta_{k,k-1}\mathbf{z}^{(k-1)}\cdots - \beta_{k,0}\mathbf{z}^{(0)}, \qquad (5.6.89)$$

where the coefficient $\gamma_{k+1}$ is chosen to normalize the new Lanczos vector, or Krylov vector, $\mathbf{z}^{(k+1)}$, and the coefficients $\beta_{k,0},\ldots,\beta_{k,k-1}$ and $\alpha_k$ are chosen to enforce the orthogonality of $\mathbf{z}^{(k+1)}$ with respect to the other Lanczos vectors,

$$\begin{cases} \mathbf{z}^{(k+1)T}\mathbf{M}\mathbf{z}^{(j)} = 0, & j < k+1 \\ \mathbf{z}^{(k+1)T}\mathbf{M}\mathbf{z}^{(k+1)} = 1. \end{cases} \qquad (5.6.90)$$

If one pre-multiplies Eq. (5.6.89) with $\mathbf{z}^{(k)T}\mathbf{M}$ and utilizes the orthonormal condition Eq. (5.6.90),

$$\gamma_{k+1}\underbrace{\mathbf{z}^{(k)T}\mathbf{M}\mathbf{z}^{(k+1)}}_{=0} = \mathbf{z}^{(k)T}\mathbf{M}\mathbf{D}\mathbf{z}^{(k)} - \alpha_k\underbrace{\mathbf{z}^{(k)T}\mathbf{M}\mathbf{z}^{(k)}}_{=1} - \beta_{k,k-1}\underbrace{\mathbf{z}^{(k)T}\mathbf{M}\mathbf{z}^{(k-1)}}_{=0}$$
$$- \cdots - \beta_{k,0}\underbrace{\mathbf{z}^{(k)T}\mathbf{M}\mathbf{z}^{(0)}}_{=0}$$

and $\alpha_k$ is found to be,

$$\alpha_k = \mathbf{z}^{(k)T}\mathbf{M}\mathbf{D}\mathbf{z}^{(k)}. \qquad (5.6.91)$$

Similarly, one can find the value of $\beta_{k,j}$ by pre-multiplication with $\mathbf{z}^{(j)T}\mathbf{M}$,

$$\beta_{k,j} = \mathbf{z}^{(j)T}\mathbf{M}\mathbf{D}\mathbf{z}^{(k)}, \quad j = 0,\ldots,k-1. \qquad (5.6.92)$$

However, not all $\beta_{k,j}$ coefficients are nonzero for symmetric matrices $\mathbf{M}$ and $\mathbf{K}$. Combining Eqs. (5.6.89) and (5.6.92) with $k = j$,

$$\begin{aligned} \beta_{k,j} &= \mathbf{z}^{(j)T}\mathbf{MD}\mathbf{z}^{(k)} = \mathbf{z}^{(j)T}\mathbf{MK}^{-1}\mathbf{M}\mathbf{z}^{(k)} = \mathbf{z}^{(k)T}\mathbf{MK}^{-1}\mathbf{M}\mathbf{z}^{(j)} \\ &= \mathbf{z}^{(k)T}\mathbf{M}(\gamma_{j+1}\mathbf{z}^{(j+1)} + \alpha_j\mathbf{z}^{(j)} + \beta_{j,j-1}\mathbf{z}^{(j-1)} \cdots + \beta_{j,0}\mathbf{z}^{(0)}) \\ &= \begin{cases} \gamma_{j+1}, & \text{when } j = k - 1 \\ 0, & \text{when } j < k - 1. \end{cases} \end{aligned}$$

(5.6.93)

To summarize, at the $k$th iteration, only one $\beta$ is nonzero,

$$\beta_{k,j} = \mathbf{z}^{(k)T}\mathbf{MD}\mathbf{z}^{(j)} = 0, \quad \text{when } j < k - 1 \tag{5.6.94}$$

$$\beta_{k,k-1} = \mathbf{z}^{(k)T}\mathbf{MD}\mathbf{z}^{(k-1)} = \gamma_k. \tag{5.6.95}$$

As a result, Eq. (5.6.89) is simplified substantially as a compact three-term recurrence relation

$$\gamma_{k+1}\mathbf{z}^{(k+1)} = \mathbf{D}\mathbf{z}^{(k)} - \alpha_k\mathbf{z}^{(k)} - \gamma_k\mathbf{z}^{(k-1)}. \tag{5.6.96}$$

Furthermore, the recurrence relation can be represented in a matrix form by

$$\mathbf{DZ}^{(k)} = \mathbf{Z}^{(k)}\mathbf{T}^{(k)} + \mathbf{S}^{(k)}. \tag{5.6.97}$$

where $\mathbf{Z}^{(k)}$ spans the $k$th Krylov space, $\mathbf{S}^{(k)}$ is the remainder matrix, and $\mathbf{T}^{(k)}$ is a **symmetric tridiagonal** matrix,

$$\mathbf{Z}^{(k)} = [\mathbf{z}^{(0)}, \mathbf{z}^{(1)}, \ldots, \mathbf{z}^{(k)}], \quad \mathbf{S}^{(k)} = [\mathbf{0}, \ldots, \mathbf{0}, \gamma_{k+1}\mathbf{z}^{(k+1)}] \tag{5.6.98}$$

$$\mathbf{T}^{(k)} = \begin{bmatrix} \alpha_0 & \gamma_1 & & & \\ \gamma_1 & \alpha_1 & \gamma_2 & & \\ & \ddots & \ddots & \ddots & \\ & & \gamma_{k-1} & \alpha_{k-1} & \gamma_k \\ & & & \gamma_k & \alpha_k \end{bmatrix}. \tag{5.6.99}$$

An elegant equality can be obtained from Eq. (5.6.97) by pre-multiplying $\mathbf{Z}^{(k)T}\mathbf{M}$, and employing the orthonormal conditions Eq. (5.6.90),

$$\mathbf{Z}^{(k)T}\mathbf{MD}\mathbf{Z}^{(k)} = \underbrace{\mathbf{Z}^{(k)T}\mathbf{M}\mathbf{Z}^{(k)}}_{=\mathbf{I}}\mathbf{T}^{(k)} + \underbrace{\mathbf{Z}^{(k)T}\mathbf{M}\mathbf{S}^{(k)}}_{=\mathbf{O}} = \mathbf{T}^{(k)}. \tag{5.6.100}$$

Finally, the Lanczos algorithm is completed by searching for approximate eigenvectors of the original eigenproblem in the subspace spanned by the Lanczos vectors. The eigenvectors are approximated as

$$\mathbf{u} = \mathbf{Z}^{(k)}\mathbf{y} \tag{5.6.101}$$

such that

$$\mathbf{Du} = \lambda\mathbf{u} + \mathbf{r}, \tag{5.6.102}$$

where a residual vector $\mathbf{r}$ is introduced, because the approximated $\mathbf{u}$ cannot satisfy the eigenproblem $\mathbf{Dx} = \lambda\mathbf{x}$ exactly. Next, pre-multiplying Eq. (5.6.102) with $\mathbf{Z}^{(k)T}\mathbf{M}$ and requiring $\mathbf{r}$ to be orthogonal to the subspace, that is, $\mathbf{Z}^{(k)T}\mathbf{Mr} = \mathbf{0}$, one can utilize Eq. (5.6.100) to obtain

$$\mathbf{Z}^{(k)T}\mathbf{MDu} = \lambda\mathbf{Z}^{(k)T}\mathbf{Mu} + \mathbf{Z}^{(k)T}\mathbf{Mr}$$
$$\mathbf{Z}^{(k)T}\mathbf{MDZ}^{(k)}\mathbf{y} = \lambda\mathbf{Z}^{(k)T}\mathbf{MZ}^{(k)}\mathbf{y}$$
$$\mathbf{T}^{(k)}\mathbf{y} = \lambda\mathbf{y}. \qquad (5.6.103)$$

This implies that one can obtain the eigenpairs of the large eigenproblem by solving a small eigenproblem associated with a symmetric tridiagonal matrix $\mathbf{T}^{(k)}$, Eq. (5.6.103). The computation of every element in $\mathbf{T}^{(k)}$ requires just a few matrix-vector multiplications and one solution of a linear system, and the eigenproblem with $\mathbf{T}^{(k)}$ can be solved efficiently using the QR algorithm discussed in Section 5.6.3. This is the reason for the success of the Lanczos algorithm for treating effectively very large systems.

In practice, the implementation of the Lanczos method requires considerable care. First of all, the Lanczos tridiagonalization procedure is prone to numerical instability. The orthogonality of the Lanczos vectors can be compromised after a few iterations due to the finite precision in a digital computer. The loss of orthogonality results in numerical errors in the eigenvalues and even spurious eigenpairs. A remedy is to perform a re-orthogonalization of the Lanczos vectors during the iterations to eliminate the numerical errors.

Furthermore, for improved numerical stability and computational efficiency, several variants of the Lanczos algorithm are typically implemented in modern finite element analysis codes. Examples include (1) preconditioned Lanczos algorithm for ill-conditioned eigenproblems (Knyazev, 1987), (2) shifted block Lanczos algorithm for eigenproblems with eigenvalue multiplicity (Grimes et al., 1994), and (3) Implicit Restarted Arnoldi/Lanczos iteration algorithm for extremely large eigenproblems (Sorensen, 1997).

The complete Lanczos algorithm is summarized in Algorithm 5.5 on page 291, where the tridiagonal eigenproblem is solved using Algorithm 5.2 on page 279.

---

**Algorithm 5.5** Lanczos algorithm

---

**Require:** Matrices $\mathbf{K}$ and $\mathbf{M}$, number of requested eigenpairs $m$, maximum number of iteration $N_{\max}$, tolerance $\epsilon$.

1: Factorize $\mathbf{K} = \mathbf{L}_K\mathbf{L}_K^T$
2: Pick a random vector $\mathbf{z}$ ▷ Generate the first Krylov vector
3: Compute $\tilde{\mathbf{y}} = \mathbf{Mz}$
4: Compute $\gamma_0 = \sqrt{\mathbf{z}^T\tilde{\mathbf{y}}}$, that is, normalization factor for $\mathbf{z}^{(0)}$
5: Normalize the vectors $\mathbf{z}^{(0)} = \mathbf{z}/\gamma_0$ and $\mathbf{y} = \tilde{\mathbf{y}}/\gamma_0$
6: Compute $\tilde{\mathbf{z}} = \mathbf{Mz}^{(0)}$ ▷ Prepare for the second Krylov vector
7: Solve $\mathbf{Kz} = \tilde{\mathbf{z}}$ using the factor $\mathbf{L}_K$

---

8: Compute $\alpha_0 = \mathbf{z}^T \mathbf{y}$, which implements Eq. (5.6.91)

9: Update the vector $\mathbf{z} \leftarrow \mathbf{z} - \alpha_0 \mathbf{z}^{(0)}$

10: **for** $i = 1, \ldots, N_{\max}$ **do**                                        ▷ Three-term recurrence

11:     Compute $\tilde{\mathbf{y}} = \mathbf{Mz}$

12:     Compute $\gamma_i = \sqrt{\mathbf{z}^T \tilde{\mathbf{y}}}$, that is, normalization factor for $\mathbf{z}^{(i)}$

13:     Normalize the vectors $\mathbf{z}^{(i)} = \mathbf{z}/\gamma_i$ and $\mathbf{y} = \tilde{\mathbf{y}}/\gamma_i$

14:     Collect Krylov vectors $\mathbf{Z}^{(i)} = [\mathbf{z}^{(0)}, \ldots, \mathbf{z}^{(i)}]$

15:     Compute $\tilde{\mathbf{z}} = \mathbf{Mz}^{(i)}$

16:     Solve $\mathbf{Kz} = \tilde{\mathbf{z}}$ using the factor $\mathbf{L}_K$

17:     Compute $\alpha_i = \mathbf{z}^T \mathbf{y}$, which implements Eq. (5.6.91)

18:     Update the vector $\mathbf{z} \leftarrow \mathbf{z} - \alpha_i \mathbf{z}^{(i)} - \gamma_i \mathbf{z}^{(i-1)}$

19:     **if** re-orthogonalization is necessary **then**

20:         Sweep out the existing Krylov vectors $\mathbf{z} \leftarrow \mathbf{z} - \mathbf{Z}^{(i)} \mathbf{Z}^{(i)T} \mathbf{Mz}$

21:     **end if**

22:     **if** $i \geq m - 1$ **then**

23:         Construct a tridiagonal matrix $\mathbf{T}^{(i)}$

24:         Solve the reduced eigenproblem $\mathbf{T}^{(i)} \mathbf{W}^{(i)} = \mathbf{W}^{(i)} \mathbf{\Lambda}^{(i)}$

25:         **if** $\max_j \left| (\Lambda_{jj}^{(i)} - \Lambda_{jj}^{(i-1)})/\Lambda_{jj}^{(i)} \right| < \epsilon$ **then**

26:             Break

27:         **end if**

28:     **end if**

29: **end for**

30: Set $\mathbf{w} = [1/\Lambda_{11}^{(i)}, \ldots, 1/\Lambda_{mm}^{(i)}]$

31: Set $\mathbf{U}$ as the first $m$ column vectors of $\mathbf{Z}^{(i)} \mathbf{W}^{(i)}$

32: **return** The requested $m$ eigenvectors $\mathbf{U}$ and the $m$ eigenvalues $\mathbf{w}$

## Examples Illustrating the Subspace Methods

In this section, a numerical example is provided for solving a generalized eigenvalue problem using the subspace iteration method described in Algorithm 5.4 on page 287 and the Lanczos method in Algorithm 5.5 on page 291. Both of the methods are designed for large-scale eigenproblems. Therefore, instead of solving the generalized eigenvalue problem of $3 \times 3$ matrices from Section 5.5, a slightly more complex problem, vibration of a cantilevered beam, is considered. The beam has a length of $L = 1$ m, a bending stiffness of $EI = 6 \times 10^3$ kg m$^3$/s$^2$, and a mass per unit length of $m = 6.0$ kg/m.

The beam is divided into 10 elements of equal length, as shown in Figure 5.8. The stiffness matrix and consistent mass matrix for each element have been derived in Eqs. (4.2.18) and (4.2.25), respectively. The element stiffness matrix is

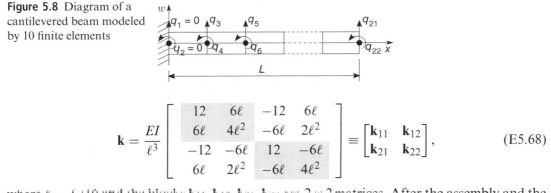

**Figure 5.8** Diagram of a cantilevered beam modeled by 10 finite elements

$$\mathbf{k} = \frac{EI}{\ell^3} \begin{bmatrix} 12 & 6\ell & -12 & 6\ell \\ 6\ell & 4\ell^2 & -6\ell & 2\ell^2 \\ -12 & -6\ell & 12 & -6\ell \\ 6\ell & 2\ell^2 & -6\ell & 4\ell^2 \end{bmatrix} \equiv \begin{bmatrix} \mathbf{k}_{11} & \mathbf{k}_{12} \\ \mathbf{k}_{21} & \mathbf{k}_{22} \end{bmatrix}, \tag{E5.68}$$

where $\ell = L/10$ and the blocks $\mathbf{k}_{11}, \mathbf{k}_{12}, \mathbf{k}_{21}, \mathbf{k}_{22}$ are $2 \times 2$ matrices. After the assembly and the application of constraints, the stiffness matrix is constructed as a $20 \times 20$ **banded** matrix

$$\mathbf{K} = \begin{bmatrix} \mathbf{k}^* & \mathbf{k}_{12} & \mathbf{O} & \mathbf{O} & \cdots & \mathbf{O} \\ \mathbf{k}_{21} & \mathbf{k}^* & \mathbf{k}_{12} & \mathbf{O} & \cdots & \mathbf{O} \\ \mathbf{O} & \mathbf{k}_{21} & \mathbf{k}^* & \ddots & \cdots & \mathbf{O} \\ \mathbf{O} & \mathbf{O} & \ddots & \ddots & \ddots & \cdots \\ \vdots & \vdots & \vdots & \ddots & \mathbf{k}^* & \mathbf{k}_{12} \\ \mathbf{O} & \mathbf{O} & \mathbf{O} & \vdots & \mathbf{k}_{21} & \mathbf{k}_{22} \end{bmatrix}, \tag{E5.69}$$

where $\mathbf{O}$ is a $2 \times 2$ zero matrix and

$$\mathbf{k}^* = \mathbf{k}_{11} + \mathbf{k}_{22} = \begin{bmatrix} 24 & 0 \\ 0 & 8\ell^2 \end{bmatrix}. \tag{E5.70}$$

The mass matrix $\mathbf{M}$ is constructed from the element mass matrix in a similar manner. It is also a $20 \times 20$ **banded** matrix having the same nonzero pattern as the stiffness matrix $\mathbf{K}$.

Next, the subspace methods are employed to solve the generalized eigenvalue problem associated with the $20 \times 20$ stiffness and mass matrices, in order to find the first two structural modes of the cantilevered beam. However, note that this example is for illustration only. There are specialized algorithms for generalized eigenvalue problems with banded matrices, which are typically more efficient than subspace methods.

**Example for Subspace Iteration Method**   To apply the subspace iteration method, the initial guess of the subspace is specified as the first two columns of a $20 \times 20$ identity matrix,

$$\mathbf{Z}^{(0)} = \begin{bmatrix} 1 & 0 & 0 & 0 & \cdots & 0 \\ 0 & 1 & 0 & 0 & \cdots & 0 \end{bmatrix}^T. \tag{E5.71}$$

The first iteration is performed in the following steps.

(1) Matrix-matrix multiplication for inverse iteration, which is obtained by solving the following linear equation,

$$\mathbf{K}\tilde{\mathbf{Z}} = \mathbf{M}\mathbf{Z}^{(0)}, \tag{E5.72}$$

where $\tilde{\mathbf{Z}}$ is a $20 \times 2$ matrix. Due to the size of $\tilde{\mathbf{Z}}$, its detailed values are not presented in this example.

(2) Construct reduced mass and stiffness matrices

$$\mathbf{M}_r = \tilde{\mathbf{Z}}^T \mathbf{M} \tilde{\mathbf{Z}} = \begin{bmatrix} 5.6613838\text{e--}13 & 6.5189971\text{e--}15 \\ 6.5189971\text{e--}15 & 7.5075403\text{e--}17 \end{bmatrix} \tag{E5.73}$$

$$\mathbf{K}_r = \tilde{\mathbf{Z}}^T \mathbf{K} \tilde{\mathbf{Z}} = \begin{bmatrix} 2.1012925\text{e--}08 & 2.0380952\text{e--}10 \\ 2.0380952\text{e--}10 & 2.1791383\text{e--}12 \end{bmatrix}. \tag{E5.74}$$

(3) Solve the reduced eigenproblem associated with $\mathbf{K}_r$ and $\mathbf{M}_r$ using the QR algorithm, Algorithm 5.2 on page 279

$$\mathbf{K}_r \mathbf{W}^{(1)} = \mathbf{M}_r \mathbf{W}^{(1)} \mathbf{\Lambda}^{(1)}. \tag{E5.75}$$

The reduced eigensolutions are

$$\mathbf{W}^{(1)} = \begin{bmatrix} 8.2266162\text{e+}05 & -1.1432465\text{e+}08 \\ -1.8684296\text{e+}08 & 9.9262831\text{e+}09 \end{bmatrix} \tag{E5.76}$$

$$\text{diag}(\mathbf{\Lambda}^{(1)}) = [\, 2.7640793\text{e+}04,\ 2.6780682\text{e+}07\,], \tag{E5.77}$$

where the first two diagonal elements of $\mathbf{\Lambda}^{(1)}$ are expected to converge to the first two eigenvalues of generalized eigenproblem.

(4) Update subspace

$$\mathbf{Z}^{(1)} = \tilde{\mathbf{Z}} \mathbf{W}^{(1)}, \tag{E5.78}$$

where the values of $\mathbf{Z}^{(1)}$ are not presented due to its size.

The procedure described by Eqs. (E5.72)–(E5.78) is repeated to generate new subspaces, $\mathbf{Z}^{(2)}$, $\mathbf{Z}^{(3)}$, ..., until the eigenvalues of the reduced problem converge. The outcomes of all the iterations are listed in Table 5.4. The first eigenvalue converges quickly within three iterations, while the second eigenvalue converges in six iterations. The modal frequencies are

$$\omega_1 = \sqrt{\lambda_1} = 1.1118626\text{e+}02 \text{ rad/s} \tag{E5.79}$$

$$\omega_2 = \sqrt{\lambda_2} = 6.9681487\text{e+}02 \text{ rad/s}. \tag{E5.80}$$

**Table 5.4 Convergence history of the subspace iteration for the first two eigenvalues**

| Iteration, $i$ | Eigenvalues, $\omega^2$ |
| --- | --- |
| 1 | [ 2.7640793e+04, 2.6780682e+07] |
| 2 | [ 1.2362508e+04, 5.1396200e+05] |
| 3 | [ 1.2362385e+04, 4.8594360e+05] |
| 4 | [ 1.2362385e+04, 4.8555728e+05] |
| 5 | [ 1.2362385e+04, 4.8555106e+05] |
| 6 | [ 1.2362385e+04, 4.8555096e+05] |
| 7 | [ 1.2362385e+04, 4.8555096e+05] |

Upon convergence at the seventh iteration, the first two columns of $\mathbf{Z}^{(7)}$ are the eigenvectors associated with the first two eigenvalues. In Figure 5.9, the numerical solutions are compared with the exact solutions for a cantilever beam, obtained in the section on continuous systems. It is clear that the numerical and analytical mode shapes are almost identical. The numerical eigenvalues are marginally higher than the exact eigenvalues. The errors are less than 0.01% and are attributed to the numerical discretization introduced by the finite element method.

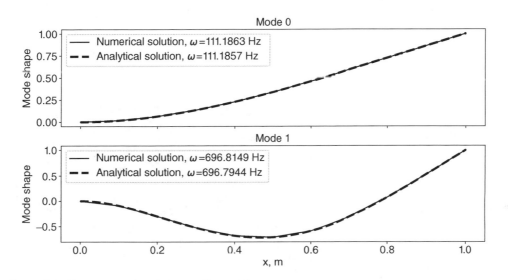

**Figure 5.9** First two modes of the cantilevered beam

**Example for Lanczos Algorithm**   The Lanczos algorithm consists of two stages. In the first stage, the zeroth and the first Krylov vectors are computed. In the second stage, the three-term recurrence formula, Eq. (5.6.96), is employed to generate new Krylov vectors, which are used to estimate the eigenpairs of interest.

**Stage 1: Initialization** An initial guess vector is specified as

$$\mathbf{z}^* = [1, 0, \ldots, 0]^T. \tag{E5.81}$$

This vector is normalized w.r.t. $\mathbf{M}$ to obtain the **zeroth** Krylov vector

$$\mathbf{z}^{(0)} = \mathbf{z}^* / \left\| \mathbf{z}^* \right\|_{\mathbf{M}}. \tag{E5.82}$$

The first Krylov vector is computed by the following three steps

(1) Perform an inverse iteration on $\mathbf{z}^{(0)}$ by solving the following static problem,

$$\mathbf{K}\tilde{\mathbf{z}} = \mathbf{M}\mathbf{z}^{(0)}. \tag{E5.83}$$

(2) The vector component associated with $\mathbf{z}^{(0)}$ is removed from $\tilde{\mathbf{z}}$,

$$\mathbf{z}^* = \tilde{\mathbf{z}} - \alpha_0 \mathbf{z}^{(0)}, \quad \alpha_0 = \tilde{\mathbf{z}}^T \mathbf{M} \mathbf{z}^{(0)} = 5.3330534\text{e–}05 \tag{E5.84}$$

(3) The vector $\mathbf{z}^*$ is normalized to obtain the first Krylov vector,

$$\mathbf{z}^{(1)} = \mathbf{z}^*/\gamma_1, \quad \gamma_1 = \|\mathbf{z}^*\|_{\mathbf{M}} = 3.7889824\text{e--}05. \tag{E5.85}$$

**Stage 2: Lanczos Iterations** With the first two Krylov vectors computed, the three-term recurrence formula, Eq. (5.6.96), is employed to compute the subsequent Krylov vectors. The second Krylov vector is computed by the following three steps

(1) Perform an inverse iteration on $\mathbf{z}^{(1)}$ by solving the following static problem

$$\mathbf{K}\tilde{\mathbf{z}} = \mathbf{M}\mathbf{z}^{(1)}. \tag{E5.86}$$

(2) The vector components associated with the previous two Krylov vectors, namely $\mathbf{z}^{(0)}$ and $\mathbf{z}^{(1)}$, are removed from $\tilde{\mathbf{z}}$

$$\mathbf{z}^* = \tilde{\mathbf{z}} - \alpha_1 \mathbf{z}^{(1)} - \gamma_1 \mathbf{z}^{(0)}, \quad \alpha_1 = \tilde{\mathbf{z}}^T \mathbf{M} \mathbf{z}^{(1)} = 2.8781754\text{e--}05, \tag{E5.87}$$

where note that $\gamma_1$ has been computed in Eq. (E5.85).

(3) The vector $\mathbf{z}^*$ is normalized to obtain the second Krylov vector

$$\mathbf{z}^{(2)} = \mathbf{z}^*/\gamma_2, \quad \gamma_2 = \|\mathbf{z}^*\|_{\mathbf{M}} = 1.1797906\text{e--}06. \tag{E5.88}$$

Since two structural modes are required, having found coefficients $\alpha_0$, $\gamma_1$, and $\alpha_1$ from Eqs. (E5.84), (E5.85), (E5.87), respectively, the initial estimates of the eigenvalues are obtained by solving the following $2 \times 2$ standard eigenproblem,

$$\mathbf{TW}^{(1)} = \mathbf{W}^{(1)}\mathbf{\Lambda}^{(1)} \tag{E5.89}$$

$$\mathbf{T} = \begin{bmatrix} \alpha_0 & \gamma_1 \\ \gamma_1 & \alpha_1 \end{bmatrix} = \begin{bmatrix} 5.3330534\text{e--}05 & 3.7889824\text{e--}05 \\ 3.7889824\text{e--}05 & 2.8781754\text{e--}05 \end{bmatrix}.$$

The reduced eigensolutions are

$$\mathbf{W}^{(1)} = \begin{bmatrix} -8.0875894\text{e--}01 & 5.8814027\text{e--}01 \\ -5.8814027\text{e--}01 & -8.0875894\text{e--}01 \end{bmatrix} \tag{E5.90}$$

$$\text{diag}(\mathbf{\Lambda}^{(1)}) = [\, 8.0884519\text{e--}05, \ 1.2277695\text{e--}06 \,]. \tag{E5.91}$$

Subsequent iterations are performed in a similar manner. The new Krylov vectors are computed using the three-term recurrence formula by repeating the procedure in Eqs. (E5.86)–(E5.88). Each time a new Krylov vector is generated, a new symmetric tridiagonal eigenproblem is solved to generate the estimates of the eigenvalues, by repeating the procedure of Eqs. (E5.89)–(E5.91). The outcomes of all the iterations are listed in Table 5.5, where only the first three eigenvalues are retained. The first eigenvalue converges at the second iteration, while the second eigenvalue converges in five iterations. The modal frequencies are

$$\omega_1 = 1/\sqrt{\lambda_1} = 1.1118626\text{e+}02 \text{ rad/s} \tag{E5.92}$$

$$\omega_2 = 1/\sqrt{\lambda_2} = 6.9681487\text{e+}02 \text{ rad/s}, \tag{E5.93}$$

**Table 5.5 Convergence history of the Lanczos method for the first three eigenvalues**

| Iteration, $i$ | Eigenvalues |
| --- | --- |
| 1 | [ 8.0884519e–05, 1.2277695e–06] |
| 2 | [ 8.0890543e–05, 2.0544089e–06, 1.2648604e–07] |
| 3 | [ 8.0890543e–05, 2.0595138e–06, 2.5798940e–07, ...] |
| 4 | [ 8.0890543e–05, 2.0595161e–06, 2.6255177e–07, ...] |
| 5 | [ 8.0890543e–05, 2.0595161e–06, 2.6257159e–07, ...] |

which are identical to the values computed by the subspace iteration method in Eqs. (E5.79)–(E5.80).

Upon convergence at the fifth iteration, a $6 \times 6$ reduced eigenproblem is solved with the solution denoted $\mathbf{W}^{(5)}$ and $\mathbf{\Lambda}^{(5)}$. The eigenvectors of the generalized eigenproblem are recovered from the first six Krylov vectors by

$$\mathbf{U} = [\mathbf{z}^{(0)}, \ldots, \mathbf{z}^{(5)}]\mathbf{W}^{(5)}. \tag{E5.94}$$

The first two columns of $\mathbf{U}$ are the eigenvectors associated with the first two eigenvalues. The eigenvectors from the Lanczos algorithm are identical to those found by the subspace iteration method that was illustrated in Figure 5.9.

**Convergence Comparison**    Comparison of the convergence of the subspace iteration method and the Lanczos algorithm is illustrated by Figure 5.10. To demonstrate their scalability, the two methods are applied to a larger generalized eigenproblem, representing the same cantilevered beam structure that is divided into 200 finite elements. However, note that the mass and stiffness matrices in this extended example are banded and this problem may be solved efficiently using some specialized algorithms for banded matrices.

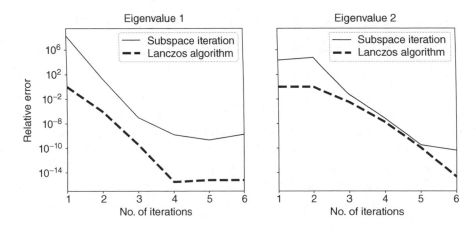

**Figure 5.10** Convergence history of subspace iteration and Lanczos algorithm

The relative error $\epsilon$ at each iteration in the subspace methods is defined as

$$\epsilon = \frac{|\omega^2_{\text{numerical}} - \omega^2_{\text{exact}}|}{\omega^2_{\text{exact}}}. \tag{E5.95}$$

Regardless of the problem size, on average, both the methods require two to four inverse iterations per converged eigenvalue. The Lanczos algorithm converges faster than the subspace iteration method. The difference is due to the treatment of the subspace. In the subspace iteration, the subspace has fixed dimensions and does not change as a result of previous iterations. In the Lanczos algorithm, the subspace is adaptively expanded at each step based on all the previous iterates. The adaptively expanded subspace provides the Lanczos algorithm with its powerful nature when solving the eigenvalue problems for large-scale structural dynamic problems.

Finally, note that in this example the exact eigenvalue is used as a reference for the computation of the relative error. In general the exact solution is unknown, and one may define the relative error in the following form, using two consecutive iterates of eigenvalues,

$$\epsilon = \frac{\left|\left(\omega^{(k+1)}\right)^2 - \left(\omega^{(k)}\right)^2\right|}{\left(\omega^{(k+1)}\right)^2}. \tag{E5.96}$$

To summarize, the examples presented in this section illustrate how the subspace iteration method and the Lanczos algorithm are employed to find the first two natural frequencies and structural modes of a cantilevered beam. In both methods, a series of reduced eigenproblems are constructed and solved to find the required eigensolutions. These solutions are found using less than 10 iterations even if the eigenproblem contains several hundreds of degrees of freedom. In this particular problem, the Lanczos algorithm converges slightly faster than the subspace iteration method.

## 5.6.5 Brief Discussion of Methods for Nonsymmetric Eigenvalue Problems

### Origin and Properties of Nonsymmetric Eigenvalue Problems

Nonsymmetric eigenvalue problems are encountered in nonconservative structural dynamics problems, where the total energy of the structure is not conserved. There are several factors that cause a system to be nonconservative. One typical factor is the damping, which produces energy dissipation over time. Another factor can be the inclusion of nonconservative loads, that is, loads that depend on the motion of the structure. The nonconservative loads are common in fluid–structure interaction problems, for example, in aeroelasticity, where energy from the flow is absorbed by the structure, causing unstable structural responses. A linear structural dynamics problem with damping and nonconservative loads can be written as

$$\mathbf{M}\ddot{\mathbf{q}} + \mathbf{C}\dot{\mathbf{q}} + \mathbf{K}\mathbf{q} = \mathbf{F}(\mathbf{q}, \dot{\mathbf{q}}, \ddot{\mathbf{q}}), \tag{5.6.104}$$

where $\mathbf{K}$ and $\mathbf{M}$ are the symmetric stiffness and mass matrices as those in Eq. (5.1.1), $\mathbf{C}$ is the damping matrix, and $\mathbf{F}$ is a nonconservative load that may depend on structural displacement, velocity, and acceleration.

To understand the stability of the nonconservative system, a common technique is the **linearized stability analysis**, when a system is linearized about an equilibrium position. The static equilibrium point of Eq. (5.6.104), $\mathbf{q}_e$, satisfies

$$\mathbf{K}\mathbf{q}_e = \mathbf{F}(\mathbf{q}_e, \dot{\mathbf{q}} = 0, \ddot{\mathbf{q}} = 0). \qquad (5.6.105)$$

At the static equilibrium point, Eq. (5.6.104) is linearized as

$$(\mathbf{M} + \mathbf{M}_N)\ddot{\mathbf{q}} + (\mathbf{C} + \mathbf{C}_N)\dot{\mathbf{q}} + (\mathbf{K} + \mathbf{K}_N)\mathbf{q} = 0, \qquad (5.6.106)$$

where the additional mass, damping, and stiffness matrices are the Jacobians of $\mathbf{F}$ computed at the equilibrium point,

$$\mathbf{M}_N = \frac{\partial \mathbf{F}}{\partial \ddot{\mathbf{q}}}, \quad \mathbf{C}_N = \frac{\partial \mathbf{F}}{\partial \dot{\mathbf{q}}}, \quad \mathbf{K}_N = \frac{\partial \mathbf{F}}{\partial \mathbf{q}}. \qquad (5.6.107)$$

It is more convenient to analyze Eq. (5.6.106) in a state-space form, with $\mathbf{x} = [\mathbf{q}^T, \dot{\mathbf{q}}^T]^T$,

$$\dot{\mathbf{x}} = \mathbf{A}\mathbf{x}, \qquad (5.6.108)$$

where $\mathbf{A}$ is a $2n \times 2n$ nonsymmetric system matrix

$$\mathbf{A} = \begin{bmatrix} \mathbf{O} & \mathbf{I} \\ -(\mathbf{M} + \mathbf{M}_N)^{-1}(\mathbf{K} + \mathbf{K}_N) & -(\mathbf{M} + \mathbf{M}_N)^{-1}(\mathbf{C} + \mathbf{C}_N) \end{bmatrix} \qquad (5.6.109)$$

The stability analysis of Eq. (5.6.108) is performed by assuming a structural motion in exponential form,

$$\mathbf{x}(t) = \mathbf{x}_0 e^{\lambda t}, \quad \lambda = \zeta + i\omega, \qquad (5.6.110)$$

where $\omega$ is the frequency as usual and $\zeta$ is the damping. When $\zeta > 0$, Eq. (5.6.110) represents a diverging and unstable motion; when $\zeta < 0$, it represents a decaying and stable motion.

Combining Eqs. (5.6.106) and (5.6.110) yields a $2n \times 2n$ standard nonsymmetric eigenvalue problem

$$\mathbf{A}\mathbf{x} = \lambda \mathbf{x}. \qquad (5.6.111)$$

Clearly, nonsymmetric eigenvalue problems are more complex to analyze and solve than the symmetric ones. For simplicity, the following discussion assumes that the nonsymmetric eigenvalue problem has $2n$ distinct eigenvalues and $2n$ independent eigenvectors

$$\mathbf{A}\mathbf{u}_i = \lambda_i \mathbf{u}_i, \quad i = 1, \ldots, 2n. \qquad (5.6.112)$$

The eigenvalues and eigenvectors may appear in conjugate pairs, which is common in nonconservative structural dynamic problems. That means, if $\lambda = \zeta + i\omega$ and $\mathbf{u} = \mathbf{u}_r + i\mathbf{u}_i$ are an eigenpair of $\mathbf{A}$, so are their conjugate,

$$\mathbf{A}\bar{\mathbf{u}} = \bar{\lambda}\bar{\mathbf{u}},\tag{5.6.113}$$

where $\bar{\lambda} = \zeta - i\omega$ and $\bar{\mathbf{u}} = \mathbf{u}_r - i\mathbf{u}_i$. A conjugate eigenpair usually indicates oscillatory structural motion with frequency $\omega$ and damping $\zeta$.

Unlike the symmetric case, $\mathbf{A}$ and $\mathbf{A}^T$ have the same eigenvalues but different eigenvectors. The eigenvectors of $\mathbf{A}$ are referred to as right eigenvectors or simply eigenvectors, while the eigenvectors of $\mathbf{A}^T$ are called left eigenvectors or adjoint eigenvectors

$$\mathbf{A}^T\mathbf{v}_i = \lambda_i\mathbf{v}_i, \quad i = 1,\ldots,2n.\tag{5.6.114}$$

In fact, when $\mathbf{A}$ is symmetric, $\mathbf{A} = \mathbf{A}^T$, the adjoint eigenvector $\mathbf{v}_i$ is the same as the eigenvector $\mathbf{u}_i$. This is one of the reasons why the symmetric eigenvalue problem is said to be self-adjoint.

Similar to the orthogonality condition discussed in Section 5.3, the two sets of eigenvectors satisfy the bi-orthogonality condition,

$$\mathbf{v}_j^T\mathbf{u}_i = \delta_{ij}\tag{5.6.115}$$
$$\mathbf{v}_j^T\mathbf{A}\mathbf{u}_i = \lambda_i\delta_{ij}.\tag{5.6.116}$$

Equation (5.6.115) implies that

$$\mathbf{V}^T\mathbf{U} = \mathbf{I}, \text{ or } \mathbf{V}^T = \mathbf{U}^{-1},\tag{5.6.117}$$

where $\mathbf{V} = [\mathbf{v}_1,\ldots,\mathbf{v}_{2n}]$ and $\mathbf{U} = [\mathbf{u}_1,\ldots,\mathbf{u}_{2n}]$. Equation (5.6.116) implies that the matrix $\mathbf{A}$ can be diagonalized, as shown in Eq. (5.6.118)

$$\mathbf{V}^T\mathbf{A}\mathbf{U} = \mathbf{\Lambda}, \text{ or } \mathbf{U}^{-1}\mathbf{A}\mathbf{U} = \mathbf{\Lambda},\tag{5.6.118}$$

where $\mathbf{\Lambda}$ is a diagonal matrix containing the eigenvalues.

In the presence of complex conjugate eigenpairs, the diagonalization in Eq. (5.6.118) involves complex matrices $\mathbf{U}$, $\mathbf{V}$, and $\mathbf{\Lambda}$. In an eigenvalue algorithm for real matrices, it is usually advisable to keep real numbers in the computation and avoid complex arithmetics in order to reduce the computational cost. Therefore, instead of the eigenvalue decomposition, the eigenvalue algorithms typically decompose a real matrix $\mathbf{A}$ into an alternative form called **Schur decomposition**. The decomposition is defined as

$$\mathbf{A} = \mathbf{Q}\mathbf{T}\mathbf{Q}^T,\tag{5.6.119}$$

where $\mathbf{Q}$ is an orthonormal matrix for similarity transformation and $\mathbf{T}$ is a **quasi-triangular** real matrix having either scalars or $2 \times 2$ blocks along its diagonal. An example of a quasi-triangular matrix is illustrated below:

$$
\mathbf{T}^* = \begin{bmatrix}
\lambda_1 & \times & \times & \times & \times & \times \\
 & \lambda_2 & \times & \times & \times & \times \\
 & & \lambda_3 & \times & \times & \times \\
 & & & \lambda_4 & \times & \times \\
 & & & & \alpha & \beta \\
 & & & & -\beta & \alpha
\end{bmatrix}.
$$

The eigenvalues of $\mathbf{T}^*$ consist of four real eigenvalues $\lambda_1$, $\lambda_2$, $\lambda_3$, and $\lambda_4$ from the first four shaded diagonal elements, and a conjugate pair of complex eigenvalues $\lambda_{5,6} = \alpha \pm i\beta$ from the shaded $2 \times 2$ block.

After the Schur decomposition, Eq. (5.6.119), $\mathbf{A}$ and $\mathbf{T}$ have the same eigenvalues. The diagonal $2 \times 2$ blocks of $\mathbf{T}$ represent the conjugate pairs of complex eigenvalues of $\mathbf{A}$, while the rest of the diagonal elements of $\mathbf{T}$ are the real eigenvalues of $\mathbf{A}$. When all the eigenvalues of $\mathbf{A}$ are real, $\mathbf{T}$ is exactly triangular with the eigenvalues of $\mathbf{A}$ on its diagonal.

### Numerical Algorithms for Nonsymmetric Eigenvalue Problems

When correctly modified, most numerical algorithms in previous sections can still be used to solve nonsymmetric eigenvalue problems. Next, the differences between the symmetric and nonsymmetric eigenvalue algorithms are highlighted but not discussed in detail. The interested readers are referred to Bai et al. (2000) and Saad (2011) for a more detailed treatment of this topic.

**Vector Iteration Methods**    These methods only require matrix-vector multiplication and hence are directly applicable to nonsymmetric problems. However, there are two caveats. First, instead of converging to the dominant conjugate complex pair of eigenvectors, the iterations converge to a real vector, which is a linear combination of the real and imaginary parts of the conjugate eigenvectors. This behavior is expected, since matrix-vector multiplication in power iteration is performed using real numbers. Therefore complex numbers cannot appear in the iteration process. Upon convergence, one needs to recover the conjugate eigenvectors from the last three successive iterates using a least squares approach. Another difference is encountered when computing multiple eigenpairs. The previous $m$ eigenvectors, including the conjugate eigenvectors, need to be swept out from an iterate $\mathbf{z}^{(k)}$ in a modified manner using both the left and right eigenvectors,

$$
\tilde{\mathbf{z}} = \mathbf{z}^{(k)} - \mathbf{u}_1 \mathbf{v}_1^T \mathbf{z}^{(k)} \cdots - \mathbf{u}_m \mathbf{v}_m^T \mathbf{z}^{(k)}. \tag{5.6.120}
$$

**Matrix Transformation Methods**    Two methods of this type have been discussed. The Jacobi's method is designed to successively reduce the magnitude of the off-diagonal elements of a symmetric matrix, and therefore it is not applicable to a nonsymmetric matrix.

However, the QR algorithm is applicable to nonsymmetric problems, but requires some modifications. First, the two-step strategy, which first tridiagonalizes and subsequently diagonalizes a symmetric matrix, has to be generalized for nonsymmetric matrices:

$$
\begin{bmatrix} \times & \times & \times & \times \\ \times & \times & \times & \times \\ \times & \times & \times & \times \\ \times & \times & \times & \times \end{bmatrix} \xrightarrow{\text{Hessenberg}} \begin{bmatrix} \times & \times & \times & \times \\ \times & \times & \times & \times \\ & \times & \times & \times \\ & & \times & \times \end{bmatrix} \xrightarrow{\text{Quasi-triangular}} \begin{bmatrix} \times & \times & \times & \times \\ & \times & \times & \times \\ & & \times & \times \\ & & \times & \times \end{bmatrix}
$$

The matrix $\mathbf{A}$ is first reduced to a form called **upper Hessenberg** matrix, which is a matrix having all zeros below the first subdiagonal. The reduction to Hessenberg form can be achieved using either Givens rotation or Householder reflection in a manner similar to the tridiagonalization in Section 5.6.3. Subsequently, the Hessenberg matrix is transformed to a quasi-triangular form via Schur decomposition, so that the eigenvalues of $\mathbf{A}$ are revealed as either the diagonal elements or the diagonal $2 \times 2$ blocks. The Schur decomposition can be accomplished using the QR algorithm.

There is one issue in the QR algorithm with shift for Schur decomposition. Since a shift $\mu$ converges to an eigenvalue of $\mathbf{A}$ during the iterations, it becomes complex when $\mathbf{A}$ has complex eigenvalues. Computing a shifted matrix $\mathbf{A} - \mu\mathbf{I}$ results in a complex matrix, which is not desirable. To avoid complex arithmetics, one needs to apply a double-shift using the shift $\mu$ and its conjugate $\bar{\mu}$, such that an iteration of QR algorithm becomes

(1) QR decomposition:

$$
\mathbf{Q}^{(k)}\mathbf{R}^{(k)} = (\mathbf{A}^{(k-1)} - \bar{\mu}\mathbf{I})(\mathbf{A}^{(k-1)} - \mu\mathbf{I})
$$

$$
= [\mathbf{A}^{(k-1)} - \text{Re}(\mu)\mathbf{I}]^2 + [\text{Im}(\mu)\mathbf{I}]^2 \qquad (5.6.121)
$$

(2) Matrix reconstruction: $\mathbf{A}^{(k)} = \mathbf{Q}^{(k)T}\mathbf{A}^{(k-1)}\mathbf{Q}^{(k)}$

**Subspace Methods**    The subspace iteration method can be directly applied to solve nonsymmetric eigenvalue problems. However, for improved computational efficiency, it is again recommended replacing the solution of the reduced eigenvalue problem by a Schur decomposition, so as to avoid complex arithmetic.

The Lanczos algorithm does not work for nonsymmetric problems, because the three-term recurrence relation Eq. (5.6.96) is no longer valid. The Arnoldi algorithm is usually employed instead for nonsymmetric problems. In Arnoldi algorithm, the new Krylov vector is generated using Eq. (5.6.89) with all the coefficients $\beta_{k,j}$ being nonzero. The original large-scale nonsymmetric eigenproblem is converted to an eigenproblem associated with a small upper Hessenberg matrix. The rest of the Arnoldi algorithm is identical to the Lanczos algorithm (Bai et al., 2000).

# BIBLIOGRAPHY

Arnoldi, W. E. (1951). The principle of minimized iterations in the solution of the matrix eigenvalue problem. *Quarterly of Applied Mathematics*, 9(1): 17–29.

Bai, Z., Demmel, J., Dongarra, J., Ruhe, A., and van der Vorst, H. (2000). *Templates for the Solution of Algebraic Eigenvalue Problems: A Practical Guide*. Society for Industrial and Applied Mathematics.

Bathe, K.-J. (2006). *Finite Element Procedures*, pages 958–978. Prentice Hall.

Bathe, K.-J. (2013). The subspace iteration method–revisited. *Computers & Structures*, 126: 177–183.

Bathe, K.-J. and Wilson, E. L. (1972). Large eigenvalue problems in dynamic analysis. *Journal of the Engineering Mechanics Division*, 98(6): 1471–1485.

Craig, R. R. Jr and Kurdila, A. J. (2006). *Fundamentals of Structural Dynamics*, pages 469–499, John Wiley & Sons.

Dong, S. B., Wolf, J. A. Jr, and Peterson, F. E. (1972). On a direct-iterative eigensolution technique. *International Journal for Numerical Methods in Engineering*, 4(2): 155–161.

Forsythe, G. E. and Henrici, P. (1960). The cyclic Jacobi method for computing the principal values of a complex matrix. *Transactions of the American Mathematical Society*, 94(1): 1–23.

Francis, J. G. (1961). The QR transformation a unitary analogue to the LR transformation–part 1. *The Computer Journal*, 4(3): 265–271.

Francis, J. G. (1962). The QR transformation–part 2. *The Computer Journal*, 4(4): 332–345.

Géradin, M. and Rixen, D. J. (2014). *Mechanical Vibrations: Theory and Application to Structural Dynamics*, pages 415–510. John Wiley & Sons.

Givens, W. (1953). A method of computing eigenvalues and eigenvectors suggested by classical results on symmetric matrices. *National Bureau of Standards, Applied Mathematics Series*, 29: 117–122.

Grimes, R. G., Lewis, J. G., and Simon, H. D. (1994). A shifted block Lanczos algorithm for solving sparse symmetric generalized eigenproblems. *SIAM Journal on Matrix Analysis and Applications*, 15(1): 228–272.

Householder, A. S. (1958). Unitary triangularization of a nonsymmetric matrix. *Journal of the ACM*, 5(4): 339–342.

Hurty W. C. and Rubinstein M. F. (1964). *Dynamics of Structures*, pages 110–140. Prentice Hall Series in Engineering of the Physical Sciences.

Knyazev, A. V. (1987). Convergence rate estimates for iterative methods for a mesh symmetric eigenvalue problem. *Soviet Journal of Numerical Analysis and Mathematical Modelling*, 2(5): 371–396.

Krylov, A. (1931). On the numerical solution of equations whose solution determine the frequency of small vibrations of material systems. *Izv. Akad. Nauk. SSSR Otd Mat. Estest*, 1: 491–539.

Lanczos, C. (1950). An iteration method for the solution of the eigenvalue problem of linear differential and integral operators. *Journal of Research of the National Bureau of Standards*, 45(4): 255–282.

Meirovitch, L. (1997). *Principles and Techniques of Vibrations*, volume 1, pages 268–360. Prentice Hall.

Rutishauser, H. (1958). Solution of eigenvalue problems with the LR-transformation. *National Bureau of Standards, Applied Mathematics Series*, 49: 47–81.

Saad, Y. (2011). *Numerical Methods for Large Eigenvalue Problems*. SIAM, revised edition.

Sorensen, D. C. (1997). Implicitly restarted Arnoldi/Lanczos methods for large scale eigenvalue calculations. In Ahmed Sameh, David E. Keyes, and V. Venkatakrishnan (eds.), *Parallel Numerical Algorithms*, pages 119–165. Springer.

Trefethen, L. N. and Bau III, D. (1997). *Numerical Linear Algebra*, volume 50, pages 179–240. SIAM.

## PROBLEMS

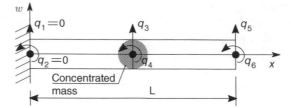

1.  Consider a cantilevered beam with a concentrated mass $M$ at the center, as shown in the figure, which resembles the wing of an aircraft with an engine. The beam has a length of $L = 1$ m, a bending stiffness of $EI = 6 \times 10^3$ kg m$^3$/s$^2$, and a mass per unit length of $m = 6.0$ kg/m. Let $M = \kappa m L$, where $\kappa$ is a constant. Divide the beam into two elements of equal length.
    (a) Follow the derivation in Section 5.6.4 and construct the stiffness matrix $\mathbf{K}$ and mass matrix $\mathbf{M}$ for this problem. Keep the constant $\kappa$ as a symbol. Hint: The matrices should be of size $4 \times 4$.
    (b) Let $\kappa = 0$, that is, remove the concentrated mass, solve the generalized eigenvalue problem using the analytical approach outlined in Section 5.5.
        • Solve the characteristic equation and find the eigenvalues, that is, the modal frequencies.
        • For lowest frequency, solve a linear system of equation to identify the mode shape.
        • Normalize the mode shape with respect to the mass matrix.
    (c) Let $\kappa = 0.01$ and $\kappa = 0.1$, and solve the two new generalized eigenvalue problems.
    (d) Plot the frequencies against $\kappa$. How do the frequencies change as $\kappa$ increases? What are the possible explanations?
    (e) Plot and compare the fundamental mode shapes associated with different $\kappa$s. How does the mode shape change as $\kappa$ increases? What are the possible explanations?

2.  In this problem, some properties of the Rayleigh's quotient are explored.
    (a) Consider the generalized eigenproblem in the numerical example in Section 5.5,

$$\mathbf{K} = 600 \begin{bmatrix} 1 & -1 & 0 \\ -1 & 3 & -2 \\ 0 & -2 & 5 \end{bmatrix}, \quad \mathbf{M} = \begin{bmatrix} 1 & 0 & 0 \\ 0 & 1.5 & 0 \\ 0 & 0 & 2.0 \end{bmatrix}. \tag{P5.1}$$

It has been found that the eigenvalues are

$$\lambda_1 = 0.35, \ \lambda_2 = 1.61, \ \lambda_3 = 3.54 \tag{P5.2}$$

and the eigenvectors are

$$[\mathbf{u}_1, \mathbf{u}_2, \mathbf{u}_3] = \begin{bmatrix} 1.000 & 1.000 & -0.388 \\ 0.647 & -0.602 & 1.000 \\ 0.301 & -0.676 & -0.965 \end{bmatrix}. \tag{P5.3}$$

Next, the Min–Max properties of Rayleigh's quotient, Eqs. (5.6.13) and (5.6.14), are visualized:

- Generate a random vector $\mathbf{r}$ of dimension $3 \times 1$, and normalize the vector such that $\mathbf{r}^T \mathbf{M} \mathbf{r} = 1$. Plot Rayleigh's quotient $R(\mathbf{u}_1 + \alpha \mathbf{r})$ for $-1 < \alpha < 1$. What is the meaning of the minimum value of this curve?
- Now plot $R(\mathbf{u}_3 + \alpha \mathbf{r})$ for $-1 < \alpha < 1$. What is the meaning of the maximum value of this curve?
- How would one construct a vector $\mathbf{t}$, such that the curve for $R(\mathbf{u}_2 + \alpha \mathbf{t})$ for $-1 < \alpha < 1$ has an extreme value of $\lambda_2$?

(b) Consider a real symmetric matrix $\mathbf{A}$ of size $n \times n$, and denote its upper-left $(n-1) \times (n-1)$ block as $\mathbf{B}$, that is,

$$\mathbf{A} = \begin{bmatrix} \mathbf{B} & \mathbf{b} \\ \mathbf{b}^T & a \end{bmatrix}, \tag{P5.4}$$

where $\mathbf{b} \in \mathbb{R}^{n-1}$ and $a \in \mathbb{R}$. Assume that $\mathbf{A}$ has distinct eigenvalues in increasing order as $\lambda_1^A, \lambda_2^A, \ldots, \lambda_n^A$, and similarly the distinct eigenvalues of $\mathbf{B}$ are denoted $\lambda_1^B, \lambda_2^B, \ldots, \lambda_{n-1}^B$. Show that the two sets of eigenvalues **interlace**

$$\lambda_1^A < \lambda_1^B < \lambda_2^A < \lambda_2^B < \cdots < \lambda_{n-1}^B < \lambda_n^A \tag{P5.5}$$

Hint: Equation (P5.5) essentially entails two inequalities $\lambda_i^B < \lambda_{i+1}^A$ and $\lambda_i^A < \lambda_i^B$. One may prove these two inequalities utilizing the Min–Max properties of Rayleigh's quotient.

3. In Section 5.2, it was mentioned that sometimes a structure may admit rigid-body motions and the stiffness matrix may be semi-positive definite. In this case, the stiffness matrix $\mathbf{K}$ cannot be inverted and it is no longer valid to compute the dynamic matrix $\mathbf{D} = \mathbf{K}^{-1}\mathbf{M}$. However, it is possible to employ the deflation technique discussed in Section 5.6.2 to develop a modified form of dynamic matrix, so that one can find eigenvectors associated with nonzero eigenvalues when the rigid-body modes are known.

(a) Suppose a $n \times n$ stiffness matrix $\mathbf{K}$ has a rank of $n - m$. It means that there exists a rank-$m$ $n \times m$ matrix $\mathbf{V}$ such that

$$\mathbf{KV} = \mathbf{0}. \tag{P5.6}$$

Suppose $\mathbf{K}$ may be partitioned as

$$\mathbf{K} = \begin{bmatrix} \mathbf{K}_{11} & \mathbf{K}_{12} \\ \mathbf{K}_{21} & \mathbf{K}_{22} \end{bmatrix}, \tag{P5.7}$$

where $\mathbf{K}_{11}$ is a $(n-m) \times (n-m)$ non-singular matrix, and $\mathbf{K}_{22}$ is a $m \times m$ singular matrix. Show that the matrix $\mathbf{V}$ can be computed as

$$\mathbf{V} = \begin{bmatrix} -\mathbf{K}_{11}^{-1}\mathbf{K}_{12} \\ \mathbf{I}_m \end{bmatrix}, \tag{P5.8}$$

where $\mathbf{I}_m$ is a $m \times m$ identity matrix.

(b) Orthogonalize $\mathbf{V}$ with respect to the mass matrix to generate a $n \times m$ matrix $\mathbf{U}$ such that $\mathbf{U}^T\mathbf{M}\mathbf{U} = \mathbf{I}$. One may define a modified dynamic matrix as

$$\tilde{\mathbf{D}} = \mathbf{P}(\mathbf{U})\mathbf{K}^+\mathbf{P}(\mathbf{U})\mathbf{M}, \tag{P5.9}$$

where $\mathbf{P}(\mathbf{U})$ is the projection matrix defined in Eq. (5.6.37), and

$$\mathbf{K}^+ = \begin{bmatrix} \mathbf{K}_{11}^{-1} & \mathbf{0} \\ \mathbf{0} & \mathbf{O} \end{bmatrix} \tag{P5.10}$$

with $\mathbf{O}$ as a $m \times m$ zero matrix. Show that applying inverse iteration to $\tilde{\mathbf{D}}$ produces the eigenvector associated with the first nonzero frequency of the generalized eigenproblem.

(c) Consider the following stiffness and mass matrices,

$$\mathbf{K} = 600 \begin{bmatrix} 1 & -1 & 0 & 0 \\ -1 & 3 & -2 & 0 \\ 0 & -2 & 5 & 0 \\ 0 & 0 & 0 & 0 \end{bmatrix}, \quad \mathbf{M} = \begin{bmatrix} 1 & 0 & 0 & 0 \\ 0 & 1.5 & 0 & 0 \\ 0 & 0 & 2.0 & 0 \\ 0 & 0 & 0 & 2.5 \end{bmatrix} \tag{P5.11}$$

- Identify the matrix $\mathbf{V}$ for the $4 \times 4$ matrix $\mathbf{K}$.
- Compute the modified dynamic matrix $\tilde{\mathbf{D}}$.
- Apply inverse iteration to $\tilde{\mathbf{D}}$ to identify the eigenpairs associated with nonzero frequencies. Are they identical to the eigenpairs in Problem 2?

4. Section 5.6.3.1 outlined the Jacobi's method for the eigenproblem of a $n \times n$ symmetric matrix $\mathbf{S}$. In this problem the Jacobi's method is examined with further detail.

(a) Show that the pivot $S_{ij}$ of $\mathbf{S}$ is driven to zero using a Givens rotation matrix with the parameter $\theta$ chosen by Eq. (5.6.49). Hint: One may focus on a $2 \times 2$ block of matrix,

$$\mathbf{S}' = \begin{bmatrix} S_{ii} & S_{ij} \\ S_{ji} & S_{jj} \end{bmatrix}. \tag{P5.12}$$

(b) The iteration of the Jacobi's method involves the multiplication between a Givens rotation matrix $\mathbf{Q}$ and $\mathbf{S}$,

$$\mathbf{S}' = \mathbf{Q}^T\mathbf{S}\mathbf{Q}. \tag{P5.13}$$

If Eq. (P5.13) is done without any simplification, the operation count will be $2n^3$ and thus the computational cost is extremely expensive. Show that the computation of Eq. (P5.13) can be done using just $O(n)$ multiplications. Hint: Most entries of $\mathbf{Q}$ are zero.

(c) Apply the Jacobi's method to solve the standard eigenproblem associated with the following tridiagonal matrix,

$$\mathbf{S} = \begin{bmatrix} 1 & -1 & 0 \\ -1 & 3 & -2 \\ 0 & -2 & 5 \end{bmatrix}. \tag{P5.14}$$

Iterate until when the eigenvalues have eight significant figures. Does the Jacobi's method preserve the tridiagonal structure of the matrix? It is called one **sweep** when the iterations are applied once to all the off-diagonal elements of $\mathbf{S}$. For the $3 \times 3$ matrix $\mathbf{S}$, one sweep consists of three iterations. How many sweeps does it take to solve the eigenproblem?

(d) Solve the standard eigenproblem associated with Eq. (P5.14) using the QR algorithm, and record the number of iterations. The computational cost of one sweep in Jacobi's method is comparable to the cost of one iteration in the QR algorithm. How does the computational costs compare between these two methods?

5. In Section 5.6.3.4 it was mentioned that the QR decomposition of a $n \times n$ tridiagonal matrix $\mathbf{T}$ requires a low operation count of $O(n)$. This special version of QR decomposition relies on the construction of a special sequence of orthonormal matrices, $\{\mathbf{Q}_k\}_{k=1}^{n-1}$, such that

- When multiplied with $\mathbf{T}$, $\mathbf{Q}_1$ eliminates the $(2, 1)$th element in $\mathbf{T}$.
- When multiplied with $\mathbf{Q}_1\mathbf{T}$, $\mathbf{Q}_2$ eliminates the $(3, 2)$th element in $\mathbf{Q}_1\mathbf{T}$.
- This procedure of elimination continues with $\mathbf{Q}_3, \mathbf{Q}_4, \cdots$
- In the end, one obtains an upper triangular matrix $\mathbf{R}$

$$\mathbf{Q}_{n-1} \cdots \mathbf{Q}_2\mathbf{Q}_1\mathbf{T} = \mathbf{R}, \tag{P5.15}$$

where $\mathbf{R}$ has only three nonzero diagonals,

$$\begin{bmatrix} \times & \times & \times & & \\ & \times & \times & \times & \\ & & \times & \times & \times \\ & & & \times & \times \\ & & & & \times \end{bmatrix}. \tag{P5.16}$$

Defining

$$\mathbf{Q} = \mathbf{Q}_1^T\mathbf{Q}_2^T \cdots \mathbf{Q}_{n-1}^T \tag{P5.17}$$

one obtains the QR decomposition $\mathbf{T} = \mathbf{QR}$.

(a) Show that $\mathbf{Q}$ must have an upper Hessenberg form,

$$\begin{bmatrix} \times & \times & \times & \times & \times \\ \times & \times & \times & \times & \times \\ & \times & \times & \times & \times \\ & & \times & \times & \times \\ & & & \times & \times \end{bmatrix} \tag{P5.18}$$

otherwise the product $\mathbf{QR}$ cannot be tridiagonal.

(b) Show that the orthonormal matrices can be constructed using Householder transformation

$$\mathbf{Q}_k = \mathbf{I} - 2\mathbf{u}_k\mathbf{u}_k^T, \tag{P5.19}$$

where $\mathbf{u}_k$ has only **two** nonzero elements.

(c) Show that using the matrix sequence based on Eq. (P5.19), Eq. (P5.15) can be computed at a computational cost of $O(n)$.

(d) Show that one may also construct the matrix sequence $\{\mathbf{Q}_k\}_{k=1}^{n-1}$ using Givens rotation, such that Eq. (P5.15) can be still evaluated at a computational cost of $O(n)$.

6. In the subspace iteration method, the buffer vector technique is introduced to accelerate the convergence of the eigensolutions. Suppose one would like to find the first $m$ eigenpairs of an eigenproblem having a dimension of $n$. Using the buffer vectors, one works with an augmented subspace of dimension $p$, $p > m$. In this case the convergence rate of the $i$th eigenvector is $\omega_i^2/\omega_{p+1}^2$. This problem explores the effect of $p$ on the computational cost of the algorithm.

(a) When $n \gg p$, show that the computational cost of one iteration of subspace iteration method is at most $O(n^2 p)$. One may assume that the cost of solving a $p \times p$ reduced eigenproblem using QR algorithm is $O(p^3)$.

(b) To achieve a desired accuracy for all the eigenvectors, show that the total computational cost of subspace iteration is

$$C(p) = c_1 n^2 p L(p), \tag{P5.20}$$

where $c_1$ is a constant and

$$L(p) = \left[\log\left(\omega_m^2/\omega_{p+1}^2\right)\right]^{-1}. \tag{P5.21}$$

Note that $C(m)$ is the computational cost of subspace iteration without the buffer vector.

(c) Suppose the natural frequencies of the eigenproblem are distributed quadratically, that is,

$$\omega_i = \omega_0 i^2, \tag{P5.22}$$

where $\omega_0$ is the fundamental frequency. Develop an expression for the ratio between $C(p)$ and $C(m)$. Let $p = km$, $k > 1$. Plot the ratio $C(p)/C(m)$ versus

$1 \leq k \leq 10$ for $m = 2, 4, 8$. When does the ratio $C(p)/C(m)$ achieve the minimum? What is the implication of this result on the choice of $p$ for accelerating the convergence?

7. The best approach to understanding a numerical algorithm is to implement it on a computer. In this and the following problems, the programming procedures are provided such that one can develop computer programs for solving generalized eigenproblems. This problem focuses on the QR algorithm as outlined in Algorithm 5.2 on page 279 and Algorithm 5.3 on page 280. It is assumed that only elementary building blocks, such as matrix multiplication and Cholesky decomposition, are available in the programming language (e.g. Python or MATLAB).

(a) Develop a function S,L=preprocess(M,K) that converts a generalized eigenproblem to a standard eigenproblem. The output S is a real symmetric matrix, and L is a lower triangular matrix such that $\mathbf{S} = \mathbf{L}^T \mathbf{L}$.

(b) Develop a function T=tridiagonalize(S) that reduces a real symmetric $n \times n$ matrix S to tridiagonal form using Householder transformation. The output T should be a symmetric and tridiagonal matrix.

(c) Develop a function Q,w=qrshift(T) that performs the QR algorithm with shift for a tridiagonal matrix T. The output Q should be a $n \times n$ orthonormal matrix containing the eigenvectors, and w should contain $n$ eigenvalues. For the step of QR decomposition, one may employ the special procedure for tridiagonal matrices that is outlined in Problem 5.

(d) Develop a function U,l=postprocess(Q,w,L) that converts the eigenpairs (Q,w) of the tridiagonal matrix to the eigenvectors U and eigenvalues l of the generalized eigenproblem.

(e) Develop a program that consecutively executes the functions preprocess, tridiagonalize, qrshift, and postprocess to solve a generalized eigenproblem. Apply this program to solve the eigenproblem in Problem 2.

8. This problem outlines the procedure to develop a computer program for Lanczos algorithm, that is, Algorithm 5.5 on page 291.

(a) Develop a function T,Z=initialize(M,K) that initializes a random vector to generate the first Krylov vector, stored as a $n \times 1$ "matrix" Z, and the first row of the tridiagonal matrix T.

(b) Develop a function Tnew,Znew=threeterm(Told,Zold) that implements the three-term recurrence formula. Given the previous Krylov vectors Zold and tridiagonal matrix Told, it computes a new Krylov vector and the new coefficients in the tridiagonal matrix and produces new matrices Znew and Tnew.

(c) Develop a function Q,w=tridiag_solve(T) that solves a standard eigenproblem of a tridiagonal matrix. One may reuse the function qrshift from Problem 7 to accomplish this task.

(d) Develop a function U,l=postprocess(Q,w,Z) that converts the eigenpairs (Q,w) of the tridiagonal matrix to the eigenvectors U and eigenvalues l of the generalized eigenproblem.

(e) Develop a program that combines the functions `initialize`, `threeterm`, `tridiag_solve`, and `postprocess` to solve a generalized eigenproblem. Apply this program to solve the numerical example presented in Section 5.6.4.3.

9. Sometimes one is not interested in the specific vibration modes of a structure, and only wants to know the natural frequencies. For example, to mitigate resonance, one only needs to tune the natural frequencies of a structure so as to avoid excitations of a certain range of frequencies; and in this process the structural mode shapes are irrelevant. Indeed there are algorithms that only compute the eigenvalues of a generalized eigenproblem, without having to compute eigenvectors. An important algorithm of this type is the **bisection method**. It is developed for a real symmetric tridiagonal matrix,

$$\mathbf{A} = \begin{bmatrix} a_1 & b_1 & & & \\ b_1 & a_2 & b_2 & & \\ & b_2 & a_3 & \ddots & \\ & & \ddots & \ddots & b_{n-1} \\ & & & b_{n-1} & a_n \end{bmatrix}, \tag{P5.23}$$

where the off-diagonal terms are all nonzero. In many finite element programs, the bisection method is combined with the tridiagonalization method in Section 5.6.3 to find the eigenvalues of a real symmetric matrix without computing the eigenvectors. The idea of bisection method is outlined in the following.

(a) Define $\mathbf{A}^{(k)}$ as the upper-left $k \times k$ block of $\mathbf{A}$. For example, $\mathbf{A}^{(n)} = \mathbf{A}$, and

$$\mathbf{A}^{(3)} = \begin{bmatrix} a_1 & b_1 & \\ b_1 & a_2 & b_2 \\ & b_2 & a_3 \end{bmatrix}. \tag{P5.24}$$

Utilizing the interlacing property of the eigenvalues of real symmetric matrices (see Problem 2), show that the number of negative eigenvalues of $\mathbf{A}$ equals to the number of sign changes in the following so-called **Sturm sequence**,

$$1, \det(\mathbf{A}^{(1)}), \det(\mathbf{A}^{(2)}), \ldots, \det(\mathbf{A}^{(n)}). \tag{P5.25}$$

In other words, when the eigenvalues of $\mathbf{A}$ are all positive, the number of eigenvalues that are less than $a$ equals to the number of sign changes in the Sturm sequence associated with $\mathbf{A} - a\mathbf{I}$.

(b) Denote the characteristic polynomial of the matrix $\mathbf{A}^{(k)}$ as

$$p^{(k)}(x) = \det(\mathbf{A}^{(k)} - x\mathbf{I}). \tag{P5.26}$$

It is clear that the eigenvalues of $\mathbf{A}^{(k)}$ are the roots of $p^{(k)}(x)$. Show that

$$p^{(k)}(x) = (a_k - x)p^{(k-1)}(x) - b_{k-1}^2 p^{(k-2)}(x). \tag{P5.27}$$

Equation (P5.27) provides a means to evaluate the characteristic polynomial $p^{(n)}(x)$ in a recursive manner at the cost of $O(n)$.

(c) Using the Sturm sequence and the recurrent relation Eq. (P5.27), describe a bisection procedure such that one can find an eigenvalue of $\mathbf{A}$ in an interval $[p, q]$ at a cost of $O(n \log \epsilon)$, where $\epsilon$ is the desired accuracy of the eigenvalue. Hint: To check if an interval $[a, b]$ contains an eigenvalue, one may examine the Sturm sequences associated with $\mathbf{A} - a\mathbf{I}$ and $\mathbf{A} - b\mathbf{I}$.

10. Consider the same $3 \times 3$ stiffness and mass matrices used in the Problem 2, but now also include a damping matrix that is proportional to the stiffness matrix,

$$\mathbf{C} = \alpha \mathbf{K}. \tag{P5.28}$$

(a) Using Eq. (5.6.108), formulate a nonsymmetric eigenvalue problem $\mathbf{Ax} = \lambda \mathbf{x}$ involving the damping matrix, where $\mathbf{A}$ is a $6 \times 6$ matrix.

(b) Modify the program for QR algorithm in Problem 7, such that it can handle complex eigenvalues and eigenvectors. Specifically, there are two functions that need modification,

- The function `tridiagonalize` should be replaced by a function `H=hessenberg(A)` that reduces a real nonsymmetric $2n \times 2n$ matrix `A` to upper Hessenberg form using Householder transformation.
- The function `Q,w=qrshift(T)` should be modified to apply the double-shift, Eq. (5.6.121), in order to handle complex eigenvalues using real arithmetics.

(c) As a sanity check, apply the modified QR program to solve the nonsymmetric eigenvalue problem with $\alpha = 0$. There should be three conjugate pairs of eigenvalues and eigenvectors. Identify the one-to-one correspondence between the these conjugate pairs and the eigenpairs listed in Problem 2.

(d) Now apply the modified QR program to solve the nonsymmetric eigenvalue problem with $\alpha = 0.01, 0.05, 0.25$. Observe how the eigenpairs change with respect to the undamped case.

# 6 Damping

*Damping is the turbulence of solid mechanics.*
– George Lesieutre, *AIAA SDM Lecture, 2014*

## 6.1 Introduction

Damping is important in structural dynamics when applied to aerospace or mechanical systems and must be considered in structural design. Typical built-up aerospace structures are lightly damped with relatively weak frequency dependence (Ungar, 1973). Structural members can exhibit extremely low damping; for instance, Min et al. (2011) report damping as low as 0.01% in Inconel 718 specimens used in high-temperature turbine blades. Damping has an important role in reducing dynamic response to avoid excessive deflection or stress. It also reduces fatigue loads and radiated noise. It contributes to aeroelastic stability. It reduces settling times following transient excitation, providing adequate margin for active structural control in the presence of uncertainty. Therefore, the system benefits from considering damping early in the design process, instead of considering it later in the design process.

Passive damping in structures is the result of numerous physical mechanisms, only some of which can be readily exploited by design (Lesieutre, 2010). The emphasis herein is on material damping that is related to damping mechanisms associated with energy dissipation within the volume of a structure.

Other sources of damping are also related to structural configurations and materials, but these are usually addressed on an empirical basis with less reliance on design models. Such sources include *nonstructural subsystems* such as cable harnesses, fluid reservoirs, and friction or fluid pumping in joints. *Friction damping* has been used successfully for some time to damp blade vibration in turbomachinery, and structural damping associated with acoustic radiation is relatively well understood. The effects of *discrete* devices employing viscous fluids or magnetic eddy currents can often be included as viscous damping elements in design models. Particle impact dampers can also be effective in damping vibration.

From a design perspective, damping represents a subset of approaches to *vibration control*, passive examples of which include vibration isolation and structural modification, including lossy struts and vibration absorbers (Mead, 1999). The most effective methods of designing material-based passive damping into structures involve the use of: high damping viscoelastic polymers in **layered damping treatments** (Nashif et al., 1985), discrete elastomeric elements, shunted piezoelectrics, damped composites, and shape memory alloys. Other approaches include eddy current damping, impact and particle dampers, and viscous damping devices (Johnson, 1995). Use of other smart materials such as electrorheological (ER) and magnetorheological (MR) fluids for damping should be regarded as semi-active, that is, requiring some activation.

Accurate models are needed in order to be able to consider damping during the design process. These models must capture the general effects of damping, while enabling the detailed design of effective damping treatments. Due to the variety of dissipation mechanisms, structural designers have not used physics-based damping models for analysis. However, advances in computing methodology and power are enabling the use of damping models with increased complexity.

This chapter focuses on damping models that are essential for the design of improved aerospace and mechanical structural configurations. We first consider **viscous damping** as a simple linear damping model and identify shortcomings that limit its practical application. An improved viscous damping model for flexural structures is presented. We explore the **complex stiffness** model as a potential improvement over viscous damping and characterize its strengths and limitations. The need for improved accuracy motivates the use of internal variable **viscoelastic** models that can accurately capture material behavior over a broad range of frequencies, making them suitable for structural finite element models. The effects of in-plane loads on the modal damping of flexural structures are explored, and the concept of the **modal strain energy method** is introduced as an approach suitable for the design of various damping treatments.

## 6.2    Modal Damping

As shown in Section 7.1, **modal analysis**, or **modal superposition**, is an effective means of obtaining the dynamic response of linear structures. This approach describes the structure as a system of parallel SDOF (modal) oscillators, where each mode can be damped differently. The time response is determined using a number of modes that is much smaller than the total number of physical degrees of freedom present in a structural dynamics model.

### 6.2.1    Viscous Damping of SDOF Modes

A common use of modal superposition involves the solution to the undamped eigenvalue problem for a discretized structural (finite element) model. The undamped matrix equation of motion has the following general form:

$$[M]\{\ddot{q}\} + [K]\{q\} = \{Q(t)\}.  \tag{6.2.1}$$

The undamped modal equations of motion have the form

$$\{\ddot{\alpha}\} + \left[\omega_m^2\right]\{\alpha\} = [\Psi]^{\mathrm{T}}\{Q(t)\}, \tag{6.2.2}$$

where $\{\alpha\}$ is a vector of modal amplitudes, $\left[\omega_m^2\right]$ is a diagonal matrix of squared undamped natural frequencies, and $[\Psi]$ is a matrix of (mass-normalized) eigenvectors. Damping may be included in Eq. (6.2.2) by adding a viscous **modal damping** term for each mode:

$$\{\ddot{\alpha}\} + [2\zeta_m\omega_m]\{\dot{\alpha}\} + \left[\omega_m^2\right]\{\alpha\} = [\Psi]^{\mathrm{T}}\{Q(t)\}, \tag{6.2.3}$$

where $\zeta_m$ is the (assigned) damping ratio for mode $m$ and $[2\zeta_m\omega_m]$ is a diagonal modal damping matrix. The equation of motion for a single mode in Eq. (6.2.3) is identical to that for an SDOF system, as can be seen by comparing it with Eq. (1.3.2).

Equation (6.2.3) and the method of modal superposition provide satisfactory accuracy for many structural dynamics response problems. Once this approach is selected for use, determining appropriate modal damping ratios becomes the main concern.

One approach is based on experience with similar structures. Although there is no assurance that this approach will yield correct damping values for any modes, it can provide a valuable starting point for analysis. For example, values of modal damping in the range from 0.003 to 0.03 might be appropriate for lightly damped, built-up aerospace structures (Johnson, 1995).

Alternate approaches to estimating modal damping for design purposes involve establishing a lower bound based on **material loss** contributions. Such approaches are especially effective when high-damping materials are used to augment nominal lightly damped structures. An example of this approach involves the use of complex stiffnesses (see Sections 6.4.2 and 6.6.3), while another is based on the distribution of **modal strain energy** among constituent parts (Section 6.10.1). If the structure considered is available for testing, then its modal damping can be characterized experimentally. Section 6.2.2 describes a method for that.

## 6.2.2 Measuring Modal Damping: The Log Decrement Method

The free response of an isolated mode of vibration representing an SDOF system can be used to estimate its modal damping ratio experimentally. The **log decrement** approach exploits the decaying exponential envelope of the free response, $e^{-\zeta\omega_n t}$.

The log decrement, $\delta$, is defined as the natural logarithm of the ratio of the amplitudes of any two successive response peaks (peak-to-next-peak):

$$\delta = \ln\left(e^{\zeta\omega_n T_d}\right) = \zeta\omega_n T_d = \frac{2\pi\zeta}{\sqrt{1-\zeta^2}}, \tag{6.2.4}$$

where $T_d = 2\pi/\omega_d$ is the period of oscillations of the damped response.

Data spanning multiple peaks (periods) can also be used to calculate the log decrement from measurements:

$$\delta = \frac{1}{N} \ln \left( \frac{u(t_1)}{u(t_1 + N\,T_d)} \right) = \frac{1}{N} \ln \left( e^{\zeta \omega_n N\,T_d} \right), \qquad (6.2.5)$$

where $u(t_1)$ is the peak amplitude (at some time $t_1$) and $u(t_1 + NT_d)$ is the peak amplitude $N$ periods later. Figure 6.1 illustrates the elements of this response.

The damping ratio is then found as

$$\zeta = \frac{(\delta/2\pi)}{\sqrt{1 + (\delta/2\pi)^2}}, \qquad (6.2.6)$$

and for small (typical) values of $\zeta$,

$$\zeta \approx \delta/2\pi. \qquad (6.2.7)$$

Note that the possible (nonlinear) dependence of damping on the response level can be assessed by considering response peaks separately at relatively high and low amplitudes.

**Figure 6.1** Rate of decay of viscously damped motion

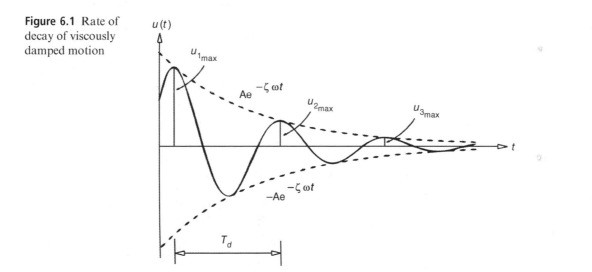

## 6.3    Viscous Damping in Continuous Beam Models

Modal damping is a special case of a more general class of damping known as **viscous damping**. This term describes models in which resisting forces or stresses proportionally oppose velocities or strain rates. Many discrete devices provide such viscous damping forces, including eddy current dampers (Sodano et al., 2006) and MR or electrorheological (ER) dampers (Wereley and Pang, 1998). Viscous damping models can be developed for use with both continuum and discretized structural dynamic models, and they are commonly used in practice in conjunction with finite element models.

As a representative problem, consider bending of an undamped Bernoulli–Euler beam vibrating in the plane of the page, as shown in Figure 6.2. The governing linear differential equation of motion, neglecting damping, is

$$\rho A(x)\frac{\partial^2 w}{\partial t^2} + \frac{\partial^2}{\partial x^2}\left(EI(x)\,\frac{\partial^2 w}{\partial x^2}\right) = p_z(x, t), \tag{6.3.1}$$

where $w(x, t)$ is the bending displacement of the neutral axis, $\rho A(x)$ is the mass per unit length, $EI(x)$ is the flexural rigidity, $p_z(x, t)$ is the distributed lateral load, and $L$ is the length of the beam.

**Figure 6.2** A simply supported beam undergoing transverse vibration

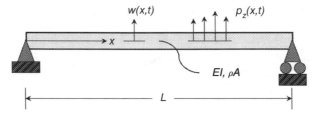

For convenience, specialize to the case of a simply-supported **uniform** beam. There is no loss of generality regarding the relationship between modal damping and the details of various viscous damping models to be considered. The governing equations consist of a partial differential equation (PDE) of motion,

$$\rho A\frac{\partial^2 w}{\partial t^2} + EI\frac{\partial^4 w}{\partial x^4} = p_z(x, t), \tag{6.3.2}$$

and a set of four boundary conditions, two at each end of the beam. In this case, the bending displacements and moments are zero at the supports

$$w(0, t) = \frac{\partial^2 w(0, t)}{\partial x^2} = 0 \quad \text{and} \quad w(L, t) = \frac{\partial^2 w(L, t)}{\partial x^2} = 0. \tag{6.3.3}$$

The solution of the structural dynamics boundary value eigenvalue problem involves mode shapes having an integer number of half-sine waves

$$\psi_m(x) = a_m \sin\left(\frac{m\pi x}{L}\right), \quad m = 1, 2, \ldots. \tag{6.3.4}$$

The natural frequencies are given by

$$\omega_m^2 = \left(\frac{m\pi}{L}\right)^4\left(\frac{EI}{\rho A}\right) \tag{6.3.5}$$

$$\omega_m = m^2\pi^2\sqrt{\frac{EI}{\rho A L^4}}. \tag{6.3.6}$$

Note that the nominal natural frequencies increase with the square of the mode number.

Next, consider two different kinds of viscous damping models: **strain-based** and **motion-based**. In all cases, "light" damping is assumed so that the vibration modes are under-damped, that is, $\zeta_m < 1$.

### 6.3.1 Strain-Based Viscous Damping

This damping model is obtained by considering a material constitutive relation where the stress depends on the strain and the strain rate:

$$\sigma(x, z) = E\,\varepsilon(x, z) + c_v\,\dot{\varepsilon}(x, z). \tag{6.3.7}$$

This is a simple example of a **viscoelastic** constitutive relation that combines elastic and viscous behaviors. When applied to this beam example, the model introduces an *internal* bending moment proportional to and opposing the time rate of change of the beam cur-vature. Differentiating this bending moment twice with respect to the spatial coordinate $x$ yields the associated effective distributed lateral force. This produces an additional term in the equation of motion (6.3.2):

$$\rho A \frac{\partial^2 w}{\partial t^2} + \underbrace{c_K \frac{\partial}{\partial t} \frac{\partial^4 w}{\partial x^4}}_{\text{strain-based viscous damping}} + EI \frac{\partial^4 w}{\partial x^4} = p_z(x, t), \tag{6.3.8}$$

where $c_K = c_v I$ and $I$ is the area moment of inertia of the beam cross-section.

Assuming free vibration in mode $m$, the following modal equation of motion is obtained:

$$\ddot{a}_m + \frac{c_K}{\rho A}\left(\frac{m\pi}{L}\right)^4 \dot{a}_m + \frac{1}{\rho A}\left[EI\left(\frac{m\pi}{L}\right)^4\right] a_m = 0. \tag{6.3.9}$$

Comparing terms with those in the canonical unforced SDOF modal equation of motion, Eq. (6.2.3), the modal damping ratio for strain-based viscous damping, $\zeta_{K\,m}$, in terms of the beam properties is found as

$$\zeta_{K\,m} = \frac{c_K\left(\frac{m\pi}{L}\right)^2}{2(\rho A\,EI)^{1/2}} = \frac{c_K\left(\frac{m\pi}{L}\right)^2\left(\frac{EI}{\rho A}\right)^{1/2}}{2EI} = \frac{c_K\,\omega_m}{2EI}. \tag{6.3.10}$$

In this case, the nominal modal damping *increases* with the square of the mode number, that is, with the nominal natural frequency. In practice, the viscous damping coefficient $c_K$ can be chosen to provide the desired modal damping for a particular mode; the modal damping of all other modes is dictated by Eq. (6.3.10). For large mode numbers, $\zeta_{K\,m}$ eventually becomes unreasonably large.

### 6.3.2 Motion-Based Viscous Damping

This damping model is obtained by considering discrete and/or distributed forces that op-pose velocities. Sometimes these damping forces oppose internal relative velocities and come in equal and opposite pairs; this is the case when a viscous strut is part of the structure. When these forces are solely external to the structure, this model is sometimes referred to as **skyhook damping** to emphasize its external nature.

When applied to the beam example, this model introduces an *external* distributed lateral force that is proportional to and opposing the local transverse velocity. This adds a term to the equation of motion (6.3.2) that is independent of the beam material:

$$\rho A \frac{\partial^2 w}{\partial t^2} + \underbrace{c_M \frac{\partial w}{\partial t}}_{\text{Motion-based viscous damping}} + EI \frac{\partial^4 w}{\partial x^4} = p_z(x, t). \tag{6.3.11}$$

Again, assuming free vibration in mode $m$, the following equation is obtained:

$$\ddot{a}_m + \frac{c_M}{\rho A}\dot{a}_m + \frac{1}{\rho A}\left[EI\left(\frac{m\pi}{L}\right)^4\right]a_m = 0. \tag{6.3.12}$$

The modal damping ratio for motion-based viscous damping, $\zeta_{M\,m}$, in terms of the beam properties is

$$\zeta_{M\,m} = \frac{c_M}{2(\rho A\,EI)^{1/2}\left(\frac{m\pi}{L}\right)^2} = \frac{c_M\left(\frac{1}{\rho A}\right)}{2\left(\frac{EI}{\rho A}\right)^{1/2}\left(\frac{m\pi}{L}\right)^2} = \frac{c_M}{2\rho A\,\omega_m}. \tag{6.3.13}$$

In this case, the damping *decreases* with the square of the mode number, that is, with the nominal natural frequency. Again, in practice, the viscous damping coefficient $c_M$ can be chosen to provide the desired modal damping for a particular mode; the modal damping of all other modes is dictated by Eq. (6.3.13). For low-frequency modes, $\zeta_{M\,m}$ may become unreasonably large.

### 6.3.3 Rayleigh Damping

**Rayleigh damping** combines the previously considered strain- and motion-based viscous damping models. For a simply supported uniform planar beam, the modal damping ratios associated with Rayleigh damping, $\zeta_{R\,m}$, are given by a combination of Eqs. (6.3.10) and (6.3.13):

$$\zeta_{R\,m} = \frac{c_K\left(\frac{m\pi}{L}\right)^2\left(\frac{EI}{\rho A}\right)^{1/2}}{2EI} + \frac{c_M\left(\frac{1}{\rho A}\right)}{2\left(\frac{EI}{\rho A}\right)^{1/2}\left(\frac{m\pi}{L}\right)^2} = \frac{c_K\,\omega_m}{2EI} + \frac{c_M}{2\rho A\,\omega_m}. \tag{6.3.14}$$

Treating the undamped natural frequencies of a beam, $\omega_m$, as a continuous variable, Figure 6.3 illustrates the general variation of modal damping with modal frequency for fixed model coefficients. A serious deficiency of this viscous damping model is that very low frequency and very high frequency modes can exhibit unreasonably high damping.

While no physical justification is provided for these viscous damping models, in practice, the Rayleigh damping model offers the advantage of two adjustable coefficients that may

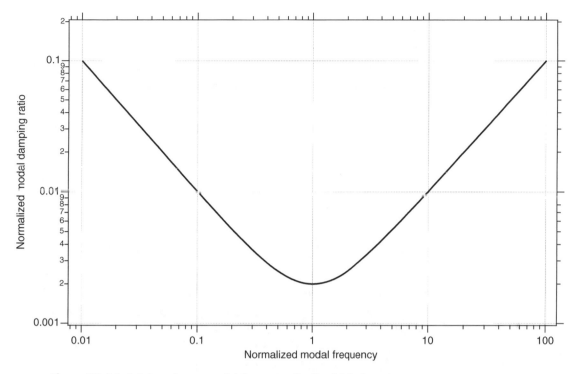

**Figure 6.3** Modal damping vs modal frequency for Rayleigh damping

be used to provide the desired modal damping for multiple modes, perhaps in a limited frequency range of interest.

### 6.3.4 Geometric Viscous Damping: Constant Modal Damping

One of the deficiencies of the preceding viscous damping models for continuous beams is the unrealistic variation of modal damping with modal frequency. Strain-based viscous damping leads to modal damping that increases with modal frequency, while motion-based viscous damping leads to modal damping that decreases with modal frequency. Consideration of the corresponding equations of motion, Eqs. (6.3.8) and (6.3.11), suggests the use of a damping model involving the time rate of change of the curvature (Lesieutre and Kauffman, 2013) as follows:

$$\rho A \frac{\partial^2 w}{\partial t^2} - \underbrace{c_G \frac{\partial}{\partial t}\frac{\partial^2 w}{\partial x^2}}_{\text{Geometric viscous damping}} + EI \frac{\partial^4 w}{\partial x^4} = p_z(x,t). \qquad (6.3.15)$$

This damping model involves an internal shear force that is proportional to the time rate of change of slope; it could be considered to be a dynamic (viscous) component of the shear force. This lateral force is analogous to that associated with a longitudinal (tensile) force carried by a deformed beam, in which the lateral (shear) component of the longitudinal force is proportional to the local slope – hence the label **geometric**. Note that this model is similar to the *square-root damping* operator model developed by Chen and Russell (1982).

Again, assuming free vibration in mode *m*, the following modal equations of motion are obtained:

$$\ddot{a}_m + \frac{c_G}{\rho A}\left(\frac{m\pi}{L}\right)^2 \dot{a}_m + \frac{1}{\rho A}\left[EI\left(\frac{m\pi}{L}\right)^4\right]a_m = 0. \tag{6.3.16}$$

For this rotation-based geometric viscous damping model, the modal damping ratio for each mode, $\zeta_{Gm}$, is found in terms of the beam properties as

$$\zeta_{Gm} = \frac{c_G}{2\rho A\,\omega_m}\left(\frac{m\pi}{L}\right)^2 = \frac{c_G}{2\,(\rho A\,EI)^{1/2}}. \tag{6.3.17}$$

In this case of simply supported boundary conditions, the nominal modal damping is *independent* of the mode number and the modal natural frequency, and the vibration mode shapes are *real*. In practice, the geometric viscous damping coefficient $c_G$ may be chosen to provide the desired nearly constant modal damping for all modes (for these and certain other boundary conditions) (Lesieutre and Kauffman, 2013). This geometric damping model can be very useful for determining the time response of damped flexural structures (e.g., beams and plates) to arbitrary forcing.

## 6.3.5    Effect of an Axial Load on Beam Modal Damping

**Membrane loads** (or *in-plane* internal *pre-stress* loads) are often encountered in thin-walled aerospace structures. Examples include tensile loads associated with spinning helicopter rotor blades, bladed disks in turbine engines, or pressurized aircraft cabins, and compressive loads due to gravitational forces and acceleration or to pressure loads in buoyant structures. The primary effects of such membrane loads on the transverse dynamics of flexural structures (beams, plates, and shells) are generally considered to be changes in the natural frequencies and perhaps mode shapes. These can be qualitatively understood by analogy with the behavior of a string in tension.

Other effects of these membrane loads, however, are not as widely appreciated. For instance, such loads can change the damping observed in various modes of structural vibration. This effect can be considerable for applications such as pressurized aircraft fuselages or spinning rotor blades (Lesieutre, 2009). The effect of a constant axial load on the modal damping of a simply supported uniform beam is considered here.

The (lateral) equation of motion for an undamped uniform planar beam with a uniform tensile axial load, $T(x) = T$, is

$$\rho A\frac{\partial^2 w}{\partial t^2} - T\,\frac{\partial^2 w}{\partial x^2} + EI\frac{\partial^4 w}{\partial x^4} = p_z(x,t). \tag{6.3.18}$$

For simply supported boundary conditions, the mode shapes of a uniform beam are not changed by a uniform axial load, and the free vibration modal equations of motion are as follows:

$$\ddot{a}_m + \frac{1}{\rho A}\left[ T\left(\frac{m\pi}{L}\right)^2 + EI\left(\frac{m\pi}{L}\right)^4 \right] a_m = 0. \qquad (6.3.19)$$

This yields an expression for the natural frequencies, $\omega_m$,

$$\omega_m^2 = \frac{1}{\rho A}\left[ EI\left(\frac{m\pi}{L}\right)^4 + T\left(\frac{m\pi}{L}\right)^2 \right] \qquad (6.3.20)$$

that can be rewritten as

$$\omega_m^2 = \frac{EI}{\rho A}\left(\frac{m\pi}{L}\right)^4 \left[ 1 + \frac{T}{EI\left(\frac{m\pi}{L}\right)^2} \right] = \omega_{m0}^2 \left( 1 + \frac{T}{m^2 \, P_{cr}} \right), \qquad (6.3.21)$$

where $\omega_{m0}$ is the undamped natural frequency of mode $m$ in the absence of axial loads and $P_{cr}$ is the critical (compressive) buckling load of a simply supported beam column:

$$P_{cr} = \frac{\pi^2 EI}{L^2}. \qquad (6.3.22)$$

Clearly, tension increases the modal frequencies, with the strongest effect on the lower modes. Compression ($T < 0$) decreases them and, as $T$ approaches $T_{cr} = -P_{cr}$, the first natural frequency ($m = 1$) approaches zero, an indication of elastic instability, or buckling.

To determine the effect of an axial load on modal damping, include a damping term in the modal equation of motion that assumes nominal constant modal damping $\zeta_{m0} = \zeta_0$ for all modes in the absence of axial loads:

$$\ddot{a}_m + 2\,\zeta_0\,\omega_{m0}\,\dot{a}_m + \omega_m^2\,a_m = 0. \qquad (6.3.23)$$

Modal damping in the presence of an axial load can be determined by equating

$$2\,\zeta_m\,\omega_m = 2\,\zeta_0\,\omega_{m0}, \qquad (6.3.24)$$

so

$$\begin{aligned} \zeta_m &= \zeta_0 \frac{\omega_{m0}}{\omega_m} \\ &= \zeta_0 \left( \frac{m^2}{m^2 + (T/P_{cr})} \right)^{1/2}. \end{aligned} \qquad (6.3.25)$$

Equation (6.3.25) indicates that tension decreases modal damping and that the effect is diminished for higher mode numbers. Conversely, compression increases modal damping to a point. In practice, the compressive load can rarely approach the critical buckling load. Figure 6.4 illustrates the general variation of modal damping with the mode number $m$ and the level of the axial load ($T/P_{cr}$).

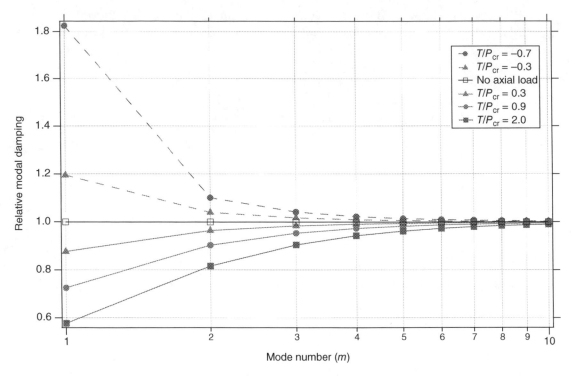

**Figure 6.4** Modal damping vs the mode number and the axial load

## 6.4    Complex Stiffness

The term **structural damping** is used broadly to refer to energy dissipation by **internal friction** in the material comprising a structure. Other terms are also used, including **solid damping** and **hysteretic damping**. In contrast with the viscous damping model, structural damping does not exhibit strong frequency dependence over a wide frequency range. A **structural damping** model can be developed for forced harmonic response based on the notion of a **complex stiffness** (Myklestad, 1952), a model that can yield frequency-independent damping. Essential assumptions include that it is proportional to stiffness (damping requires deformation) and forced harmonic response.

### 6.4.1    SDOF System with Structural Damping

To incorporate structural damping in the equation of motion for a **harmonically forced** SDOF system, consider a damping force that is in phase with and opposing the velocity, with a viscous damping coefficient $c$ that is inversely proportional to the forcing frequency $\Omega$:

$$c = h/\pi\Omega, \tag{6.4.1}$$

where $h$ is called the **hysteretic damping coefficient**.

**Harmonically Forced Vibration of an SDOF System with Structural Damping**

For harmonic forcing, $P(t) = P_0 \sin(\Omega t)$, the SDOF equation of motion, Eq. (1.3.2), takes the following form:

$$m\ddot{u}(t) + \frac{h}{\pi\Omega}\dot{u}(t) + ku(t) = P_0 \sin(\Omega t). \tag{6.4.2}$$

Recognizing that in steady-state harmonic response the velocity leads the displacement by $\pi/2$ radians in phase, the preceding equation can be recast as

$$m\ddot{u} + k(1 + i\gamma)u = P_0 \sin(\Omega t), \quad \text{with} \quad \gamma = h/\pi k, \tag{6.4.3}$$

where the quantity $k(1 + i\gamma)$ is the **complex stiffness** and $\gamma$ is the **structural damping factor**. Despite the presence of time derivatives in the equation of motion, the response is understood to be harmonic.

With this damping model, the magnification factor in the steady-state solution to (6.4.3) can be written as

$$A_d = \frac{1}{\sqrt{(1 - \alpha^2)^2 + (h/\pi k)^2}} = \frac{1}{\sqrt{(1 - \alpha^2)^2 + \gamma^2}}. \tag{6.4.4}$$

The energy dissipation per cycle $\Delta W$ for structural damping is

$$\Delta W = hA^2, \tag{6.4.5}$$

where $A$ is the maximum displacement.

The graphical representation of structural damping depicted in Figure 6.5 involves a relationship between the internal force and the displacement of the form

$$\left(\frac{u}{A}\right)^2 + \left(\frac{F - ku}{hA/\pi}\right)^2 = 1. \tag{6.4.6}$$

**Figure 6.5** Force–displacement hysteresis loop for structural damping

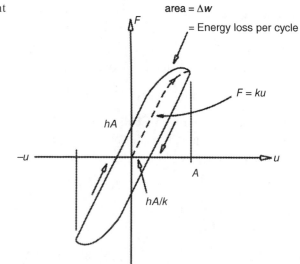

For small damping, the slope of this hysteresis loop is approximately $k$, and the intercept values on the force and displacement axes are $F = hA$ and $u = hA/k$, respectively.

The fractional energy loss per cycle, the **specific damping capacity**, $\psi = \frac{\Delta W}{U}$, is sometimes used as a measure of intrinsic *material* damping. The peak potential energy, here $U = \frac{1}{2}kA^2$, is the reference for determining fractional loss. An alternate measure of material damping is the **loss factor**, defined as the fractional energy loss per radian:

$$\eta = \frac{\Delta W}{2\pi U}. \tag{6.4.7}$$

For structural damping, the loss factor is

$$\eta_{\text{structural}} = h/\pi k = \gamma, \tag{6.4.8}$$

and for comparison, the *effective* loss factor for viscous damping is

$$\eta_{\text{viscous}} = c\,\Omega/k = 2\zeta \quad \text{at resonance.} \tag{6.4.9}$$

## 6.4.2  Material Complex Modulus

In a discretized continuum model of a structural member, the material constitutive relations provide the basis for a **structural** or **hysteretic** damping matrix. In this case, the response to harmonic forcing is considered, and the single-modulus linear elastic constitutive equations are modified to include a **material loss factor**, $\eta$:

$$\sigma(\Omega) = (1 + i\eta)\,E\,\varepsilon(\Omega). \tag{6.4.10}$$

Materials scientists routinely measure damping loss factors, often as functions of frequency, temperature, and amplitude. In addition to representing the fractional energy dissipation per radian of motion, the material loss factor may be regarded as the phase difference in forced harmonic response between an applied stress and the resulting strain. For materials with relatively high damping, these phase differences are often measured in a laboratory.

Equation (6.4.10) may also be expressed in a form that relates multiple stresses and strains in terms of multiple material moduli and loss factors:

$$\{\sigma(\Omega)\} = \big[[E'] + i[E'']\big]\{\varepsilon(\Omega)\}, \tag{6.4.11}$$

where $[E']$ is associated with **storage moduli**, or the real parts of complex moduli, and $[E'']$ with **loss moduli**, or the imaginary parts of complex moduli. An elemental stiffness matrix may then be constructed having the form

$$[K^*] = [K'] + i[K'']. \tag{6.4.12}$$

When the use of a single-material modulus suffices, or all moduli have the same loss factor, Eq. (6.4.12) may be expressed as

$$[K^*] = (1 + i\eta)[K']. \tag{6.4.13}$$

## 6.5   Viscoelastic Material Models

For some applications that require high model fidelity, the damping models described in the preceding subsections may be inadequate. An example of such an application might be a structural dynamics model that is to be used as the basis for the design of a high-performance feedback controller, in which good phase accuracy is required (Gueler et al., 1993).

While a (viscous) modal damping model used in combination with the MSE method may capture essential dissipative properties, it might not accurately capture the relative phase differences in response, as well as potential frequency-dependent properties. Several damping models suitable for use with finite element analysis have been developed to address such shortcomings. The emphasis here is on models that are compatible with linear analysis, thus neglecting friction and other nonlinear mechanisms.

Materials that exhibit a combination of elastic and viscous behavior are said to be **viscoelastic**. Both the viscous damping and complex stiffness models may be considered special cases of more general viscoelastic material models. Of particular interest for engineering structures that must maintain their shape over time are materials that fully recover their original state upon removal of loads; these are sometimes called **anelastic** materials.

Idealized types of dynamic behavior are often of interest, especially for materials characterization. These behaviors include stress relaxation at constant strain, strain creep under constant stress, and harmonic response. In general, polymeric viscoelastic materials exhibit frequency- and temperature-dependent behavior. At low temperatures or high frequencies, such materials tend to be relatively stiff; conversely, at high temperatures or low frequencies, they tend to be relatively soft. The loss factors for such materials peak in an intermediate range. Frequency and temperature dependence can often be related through the so-called principle of **time–temperature superposition**.

General forms of constitutive models for linear viscoelastic materials include **hereditary integrals** and **differential operators**, and are described in the following sections.

### 6.5.1   Hereditary Integrals

**Hereditary integrals** can capture the characteristic time-dependent behavior observed in many materials, including stress relaxation and creep, in terms of strain or stress histories. Constitutive relations involving convolution integrals are sometimes called **Volterra equations**.

In one such model, the current stress is expressed in terms of the strain history:

$$\{\sigma(t)\} = \int_0^t [E(t-\tau)]\left\{\frac{d\varepsilon(\tau)}{d\tau}\right\} d\tau, \tag{6.5.1}$$

where $E(t-\tau)$ is known as the **relaxation function**. It has the property of **fading memory**, such that $E(t-\tau)$ decreases monotonically with increasing values of the argument, $t-\tau$. As a result, the instantaneous (or unrelaxed) modulus, $E_0$ or $E_u$, is greater than the long-term (relaxed) modulus, $E_\infty$ or $E_r$. A step change in strain results in stress that gradually relaxes over time to a long-term lower value – a characteristic of **stress relaxation**.

A 1-D mechanical analogy that captures certain aspects of stress relaxation is known as the **Maxwell model**. It consists of a spring and dashpot in series, as shown in Figure 6.6. With the instantaneous imposition of an initial displacement, the dashpot is initially locked, the spring experiences the entire deformation, and the associated force is high: $f_0 = k u_0$. The force in the dashpot gradually decreases to zero over time and, since they are in series, so does the force in the spring. Shortcomings of this Maxwell model for structural materials are that it relaxes to zero stress and does not return to its initial state when unloaded.

**Figure 6.6** Maxwell model for stress relaxation

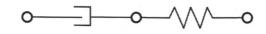

Another version of a viscoelastic model describes the current strain in terms of the stress history:

$$\{\varepsilon(t)\} = \int_0^t [C(t-\tau)] \left\{ \frac{d\sigma(\tau)}{d\tau} \right\} d\tau, \qquad (6.5.2)$$

where $C(t-\tau)$ is known as the **creep function**. A step change in stress results in strain that gradually increases over time to a long-term higher value – a characteristic of **creep**.

A 1-D mechanical analogy that captures certain aspects of creep is known as the **Kelvin–Voigt model**. It consists of a spring and dashpot in parallel, as shown in Figure 6.7. With the instantaneous imposition of an initial force, the dashpot is initially locked, carrying the entire force, and the spring experiences no initial deformation. The force in the dashpot gradually decreases to zero over time and, since they are in parallel, the spring takes up the entire force and extends to a value $u_\infty = f_0/k$. A shortcoming of this Kelvin–Voigt model for structural materials is that its initial stiffness is too high, but it does return to its initial state when unloaded.

**Figure 6.7** Kelvin–Voigt model for creep

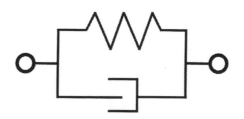

Considerable effort may be expended to determine the form of, and parameters associated with, the kernel(s) of these hereditary integral(s), $[E(t-\tau)]$ and $[C(t-\tau)]$. General relaxation and creep functions may be approximated in practice by relatively simple functions, and the mechanical analogies motivate useful viscoelastic damping models.

## 6.5.2   Differential Operators

**Differential operator** models represent the characteristic time-dependent behavior observed in many viscoelastic materials in a (linear) operator form, relating stress and strain and their time derivatives at any time. Their general form is

$$
P\left(\sigma(t)\right) = p_0\sigma + p_1\dot{\sigma} + p_2\ddot{\sigma} + \cdots + p_{Np}\frac{\partial^{Np}\sigma}{\partial t^{Np}}
$$
$$
= q_0\varepsilon + q_1\dot{\varepsilon} + q_2\ddot{\varepsilon} + \cdots + q_{Nq}\frac{\partial^{Nq}\varepsilon}{\partial t^{Nq}} = Q\left(\varepsilon(t)\right). \tag{6.5.3}
$$

In practice, the coefficients of such an operator model must be based on experimental measurements of the behavior of specific materials. Such measurements might include, in order of ease of obtaining them: creep, frequency response, and stress relaxation. And the number of coefficients should be limited to the minimum needed to represent material behavior with adequate accuracy in an operating range of interest. An example of such a simple model is

$$
p_0\sigma + p_1\dot{\sigma} = q_0\varepsilon + q_1\dot{\varepsilon}. \tag{6.5.4}
$$

In this model, without loss of generality, $p_0$ may be taken equal to 1, so that the model has three free coefficients.

## 6.5.3   Internal Variable Models

One approach to capturing viscoelastic (frequency- and temperature-dependent) material behavior in a time-domain model involves the introduction of internal dynamic coordinates. These are effectively based on various forms of Eq. (6.5.3) posed in terms of internal variables.

Several such models are available and, although they differ in some respects, they share many common features. Examples include the Golla–Hughes–McTavish (McTavish and Hughes, 1993) and **anelastic displacement fields (ADF)** (Lesieutre and Bianchini, 1995) models. Such models can capture the frequency dependence of damping and stiffness exhibited by the high-loss viscoelastic materials that are frequently used in layered damping treatments.

### The ADF Model

Figure 6.8 shows a 1-D mechanical analogy of material behavior – the **standard anelastic solid** (Nowick and Berry, 1972) – that also illustrates the general structure of the ADF model. Note that this resembles a Kelvin–Voigt model with an added series stiffness that remedies the main deficiency of that model.

The deformation of this system is described primarily by the stress, $\sigma$, and the total strain, $\varepsilon$, but its apparent instantaneous stiffness is affected by the dynamics of an internal **anelastic strain**, $\varepsilon^A$ (Lesieutre and Bianchini, 1995). If this system is subjected to harmonic forcing, its apparent stiffness and damping vary with frequency. At very high frequencies, the internal

**Figure 6.8** A viscoelastic model
with one internal coordinate

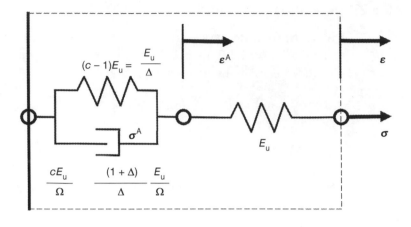

dashpot is essentially locked, the (*unrelaxed or instantaneous*) material modulus is high ($E_u$), and the damping is low. At very low frequencies, the internal dashpot slides freely, the (*relaxed or static*) modulus is lower ($E_r = E_u/(1 + \Delta)$), and the damping remains low. At some intermediate frequency, the damping attains a peak value (loss factor $\eta \approx \Delta/2$ for small damping). The peak loss factor and change in modulus both depend on the strength of the coupling between the total strain and the internal strain. The **relaxation strength** $\Delta$ describes the strength of this coupling, such that $E_u = E_r(1 + \Delta)$. The frequency at which peak damping is observed ($\approx \Omega$) is related to the inverse of the relaxation time for the internal strain.

For uniaxial stress, the ADF material constitutive equations are, for the **stress**, $\sigma$,

$$\sigma = E_u(\varepsilon - \varepsilon^A) \tag{6.5.5}$$

and, for the **anelastic stress**, $\sigma^A$,

$$\sigma^A = E_u\varepsilon - E^A\varepsilon^A = E_u\varepsilon - c\,E\varepsilon^A = E_u(\varepsilon - c\varepsilon^A). \tag{6.5.6}$$

where $c$ is a parameter that relates the anelastic modulus, $E^A$, to the elastic material modulus, $E_u$. The anelastic stress is zero when the anelastic strain is equal to its equilibrium value and is related to the force in the dashpot of the 1-D mechanical analogy.

The corresponding 1-D equation of motion (for a rod) is

$$\rho A(x)\frac{\partial^2 u(x,t)}{\partial t^2} - \frac{\partial}{\partial x}\left(A(x)\,\sigma(x,t)\right) = p_x(x,t). \tag{6.5.7}$$

Using the constitutive equations, and specializing to the case of a *uniform* rod, this becomes

$$\rho A\frac{\partial^2 u}{\partial t^2} - E_u A\left(\frac{\partial^2 u}{\partial x^2} - \frac{\partial^2 u^A}{\partial x^2}\right) = p_x(x,t), \tag{6.5.8}$$

where the (internal) **anelastic displacement**, $u^A(x,t)$, is that part of the total displacement which is not proportional to the instantaneous stress.

The ADF viscoelastic model includes an additional **relaxation equation** that describes the time evolution of the internal anelastic displacement or strain. It states that the time evolution of the anelastic strain is proportional to the anelastic stress:

$$E^A \dot{\varepsilon}^A = \Omega \, \sigma^A = \Omega \left( E_u \varepsilon - E^A \varepsilon^A \right) \tag{6.5.9}$$

or, in a form similar to the equation of motion, Eq. (6.5.8),

$$\frac{c}{\Omega} E_u A \frac{\partial}{\partial t} \left( \frac{\partial^2 u^A}{\partial x^2} \right) - E_u A \left( \frac{\partial^2 u}{\partial x^2} - c \frac{\partial^2 u^A}{\partial x^2} \right) = 0. \tag{6.5.10}$$

Note that there is no external forcing of this ADF relaxation equation; the internal displacement field is driven only through coupling to the total displacement.

With a single internal ADF, the loss factor is proportional to frequency at low frequencies and inversely proportional at high frequencies. Weaker frequency dependence can be approximated by using multiple internal fields as "building blocks," each with its own relaxation strength ($\Delta_n$) and dynamics, distributed over a range of relaxation times (or $\Omega_n$). The effective complex modulus of a material modeled using multiple ADFs is given by (Lesieutre and Bianchini, 1995)

$$E^*(\omega) = E_r \left( 1 + \sum_{n=1}^{N} \Delta_n \frac{(\omega/\Omega_n)^2}{1 + (\omega/\Omega_n)^2} \right) + iE_r \left( \sum_{n=1}^{N} \Delta_n \frac{(\omega/\Omega_n)}{1 + (\omega/\Omega_n)^2} \right). \tag{6.5.11}$$

Note that the unrelaxed material modulus, $E_u$, is related to the relaxed or static modulus, $E_r$, through the **total relaxation strength**, $\Delta \geq 0$,

$$E_u = E_r(1 + \Delta), \tag{6.5.12}$$

and that the total relaxation is the sum of relaxation strengths, $\Delta_n$, associated with individual ADFs:

$$\Delta = \sum_{1}^{N} \Delta_n. \tag{6.5.13}$$

The anelastic parameter $c_n$ for any internal field is related to its relaxation strength and the total relaxation strength by

$$c_n = \frac{1 + \Delta}{\Delta_n}. \tag{6.5.14}$$

When the relaxation strength, $\Delta_n$, is small, $c_n$ is large. Conversely, when the relaxation strength is large, $c_n$ approaches a minimum value of 1.

Temperature effects can be included by introducing a **time–temperature shift factor**, $\alpha_T(T)$, that increases the relaxation rate of the internal fields with increasing temperature (Lesieutre and Govindswamy, 1996). This modifies the relaxation equation, Eq. (6.5.10), as

$$\alpha_T \frac{c}{\Omega} E_u A \frac{\partial}{\partial t}\left(\frac{\partial^2 u^A}{\partial x^2}\right) - E_u A \left(\frac{\partial^2 u}{\partial x^2} - c\frac{\partial^2 u^A}{\partial x^2}\right) = 0. \tag{6.5.15}$$

Note that $\alpha_T(T)$ generally decreases with temperature, which has the effect of increasing the relaxation rate.

This approach is consistent with the **time–temperature superposition principle**, which is used in practice to determine the mechanical properties of linear viscoelastic materials at some operating condition from those known at a reference condition. These properties are often expressed in terms of a **reduced frequency** (Ferry, 1980), which combines frequency and temperature into a single independent parameter.

Internal variable viscoelastic models such as the ADF model outlined in this section are capable of accurately capturing the essential elastic and dissipative aspects of real structural behavior and are finding increased use in practice. Importantly, they are readily implemented in finite element models, as the internal displacement field(s) can be approximated in precisely the same way as the total displacement field.

### 6.5.4 Fractional Derivative Models

Mathematically, a fractional derivative is a derivative of arbitrary, not necessarily integer, order. Fractional derivative models provide a compact means of representing relatively weak frequency-dependent properties in the frequency domain (Bagley and Torvik, 1983). For example, a single-material complex modulus might be represented as

$$E^*(\omega) = \frac{q_0 + q_1(i\omega)^\alpha}{1 + p_1(i\omega)^\alpha} \tag{6.5.16}$$

or:

$$E^*(\omega) = E_r\left(1 + \Delta\frac{(\omega/\Omega)^{2\alpha}}{1 + (\omega/\Omega)^{2\alpha}}\right) + i\,E_r\left(\Delta\frac{(\omega/\Omega)^\alpha}{1 + (\omega/\Omega)^{2\alpha}}\right), \tag{6.5.17}$$

where $E_r$, $\Delta$, and $\alpha$ are material properties, with $\alpha$ being the slope of the loss factor on a log–log plot versus frequency, at frequencies below that at which peak damping is observed. Values of $\alpha$ for real polymeric materials range from nearly 0 to 1, with $\alpha \approx 1/2$ being typical.

This method was developed as a time-domain model, effectively from a version of Eq. (6.5.4) modified to employ fractional time derivatives in the relaxation of the internal variables. Note that a noninteger fractional derivative of a function $f(x)$ at $x = a$ is nonlocal, that is, it depends to some extent on all values of $f(x)$. As a result, to determine dynamic response to general forcing, specialized numerical methods are needed for time integration (Enelund and Lesieutre, 1999).

### 6.5.5 Nonlinear Viscoelastic Models

Elastomeric materials are commonly used in helicopter blade lag dampers, engine mounts, and motion-accommodating bearings. Their performance exhibits some dependence on frequency and temperature as well as on response amplitude, an indication of nonlinear behavior. As components of dampers, these elastomers can experience substantial shear strains in service, up to several tens of percent. Experimental data for representative materials indicate that there is a peak in the loss modulus at strain amplitudes just below 1% and that the material storage modulus is nearly independent of amplitude at very high strain amplitudes (Brackbill et al., 2002). Both the storage and loss moduli, defined in Eq. (6.4.11), exhibit relatively weak frequency dependence.

As described in Section 6.5.3 and Eq. (6.5.11), a linear multiple-ADF model can be used to capture the frequency dependence of behavior. A softening (cubic) nonlinear dependence of the relaxation strain rate on internal stress can be added to capture the high-amplitude behavior. And a rate-independent nonlinearity provided by a yielding parallel spring–slider arrangement can capture the nonlinear behavior at low amplitude. Figure 6.9 shows a 1-D mechanical analogy of a nonlinear viscoelastic material model having such features (Austrell, 1997; Brackbill et al., 2002).

**Figure 6.9** A nonlinear viscoelastic model with internal coordinates, softening nonlinear viscosity, and yielding

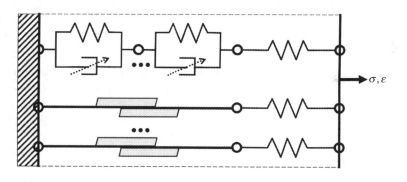

Ni–Ti shape memory alloys (SMA) also exhibit nonlinear viscoelastic behavior (Wolons et al., 1998). **Pseudoelastic hysteresis** is associated with transformation between austenite and martensite phases. The damping observed is potentially very high and tends to diminish at higher frequencies, temperatures, and static strain offset.

## 6.6 Viscous and Viscoelastic Damping in Finite Element Models

This section addresses the incorporation of various damping models in structural finite element models. Consider first an elastic (undamped) linear structure. Application of the finite element modeling method yields a discretized matrix equation of motion having the following general form:

$$[M]\{\ddot{q}\} + [K]\{q\} = \{Q(t)\}, \tag{6.6.1}$$

where $[M]$ and $[K]$ are the system mass and stiffness matrices, and $\{q\}$ and $\{Q\}$ are vectors of global nodal displacements and forces, respectively. Analysis of the free vibration eigenvalue problem associated with the matrix equation of motion yields the normal modes of vibration, each consisting of an undamped natural frequency $\omega_m$ and a real eigenvector (discrete mode shape) $\{\psi\}_m$.

The natural extension of this linear matrix equation of motion, Eq. (6.6.1), to include damping involves the addition of a matrix term multiplying the vector of nodal velocities:

$$[M]\{\ddot{q}\} + [C]\{\dot{q}\} + [K]\{q\} = \{Q(t)\},\qquad(6.6.2)$$

where $[C]$ denotes the **viscous damping matrix**.

This general second-order damped matrix equation of motion may also be expressed (nonuniquely) in a first-order form as follows:

$$\begin{bmatrix}[M] & 0 \\ 0 & [I]\end{bmatrix}\begin{Bmatrix}\{\ddot{q}\} \\ \{\dot{q}\}\end{Bmatrix} + \begin{bmatrix}[C] & [K] \\ [-I] & 0\end{bmatrix}\begin{Bmatrix}\{\dot{q}\} \\ \{q\}\end{Bmatrix} = \begin{Bmatrix}\{Q(t)\} \\ 0\end{Bmatrix}.\qquad(6.6.3)$$

This form is nonunique because the identity matrices in the second matrix equation could be replaced by different matrices. Such matrices should be positive definite (so that the product with $\{\dot{q}\}$ cannot be zero unless $\{\dot{q}\} = 0$).

Analysis of the eigenvalue problem associated with the free vibration form of Eq. (6.6.2) or (6.6.3) yields **complex modes of vibration**, each consisting of a **complex natural frequency** (corresponding to damped oscillatory vibration) $p_m$, and a **complex eigenvector** (discrete mode shape) $\{\psi\}_m$. The associated (discretized) characteristic structural response has the following form:

$$\{\psi\}_m e^{p_m t},\qquad(6.6.4)$$

where, assuming underdamped oscillatory motion, $p_m$ is often expressed as

$$p_m = -\zeta_m \omega_m \pm i\omega_m \sqrt{1 - \zeta_m^2}.\qquad(6.6.5)$$

In Eq. (6.6.5), $\zeta_m$ is the *modal damping ratio* and $\omega_m$ is the *natural frequency*, the magnitude of the complex natural frequency. Eigenvalues corresponding to oscillatory motion $p_m$ appear in complex conjugate pairs, and the associated eigenvectors are also complex conjugates.

Note that if the eigenvalue problem is formulated for the first-order equations of motion, Eq. (6.6.3), the eigenvectors contain information about the global nodal velocities $\{\dot{q}\}$ as well as the displacements $\{q\}$, such that the characteristic velocity response is $p_m\{\psi\}_m e^{p_m t}$.

## 6.6.1 Elemental Viscous Damping

In a finite element context, the global damping matrix, $[C]$, may consist of assembled elemental viscous damping matrices and discrete viscous devices. As such, it can be expected to have symmetry and definiteness properties like those of $[K]$ or, perhaps, $[M]$. Furthermore, that part of $[C]$ associated with elemental viscous damping has a matrix structure (connectivity) like that of $[K]$ or $[M]$.

The basis for a strain-based elemental viscous damping matrix must necessarily be the constitutive relations of the material from which the structural member is made. As noted in Eq. (6.3.7), the linear elastic constitutive equations are augmented such that part of the instantaneous stress is proportional to the strain rate:

$$\{\sigma\} = [E]\{\varepsilon\} + [F]\{\dot{\varepsilon}\}.\tag{6.6.6}$$

One of the deficiencies of the viscous damping model is that the constitutive behavior described by Eq. (6.6.6) does not represent the experimentally observed behavior of structural or damping materials very well. As a result, structural analysts have difficulty obtaining material strain rate coefficients $[F]$ for most materials – materials scientists typically do not try to determine such properties because the basic structure of the model is inadequate.

### 6.6.2 Proportional Damping

**Proportional damping** is a subset of general viscous damping in which the global viscous damping matrix $[C]$ is a linear combination of the global mass and stiffness matrices. It is closely related to the Rayleigh damping model:

$$[C] = \beta\,[M] + \gamma\,[K].\tag{6.6.7}$$

The proportional viscous damping model does not represent material behavior well, and its deficiencies are observed at the structural level as well.

Substitute the viscous damping matrix given in Eq. (6.6.7) into the matrix equation of motion, Eq. (6.6.2). Now note that the same real eigenvectors that diagonalize the mass and stiffness matrices of the undamped problem (Eq. (6.6.1)) will also diagonalize this form of the viscous damping matrix $[C]$, allowing a straightforward calculation of the modal damping. This feature is perhaps the greatest attraction of the proportional damping model. For underdamped modes, modal damping is found as

$$\zeta_m = \frac{\beta}{2\,\omega_m} + \frac{\gamma}{2}\,\omega_m.\tag{6.6.8}$$

Note that the proportional damping model can be generalized by considering additional combinations of powers of the mass and stiffness matrices that diagonalize the system damping matrix (Adhikari, 2006).

In the case of proportionality to the stiffness matrix, modal damping increases monotonically with modal frequency:

$$\zeta_m = \min\left(\frac{\gamma}{2}\,\omega_m,\ 1\right).\tag{6.6.9}$$

One result is that all modes having a natural frequency greater than $2/\gamma$ are overdamped; that is, in free vibration, they do not respond in an oscillatory manner. Such unreasonably high damping is a deficiency of this model. Another deficiency is associated with proportionality to the mass matrix. In the event that a structure of interest possesses a rigid body (zero frequency) mode, that mode will have unreasonably high damping. Many structures exhibit modal damping that depends only weakly on frequency and, when the source is material based, it generally increases, then decreases with frequency over a broad frequency range.

Despite its drawbacks, viscous damping is a simple way of introducing damping in a structural model and may be adequate under some applications, for example, when accuracy is only needed over a limited frequency range. Perhaps the greatest utility of the viscous damping model is the possibility of identifying a viscous damping matrix from experiments. Although such a damping matrix would not necessarily be *element based*, a desirable feature, it might represent damping adequately for certain applications.

### 6.6.3    Structural Complex Stiffness

For stiffness-proportional damping, modal damping increases monotonically with modal frequency, as indicated in Eq. (6.6.9). Motivated by this observation, one can modify a single-modulus elemental viscous damping matrix by dividing it by a frequency:

$$[C]_{\text{hysteretic}} = \frac{1}{\Omega}[C]_{\text{viscous}} = \frac{\gamma}{\Omega}[K], \qquad (6.6.10)$$

where $\Omega$ is interpreted as a harmonic forcing frequency. In this case, Eq. (6.6.2) is modified to yield

$$[M]\{\ddot{q}\} + \frac{\gamma}{\Omega}[K]\{\dot{q}\} + [K]\{q\} = \{Q\}\,e^{i\Omega t}. \qquad (6.6.11)$$

Then, considering the relationship of velocity to displacement in forced harmonic response, one obtains

$$[M]\{\ddot{q}\} + (1 + i\gamma)[K]\{q(t)\} = \{Q\}\,e^{i\Omega t}. \qquad (6.6.12)$$

Finally, defining a **complex stiffness matrix** $[K^*] = (1 + i\gamma)[K]$ and using the result that the response $\{q(t)\}$ is harmonic at the forcing frequency yields

$$\left[ -\Omega^2[M] + [K^*] \right]\{q(\Omega)\} = \{Q\}. \qquad (6.6.13)$$

Note that the response vector, $\{q(\Omega)\}$, is generally complex, indicating possible phase differences between the response and the forcing, as well as between the responses at different points on the structure.

The damped structural model described by Eq. (6.6.13) has some deficiencies but also has some practical applications. It is not useful for obtaining time response directly, as it describes a frequency response. However, when the time-domain forcing function in Eq. (6.6.12) is not harmonic, but nevertheless has a Laplace transform, it may be possible to find the response via inverse transformation using the **elastic–viscoelastic correspondence principle** (Ferry, 1980).

This complex stiffness damping model is ideal for frequency response analysis, as loss factors and stiffnesses can, in principle, be functions of frequency:

$$\begin{aligned}
\left[ -\Omega^2[M] + (1 + i\eta(\Omega))[K(\Omega)] \right]\{q(\Omega)\} &= \{Q\} \\
\left[ -\Omega^2[M] + [[K'(\Omega)] + i[K''(\Omega)]] \right]\{q(\Omega)\} &= \{Q\}.
\end{aligned} \qquad (6.6.14)$$

Because the high-damping viscoelastic materials that are sometimes used to enhance structural damping exhibit significant frequency-dependent stiffness and damping, this feature is particularly useful in practice.

### 6.6.4 Viscoelastic Finite Elements

The internal variable models addressed in Section 6.5.3 are quite compatible with finite element structural analysis methods. In the ADF approach, for instance, additional nodal displacement coordinates, identical to those of an elastic element, are introduced to model the internal displacement field(s). The internal fields are interpolated in the same way as the total displacement field. Additional first-order equations of motion are developed to describe the relaxation/creep dynamics of the internal displacement fields and their coupling to the total displacement; note that the structure of these equations does not directly reflect the 1-D mechanical analogies. The boundary conditions for the internal coordinates are implemented just as those for the corresponding total displacements.

Equation (6.6.15) shows the general structure of the finite-element-based equations of motion with a single set of internal ADF coordinates, $\{q_A\}$. In this form, $[K]$ is determined using the high-frequency (unrelaxed) material stiffness, $E_u$. Evidently, higher accuracy comes at a cost of additional coordinates and material properties:

$$\begin{bmatrix} [M] & \\ & \end{bmatrix}\begin{Bmatrix} \{\ddot{q}\} \\ \end{Bmatrix} + \begin{bmatrix} & \\ & \frac{c}{\Omega}[K] \end{bmatrix}\begin{Bmatrix} \\ \{\dot{q}_A\} \end{Bmatrix} + \begin{bmatrix} [K] & -[K] \\ -[K] & c[K] \end{bmatrix}\begin{Bmatrix} \{q\} \\ \{q_A\} \end{Bmatrix} = \begin{Bmatrix} \{Q(t)\} \\ 0 \end{Bmatrix}, \quad (6.6.15)$$

Such internal variable approaches capture the frequency-dependent elastic and dissipative aspects of structural behavior in a linear time-domain model with fixed (not frequency-dependent) system matrices.

### 6.6.5 Damping Matrices

For reference, this section provides elemental viscous and viscoelastic damping matrices for planar rod- and beam-type finite elements. These may be assembled into global damping matrices using the usual matrix assembly process of the finite element method.

#### Rods

A proportional **viscous damping matrix** for a two-noded axial rod finite element of length $L$ can be constructed from the element mass and stiffness matrices developed in Section 4.1 (Eqs. (4.1.21) and (4.1.9)) as follows:

$$[C_{\mathrm{rod}}] = \beta_{\mathrm{rod}}\,[M_{\mathrm{rod}}] + \gamma_{\mathrm{rod}}\,[K_{\mathrm{rod}}]$$

$$= \beta_{\mathrm{rod}}\,\frac{\rho AL}{6}\begin{bmatrix} 2 & 1 \\ 1 & 1 \end{bmatrix} + \gamma_{\mathrm{rod}}\,\frac{EA}{L}\begin{bmatrix} 1 & -1 \\ -1 & 1 \end{bmatrix}. \qquad (6.6.16)$$

Note that even if a proportional damping model is used at the elemental level, the assembly of elements having different proportional damping models can lead to a nonproportional global viscous damping matrix. For illustration, consider the case of a rod modeled using two finite elements, each of length $L$. Let the damping matrix for each element be proportional to the stiffness matrix for that element, and let the coefficient $\gamma_2$ for the second rod element be twice that of the first element, $\gamma_1 = \gamma$. The assembled stiffness matrix is

$$[K_{2r}] = \frac{EA}{L}\begin{bmatrix} 1 & -1 & 0 \\ -1 & 2 & -1 \\ 0 & -1 & 1 \end{bmatrix}, \qquad (6.6.17)$$

while the assembled damping matrix is

$$[C_{2r}] = \gamma\,\frac{EA}{L}\begin{bmatrix} 1 & -1 & 0 \\ -1 & 3 & -2 \\ 0 & -2 & 2 \end{bmatrix}. \qquad (6.6.18)$$

In this case, the global viscous damping matrix is clearly not proportional to the global stiffness matrix, even though there is proportional damping at the element level.

The elemental matrices and structure of the **viscoelastic** equations of motion for a two-noded axial rod finite element can be constructed from the element mass and stiffness matrices following Eq. (6.6.15). Note that this form uses a single set of internal ADF coordinates $\{q_A\}$ and that $E = E_u$ represents the *unrelaxed* (high-frequency, instantaneous) Young's modulus. The material parameter $c$ can be determined using Eqs. (6.5.12) and (6.5.14), first finding the total relaxation as $\Delta = \frac{E_u}{E_r} - 1$ and then $c = \frac{1-\Delta}{\Delta}$:

$$\begin{bmatrix} \begin{bmatrix} \frac{\rho AL}{3} & \frac{\rho AL}{6} \\ \frac{\rho AL}{6} & \frac{\rho AL}{3} \end{bmatrix} \end{bmatrix} \left\{ \begin{matrix} \ddot{q}_1 \\ \ddot{q}_2 \end{matrix} \right\} + \begin{bmatrix} \begin{bmatrix} 0 & 0 \\ 0 & 0 \end{bmatrix} & \begin{bmatrix} 0 & 0 \\ 0 & 0 \end{bmatrix} \\ \begin{bmatrix} 0 & 0 \\ 0 & 0 \end{bmatrix} & \begin{bmatrix} \frac{cEA}{\Omega L} & -\frac{cEA}{\Omega L} \\ -\frac{cEA}{\Omega L} & \frac{cEA}{\Omega L} \end{bmatrix} \end{bmatrix} \left\{ \begin{matrix} \dot{q}_1 \\ \dot{q}_2 \\ \dot{q}_{A1} \\ \dot{q}_{A2} \end{matrix} \right\}$$

$$+ \begin{bmatrix} \begin{bmatrix} \frac{EA}{L} & -\frac{EA}{L} \\ -\frac{EA}{L} & \frac{EA}{L} \end{bmatrix} & \begin{bmatrix} -\frac{EA}{L} & \frac{EA}{L} \\ \frac{EA}{L} & -\frac{EA}{L} \end{bmatrix} \\ \begin{bmatrix} -\frac{EA}{L} & \frac{EA}{L} \\ \frac{EA}{L} & -\frac{EA}{L} \end{bmatrix} & \begin{bmatrix} \frac{cEA}{L} & -\frac{cEA}{L} \\ -\frac{cEA}{L} & \frac{cEA}{L} \end{bmatrix} \end{bmatrix} \left\{ \begin{matrix} q_1 \\ q_2 \\ q_{A1} \\ q_{A2} \end{matrix} \right\} = \left\{ \begin{matrix} Q_1(t) \\ Q_2(t) \\ 0 \\ 0 \end{matrix} \right\}. \qquad (6.6.19)$$

**Beams**

A proportional **viscous damping matrix** for a uniform conventional beam finite element with four nodal displacements can be constructed from the element mass and stiffness matrices developed in Section 4.2 (Eqs. (4.2.25) and (4.2.18)) as follows:

$$[C_{\text{beam}}] = \beta_{\text{beam}} [M_{\text{beam}}] + \gamma_{\text{beam}} [K_{\text{beam}}]$$

$$[C_{\text{beam}}] = \beta_{\text{beam}} \frac{\rho A L}{420} \begin{bmatrix} 156 & 22L & 54 & -13L \\ 22L & 4L^2 & 13L & -3L^2 \\ 54 & 13L & 156 & -22L \\ -13L & -3L^2 & -22L & 4L^2 \end{bmatrix}$$

$$+ \gamma_{\text{beam}} \frac{EI}{L^3} \begin{bmatrix} 12 & 6L & -12 & 6L \\ 6L & 4L^2 & -6L & 2L^2 \\ -12 & -6L & 12 & -6L \\ 6L & 2L^2 & -6L & 4L^2 \end{bmatrix}. \quad (6.6.20)$$

The structure of the **viscoelastic** equations of motion for a conventional beam finite element is given by Eq. (6.6.15). Note that a single set of internal ADF coordinates $\{q_A\}$ is used and that $E = E_u$ represents the unrelaxed (high-frequency, or instantaneous) Young's modulus:

$$\begin{bmatrix} [M_{\text{beam}}] & \\ & \end{bmatrix} \begin{Bmatrix} \{\ddot{q}\} \\ \end{Bmatrix} + \begin{bmatrix} & \\ & \frac{c}{\Omega}[K_{\text{beam}}] \end{bmatrix} \begin{Bmatrix} \\ \{\dot{q}_A\} \end{Bmatrix}$$

$$+ \begin{bmatrix} [K_{\text{beam}}] & -[K_{\text{beam}}] \\ -[K_{\text{beam}}] & c[K_{\text{beam}}] \end{bmatrix} \begin{Bmatrix} \{q\} \\ \{q_A\} \end{Bmatrix} = \begin{Bmatrix} \{Q(t)\} \\ \end{Bmatrix}, \quad (6.6.21)$$

where $[M_{beam}]$ and $[K_{beam}]$ are given as parts of Eq. (6.6.20).

When beam damping is nearly constant or only weakly dependent on frequency, the **geometric viscous damping matrix** may be used, as described in Section 6.3.4. In the case of a conventional beam finite element, the geometric damping matrix can be found using the element geometric stiffness matrix developed in Section 4.6 (Eq. (4.6.24)) and is given by

$$[C_{\text{geom}}] = \frac{c_G}{30\,L} \begin{bmatrix} 36 & 3L & -36 & 3L \\ 3L & 4L^2 & -3L & -L^2 \\ -36 & -3L & 36 & -3L \\ 3L & -L^2 & -3L & 4L^2 \end{bmatrix}. \quad (6.6.22)$$

## 6.7    Natural Modes of Damped Systems

The structural dynamics eigenvalue problem can be defined for linear models of damped systems. The eigenvalues themselves describe the characteristic time responses of the the system, including not only natural frequencies, but damping and, in the case of viscoelastic models, first-order inverse relaxation time constants. The eigenfunctions/eigenvectors describe the characteristic displacement patterns associated with each eigenvalue and, unlike those of undamped systems, can be complex, indicating phase differences between the times of peak excursions at different points on the structure. This section focuses on the eigenvalue problem for discretized structural models.

### 6.7.1  The Eigenvalue Problem for Viscous Damping

Consider free vibration described by Eq. (6.6.2), the discretized matrix equation of motion for a viscously damped structure,

$$[M]\{\ddot{q}\} + [C]\{\dot{q}\} + [K]\{q\} = \{0\}. \tag{6.7.1}$$

Assume that $[C]$ is is symmetric and positive semidefinite, like $[K]$, and not necessarily a proportional damping matrix. Like Eq. (6.6.3), Eq. (6.7.1) may be expressed (nonuniquely) in a *symmetric* first-order form as follows:

$$\begin{bmatrix} [M] & 0 \\ 0 & [-K] \end{bmatrix} \begin{Bmatrix} \{\ddot{q}\} \\ \{\dot{q}\} \end{Bmatrix} + \begin{bmatrix} [C] & [K] \\ [K] & 0 \end{bmatrix} \begin{Bmatrix} \{\dot{q}\} \\ \{q\} \end{Bmatrix} = \begin{Bmatrix} \{0\} \\ \{0\} \end{Bmatrix}. \tag{6.7.2}$$

In this symmetric form, $[K]$ is ideally positive definite so that $[K]\{\dot{q}\} \neq 0$ unless $\{\dot{q}\} = 0$. If $[K]$ is not positive definite, Eq. (6.6.3) should be used instead.

Eigenvalue problems associated with Eq. (6.7.1) or (6.7.2) may be defined by considering a complex exponential response of the form

$$\{q(t)\} = \{\psi\}_m e^{p_m t}. \tag{6.7.3}$$

Substituting Eq. (6.7.3) into Eq. (6.7.1) and eliminating the common time dependence yields

$$\left[ [M] p_m^2 + [C] p_m + [K] \right] \{\psi\}_m = \{0\}. \tag{6.7.4}$$

Solutions to this eigenvalue problem are only available for certain (complex) values of $p_m$, the eigenvalues, and corresponding $\{\psi\}_m$, the eigenvectors. The eigenvectors are also generally complex, reflecting potential phase differences in the motion at different points on the structure.

This eigenvalue problem is best addressed in a first-order form, based on Eq. (6.7.2). It can be expressed compactly as

$$[[B] p_m - [A]] \{\Psi\}_m = \{0\}, \tag{6.7.5}$$

where $[B]$ and $[A]$ are symmetric, and the eigenvectors for the eigenvalue problem in first-order form are related to the eigenvectors for the eigenvalue problem in second-order form:

$$\{\Psi\}_m = \begin{Bmatrix} p_m \{\psi\}_m \\ \{\psi\}_m \end{Bmatrix}. \tag{6.7.6}$$

Note that the eigenvalue problem in first-order form yields twice as many eigenvectors as the problem in second-order form.

Usually, the characteristic response is decaying oscillatory motion, and $p_m$ can be expressed as

$$p_{m1} = -\zeta_m \omega_m + i\omega_m \sqrt{1 - \zeta_m^2} \tag{6.7.7}$$

or

$$p_{m2} = -\zeta_m \omega_m - i\omega_m \sqrt{1 - \zeta_m^2}, \tag{6.7.8}$$

where $\zeta_m$ is the modal damping ratio and $\omega_m$ is the magnitude of the complex natural frequency, the **natural frequency**. Eigenvalues corresponding to the oscillatory motion $p_m$ appear in complex conjugate pairs, and the associated eigenvectors are also complex conjugates.

### 6.7.2 Complex Modes and Orthogonality

As in the undamped structural dynamics eigenvalue problem, the eigenvectors of the analogous damped problem, even in first-order form, exhibit certain orthogonality properties. Consider the eigenvalue problem of Eq. (6.7.5) for mode $r$,

$$p_r [B]\{\Psi\}_r = [A]\{\Psi\}_r. \tag{6.7.9}$$

Premultiply both sides by the transpose of the $s$th eigenvector:

$$p_r \{\Psi\}_s^T [B]\{\Psi\}_r = \{\Psi\}_s^T [A]\{\Psi\}_r. \tag{6.7.10}$$

Now, consider the eigenvalue problem for mode $s$, and premultiply both sides by the transpose of the $r$th eigenvector,

$$p_s \{\Psi\}_r^T [B]\{\Psi\}_s = \{\Psi\}_r^T [A]\{\Psi\}_s. \tag{6.7.11}$$

With $[A]$ and $[B]$ both being symmetric, transposing the result yields

$$p_s \{\Psi\}_s^T [B]\{\Psi\}_r = \{\Psi\}_s^T [A]\{\Psi\}_r. \tag{6.7.12}$$

Now, subtract Eq. (6.7.12) from Eq. (6.7.10) to find

$$(p_r - p_s) \{\Psi\}_s^T [B]\{\Psi\}_r = \{0\}. \tag{6.7.13}$$

For distinct modes $r$ and $s$, such that $p_r \neq p_s$, the corresponding eigenvectors are orthogonal with respect to the $[B]$ matrix:

$$\{\Psi\}_s^T [B]\{\Psi\}_r = \{0\}, \quad \text{for} \quad p_r \neq p_s. \tag{6.7.14}$$

Furthermore, considering Eq. (6.7.10) or Eq. (6.7.12), the corresponding eigenvectors must also be orthogonal with respect to the $[A]$ matrix:

$$\{\Psi\}_s^T [A]\{\Psi\}_r = \{0\}, \quad \text{for} \quad p_r \neq p_s. \tag{6.7.15}$$

Note that Eqs. (6.7.14) and (6.7.15) apply to both the real and imaginary parts of complex eigenvectors.

Finally, consider Eq. (6.7.10) in the case where $s = r$:

$$p_r \{\Psi\}_r^T [B]\{\Psi\}_r = \{\Psi\}_r^T [A]\{\Psi\}_r. \tag{6.7.16}$$

If the eigenvectors $\{\Psi\}_r$ are scaled such that

$$\{\Psi\}_r^T [B]\{\Psi\}_r = 1, \tag{6.7.17}$$

then

$$\{\Psi\}_r^{\mathrm{T}}[A]\{\Psi\}_r = p_r. \qquad (6.7.18)$$

The results in this subsection are similar to those found for the real eigenvectors of un-damped structures. An important distinction is that a symmetric form of the first-order equations of motion is used, so that the system matrices and eigenvectors have twice the dimension of those of the undamped system.

### 6.7.3  Natural Modes of Viscoelastic Systems

Consider free vibration described by Eq. (6.6.15), which shows the general structure of the viscoelastic finite-element-based equations of motion using a single set of internal ADF coordinates $\{q_A\}$. In this form, $[K]$ represents the elastic stiffness matrix for any element or assemblage, calculated using a common unrelaxed (high-frequency or instantaneous) material stiffness:

$$\begin{bmatrix} [M] & \\ & \frac{c}{\Omega}[K] \end{bmatrix} \begin{Bmatrix} \{\ddot{q}\} \\ \{\dot{q}_A\} \end{Bmatrix} + \begin{bmatrix} [K] & -[K] \\ -[K] & c[K] \end{bmatrix} \begin{Bmatrix} \{q\} \\ \{q_A\} \end{Bmatrix} = \begin{Bmatrix} \{0\} \\ \{0\} \end{Bmatrix}. \qquad (6.7.19)$$

Equation (6.7.19) can be expressed (nonuniquely) in first-order form as

$$\begin{bmatrix} [M] & & \\ & [I] & \\ & & \frac{c}{\Omega}[K] \end{bmatrix} \begin{Bmatrix} \{\ddot{q}\} \\ \{\dot{q}\} \\ \{\dot{q}^A\} \end{Bmatrix} + \begin{bmatrix} & [K] & -[K] \\ -[I] & & \\ & -[K] & c[K] \end{bmatrix} \begin{Bmatrix} \{\dot{q}\} \\ \{q\} \\ \{q_A\} \end{Bmatrix} = \begin{Bmatrix} \{0\} \\ \{0\} \\ \{0\} \end{Bmatrix}. \qquad (6.7.20)$$

An alternative *symmetric* form is

$$\begin{bmatrix} & [M] & \\ [M] & & \\ & & \frac{c}{\Omega}[K] \end{bmatrix} \begin{Bmatrix} \{\dot{q}\} \\ \{\ddot{q}\} \\ \{\dot{q}^A\} \end{Bmatrix} + \begin{bmatrix} [K] & & -[K] \\ & -[M] & \\ -[K] & & c[K] \end{bmatrix} \begin{Bmatrix} \{q\} \\ \{\dot{q}\} \\ \{q_A\} \end{Bmatrix} = \begin{Bmatrix} \{0\} \\ \{0\} \\ \{0\} \end{Bmatrix}. \qquad (6.7.21)$$

The eigenvalue problem for this system takes the same general form as Eq. (6.7.5). Note that in addition to damped oscillatory modes, the spectrum of eigenvalues includes first-order **relaxation modes** of the form

$$p_m = -\alpha_m = -1/\tau_m, \qquad (6.7.22)$$

where $\tau_m > 0$ is a **relaxation time constant**. Plotted in the complex plane, these eigenvalues lie on the negative real axis.

Figure 6.10 shows representative damping-vs-frequency results for the vibration modes of two 1-D rods modeled using a single ADF and 20 elements. Two materials are considered, each having the same static (unrelaxed) modulus $E_r$ and the same inverse relaxation time constant $\Omega = 1$. One has a modest relaxation strength of $\Delta = 0.125$, and the other has a stronger relaxation strength of $\Delta = 1.250$. The modal damping ratios are plotted versus the natural frequencies for the two rods. Note that the fundamental natural frequencies are the same for both, consistent with having the same low-frequency modulus. For higher mode

**Figure 6.10** Modal damping vs. modal frequency for a viscoelastic rod modeled using a single ADF

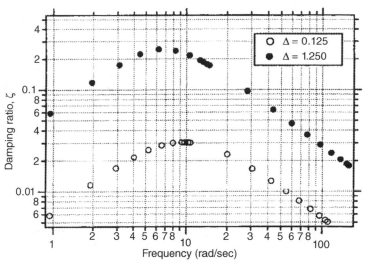

numbers, however, the rod made from the material with the larger relaxation strength also has higher natural frequencies. At all corresponding mode numbers, the rod made from the material with the larger relaxation strength has higher damping. Peak damping for the smaller relaxation strength is $\zeta \approx \Lambda/4$, and it is observed at $\omega \approx \Omega$. Peak damping for the larger relaxation strength is $\zeta < \Delta/4$, and it is observed at $\omega < \Omega$. In both cases, the damping tends to small values at both low and high frequencies.

Figures 6.11 and 6.12 show results for the damping and frequency of the vibration modes of a 1-D rod modeled using three ADFs and 35 elements. In Figure 6.11, the ADF parameters are selected to provide approximately constant damping over two decades of frequency.

**Figure 6.11** Modal damping vs. modal frequency for a viscoelastic rod modeled using three ADFs to approximate constant damping

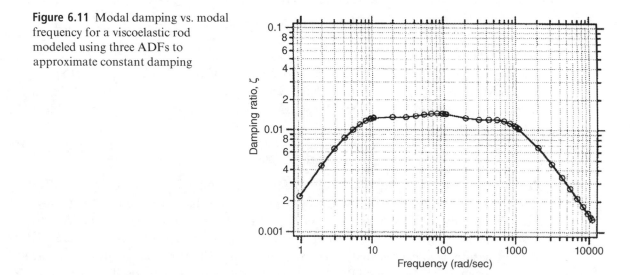

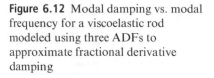

**Figure 6.12** Modal damping vs. modal frequency for a viscoelastic rod modeled using three ADFs to approximate fractional derivative damping

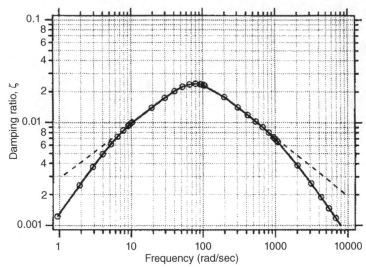

While there are some variations around the nominal value, and the contributions of the individual ADFs are very clear, the damping is much closer to constant than is possible with a conventional proportional damping model. In Figure 6.12, the ADF parameters are selected to approximate a fractional derivative model with $\alpha = 1/2$ over a similar frequency range. Over the target frequency range, the approximation is very good.

### 6.7.4   The Eigenvalue Problem for Complex Stiffness

Consider the free vibration version of Eq. (6.6.12):

$$[M]\{\ddot{q}\} + \left[K^*\right]\{q(t)\} = \{0\},\tag{6.7.23}$$

with $[K^*] = (1 + i\gamma)[K]$.

This equation was developed under the assumption of forced harmonic motion, so it is inconsistent to use it to explore characteristic free vibration. Nevertheless, this is sometimes done in practice. Assuming exponential time response, an eigenvalue problem may be expressed as

$$\left[\left[K^*\right] + p_m^2[M]\right]\{\psi_m\} = \{0\}.\tag{6.7.24}$$

Because the stiffness matrix is complex, the damped natural frequencies are related to square roots of complex eigenvalues $p_m^2$, and the results must be interpreted very carefully. These will typically be of the form $p_m = \pm(-\zeta\omega_n + i\sqrt{1 - \zeta^2}\omega_n)$, so a good practice is to choose the roots $p_m$ that have a positive imaginary part. (If the eigenvalues are treated as squares of natural frequencies, that is, as $-\Omega_m^2$, the results will be rotated 90 degrees in the complex plane and will be more difficult to interpret.) Because damped oscillatory motion is inconsistent with the basic assumption of the complex stiffness model, the results will be more accurate if the free vibration is nearly harmonic, that is, lightly damped.

If a single global loss factor can be isolated as in Eq. (6.6.13), real eigenvectors result. For light damping, the modal damping ratio is approximately half of the loss factor, $\zeta \approx \eta/2$. In the more general case, in which different elements of a structure have different loss factors, complex eigenvectors result. The accuracy of natural frequencies and eigenvectors determined using this method will decrease with increasing loss factor(s).

### 6.7.5 Effects of Axial Loads on Modal Damping

As discussed in Section 6.3.5, the presence of tensile or compressive loads in flexural structures like beams or plates affects their modal damping: Tensile loads decrease it, while compressive loads increase it. Consider the free vibration of the structure described by Eq. (6.6.2). This discretized matrix equation of motion for a viscously damped structure is modified to include the effects of axial loads:

$$[M]\{\ddot{q}\} + [C]\{\dot{q}\} + \big[[K] + \big[K_g(T)\big]\big]\{q\} = \{0\}. \tag{6.7.25}$$

Here, $T$ is a parameter representing the level of a tensile axial load and $[K_g]$ is the associated global **geometric stiffness matrix**. In the case of a uniform two-noded planar beam finite element, the element geometric stiffness matrix developed in Section 4.6 (Eq. (4.6.24)) is

$$[K_g] = \frac{T}{30\,L} \begin{bmatrix} 36 & 3L & -36 & 3L \\ 3L & 4L^2 & -3L & -L^2 \\ -36 & -3L & 36 & -3L \\ 3L & -L^2 & -3L & 4L^2 \end{bmatrix}. \tag{6.7.26}$$

In Eq. (6.7.25), $[C]$ is a general symmetric, positive semidefinite viscous damping matrix; a good choice for approximately constant modal damping in the absence of axial loads would be the geometric damping matrix defined in Eq. (6.6.22).

Recognizing that the global stiffness matrix is a combination of the elastic and geometric stiffness matrices, Eq. (6.7.25) has the same structure as the eigenvalue problem for viscously damped structures addressed in Section 6.7.1. Changing the level of axial loads in the structure will change the natural frequencies, damping, and eigenvectors (discrete mode shapes).

## 6.8   Modal Superposition Using Complex Modes

To obtain the dynamic time or frequency response of a damped system, the computational methods described in Chapter 7 can be used directly with the system mass, stiffness, and damping matrices. In cases where computational efficiency is a concern, however, it is appropriate to use **modal analysis** to obtain dynamic response. Section 7.1 addresses the **normal mode method** for undamped structures.

As noted previously in this chapter, some eigenvalue problems for damped structures yield complex eigenvalues and eigenvectors as solutions. These complex eigenvalues and eigenvectors can be used as the basis for modal analysis, similar to the way that normal (real) modes are used to obtain the response of undamped systems.

If the eigenvectors of the modes of interest are approximately real, that is, the motion at all points on the structure are almost in phase and reach their extreme excursions at approximately the same time, then the methods described in Sections 6.2.1 and 1.3 can be used. Each mode is treated as an SDOF system, and the modal natural frequencies and damping are determined from the eigenvalues.

If the eigenvectors of the modes of interest exhibit significant phase differences, then modal analysis proceeds as it does in normal modes analysis, but the *complex eigenvectors* are used. Because the eigenvalues and eigenvectors come in complex conjugate pairs, twice as many modes are available when compared to normal modes analysis. Only the real part of the total complex response contributes to the physical response.

## 6.9     Friction or Coulomb Damping

Built-up structures are frequently assembled by mechanically joining structural members together using, for example, rivets or bolts. The relative motion of the loaded interface between members when the structure deforms is a potential source of energy dissipation in addition to that which is intrinsic to the material. Friction damping, or **Coulomb damping**, is due to the frictional force between sliding contact surfaces. This frictional force always acts to oppose the motion (velocity), and its magnitude is taken to be proportional to the normal force $N$ between the contact surfaces:

$$F_d = \mu N, \tag{6.9.1}$$

where $\mu$ is the kinetic coefficient of friction.

Consider the SDOF system with Coulomb damping shown in Figure 6.13. Assuming the net force on the mass is sufficient to overcome static friction – that is, the mass is sliding in the $+u$ direction – the equation of motion for velocity in the $(+u)$ horizontal direction is

$$m\ddot{u}(t) + \mu N + ku(t) = P(t) \quad \text{or} \quad \ddot{u} + \omega_n^2(u + a) = \frac{\omega_n^2}{k}P(t), \tag{6.9.2}$$

where $a = \mu N/k$ is the minimum static deflection of the spring needed to overcome the friction force. Defining $u_1 = u + a$ and noting that $\ddot{u}_1 = \ddot{u}$, Eq. (6.9.2) becomes

$$\ddot{u}_1(t) + \omega_n^2 u_1(t) = \frac{\omega_n^2}{k}P(t). \tag{6.9.3}$$

Note that Eq. (6.9.3) has the same form as that for undamped motion, with the frequency $\omega_n$ unchanged by the presence of Coulomb damping.

For velocity in the $(-u)$ direction, Eq. (6.9.3) also holds, but with $u_2 = u - a$ replacing $u_1 = u + a$.

**Figure 6.13**
SDOF system
with Coulomb
damping

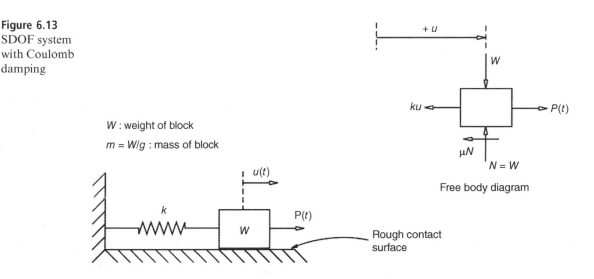

$W$ : weight of block

$m = W/g$ : mass of block

Free body diagram

Rough contact surface

Thus, a similar undamped equation of motion describes motion in both directions, but the direction of motion must be tracked in order to use the appropriate one at any given time. The motion of this SDOF system thus consists of a sequence of piecewise linear trajectories.

### 6.9.1    Free Response of an SDOF System with Coulomb Damping

The following example illustrates the free vibration response for a system with Coulomb damping.

### Example

Consider an SDOF system with Coulomb damping, given an initial displacement $u(0) = u_0$ with zero velocity $\dot{u}(0) = v_0 = 0$. Here $P(t) = 0$. The first portion of the damped free response will be in the $(-u)$ direction. In terms of $u_2$, the initial conditions take the form

$$u_2(0) = u_0 - a \quad \text{and} \quad \dot{u}_2(0) = 0.$$

The solution for the first excursion in the $(-u)$ direction is given by

$$u_2(t) = (u_0 - a)\cos(\omega t).$$

The time at the end of the first excursion is

$$t = \pi/\omega_n = T_n/2,$$

and the displacement and velocity in terms of $u_2$ and $u$ are

$$u_2(T_n/2) = -(u_0 - a) \quad \rightarrow u(T_n/2) = -(u_0 - 2a)$$
$$\dot{u}_2(T_n/2) = 0 \qquad\qquad \rightarrow \dot{u}(T_n/2) = 0,$$

where $T_n = 2\pi/\omega_n$.

From this result, the initial conditions for the subsequent excursion in the $(+u)$ direction can be expressed in terms of $u_1$:

$$u_1(T_n/2) = -(u_0 - 3a) \quad \text{and} \quad \dot{u}_1(T_n/2) = 0,$$

and the solution for the second excursion is

$$u_1(t') = -(u_0 - 3a)\cos(\omega t'), \quad \text{where} \quad t' = t - T_n/2.$$

This solution procedure can be repeated until the SDOF system excursions diminish to a point at which the spring force cannot overcome the (static) Coulomb friction force, with $u \leq a$. Figure 6.14 shows a plot of representative motion.

**Figure 6.14** Free vibration with Coulomb damping

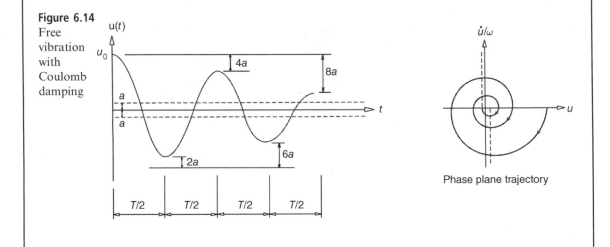

### 6.9.2 Forced Response of an SDOF System with Coulomb Damping

Forced response with Coulomb damping may be obtained using an approach similar to that for undamped forced vibration analysis. As in the analysis of free vibration with Coulomb damping, the appropriate equation of motion depends on the direction of the motion. A sequence of directed piecewise linear motions combine to represent the time evolution of this nonlinear system.

## 6.10 Damping Treatments

Several approaches to designing damping into structures are available, including layered damping treatments, shunted piezoelectric damping, damped composite materials, and particle damping. One or more of the damping modeling methods described in the preceding may often be used to predict the performance of one of these damping treatments.

### 6.10.1 The Modal Strain Energy Method

One materials-based approach to design of damping is the **MSE** method. In this approach, a **modal loss factor** is considered to be a weighted sum of material (or modulus/component/subsystem) loss factors (Ungar and Kerwin Jr., 1962). The weighting terms are the fraction of strain energy $U_{mi}/U_m$ stored in each material (or component) when the structure deforms into the mode shape of interest; this is usually estimated from the analysis of an undamped model. For light damping, the modal damping of mode $m$ is approximately half of the modal loss factor:

$$\zeta_m \cong \tfrac{1}{2}\eta_m = 1/2 \sum_{\substack{i \\ \text{materials}}} \eta_i \frac{U_{mi}}{U_m}. \qquad (6.10.1)$$

The MSE method is typically implemented in a finite element context and is used to predict the damping performance of layered damping treatments as well as damped composite materials. The complex stiffness method usually predicts damping values similar to those predicted by this MSE method.

### 6.10.2 Layered Damping Treatments

One of the most effective and widely used additive damping methods is the constrained-layer damping treatment (DiTaranto, 1965). In this approach, a lossy viscoelastic material is sandwiched between a flexural base structure and a constraining layer. As the base structure deforms, the constraining layer induces shear in the viscoelastic layer, as illustrated in Figure 6.15, and some of the associated strain energy is dissipated. In design practice, a finite element analysis is used to determine the strain energy distribution in the built-up structure, and modal damping is estimated using the modal strain energy method (Johnson and Kienholz, 1982) or the complex stiffness method.

**Figure 6.15** Constrained layer damping treatments are popular and effective

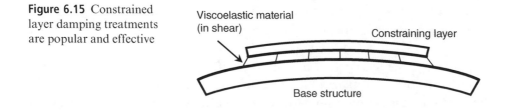

Viscoelastic material (in shear)

Constraining layer

Base structure

An effective design requires careful consideration of the operational environment, especially temperature and frequency, in the selection of a damping material. Frequency- and temperature dependence can often be related through the so-called principle of time–temperature superposition; in practice, the relationship is often displayed in a reduced frequency nomogram, as shown in Figure 6.16 (Soovere and Drake, 1984). To determine

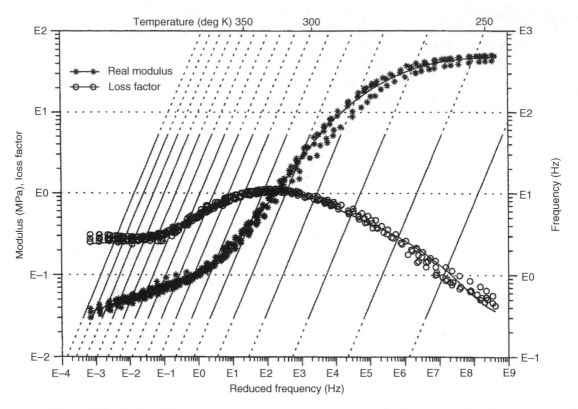

**Figure 6.16** Reduced-frequency nomograms can capture simultaneously the frequency- and temperature-dependent behavior of many viscoelastic polymers (Soovere and Drake, 1984)

a material storage modulus and loss factor using such a nomogram, one first locates the operating temperature at the top and the frequency on the right. Then, one finds their intersection in the interior of the nomogram and reads the **reduced frequency** directly below at the bottom. The storage modulus and loss factor are presented as functions of this reduced frequency.

A related approach is the unconstrained or free-layer damping treatment. In this treatment, a relatively stiff, but lossy material is added to a surface of the base structure. This layer must be relatively thick in order to participate significantly in terms of strain energy and is generally less effective than the constrained layer approach on a weight basis.

Several features of such layered damping treatments may be adjusted in the pursuit of high performance designs. These include: the materials and thicknesses of the viscoelastic and constraining layers; the lengths, segmentation, and placement of the treatments; and the possibility of increasing the distance from the mid-plane of the base structure in a standoff configuration. These layered damping treatments may also be modified for use in laminated composite shells, sandwich panels, and adhesive joints.

### 6.10.3 Damped Composite Materials

Fiber-reinforced composite materials are well suited for use in precision aerospace structures because of the potential for superior mechanical and thermal properties, as well as for reduced parts count in manufacturing. Damping measured for monolithic composite structures, however, is often remarkably low compared to that of built-up metallic structures, despite the common use of polymeric matrix materials. For instance, Abbruzzese et al. (2008) measured the modal damping of the bare composite structure of the James Webb Space Telescope at 35K and found it to be as low as 0.017%. Because of its practical importance, continuing effort is being directed toward understanding and increasing composite damping.

From a design perspective, the damping of fiber-reinforced composites may be estimated using the (modal) strain energy method or the complex stiffness method; this requires knowledge (or assumption) of the damping of the constituent fiber and matrix materials or of a composite lamina. The loss factors of representative high-modulus carbon fibers tend to be quite low, less than 0.1%. For composites made from such fibers, the damping associated with **fiber-dominated** composite moduli is low; most damping of significance is provided by the polymeric matrix material, particularly when it is loaded in shear. This may be exploited by control of lamina orientation in a continuous fiber-reinforced laminate or by using fibers having relatively low aspect ratios (Lin et al., 1984; Suarez et al., 1986).

Composites with high damping fibers offer an interesting design alternative. Because of the strain energy distribution associated with fiber-dominated composite moduli, even a small increase in fiber damping can have a significant impact on structural damping. Some efforts have addressed the damping of carbon fibers; measured damping in bromine-intercalated P100 fibers was 10 times higher (peak loss factor 0.5%) than that of pristine fibers (Lesieutre et al., 1991). In addition, piezoelectric ceramic fibers are commercially available, and resistive shunting offers the potential to achieve significant damping – loss factors of up to several tens of percent – in addition to good stiffness (Lesieutre et al., 1993). The development of damped fibers provides potential for tailoring damping as well as stiffness during composite structural design.

### 6.10.4 Shunted Piezoelectric Materials

Piezoelectric materials exhibit coupled elastic and electric behavior and are frequently used as sensors and sometimes as actuators. Quartz is a naturally occurring piezoelectric material, and many ceramic materials such as lead zirconate titanate have been engineered to exhibit stronger piezoelectric behavior.

Shunted piezoelectric materials provide new options to engineers who work on solutions to structural vibration control problems. Modern piezoelectric materials exhibit relatively strong electromechanical coupling. If such a piezoelectric element is attached to a structure, it is strained as the structure deforms and converts a portion of the energy associated with vibration into electrical energy. The piezoelectric element (which behaves electrically as a capacitor), in combination with a network of electrical elements connected to it (a **shunt** network), comprises an electrical system that can be configured to accomplish vibration control through its treatment of this electrical energy (Forward, 1979; Hagood and von Flotow, 1991).

Four basic kinds of shunt circuits are typically used: resistive, inductive, capacitive, and switched (Lesieutre, 1998). Each of these kinds of shunts results in a different kind of dynamic behavior. A resistive shunt dissipates energy through Joule heating, which has the effect of structural damping. An inductive shunt results in a resonant LC circuit, the behavior of which is analogous to that of a mechanical vibration absorber (tuned mass damper). A capacitive shunt changes the effective stiffness of the piezoelectric element, which can be used to advantage in, for example, a tunable mechanical vibration absorber. A switched shunt offers the possibilities of controlling the energy transfer to reduce frequency-dependent behavior or for energy harvesting, which also has the effect of structural damping (Lesieutre et al., 2004).

**Resistive shunt damping** exhibits frequency dependence for both the loss factor and modulus that is relatively strong compared to that of most viscoelastic materials. Peak damping depends on the strength of the modal electromechanical coupling, and the frequency at which peak damping is observed is related to the inverse of the $RC$ time constant for the shunt circuit. In order to achieve maximum damping in practice, a key concept involves matching a characteristic time constant of a shunted piezoelectric system to the period of a design vibration or forcing frequency.

Figure 6.17 shows the damping loss factor and relative modulus for two piezoelectric materials: one having an electromechanical coupling coefficient of $k = 0.70$ and the other $k = 0.35$. The material having the higher coupling coefficient exhibits higher peak damping as well as a larger change in modulus with frequency. A coupling coefficient of 0.70 is typical for polycrystalline piezoelectric ceramics under uniaxial loads in the poling direction, and the effective peak loss factor is 0.34. This is remarkably high for a potential structural material that is approximately as stiff as aluminum. A coupling coefficient of 0.35 is typical for uniaxial loads perpendicular to the poling direction, and the effective peak loss factor is 0.065, still quite high.

Energy methods may be used to develop approximate equations of motion for piezoelectric structures. In a typical approach, the mechanical displacement vector field and the scalar electric potential field are interpolated in terms of nodal variables, while the dual quantities, force and charge, are considered as forcing terms. This leads to discretized finite element equations of motion having the following structure:

$$\begin{bmatrix}[M] & \\ & \end{bmatrix}\begin{Bmatrix}\{\ddot{d}\}\\ \end{Bmatrix} + \begin{bmatrix}[K+k^E] & -[p]^t\\ [p] & [C^S]\end{bmatrix}\begin{Bmatrix}\{d\}\\ \{V\}\end{Bmatrix} = \begin{Bmatrix}\{F\}\\ \{Q\}\end{Bmatrix},\qquad (6.10.2)$$

where $d$ are the nodal displacements, $V$ are the voltages (potentials), $F$ are the forces, and $Q$ are the charges. $M$ and $K$ are the mass and base stiffness matrices, while $k^E$ is the piezoelectric stiffness at constant electric field, $C^S$ is the piezoelectric capacitance at constant strain,

**Figure 6.17** Resistively shunted piezoelectric ceramics can provide relatively high damping, albeit with significant frequency dependence

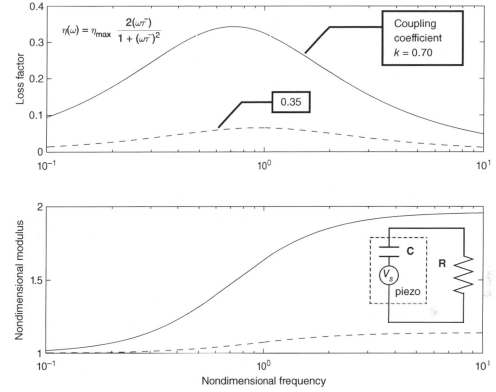

and $p$ is the piezoelectric coupling. The roles of nodal voltage and charge may be interchanged as needed to facilitate the introduction of discrete shunting elements and associated dynamics to these equations.

### 6.10.5 Particle Damping

Particle impact dampers (PIDs) are enclosures partially filled with particles. When attached to or embedded in a vibrating structure, they provide damping by dissipating energy through inelastic particle–enclosure and particle–particle collisions. Damping performance is affected by the (effective) mass of the particle(s), effective coefficient of restitution, vibration amplitude, excitation frequency, and gap size. PIDs might be considered for use when operating conditions, say high temperatures, preclude the use of alternative damping approaches (Panossian, 1992).

The simplest way to model a group of contained particles is to assume that it behaves as a single effective particle. Masri (1970) used this approach to show that, in the absence of gravity, with a single coefficient of restitution and optimal gap size, symmetric equi-spaced impacts provide maximum damping when the system is excited near resonance.

A more general approach is to track the motions of the enclosure and the particle bed through time. Friend and Kinra (2000) used this approach to analyze the dynamic response of a beam subjected to a given initial displacement and release in the presence of gravity. They found an effective coefficient of restitution for the particle bed, and then successfully predicted the behavior of the beam with differing initial displacements. A more general modeling approach based on **particle dynamics** tracks the movements and interactions of individual particles in the enclosure; this is relatively expensive computationally. Using a particle dynamics approach, Saluena et al. (1999) found dissipation to be relatively small in the "solid" phase, large around the fluidization point (which occurs around $1g$ of acceleration), diminishing in the "liquid" regime, and rising again in the "gas" regime.

Figure 6.18 shows a simple model of a particle damper. It consists of a primary (base) structure and a single effective impacting particle, which is contained inside an enclosure. The base and the enclosure move together, while the particle, when it is not in contact with either the floor or the ceiling, moves solely under the influence of gravity. For simplicity, the base is assumed to undergo prescribed harmonic motion parallel to gravity, and normal impacts are the primary energy dissipation mechanism.

**Figure 6.18** Particle damping can provide good performance when engineered for two impacts per cycle

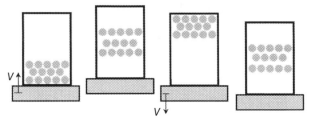

The dynamics described by this model depend on three nondimensional parameters: (base) acceleration $\beta$; gap clearance $d$; and the coefficient of restitution $e$. Tracking the steady states of the PID reveals many interesting dynamic features. The most significant of these is the existence of solutions that correspond to two impacts per cycle; the phases of the impacts are very close to those corresponding to the maximum upward and downward velocities of the base. These solutions are observed over a wide and continuous range of $\beta$. Within this range, one can obtain relatively high loss factors by choosing $d$ and $e$ as described in Ramachandran and Lesieutre (2008). The relative insensitivity of damping to operating conditions and tuning makes the two-impacts-per-cycle motions desirable from a design perspective.

Under harmonic excitation and at relatively low acceleration (but $>1g$), a high coefficient of restitution in combination with a relatively large gap yields high effective loss. In practice, a large $e$ might be realized using a single particle. More typically, for relatively high operating acceleration, a smaller gap (and thus device size) is enabled by a low effective coefficient of restitution, $e$, and results in modestly lower peak loss. In practice, a small $e$ can be realized using 4–5 layers of smaller particles.

To effectively damp a vibrating structure, a PID is best placed at locations of high motion. Given the design base acceleration level, the coefficient of restitution $e$ can be established roughly, and then the gap size $d$ can be set. Increasing total particle mass increases the dissipation up to a point beyond which mass-loading effects reduce the base motion and the resulting dissipation. Iteration may be needed to converge on a suitable design (Yang et al., 2005).

Although considerable experience has shown PIDs to be an effective means of energy dissipation, modeling and analysis continues to be an active area of research, especially for nonharmonic excitation over a broad frequency range.

Passive damping is an important aspect of aerospace structural dynamics that is best considered early in the design of new vehicles and systems. In order to confidently consider damping during the design process, efficient, accurate models are needed; these must capture the general dissipative effects of damping, while also supporting the detailed design of damping treatments. A student who has mastered the material in this section will be well prepared to address vibration problems by conceiving, modeling, and designing appropriate damping solutions.

## BIBLIOGRAPHY

Abbruzzese, N., Lin, P., and Innis, J. (2008). Measurement of damping in composite structure assembly at cryogenic temperatures. In *49th AIAA/ASME/ASCE/AHS/ASC Structures, Structural Dynamics, and Materials Conference*. doi: 10.2514/6.2008-2192.

Adhikari, S. (2006). Damping modelling using generalized proportional damping. *Journal of Sound and Vibration*, 293(1–2): 156–170. doi: 10.1016/j.jsv.2005.09.034.

Austrell, P.-E. (1997). *Modeling of elasticity and damping for filled elastomers*. PhD thesis, Lund University (Sweden).

Bagley, R. L. and Torvik, P. J. (1983). Fractional calculus – A different approach to the analysis of viscoelastically damped structures. *AIAA Journal*, 21(5): 741–748. doi: 10.2514/3.8142.

Brackbill, C. R., Smith, E. C., and Lesieutre, E. C. (2002). Application of a refined time domain elastomeric damper model to helicopter rotor aeroelastic response and stability. *Journal of the American Helicopter Society*, 47(3): 186–197. doi: 10.4050/JAHS.47.186.

Chen, G. and Russell, D. L. (1982). A mathematical model for linear elastic systems with structural damping. *Journal of Applied Mechanics*, 39(4): 433–454. doi: 10.1090/qam/644099.

DiTaranto, R. (1965). Theory of vibratory bending for elastic and viscoelastic layered finite-length beams. *Journal of Applied Mechanics*, 32(4): 881–886. doi: 10.1115/1.3627330.

Enelund, M. and Lesieutre, G. A. (1999). Time domain modeling of damping using anelastic displacement fields and fractional calculus. *International Journal of Solids and Structures*, 36(29): 4447–4472. doi: 10.1016/S0020-7683(98)00194-2.

Ferry, J. D. (1980). *Viscoelastic Properties of Polymers*, 3rd ed. John Wiley & Sons.

Forward, R. L. (1979). Electronic damping of vibrations in optical structures. *Applied Optics*, 18(5): 690–697. doi: 10.1364/AO.18.000690.

Friend, R. D. and Kinra, V. K. (2000). Particle impact damping. *Journal of Sound and Vibration*, 233(1): 93–118. doi: 10.1006/jsvi.1999.2795.

Gueler, R., von Flotow, A. H., and Vos, D. W. (1993). Passive damping for robust feedback control of flexible structures. *Journal of Guidance, Control, and Dynamics*, 16(4): 662–667. doi: 10.2514/3.21065.

Hagood, N. and von Flotow, A. (1991). Damping of structural vibrations with piezoelectric materials and passive electrical networks. *Journal of Sound and Vibration*, 146(2): 243–268. doi: 10.1016/0022-460X(91)90762-9.

Johnson, C. D. (1995). Design of passive damping systems. *Journal of Mechanical Design*, 117(B): 171–176.

Johnson, C. D. and Kienholz, D. A. (1982). Finite element prediction of damping in structures with constrained viscoelastic layers. *AIAA Journal*, 20(9): 1284  1290. doi: 10.2514/3.51190.

Lesieutre, G., Ottman, G., and Hofmann, H. (2004). Damping as a result of piezoelectric energy harvesting. *Journal of Sound and Vibration*, 269(3–5): 991–1001. doi: 10.1016/S0022-460X(03)00210-4.

Lesieutre, G. A. (1998). Vibration damping and control using shunted piezoelectric materials. *The Shock and Vibration Digest*, 30(3): 187–195. doi: 10.1177/058310249803000301.

Lesieutre, G. A. (2009). How membrane loads influence the modal damping of flexural structures. *AIAA Journal*, 47(7): 1642–1646. doi: 10.2514/1.37618.

Lesieutre, G. A. (2010). Damping in structural dynamics. In *Encyclopedia of Aerospace Engineering*. John Wiley & Sons. doi: 10.1002/9780470686652.eae146.

Lesieutre, G. A. and Bianchini, E. (1995). Time domain modeling of linear viscoelasticity using anelastic displacement fields. *Journal of Vibration and Acoustics*, 117(4): 424–430. doi: 10.1115/1.2874474.

Lesieutre, G. A. and Govindswamy, K. (1996). Finite element modeling of frequency dependent and temperature-dependent dynamic behavior of viscoelastic materials in simple shear. *International Journal of Solids and Structures*, 33(3): 419–432. doi: 10.1016/0020-7683(95)00048-F.

Lesieutre, G. A. and Kauffman, J. L. (2013). A 'geometric' viscous damping model for nearly constant beam modal damping. *AIAA Journal*, 51(7): 1688–1694. doi: 10.2514/1.J052174.

Lesieutre, G. A., Eckel, A. J., and DiCarlo, J. A. (1991). Damping of bromine-intercalated P-100 graphite fibers. *Carbon*, 29(7): 1025–1032. doi: 10.1016/0008-6223(91)90182-I.

Lesieutre, G. A., Yarlagadda, S., Yoshikawa, S., Kurtz, S. K., and Xu, Q. C. (1993). Passively damped structural composite materials using resistively-shunted piezoceramic fibers. *Journal of Materials Engineering and Performance*, 2(6): 887–892. doi: 10.1007/BF02645690.

Lin, D. X., Ni, R. G., and Adams, R. D. (1984). Prediction and measurement of the vibrational damping parameters of carbon and glass fibre-reinforced plastics plates. *Journal of Composite Materials*, 18(2): 132–152. doi: 10.1177/002199838401800204.

Masri, S. (1970). General motion of impact dampers. *Journal of the Acoustical Society of America*, 47 (1P2): 229. doi: 10.1121/1.1911470.

McTavish, D. J. and Hughes, P. C. (1993). Modeling of linear viscoelastic space structures. *Journal of Vibration and Acoustics*, 115(1): 103–110. doi: 10.1115/1.2930302.

Mead, D. J. (1999). *Passive Vibration Control*. John Wiley & Sons.

Min, J., Harris, D., and Ting, J. (2011). Advances in ceramic matrix composite blade damping characteristics for aerospace turbomachinery applications. In *52nd AIAA/ASME/ASCE/AHS/ASC Structures, Structural Dynamics and Materials Conference*. doi: 10.2514/6.2011-1784.

Myklestad, N. (1952). The concept of complex damping. *Journal of Applied Mechanics*, 19(3): 284–286. doi: 10.1115/1.4010499.

Nashif, A. D., Jones, D. I., and Henderson, J. P. (1985). *Vibration Damping*. John Wiley & Sons.

Nowick, A. and Berry, B. (1972). *Anelastic Relaxation in Crystalline Solids*. Academic Press.

Panossian, H. V. (1992). Structural damping enhancement via non-obstructive particle damping technique. *Journal of Vibration and Acoustics*, 114(1): 101–105. doi: 10.1115/1.2930221.

Ramachandran, S. and Lesieutre, G. (2008). Dynamics and performance of a harmonically excited vertical impact damper. *Journal of Vibration and Acoustics*, 130(2). doi: 10.1115/1.2827364.

Saluena, C., Poschel, T., and Esipov, S. E. (1999). Dissipative properties of vibrated granular materials. *Physical Review* E, 59(4): 4422–4425. doi: 10.1103/PhysRevE.59.4422.

Sodano, H., Bae, J.-S., Inman, D., and Belvin, W. (2006). Improved concept and model of eddy current damper. *Journal of Vibration and Acoustics*, 128(3): 294–302. doi: 10.1115/1.2172256.

Soovere, J. and Drake, M. L. (1984). Aerospace structures technology damping design guide. *Technical Report AFWAL-TR-84-3089*, US Air Force Wright Aeronautical Laboratories.

Suarez, S., Gibson, R., Sun, C., and Chaturvedi, S. (1986). The influence of fiber length and fiber orientation on damping and stiffness of polymer composite materials. *Experimental Mechanics*, 26 (2): 175–184. doi: 10.1007/BF02320012.

Ungar, E. (1973). The status of engineering knowledge concerning the damping of built-up structures. *Journal of Sound and Vibration*, 26(1): 141–154. doi: 10.1016/S0022-460X(73)80210-X.

Ungar, E. E., and Kerwin Jr, E. M. (1962). Loss factors of viscoelastic systems in terms of energy concepts. *Journal of the Acoustical Society of America*, 34(7): 954–957. doi: 10.1121/1.1918227.

Wereley, N. M. and Pang, L. (1998). Nondimensional analysis of semi-active electrorheological and magnetorheological dampers using approximate parallel plate models. *Smart Materials and Structures*, 7(5): 732–743. doi: 10.1088/0964-1726/7/5/015.

Wolons, D., Gandhi, F., and Malovrh, B. (1998). Experimental investigation of the pseudoelastic hysteresis damping characteristics of shape memory alloy wires. *Journal of Intelligent Material Systems and Structures*, 9(2): 116–126.

Yang, M., Lesieutre, G., Hambric, S., and Koopmann, G. (2005). Development of a design curve for particle impact dampers. *Noise Control Engineering Journal*, 53(1): 5–13. doi: 10.3397/1.2839240.

## PROBLEMS

1. Consider an SDOF system comprising a mass and a lossy spring. The spring has a loss factor $\eta = 0.05$. Determine the modal damping ratio $\zeta$ and comment on the relationship between the loss factor and the damping ratio.

2. Shortcomings of the Maxwell model shown in Figure 6.6 are that it relaxes to zero stress and does not return to its initial state when unloaded. By what addition might these deficiencies be remedied? Compare the result to the standard anelastic solid shown in Figure 6.8.

3. Consider the fractional derivative damping model described in Section 6.5.4. Compare the expressions for the material complex modulus in Eqs. (6.5.16) and (6.5.17). Develop expressions for $E_r$, $\Delta$, $\Omega$, and $E_u$ in terms of $p_1$, $q_0$, and $q_1$.

4. Show that the real and imaginary parts of complex eigenvectors found as solutions to Eq. (6.7.9) respect orthogonality conditions, even for complex conjugate modes.

5. Consider a simply-supported planar beam. The beam is made from two materials that have comparable elastic properties, but different damping properties.

    thickness, $h = 0.002$m

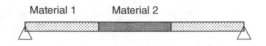

Material 1          Material 2

width, $b = 0.02$m
total length, $L = 0.300$m (each section is one-third of this).

The properties of the two materials are as follows:
**Material 1**

density, $\rho_1 = 3,000$kg/m$^3$
Young's modulus, $E_1 = 100 \times 10^9$ Pa
Loss factor, $\eta_1 = 0.02$

**Material 2**

density, $\rho_2 = 2000kg/m^3$
Young's modulus, $E_2 = 50 \times 10^9$ Pa
Loss factor, $\eta_2 = 0.20$

A planar beam finite element model is suggested for use in the following problems. Neglect transverse shear.

(a) Modal strain energy method
  - Estimate the natural frequencies and mode shapes (eigenvectors) of the five lowest frequency modes of the structure. (How many elements are needed to capture these accurately?)
  - Using the modal strain energy method, estimate the modal damping ratios (not loss factors) for each of these modes.
  - Discuss the results.

(b) Complex stiffness method
  - Using the complex stiffness method, estimate the natural frequencies, modal damping ratios, and mode shapes for the five lowest frequency modes of the structure.
  - Discuss the results.

(c) Comparison
  - Compare the results obtained using the two different methods. (Address at least accuracy, ease of use, and insight.)
  - How would the comparison change if the loss factors were increased by a factor of 5?

6. Consider the same beam addressed in Problem 5. Implement a geometric viscous damping model and estimate the natural frequencies, modal damping ratios, and mode shapes for the five lowest frequency modes of the structure. Compare the results to those obtained in Problem 5.

7. Consider the same beam addressed in Problem 5. Implement a proportional viscous damping model to match the modal damping ratio for at least the first mode. Estimate

the natural frequencies, modal damping ratios, and mode shapes for the five lowest frequency modes of the structure. Compare the results to those obtained in Problem 5

8. Consider the same structure addressed in Problem 5, but consider it to be a uniform *rod* instead of a beam. Assume that it is made only of Material 2, that the loss factor is the peak loss factor, and that the Young's modulus is the static (low-frequency) value $E_r$.

   Implement an ADF viscoelastic damping model that exhibits the peak loss factor at the natural frequency of mode 3. Estimate the longitudinal natural frequencies, modal damping ratios, and mode shapes for the five lowest frequency modes of the structure. Are any relaxation modes observed in the frequency range of interest?

9. Consider the same structure addressed in Problem 5, but consider it to be a uniform beam made only of Material 1. Implement a proportional viscous damping model to match the modal damping ratio for at least the first mode.

   Consider the eigenvalue problem for this structure in second-order form, and in first-order form. Compare the eigenvalues (natural frequencies and damping) and eigenvectors that result in each case.

10. Consider the same beam addressed in Problem 6. Add an axial load to the beam and include its effect in a finite element model by adding a geometric stiffness matrix. Consider several axial loads in the range from −0.9 times the buckling load $P_{cr}$ (compressive) to 2.0 times the buckling load (tension). How are the natural frequencies and damping of the first three modes affected?

# 7     Dynamic Response of Structures

## 7.1   Dynamic Response Using the Normal Mode Method

The normal mode method takes advantage of the orthogonality conditions of a structural dynamic system, with respect to the mass and stiffness matrices, in order to decouple the system.

Consider first an $n$-degree-of-freedom (DOF) undamped system for this case the equations governing the forced response problem are given by

$$[m]\{\ddot{q}\} + [k]\{q\} = \{F(t)\}, \tag{7.1.1}$$

and the corresponding free vibration problem is governed by the following eigenvalue problem:

$$\omega^2[m]\{q\} = [k]\{q\}. \tag{7.1.2}$$

Solution of Eq. (7.1.2) yields the modal matrix $[\Phi]$, which has the property of diagonalizing the mass matrix $[m]$ and the stiffness matrix $[k]$, due to the orthogonality conditions with respect to $[m]$ and $[k]$, thus

$$[\Phi]^T[m][\Phi] = [M_{rr}] \tag{7.1.3}$$

$$[\Phi]^T[k][\Phi] = [K_{rr}], \tag{7.1.4}$$

where the elements $M_{rr}$ and $K_{rr}$ in the diagonal matrices are known as the generalized mass and stiffness matrices. Obviously from Eqs. (7.1.3), (7.1.4), and (7.1.2), it is clear that

$$\omega_r^2 M_{rr} = K_{rr}. \tag{7.1.5}$$

Based on these equations, the forced response can be conveniently obtained by using the normal modes as the appropriate coordinate system. Introducing the transformation

$$\{q\} = [\Phi]\{\eta\}, \tag{7.1.6}$$

where it is understood that the columns of $[\Phi]$ contain the $n$ eigenvectors representing the solution of the eigenvalue problem given by Eq. (7.1.2). Thus, from Eqs. (7.1.1) and (7.1.6),

$$[m][\Phi]\{\ddot{\eta}\} + [k][\Phi]\{\eta\} = [F(t)].$$

Pre-multiplying by $[\Phi]^T$

$$[\Phi]^T[m][\Phi]\{\ddot{\eta}\} + [\Phi]^T[k][\Phi]\{\eta\} = [\Phi]^T\{F(t)\}. \tag{7.1.7}$$

Define

$$[\Phi]^T\{F(t)\} = \{f(t)\}. \tag{7.1.8}$$

From Eqs. (7.1.3), (7.1.4), (7.1.5), (7.1.7), and (7.1.8),

$$[M_{rr}]\{\ddot{\eta}\} + [\omega_r^2 M_{rr}]\{\eta\} = \{f(t)\}, \tag{7.1.9}$$

which means that the response of the system is governed by an uncoupled system of linear, second-order differential equations with constant coefficients, which is a simple problem from a mathematical point of view. Thus, the response is obtained by solving the following set of equations:

$$M_{rr}\ddot{\eta}_r + \omega_r^2 M_{rr}\eta = f_r(t) \qquad r = 1, 2, \ldots, n,$$

where $f_r(t)$ is the generalized excitation, in any mode $r$.

Since the system is linear and the principle of superposition applies, knowledge of $\eta_r(t)$, $r = 1, 2, \ldots, n$, enables one to obtain the response from Eq. (7.1.6):

$$\{q(t)\} = [\Phi]\{\eta(t)\}.$$

Apply the same procedure to a damped $n$-DOF system:

$$[m]\{\ddot{q}\} + [c]\{\dot{q}\} + [k]\{q\} = \{F(t)\}. \tag{7.1.10}$$

The normal mode transformation can be used again; however, the success of the method in decoupling the system depends now on the properties of the damping matrix:

$$[C] = [\Phi]^T[c][\Phi].$$

If $[C]$ is a diagonal matrix or can be assumed to be such the response problem decouples again. A nondiagonal $[C]$ matrix requires a somewhat more complicated treatment of the dynamic response problem. In order to understand better the damping matrix, the role of damping in structural dynamics is reviewed in the next section.

### 7.1.1 Illustration of Generalized Loads in the Normal Mode Method

It is evident from Eq. (7.1.8) that in the application of the normal mode method, one always encounters the loading vector $\{f(t)\}$ given by

$$\{f(t)\} = [\Phi]^T\{F(t)\}.$$

The precise meaning of this vector can be best illustrated by examples. Consider the three-story building used earlier when discussing eigenvalue solution methods for calculating free vibration modes and frequencies in a discrete system (Section 5.5). Assume that at each DOF a time-dependent excitation is applied, that is,

$$\{F(t)\} = \left\{ \begin{array}{c} F_a(t) \\ F_b(t) \\ F_c(t) \end{array} \right\}.$$
(7.1.11)

Thus, the structure with the time-dependent applied loads is shown in Figure 7.1 and the equation of motion is given by

$$600 \begin{bmatrix} 1 & -1 & 0 \\ -1 & 3 & -2 \\ 0 & -2 & 5 \end{bmatrix} \left\{ \begin{array}{c} v_a \\ v_b \\ v_c \end{array} \right\} + \begin{bmatrix} 1.0 & 0 & 0 \\ 0 & 1.5 & 0 \\ 0 & 0 & 2.0 \end{bmatrix} \left\{ \begin{array}{c} \ddot{v}_a \\ \ddot{v}_b \\ \ddot{v}_c \end{array} \right\} = \left\{ \begin{array}{c} F_a(t) \\ F_b(t) \\ F_c(t) \end{array} \right\}.$$
(7.1.12)

The normal mode transformation for this case is given by

$$\left\{ \begin{array}{c} v_a \\ v_b \\ v_c \end{array} \right\} = \begin{bmatrix} 0.742 & -0.636 & 0.210 \\ 0.482 & 0.385 & -0.535 \\ 0.224 & 0.431 & 0.514 \end{bmatrix} \left\{ \begin{array}{c} \eta_1 \\ \eta_2 \\ \eta_3 \end{array} \right\} = [\Phi]\{\eta\}.$$
(7.1.13)

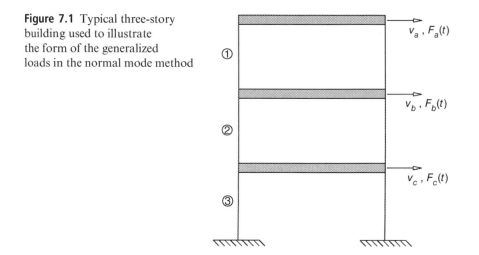

**Figure 7.1** Typical three-story building used to illustrate the form of the generalized loads in the normal mode method

After applying the normal mode transformation, the response problem is

$$[I]\{\ddot{\eta}\} + [\omega_i^2]\{\eta\} = [\Phi]^T\{F(t)\} = \{f(t)\}$$
(7.1.14)

or for our case

$$\left\{ \begin{array}{c} \ddot{\eta}_1 \\ \ddot{\eta}_2 \\ \ddot{\eta}_3 \end{array} \right\} + \begin{bmatrix} 210.096 & 0.0 & 0.0 \\ 0.0 & 963.36 & 0.0 \\ 0.0 & 0.0 & 2126.58 \end{bmatrix} \left\{ \begin{array}{c} \eta_1 \\ \eta_2 \\ \eta_3 \end{array} \right\} = \left\{ \begin{array}{c} f_1(t) \\ f_2(t) \\ f_3(t) \end{array} \right\}.$$
(7.1.15)

A number of different cases can be encountered depending on the *spatial* distribution of the applied excitation.

Case 1: The time-dependent forcing functions are independent, that is,

$$\{F(t)\} = \left\{ \begin{array}{c} F_a(t) \\ F_b(t) \\ F_c(t) \end{array} \right\} = \left[ \begin{array}{ccc} 1 & 0 & 0 \\ 0 & 1 & 0 \\ 0 & 0 & 1 \end{array} \right] \left\{ \begin{array}{c} F_a(t) \\ F_b(t) \\ F_c(t) \end{array} \right\} \tag{7.1.16}$$

$$\{f(t)\} = [\Phi]^T\{F(t)\} = [\Phi]^T[I]\{F(t)\} = [P]\{F(t)\}. \tag{7.1.17}$$

Case 2: A particular loading configuration, having the same time dependence at each DOF

$$[F(t)] = \left\{ \begin{array}{c} F_a \\ F_b \\ F_c \end{array} \right\} = \left\{ \begin{array}{c} 4 \\ 2 \\ 2 \end{array} \right\} T(t)$$

$$\{f(t)\} = [\Phi]^T\{F(t)\} = [\Phi]^T \left\{ \begin{array}{c} 4 \\ 2 \\ 2 \end{array} \right\} = \{P\}T(t) =$$

$$= \left\{ \begin{array}{c} 4.380 \\ -0.912 \\ 0.798 \end{array} \right\} T(t). \tag{7.1.18}$$

Case 3: Two particular loading configurations, for example,

$$\{F(t)\} = \left\{ \begin{array}{c} 4 \\ 2 \\ 2 \end{array} \right\} T_1(t) + \left\{ \begin{array}{c} 1 \\ 0 \\ 0 \end{array} \right\} T_2(t) = \left[ \begin{array}{cc} 4 & 1 \\ 2 & 0 \\ 2 & 0 \end{array} \right] \left\{ \begin{array}{c} T_1(t) \\ T_2(t) \end{array} \right\} \tag{7.1.19}$$

$$= [F_s]\{T(t)\}.$$

For this case,

$$\{f(t)\} = [\Phi]^T[F_s]\{T(t)\} = [P]\{T(t)\} =$$

$$= [\Phi]^T \left[ \begin{array}{cc} 4 & 1 \\ 2 & 0 \\ 2 & 0 \end{array} \right] \left\{ \begin{array}{c} T_1(t) \\ T_2(t) \end{array} \right\} = \left[ \begin{array}{cc} 4.380 & 0.742 \\ -0.912 & -0.636 \\ 0.798 & 0.210 \end{array} \right] \left\{ \begin{array}{c} T_1(t) \\ T_2(t) \end{array} \right\}, \tag{7.1.20}$$

where in these equations $[F_s]$ is a matrix that describes the spatial distribution of the time-dependent applied load.

## 7.1.2 Damped $N$-DOF Systems

Introduction of damping into the equations of motion is straightforward. However, one should remember that in reality, the correct damping characterization of a built-up structure is a complicated task which in most cases is best accomplished by experimental testing.

Introduction of structural damping for an $n$-DOF system can be accomplished by operating on the structural part of the equations (see also Chapter 6).

Thus, if the undamped system is given by

$$[m]\{\ddot{q}\} + [k]\{q\} = \{F(t)\}, \tag{7.1.21}$$

then the same system with structural damping can be written as

$$[m]\{\ddot{q}\} + (1 + ig)[k]\{q\} = \{F(t)\}. \tag{7.1.22}$$

From the treatment shown in Section 7.1, it is clear that the normal mode transformation

$$\{q\} = [\Phi]\{\eta\} \tag{7.1.23}$$

will decouple this system, since

$$[\Phi]^T[m][\Phi]\{\ddot{\eta}\} + (1 + ig)[\Phi]^T[k][\Phi]\{\eta\} = [\Phi]^T\{F(t)\}$$
$$[M_{rr}]\{\ddot{\eta}\} + (1 + ig)[K_{rr}]\{\eta\} = \{F(t)\}. \tag{7.1.24}$$

The normal modes are orthogonal with respect to the stiffness matrix.

In many applications, it is convenient to replace the various types of damping by equivalent viscous damping (see Chapter 6); for this case, the equations of motion are

$$[m]\{\ddot{q}\} + [c]\{\dot{q}\} + [k]\{q\} = \{F(t)\}. \tag{7.1.25}$$

### 7.1.3    Rayleigh's Dissipation Function

The kinetic energy of an $n$-DOF system is

$$T = \frac{1}{2}\sum_{i=1}^{n}\sum_{j=1}^{n} m_{ij}\dot{q}_i\dot{q}_j. \tag{7.1.26}$$

Similarly, the potential or strain energy is given by

$$U = \frac{1}{2}\sum_{i=1}^{n}\sum_{j=1}^{n} k_{ij}q_iq_j. \tag{7.1.27}$$

Next, assume that the damping forces in the structural dynamic system can be represented by viscous damping. For this case, one can define Rayleigh's dissipation function from which the damping forces can be obtained; this function which has a functional form similar to Eqs. (7.1.26) and (7.1.27) is given by

$$R = \frac{1}{2}\sum_{i=1}^{n}\sum_{j=1}^{n} c_{ij}\dot{q}_i\dot{q}_j. \tag{7.1.28}$$

The equations of motion can be obtained by using Lagrange's equations:

$$\frac{d}{dt}\left(\frac{\partial T}{\partial \dot{q}_r}\right) - \frac{\partial T}{\partial q_r} + \frac{\partial R}{\partial \dot{q}_r} + \frac{\partial U}{\partial q_r} = Q_{Ar} \quad r = 1, 2, \ldots, n, \tag{7.1.29}$$

where $Q_{Ar}$ are the generalized forces acting on the system. Equations (7.1.26) through (7.1.29) can be combined to give

$$[m]\{\ddot{q}\} + [c]\{\dot{q}\} + [k]\{q\} = \{Q\}.$$   (7.1.30)

Applying the normal mode transformation on this system

$$\{q\} = [\Phi]\{\eta\}$$   (7.1.31)

and assuming the normal mode matrix to be normalized with respect to the mass matrix yields

$$[\Phi]^T[m][\Phi] = [I].$$   (7.1.32)

A combination of Eqs. (7.1.30) through (7.1.32) results in

$$\{\ddot{\eta}\} + [C]\{\dot{\eta}\} + [\omega_n^2]\{\eta\} = \{F_A\},$$   (7.1.33)

where

$$\{F_A\} = [\Phi]^T\{Q\}$$   (7.1.34)

and

$$[C] = [\Phi]^T[c][\Phi].$$   (7.1.35)

In general, the viscous damping matrix $[C]$ will not be diagonal. In this case, the damping is said to be nonproportional. The equation of motion corresponding to the $r$th DOF, Eq. (7.1.33), is

$$\ddot{\eta}_r + C_{rr}\dot{\eta}_r + \underbrace{\sum_{j=1, j\neq r}^{n} C_{rj}\dot{\eta}_j} + \omega_r^2\eta_r = F_{Ar}(t).$$   (7.1.36)

The terms underlined by the curly braces represent coupling between the normal modes as defined by the undamped eigenvalue problem. Thus, it is clear that a nondiagonal damping matrix complicates the response problem and requires the treatment of an $n$-DOF coupled system for response calculations.

### 7.1.4 Proportional Damping

The general equation of motion represented by Eq. (7.1.33) can be decoupled provided that an assumption is made regarding the properties of the $[c]$-matrix in Eq. (7.1.35). A convenient assumption is one where the damping matrix is assumed to be a linear combination of the mass and stiffness matrix, that is,

$$[c] \equiv \alpha[m] + \beta[k].$$   (7.1.37)

A damping matrix of this form is called **proportional damping**. This assumption was first introduced by Rayleigh. A combination of Eqs. (7.1.35) through (7.1.37) yields

$$[C] = [\Phi]^T [c][\Phi] = [\Phi]^T \left[ \alpha[m] + \beta[k] \right][\Phi] = \left( \alpha[I] + \beta \left[ \ \omega_r^2 \ \right] \right) \tag{7.1.38}$$

and Eq. (7.1.38) decouples into $n$ equations of the form

$$\ddot{\eta} + \left( \alpha + \beta\omega_r^2 \right) \dot{\eta}_r + \omega_r^2 \eta_r = F_{Ar}(t). \tag{7.1.39}$$

Thus, it is again possible to consider the response in each DOF independently.

## Example

Consider a 2-DOF system with a given amount of modal critical damping associated with each mode for this case:

$$\ddot{\eta}_1 + 2\zeta_1\omega_1\dot{\eta}_1 + \omega_1^2\eta_1 = F_{A1}(t)$$
$$\ddot{\eta}_2 + 2\zeta_2\omega_2\dot{\eta}_2 + \omega_2^2\eta_2 = F_{A2}(t). \tag{7.1.40}$$

Comparing Eqs. (7.1.39) and (7.1.40), it is clear that

$$2\zeta_1\omega_1 = \alpha + \beta\omega_1^2$$
$$2\zeta_2\omega_2 = \alpha + \beta\omega_2^2. \tag{7.1.41}$$

Solving for $\alpha$ and $\beta$, one has

$$\alpha = \frac{2\omega_1\omega_2 (\zeta_2\omega_1 - \zeta_1\omega_2)}{\left(\omega_1^2 - \omega_2^2\right)} \tag{7.1.42}$$

$$\beta = \frac{2(\zeta_1\omega_1 - \zeta_2\omega_2)}{\omega_1^2 - \omega_2^2}. \tag{7.1.43}$$

Or in general

$$\zeta_j = \frac{\left(\alpha + \beta\omega_j^2\right)}{2\omega_j}. \tag{7.1.44}$$

Note that while this one-to-one correspondence between modal critical damping ratios and the constants of proportionality $\alpha$ and $\beta$ can be obtained for the 2-DOF case, in a unique manner, it can be difficult for an $n$-DOF case.

If modal damping $\zeta_1, \zeta_2, \ldots, \zeta_n$ is obtained by an experimental method, one will have in general $n$-quantities (or equations) from which two parameters $\alpha$ and $\beta$ have to be determined. This is an overdetermined problem and more sophisticated mathematical methods such as least squares techniques have to be used to determine $\alpha$ and $\beta$. Additional material on damping can be found in Nashif et al. (1985), at the end of this chapter as well as in Chapter 6.

## 7.2    Response Analysis in the Time and Frequency Domain

### 7.2.1    Solution of the Response Problem Using the Laplace Transform Method

Consider the single-DOF (SDOF) equation representing motion (response) in a particular mode, obtained after decoupling the dynamic system using the normal mode method, that is,

$$\ddot{\eta}_r + 2\zeta\dot{\eta}_r + \omega_r^2\eta_r = \frac{\Gamma_r f(t)}{M_r}. \tag{7.2.1}$$

Furthermore, note that the problem is linear, hence when the right-hand side of Eq. (7.2.1) is replaced by

$$\ddot{\eta}_r + 2\zeta\dot{\eta}_r + \omega_r^2\eta_r = \sum_{J=1}^{m} \frac{\Gamma_{rJ} f_J(t)}{M_r} \tag{7.2.2}$$

using the principle of superposition, the total response of the system is obtained by summing up the response in the individual DOFs.

**The Laplace Transform**

One convenient method for solving Eq. (7.2.1) is the Laplace Transform method (L-T). The L-T of a function in the time domain $t$ is defined as (see Appendix B)

$$\mathscr{L}[f(t)] = \int_0^\infty e^{-st} f(t)dt = \bar{f}(s) = \bar{f}. \tag{7.2.3}$$

The main advantage of this method is that it changes an ordinary differential equation into an algebraic equation, so that the solution is obtained in a straightforward manner.

In general for the $n$th derivative it can be shown that (Appendix B)

$$\mathscr{L}\frac{d^n f(t)}{dt^n} = -f^{(n-1)}(0) - st^{(n-2)}(0) - \cdots - s^{n-1}f(0) + s^n\bar{f}(s). \tag{7.2.4}$$

Consider the solution to Eq. (7.2.1) with initial conditions given by $\eta_r(0) = \dot{\eta}_r(0) = 0, r = 1, 2, \ldots, n$, corresponding to $w(x,0) = \dot{w}(x,0) = 0$, applying the Laplace transform

$$(s^2 + 2\zeta s + \omega_r^2)\bar{\eta}_r(s) = \Gamma_r\frac{\bar{f}(s)}{M_r} \tag{7.2.5}$$

$$\begin{aligned}
\bar{\eta}_r(s) &= \frac{\Gamma_r\bar{f}(s)}{M_r(s^2 + 2\zeta s + \omega_r^2)} = \frac{\Gamma_r\omega_r^2}{M_r\omega_r^2}\frac{\bar{f}(s)}{(\omega_r^2 + s^2 + 2\zeta s)} \\
&= \frac{\Gamma_r}{\omega_r^2 M_r}H(s)\bar{f}(s),
\end{aligned} \tag{7.2.6}$$

where

$$H(s) = \frac{\omega_r^2}{(s^2 + 2\zeta s + \omega_r^2)} = \frac{\omega_r^2}{(s + \zeta)^2 + (\omega_r^2 - \zeta^2)}. \tag{7.2.7}$$

The response in the time domain is required, that is, $\eta_r(t)$ is the expression required, thus

$$\eta_r(t) = \frac{\Gamma_r}{\omega_r^2 M_r} \mathscr{L}^{-1}[H(s)\bar{f}(s)]. \tag{7.2.8}$$

The inverse transform of a product of two functions is given by a *convolution integral*, given below:

$$\mathscr{L}^{-1}[H(s)\bar{f}(s)] = \underbrace{\int_0^t h(t-\tau)f(\tau)d\tau}_{D_r(t)}. \tag{7.2.9}$$

For the case when $H(s)$ is defined by Eq. (7.2.7) from the table of L-T, one has

$$h(t-\tau) = \mathscr{L}^{-1}[H(s)] = \frac{\omega_r^2}{\sqrt{\omega_r^2 - \zeta^2}} e^{-\zeta(t-\tau)} \sin\left[\sqrt{\omega_r^2 - \zeta_r^2}\,(t-\tau)\right]. \tag{7.2.10}$$

Combining Eqs. (7.2.8), (7.2.9), and (7.2.10), one has

$$\eta_r(t) = \frac{\Gamma_r D_r(t)}{\omega_r^2 M_r}, \tag{7.2.11}$$

where

$$D_r(t) = \int_0^t h(t-\tau)f(\tau)d\tau = \int_0^t \frac{\omega_r^2}{\sqrt{\omega_r^2 - \zeta_r^2}} \bar{e}^{\zeta(t-\tau)} \sin\left[\sqrt{\omega_r^2 - \zeta_r^2}(t-\tau)\right]f(\tau)d\tau. \tag{7.2.12}$$

$D_r(t)$ is the *dynamic load factor*, which is a *dimensionless* quantity and is a function of time, depends upon the following quantities:

(a) Natural frequency $\omega_r$, (b) damping $\zeta_r$, and (c) forcing $f(t)$.
The total response of a structure in which

$$w(x, t) = \sum_{i-1}^{n} \Phi_i(x)\eta_i(t)$$

can be written as

$$w(x, t) = \sum_{i=1}^{n} \frac{\Gamma_i}{\omega_i^2 M_i} \Phi_i(x)D_i(t). \tag{7.2.13}$$

## Some Special Cases

(1) *Unit Step Function*
*Indicial Response* [$\equiv g(t)$] is defined as the response of a system with zero initial condition to a step function, Figure 7.2:

$$\mathscr{L}[u(t)] = 1/s$$
$$\mathscr{L}[g(t)] = \bar{g}(s) = G(s)\mathscr{L}[u(t)] = G(s)/s. \tag{7.2.14}$$

$\bar{x}_0(s) = G(s)F_{in}(s) \equiv$ this is the definition of the transfer function $G(s)$

$$g(t) = \mathscr{L}^{-1}[\bar{g}(s)] = \mathscr{L}^{-1}[G(s)/s].$$

**Figure 7.2** Unit step function

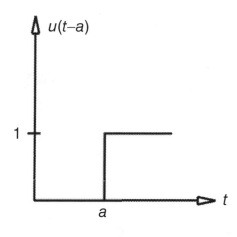

(2) *Impulsive Response, Unit Impulse*
The unit impulse function or Dirac's delta function is defined by

$$\delta(t-a) = 0 \quad t \neq 0$$

and shown in Figure 7.3. For this case,

$$\int_0^\infty \delta(t-a)dt = 1 \qquad \text{[dimensions 1/s]}$$

**Figure 7.3** Unit impulse
function or
Dirac delta function

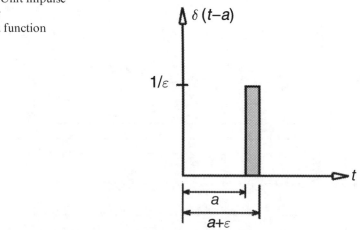

*Impulsive response* denoted by $h(t)$ is the response of a system with zero initial conditions to unit impulse applied at $t = 0$:

$$\mathscr{L}[\delta(t)] = 1.0$$
$$\mathscr{L}[h(t)] = \bar{h}(s) = G(s)\mathscr{L}[\delta(t)] = G(s) \qquad (7.2.15)$$
$$h(t) = \mathscr{L}^{-1}[\bar{h}(s)] = \mathscr{L}^{-1}[G(s)].$$

From comparison of Eqs. (7.2.14) and (7.2.15), it is evident that the unit impulse is the time derivative of the indicial response function, that is,

$$\bar{h}(s) = s\bar{g}(s). \qquad (7.2.16)$$

## 7.2.2  Example of a Response Calculation

Consider a beam on an elastic foundation, having a spring stiffness $\beta_f$ per unit length, under some arbitrary distributed time-dependent loading $p_z(x, t)$ as shown in Figure 7.4. The differential equation of motion is given by

$$\frac{m\partial^2 w}{\partial t^2} + EI\frac{\partial^4 w}{\partial x^4} + \beta_f w = p_z(x, t). \qquad (7.2.17)$$

and the boundary conditions are

$$EI\frac{\partial^2 w}{\partial x^2}\Big|_{x=0,\ell} = 0 \; ; \; w|_{x=0,\ell}. \qquad (7.2.18)$$

Assume solution in the form

$$w(x, t) = W(x)e^{i\omega t}. \qquad (7.2.19)$$

**Figure 7.4**
Beam on a spring foundation used for response calculation example

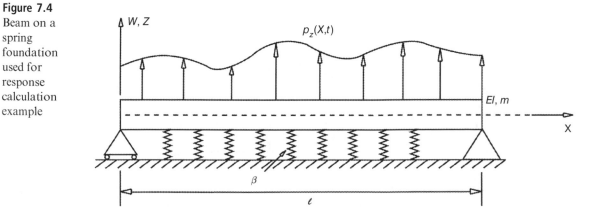

The free vibration problem must be first solved to obtain the normal modes, from Eqs. (7.2.17) and (7.2.19):

$$-m\omega^2 W + EI\frac{d^4 W}{dx^4} + \beta_f W = 0$$

$$\frac{d^4 W}{dx^4} - \frac{m\omega^2 - \beta_f}{EI}W = 0; \qquad \text{define} \quad k^2 = \frac{\omega^2 m - \beta_f}{EI} \qquad (7.2.20)$$

$$\frac{d^4 W}{dx^4} - k^2 W = 0. \qquad (7.2.21)$$

Solution is given by

$$W(x) = C\sinh\sqrt{k}x + D\cosh\sqrt{k}x + E\sin\sqrt{k}x + F\cos\sqrt{k}x. \qquad (7.2.22)$$

Boundary conditions yield

$$W(0) = W(\ell) = W'(0) = W''(\ell) = 0, \qquad \text{where } (\ )' = \frac{d}{dx}$$

from which the following four equations are obtained:

$$D + F = 0 \qquad (7.2.23)$$
$$C\sinh\sqrt{k}\ell + D\cosh\sqrt{k}\ell + E\sin\sqrt{k}\ell + F\cos\sqrt{k}\ell = 0 \qquad (7.2.24)$$
$$D - F = 0 \qquad (7.2.25)$$
$$\sinh\sqrt{k}\ell + D\cosh\sqrt{k}\ell - E\sin\sqrt{k}\ell - F\cos\sqrt{k}\ell = 0. \qquad (7.2.26)$$

From Eqs. (7.2.23) and (7.2.25) $D = F = 0$.

Adding and subtracting Eqs. (7.2.24) and (7.2.26) yields

$$C\ \sinh\sqrt{k}\ell = 0$$

$$E\ \sin\sqrt{k}\ell = 0.$$

Recall that $\sinh x = \frac{1}{2}(e^x - e^{-x})$ and $\cosh x = \frac{1}{2}(e^x + e^{-x})$.

Since $\sinh\sqrt{k}\ell$ cannot be zero for a finite value of $\sqrt{k}\ell$, then $C = 0$ and $\sin\sqrt{k}\ell = 0$:

$$\sqrt{k}\ \ell = n\pi \qquad (7.2.27)$$

$$k_n^2 = \frac{n^4\pi^4}{\ell^4} = \frac{m\omega_n^2 - \beta_f}{EI}. \qquad (7.2.28)$$

Thus

$$\omega_n^2 = \frac{\beta_f}{m} + \frac{EI}{m}\frac{n^4\pi^4}{\ell^4}. \qquad (7.2.29)$$

Also $A = B = C = D = 0$. Thus

$$W_n = W_n(x) = E\sin\sqrt{k_n}x = E\sin\frac{n\pi x}{\ell}.$$

An interesting observation can be made from Eq. (7.2.28), which indicates that a solution to the free vibration problem exists only when $m\omega_n^2 > \beta_f$ or when

$$\omega_n^2 > \frac{\beta_f \ell}{m\ell} = \frac{K}{M} = \Omega^2,$$

where $\Omega^2$ would be the square of the frequency of the oscillation of the whole beam on its spring foundation as a rigid body, when not restrained by its simple supports.

Using the normal mode approach, the equation of motion can be decoupled:

$$w(x, t) = \sum_{i=1}^{\infty} \phi_i(x)\eta_i(t)$$

$$m \sum_{i=1}^{\infty} \phi_i \ddot{\eta}_i + EI \sum_{i=1}^{\infty} \phi_i^{IV} \eta_i + \beta_f \sum_{i=1}^{\infty} \phi_i \eta_i = p_z(x, t).$$

Multiply by $\phi_j$ and integrate between 0 and $\ell$:

$$m \sum_{i=1}^{\infty} \ddot{\eta}_i \int_0^\ell \phi_i \phi_j dx + EI \sum_{i=1}^{\infty} \int_0^\ell \phi_i^{IV} \phi_j \eta_i dx + \beta_f \sum_{i=1}^{\infty} \eta_i \int_0^\ell \phi_i \phi_j dx$$

$$= \int_0^\ell p_z(x, t)\phi_j dx. \quad (7.2.30)$$

Note that in evaluating the second term in Eq. (7.2.30), one can use the original differential equation, that is,

$$\phi_i^{IV} - k_i^2 \phi_i = 0$$

to yield

$$\sum_{i=1}^{\infty} \ddot{\eta}_i \int_0^\ell m\phi_i \phi_j dx + \sum_{i=1}^{\infty} \eta_i k_i^2 \int_0^\ell EI\phi_i \phi_j dx +$$

$$+ \beta_f \sum_{i=1}^{\infty} \eta_i \int_0^\ell \phi_i \phi_j dx = \int_0^\ell p_z(x, t)\phi_j dx. \quad (7.2.31)$$

Using Eqs. (7.2.28) and (7.2.29),

$$\sum_{i=1}^{\infty} \ddot{\eta}_i \int_0^\ell m\phi_i \phi_j dx = \omega_i^2 \sum_{i=1}^{\infty} \eta_i \int_0^\ell m\phi_i \phi_j dx - \sum_{i=1}^{\infty} \eta_i \beta_f \int_0^\ell \phi_i \phi_j dx$$

$$+ \sum_{i=1}^{\infty} \beta_f \eta_i \int_0^\ell \phi_i \phi_j dx = \sum_{i=1}^{\infty} p_z(x, t)\phi_j dx. \quad (7.2.32)$$

Due to the mode shapes used,

$$\int_0^\ell \phi_i \phi_j dx = \delta_{ij},$$

where $\delta_{ij}$ is the Kroenecker delta.

Thus

$$M_j \ddot{\eta}_j + M_j \omega_j^2 \eta_j = P_j, \qquad (7.2.33)$$

where

$$M_j = \int_0^\ell m\phi_j^2 dx$$

$$P_j = \int_0^\ell p_z(x,t)\phi_j dx.$$

Using normalized mode shapes,

$$M_j = m \int_0^\ell \sin^2 \frac{\pi j x}{\ell} dx = \frac{m\ell}{2}. \qquad (7.2.34)$$

From Eqs. (7.2.33) and (7.2.34),

$$\ddot{\eta}_j + \omega_j^2 \eta_j = \frac{2}{m\ell} \int_0^\ell p_z(x,t) \sin \frac{\pi j x}{\ell} dx. \qquad (7.2.35)$$

Suppose a concentrated load is suddenly applied as a step function at the middle of the beam, that is,

$$p_z(x,t) = Pu(t)\delta(x - \xi), \qquad (7.2.36)$$

where $\delta(x - \xi)$ is the Dirac delta function, which is zero everywhere except at $x = \xi$. Its integral properties are

$$\left. \begin{array}{l} \int_0^\ell \delta(x - \xi)dx = 1 \\ \int_0^\ell g(x)\delta(x - \xi)dx = g(\xi) \end{array} \right\}, \qquad (7.2.37)$$

where $g(x)$ is an arbitrary function.

The loading is symmetric; thus, only symmetric response is of interest, that is, $j = 1, 3, 5$:

$$\Gamma_r f(t) = \int_0^\ell u(t)P \sin \frac{\pi x_j}{\ell} \delta\left(x - \frac{\ell}{2}\right) dx = u(t)P \sin \frac{\pi j}{2}. \qquad (7.2.38)$$

And Eq. (7.2.35) becomes

$$\ddot{\eta}_j + \omega_j^2 \eta_j = \frac{2Pu(t)}{m\ell} \sin \frac{\pi j}{2}. \qquad (7.2.39)$$

Initial conditions are zero, that is, $\eta_j(0) = \dot{\eta}_j(0) = 0$

Take Laplace transform of Eq. (7.2.39):

$$(s^2 + \omega_j^2)\bar{\eta}_j(s) = \frac{2P}{m\ell}\frac{1}{s} \sin \frac{\pi j}{2}$$

$$\bar{\eta}_j(s) = \frac{2P/m\ell \sin \pi j/2}{s(s^2 + \omega_j^2)}$$

$$\eta_j(t) = \mathscr{L}^{-1}[\bar{\eta}_j(s)] = \frac{2P}{m\ell}\frac{1}{\omega_j^2}(1 - \cos\omega_j t)\sin\frac{\pi j}{2}.$$

Thus, the final response will be

$$w(x, t) = \sum_{i=1,3,5}^{\infty}\sin\left(\frac{\pi i}{2}\right)\frac{2P}{m\ell}\sin\left(\frac{i\pi x}{\ell}\right)\frac{1}{\omega_i^2}(1 - \cos\omega_i t), \qquad (7.2.40)$$

where $\omega_i^2$ is given by Eq. (7.2.29).

Assume next that the system had constant viscous damping. Then the equations of motion can be rewritten as

$$M_j\ddot{\eta}_j + \sum_{j=1}^{\infty}\int_0^{\ell}c\phi_i\phi_j dx\,\dot{\eta}_i + M_j\omega_j^2\eta_j = \Gamma_j f(t), \qquad (7.2.41)$$

where $c$ is the distributed viscous damping per unit span.

The term $\sum_{j=1}^{\infty}\dot{\eta}_i\int_0^{\ell}c\phi_i\phi_j dx$ would usually couple the equations of motion, however, due to our fortunate selection for the mode shapes $\int_0^{\ell}\phi_i\phi_j dx = 0$ and $\int_0^{\ell}\phi_j^2 dx = \ell/2$. Thus, Eq. (7.2.41) becomes

$$M_j\ddot{\eta}_j + \frac{c\ell}{2}\dot{\eta}_j + M_j\omega_j^2\eta_j = \Gamma_j f(t). \qquad (7.2.42)$$

Thus, the response problem decouples and is convenient to solve.

Representing the damping term by

$$\frac{c\ell}{2} = 2\zeta$$

enables one to write Eq. (7.2.42) in the form

$$\ddot{\eta}_j + 2\zeta\dot{\eta}_j + \omega_j^2\eta_j = \Gamma_j\frac{f(t)}{M_j}. \qquad (7.2.43)$$

Assume a step loading in the time domain

$$\eta_j = \frac{1}{s}\frac{\Gamma_j}{M_j}\frac{1}{(s^2 + 2\zeta s + \omega_j^2)} = \frac{1}{s}\frac{\Gamma_j}{M_j}\frac{1}{[(s + \zeta)^2 + (\omega_j^2 - \zeta^2)]}.$$

From the table of Laplace transforms, the inverse transform is given by

$$\eta_j(t) = \frac{\Gamma_j}{M_j}\left\{\frac{1}{\omega_j^2} - \frac{1}{\omega_j^2\sqrt{1 - \zeta^2}}e^{-\zeta\omega_j t}\sin\left[(\omega_j\sqrt{1 - \zeta^2})t + \phi\right]\right\}$$

$$\eta_i(t) = \frac{\Gamma_j}{\omega_j^2 M_j}\left\{1 - \frac{1}{\sqrt{1 - \xi^2}}e^{-\zeta\omega_j t}\sin\left[(\omega_j\sqrt{1 - \zeta^2})t + \phi\right]\right\}. \qquad (7.2.44)$$

Thus, for this case, the response is

$$w(x, t) = \sum_{i=1,3,5}^{\infty} \sin\left(\frac{\pi i}{2}\right) \frac{2P}{m\ell} \sin\frac{i\pi x}{\ell} \frac{1}{\omega_i^2} \{1-$$

$$- \frac{1}{\sqrt{1-\zeta^2}} e^{-\zeta\omega_i t} \sin\left[(\omega_i\sqrt{1-\zeta^2})\, t + \phi\right]\} . \qquad (7.2.45)$$

## 7.3  Response of Displacement-Driven Structures

The response of displacement-driven structures is an important problem in structural dynamics. Typical examples that belong to this class of problems are a structure mounted on a shaker table, a building excited by ground motion or a satellite or payload on the top of a spacecraft or launcher. Consider a general example shown in Figure 7.5.

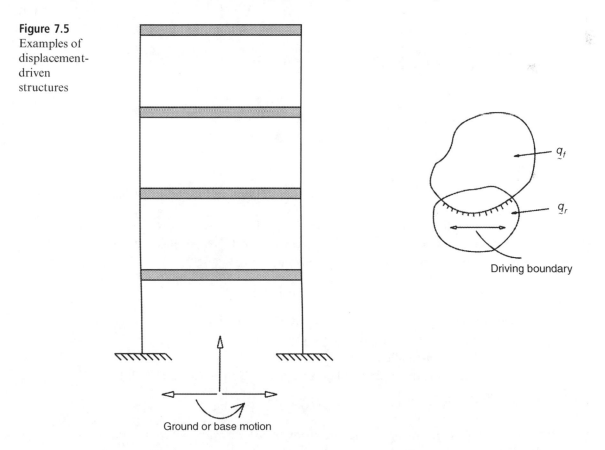

**Figure 7.5**
Examples of displacement-driven structures

Ground or base motion

$q_f$

$q_r$

Driving boundary

Consider the symbolic structure shown in the right-hand side of Figure 7.5 which can be represented by a generalized displacement vector $\{q\}$. The generalized displacement vector can be partitioned in the following manner:

$$\{q\} = \left\{ \frac{\mathbf{q}_r}{\mathbf{q}_f} \right\},$$

where $\{q_r\}$ are the nodal DOFs, including rotations that are prescribed (given) as a function of time. These are the ones that "drive" the structure. $\{q_f\}$ are the remaining nodal DOFs. Furthermore, it is important to realize that in this problem the number of $\{q_r\}$s are determined from the physics of the problem and therefore there may be **more** $\{q_r\}$s than are absolutely necessary in order to **prevent rigid body motion**. It is also important to note that the only impressed forces are the reactions $Q_r$ applied to the structure. Thus, the equations of motion can be written in the following form:

$$\begin{bmatrix} \mathbf{m}_{rr} & \mathbf{m}_{rf} \\ \mathbf{m}_{fr} & \mathbf{m}_{ff} \end{bmatrix} \left\{ \begin{array}{c} \ddot{\mathbf{q}}_r \\ \ddot{\mathbf{q}}_f \end{array} \right\} + \begin{bmatrix} \mathbf{k}_{rr} & \mathbf{k}_{rf} \\ \mathbf{k}_{fr} & \mathbf{k}_{ff} \end{bmatrix} \left\{ \begin{array}{c} \mathbf{q}_r \\ \mathbf{q}_f \end{array} \right\} = \left\{ \begin{array}{c} \mathbf{Q}_r \\ 0 \end{array} \right\}. \tag{7.3.1}$$

Consider a simpler static problem associated with Eq. (7.3.1) for this case one can write:

$$[k] \left\{ \begin{array}{c} \mathbf{q}_r \\ \mathbf{q}_f \end{array} \right\} = \left\{ \begin{array}{c} Q_r \\ 0 \end{array} \right\}. \tag{7.3.2}$$

In Eq. (7.3.2), the reactions are prescribed; therefore, in a partitioned form, Eq. (7.3.2) is equivalent to the following

$$[k_{rr}]\{q_r\} + [k_{rf}]\{q_f\} = \{Q_r\} \tag{7.3.3}$$

$$[k_{fr}]\{q_r\} + [k_{ff}]\{q_f\} = \{0\}. \tag{7.3.4}$$

From Eq. (7.3.4)

$$\{q_f\} = -[k_{ff}]^{-1}[k_{fr}]\{q_r\}. \tag{7.3.5}$$

$\{q_f\} = $ here are the displacements for a prescribed $\{q_r\}$.

Equation (7.3.3) gives the reaction forces due to the prescribed $\{q_r\}$, while Eq. (7.3.5) represents a transformation that can be also written as

$$\{q_f\} = [T]\{q_r\}. \tag{7.3.6}$$

If $\{q_r\}$ is exactly equal to the number of DOFs required to prevent rigid body motion, then the columns of $[T]$ are the rigid body modes; however, when this statement is not satisfied the columns of $[T]$ are called *constraint* modes.

Next, consider the dynamic problem and define relative coordinates $\{u\}$ by

$$\{q_f\} = -[k_{ff}]^{-1}[k_{fr}]\{q_r\} + \{u\}. \tag{7.3.7}$$

This is the *dynamic* $\{q_f\}$ different from the *static value* given by Eq. (7.3.5). Using Eq. (7.3.1) one can write

$$[m_{rr}]\{\ddot{q}_r\} + [m_{rf}]\{\ddot{q}_f\} + [k_{rr}]\{q_r\} + [k_{rf}]\{q_f\} = \{Q_r\}$$
$$[m_{fr}]\{\ddot{q}_r\} + [m_{ff}]\{\ddot{q}_f\} + [k_{fr}]\{q_r\} + [k_{ff}]\{q_f\} = \{0\}. \tag{7.3.8}$$

Introduce the change of variable corresponding to Eq. (7.3.7) into Eq. (7.3.8):

$$[m_{fr}]\{\ddot{q}_r\} + [m_{ff}](-[k_{ff}]^{-1}[k_{fr}]\{\ddot{q}_r\} + \{\ddot{u}\})$$
$$+[k_{fr}]\{q_r\} + [k_{ff}](-[k_{ff}]^{-1}[k_{fr}]\{q_r\} + \{u\}) = 0.$$

From the last equation,

$$[m_{ff}]\{\ddot{u}\} + [k_{ff}]\{u\} = [m_{ff}][k_{ff}]^{-1}[k_{fr}]\{\ddot{q}_r\}$$
$$-[k_{fr}]\{q_r\} + [k_{ff}][k_{ff}]^{-1}[k_{fr}]\{q_r\} - [m_{fr}]\{\ddot{q}_r\}. \tag{7.3.9}$$

Thus, finally one has

$$[m_{ff}]\{\ddot{u}\} + [k_{ff}]\{u\} = ([m_{ff}][k_{ff}]^{-1}[k_{fr}] - [m_{fr}])\{\ddot{q}_r\}. \tag{7.3.10}$$

Note that Eq. (7.3.10) is written in terms of the relative coordinate $\{u\}$. After solving Eq. (7.3.10), Eq. (7.3.8) can be used to determine the forced reactions $\{Q_r\}$.

Equation (7.3.10) also indicates that the **acceleration records** of an earthquake, a shaker table motion, or a missile launch are the important quantities that should be measured and not the displacements themselves that drive the structure.

## 7.4   Time Integration for the Solution of the Structural Dynamics Response

### 7.4.1   Introduction

Structural behavior under dynamic loads is a central problem in structural dynamics. Dynamic loads may take several forms consisting of external or internal forces and pressures, as well as accelerations due to base motion. In some cases, these loads may depend on the motion, as is the case for aerodynamic loads, or actively controlled responses.

In the majority of the cases, determination of structural dynamic response requires **numerical solution** of an initial-value problem (IVP) corresponding to the solution of a system of ordinary differential equations (ODEs) in the time domain. A typical system representing this case can be written as

$$\mathbf{M\ddot{q}} + \mathbf{C\dot{q}} + \mathbf{Kq} = \mathbf{f}_E(\mathbf{q}, \dot{\mathbf{q}}, t) \tag{7.4.1}$$

with initial condition

$$\mathbf{q}(0) = \mathbf{q}_0, \quad \dot{\mathbf{q}}(0) = \dot{\mathbf{q}}_0, \tag{7.4.2}$$

where $\mathbf{M}$, $\mathbf{C}$, and $\mathbf{K}$ are, respectively, the mass, damping, and stiffness matrices; $\mathbf{f}_E$ is the external loading vector; and $\mathbf{q}$ is the displacement vector. The $\mathbf{M}$, $\mathbf{C}$, and $\mathbf{K}$ matrices are usually constant matrices for a given system. However, these matrices can be nonlinear functions of time and displacements such as for the cases of large structural deformation, viscoelastic material properties, or other effects. Equation (7.4.1) can be obtained from a finite element analysis of a structural dynamic system, or from global approximation models obtained by the Rayleigh–Ritz or Galerkin methods.

The time integration of Eq. (7.4.1) is accomplished by a suitable, numerical, step-by-step procedure. Structural dynamic systems are frequently represented by $\mathbf{M}$, $\mathbf{C}$, and $\mathbf{K}$ matrices of considerable size, particularly when they are obtained from the finite element analysis. Therefore, the methods of particular interest to structural dynamicists are those that take advantage of the special characteristics of the $\mathbf{M}$, $\mathbf{C}$, and $\mathbf{K}$ matrices as well as the specific second-order form of the ODEs. The time integration methods for second-order equations of motion are referred to as **direct integration** methods. The term "direct" implies that prior to numerical integration, no transformation of the equations into a different form, such as the normal mode type transformation or transformation into first-order form, is performed.

When the structural dynamic problem is nonlinear, or coupled with another physical system, for example, an aerodynamic loading governed by a first-order ODE system, then it is not uncommon to solve Eq. (7.4.1) numerically using conventional methods of numerical integration such as the Runge–Kutta methods (Zienkiewicz and Taylor, 2000; Felippa et al. 2001).

Time integration implies two things: (1) instead of satisfying Eq. (7.4.1) at any instant in time $t$, the equations are satisfied at discrete time intervals $\Delta t$; (2) to perform the integration some variation of the displacements, velocities, and accelerations inside each interval $\Delta t$ has to be assumed. The nature of these assumptions governs the accuracy, stability, and cost of the solution procedure.

## General Considerations

The objective of a structural dynamic response computation is to obtain a reliable and accurate solution at minimal computational cost. By minimizing the computational cost of an individual analysis, a larger combination of cases can be explored, leading to a better overall design.

General considerations in the selection of a numerical integration method include convergence, accuracy, and stability.

(1) **Convergence** provides assurance that the numerical solution converges to the exact solution of the underlying mathematical model as a time step size is reduced or an integration order is increased.
(2) **Accuracy** is the quality of the solution indicating how well the numerical solution matches a Taylor series expansion of the solution in the neighborhood of a particular time.
(3) **Stability** implies how a small error in the solution propagates through the solution as a function of time.

Accuracy and stability are both affected by the step size in time. Some methods may require a minimal step size for conditional stability, especially for nonlinear problems.

Another important consideration for the numerical methods is **consistency**. This implies that the temporal discretization and spatial discretization are not independent. For example, in wave propagation in solids, the time step size should be on the order of the speed of propagation of a disturbance times a characteristic length, that is, the step size for spatial discretization. This requirement on the time and spatial step sizes is known as the Courant–Friedrichs–Lewy (CFL) condition, which is also frequently used in computational fluid dynamics (CFD).

## Types of Integration Methods

Some numerical integration approaches employ a single time step called **one-step** methods; higher-order accuracy can be obtained using values of the response and derivatives at multiple times, which are known as **multistep** methods. Other approaches, such as the well-known Runge–Kutta method, use several evaluation points in the nominal time interval to obtain improved accuracy at the end of the interval. The intrinsic error in a response calculation can be estimated by implementing parallel schemes of different order, and a similar measure can indicate the presence of local stiffness in the response. Knowledge of the error can be used to guide selection of the time step in so-called **variable time-step** methods. Such knowledge can alternatively be used to modify the order of the integration scheme in **variable-order** methods. Section 7.4.2 focuses on the discussion of the variable time-step methods and multistep methods.

Some approaches use time derivatives calculated from the equations of motion evaluated at the current time to approximate the response at some future time. Such approaches are known as **explicit**, as the approximate solution at the future time is obtained directly using information that is already available. While often suitable, explicit approaches can have difficulties for problems that exhibit so-called numerical **stiffness**. This implies that the response of a stiff system involves a broad range of time scales. An example of a situation in which this kind of stiffness arises is fluid-structure interaction. In these cases, approaches other than explicit might yield more stable and accurate results. An important example of an explicit method in structural dynamics is the central difference method, which is discussed in detail in Section 7.4.3.

Approaches that use the equations of motion written at some time in the future to solve for the response at that future time are known as **implicit**. Because the derivatives at that future time depend on the response at that time, iteration is needed to determine a self-consistent solution. For a given time step size, implicit schemes tend to be more stable, even if they are more computationally intensive. The implicit methods are frequently used in structural dynamic analysis. Two instances are the Newmark-$\beta$ methods and the generalized-$\alpha$ methods, which are discussed in detail in Section 7.4.4.

The reader is not expected to engage in the development of new numerical integration methods, but complement and use existing methods. The purpose of this chapter is to describe useful aspects of the numerical solution of structural dynamic equations of motion, so that the reader can select appropriate methods for a variety of applications.

### 7.4.2   Numerical Methods for First-Order Equations

#### Transformation of Equations of Motion

Since most general numerical integration schemes assume the system of interest is described as first-order ODEs, Eq. (7.4.1) may be expressed in a general form as

$$\dot{\mathbf{x}} = \mathbf{A}\mathbf{x} + \mathbf{B}\mathbf{u}(\mathbf{x}, t) \equiv \mathbf{f}(\mathbf{x}, t), \tag{7.4.3}$$

where $\mathbf{f}$ denotes the system equation, and the state vector $\mathbf{x}$ and input vector $\mathbf{u}$ are, respectively,

$$\mathbf{x} = \begin{bmatrix} \mathbf{q} \\ \dot{\mathbf{q}} \end{bmatrix}, \quad \mathbf{u}(\mathbf{x}, t) = \mathbf{f}_E(\mathbf{q}, \dot{\mathbf{q}}, t) \tag{7.4.4}$$

and the system matrices are

$$\mathbf{A} = \begin{bmatrix} 0 & \mathbf{I} \\ -\mathbf{M}^{-1}\mathbf{K} & -\mathbf{M}^{-1}\mathbf{C} \end{bmatrix}, \quad \mathbf{B} = \begin{bmatrix} 0 \\ \mathbf{M}^{-1} \end{bmatrix}. \tag{7.4.5}$$

The initial conditions are

$$\mathbf{x}(0) = \mathbf{x}_0 = \begin{bmatrix} \mathbf{q}_0 \\ \dot{\mathbf{q}}_0 \end{bmatrix}. \tag{7.4.6}$$

The solution to Eq. (7.4.3) is written as

$$\mathbf{x}(t) = \mathbf{x}_0 + \int_0^t \mathbf{f}(\mathbf{x}, t)dt. \tag{7.4.7}$$

The problem of finding the dynamic response of a structure usually starts from specified initial conditions of the motion as well as the subsequent dynamic loading; however, the loading may also depend on the response. In the simplest approach, the response solution is sought at a sequence of times separated by a uniform time interval, denoted the **time step size**, $\Delta t$

$$\mathbf{x}(0), \mathbf{x}(\Delta t), \mathbf{x}(2\Delta t), \ldots, \mathbf{x}(i\Delta t), \ldots, \mathbf{x}(n\Delta t) \tag{7.4.8}$$

or alternatively

$$\mathbf{x}_0, \mathbf{x}_1, \mathbf{x}_2, \ldots, \mathbf{x}_i, \ldots, \mathbf{x}_n,$$

where the index $i$, $0 \leq i \leq n$, denotes the end of the $i$th time interval, that is, at time $t = t_i = i\Delta t$. The solutions Eq. (7.4.8) can be transformed into a recursive expression using an integral form

$$\mathbf{x}_{i+1} = \mathbf{x}_i + \int_{t_i}^{t_{i+1}} \mathbf{f}(\mathbf{x}, t)dt, \quad i = 0, 1, 2, \ldots, (n-1) \tag{7.4.9}$$

or in a differential form

$$\mathbf{x}_{i+1} = \mathbf{x}_i + \bar{\mathbf{v}}_i \Delta t, \quad i = 0, 1, 2, \ldots, (n-1), \tag{7.4.10}$$

where $\bar{\mathbf{v}}_i$ is an average rate of change over the time interval $[t_i, t_{i+1}]$:

$$\bar{\mathbf{v}}_i = \frac{1}{\Delta t} \int_{t_i}^{t_{i+1}} \mathbf{f}(\mathbf{x}, t) dt. \tag{7.4.11}$$

The goal of a time integration method is therefore to develop an accurate and efficient approximation of either the time integral in Eq. (7.4.9) or the time derivative in Eq. (7.4.10), using the numerical solutions $\mathbf{x}_0, \mathbf{x}_1, \ldots, \mathbf{x}_i, \mathbf{x}_{i+1}, \ldots$.

## The Euler Methods

This section introduces several simple time integration methods related to the classical forward Euler method, which dates back to the eighteenth century. Despite their simplicity, they serve as the basis for formulating more advanced and practical numerical methods in structural dynamic problems.

**Forward Euler Method**    The integral in Eq. (7.4.9) is approximated by assuming that the function $\mathbf{f}(\mathbf{x}, t)$ is a constant $\mathbf{f}(\mathbf{x}_i, t_i)$ over the time interval $[t_i, t_{i+1}]$, as illustrated in Figure 7.6a:

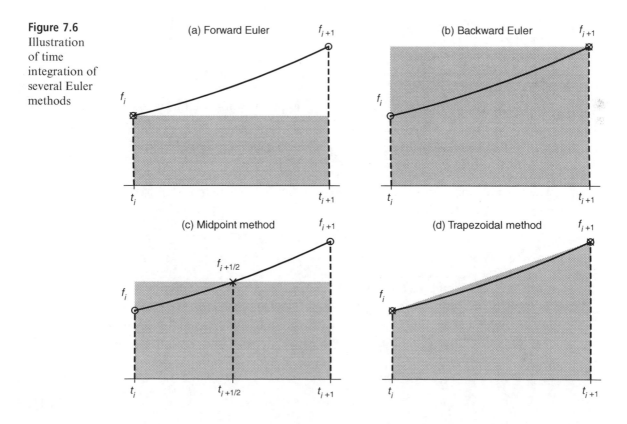

**Figure 7.6**
Illustration of time integration of several Euler methods

$$\int_{t_i}^{t_{i+1}} \mathbf{f}(\mathbf{x}, t)dt = \mathbf{f}(\mathbf{x}_i, t_i)\Delta t + O(\Delta t^2), \qquad (7.4.12)$$

where the term $O(\Delta t^2)$ represents an error term whose magnitude is proportional to $\Delta t^2$. Equation (7.4.9) is rearranged to yield an explicit expression for the solution at the end of the time interval in terms of quantities known at the beginning of the time interval:

$$\mathbf{x}_{i+1} = \mathbf{x}_i + \mathbf{f}(\mathbf{x}_i, t_i)\Delta t + O(\Delta t^2) \qquad (7.4.13a)$$
$$\approx \mathbf{x}_i + \mathbf{f}(\mathbf{x}_i, t_i)\Delta t. \qquad (7.4.13b)$$

Starting from specified initial conditions $\mathbf{x}(0) = \mathbf{x}_0$, Eq. (7.4.13b) provides a recursive, time-marching scheme for the approximate numerical solution of Eq. (7.4.3).

Equation (7.4.13a) is essentially a Taylor series expansion of $\mathbf{x}_{i+1}$ at $t = t_i$, and the second-order error term $O(\Delta t^2)$ is called a **local truncation error**, defining the error introduced by the numerical scheme *at each step*. For a fixed time interval $[0; T]$, required number of steps is proportional to $\Delta t^{-1}$. Hence, the **global error** is proportional to $(\Delta t^2)(\Delta t^{-1}) = \Delta t^1$. Therefore, the forward Euler method is called **first-order** time accurate.

Finally, comparing Eqs. (7.4.10) and (7.4.13b), it is clear that the average rate of change is approximated as

$$\bar{\mathbf{v}}_i = \mathbf{f}(\mathbf{x}_i, t_i) = \dot{\mathbf{x}}_i, \qquad (7.4.14)$$

which indicates that the forward Euler method can be obtained also by approximating the time derivative in Eq. (7.4.3) using a first-order forward finite-difference (FD) scheme:

$$\dot{\mathbf{x}}_i = \frac{\mathbf{x}_{i+1} - \mathbf{x}_i}{\Delta t} + O(\Delta t). \qquad (7.4.15)$$

**Backward Euler Method**    An alternative approximation of the integral in Eq. (7.4.9), illustrated in Figure 7.6b, is

$$\int_{t_i}^{t_{i+1}} \mathbf{f}(\mathbf{x}, t)dt = \mathbf{f}(\mathbf{x}_{i+1}, t_{i+1})\Delta t + O(\Delta t^2). \qquad (7.4.16)$$

The rearrangement of Eq. (7.4.9) yields another recursive expression for the solution at the end of the time interval:

$$\mathbf{x}_{i+1} = \mathbf{x}_i + \mathbf{f}(\mathbf{x}_{i+1}, t_{i+1})\Delta t + O(\Delta t^2) \qquad (7.4.17a)$$
$$\approx \mathbf{x}_i + \mathbf{f}(\mathbf{x}_{i+1}, t_{i+1})\Delta t. \qquad (7.4.17b)$$

The new expression Eq. (7.4.17b) is implicit, because the unknown solution $\mathbf{x}_{i+1}$ is expressed in terms of not only the known solution $\mathbf{x}_i$ but also the unknown $\mathbf{x}_{i+1}$ itself, and one needs to solve a nonlinear equation in order to find $\mathbf{x}_{i+1}$. From Eq. (7.4.17a), it is clear that the backward Euler method also has a second-order local truncation error and hence is first-order time accurate.

In the backward Euler method, the average rate of change is

$$\bar{\mathbf{v}}_i = \mathbf{f}(\mathbf{x}_{i+1}, t_{i+1}) = \dot{\mathbf{x}}_{i+1} \qquad (7.4.18)$$

and the time derivative in Eq. (7.4.10) is effectively approximated using a first-order backward FD scheme:

$$\bar{\mathbf{v}}_i = \frac{\mathbf{x}_{i+1} - \mathbf{x}_i}{\Delta t} + O(\Delta t). \tag{7.4.19}$$

**Midpoint and Trapezoidal Methods**    The order of accuracy of a time integration method is improved using higher-order approximations of the integral in Eq. (7.4.9). Two possible approximations are illustrated in Figures 7.6c and 7.6d. The first approximation is based on the midpoint rule:

$$\int_{t_i}^{t_{i+1}} \mathbf{f}(\mathbf{x}, t) dt = \mathbf{f}(\mathbf{x}_{i+1/2}, t_{i+1/2}) \Delta t + O(\Delta t^3), \tag{7.4.20}$$

where $\mathbf{x}_{i+1/2} = \frac{1}{2}(\mathbf{x}_i + \mathbf{x}_{i+1})$. It leads to the **midpoint method**

$$\mathbf{x}_{i+1} = \mathbf{x}_i + \mathbf{f}(\mathbf{x}_{i+1/2}, t_{i+1/2}) \Delta t \tag{7.4.21}$$

having an average rate of change $\bar{\mathbf{v}}_i = \mathbf{f}(\mathbf{x}_{i+1/2}, t_{i+1/2})$.

The second approximation is based on the trapezoidal rule

$$\int_{t_i}^{t_{i+1}} \mathbf{f}(\mathbf{x}, t) dt = \frac{\Delta t}{2} [\mathbf{f}(\mathbf{x}_i, t_i) + \mathbf{f}(\mathbf{x}_{i+1}, t_{i+1})] + O(\Delta t^3), \tag{7.4.22}$$

which leads to the **trapezoidal method**

$$\mathbf{x}_{i+1} = \mathbf{x}_i + \frac{\Delta t}{2} [\mathbf{f}(\mathbf{x}_i, t_i) + \mathbf{f}(\mathbf{x}_{i+1}, t_{i+1})] \tag{7.4.23}$$

having an average rate of change $\bar{\mathbf{v}}_i = \frac{1}{2}[\mathbf{f}(\mathbf{x}_i, t_i) + \mathbf{f}(\mathbf{x}_{i+1}, t_{i+1})]$.

Both midpoint and trapezoidal methods are implicit and second-order time accurate; and the two methods are equivalent when the function $\mathbf{f}(\mathbf{x}, t)$ is linear.

**Improved Euler Method**    It is also possible to employ a **predictor-corrector** approach to convert an implicit higher-order method to an explicit one by estimating the unknown solution $\mathbf{x}_{i+1}$ using a predictor.

For example, the trapezoidal method is converted to an explicit method, called improved Euler method, which consists of two steps. The predictor step estimates $\mathbf{x}_{i+1}$ using the forward Euler method:

$$\mathbf{x}_{i+1}^* = \mathbf{x}_i + \mathbf{f}(\mathbf{x}_i, t) \Delta t \tag{7.4.24}$$

and subsequently the corrector step computes $\mathbf{x}_{i+1}$ using a modified form of Eq. (7.4.23):

$$\mathbf{x}_{i+1} = \mathbf{x}_i + \frac{\Delta t}{2} \left[ \mathbf{f}(\mathbf{x}_i, t_i) + \mathbf{f}(\mathbf{x}_{i+1}^*, t_{i+1}) \right]. \tag{7.4.25}$$

Using Eqs. (7.4.24) and (7.4.25), the new solution $\mathbf{x}_{i+1}$ is computed without solving a coupled equation, and hence the method is explicit. Furthermore, applying Taylor series expansion to Eqs. (7.4.24) and (7.4.25), one can show that the improved Euler method is second-order time accurate; see Ascher and Petzold (1998) for a detailed proof.

### Absolute Stability and Its Analysis

When a numerical method is stable, it is guaranteed that small errors introduced into the time integration process will not be amplified by the subsequent computations. Numerical errors are unavoidable in time integration, and therefore it is imperative to employ a stable numerical method to obtain the response of structural dynamic system. There are multiple definitions of the stability, and this section focuses on the **absolute stability**, or A-stability, which characterizes the behavior of a numerical solution with a fixed $\Delta t$ as the time goes to infinity. In particular, if the exact solution is expected to decay to zero, so should the numerical solution.

The absolute stability is studied by considering a scalar linear test equation

$$\dot{x}(t) = \lambda x(t), \quad x(0) = x_0, \tag{7.4.26}$$

where $\lambda$ is a complex number and the real part of $\lambda$ is negative, $\mathrm{Re}(\lambda) < 0$. The exact solution to Eq. (7.4.26) is

$$x(t) = \exp(\lambda t)x_0. \tag{7.4.27}$$

Since $\mathrm{Re}(\lambda) < 0$, $x(t) \to 0$ as $t \to \infty$.

When a numerical method is applied to solve Eq. (7.4.26), a recursive equation is formed:

$$x_{n+1} = R(z)x_n, \quad z = \lambda \Delta t, \tag{7.4.28}$$

where $R(z)$ is called **stability polynomial**, a rational polynomial of a complex variable. For example,

(1) Forward Euler method: $R(z) = 1 + z$
(2) Backward Euler method: $R(z) = (1 - z)^{-1}$
(3) Midpoint/trapezoidal method: $R(z) = \dfrac{1 + z/2}{1 - z/2}$
(4) Improved Euler Method: $R(z) = 1 + z + \frac{1}{2}z^2$

The absolute value of stability polynomial, $|R(z)|$, characterizes the growth rate of the numerical solution. To ensure that the numerical solution behaves in a similar manner as the exact solution, that is, $x_{n+1}$ decays to zero as $n \to \infty$, it is required that

$$|R(z)| < 1. \tag{7.4.29}$$

Equation (7.4.29) defines a **stability region**, from which one can choose a time step size to ensure the absolute stability of a numerical method.

The stability regions of the four methods discussed in the previous section are illustrated in Figure 7.7. For the forward and improved Euler methods, the stability regions are finite: One is a disk of radius 1 centered at the point $z = -1$, and the other is a region of stretched disk. For both methods, the most negative intersection of the stability region with the negative real axis is $z_b = -2$; therefore, when $\lambda$ is real, to maintain A-stability, the time step size should satisfy

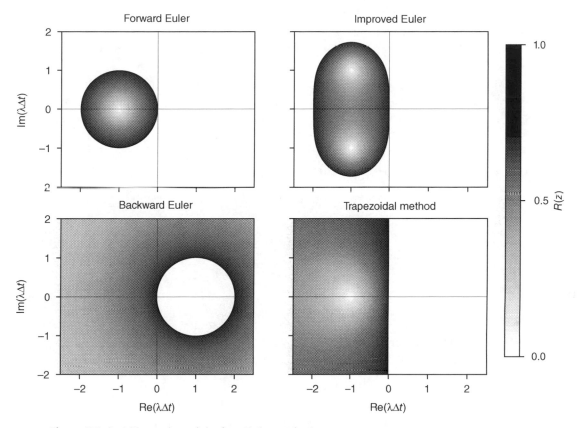

**Figure 7.7** Stability region of the four Euler methods

$$\Delta t < \Delta t_{\mathrm{cr}} = \frac{z_b}{\lambda}. \tag{7.4.30}$$

As for the backward Euler and midpoint/trapezoidal methods, the stability regions are infinite: One is the outer region of a disk of radius 1 centered at the point $z = 1$, and the other is the entire left half of the complex plane. Therefore, as long as $\mathrm{Re}(\lambda) < 0$, both methods are A-stable for any step size.

Integration schemes that require $\Delta t \leq \Delta t_{\mathrm{cr}}$ are said to be **conditionally stable**, which is the case for explicit methods such as the forward and improved Euler methods. If a time step is used with $\Delta t > \Delta t_{\mathrm{cr}}$, the integration fails, indicating that errors resulting from the numerical integration or round-off in the computer grow and make the response calculations unacceptable in most cases. Integration schemes that do not have a constraint on $\Delta t$ are said to be **unconditionally stable**, which is the case for implicit methods such as the backward Euler and midpoint/trapezoidal methods.

The main advantage of unconditionally stable integration schemes is due to the fact that to obtain a desired accuracy in integration, the time step $\Delta t$ can be selected without limitations such as Eq. (7.4.30), and therefore $\Delta t$ can be much larger resulting in significant savings of computer time. However, the price paid for the unconditional stability is the need to solve systems of equations, possibly nonlinear, at each time step, while the conditionally stable schemes are computationally efficient since solving systems of equations is unnecessary.

Finally, note that the methods discussed previously can be applied to solve nonlinear problems; however, the statement made regarding stability and accuracy of these methods is strictly restricted to linear systems, such as the model problem, Eq. (7.4.26), and linear structural dynamic problems.

## Numerical Examples for Euler Methods

In this example, the forward, backward, and improved Euler methods are applied to solve an SDOF undamped oscillator problem:

$$\ddot{q} + q = 0, \quad q(0) = 1, \quad \dot{q}(0) = 0. \tag{E7.97}$$

The analytical solution for the SDOF oscillator is

$$q(t) = \cos(t) \tag{E7.98}$$

with a time period $T = 2\pi$ and an amplitude of 1. An ideal numerical solution should capture accurately both the time period and the amplitude. The numerical solution also needs to be stable; for the SDOF oscillator, the stability implies that the amplitude does not grow over time.

**Conversion to First-Order System**    To apply the Euler methods, the second-order equation of motion (Eq. (E7.97)) needs to be converted to a first-order form:

$$\dot{\mathbf{x}} = \mathbf{f}(\mathbf{x}, t) \equiv \begin{bmatrix} 0 & 1 \\ -1 & 0 \end{bmatrix} \mathbf{x} \tag{E7.99a}$$

$$\mathbf{x}(0) = \begin{bmatrix} 1 \\ 0 \end{bmatrix}, \tag{E7.99b}$$

where the state vector $\mathbf{x} = [q, \dot{q}]^T$.

**Forward Euler Method**    To solve Eq. (E7.99) using the forward Euler method, one has to construct an iterative formulation using Eq. (7.4.13b):

$$\mathbf{x}_{k+1} = \mathbf{x}_k + \Delta t \mathbf{f}(\mathbf{x}_k, t_k) = \begin{bmatrix} 1 & \Delta t \\ -\Delta t & 1 \end{bmatrix} \mathbf{x}_k. \tag{E7.100}$$

Using a time step size $\Delta t = 0.1$ and the initial condition $\mathbf{x}_0 = [1, 0]^T$, one can compute the numerical solutions at the first time step:

$$\mathbf{x}_1 = \begin{bmatrix} 1 & 0.1 \\ -0.1 & 1 \end{bmatrix} \mathbf{x}_0 = \begin{bmatrix} 1.0000 \\ -0.1000 \end{bmatrix}. \tag{E7.101}$$

**Table 7.1 Comparison of numerical solutions using Euler methods**

| Step | Exact solution | Error (%) | Forward Euler | Error (%) |
|------|----------------|-----------|---------------|-----------|
| 1 | [0.9950, −0.0998] | N/A | [1.0000, −0.1000] | 0.4999 |
| 2 | [0.9801, −0.1987] | N/A | [0.9900, −0.2000] | 1.0022 |
| 3 | [0.9553, −0.2955] | N/A | [0.9700, −0.2990] | 1.5071 |
| 4 | [0.9211, −0.3894] | N/A | [0.9401, −0.3960] | 2.0145 |
| 5 | [0.8776, −0.4794] | N/A | [0.9005, −0.4900] | 2.5244 |

| Step | Backward Euler | Error (%) | Improved Euler | Error (%) |
|------|----------------|-----------|----------------|-----------|
| 1 | [0.9901, −0.0990] | 0.4974 | [0.9950, −0.1000] | 0.0167 |
| 2 | [0.9705, −0.1961] | 0.9923 | [0.9800, −0.1990] | 0.0333 |
| 3 | [0.9415, −0.2902] | 1.4847 | [0.9552, −0.2960] | 0.0500 |
| 4 | [0.9034, −0.3805] | 1.9748 | [0.9208, −0.3900] | 0.0667 |
| 5 | [0.8568, −0.4662] | 2.4623 | [0.8772, −0.4802] | 0.0833 |

The second time step is

$$\mathbf{x}_2 = \begin{bmatrix} 1 & 0.1 \\ -0.1 & 1 \end{bmatrix} \mathbf{x}_1 = \begin{bmatrix} 0.9900 \\ -0.2000 \end{bmatrix} \tag{E7.102}$$

and the subsequent steps are computed in a similar manner.

The first five steps of the forward Euler solutions are listed in Table 7.1 and compared with the exact solution. The forward Euler method produces an average of 0.5% numerical error per time step and therefore one can expect rapid accumulation of error in the forward Euler solution.

**Backward Euler Method**    To solve Eq. (E7.99) using the backward Euler method, an iterative formulation using Eq. (7.4.17b) is constructed:

$$\mathbf{x}_{k+1} = \mathbf{x}_k + \Delta t \mathbf{f}(\mathbf{x}_{k+1}, t_{k+1}) = \mathbf{x}_k + \begin{bmatrix} 0 & \Delta t \\ -\Delta t & 0 \end{bmatrix} \mathbf{x}_{k+1} \tag{E7.103}$$

and rearranging

$$\begin{bmatrix} 1 & -\Delta t \\ \Delta t & 1 \end{bmatrix} \mathbf{x}_{k+1} = \mathbf{x}_k. \tag{E7.104}$$

As shown in Eq. (E7.104), the backward Euler method requires solving a system of equations at every time step.

Using a time step size $\Delta t = 0.1$ and the initial condition $\mathbf{x}_0 = [1, 0]^T$, the solution at the first time step $\mathbf{x}_1$ is found by solving

$$\begin{bmatrix} 1 & -0.1 \\ 0.1 & 1 \end{bmatrix} \mathbf{x}_1 = \mathbf{x}_0 = \begin{bmatrix} 1 \\ 0 \end{bmatrix}, \tag{E7.105}$$

which leads to $\mathbf{x}_1 = [0.9901, -0.0990]^T$. Similarly, the next solution $\mathbf{x}_2$ is found by solving

$$\begin{bmatrix} 1 & -0.1 \\ 0.1 & 1 \end{bmatrix} \mathbf{x}_2 = \mathbf{x}_1 = \begin{bmatrix} 0.9901 \\ -0.0990 \end{bmatrix}, \tag{E7.106}$$

which leads to $\mathbf{x}_2 = [0.9705, -0.1961]^T$. Continuing to solve Eq. (E7.104) iteratively to find the solutions at the subsequent time steps, the first five steps of the backward Euler solutions are listed in Table 7.1, in which a rapid growth of 0.5% error per time step is observed. The forward and backward Euler methods have a fast growth of error, since both are only first-order time accurate.

**Improved Euler Method**    Next, Eq. (E7.99) is solved using the improved Euler method. An iterative predictor–corrector formulation is constructed using Eqs. (7.4.24) and (7.4.25). The predictor step is

$$\mathbf{x}_{k+1}^* = \mathbf{x}_k + \mathbf{f}(\mathbf{x}_k, t_k)\Delta t = \begin{bmatrix} 1 & \Delta t \\ -\Delta t & 1 \end{bmatrix} \mathbf{x}_k. \tag{E7.107}$$

Note that Eq. (E7.107) is identical to Eq. (E7.100), since the forward Euler method is used as the predictor in the improved Euler method. The corrector step is

$$\begin{aligned} \mathbf{x}_{k+1} &= \mathbf{x}_k + \frac{\Delta t}{2}\left[\mathbf{f}(\mathbf{x}_k, t_k) + \mathbf{f}(\mathbf{x}_{k+1}^*, t_{k+1})\right] \\ &= \mathbf{x}_k + \frac{\Delta t}{2}\left(\begin{bmatrix} 0 & 1 \\ -1 & 0 \end{bmatrix} \mathbf{x}_k + \begin{bmatrix} 0 & 1 \\ -1 & 0 \end{bmatrix} \mathbf{x}_{k+1}^*\right). \end{aligned} \tag{E7.108}$$

Combining the predictor step, Eq. (E7.107), and the corrector step, (Eq. (E7.108)), the iterative formula for the improved Euler method is

$$\begin{aligned} \mathbf{x}_{k+1} &= \mathbf{x}_k + \frac{\Delta t}{2}\left(\begin{bmatrix} 0 & 1 \\ -1 & 0 \end{bmatrix} \mathbf{x}_k + \begin{bmatrix} 0 & 1 \\ -1 & 0 \end{bmatrix}\begin{bmatrix} 1 & \Delta t \\ -\Delta t & 1 \end{bmatrix} \mathbf{x}_k\right) \\ &= \begin{bmatrix} 1 - \Delta t^2/2 & \Delta t \\ -\Delta t & 1 - \Delta t^2/2 \end{bmatrix} \mathbf{x}_k. \end{aligned} \tag{E7.109}$$

Using a time step size $\Delta t = 0.1$ and the initial condition $\mathbf{x}_0 = [1, 0]^T$, one can compute the numerical solutions at the subsequent steps:

$$\mathbf{x}_1 = \begin{bmatrix} 0.995 & 0.1 \\ -0.1 & 0.995 \end{bmatrix} \mathbf{x}_0 = \begin{bmatrix} 0.9950 \\ -0.1000 \end{bmatrix} \tag{E7.110}$$

$$\mathbf{x}_2 = \begin{bmatrix} 0.995 & 0.1 \\ -0.1 & 0.995 \end{bmatrix} \mathbf{x}_1 = \begin{bmatrix} 0.9800 \\ -0.1990 \end{bmatrix} \tag{E7.111}$$

$\mathbf{x}_3, \mathbf{x}_4$, and so on.

The first five steps of the improved Euler solutions are listed in Table 7.1. Due to the second-order time accuracy of the improved Euler method, the numerical error is one order of magnitude smaller than those in the forward and backward Euler solutions.

**Absolute Stability Analysis**    Subsequently, using the stability polynomial $R(z)$ in Eq. (7.4.28), one can estimate the absolute stability of the three Euler methods. The eigenvalues of the equation of motion (Eq. (E7.99)) are $\lambda_{1,2} = \pm i$, where $i$ is the unit imaginary number; therefore in the stability polynomial $z = \lambda\Delta t = \pm i\Delta t$.

For the forward Euler method, the stability polynomial is

$$R(z) = 1 + \lambda \Delta t = 1 \pm i \Delta t. \qquad (E7.112)$$

The absolute stability requires the growth rate $|R(z)| < 1$; however, for any time step size $\Delta t$,

$$|R(z)| = |1 \pm i\Delta t| = \sqrt{1 + \Delta t^2} > 1. \qquad (E7.113)$$

Therefore, forward Euler method is unstable for the SDOF oscillator, and one can expect the amplitude of oscillation in the numerical solution to increase over time.

For the backward Euler method, the growth rate is

$$|R(z)| = |(1 - \lambda \Delta t)^{-1}| = |1 \pm i\Delta t|^{-1} = \frac{1}{\sqrt{1 + \Delta t^2}} < 1 \qquad (E7.114)$$

for any time step size $\Delta t$. Therefore, backward Euler method is unconditionally stable for the SDOF oscillator.

For the improved Euler method, the growth rate is

$$|R(z)| = \left| 1 + \lambda \Delta t + \frac{1}{2} \lambda^2 \Delta t^2 \right| = \left| 1 - \frac{1}{2} \Delta t^2 \pm i\Delta t \right|$$
$$= \sqrt{1 + \Delta t^4 / 4} > 1. \qquad (E7.115)$$

Therefore, improved Euler method is unstable for the SDOF oscillator for *any* time step sizes. However, when $\Delta t \ll 1$, the growth rate in Eq. (E7.115) is much smaller than that in Eq. (E7.113), indicating that the oscillation amplitude in the improved Euler solution grows at a slower rate than that in the forward Euler solution.

**Graphical Comparison of the Euler Methods**    Finally, the numerical solutions generated by the forward, backward, and improved Euler methods are compared in Figure 7.8. As a second-order time accurate method, the improved Euler method produces a solution that has superior accuracy compared to those generated by the forward and backward Euler methods, which are first-order time accurate. While the improved Euler method is found to be unstable for any time step size in Eq. (E7.115), for the first two periods of oscillation, the improved Euler solution captures the period and amplitude of the harmonic oscillation with an error less than 1%.

The forward Euler method produces an unstable growing oscillatory response, even though the equation of motion (Eq. (E7.97)) is undamped and neutrally stable. Using a time step size of 0.1, after two periods of oscillation, the amplitude error of forward Euler solution is as high as 86%. The numerical error is reduced by decreasing the time step size by a factor of 10. However, the forward Euler solution is still unstable and less accurate than the improved Euler method. The unstable characteristics of the forward Euler method relate to the absolute stability analysis in Eq. (E7.113). The comparison between forward and improved Euler methods also highlights the importance of using a numerical method with high-order time accuracy.

The backward Euler method produces a stable and damped oscillatory response, as predicted by the stability analysis in Eq. (E7.114). The damped response indicates that the total energy of the oscillator, that is, the sum of kinetic energy and potential energy, dissipates over

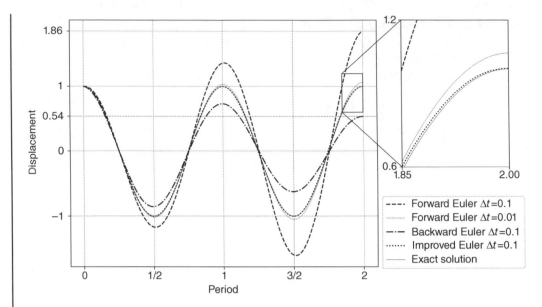

**Figure 7.8** Comparison of three Euler methods

time. However, such energy dissipation is not physical, because the total energy is expected to be conserved in the SDOF undamped oscillator, Eq. (E7.97). The non-physical energy dissipation in the numerical solution is referred to as **numerical dissipation** and is attributed to the time discretization error of a time integration method.

Due to the higher order of time accuracy, the unstable characteristics of the improved Euler method are visible only over long-term numerical simulation. The simulation performed by the improved Euler method is continued until the 20th period of oscillation, and the last two periods are compared with the analytical solution in Figure 7.9. The improved Euler solution shows some numerical errors in both the amplitude and the frequency: The 20th oscillation ends at $t = 19.97T$ with an amplitude 1.01, as marked in circle in Figure 7.9. Therefore, over the 20 periods of simulation, the improved Euler method results in an error of 0.15% in the period and an error of 1% in amplitude. These errors appear small for this particular SDOF problem, but the **periodicity error** and the **amplitude error** can become larger when the improved Euler method is applied to more complex problems. The periodicity and amplitude errors are closely related to numerical dissipation and are important characteristics of a numerical method, which will be analyzed in Sections 7.4.3 and 7.4.4.

**Summary**    In this example, the forward, backward, and improved Euler methods are employed to generate the numerical solution of an SDOF undamped oscillator, one of the simplest problems in structural dynamics. It is demonstrated that a method having higher-order time accuracy, such as the improved Euler method, produces a solution that is more accurate than the lower-order methods when the same time step size is used. The example also

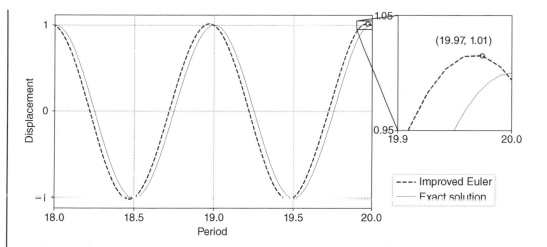

**Figure 7.9** Improved Euler solution over the 19th to 20th period of SDOF oscillation

shows that an unstable time integration method can result in a spurious growth of oscillation in the numerical solution even if the actual system is neutrally stable; and such unstable characteristics have to be avoided in practice. On the other hand, a stable method, such as the backward Euler method, can possess a high amount of numerical dissipation and cause a rapid non-physical dissipation of energy in the numerical solution, leading to a large numerical error. Overall, this numerical example underlines the importance of using high-order stable time integration methods with low numerical dissipation for solving structural dynamic problems. Such methods are discussed in the following sections of this chapter.

### Other Applicable Methods for First-Order Equations

In this section, two additional time integration methods for first-order equations of motion are introduced. These methods are used in some specialized structural dynamic applications.

**Exponential Integrator Method**    When a linear structural dynamic problem is considered, the exponential integrator method can be used to mitigate the stiffness issue in the problem. The equations of motion, Eq. (7.4.3), are linear

$$\dot{\mathbf{x}} = \mathbf{A}\mathbf{x} + \mathbf{B}\mathbf{u}(t), \quad \mathbf{x}(0) = \mathbf{x}_0. \tag{7.4.31}$$

The solution of Eq. (7.4.9) is

$$\mathbf{x}(t) = \int_{t_0}^{t} [\mathbf{A}\mathbf{x} + \mathbf{B}\mathbf{u}(\tau)] \, d\tau$$

$$= \mathbf{\Phi}(t, t_0)\mathbf{x}(t_0) + \int_{t_0}^{t} \mathbf{\Phi}(t, \tau)\mathbf{B}\mathbf{u}(\tau) d\tau, \tag{7.4.32}$$

where $\Phi(t, t_0)$ transforms the state at time $t_0$ to the state at time $t$ and therefore is called the **state transition matrix**. In the time-invariant case, that is, when $\mathbf{A}$ is constant, the state transition matrix is computed exactly as

$$\Phi(t, t_0) = \exp[\mathbf{A}(t - t_0)], \tag{7.4.33}$$

where the definition of a matrix exponential is provided in Appendix A.4.5. Subsequently, the recursive formula for the numerical solution is found to be

$$\mathbf{x}_{i+1} = \Phi(\Delta t)\mathbf{x}_i + \int_{t_i}^{t_{i+1}} \Phi(t_{i+1}, \tau)\mathbf{B}\mathbf{u}(\tau)d\tau$$

$$= \exp(\mathbf{A}\Delta t)\mathbf{x}_i + \int_{t_i}^{t_{i+1}} \exp[\mathbf{A}(t_{i+1} - \tau)]\mathbf{B}\mathbf{u}(\tau)d\tau. \tag{7.4.34}$$

The state transition matrix for the time-varying case is discussed in Chapter 8.

When the forcing term $\mathbf{u}(t)$ is time dependent, it can be approximated using a **zero-order hold**, which implies holding the value of the forcing term constant over a time step. The solution at the end of the time interval can be expressed in terms of the force at the beginning of the interval:

$$\mathbf{x}_{i+1} = \exp(\mathbf{A}\Delta t)\mathbf{x}_i + \mathbf{A}^{-1}[\exp(\mathbf{A}\Delta t) - \mathbf{I}]\mathbf{B}\mathbf{u}_i. \tag{7.4.35}$$

Starting from specified initial conditions, Eq. (7.4.35) provides a recursive, time-marching scheme for the approximate numerical solution of Eq. (7.4.3).

An interesting connection can be made when comparing Eq. (7.4.35) to the various time integration methods discussed in the Euler methods. For example, assuming that the input vector $\mathbf{u}$ is constant, the recursive formulae associated with improved and forward Euler methods are, respectively,

$$\mathbf{x}_{i+1} = \left(\mathbf{I} + \Delta t\mathbf{A} + \frac{\Delta t^2}{2}\mathbf{A}^2\right)\mathbf{x}_i + \Delta t\left(\mathbf{I} + \frac{\Delta t}{2}\mathbf{A}\right)\mathbf{B}\mathbf{u} \tag{7.4.36a}$$

$$\mathbf{x}_{i+1} = (\mathbf{I} + \Delta t\mathbf{A})\mathbf{x}_i + \Delta t\mathbf{B}\mathbf{u}. \tag{7.4.36b}$$

The Taylor series expansion of the matrix exponential is

$$\exp(\mathbf{A}\Delta t) = \mathbf{I} + \Delta t\mathbf{A} + \frac{\Delta t^2}{2}\mathbf{A}^2 + \frac{\Delta t^3}{6}\mathbf{A}^3 + O(\Delta t^4) \tag{7.4.37}$$

and therefore the series expansion of the exponential integrator formula, Eq. (7.4.35), with respect to $\Delta t$ is

$$\mathbf{x}_{i+1} = \left(\mathbf{I} + \Delta t\mathbf{A} + \frac{\Delta t^2}{2}\mathbf{A}^2\right)\mathbf{x}_i + \Delta t\left(\mathbf{I} + \frac{\Delta t}{2}\mathbf{A}\right)\mathbf{B}\mathbf{u} + O(\Delta t^3) \tag{7.4.38a}$$

$$= (\mathbf{I} + \Delta t\mathbf{A})\mathbf{x}_i + \Delta t\mathbf{B}\mathbf{u} + O(\Delta t^2). \tag{7.4.38b}$$

Comparing Eqs. (7.4.36) and (7.4.38), it is evident that the improved and forward Euler methods approximate the exponential integrator with local truncation errors of third order and second order, respectively. Similar conclusions can be drawn for the backward Euler and trapezoidal methods. The local truncation errors can result in the inaccurate prediction of the high-frequency components in a structural dynamic response, represented by the larger eigenvalues of the system matrix $\mathbf{A}$, which can cause a stiffness issue. The exact integration of the system matrix $\mathbf{A}$ in the exponential integrator helps mitigate this stiffness issue.

In the nonlinear case where the forcing term depends on the response, the effects of the forcing term can also be approximated using a zero-order hold. If better performance is needed, the forcing term can be evaluated using the solution at the end of the time step, requiring iteration to find the solution implicitly. If linear variation of $\mathbf{Bu}(\tau)$ is assumed over the time step, the solution at the end of the time step can be expressed as

$$\mathbf{x}_{i+1} = \exp(\mathbf{A}\Delta t)\mathbf{x}_i + \frac{1}{\Delta t}\mathbf{A}^{-2}\left[\mathbf{I} + \exp(\mathbf{A}\Delta t)(\mathbf{A}\Delta t - \mathbf{I})\right]\mathbf{Bu}_i$$
$$+ \frac{1}{\Delta t}\mathbf{A}^{-2}\left[\exp(\mathbf{A}\Delta t) - (\mathbf{A}\Delta t + \mathbf{I})\right]\mathbf{Bu}_{i+1}. \tag{7.4.39}$$

If $\mathbf{u}_{i+1}$ depends on the solution $\mathbf{x}_{i+1}$, then a consistent solution can be sought by fixed-point iteration.

**Linear Multistep Methods**    The improved Euler method uses intermediate Euler steps in a time step to achieve increased accuracy but discards most information when taking the next step. Multistep methods seek efficiency by using previous information rather than discarding it. Most useful multistep methods are linear, with a form that relates linear combinations of previous solutions and time derivatives:

$$\mathbf{x}_{i+s} + a_{s-1}\mathbf{x}_{i+s-1} + \cdots + a_0\mathbf{x}_i = \Delta t\left(b_s\mathbf{d}_{i+s} + b_{s-1}\mathbf{d}_{i+s-1} + \cdots + b_0\mathbf{d}_i\right). \tag{7.4.40}$$

Note that such a scheme is not **self-starting**, since it requires information from prior to the initial time. The first few points are usually obtained using an alternative integration method, such as the well-known Runge–Kutta method. Some details of the Runge–Kutta method are provided in Chapter 8.

An example of an explicit linear multistep method is the **Adams–Bashforth** method of order 4 (AB4),

$$\mathbf{x}_{i+1} = \mathbf{x}_i + \frac{\Delta t}{24}\left(55\mathbf{d}_i - 59\mathbf{d}_{i-1} + 37\mathbf{d}_{i-2} - 9\mathbf{d}_{i-3}\right) \tag{7.4.41}$$

with a local truncation error

$$\mathbf{e}_{AB4} \approx \frac{251}{720}\Delta t^5\frac{d^5\mathbf{x}}{dt^5}. \tag{7.4.42}$$

An implicit example is the **Adams–Moulton** method of order 4 (AM4),

$$\mathbf{x}_{i+1} = \mathbf{x}_i + \frac{\Delta t}{24}\left(9\mathbf{d}_{i+1} + 19\mathbf{d}_i - 5\mathbf{d}_{i-1} + \mathbf{d}_{i-2}\right) \tag{7.4.43}$$

with a local truncation error

$$\mathbf{e}_{AM4} \approx -\frac{19}{720}\Delta t^5\frac{d^5\mathbf{x}}{dt^5}. \tag{7.4.44}$$

The explicit and implicit linear multistep methods are frequently combined to construct predictor–corrector methods. A well-known multistep implementation is the

Adams–Bashforth–Moulton method. For the case of order 4, the explicit AB4 method, Eq. (7.4.41), is used as the predictor to estimate the new solution at $t = t_{i+1}$; subsequently, the implicit AM4 method, Eq. (7.4.43), is used as the corrector to improve the estimate:

$$\mathbf{x}_{i+1}^{(0)} = \mathbf{x}_i + \frac{\Delta t}{24} \left( 55\mathbf{d}_i - 59\mathbf{d}_{i-1} + 37\mathbf{d}_{i-2} - 9\mathbf{d}_{i-3} \right) \tag{7.4.45a}$$

$$\hat{\mathbf{d}}_{i+1} = \mathbf{f}(\mathbf{x}_{i+1}^{(0)}, t_{i+1}) \tag{7.4.45b}$$

$$\mathbf{x}_{i+1}^{(1)} = \mathbf{x}_i + \frac{\Delta t}{24} \left( 9\hat{\mathbf{d}}_{i+1} + 19\mathbf{d}_i - 5\mathbf{d}_{i-1} + \mathbf{d}_{i-2} \right). \tag{7.4.45c}$$

The solution $\mathbf{x}_{i+1}^{(1)}$ can be considered the new solution $\mathbf{x}_{i+1}$ at $t = t_{i+1}$.

Since the local truncation error for each of the AB4 and AM4 methods is known, using both sequentially enables estimation of the error at each time step. For small time step sizes, this is given approximately by

$$\mathbf{e}_{i+1} \approx \frac{\|\mathbf{e}_{AM4}\|}{\|\mathbf{e}_{AM4} - \mathbf{e}_{AB4}\|} \left( \mathbf{e}_{AM4} - \mathbf{e}_{AB4} \right) \approx \frac{1}{14} \left( \mathbf{x}_{i+1}^{(1)} - \mathbf{x}_{i+1}^{(0)} \right). \tag{7.4.46}$$

Finally, Eqs. (7.4.45b) and (7.4.45c) can be iterated to compute $\mathbf{x}_{i+1}^{(2)}, \mathbf{x}_{i+1}^{(3)}, \ldots$, until the difference between the successive solutions converges to within a specified tolerance; this is equivalent to solving the implicit AM4 formula, Eq. (7.4.43), using the fixed-point iteration method.

## 7.4.3   Explicit Methods for Second-Order Equations

This section presents the explicit methods that are directly applicable to second-order equations of motion, appearing in structural dynamics:

$$\mathbf{M}\ddot{\mathbf{q}} + \mathbf{C}\dot{\mathbf{q}} + \mathbf{K}\mathbf{q} = \mathbf{f}_E(\mathbf{q}, \dot{\mathbf{q}}, t) \tag{7.4.47}$$

$$\mathbf{q}(0) = \mathbf{q}_0, \quad \dot{\mathbf{q}}(0) = \dot{\mathbf{q}}_0. \tag{7.4.48}$$

First, we focus on the **central difference method**, a second-order time accurate algorithm that is implemented in many explicit finite element codes for structural dynamics. Along with the formulation of the central difference method, its numerical stability characteristics are examined in detail.

### Formulation the Central Difference Method

In the central difference method, the acceleration and velocity vectors are both approximated using second-order central FD schemes:

$$\ddot{\mathbf{q}}_i = \frac{\mathbf{q}_{i-1} - 2\mathbf{q}_i + \mathbf{q}_{i+1}}{\Delta t^2} \tag{7.4.49a}$$

$$\dot{\mathbf{q}}_i = \frac{-\mathbf{q}_{i-1} + \mathbf{q}_{i+1}}{2\Delta t}. \tag{7.4.49b}$$

In these approximations, the error is of order $\Delta t^2$.

The new displacement solution $\mathbf{q}_{i+1}$ is obtained by using Eq. (7.4.47) at time $t = t_i$:

$$\mathbf{M}\ddot{\mathbf{q}}_i + \mathbf{C}\dot{\mathbf{q}}_i + \mathbf{K}\mathbf{q}_i = \mathbf{f}_i, \tag{7.4.50}$$

where $\mathbf{f}_i$ is the external loading vector $\mathbf{f}_E$ at $t = t_i$ and the subscript $E$ is omitted for clarity. Substituting Eqs. (7.4.49a) and (7.4.49b) into Eq. (7.4.50) yields

$$\mathbf{M}\left(\frac{\mathbf{q}_{i-1} - 2\mathbf{q}_i + \mathbf{q}_{i+1}}{\Delta t^2}\right) + \mathbf{C}\left(\frac{-\mathbf{q}_{i-1} + \mathbf{q}_{i+1}}{2\Delta t}\right) + \mathbf{K}\mathbf{q}_i = \mathbf{f}_i \tag{7.4.51}$$

and rearranging

$$\left(\frac{1}{\Delta t^2}\mathbf{M} + \frac{1}{2\Delta t}\mathbf{C}\right)\mathbf{q}_{i+1} = \mathbf{f}_i - \left(\mathbf{K} - \frac{2}{\Delta t^2}\mathbf{M}\right)\mathbf{q}_i$$
$$- \left(\frac{1}{\Delta t^2}\mathbf{M} - \frac{1}{2\Delta t}\mathbf{C}\right)\mathbf{q}_{i-1}. \tag{7.4.52}$$

Equation (7.4.52) can be solved to obtain $\mathbf{q}_{i+1}$. It is important to note that the solution for $\mathbf{q}_{i+1}$ is based upon a dynamic equilibrium condition at time $t = t_i$, Eq. (7.4.50). Therefore, the central difference method is an explicit integration method.

The central difference method is computationally efficient when the system has no damping $\mathbf{C} = 0$. In this case, Eq. (7.4.52) reduces to

$$\frac{1}{\Delta t^2}\mathbf{M}\mathbf{q}_{i+1} = \hat{\mathbf{f}}_i, \tag{7.4.53}$$

where

$$\hat{\mathbf{f}}_i = \mathbf{f}_i - \left(\mathbf{K} - \frac{2}{\Delta t^2}\mathbf{M}\right)\mathbf{q}_i - \frac{1}{\Delta t^2}\mathbf{M}\mathbf{q}_{i-1}. \tag{7.4.54}$$

Thus, if the mass matrix is diagonal (i.e., lumped mass matrix), the system equations, Eq. (7.4.47), can be solved without factorizing a matrix or solving a linear system, which implies that only matrix multiplications are required to obtain the right-hand side effective load vector $\hat{\mathbf{f}}_i$. The use of the lumped mass matrix is the main reason for the high computational efficiency of the central difference method, or the explicit methods in general.

**Starting Procedure** When using the central difference method, the calculation of $\mathbf{q}_{i+1}$ involved $\mathbf{q}_{i-1}$. Therefore, to calculate the solution $\mathbf{q}_1$ at time $t = t_1 = \Delta t$, the solutions at $t = 0$ and $t = -\Delta t$ (i.e., $\mathbf{q}_0$ and $\mathbf{q}_{-1}$) are required. However, in a structural dynamic problem, typically only the regular initial conditions $\mathbf{q}_0$ and $\dot{\mathbf{q}}_0$ are known. Therefore, a special starting procedure has to be used to find $\mathbf{q}_{-1}$ for the central difference method.

First, the acceleration at time $t = 0$, $\ddot{\mathbf{q}}_0$, is computed from Eq. (7.4.50):

$$\mathbf{M}\ddot{\mathbf{q}}_0 + \mathbf{C}\dot{\mathbf{q}}_0 + \mathbf{K}\mathbf{q}_0 = \mathbf{f}_0 \tag{7.4.55}$$

and rearranging

$$\ddot{\mathbf{q}}_0 = \mathbf{M}^{-1}\left(\mathbf{f}_0 - \mathbf{C}\dot{\mathbf{q}}_0 - \mathbf{K}\mathbf{q}_0\right). \tag{7.4.56}$$

Next, with $\mathbf{q}_0$, $\dot{\mathbf{q}}_0$, and $\ddot{\mathbf{q}}_0$ known, using Eqs. (7.4.49a) and (7.4.49b) for time $t = 0$,

$$\ddot{\mathbf{q}}_0 = \frac{\mathbf{q}_{-1} - 2\mathbf{q}_0 + \mathbf{q}_1}{\Delta t^2}, \qquad \text{or} \quad \Delta t^2 \ddot{\mathbf{q}}_0 - \mathbf{q}_{-1} + 2\mathbf{q}_0 = \mathbf{q}_1 \qquad (7.4.57a)$$

$$\dot{\mathbf{q}}_0 = \frac{-\mathbf{q}_{-1} + \mathbf{q}_1}{2\Delta t}, \qquad \text{or} \quad 2\Delta t \dot{\mathbf{q}}_0 + \mathbf{q}_{-1} = \mathbf{q}_1. \qquad (7.4.57b)$$

Eliminating $\mathbf{q}_1$ from Eqs. (7.4.57a) and (7.4.57b), and solving for $\mathbf{q}_{-1}$, one obtains

$$\mathbf{q}_{-1} = \mathbf{q}_0 - \Delta t \dot{\mathbf{q}}_0 + \frac{\Delta t^2}{2} \ddot{\mathbf{q}}_0. \qquad (7.4.58)$$

With $\mathbf{q}_{-1}$ from Eq. (7.4.58) and $\mathbf{q}_0$ from the initial condition, the central difference method can be started to solve a structural dynamic problem.

The central difference method is defined completely using Eqs. (7.4.49), (7.4.53), (7.4.54), and (7.4.58). However, for computational efficiency, the central difference method is usually implemented using an alternative **velocity-based** formulation, which also makes the implementation of **adaptive time stepping** easier. The velocity-based formulation will be presented after the discussion of the stability analysis.

### Spectral Stability and Oscillatory Characteristics

In Section 7.4.2, the absolute stability was developed based on a scalar linear first-order differential equation, and employed to characterize the stability of numerical methods for first-order equations of motion. This section presents a framework for the stability analysis of numerical methods for second-order equations of motion based on the concept of **spectral stability**, an extension of the absolute stability. Furthermore, since structural dynamic problems involve oscillations, the analysis of spectral stability is used to characterize the ability of the numerical methods to reproduce oscillatory responses, in terms of **amplitude** and **periodicity errors**.

**Unit Structural Dynamic Problem**    For stability analysis, it is mathematically inconvenient to directly work with the equations of motion for multiple DOFs. Therefore, the first step is to develop a unit problem that is sufficiently simple yet captures the main oscillatory characteristics of the structural dynamic response. Starting from the linear structural dynamic problem,

$$\mathbf{M}\ddot{\mathbf{q}} + \mathbf{C}\dot{\mathbf{q}} + \mathbf{K}\mathbf{q} = \mathbf{f}(t) \qquad (7.4.59a)$$

$$\mathbf{q}(0) = \mathbf{q}_0, \quad \dot{\mathbf{q}}(0) = \dot{\mathbf{q}}_0, \qquad (7.4.59b)$$

where $\mathbf{M}$, $\mathbf{C}$, and $\mathbf{K}$ are assumed to be constant matrices.

Recall the structural dynamics free vibration problem

$$\mathbf{K}\mathbf{u} = \omega^2 \mathbf{M}\mathbf{u} \qquad (7.4.60)$$

and introduce the normal mode transformation

$$\mathbf{q}(t) = \mathbf{\Phi}\mathbf{x}(t), \qquad (7.4.61)$$

where the columns of $\boldsymbol{\Phi}$ are the orthonormalized eigenvectors of the undamped free vibration problem and it is understood that the normalization is with respect to the mass matrix. Furthermore, assume that the damping matrix in Eq. (7.4.59) is a proportional type of damping matrix, as discussed in Chapter 6. Combining Eqs. (7.4.59) and (7.4.61) yields

$$\ddot{\mathbf{x}} + \mathbf{C}_D \dot{\mathbf{x}} + \boldsymbol{\Lambda}\mathbf{x} = \boldsymbol{\Phi}^T \mathbf{f}(t), \tag{7.4.62}$$

where

$$\mathbf{C}_D = \boldsymbol{\Phi}^T \mathbf{C}\boldsymbol{\Phi}, \quad \boldsymbol{\Lambda} = \boldsymbol{\Phi}^T \mathbf{K}\boldsymbol{\Phi}. \tag{7.4.63}$$

The new damping matrix $\mathbf{C}_D$ is diagonal, whose elements are given by $C_{Di} = 2\zeta_i\omega_i$ and $\zeta_i$ is the damping ratio in the $i$th mode, while $\boldsymbol{\Lambda}$ is the diagonal matrix containing the free vibration frequencies of the undamped system, $\omega_1^2, \dots, \omega_n^2$. The DOFs in Eq. (7.4.62) are obviously uncoupled due to the orthogonality of normal modes. In solving Eq. (7.4.61) using numerical integrations, one can choose for each decoupled equation a time increment or step size, compatible with the natural period of the system for that particular normal mode. On the other hand, when Eq. (7.4.59) is used as the basis of direct numerical integration, a common step size $\Delta t$ is used for all the equations. Therefore, in dealing with the stability analysis of a numerical integration scheme used for the structural dynamics response problem represented by Eq. (7.4.59), one can instead deal with the decoupled system, Eq. (7.4.62), which is represented by an SDOF damped oscillator:

$$\ddot{x} + 2\zeta\omega\dot{x} + \omega^2 x = f(t) \tag{7.4.64a}$$

$$x(0) = x_0 = \boldsymbol{\Phi}^T \mathbf{q}_0, \quad \dot{x}(0) = \dot{x}_0 = \boldsymbol{\Phi}^T \dot{\mathbf{q}}_0. \tag{7.4.64b}$$

**Formulation of a Recursive Relation**   When a numerical method is applied to solve Eq. (7.4.64), a recursive equation can be established:

$$\boldsymbol{\chi}_{i+1} = \mathbf{A}\boldsymbol{\chi}_i + \mathbf{L}\mathbf{f}_{i+s}, \tag{7.4.65}$$

where $\boldsymbol{\chi}_{i+1}$ and $\boldsymbol{\chi}_i$ are vectors in which the solutions for times $t = (i+1)\Delta t$ and $t = i\Delta t$ are stored, while $\mathbf{f}_{i+s}$ is the load at time $t = (i+s)\Delta t$ where the value of $s$ depends on the type of integration being employed, and $0 \le s \le 1$. The matrixes $\mathbf{A}$ and $\mathbf{L}$ are denoted as integration approximation and load operators, respectively. The matrix $\mathbf{A}$ is also called an **amplification matrix**. Equation (7.4.65) can be expanded to establish additional recursive relations:

$$\begin{aligned}
\boldsymbol{\chi}_{i+1} &= \mathbf{A}\boldsymbol{\chi}_i + \mathbf{L}\mathbf{f}_{i+s} \\
&= \mathbf{A}\left(\mathbf{A}\boldsymbol{\chi}_{i-1} + \mathbf{L}\mathbf{f}_{i-1+s}\right) + \mathbf{L}\mathbf{f}_{i+s} \\
&= \mathbf{A}^2\boldsymbol{\chi}_{i-1} + \mathbf{A}\mathbf{L}\mathbf{f}_{i-1+s} + \mathbf{L}\mathbf{f}_{i+s}
\end{aligned} \tag{7.4.66}$$

and more generally

$$\boldsymbol{\chi}_n = \mathbf{A}^n\boldsymbol{\chi}_0 + \mathbf{A}^{n-1}\mathbf{L}\mathbf{f}_s + \mathbf{A}^{n-2}\mathbf{L}\mathbf{f}_{s+1} + \cdots + \mathbf{L}\mathbf{f}_{s+n-1}. \tag{7.4.67}$$

**Spectral Stability**   The stability of an integration method is determined by examining the behavior of the numerical solution for arbitrary initial conditions. Stability problems are usually a fundamental property of the homogeneous system and therefore one is usually interested in the integration of Eq. (7.4.64) in the absence of the loading term. For this case, $\mathbf{f}_i = 0$, and Eq. (7.4.67) reduces to

$$\chi_n = \mathbf{A}^n \chi_0. \tag{7.4.68}$$

As indicated previously, time integration methods are classified as unconditionally stable or as merely conditionally stable. A method is unconditionally stable if the solution for any set of initial conditions does not grow without bound for any time step size $\Delta t$, in particular when the ratio of $\Delta t$ and the period of oscillation $T$ is large. The method is only conditionally stable if the above statement is true only provided that the ratio $\Delta t / T$ is smaller than a certain value, and this value is denoted as the critical step size ratio or stability limit.

Therefore, the stability analysis of Eq. (7.4.68) is reduced to determination of the conditions when the matrix power $\mathbf{A}^n$ is bounded. The detailed discussion of the matrix power and its boundedness is provided in Appendix A.4.5. The boundedness of $\mathbf{A}^n$ is characterized using the **spectral radius** of $\mathbf{A}$, defined as the largest magnitude of the eigenvalues of $\mathbf{A}$:

$$\rho(\mathbf{A}) = \max_i |\lambda_i|, \quad i = 1, 2, \dots. \tag{7.4.69}$$

The matrix power $\mathbf{A}^n$ is bounded for $n \to \infty$ if and only if

$$\begin{cases} \rho(\mathbf{A}) < 1, & \mathbf{A} \text{ defective} \\ \rho(\mathbf{A}) \le 1, & \text{otherwise} \end{cases} \tag{7.4.70}$$

where a defective matrix implies a matrix that cannot be diagonalized, see Appendix A.4.4 for more details. Equation (7.4.70) serves as the **stability criterion** for the numerical method. Furthermore, if $\rho(\mathbf{A}) < 1$, $\mathbf{A}^n$ converges to zero when $n \to \infty$, and a smaller spectral radius results in an even faster convergence.

A numerical method is **spectrally stable** if the amplification matrix $\mathbf{A}$ satisfies the stability criterion Eq. (7.4.70). A numerical method is **conditionally (spectrally) stable** if the spectral stability is satisfied for $\Delta t \le \Delta t_{\mathrm{cr}}$; otherwise, the method is **unconditionally (spectrally) stable**. Note that these definitions associated with spectral stability are analogous to those related to absolute stability.

**Oscillatory Characteristics**   The time accuracy and spectral stability are two simple indicators used to characterize the time integration methods. For structural dynamics applications, a more quantitative understanding of the performance of these methods is obtained from two additional indicators, amplitude error and periodicity error, which were first observed and introduced in the numerical examples in section "Numerical Examples for Euler Methods," pp. 384–389. These two indicators characterize how well the numerical method captures the oscillatory response of a structural dynamic system.

The oscillatory characteristics of the numerical solution are characterized by the eigenvalues of the amplification matrix:

$$\lambda = \exp\left(-\bar{\zeta}\bar{\omega}\Delta t\right)\exp\left(i\bar{\omega}_d\Delta t\right),$$   (7.4.71)

where $\bar{\omega}$, $\bar{\zeta}$, and $\bar{\omega}_d$ are the numerical counterparts of the physical parameters $\omega$, $\zeta$, and $\omega_d$ characterizing the SDOF damped oscillator, and

$$\bar{\omega}_d = \bar{\omega}\sqrt{1-\bar{\zeta}^2}.$$   (7.4.72)

From an eigenvalue $\lambda$, the numerical natural frequency and damping ratio are computed as

$$\bar{\omega} = \frac{1}{\Delta t}\sqrt{a^2+b^2}$$   (7.4.73a)

$$\bar{\zeta} = -\frac{a}{\bar{\omega}\Delta t},$$   (7.4.73b)

where

$$a = \mathrm{Log}|\lambda|, \quad b = \mathrm{Arg}(\lambda).$$   (7.4.74)

From $\bar{\omega}$, the oscillation period of a numerical solution is computed as

$$\bar{T} = \frac{2\pi}{\bar{\omega}}.$$   (7.4.75)

The difference between the numerical and exact damping ratios,

$$\Delta\zeta = \bar{\zeta} - \zeta,$$   (7.4.76)

characterizes the numerical dissipation introduced by the numerical method. A positive value of $\Delta\zeta$ indicates that the oscillation in the numerical solution decays faster than the actual oscillation and vice versa. A high numerical dissipation causes a large amplitude error. The effects of numerical dissipation have been observed in the numerical examples in section "Numerical Examples for Euler Methods," pp. 384–389. The numerical dissipation is closely related to the spectral radius and the spectral stability, and the relation is summarized for the case of an undamped oscillator, $\zeta = 0$, in Table 7.2.

The relative difference between the numerical and exact oscillation periods,

$$\frac{\Delta T}{T} = \frac{\bar{T}}{T} - 1 = \frac{\omega}{\bar{\omega}} - 1,$$   (7.4.77)

**Table 7.2 Numerical dissipation versus spectral stability ($\zeta = 0$)**

| $\Delta\zeta$ | $\rho(\mathbf{A})$ | Stability |
|---|---|---|
| $> 0$ | $< 1$ | Stable |
| $= 0$ | $= 1$ | Neutrally stable |
| $< 0$ | $> 1$ | Unstable |

characterizes the periodicity error introduced by the numerical method. A positive value of $\frac{\Delta T}{T}$ indicates that the oscillation period in the numerical solution is elongated when compared to the actual oscillation and vice versa. Typically the periodicity error is a function of the frequency, which causes a **frequency dispersion** in the numerical solution.

### Stability Analysis of the Central Difference Method

Consider next the central difference method that has been discussed earlier. Recall that for this method the equilibrium at time $t = t_i$ is satisfied:

$$\ddot{x}_i + 2\zeta\omega\dot{x}_i + \omega^2 x_i = f_i \tag{7.4.78}$$

and furthermore

$$\ddot{x}_i = \frac{x_{i-1} - 2x_i + x_{i+1}}{\Delta t^2} \tag{7.4.79a}$$

$$\dot{x}_i = \frac{-x_{i-1} + x_{i+1}}{2\Delta t}. \tag{7.4.79b}$$

Similar to the procedure in Eqs. (7.4.51) and (7.4.52), substituting Eqs. (7.4.79a) and (7.4.79b) into Eq. (7.4.78) yields an expression for $x_{i+1}$:

$$\left(\frac{1}{\Delta t^2} + \frac{\zeta\omega}{\Delta t}\right) x_{i+1} = f_i - \left(\omega^2 - \frac{2}{\Delta t^2}\right) x_i - \left(\frac{1}{\Delta t^2} - \frac{\zeta\omega}{\Delta t}\right) x_{i-1}$$

or

$$x_{i+1} = \frac{2 - \omega^2\Delta t^2}{1 + \zeta\omega\Delta t} x_i - \frac{1 - \zeta\omega\Delta t}{1 + \zeta\omega\Delta t} x_{i-1} + \frac{\Delta t^2}{1 + \zeta\omega\Delta t} f_i. \tag{7.4.80}$$

Equation (7.4.80) can be written in the recursive form of Eq. (7.4.65) by defining

$$\boldsymbol{\chi}_i = [x_i, \ x_{i-1}]^T, \quad \mathbf{f}_{i+s} = [f_i] \tag{7.4.81}$$

and so that

$$\boldsymbol{\chi}_{i+1} = \mathbf{A}\boldsymbol{\chi}_i + \mathbf{L}\mathbf{f}_{i+s}$$

$$\begin{bmatrix} x_{i+1} \\ x_i \end{bmatrix} = \mathbf{A} \begin{bmatrix} x_i \\ x_{i-1} \end{bmatrix} + \mathbf{L}[f_i]$$

$$= \begin{bmatrix} \dfrac{2 - \omega^2\Delta t^2}{1 + \zeta\omega\Delta t} & -\dfrac{1 - \zeta\omega\Delta t}{1 + \zeta\omega\Delta t} \\ 1 & 0 \end{bmatrix} \begin{bmatrix} x_i \\ x_{i-1} \end{bmatrix} + \begin{bmatrix} \dfrac{\Delta t^2}{1 + \zeta\omega\Delta t} \\ 0 \end{bmatrix} f_i. \tag{7.4.82}$$

Now analyze the central difference method to determine the critical step size for stability. For simplicity, consider the undamped case $\zeta = 0$.

First, calculate the spectral radius of the amplification matrix assuming the simplification of $\zeta = 0$:

$$\mathbf{A} = \begin{bmatrix} 2 - \omega^2\Delta t^2 & -1 \\ 1 & 0 \end{bmatrix}. \tag{7.4.83}$$

To find the eigenvalues,

$$\begin{vmatrix} 2 - \omega^2 \Delta t^2 - \lambda & -1 \\ 1 & -\lambda \end{vmatrix} = 0 \tag{7.4.84}$$

and the roots are found to be

$$\lambda_{1,2} = \alpha \pm \sqrt{\alpha^2 - 1}, \tag{7.4.85}$$

where $\alpha = \dfrac{2 - \omega^2 \Delta t^2}{2}$.

Next, apply the stability criteria, Eq. (7.4.70):

$$\left| \lambda_{1,2} \right| = \left| \alpha \pm \sqrt{\alpha^2 - 1} \right| \leq 1. \tag{7.4.86}$$

When $|\alpha| > 1$, both eigenvalues are real and positive, and Eq. (7.4.86) simplifies as

$$\alpha \pm \sqrt{\alpha^2 - 1} \leq 1$$
$$\alpha^2 - 1 \leq (1 - \alpha)^2$$
$$\alpha \leq 1, \tag{7.4.87}$$

which implies that the stability criteria are not satisfied.

When $|\alpha| < 1$, the eigenvalues are written in the form of a complex conjugate pair:

$$\lambda_{1,2} = \alpha \pm i\sqrt{1 - \alpha^2} \tag{7.4.88}$$

and

$$\left| \lambda_{1,2} \right| = \sqrt{\alpha^2 + \left( \sqrt{1 - \alpha^2} \right)^2} = 1, \tag{7.4.89}$$

which means the stability criteria is satisfied.

Combining Eqs. (7.4.87) and (7.4.89), the central difference method is stable when $|\alpha| \leq 1$, or

$$-1 \leq \frac{2 - \omega^2 \Delta t^2}{2} \leq 1 \quad \text{or} \quad \omega \Delta t \leq 2. \tag{7.4.90}$$

Therefore,

$$\Delta t \leq \frac{2}{\omega_{\max}}. \tag{7.4.91}$$

Let $\omega_{\max} = \frac{2\pi}{T_{\min}}$, where $T_{\min}$ is the smallest period of natural vibration in the system; Eq. (7.4.91) is rewritten as

$$\Delta t \leq \frac{T_{\min}}{\pi} \approx 0.318 T_{\min}. \tag{7.4.92}$$

Therefore, for the central difference method to produce stable solutions, the time step size needs to be not larger than 31.8% of the minimum oscillation period of interest.

Finally, the amplitude and period accuracy are characterized using the eigenvalues found in Eq. (7.4.88). Using Eq. (7.4.73), the numerical damping ratio and frequency are found to be

$$\bar{\zeta} = 0, \quad \bar{\omega} = \frac{1}{\Delta t} \arctan \frac{\sqrt{1 - \alpha^2}}{\alpha}. \tag{7.4.93}$$

It is clear that the central difference method does not introduce any numerical dissipation in the pure oscillation case. However, it introduces a periodicity error:

$$\frac{\Delta T}{T} \approx -\frac{\omega^2 \Delta t^2}{24} + O(\omega^3 \Delta t^3). \tag{7.4.94}$$

This implies that the oscillation period of the numerical solution obtained by the central difference method is smaller than the actual period with a difference of second-order accuracy. The difference is proportional to the square of the frequency, which introduces a larger error in the higher-frequency components of the numerical solution.

While Eq. (7.4.92) provides an upper bound of the time step size for the central difference method, usually a smaller $\Delta t$ is required to resolve the oscillation of interest with sufficient accuracy. The expression for periodicity error, Eq. (7.4.94), provides an estimate allowing the user to select the time step size. Suppose a maximum periodicity error of $p$ is required for the minimum oscillation period $T_{\min}$:

$$\left| \frac{\Delta T}{T} \right| = \frac{\omega^2 \Delta t^2}{24} \leq p \quad \text{or} \quad \Delta t \leq \frac{\sqrt{6p}}{\pi} T_{\min}. \tag{7.4.95}$$

For example, when $p = 5\%$, $\Delta t \leq 0.174 T_{\min}$, which is a tighter bound for $\Delta t$ than the stability requirement, Eq. (7.4.92).

## Adaptive Time Stepping and Velocity-Based Formulation

In many nonlinear structural dynamic problems, the frequency characteristics often depend on the structural response. For example, for a structure undergoing large deformation with finite strain, the structural frequencies can increase due to the geometric stiffening effect, and thus a smaller time step size is required to resolve the oscillations. Another example is when the structural load undergoes discontinuity, which can cause transient high-frequency components in the structural response; and smaller time step sizes should be used after the introduction of the load discontinuity to maintain the accuracy of the solution. Therefore, it makes sense to use adaptive time stepping for nonlinear structural dynamic problems, that is, to employ different time step sizes over different stages of the numerical solution.

In adaptive time stepping, the solutions are found for a series of **non-uniformly** distributed time stations $\mathbf{q}_i = \mathbf{q}(t_i)$, $i = 1, 2, \ldots, n$, with the time step size $\Delta t_i = t_i - t_{i-1}$. With the non-uniform time step sizes, it is computationally more efficient to implement the central difference method in a velocity-based formulation, where the displacements, velocities, and accelerations are computed at regular time stations $t_1, t_2, \ldots, t_n$, but an extra set of velocities are computed at the middle points $t_{1/2}, t_{3/2}, \ldots, t_{n-1/2}$:

$$t_{i+1/2} = t_i + \frac{1}{2}\Delta t_{i+1}. \tag{7.4.96}$$

The velocity-based formulation relies on the second-order central difference scheme for nonuniform step sizes:

$$\dot{\mathbf{q}}_{i+1/2} = \frac{\mathbf{q}_{i+1} - \mathbf{q}_i}{\Delta t_{i+1}} \tag{7.4.97a}$$

$$\dot{\mathbf{q}}_{i-1/2} = \frac{\mathbf{q}_i - \mathbf{q}_{i-1}}{\Delta t_i} \tag{7.4.97b}$$

$$\ddot{\mathbf{q}}_i = \frac{\dot{\mathbf{q}}_{i+1/2} - \dot{\mathbf{q}}_{i-1/2}}{\Delta t_{i+1/2}}, \tag{7.4.97c}$$

where $\Delta t_{i+1/2} = (\Delta t_{i+1} + \Delta t_i)/2$.

Consider a general nonlinear structural dynamic problem:

$$\mathbf{M}\ddot{\mathbf{q}} + \mathbf{f}_I(\mathbf{q}, \dot{\mathbf{q}}, t) = \mathbf{f}_E(\mathbf{q}, \dot{\mathbf{q}}, t) \tag{7.4.98a}$$

$$\mathbf{q}(0) = \mathbf{q}_0, \quad \dot{\mathbf{q}}(0) = \dot{\mathbf{q}}_0, \tag{7.4.98b}$$

where $\mathbf{f}_I$ and $\mathbf{f}_E$ are internal and external forcing vectors, respectively. To initiate the velocity-based central difference method at time $t = t_0 = 0$, one first needs to compute the acceleration $\ddot{\mathbf{q}}_0$ at $t = 0$ and the middle-point velocity $\dot{\mathbf{q}}_{-1/2}$ at $t = -\Delta t_0/2$. The acceleration is computed using Eq. (7.4.98a) and the initial conditions:

$$\ddot{\mathbf{q}}_0 = \mathbf{M}^{-1}\left(\mathbf{f}_E(\mathbf{q}_0, \dot{\mathbf{q}}_0, t_0) - \mathbf{f}_I(\mathbf{q}_0, \dot{\mathbf{q}}_0, t_0)\right). \tag{7.4.99}$$

The middle-point velocity is computed as

$$\dot{\mathbf{q}}_{-1/2} = \dot{\mathbf{q}}_0 - \frac{\Delta t_0}{2}\ddot{\mathbf{q}}_0. \tag{7.4.100}$$

Next, with the solutions at time $t = t_i$ known, the central difference method determines the middle-point velocity $\dot{\mathbf{q}}_{i+1/2}$ and the displacement $\mathbf{q}_{i+1}$. The new middle-point velocity is computed using Eq. (7.4.97c):

$$\dot{\mathbf{q}}_{i+1/2} = \dot{\mathbf{q}}_{i-1/2} + \Delta t_{i+1/2}\ddot{\mathbf{q}}_i. \tag{7.4.101}$$

The new displacement is computed using Eq. (7.4.97a):

$$\mathbf{q}_{i+1} = \mathbf{q}_i + \Delta t_{i+1}\dot{\mathbf{q}}_{i+1/2}. \tag{7.4.102}$$

Subsequently, the velocity $\dot{\mathbf{q}}_{i+1}$ and acceleration $\ddot{\mathbf{q}}_{i+1}$ are computed. If the external and internal force vectors are independent of the velocity, the acceleration and velocity can be evaluated explicitly in a sequential manner, without having to solve an equation:

$$\ddot{\mathbf{q}}_{i+1} = \mathbf{M}^{-1}\left[\mathbf{f}_E(\mathbf{q}_{i+1}, t_{i+1}) - \mathbf{f}_I(\mathbf{q}_{i+1}, t_{i+1})\right] \tag{7.4.103a}$$

$$\dot{\mathbf{q}}_{i+1} = \dot{\mathbf{q}}_{i+1/2} + \frac{\Delta t_{i+1}}{2}\ddot{\mathbf{q}}_{i+1}. \tag{7.4.103b}$$

When $\mathbf{f}_I$ and $\mathbf{f}_E$ are functions of velocity $\dot{\mathbf{q}}$, such as cases when the structural damping is present, Eqs. (7.4.103a) and (7.4.103b) become a coupled system of nonlinear equations with unknowns $\ddot{\mathbf{q}}_{i+1}$ and $\dot{\mathbf{q}}_{i+1}$. The coupled equations can be solved using a fixed-point iteration approach. Starting from $j = 0$ and an initial guess $\dot{\mathbf{q}}_{i+1}^{(0)} = \dot{\mathbf{q}}_{i+1/2}$, the iterative solution is given by

$$\ddot{\mathbf{q}}_{i+1}^{(j)} = \mathbf{M}^{-1}\left[\mathbf{f}_E(\mathbf{q}_{i+1}, \dot{\mathbf{q}}_{i+1}^{(j)}, t_{i+1}) - \mathbf{f}_I(\mathbf{q}_{i+1}, \dot{\mathbf{q}}_{i+1}^{(j)}, t_{i+1})\right] \tag{7.4.104a}$$

$$\dot{\mathbf{q}}_{i+1}^{(j+1)} = \dot{\mathbf{q}}_{i+1/2} + \frac{\Delta t_{i+1}}{2}\ddot{\mathbf{q}}_{i+1}^{(j)}. \tag{7.4.104b}$$

Equations (7.4.104a) and (7.4.104b) can be evaluated to generate $\left(\dot{\mathbf{q}}_{i+1}^{(1)}, \ddot{\mathbf{q}}_{i+1}^{(1)}\right)$, $\left(\dot{\mathbf{q}}_{i+1}^{(2)}, \ddot{\mathbf{q}}_{i+1}^{(2)}\right)$, ..., until the difference between two subsequent iterates is sufficiently small. Typically only a few iterations are required to achieve convergence when the damping in the system is small.

The velocity-based formulation of the central difference method for Eq. (7.4.98) is summarized and implemented in Algorithm 7.1. The algorithm is presented in the form of pseudocode. The notation used in the pseudocode is explained in Appendix D. In addition, a flow chart outlining the central difference method is shown in Figure 7.10.

---

**Algorithm 7.1** Velocity-based formulation of the central difference method

---

**Require:** Lumped mass matrix $\mathbf{M}$, internal and external forcing vectors $\mathbf{f}_I(\mathbf{q}, \dot{\mathbf{q}}, t)$ and $\mathbf{f}_E(\mathbf{q}, \dot{\mathbf{q}}, t)$, initial conditions $\mathbf{q}_0$ and $\dot{\mathbf{q}}_0$, initial time step size $\Delta t_0$, total time period $T$, maximum number of fixed-point iterations $M$, and tolerance $\epsilon$.

1: Initialize $i = 0$ and $t_0 = 0$.
2: Compute the initial acceleration $\ddot{\mathbf{q}}_0$ at $t = 0$ using Eq. (7.4.99).
3: Compute the velocity $\dot{\mathbf{q}}_{-1/2}$ at $t = -\Delta t_0/2$ using Eq. (7.4.100).
4: **while** $t_i \leq T$ **do**
5:     Compute the time step sizes $\Delta t_{i+1}$ and $\Delta t_{i+1/2} = \frac{1}{2}\left(\Delta t_i + \Delta t_{i+1}\right)$
6:     Update the time stations:

$$t_{i+1/2} = t_i + \Delta t_{i+1}/2, \quad t_{i+1} = t_i + \Delta t_{i+1}$$

7:     Update velocity $\dot{\mathbf{q}}_{i+1/2}$ at $t = t_{i+1/2}$ using Eq. (7.4.101).
8:     Update displacement $\mathbf{q}_{i+1}$ at $t = t_{i+1}$ using Eq. (7.4.102).
9:     Set initial guess for the velocity at $t = t_{i+1}$, $\dot{\mathbf{q}}_{i+1}^{(0)} = \dot{\mathbf{q}}_{i+1/2}$
10:     **for** $j = 0, \ldots, M - 1$ **do**    ▷ Perform at most $M$ fixed-point iterations
11:         Update acceleration $\ddot{\mathbf{q}}_{i+1}^{(j)}$ using Eq. (7.4.104a)
12:         Update velocity $\dot{\mathbf{q}}_{i+1}^{(j+1)}$ using Eq. (7.4.104b)
13:         **if** $||\dot{\mathbf{q}}_{i+1}^{(j+1)} - \dot{\mathbf{q}}_{i+1}^{(j)}|| < \epsilon$ or $j > M$ **then**    ▷ Check convergence
14:             Break
15:         **end if**
16:     **end for**
17:     Let $\dot{\mathbf{q}}_{i+1}^{(j+1)}$ be the velocity $\dot{\mathbf{q}}_{i+1}$ at $t = t_{i+1}$.
18:     Compute acceleration $\ddot{\mathbf{q}}_{i+1}$ using $\dot{\mathbf{q}}_{i+1}$ and Eq. (7.4.104a)
19:     Update index $i = i + 1$
20: **end while**
21: **return** The solutions $\mathbf{q}, \dot{\mathbf{q}}, \ddot{\mathbf{q}}$ at the computed time stations $t_0, t_1, \ldots, T$.

---

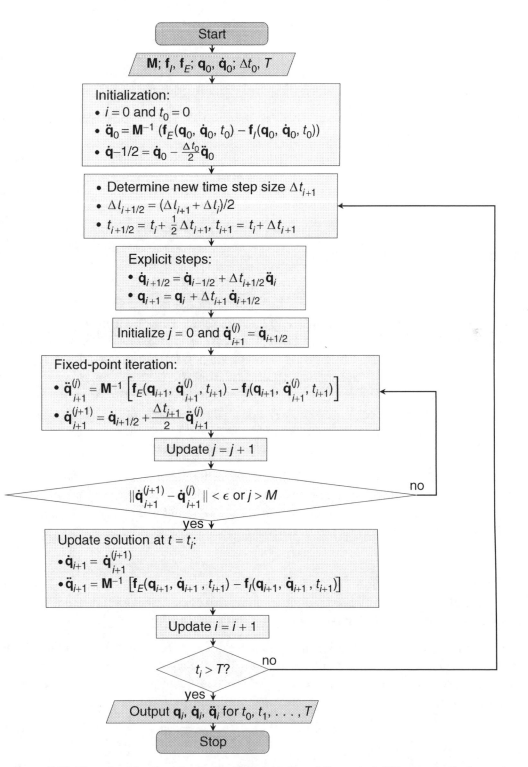

**Figure 7.10** Flow chart for the velocity-based formulation of the central difference method

## Numerical Example for the Central Difference Method

In this section, the solution procedure of the central difference method is illustrated using a numerical example, and the convergence and stability characteristics of this method are also examined.

**Problem Statement** Consider the forced response of a 2-DOF undamped oscillator, in a nondimensional form:

$$\mathbf{M\ddot{q}} + \mathbf{Kq} = \mathbf{f}_0, \quad \mathbf{q}(0) = [0, 0]^T, \quad \mathbf{\dot{q}}(0) = [0, 0]^T, \quad \text{(E7.116)}$$

where $\mathbf{f}_0 = [0.0160, 1.0000]^T$ and the mass and stiffness matrices are intentionally diagonal:

$$\mathbf{M} = \begin{bmatrix} 1 & 0 \\ 0 & 4 \end{bmatrix}, \quad \mathbf{K} = \begin{bmatrix} 16 & 0 \\ 0 & 1 \end{bmatrix}. \quad \text{(E7.117)}$$

The exact solution to Eq. (E7.116) is

$$\mathbf{q}(t) = \begin{bmatrix} 10^{-3} [1 - \cos(4t)] \\ 1 - \cos(t/2) \end{bmatrix}. \quad \text{(E7.118)}$$

Due to the diagonal form of $\mathbf{M}$ and $\mathbf{K}$, the response of higher frequency $\omega_1 = 4$ only appears in the first DOF, while the response of lower frequency $\omega_2 = 1/2$ only appears in the second DOF. Furthermore, the external force for the first DOF is much lower than the force for the second DOF. As a result, the 2-DOF oscillator example emulates the scenario that is frequently encountered in structural dynamic applications, where the structural response is dominated by the low-frequency component while the amplitude of the high-frequency component is negligible. In practice, one would be interested in obtaining and analyzing *only* the low-frequency response.

**Choice of Time Step Size** The central difference method is applied to generate the response of the 2-DOF oscillator. To start with, one needs to determine the feasible size of the time step. The maximum frequency of the system is $\omega_{\max} = 4$, and using the stability criterion Eq. (7.4.91):

$$\Delta t \le \frac{2}{\omega_{\max}} = 0.5. \quad \text{(E7.119)}$$

For a sufficient resolution of both frequency components, a time step size $\Delta t = 0.1$ is chosen. Since this problem is linear and the natural frequencies are constant, the adaptive time stepping is not employed, that is, the time step size is fixed.

**Starting Procedure** Next, the dynamical response of the 2-DOF oscillator is obtained following the velocity-based formulation in Algorithm 7.1. The algorithm is presented for a general nonlinear case. In this problem, however, the internal and external force vectors are linear and defined, respectively, as

$$\mathbf{f}_E(\mathbf{q}, \mathbf{\dot{q}}, t) = \mathbf{f}_0, \quad \mathbf{f}_I(\mathbf{q}, \mathbf{\dot{q}}, t) = \mathbf{Kq}. \quad \text{(E7.120)}$$

To initiate the loop for time stepping, one first needs to prepare two quantities given the initial displacement and velocity. The first quantity is the initial acceleration. Using Eq. (7.4.99),

$$\ddot{\mathbf{q}}_0 = \mathbf{M}^{-1}\left(\mathbf{f}_{E,0} - \mathbf{f}_{I,0}\right) = \mathbf{M}^{-1}(\mathbf{f}_0 - \mathbf{K}\mathbf{q}_0)$$
$$= \begin{bmatrix}1.6000\text{e--}02 & 2.5000\text{e--}01\end{bmatrix}^T. \tag{E7.121}$$

The second quantity is the velocity at $t = -\Delta t/2 = -0.05$. Using Eq. (7.4.100),

$$\dot{\mathbf{q}}_{-1/2} = \dot{\mathbf{q}}_0 - \frac{\Delta t}{2}\ddot{\mathbf{q}}_0 = \begin{bmatrix}-8.0000\text{e--}04 & -1.2500\text{e--}02\end{bmatrix}^T. \tag{E7.122}$$

**Time Stepping**   With $\ddot{\mathbf{q}}_0$ and $\dot{\mathbf{q}}_{-1/2}$ available, one can proceed to the computation of the numerical solution at the first time step, $t = t_1 = \Delta t = 0.1$. First, update velocity at $t = t_{1/2} = \Delta t/2 = 0.05$ using Eqs. (7.4.101), (E7.121), and (E7.122):

$$\dot{\mathbf{q}}_{1/2} = \dot{\mathbf{q}}_{-1/2} + \Delta t\ddot{\mathbf{q}}_0 = \begin{bmatrix}8.0000\text{e--}04 & 1.2500\text{e--}02\end{bmatrix}^T. \tag{E7.123}$$

Next, update the displacement at $t = t_1$ using Eqs. (7.4.102) and (E7.123):

$$\mathbf{q}_1 = \mathbf{q}_0 + \Delta t\dot{\mathbf{q}}_{1/2} = \begin{bmatrix}8.0000\text{e--}05 & 1.2500\text{e--}03\end{bmatrix}^T. \tag{E7.124}$$

Subsequently, update the acceleration at $t = t_1$ using Eqs. (7.4.103a) and (E7.124):

$$\ddot{\mathbf{q}}_1 = \mathbf{M}^{-1}(\mathbf{f}_0 - \mathbf{K}\mathbf{q}_1) = \begin{bmatrix}1.4720\text{e--}02 & 2.4969\text{e--}01\end{bmatrix}^T. \tag{E7.125}$$

Finally, update the velocity at $t = t_1$ using Eqs. (7.4.103b), (E7.123), and (E7.125):

$$\dot{\mathbf{q}}_1 = \dot{\mathbf{q}}_{1/2} + \frac{\Delta t}{2}\ddot{\mathbf{q}}_1 = \begin{bmatrix}1.5360\text{e--}03 & 2.4984\text{e--}02\end{bmatrix}^T. \tag{E7.126}$$

Repeating the procedure in Eqs. (E7.123) through (E7.126), one can compute the numerical solution at the second time step, $t = t_2 = 2\Delta t = 0.2$:

$$\dot{\mathbf{q}}_{3/2} = \dot{\mathbf{q}}_{1/2} + \Delta t\ddot{\mathbf{q}}_1 = \begin{bmatrix}2.2720\text{e--}03 & 3.7469\text{e--}02\end{bmatrix}^T \tag{E7.127a}$$

$$\mathbf{q}_2 = \mathbf{q}_1 + \Delta t\dot{\mathbf{q}}_{3/2} = \begin{bmatrix}3.0720\text{e--}04 & 4.9969\text{e--}03\end{bmatrix}^T \tag{E7.127b}$$

$$\ddot{\mathbf{q}}_2 = \mathbf{M}^{-1}(\mathbf{f}_0 - \mathbf{K}\mathbf{q}_2) = \begin{bmatrix}1.1085\text{e--}02 & 2.4875\text{e--}01\end{bmatrix}^T \tag{E7.127c}$$

$$\dot{\mathbf{q}}_2 = \dot{\mathbf{q}}_{3/2} + \frac{\Delta t}{2}\ddot{\mathbf{q}}_2 = \begin{bmatrix}2.8262\text{e--}03 & 4.9906\text{e--}02\end{bmatrix}^T. \tag{E7.127d}$$

The subsequent time steps are computed in a similar manner. The first five steps of displacements in the numerical solution are compared with the exact solution in Table 7.3. The central difference method achieves an average error as low as 0.07%.

**Effect of Time Step Size**   The central difference solution of $\Delta t = 0.1$ is compared with the exact solution in Figure 7.11. The central difference method produces no amplitude error for both low- and high-frequency responses, as expected based on the analysis of numerical dissipation in Eq. (7.4.93). The central difference method captures the period of the low-frequency

**Table 7.3 Comparison of the exact and central difference solutions**

| Step | Exact solution | Central difference | Error (%) |
|---|---|---|---|
| 1 | [7.8939e–05,1.2497e–03] | [8.0000e–05,1.2500e–03] | 0.0872 |
| 2 | [3.0329e–04,4.9958e–03] | [3.0720e–04,4.9969e–03] | 0.0808 |
| 3 | [6.3764e–04,1.1229e–02] | [6.4525e–04,1.1231e–02] | 0.0707 |
| 4 | [1.0292e–03,1.9933e–02] | [1.0401e–03,1.9938e–02] | 0.0582 |
| 5 | [1.4161e–03,3.1088e–02] | [1.4285e–03,3.1094e–02] | 0.0446 |

response well; however, there is evidence of a periodicity error in the high-frequency response. Indeed, according to Eq. (7.4.94), for a fixed time step size, the periodicity error of the central difference method is proportional to the square of frequency.

Two additional central difference solutions are also illustrated in Figure 7.11. One solution is generated using the critical time step size $\Delta t = 0.5$, as obtained from Eq. (E7.119), and the other solution employs a time step size $\Delta t = 0.5001$ that is slightly higher than the critical value. For both cases, the low-frequency response is resolved well, without noticeable amplitude and periodicity error. However, the high-frequency responses of both solutions are completely incorrect. In the solution with $\Delta t = 0.5$, the high-frequency response becomes a series of triangular waves with a fixed amplitude. As for the solution with $\Delta t = 0.5001$, the high-frequency response becomes unstable.

**Figure 7.11** Numerical solution of a 2-DOF oscillator using central difference method

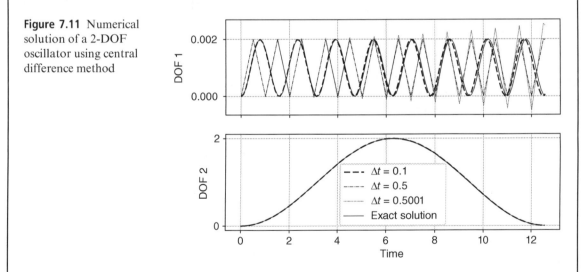

The comparison of the solutions generated with $\Delta t = 0.1$, $\Delta t = 0.5$, and $\Delta t = 0.5001$ underlines the fact that, in the central difference method, the time step size is limited by the maximum frequency in the structural response. The limitation of time step size renders the central difference method inappropriate for structural dynamic applications where only the lower-frequency component is of interest, because the time step size required to resolve the high-frequency response is excessively small for the low-frequency component and can incur unnecessary computational cost.

**Summary** This example demonstrates the solution procedure of the central difference method. It has a remarkable feature of zero amplitude error but does introduce periodicity error in the numerical solution when the time step size is excessive for the resolution of the oscillation period. A major limitation of the central difference method is that the time step size is constrained by the smallest period of oscillation in the structural response. Therefore, the central difference method is more appropriate for structural dynamic response problems where the high-frequency component is of major interest. Such an example is the structural analysis involving shocks, impacts or explosions, where high-frequency components in having a wave propagation characteristic are expected to be dominant.

### 7.4.4 Implicit Methods for Second-Order Equations

The implicit time integration methods overcome the limitation in the time step size encountered in the explicit methods, such as the central difference method. Therefore, they enable the simulation of structural dynamic response in a computationally efficient manner. Historically, various methods were developed and implemented; however, these are not in use anymore. For example, the Houbolt method (Houbolt, 1950) is one of the pioneering methods of time integration, which employs a multistep formulation to approximate the acceleration and velocity. However, due to the lack of a convenient starting procedure, the Houbolt method does not enjoy a widespread use anymore. Another example is a one-step method called the Wilson-$\theta$ method (Wilson and Clough, 1962), which has unfavorable high periodicity and amplitude errors for large time step sizes and thus is no longer adopted in dynamic analysis.

This section introduces two families of implicit time integration methods for the second-order structural dynamics equations. The first implicit algorithm is the Newmark-$\beta$ method (Newmark, 1959). It has become one of the most popular families of algorithms for the numerical structural dynamic solutions since its introduction in the late 1950s. The second one is the generalized-$\alpha$ method (Chung and Hulbert, 1993), a more recent extension of the Newmark-$\beta$ method. It has improved various aspects of the Newmark-$\beta$ method and is implemented in many of the implicit finite element codes employed for structural dynamic applications.

#### The Newmark-$\beta$ Method

The Newmark-$\beta$ method is based on the simple idea of integrating the acceleration $\ddot{\mathbf{q}}$ twice over a time interval to find the new displacement. First, starting from $t = t_i$, the velocity as a function of time is represented by the integration of acceleration:

$$\dot{\mathbf{q}}(t) = \dot{\mathbf{q}}_i + \int_{t_i}^{t} \ddot{\mathbf{q}}(s)ds, \tag{7.4.105}$$

where the variation of the acceleration $\ddot{\mathbf{q}}(s)$ over time is still unknown.

Next, based on Eq. (7.4.105), the velocity and displacement at the new time step $t = t_{i+1}$ are found to be

$$\dot{\mathbf{q}}_{i+1} = \dot{\mathbf{q}}_i + \int_{t_i}^{t_{i+1}} \ddot{\mathbf{q}}(s)ds \tag{7.4.106a}$$

$$\mathbf{q}_{i+1} = \mathbf{q}_i + \int_{t_i}^{t_{i+1}} \dot{\mathbf{q}}(s)ds$$

$$= \mathbf{q}_i + \dot{\mathbf{q}}_i \Delta t + \int_{t_i}^{t_{i+1}} (t_{i+1} - s)\ddot{\mathbf{q}}(s)ds. \tag{7.4.106b}$$

The Newmark-$\beta$ method essentially approximates the integrals of acceleration in Eqs. (7.4.106) by a weighted sum of the accelerations at $t = t_i$ and $t = t_{i+1}$, as follows:

$$\dot{\mathbf{q}}_{i+1} = \dot{\mathbf{q}}_i + [(1 - \gamma)\ddot{\mathbf{q}}_i + \gamma\ddot{\mathbf{q}}_{i+1}]\,\Delta t \tag{7.4.107a}$$

$$\mathbf{q}_{i+1} = \mathbf{q}_i + \dot{\mathbf{q}}_i \Delta t + \left[\left(\frac{1}{2} - \beta\right)\ddot{\mathbf{q}}_i + \beta\ddot{\mathbf{q}}_{i+1}\right]\Delta t^2. \tag{7.4.107b}$$

In these equations, $\beta$ and $\gamma$ are weighting parameters that can be adjusted to modify the stability and accuracy characteristics of the algorithm.

**The Newmark-$\beta$ Family of Algorithms**    The Newmark-$\beta$ method by default usually refers to the case where $\gamma = 1/2$ and $\beta = 1/4$, which is based on the assumption that the acceleration is constant over the time interval $[t_i, t_{i+1}]$ and equals to the average of the accelerations $\ddot{\mathbf{q}}_i$ and $\ddot{\mathbf{q}}_{i+1}$. Therefore, this version is also known as the **average acceleration** method, or the **trapezoidal** method. Note the similarity with the trapezoidal method developed for first-order equations of motion in Section 7.4.2.

It is also possible to employ a more sophisticated approximation of the acceleration over $[t_i, t_{i+1}]$, for example, using a linear interpolation between $\ddot{\mathbf{q}}_i$ and $\ddot{\mathbf{q}}_{i+1}$:

$$\ddot{\mathbf{q}}(t) = \ddot{\mathbf{q}}_i + \frac{t - t_i}{\Delta t}(\ddot{\mathbf{q}}_{i+1} - \ddot{\mathbf{q}}_i). \tag{7.4.108}$$

Substituting Eq. (7.4.108) into Eq. (7.4.106) and performing the integration, the coefficients are found to be $\gamma = 1/2$ and $\beta = 1/6$. This version is called the **linear acceleration** method. While linearly interpolating acceleration appears to be more accurate than using a constant value of acceleration, it turns out that this method is less accurate and stable than the average acceleration method.

A variant of the average acceleration method is called **average acceleration with damping**, where the coefficients are chosen to be

$$\gamma = \frac{1}{2} + \alpha, \quad \beta = \frac{(1 + \alpha)^2}{4}. \tag{7.4.109}$$

The free parameter $\alpha$ is introduced to tune the numerical dissipation of the average acceleration method. Typically $0 < \alpha \le 1$.

The stability and accuracy characteristics of the above three versions of Newmark-$\beta$ methods are discussed in detail later in this section.

Finally, when $\gamma = 1/2$ and $\beta = 0$, the approximations of velocity and displacement, Eq. (7.4.107), are simplified as shown below:

$$\dot{\mathbf{q}}_{i+1} = \dot{\mathbf{q}}_i + (\ddot{\mathbf{q}}_i + \ddot{\mathbf{q}}_{i+1})\frac{\Delta t}{2} \tag{7.4.110a}$$

$$\mathbf{q}_{i+1} = \mathbf{q}_i + \dot{\mathbf{q}}_i\Delta t + \ddot{\mathbf{q}}_i\frac{\Delta t^2}{2}. \tag{7.4.110b}$$

This step reduces the Newmark-$\beta$ method to the central difference method. To show the equivalence, first note that from Eq. (7.4.110a), $\ddot{\mathbf{q}}_i$ can be expressed in terms of other variables:

$$\ddot{\mathbf{q}}_i = \frac{2}{\Delta t}(\dot{\mathbf{q}}_{i+1} - \dot{\mathbf{q}}_i) - \ddot{\mathbf{q}}_{i+1}. \tag{7.4.111}$$

Next, substitute Eq. (7.4.111) into Eq. (7.4.110b) to eliminate $\ddot{\mathbf{q}}_i$:

$$\mathbf{q}_{i+1} = \mathbf{q}_i + \dot{\mathbf{q}}_{i+1}\Delta t - \ddot{\mathbf{q}}_{i+1}\frac{\Delta t^2}{2}. \tag{7.4.112}$$

Furthermore, replace $i$ and $i+1$ in Eq. (7.4.112) with $i-1$ and $i$, respectively:

$$\mathbf{q}_i = \mathbf{q}_{i-1} + \dot{\mathbf{q}}_i\Delta t - \ddot{\mathbf{q}}_i\frac{\Delta t^2}{2}. \tag{7.4.113}$$

Subsequently, combine Eqs. (7.4.110b) and (7.4.113):

$$\mathbf{q}_{i+1} = \mathbf{q}_{i-1} + 2\dot{\mathbf{q}}_i\Delta t \tag{7.4.114}$$

or equivalently

$$\dot{\mathbf{q}}_i = \frac{\mathbf{q}_{i+1} - \mathbf{q}_{i-1}}{2\Delta t}. \tag{7.4.115}$$

Finally, substitute Eq. (7.4.115) into Eq. (7.4.113)

$$\mathbf{q}_i = \frac{1}{2}(\mathbf{q}_{i+1} + \mathbf{q}_{i-1}) - \ddot{\mathbf{q}}_i\frac{\Delta t^2}{2} \tag{7.4.116}$$

and rearrange

$$\ddot{\mathbf{q}}_i = \frac{\mathbf{q}_{i-1} - 2\mathbf{q}_i + \mathbf{q}_{i+1}}{\Delta t^2}. \tag{7.4.117}$$

Equations (7.4.115) and (7.4.117) are identical to the velocity relation, Eq. (7.4.49b), and acceleration relation, Eq. (7.4.49a), of the central difference method, respectively. Therefore, the Newmark-$\beta$ method with $\gamma = 1/2$ and $\beta = 0$ is equivalent to the central difference method.

**Computer Implementation of Newmark-$\beta$ Method**   To solve a structural dynamic problem, Newmark-$\beta$ method requires the initial conditions $\mathbf{q}_0$ and $\dot{\mathbf{q}}_0$ as well as the initial acceleration $\ddot{\mathbf{q}}_0$. The initial acceleration is typically not given in an IVP, and needs to be computed using the equations of motion at $t = 0$:

$$\mathbf{M}\ddot{\mathbf{q}}_0 + \mathbf{C}\dot{\mathbf{q}}_0 + \mathbf{K}\mathbf{q}_0 = \mathbf{f}_0 \tag{7.4.118}$$

or equivalently

$$\ddot{\mathbf{q}}_0 = \mathbf{M}^{-1} \left( \mathbf{f}_0 - \mathbf{C}\dot{\mathbf{q}}_0 - \mathbf{K}\mathbf{q}_0 \right). \tag{7.4.119}$$

Next, at the beginning of the $(i+1)$th time step, the quantities $\mathbf{q}_i$, $\dot{\mathbf{q}}_i$, and $\ddot{\mathbf{q}}_i$ are known from the previous time step. The equations of motion at the time $t = t_{i+1}$ are used:

$$\mathbf{M}\ddot{\mathbf{q}}_{i+1} + \mathbf{C}\dot{\mathbf{q}}_{i+1} + \mathbf{K}\mathbf{q}_{i+1} = \mathbf{f}_{i+1}. \tag{7.4.120}$$

It is possible to substitute Eq. (7.4.107) into Eq. (7.4.120), solve for $\ddot{\mathbf{q}}_{i+1}$, and subsequently find $\dot{\mathbf{q}}_{i+1}$ and $\mathbf{q}_{i+1}$ from Eq. (7.4.107). However, it is usually more convenient to treat the displacement $\mathbf{q}_{i+1}$ as the principal unknown variable, as described in the following text.

Using Eq. (7.4.107), the acceleration $\ddot{\mathbf{q}}_{i+1}$ and the velocity $\dot{\mathbf{q}}_{i+1}$ are expressed in terms of the displacement $\mathbf{q}_{i+1}$:

$$\ddot{\mathbf{q}}_{i+1} = \frac{2\beta - 1}{2\beta}\ddot{\mathbf{q}}_i - \frac{1}{\beta \Delta t}\dot{\mathbf{q}}_i - \frac{1}{\beta \Delta t^2}\mathbf{q}_i + \frac{1}{\beta \Delta t^2}\mathbf{q}_{i+1} \tag{7.4.121a}$$

$$\dot{\mathbf{q}}_{i+1} = \frac{(2\beta - \gamma)\Delta t}{2\beta}\ddot{\mathbf{q}}_i + \frac{\beta - \gamma}{\beta}\dot{\mathbf{q}}_i - \frac{\gamma}{\beta \Delta t}\mathbf{q}_i + \frac{\gamma}{\beta \Delta t}\mathbf{q}_{i+1}. \tag{7.4.121b}$$

Subsequently, substitute Eq. (7.4.121) into Eq. (7.4.120), and rearrange

$$\mathbf{P}_1 \mathbf{q}_{i+1} = \mathbf{f}_{i+1} + \mathbf{P}_2\ddot{\mathbf{q}}_i + \mathbf{P}_3\dot{\mathbf{q}}_i + \mathbf{P}_4\mathbf{q}_i, \tag{7.4.122}$$

where

$$\mathbf{P}_1 = \frac{1}{\beta \Delta t^2}\mathbf{M} + \frac{\gamma}{\beta \Delta t}\mathbf{C} + \mathbf{K} \tag{7.4.123a}$$

$$\mathbf{P}_2 = \frac{1 - 2\beta}{2\beta}\mathbf{M} + \frac{\gamma - 2\beta}{2\beta}\Delta t\mathbf{C} \tag{7.4.123b}$$

$$\mathbf{P}_3 = \frac{1}{\beta \Delta t}\mathbf{M} + \frac{\gamma - \beta}{\beta}\mathbf{C} \tag{7.4.123c}$$

$$\mathbf{P}_4 = \frac{1}{\beta \Delta t^2}\mathbf{M} + \frac{\gamma}{\beta \Delta t}\mathbf{C}. \tag{7.4.123d}$$

For constant system matrices $\mathbf{M}$, $\mathbf{C}$, $\mathbf{K}$, and a fixed integration step size $\Delta t$, the matrices $\mathbf{P}_i$, $i = 1, 2, 3, 4$, in Eq. (7.4.122) need be calculated only once. So, at each time step, one needs only to solve a system of linear equations for $\mathbf{q}_{i+1}$ associated with the system matrix $\mathbf{P}_1$. The matrix $\mathbf{P}_1$ is positive definite and factorized once before the start of the time integration. An alternative version of the Newmark-$\beta$ method for nonlinear structural dynamic problems is provided in Section 7.4.4.

The solution procedure of the Newmark-$\beta$ method is summarized in Algorithm 7.2 for a linear structural dynamic problem. The flow chart illustrating Algorithm 7.2 is provided in Figure 7.12.

**Spectral Stability and Oscillatory Characteristics**   The stability and accuracy characteristics of the Newmark-$\beta$ method are examined using the concept of spectral stability, as done for the central difference method in Section 7.4.3.

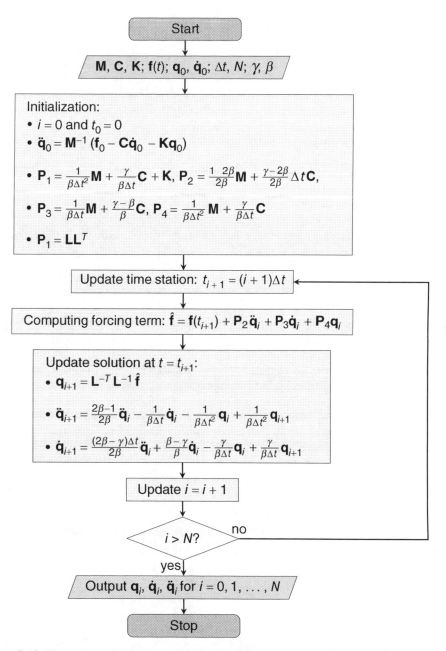

**Figure 7.12** Flow chart of the Newmark-$\beta$ method for linear structural dynamics

The spectral stability analysis is again performed using the SDOF damped oscillator:

$$\ddot{x} + 2\zeta\omega\dot{x} + \omega^2 x = f(t)$$
$$x(0) = x_0, \quad \dot{x}(0) = \dot{x}_0$$

---

**Algorithm 7.2** The Newmark-$\beta$ method (linear case)

---

**Require:** Mass, damping, and stiffness matrices $\mathbf{M}, \mathbf{C}, \mathbf{K}$, external force vector $\mathbf{f}(t)$, initial conditions $\mathbf{q}_0$ and $\dot{\mathbf{q}}_0$, time step size $\Delta t$, number of steps $N$, parameters $\gamma$ and $\beta$.

1: Initialize $i = 0$ and $t_0 = 0$.
2: Compute the initial acceleration using Eq. (7.4.119).
3: Compute the matrices $\mathbf{P}_i$, $i = 1, 2, 3, 4$, using Eq. (7.4.123).
4: Perform Cholesky decomposition $\mathbf{P}_1 = \mathbf{L}\mathbf{L}^T$.
5: **for** $i = 0, \ldots, N - 1$ **do**                                    ▷ Compute the solution at $t = t_{i+1}$
6:     Compute current time $t_{i+1} = (i + 1)\Delta t$
7:     Compute the right-hand side term in Eq. (7.4.122)

$$\hat{\mathbf{f}} = \mathbf{f}(t_{i+1}) + \mathbf{P}_2\ddot{\mathbf{q}}_i + \mathbf{P}_3\dot{\mathbf{q}}_i + \mathbf{P}_4\mathbf{q}_i \tag{7.4.124}$$

8:     Solve the linear system $\mathbf{P}_1\mathbf{q}_{i+1} = \hat{\mathbf{f}}$ by back-substitution using the factor $\mathbf{L}$.
9:     Compute the new velocity and acceleration $\dot{\mathbf{q}}_{i+1}$ and $\ddot{\mathbf{q}}_{i+1}$ using Eq. (7.4.121)
10: **end for**
11: **return** The solutions $\mathbf{q}, \dot{\mathbf{q}}, \ddot{\mathbf{q}}$ at the time stations $t_0, t_1, \ldots, t_N$.

---

The first step is to create a recursive relation and find the amplification matrix. This is achieved by considering the approximated velocity and displacement, Eq. (7.4.107), in scalar form,

$$\dot{x}_{i+1} = \dot{x}_i + [(1 - \gamma)\ddot{x}_i + \gamma\ddot{x}_{i+1}]\,\Delta t \tag{7.4.125a}$$

$$x_{i+1} = x_i + \dot{x}_i\Delta t + \left[\left(\frac{1}{2} - \beta\right)\ddot{x}_i + \beta\ddot{x}_{i+1}\right]\Delta t^2 \tag{7.4.125b}$$

and the SDOF damped oscillator equation at times $t = t_i$ and $t = t_{i+1}$:

$$\ddot{x}_i + 2\zeta\omega\dot{x}_i + \omega^2 x_i = 0 \tag{7.4.126a}$$

$$\ddot{x}_{i+1} + 2\zeta\omega\dot{x}_{i+1} + \omega^2 x_{i+1} = 0, \tag{7.4.126b}$$

where the forcing function $f(t)$ is ignored since only the amplification matrix is of interest.

Define the recursion variable as

$$\boldsymbol{\chi}_i = [x_i, \dot{x}_i\Delta t]^T \tag{7.4.127}$$

and Eq. (7.4.125) is written as

$$\boldsymbol{\chi}_{i+1} = \begin{bmatrix} 1 & 1 \\ 0 & 1 \end{bmatrix}\boldsymbol{\chi}_i + \begin{bmatrix} \frac{1}{2} - \beta & \beta \\ 1 - \gamma & \gamma \end{bmatrix}\begin{bmatrix} \ddot{x}_i \\ \ddot{x}_{i+1} \end{bmatrix}\Delta t^2. \tag{7.4.128}$$

Equation (7.4.126) is also written in a matrix form as

$$\begin{bmatrix} \ddot{x}_i \\ \ddot{x}_{i+1} \end{bmatrix} = -\begin{bmatrix} 2\zeta\omega\dot{x}_i + \omega^2 x_i \\ 2\zeta\omega\dot{x}_{i+1} + \omega^2 x_{i+1} \end{bmatrix}$$

$$= -\begin{bmatrix} 0 & 0 \\ \omega^2 & 2\zeta\omega/\Delta t \end{bmatrix}\boldsymbol{\chi}_{i+1} - \begin{bmatrix} \omega^2 & 2\zeta\omega/\Delta t \\ 0 & 0 \end{bmatrix}\boldsymbol{\chi}_i \tag{7.4.129}$$

and the accelerations in Eq. (7.4.128) are eliminated using Eq. (7.4.129):

$$
\begin{aligned}
\chi_{i+1} &= \begin{bmatrix} 1 & 1 \\ 0 & 1 \end{bmatrix} \chi_i \\
&\quad - \begin{bmatrix} \tfrac{1}{2} - \beta & \beta \\ 1 - \gamma & \gamma \end{bmatrix} \left( \begin{bmatrix} 0 & 0 \\ \omega^2 \Delta t^2 & 2\zeta\omega\Delta t \end{bmatrix} \chi_{i+1} + \begin{bmatrix} \omega^2 \Delta t^2 & 2\zeta\omega\Delta t \\ 0 & 0 \end{bmatrix} \chi_i \right) \\
&= \begin{bmatrix} 1 & 1 \\ 0 & 1 \end{bmatrix} \chi_i - \begin{bmatrix} \beta\omega^2 \Delta t^2 & 2\beta\zeta\omega\Delta t \\ \gamma\omega^2 \Delta t^2 & 2\gamma\zeta\omega\Delta t \end{bmatrix} \chi_{i+1} \\
&\quad - \begin{bmatrix} \tfrac{1}{2}(1 - 2\beta)\omega^2 \Delta t^2 & (1 - 2\beta)\zeta\omega\Delta t \\ (1 - \gamma)\omega^2 \Delta t^2 & 2(1 - \gamma)\zeta\omega\Delta t \end{bmatrix} \chi_i.
\end{aligned}
\tag{7.4.130}
$$

Subsequently, Eq. (7.4.130) is rearranged as

$$
\mathbf{A}_1 \chi_{i+1} = \mathbf{A}_2 \chi_i,
\tag{7.4.131}
$$

where

$$
\begin{aligned}
\mathbf{A}_1 &= \begin{bmatrix} 1 + \beta\omega^2 \Delta t^2 & 2\beta\zeta\omega\Delta t \\ \gamma\omega^2 \Delta t^2 & 1 + 2\gamma\zeta\omega\Delta t \end{bmatrix} \\
\mathbf{A}_2 &= \begin{bmatrix} 1 - \tfrac{1}{2}(1 - 2\beta)\omega^2 \Delta t^2 & 1 - (1 - 2\beta)\zeta\omega\Delta t \\ -(1 - \gamma)\omega^2 \Delta t^2 & 1 - 2(1 - \gamma)\zeta\omega\Delta t \end{bmatrix}.
\end{aligned}
$$

Finally, the recursion relation for the Newmark-$\beta$ method is

$$
\chi_{i+1} = \mathbf{A}_1^{-1} \mathbf{A}_2 \chi_i \equiv \mathbf{A}\chi_i.
\tag{7.4.132}
$$

Consider the undamped case, $\zeta = 0$; the amplification matrix $\mathbf{A}$ is

$$
\mathbf{A} = \begin{bmatrix} 1 - \tfrac{1}{2}(1 - 2\beta)\omega^2 \Delta t^2 & 1 \\ -\omega^2 \Delta t^2 - \tfrac{1}{2}(2\beta - \gamma)\omega^4 \Delta t^4 & 1 + (\beta - \gamma)\omega^2 \Delta t^2 \end{bmatrix}.
\tag{7.4.133}
$$

Subsequently, the eigenvalues of $\mathbf{A}$ can be examined to determine the spectral stability of the Newmark-$\beta$ method. However, the mathematical derivations are lengthy and hence omitted; a detailed derivation can be found in Hughes (2000). The key conclusions are listed as follows:

(1) When $\gamma < 1/2$, the method is unstable.
(2) When $\gamma \geq 1/2$ and $\beta < \gamma/2$, the method is conditionally stable, with the time step size requirement (for $\zeta = 0$):

$$
\Delta t < \Delta t_{cr} = \frac{T_{\min}}{\pi\sqrt{2\gamma - 4\beta}}
\tag{7.4.134}
$$

where $T_{\min}$ is the smallest period of oscillation in the structural dynamic system.
(3) When $\gamma \geq 1/2$ and $\beta \geq \gamma/2$, the method is unconditionally stable. However,

(a) when $\gamma/2 \leq \beta < (1/2 + \gamma)^2/4$, the method *cannot* reproduce the oscillatory responses of a frequency higher than a critical value, which leads to the time step size requirement

$$\Delta t < \Delta t_b = \frac{T_{\min}}{\pi}\left[\left(\gamma + \frac{1}{2}\right)^2 - 4\beta\right]^{-1/2};\qquad (7.4.135)$$

(b) when $\beta \geq (1/2 + \gamma)^2/4$, the method can capture the entire range of frequency components in a structural dynamic response.

(4) Finally, the method is second-order time accurate only when $\gamma = 1/2$, which explains why typically $\gamma = 1/2$ is used for the Newmark-$\beta$ method.

The stability characteristics are also summarized in Figure 7.13, with the markers representing the four special cases of Newmark-$\beta$ methods discussed earlier in this section. It is clear that the methods of central difference and linear acceleration are both second-order time accurate and conditionally stable, while the average acceleration method, that is, the standard Newmark-$\beta$ method, is second-order time accurate and unconditionally stable. Any combination of $(\gamma, \beta)$ on the parabolic curve $\beta = (1/2 + \gamma)^2/4$, $\gamma > 1/2$, results in the method of average acceleration with damping, and this method is first-order time accurate and unconditionally stable. In Figure 7.13, the case of $\alpha = \gamma - 1/2 = 0.1$ is illustrated.

The oscillation characteristics of the Newmark-$\beta$ methods are examined using the amplitude and periodicity errors, computed from the eigenvalues of the amplification matrix

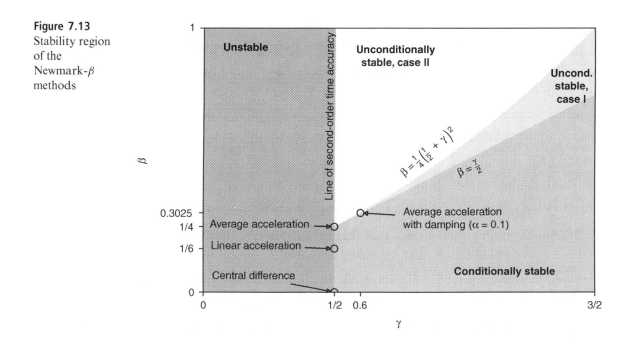

**Figure 7.13**
Stability region of the Newmark-$\beta$ methods

using Eq. (7.4.73). The exact expressions are complex and need to be examined via Taylor series expansion:

$$\Delta\zeta = \frac{1}{4}(2\gamma - 1)\omega\Delta t - \frac{1}{48}(6\beta - 3\gamma + 2)(2\gamma - 1)\omega^3\Delta t^3$$
$$+ O(\omega^5\Delta t^5) \tag{7.4.136a}$$

$$\frac{\Delta T}{T} = \frac{1}{12}(1 + 6\beta - 3\gamma)\omega^2\Delta t^2 + O(\omega^4\Delta t^4). \tag{7.4.136b}$$

Note that choosing $\gamma = 1/2$ eliminates all the terms in the expansion Eq. (7.4.136a), making $\Delta\zeta = 0$. This means that the Newmark-$\beta$ method does not introduce any numerical dissipation when it is second-order time accurate. In addition, when $\gamma \neq 1/2$, $\Delta\zeta$ and hence the numerical solution are only first-order time accurate.

For most of the practical choices of $\gamma$ and $\beta$, the periodicity error is second-order time accurate. The expression for periodicity error, Eq. (7.4.136b), provides an approach to estimating the required time step size. To have $\frac{\Delta T}{T}$ smaller than an error tolerance $p$,

$$\Delta t \leq \frac{T_{\min}}{\pi}\sqrt{\frac{3p}{1 + 6\beta - 3\gamma}}. \tag{7.4.137}$$

When $p = 5\%$, $\gamma = 1/2$, and $\beta = 1/4$, $\Delta t \leq 0.123 T_{\min}$. The choice of $p$ depends on the specific application.

Furthermore, the magnitudes of $\Delta\zeta$ and $\Delta T/T$ grow as $\beta$ increases. Implying that for any specific choice of $\gamma$, the minimum possible value of $\beta$ should be used to minimize the the amplitude and periodicity errors of the numerical solution. Therefore, the optimal combinations of $(\gamma, \beta)$ lie on the lower boundary of the unconditional stability (case II) region in Figure 7.13. This requirement explains the choice of $\gamma = 1/2$ and $\beta = 1/4$ for the average acceleration method, as well as the choice of

$$\beta = \frac{(1/2 + \gamma)^2}{4}, \quad \gamma > \frac{1}{2} \tag{7.4.138}$$

for the method of average acceleration with damping.

Using the stability characteristics and the amplitude and periodicity errors in Eq. (7.4.136), the four special cases of Newmark-$\beta$ methods shown in Figure 7.13 are characterized in Table 7.4. Note how the change in $\beta$ affects the stability of these methods, and how the deviation of $\gamma$ from $1/2$ affects the numerical dissipation of the solution.

**Spectral Radius Plot**    Another useful tool allowing examination of the stability and accuracy characteristics of numerical methods is the plot of spectral radius versus the variable $\omega\Delta t$. Examples of spectral radius plots are shown in Figure 7.14 for the central difference method, average acceleration method, and average acceleration with damping ($\alpha = 0.1$).

Explanation of the horizontal axis in Figure 7.14 is provided next. When $\omega$ is fixed, a small value of $\omega\Delta t$ means that the time step size is small when compared to the period of oscillation and therefore provides sufficient time resolution of the oscillatory response; while a large value of $\omega\Delta t$ indicates insufficient time resolution of the frequency component.

**Table 7.4 Stability characteristics of several Newmark-$\beta$ methods**

| Method | $\gamma$ | $\beta$ | $\omega\Delta t_{cr}$ | $\Delta\zeta$ | $\Delta T/T$ |
|---|---|---|---|---|---|
| Central difference | $\dfrac{1}{2}$ | $0$ | $2$ | $0$ | $-\dfrac{1}{24}\omega^2\Delta t^2$ |
| Linear acceleration | $\dfrac{1}{2}$ | $\dfrac{1}{6}$ | $2\sqrt{3}$ | $0$ | $\dfrac{1}{24}\omega^2\Delta t^2$ |
| Average acceleration | $\dfrac{1}{2}$ | $\dfrac{1}{4}$ | $\infty$ | $0$ | $-\dfrac{1}{12}\omega^2\Delta t^2$ |
| Average acceleration with damping | $\dfrac{1}{2}+\alpha$ | $\dfrac{(1+\alpha)^2}{4}$ | $\infty$ | $\dfrac{\alpha}{2}\omega\Delta t$ | $\dfrac{2+3\alpha^2}{24}\omega^2\Delta t^2$ |

**Figure 7.14**
Spectral radius plots of three cases of Newmark-$\beta$ methods

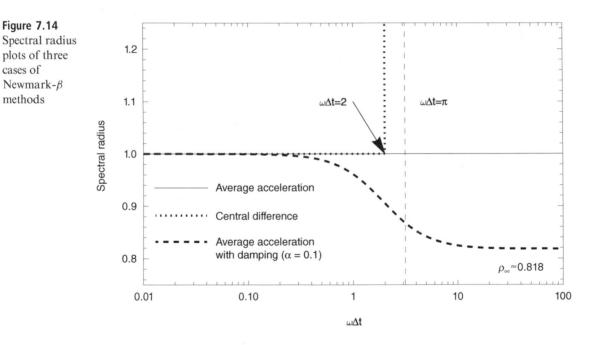

When $\Delta t$ is fixed, a small value of $\omega\Delta t$ corresponds to a frequency component that can be well resolved by this time step size. A large $\omega\Delta t$ represents a frequency component that is beyond the resolution capability of the given $\Delta t$.

The choice of $\Delta t$ for a response of frequency $\omega$ is guided by the Nyquist sampling theorem, which is a fundamental theorem in signal processing theory. The theorem states that all the information from a continuous-time signal with a maximum frequency of $f$ is completely captured using a discrete-time signal with a sampling rate of $2f$. The implication of Nyquist sampling theorem for structural dynamic response problem is that a minimum of two points per period in the numerical solution (i.e., the discrete-time signal) are needed to resolve an oscillatory structural response of period $T$ (i.e., the continuous-time signal). Therefore, the time step size for time integration should be no larger than half of the oscillation period of interest:

$$\Delta t \leq \frac{T}{2} \quad \text{or} \quad \omega \Delta t \leq \pi. \tag{7.4.139}$$

Therefore, a special range of $\omega \Delta t$ that has to be considered in the spectral radius plot is $[0, \pi]$.

Examine next the three curves in Figure 7.14. The spectral radius of the central difference method behaves as one would expect from the previous stability analysis. The spectral radius diverges quickly to infinity once $\omega \Delta t$ exceeds the critical value of 2, indicating conditional stability for this method. Furthermore, spectral radius stays as unity when $\omega \Delta t < 2$, which implies that the method does not introduce any numerical dissipation. Next, for average acceleration method, the spectral radius is unity over the entire domain of $\omega \Delta t$, which indicates unconditional stability and zero numerical dissipation. Finally, for average acceleration with damping, the spectral radius is close to unity when $\omega \Delta t$ is small but decreases to $\rho_\infty = 9/11 \approx 0.818$ when $\omega \Delta t \to \infty$. This trend indicates that the method is unconditionally stable and its numerical dissipation is frequency dependent: higher dissipation for oscillations at higher frequency.

Figure 7.14 contains another important parameter for characterizing the numerical dissipation besides $\Delta \zeta$. The parameter is the spectral radius at the high-frequency limit, or the **asymptotic spectral radius** $\rho_\infty$. For the Newmark-$\beta$ family, it is given by

$$\rho_\infty = \lim_{\omega \Delta t \to \infty} \rho(\mathbf{A}) = \sqrt{1 + \frac{1 - 2\gamma}{2\beta}}. \tag{7.4.140}$$

The examination of $\rho_\infty$ reveals some practical issues related to the Newmark-$\beta$ methods. When the second-order time accuracy is required, $\gamma = 1/2$, and the asymptotic spectral radius $\rho_\infty = 1$. This means that all frequency components of the structural dynamic system are captured without amplitude error. This property appears to be remarkable and useful, but in fact is undesirable in certain problems. One case is a structural dynamic problem based on the finite element formulation that can contain spurious high-frequency content due to numerical discretization errors. Such high-frequency content is erroneous and can contaminate the low-frequency content. Some numerical dissipation, that is, an asymptotic spectral radius of less than unity, is desirable to eliminate the high-frequency components. Another situation where $\rho_\infty < 1$ is desirable is the nonlinear problem when unconditional stability cannot be assured. Some numerical dissipation is beneficial for improving the numerical stability and achieving convergence of the numerical solution.

The lack of numerical dissipation can be remedied by employing the method of average acceleration with damping, such that

$$\gamma = \frac{1}{2} + \alpha, \quad \beta = \frac{(1 + \alpha)^2}{4}, \quad \rho_\infty = \frac{1 - \alpha}{1 + \alpha}, \tag{7.4.141}$$

where $0 < \alpha \le 1$ is used to tune $\rho_\infty$ to be a value between 0 and 1. In Figure 7.14, $\alpha = 0.1$ and hence $\rho_\infty = \frac{9}{11} \approx 0.818$. However, for Newmark-$\beta$ family, introducing the numerical dissipation *inevitably* reduces the order of time accuracy from two to one, which is undesirable in practical structural dynamic analysis.

## Numerical Example for the Newmark-$\beta$ Method

In this section, the Newmark-$\beta$ method is applied to the solution of the 2-DOF oscillator problem that was presented in section "Numerical Example for the Central Difference Method," pp. 404–407.

$$\mathbf{M}\ddot{\mathbf{q}} + \mathbf{K}\mathbf{q} = \mathbf{f}_0, \quad \mathbf{q}(0) = [0,0]^T, \quad \dot{\mathbf{q}}(0) = [0,0]^T, \tag{E7.128}$$

where

$$\mathbf{f}_0 = \begin{bmatrix} 0.0160 \\ 1.0000 \end{bmatrix}, \quad \mathbf{M} = \begin{bmatrix} 1 & 0 \\ 0 & 4 \end{bmatrix}, \quad \mathbf{K} = \begin{bmatrix} 16 & 0 \\ 0 & 1 \end{bmatrix}.$$

Equation (E7.128) is solved using the Newmark-$\beta$ method, that is, $\gamma = \frac{1}{2}$ and $\beta = \frac{1}{4}$, with a time step size of $\Delta t = \frac{1}{10}$.

**Starting Procedure**   Following Algorithm 7.2, discussed earlier, first compute the initial acceleration using Eq. (7.4.119):

$$\ddot{\mathbf{q}}_0 = \mathbf{M}^{-1}(\mathbf{f}_0 - \mathbf{C}\dot{\mathbf{q}}_0 - \mathbf{K}\mathbf{q}_0)$$

$$= \begin{bmatrix} 1 & 0 \\ 0 & 4 \end{bmatrix}^{-1} \left( \begin{bmatrix} 0.0160 \\ 1.0000 \end{bmatrix} - \begin{bmatrix} 16 & 0 \\ 0 & 1 \end{bmatrix} \begin{bmatrix} 0.0000 \\ 0.0000 \end{bmatrix} \right) = \begin{bmatrix} 1.6000e{-}02 \\ 2.5000e{-}01 \end{bmatrix}. \tag{E7.129}$$

Four coefficient matrices also have to be prepared, using Eqs. (7.4.123a) through (7.4.123d):

$$\mathbf{P}_1 = \frac{1}{\beta \Delta t^2}\mathbf{M} + \frac{\gamma}{\beta \Delta t}\mathbf{C} + \mathbf{K}$$

$$= 400 \begin{bmatrix} 1 & 0 \\ 0 & 4 \end{bmatrix} + \begin{bmatrix} 16 & 0 \\ 0 & 1 \end{bmatrix} = \begin{bmatrix} 416 & 0 \\ 0 & 1601 \end{bmatrix} \tag{E7.130}$$

$$\mathbf{P}_2 = \frac{1-2\beta}{2\beta}\mathbf{M} + \frac{\gamma - 2\beta}{2\beta}\Delta t \mathbf{C} = \begin{bmatrix} 1 & 0 \\ 0 & 4 \end{bmatrix} \tag{E7.131}$$

$$\mathbf{P}_3 = \frac{1}{\beta \Delta t}\mathbf{M} + \frac{\gamma - \beta}{\beta}\mathbf{C} = \begin{bmatrix} 40 & 0 \\ 0 & 160 \end{bmatrix} \tag{E7.132}$$

$$\mathbf{P}_4 = \frac{1}{\beta \Delta t^2}\mathbf{M} + \frac{\gamma}{\beta \Delta t}\mathbf{C} = \begin{bmatrix} 400 & 0 \\ 0 & 1600 \end{bmatrix}. \tag{E7.133}$$

**Time Stepping**   Starting from the given initial displacement $\mathbf{q}_0$ and velocity $\dot{\mathbf{q}}_0$, and the initial acceleration $\ddot{\mathbf{q}}_0$ found from the starting procedure, the displacement at the first time step is computed using Eq. (7.4.122):

$$\mathbf{P}_1\mathbf{q}_1 = \mathbf{f}_1 + \mathbf{P}_2\ddot{\mathbf{q}}_0 + \mathbf{P}_3\dot{\mathbf{q}}_0 + \mathbf{P}_4\mathbf{q}_0$$

$$\begin{bmatrix} 416 & 0 \\ 0 & 1601 \end{bmatrix}\mathbf{q}_1 = \begin{bmatrix} 0.0160 \\ 1.0000 \end{bmatrix} + \begin{bmatrix} 1 & 0 \\ 0 & 4 \end{bmatrix}\begin{bmatrix} 1.6000\text{e--}02 \\ 2.5000\text{e--}01 \end{bmatrix} + \begin{bmatrix} 40 & 0 \\ 0 & 160 \end{bmatrix}\begin{bmatrix} 0.0000 \\ 0.0000 \end{bmatrix}$$

$$+ \begin{bmatrix} 400 & 0 \\ 0 & 1600 \end{bmatrix}\begin{bmatrix} 0.0000 \\ 0.0000 \end{bmatrix}$$

$$\begin{bmatrix} 416 & 0 \\ 0 & 1601 \end{bmatrix}\mathbf{q}_1 = \begin{bmatrix} 3.2000\text{e--}02 \\ 2.0000\text{e+}00 \end{bmatrix}. \tag{E7.134}$$

Solving the linear system Eq. (E7.134), one finds the displacement

$$\mathbf{q}_1 = [7.6923\text{e--}05, 1.2492\text{e--}03]^T. \tag{E7.135}$$

Next, with the new displacement $\mathbf{q}_1$ available, one can compute the acceleration $\ddot{\mathbf{q}}_1$ and the velocity $\dot{\mathbf{q}}_1$ at the first time step using Eqs. (7.4.121a) and (7.4.121b), respectively:

$$\ddot{\mathbf{q}}_1 = \frac{2\beta - 1}{2\beta}\ddot{\mathbf{q}}_0 - \frac{1}{\beta\Delta t}\dot{\mathbf{q}}_0 - \frac{1}{\beta\Delta t^2}\mathbf{q}_0 + \frac{1}{\beta\Delta t^2}\mathbf{q}_1$$

$$= -\begin{bmatrix} 1.6000\text{e--}02 \\ 2.5000\text{e--}01 \end{bmatrix} - 40\begin{bmatrix} 0.0000 \\ 0.0000 \end{bmatrix} - 400\begin{bmatrix} 0.0000 \\ 0.0000 \end{bmatrix} + 400\begin{bmatrix} 7.6923\text{e--}05 \\ 1.2492\text{e--}03 \end{bmatrix}$$

$$= \begin{bmatrix} 1.4769\text{e--}02 \\ 2.4969\text{e--}01 \end{bmatrix} \tag{E7.136}$$

$$\dot{\mathbf{q}}_1 = \frac{(2\beta - \gamma)\Delta t}{2\beta}\ddot{\mathbf{q}}_0 + \frac{\beta - \gamma}{\beta}\dot{\mathbf{q}}_0 - -\frac{\gamma}{\beta\Delta t}\mathbf{q}_0 + \frac{\gamma}{\beta\Delta t}\mathbf{q}_1$$

$$= 0\begin{bmatrix} 1.6000\text{e--}02 \\ 2.5000\text{e--}01 \end{bmatrix} - \begin{bmatrix} 0.0000 \\ 0.0000 \end{bmatrix} - 20\begin{bmatrix} 0.0000 \\ 0.0000 \end{bmatrix} + 20\begin{bmatrix} 7.6923\text{e--}05 \\ 1.2492\text{e--}03 \end{bmatrix}$$

$$= \begin{bmatrix} 1.5385\text{e--}03 \\ 2.4984\text{e--}02 \end{bmatrix}. \tag{E7.137}$$

Following the procedure outlined in Eqs. (E7.134) through (E7.137), one can compute the structural dynamic solutions at the subsequent time steps. The numerical solution for the first five steps are compared with the exact solution in Table 7.5. The average error of the Newmark-$\beta$ method is as low as 0.13%, which is comparable to the accuracy of the central difference method as illustrated in section "Numerical Example for the Central Difference Method," pp. 404–407.

**Effects of Time Step Size and Numerical Dissipation**    The Newmark-$\beta$ solutions are compared with the exact solution in Figure 7.15.

**Table 7.5 Comparison of the exact and Newmark-$\beta$ solutions**

| Step | Exact solution | Newmark-$\beta$ | Error (%) |
|------|----------------|------------------|-----------|
| 1 | [7.8939e–05, 1.2497e–03] | [7.6923e–05, 1.2492e–03] | 0.1663 |
| 2 | [3.0329e–04, 4.9958e–03] | [2.9586e–04, 4.9938e–03] | 0.1543 |
| 3 | [6.3764e–04, 1.1229e–02] | [6.2312e–04, 1.1224e–02] | 0.1356 |
| 4 | [1.0292e–03, 1.9933e–02] | [1.0084e–03, 1.9925e–02] | 0.1123 |
| 5 | [1.4161e–03, 3.1088e–02] | [1.3923e–03, 3.1075e–02] | 0.0870 |

**Figure 7.15**
Numerical
solutions of
a linear
2-DOF
oscillator
using the
Newmark-$\beta$
method

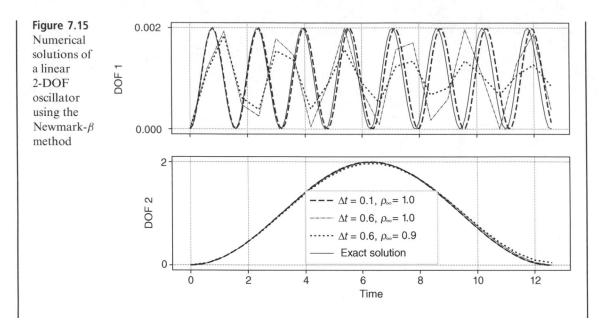

First, consider the solution generated by the Newmark-$\beta$ method, that is, $\gamma = \frac{1}{2}$ and $\beta = \frac{1}{4}$, with $\Delta t = 0.1$. There is no amplitude error in the numerical solution for both the low- and high-frequency responses, which is consistent with the theoretical analysis in Eq. (7.4.136a). The periodicity error is evident in the high-frequency response, but not in the low-frequency response. The difference is explained by Eq. (7.4.136b), where the periodicity error is shown to be proportional to the square of frequency, when $\gamma = \frac{1}{2}$ and $\beta = \frac{1}{4}$:

$$\frac{\Delta T}{T} = \frac{1}{12}\omega^2 \Delta t^2. \tag{E7.138}$$

Next, consider the solution generated by the Newmark-$\beta$ method with $\Delta t = 0.6$. Note that when the same 2-DOF oscillator problem was solved using the central difference method in section "Numerical Example for the Central Difference Method," pp. 404–407, the numerical solution becomes unstable once $\Delta t > 0.5$. However, the Newmark-$\beta$ method produces a stable solution with $\Delta t = 0.6$, where the low-frequency response is well resolved while the high-frequency response is inaccurate but does not diverge. The solution with $\Delta t = 0.6$ is sufficient for analysis purposes if one is only interested in the low-frequency response, which is the case for many structural dynamic applications. This solution illustrates that an implicit method such as the Newmark-$\beta$ method can be employed to resolve low-frequency responses with a relatively large time step size, even in the presence of high-frequency responses.

Finally, consider a third solution that is generated by the method of average acceleration with damping, that is, a modified Newmark-$\beta$ method. The method introduces damping to eliminate the high-frequency response while maintaining the amplitude of the low-frequency response. This method is useful, for example, when the high-frequency response is spurious due to numerical discretization of the finite element method and thus is *not required*. The high-frequency damping is obtained by adjusting the asymptotic spectral radius $\rho_\infty$ that determines the numerical dissipation at the high-frequency limit. For this example, $\rho_\infty$ is chosen to be 0.9. Using Eq. (7.4.141), the coefficients for the method of average acceleration with damping are determined as shown below:

$$\alpha = \frac{1 - \rho_\infty}{1 + \rho_\infty} \approx 0.0526, \quad \gamma = \frac{1}{2} + \alpha \approx 0.553, \quad \beta = \frac{(1 + \alpha)^2}{4} \approx 0.277. \quad \text{(E7.139)}$$

As shown in Figure 7.15, a strong numerical damping effect is evident in the high-frequency response, and the high-frequency response is expected to be completely eliminated if the simulation were to continue for a longer time. However, the numerical damping also introduces amplitude error in the low-frequency response and results in a lightly damped oscillation, which is undesirable. The amplitude error is first-order time accurate, based on Eq. (7.4.136a):

$$\Delta\zeta \approx \frac{1}{4}(2\gamma - 1)\omega\Delta t = \frac{1}{2}\alpha\omega\Delta t. \quad \text{(E7.140)}$$

As a result, the method of average acceleration with damping is only first-order time accurate.

**Summary**    This example has illustrated the solution procedure of the Newmark-$\beta$ method and highlighted the main advantage and disadvantage of this method. The advantage is the unconditional stability. The practical implication is that the Newmark-$\beta$ method is capable of generating an accurate low-frequency response of a structural dynamic problem without the need to accurately resolve the high-frequency response. The unconditional stability can save a considerable amount of computer time, since one does not need to use a small step size to resolve all the frequency components in a structural response. The disadvantage of the Newmark-$\beta$ method is that its second-order time accuracy and the numerical dissipation for high-frequency response are mutually exclusive. Any attempt to completely eliminate the high-frequency response reduces the overall accuracy of the numerical solution. The mutual exclusivity motivates an extension of the Newmark-$\beta$ method, which leads to the generalized-$\alpha$ method that is introduced next.

### Generalized-$\alpha$ Method

The dilemma in the Newmark-$\beta$ method is that its time accuracy and numerical dissipation are governed by different requirements of the same parameter $\gamma$. Second-order time accuracy requires $\gamma = \frac{1}{2}$ but numerical dissipation for high-frequency components requires $\gamma > \frac{1}{2}$. A resolution to this dilemma consists of introducing one or more new parameters in the numerical method to create additional flexibility in the tuning of the time accuracy and the numerical dissipation. Following this strategy, various extended versions of the Newmark-$\beta$ method were developed in the literature. Two of the representative methods are the $\alpha$-method of Hilber, Hughes, and Taylor (HHT-$\alpha$) (Hilber et al., 1977) and the generalized-$\alpha$ method (Chung and Hulbert, 1993). Both methods allow for tunable numerical dissipation of high-frequency responses, while minimizing the dissipation of low-frequency responses and maintaining second-order time accuracy. The HHT-$\alpha$ method is a special case of the generalized-$\alpha$ method, and therefore this section focuses on the latter.

In the generalized-$\alpha$ method, the velocity and displacement at the new time step are the same as those in Newmark-$\beta$ method:

$$\dot{\mathbf{q}}_{i+1} = \dot{\mathbf{q}}_i + [(1-\gamma)\ddot{\mathbf{q}}_i + \gamma\ddot{\mathbf{q}}_{i+1}]\,\Delta t \tag{7.4.142}$$

$$\mathbf{q}_{i+1} = \mathbf{q}_i + \dot{\mathbf{q}}_i\Delta t + \left[\left(\frac{1}{2}-\beta\right)\ddot{\mathbf{q}}_i + \beta\ddot{\mathbf{q}}_{i+1}\right]\Delta t^2 \tag{7.4.143}$$

but the equations of motion are written in an alternative implicit form as

$$\mathbf{M}\ddot{\mathbf{q}}_{i+1-\alpha_m} + \mathbf{C}\dot{\mathbf{q}}_{i+1-\alpha_f} + \mathbf{K}\mathbf{q}_{i+1-\alpha_f} = \mathbf{f}_{i+1-\alpha_f}, \tag{7.4.144}$$

where

$$\ddot{\mathbf{q}}_{i+1-\alpha_m} = (1-\alpha_m)\ddot{\mathbf{q}}_{i+1} + \alpha_m\ddot{\mathbf{q}}_i \tag{7.4.145a}$$

$$\dot{\mathbf{q}}_{i+1-\alpha_f} = (1-\alpha_f)\dot{\mathbf{q}}_{i+1} + \alpha_f\dot{\mathbf{q}}_i \tag{7.4.145b}$$

$$\mathbf{q}_{i+1-\alpha_f} = (1-\alpha_f)\mathbf{q}_{i+1} + \alpha_f\mathbf{q}_i \tag{7.4.145c}$$

$$\mathbf{f}_{i+1-\alpha_f} = (1-\alpha_f)\mathbf{f}_{i+1} + \alpha_f\mathbf{f}_i. \tag{7.4.145d}$$

The parameters $\alpha_m$ and $\alpha_f$ are introduced to tune the asymptotic spectral radius $\rho_\infty$ of the numerical method so as to control the numerical dissipation for the high-frequency responses. The generalized-$\alpha$ method reduces to the Newmark-$\beta$ method when $\alpha_m = \alpha_f = 0$ and the HHT-$\alpha$ method when $\alpha_m = 0$.

Like in the Newmark-$\beta$ method, it is possible to substitute Eq. (7.4.145) into Eq. (7.4.144) so as to form a compact recursive relation that expresses the new solution $\mathbf{q}_{i+1}$ in terms of the forcing vectors and the solutions at the current time $t = t_i$:

$$\mathbf{P}_1^*\mathbf{q}_{i+1} = (1-\alpha_f)\mathbf{f}_{i+1} + \mathbf{R} + \mathbf{P}_2^*\ddot{\mathbf{q}}_i + \mathbf{P}_3^*\dot{\mathbf{q}}_i + \mathbf{P}_4^*\mathbf{q}_i, \tag{7.4.146}$$

where

$$\mathbf{R} = \alpha_f\mathbf{f}_i - \alpha_m\mathbf{M}\ddot{\mathbf{q}}_i - \alpha_f\mathbf{C}\dot{\mathbf{q}}_i - \alpha_f\mathbf{K}\mathbf{q}_i \tag{7.4.147a}$$

$$\mathbf{P}_1^* = (1-\alpha_m)\frac{1}{\beta\Delta t^2}\mathbf{M} + (1-\alpha_f)\frac{\gamma}{\beta\Delta t}\mathbf{C} + (1-\alpha_f)\mathbf{K} \tag{7.4.147b}$$

$$\mathbf{P}_2^* = (1-\alpha_m)\frac{1-2\beta}{2\beta}\mathbf{M} + (1-\alpha_f)\frac{\gamma-2\beta}{2\beta}\Delta t\mathbf{C} \tag{7.4.147c}$$

$$\mathbf{P}_3^* = (1-\alpha_m)\frac{1}{\beta\Delta t}\mathbf{M} + (1-\alpha_f)\frac{\gamma-\beta}{\beta}\mathbf{C} \tag{7.4.147d}$$

$$\mathbf{P}_4^* = (1-\alpha_m)\frac{1}{\beta\Delta t^2}\mathbf{M} + (1-\alpha_f)\frac{\gamma}{\beta\Delta t}\mathbf{C}. \tag{7.4.147e}$$

The starting procedure of the generalized-$\alpha$ method is the same as that of the Newmark-$\beta$ method, Eq. (7.4.119). For linear structural dynamic problems, the solution procedure of the generalized-$\alpha$ method is similar to the one used for the Newmark-$\beta$ method and is summarized in Algorithm 7.3.

---

**Algorithm 7.3** Generalized-$\alpha$ method (linear case)

---

**Require:** Mass, damping, and stiffness matrices $\mathbf{M}, \mathbf{C}, \mathbf{K}$, external force vector $\mathbf{f}(t)$, initial conditions $\mathbf{q}_0$ and $\dot{\mathbf{q}}_0$, time step size $\Delta t$, number of steps $N$, parameters $\alpha_f$, $\alpha_m$, $\gamma$, and $\beta$.

1: Initialize $i = 0$ and $t_0 = 0$.
2: Compute the initial acceleration using Eq. (7.4.119)

$$\ddot{\mathbf{q}}_0 = \mathbf{M}^{-1}\left(\mathbf{f}_0 - \mathbf{C}\dot{\mathbf{q}}_0 - \mathbf{K}\mathbf{q}_0\right)$$

3: Compute the matrices $\mathbf{P}_i^*$, $i = 1, 2, 3, 4$, using Eq. (7.4.147).
4: Perform Cholesky decomposition $\mathbf{P}_1^* = \mathbf{L}\mathbf{L}^T$.
5: **for** $i = 0, \ldots, N - 1$ **do**                    ▷ Compute the solution at $t = t_i$
6:      Compute current time $t_i = i\Delta t$
7:      Compute the right-hand side term in Eq. (7.4.146)

$$\hat{\mathbf{f}} = (1 - \alpha_f)\mathbf{f}_{i+1} + \mathbf{R} + \mathbf{P}_2^*\ddot{\mathbf{q}}_i + \mathbf{P}_3^*\dot{\mathbf{q}}_i + \mathbf{P}_4^*\mathbf{q}_i$$

8:      Solve the linear system $\mathbf{P}_1^*\mathbf{q}_{i+1} = \hat{\mathbf{f}}$ by back-substitution using the factor $\mathbf{L}$.
9:      Compute the new velocity and acceleration $\dot{\mathbf{q}}_{i+1}$ and $\ddot{\mathbf{q}}_{i+1}$ using Eq. (7.4.121)
10: **end for**
11: **return** The solutions $\mathbf{q}, \dot{\mathbf{q}}, \ddot{\mathbf{q}}$ at the time stations $t_0, t_1, \ldots, t_N$.

---

**Spectral Stability and Oscillatory Characteristics**    The stability analysis of the generalized-$\alpha$ method starts with the formulation of the recursive relation for the SDOF structural dynamic problem. The recursive relation is constructed from the scalar form of Eq. (7.4.144) for the SDOF damped oscillator without the forcing term, and the approximated velocity and displacement, Eq. (7.4.107):

$$0 = \ddot{x}_{i+1-\alpha_m} + 2\zeta\omega\dot{x}_{i+1-\alpha_f} + \omega^2 x_{i+1-\alpha_f} \tag{7.4.148a}$$

$$\dot{x}_{i+1} = \dot{x}_i + [(1 - \gamma)\ddot{x}_i + \gamma\ddot{x}_{i+1}]\Delta t \tag{7.4.148b}$$

$$x_{i+1} = x_i + \dot{x}_i\Delta t + \left[\left(\frac{1}{2} - \beta\right)\ddot{x}_i + \beta\ddot{x}_{i+1}\right]\Delta t^2. \tag{7.4.148c}$$

Define the recursion variable as

$$\boldsymbol{\chi}_i = [x_i, \dot{x}_i\Delta t, \ddot{x}_i\Delta t^2]^T \tag{7.4.149}$$

and Eq. (7.4.148) is written as

$$\mathbf{A}_1^*\boldsymbol{\chi}_{i+1} = \mathbf{A}_2^*\boldsymbol{\chi}_i, \tag{7.4.150}$$

where

$$\mathbf{A}_1^* = \begin{bmatrix} (1-\alpha_f)\omega^2\Delta t^2 & 2(1-\alpha_f)\zeta\omega\Delta t & 1-\alpha_m \\ 1 & 0 & -\beta \\ 0 & 1 & -\gamma \end{bmatrix} \quad (7.4.151)$$

$$\mathbf{A}_2^* = \begin{bmatrix} -\alpha_f\omega^2\Delta t^2 & -2\alpha_f\zeta\omega\Delta t & -\alpha_m \\ 1 & 1 & 1/2-\beta \\ 0 & 1 & 1-\gamma \end{bmatrix}. \quad (7.4.152)$$

The amplification matrix is therefore

$$\mathbf{A} = \left(\mathbf{A}_1^*\right)^{-1}\mathbf{A}_2^*. \quad (7.4.153)$$

Next, the three eigenvalues of $\mathbf{A}$ are examined to determine the spectral stability of the generalized-$\alpha$ method. However, the mathematical derivations are lengthy and tedious, and hence are omitted; a detailed derivation can be found in Géradin and Rixen (2014). For the generalized-$\alpha$ method to be second-order time accurate and unconditionally stable, the parameters need to satisfy

$$\gamma = \frac{1}{2} + \alpha_f - \alpha_m, \quad \beta = \frac{1}{4}(1+\alpha_f-\alpha_m)^2, \quad \frac{1}{2} \geq \alpha_f \geq \alpha_m \geq 0. \quad (7.4.154)$$

The free parameters $\alpha_f$ and $\alpha_m$ can be determined using the asymptotic spectral radius $\rho_\infty$ as

$$\alpha_f = \frac{\rho_\infty}{\rho_\infty+1}, \quad \alpha_m = \frac{2\rho_\infty-1}{\rho_\infty+1}, \quad (7.4.155)$$

where $0 \leq \rho_\infty \leq 1$, and $\gamma$ and $\beta$ become

$$\gamma = \frac{1}{2} + \frac{1-\rho_\infty}{1+\rho_\infty}, \quad \beta = \frac{1}{(1+\rho_\infty)^2}. \quad (7.4.156)$$

The choice of $\alpha_f$ and $\alpha_m$ in Eq. (7.4.155) is optimal in the sense that the numerical dissipation is minimized for the low-frequency component while keeping the dissipation in the high-frequency range close to $\rho_\infty$. A typical value of $\rho_\infty = 0.95$ is chosen to obtain transient structural response while suppressing the high-frequency components of the structural dynamic system that are not of major interest.

The four variants of the generalized-$\alpha$ method are compared in Table 7.6, with the parameters expressed in terms of the asymptotic spectral radius. The variants include (1) one of the Newmark-$\beta$ methods, average acceleration with damping, (2) HHT-$\alpha$ method, a popular method used in implicit structural dynamic codes, (3) generalized-$\alpha$ method, using the optimal parameters, Eqs. (7.4.155) and (7.4.156), and (4) a special case of generalized-$\alpha$ method called the midpoint method.

The first three variants all employed the same choice of $\gamma$ and $\beta$ that are functions of $\rho_\infty$. For the Newmark-$\beta$ method, as discussed in the previous section, choosing $\gamma \neq 1/2$ results in the reduction of the order of time accuracy. The HHT-$\alpha$ method utilizes the $\alpha_f$ parameter and introduces numerical dissipation while maintaining the second-order time accuracy; but this method has a limitation due to $\rho_\infty \geq 1/2$ for the sake of stability. Using both $\alpha_f$ and $\alpha_m$,

**Table 7.6 Comparison of various forms of the generalized-$\alpha$ method**

| Method | $\rho_\infty$ | $\alpha_f$ | $\alpha_m$ | $\gamma$ | $\beta$ |
|---|---|---|---|---|---|
| Newmark-$\beta$ with damping | $[0,1]$ | $0$ | $0$ | $\dfrac{1}{2} + \dfrac{1 - \rho_\infty}{1 + \rho_\infty}$ | $\dfrac{1}{(1 + \rho_\infty)^2}$ |
| HHT-$\alpha$ | $\left[\dfrac{1}{2},1\right]$ | $\dfrac{1 - \rho_\infty}{1 + \rho_\infty}$ | $0$ | $\dfrac{1}{2} + \dfrac{1 - \rho_\infty}{1 + \rho_\infty}$ | $\dfrac{1}{(1 + \rho_\infty)^2}$ |
| Generalized-$\alpha$ | $[0,1]$ | $\dfrac{\rho_\infty}{\rho_\infty + 1}$ | $\dfrac{2\rho_\infty - 1}{\rho_\infty + 1}$ | $\dfrac{1}{2} + \dfrac{1 - \rho_\infty}{1 + \rho_\infty}$ | $\dfrac{1}{(1 + \rho_\infty)^2}$ |
| Midpoint | $1$ | $\dfrac{1}{2}$ | $\dfrac{1}{2}$ | $\dfrac{1}{2}$ | $\dfrac{1}{4}$ |

the "optimal" generalized-$\alpha$ method enables the arbitrary choice of numerical dissipation (i.e., $0 \le \rho_\infty \le 1$) without the loss of second-order time accuracy and numerical stability.

The fourth variant in Table 7.6, the midpoint method, is essentially the generalized-$\alpha$ method with $\rho_\infty = 1$. It appears that the midpoint method is the same as the average acceleration method with $\gamma = 1/2$ and $\beta = 1/4$. The only difference is that the midpoint method considers the dynamic equilibrium at time $t = t_{i+1/2}$:

$$\mathbf{M}\ddot{\mathbf{q}}_{i+1/2} + \mathbf{C}\dot{\mathbf{q}}_{i+1/2} + \mathbf{K}\mathbf{q}_{i+1/2} = \mathbf{f}_{i+1/2}, \tag{7.4.157}$$

which is different from the Newmark-$\beta$ methods that consider the dynamic equilibrium at time $t = t_{i+1}$. When considering linear problems, the midpoint and average acceleration methods generate the same solutions. But for nonlinear problems, especially coupled problems such as fluid-structural interaction, the midpoint method has a beneficial property of energy conservation that tends to result in a more accurate numerical solution. See Zienkiewicz and Taylor (2000) for a more detailed discussion.

The four variants in Table 7.6 are compared further by using the spectral radius plots in Figure 7.16, assuming $\rho_\infty = 0.9$. The midpoint method has a constant spectral radius of unity, which is the same as the average acceleration method. The spectral radii of the other three variants converge to $\rho_\infty = 0.9$ as $\omega\Delta t \to \infty$. The difference is in the range of low-frequency components. The introduction of the $\alpha_f$ and $\alpha_m$ parameters delays significantly the decrease of the spectral radius and expands the frequency range that has low or negligible numerical dissipation. Furthermore, the average acceleration with damping, the HHT-$\alpha$ method, and the generalized-$\alpha$ method all demonstrate a **low-pass filtering** characteristic, implying that all these methods keep the low-frequency components and filter out the high-frequency components in the numerical solution. As explained earlier, such low-pass filtering characteristic is desirable in structural dynamic analysis.

Finally, the oscillation characteristics of the four variants are compared in Figures 7.17 and 7.18, considering an undamped case $\zeta = 0$. Figure 7.17 shows the error in damping ratio $\Delta\zeta$ versus $\omega\Delta t$. The midpoint method has zero damping ratio error because of its constant spectral radius of unity. In the low-frequency range, $\Delta\zeta$ of the Newmark-$\beta$ method grows linearly because of its first-order accuracy in damping ratio. On the contrary, $\Delta\zeta$ of the two $\alpha$ methods grows quadratically and are much lower than that of the

**Figure 7.16**
Comparison of
spectral radius
with $\rho_\infty = 0.9$
and $\zeta = 0$

**Figure 7.17**
Comparison of
numerical
damping ratio
with $\rho_\infty = 0.9$
and $\zeta = 0$

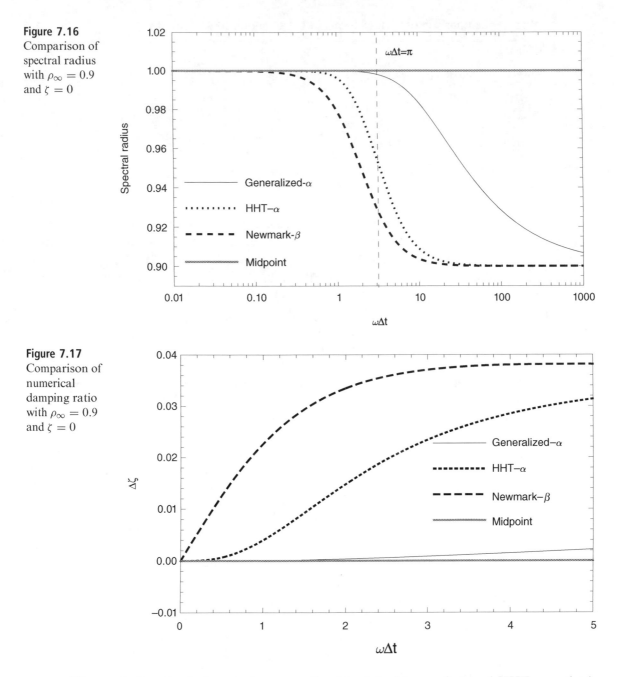

Newmark-$\beta$ method. As $\omega\Delta t$ increases, the $\Delta\zeta$ of the Newmark-$\beta$ and HHT-$\alpha$ methods gradually reach the same magnitude; however, as a remarkable feature of the generalized-$\alpha$ method, its damping ratio error remains as low as <10% of the $\Delta\zeta$ of the other two methods. The trend of the damping ratio error in Figure 7.17 agrees well with the findings from Figure 7.16 that the generalized-$\alpha$ method has the best numerical dissipation characteristics among the four variants.

The periodicity errors of the four variants are compared in Figure 7.18. All the methods have a second-order accuracy of $\Delta T/T$, which explains the quadratic trend. Despite the differences in numerical damping errors, the methods do not have significant differences in terms of $\Delta T/T$. The HHT-$\alpha$ method is slightly inferior when compared to the other three methods.

**Figure 7.18**
Comparison of periodicity error with $\rho_\infty = 0.9$ and $\zeta = 0$

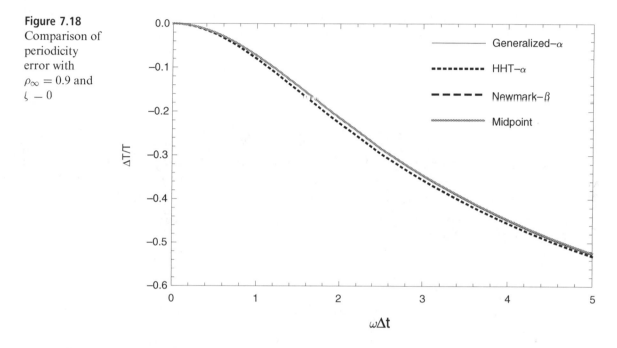

## Numerical Example for the Generalized-$\alpha$ Method

The solution procedure of the generalized-$\alpha$ method is similar to that of the Newmark-$\beta$ method. As one can recall from the numerical example in section "Numerical Example for the Newmark-$\beta$ Method," pp. 418–421, the Newmark-$\beta$ method involves computing four coefficient matrices $\mathbf{P}_1, \ldots, \mathbf{P}_4$ using Eqs. (7.4.123a) through (7.4.123d) and using $\mathbf{P}_1, \ldots, \mathbf{P}_4$ and a recursive formula, Eq. (7.4.122), to find the displacements at each time step sequentially. In the generalized-$\alpha$ method, the recursive procedure is the same, except that the four coefficient matrices are defined differently using Eqs. (7.4.147b) through (7.4.147e) and a new residual vector as defined in Eq. (7.4.147a) is involved in the recursive formula Eq. (7.4.146). Therefore, this section does not focus on the detailed solution procedure for the generalized-$\alpha$ method for, for example, a simple 2-DOF oscillator problem. Instead, the generalized-$\alpha$ method is applied to a relatively sophisticated beam vibration problem so as to illustrate the numerical characteristics of the method.

**Problem Statement**  The cantilevered beam configuration is adapted from the configuration considered in Section 5.6.4, as shown in Figure 7.19. The beam has a length of $L = 1$ m, a bending stiffness of $EI = 6 \times 10^3$ kg m$^3$/s$^2$, and a mass per unit length of

**Figure 7.19** Cantilevered beam modeled with 10 finite elements and a point mass at its end

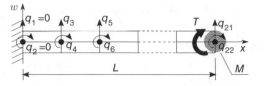

$m = 6.0$ kg/m. In this example, a point mass $M$ is added to the free end of the beam, so that the whole beam structure resembles a wing with a fuel tank at the wing tip. The point mass is assumed to be 10% of the beam mass, that is, $M = 0.1\ mL$. A constant torque $T = 10$ Nm is added to the free end of the beam. Initially the beam has zero displacement and velocity.

The beam is divided into 10 elements of equal length and provides a structural dynamic system having 20 DOFs:

$$\mathbf{M\ddot{q}} + \mathbf{Kq} = \mathbf{f}_0, \quad \mathbf{q}(0) = 0, \quad \dot{\mathbf{q}}(0) = 0. \tag{E7.141}$$

The stiffness matrix $\mathbf{K}$ and mass matrix $\mathbf{M}$ have been developed for the beam structure without the point mass in Section 5.6.4. The point mass does not modify the stiffness matrix. It is accounted for by adding $M$ to the second last diagonal entry of the mass matrix, which corresponds to the DOF of vertical displacement at the beam tip, that is, $q_{21}$ in Figure 7.19. The load vector is a 20-dimensional vector of zeros except the last entry:

$$\mathbf{f}_0 = [\underbrace{0, 0, \ldots, 0}_{19 \text{ zeros}}, T]^T. \tag{E7.142}$$

**Characteristics of the Dynamic Response**   Using the normal mode method described in Section 7.1, the solution to Eq. (E7.141) is

$$\mathbf{q}(t) = \mathbf{K}^{-1}\mathbf{f}_0 - \mathbf{\Phi}\cos{(\mathbf{\Lambda}t)}\mathbf{\Lambda}^{-2}\mathbf{\Phi}^T\mathbf{f}_0, \tag{E7.143}$$

where $\mathbf{\Lambda}$ is the diagonal matrix of eigenfrequencies and $\mathbf{\Phi}$ is the matrix of eigenvectors such that

$$\mathbf{\Phi}^T\mathbf{M\Phi} = \mathbf{I}, \quad \mathbf{\Phi}^T\mathbf{K\Phi} = \mathbf{\Lambda}^2. \tag{E7.144}$$

The term $\cos{(\mathbf{\Lambda}t)}\mathbf{\Lambda}^{-2}$ in Eq. (E7.143) is also a diagonal matrix, whose $i$th diagonal element is $\cos{(\omega_i t)}/\omega_i^2$.

Equation (E7.143) is a superposition of two components of responses, representing the beam vibration about a static equilibrium. The first term $\mathbf{K}^{-1}\mathbf{f}_0$ represents a static equilibrium position where the beam deforms given the load vector $\mathbf{f}_0$, while the second term represents the beam oscillation around the static equilibrium position.

The solution Eq. (E7.143) contains all the natural frequencies of the system, indicating that the load vector excites all the structural modes in the system. However, the term $\cos{(\mathbf{\Lambda}t)}\mathbf{\Lambda}^{-2}$ shows that the amplitude of oscillation associated with the $i$th frequency $\omega_i$ is inversely proportional to $\omega_i^2$, which indicates that the structural response is dominated by the low-frequency components. In fact, using the 10-element discretization in this example, the fundamental frequency of the beam configuration is 93.85 Hz, while the maximum frequency, that is, the

20th frequency, is 158.2 kHz. Therefore, the oscillation amplitude associated with the 20th frequency is expected to be *seven* orders of magnitude smaller than the oscillation amplitude associated with the fundamental frequency. In engineering practice, the structural responses associated with the highest 10 frequencies are negligible.

There are also additional considerations for eliminating the high-frequency responses. One can compute accurately natural frequencies of the beam configuration by increasing the number of finite elements in the discretization. A convergence study shows that the 10-element model captures only the first 6 frequencies with a satisfactory error of less than 5%. The error increases along with the order of mode. For the 20th natural frequency, the accurate value is 115.9 kHz and approximately 36% lower than the value predicted in 10-element model. The errors in the predicted frequencies indicate that the high-frequency component of solution to Eq. (E7.141) is not very significant even if it is captured correctly in the numerical solution, since it is not a physically significant and accurate representation of the beam structure.

**Unconditional Stability**    First, the numerical stability characteristics of the generalized-$\alpha$ method are highlighted by comparison to the central difference method. The central difference method has been shown to be conditionally stable for a particular time step size requirement. For this case,

$$\Delta t \leq \Delta t_{cr} = \frac{2}{\omega_{\max}} = \frac{2}{158.2\text{kHz}} \approx 1.264 \times 10^{-5} \text{ s.} \tag{E7.145}$$

Figure 7.20 illustrates the numerical solutions for the transient response of the beam tip for $0 \leq t \leq 66.95$ ms, which corresponds to the first half of the period associated with the fundamental natural frequency. The central difference solution employs $\Delta t = \Delta t_{cr}$ with 2648 time steps, while the generalized-$\alpha$ solution employs $\Delta t = 100 \Delta t_{cr}$ requiring just 26 time steps.

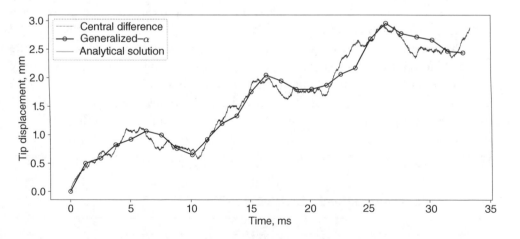

**Figure 7.20** Comparison of the central difference and generalized-$\alpha$ methods

The central difference solution resolves all the frequency components and overlaps with the exact solution computed using Eq. (E7.143). However, such accuracy is obtained at the cost of using an extremely small time step size $\Delta t_{cr}$, even though $\Delta t_{cr}$ is the maximum possible time step size for the central difference method. Furthermore, it is unnecessary to resolve the high-frequency components, which are insignificant as discussed in the analytical solution. The generalized-$\alpha$ solution removes all the issues associated with the central difference solution, employing a 100 times larger time step size. It captures well the low-frequency structural dynamic response and filters out the insignificant high-frequency components.

The comparison between the central difference and generalized-$\alpha$ methods highlights again the importance of using an implicit method for obtaining low-frequency structural dynamic responses in the presence of high-frequency components, while emphasizing the computational cost issue. Even if the cost per time step of an implicit method is higher than the cost of an explicit method, with the significant reduction in the number of time steps, the overall computational cost of an implicit method is far less than that of an explicit method.

**Numerical Dissipation**    Next, the generalized-$\alpha$ method is compared with the Newmark-$\beta$ methods. Specifically, three methods are considered: (1) generalized-$\alpha$ method with $\rho_\infty = 0.9$, (2) Newmark-$\beta$ method, and (3) average acceleration method with damping, choosing $\rho_\infty = 0.9$. The coefficients of the three methods are compared in Table 7.7. Note that the latter two methods have been compared in section "Numerical Example for the Newmark-$\beta$ Method," pp. 418–421. It was found that the Newmark-$\beta$ method lacks numerical dissipation and cannot filter out the high-frequency responses, while the average acceleration method with damping introduces too much numerical dissipation and reduces the time accuracy of the solution.

The numerical solutions are generated for 5000 time steps using a step size $\Delta t \approx 6.416 \times 10^{-4}$ s, which corresponds to 1/16 of the period associated with the second natural frequency. The three numerical solutions are compared in Figure 7.21 for two time intervals: the first 200 steps and the last 200 steps. The tip displacement appears smooth and is dominated by low-frequency responses; the tip rotation appears noisy as it contains some high-frequency responses, which can be spurious and should be filtered out.

In the first 200 steps, the generalized-$\alpha$ and Newmark-$\beta$ solutions are identical, while the average acceleration solution has started to introduce damping in the high-frequency response, as manifested by the smoothing of responses in tip rotation. As the simulation progresses, in the last 200 steps, while the generalized-$\alpha$ and Newmark-$\beta$ solutions remain identical for the tip displacement, the generalized-$\alpha$ solution is smoother than the Newmark-$\beta$ solution of the tip rotation. The smoothing effect in the generalized-$\alpha$ solution removes the high-frequency component and is due to its numerical dissipation property.

**Table 7.7 Coefficients of numerical methods in this example**

| Method | $\rho_\infty$ | $\alpha_f$ | $\alpha_m$ | $\gamma$ | $\beta$ |
|---|---|---|---|---|---|
| Generalized-$\alpha$ method | 0.9 | 0.474 | 0.421 | 0.553 | 0.277 |
| Newmark-$\beta$ method | 1.0 | 0 | 0 | 0.500 | 0.250 |
| Average acceleration with damping | 0.9 | 0 | 0 | 0.553 | 0.277 |

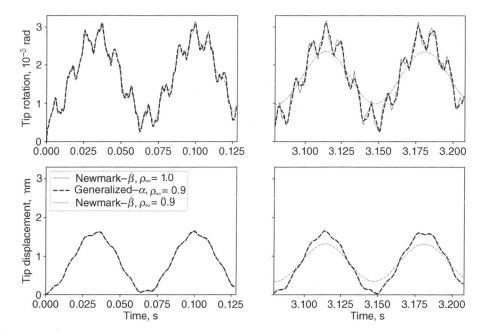

**Figure 7.21** Comparison of the Newmark-$\beta$ and generalized-$\alpha$ methods

On the other hand, most of the frequency components are lost in the average acceleration solution due to its low-order time accuracy.

The comparison in Figure 7.21 illustrates the advantage of the generalized-$\alpha$ method over the Newmark-$\beta$ methods. The generalized-$\alpha$ method achieves good numerical dissipation for high-frequency components while maintaining the time accuracy of the overall solution.

**Low-Pass Filtering**    The generalized-$\alpha$ method permits tuning of numerical dissipation for different spectral contents without sacrificing the time accuracy of the solution. The tunable numerical dissipation enables the low-pass filtering of the structural response, where the responses associated with frequencies below a threshold remain in the solution and the other higher-frequency responses are eliminated.

The low-pass filtering characteristics are useful for coupled multi-physics analysis, for example, fluid-structural interaction problems, which are commonly encountered in applications such as aeroelasticity and turbomachinery. Sometimes the characteristic frequency of the fluid system is low, which means that the fluid is receptive only to low-frequency structural responses and produces low-frequency reaction forces on the structure. As a result, it is sufficient to resolve the structural response up to the frequency of relevance to the fluid system.

To illustrate the low-pass filtering characteristics of the generalized-$\alpha$ method, consider the following three combinations of coefficients: (1) $\rho_\infty = 0.9$ and $\Delta t = \Delta t_0 \approx 6.416 \times 10^{-4}$ s, (2) $\rho_\infty = 0.0$ and $\Delta t = \Delta t_0$, and (3) $\rho_\infty = 0.0$ and $\Delta t = 80 \Delta t_0$. The first case corresponds to the coefficients used in the previous example.

The numerical solutions are shown in Figure 7.22. The first case is identical to the previous example where all the low-frequency components are resolved. The second case involves additional numerical dissipation as $\rho_\infty$ decreases from 0.9 to 0.0. Most of the spectral components are filtered out and only the fundamental frequency component is retained. For the third case, time step size $\Delta t$ is comparable to the period of the fundamental natural frequency, which introduces high numerical dissipation for *all* the frequency components. As a result the numerical solution eliminates all the oscillations and keeps only the average trend in the structural dynamic response, which corresponds to the static equilibrium position of the beam configuration.

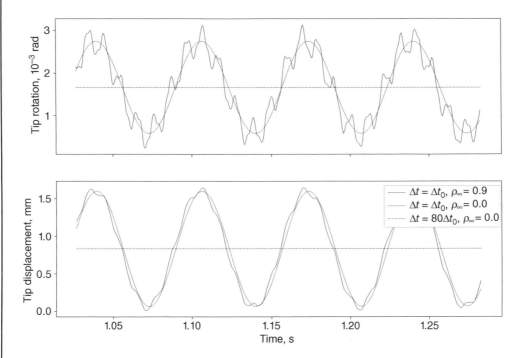

**Figure 7.22** Low-pass filtering characteristics of the generalized-$\alpha$ method

**Summary**   This example represents a cantilevered beam configuration modeled using a discrete system with 10 finite elements. In the exact solution of the discrete system, the structural response contains a wide range of spectral contents and is dominated by the low-frequency components. The high-frequency components are found to be insignificant as well as possibly spurious due to errors in the numerical discretization of the finite element model.

Three numerical examples are presented to illustrate the advantages of the generalized-$\alpha$ method. First, when compared to the central difference method, the generalized-$\alpha$ method overcomes the constraint on the time step size placed by the high-frequency components, and enables efficient computation of the low-frequency structural responses without the need to resolve the high-frequency responses. Second, when compared to the Newmark-$\beta$ methods, the generalized-$\alpha$ method possesses favorable numerical dissipation that eliminates the high-frequency response while maintaining a high time accuracy. Finally, the generalized-$\alpha$ method has a low-pass filtering capability that enables the trimming of the spectral content in the structural response.

All the numerical characteristics discussed above make the generalized-$\alpha$ method, or its special case the HHT-$\alpha$ method, the *de facto* time integration method for structural dynamic analysis.

### Application to Nonlinear Problems

Up to this point the time integration methods are discussed in the context of linear problems. However, these methods are also applicable to the solution of nonlinear structural dynamic problems. For a general nonlinear problem, it is frequent that one converts the governing equations to a first-order form and time integrate the resulting equations of motion using, for example, the Runge–Kutta method. However, since the focus of this chapter is the direct integration method for second-order equations of motion, the procedure for solving nonlinear structural dynamic problems using the generalized-$\alpha$ method is presented next.

Consider a general nonlinear structural dynamic problem:

$$\mathbf{M}\ddot{\mathbf{q}} + \mathbf{f}_I(\mathbf{q}, \dot{\mathbf{q}}, t) = \mathbf{f}_E(\mathbf{q}, \dot{\mathbf{q}}, t) \tag{7.4.158a}$$

$$\mathbf{q}(0) = \mathbf{q}_0, \quad \dot{\mathbf{q}}(0) = \dot{\mathbf{q}}_0, \tag{7.4.158b}$$

where the internal force vector $\mathbf{f}_I$ and external force vector $\mathbf{f}_E$ are both nonlinear functions of displacement $\mathbf{q}$, velocity $\dot{\mathbf{q}}$, and time. The generalized-$\alpha$ method for the nonlinear problem is implemented as

$$\mathbf{M}\ddot{\mathbf{q}}_{i+1-\alpha_m} + \mathbf{f}_{I,i+1-\alpha_f} = \mathbf{f}_{E,i+1-\alpha_f}, \tag{7.4.159}$$

where

$$\ddot{\mathbf{q}}_{i+1-\alpha_m} = (1 - \alpha_m)\ddot{\mathbf{q}}_{i+1} + \alpha_m\ddot{\mathbf{q}}_i \tag{7.4.160a}$$

$$\mathbf{f}_{I,i+1-\alpha_f} = (1 - \alpha_f)\mathbf{f}_I(\mathbf{q}_{i+1}, \dot{\mathbf{q}}_{i+1}, t_{i+1}) + \alpha_f\mathbf{f}_I(\mathbf{q}_i, \dot{\mathbf{q}}_i, t_i) \tag{7.4.160b}$$

$$\mathbf{f}_{E,i+1-\alpha_f} = (1 - \alpha_f)\mathbf{f}_E(\mathbf{q}_{i+1}, \dot{\mathbf{q}}_{i+1}, t_{i+1}) + \alpha_f\mathbf{f}_E(\mathbf{q}_i, \dot{\mathbf{q}}_i, t_i) \tag{7.4.160c}$$

and $\ddot{\mathbf{q}}_{i+1}$ and $\dot{\mathbf{q}}_{i+1}$ are computed using Eq. (7.4.121):

$$\ddot{\mathbf{q}}_{i+1} = \frac{2\beta - 1}{2\beta}\ddot{\mathbf{q}}_i - \frac{1}{\beta\Delta t}\dot{\mathbf{q}}_i - \frac{1}{\beta\Delta t^2}\mathbf{q}_i + \frac{1}{\beta\Delta t^2}\mathbf{q}_{i+1}$$

$$\dot{\mathbf{q}}_{i+1} = \frac{(2\beta - \gamma)\Delta t}{2\beta}\ddot{\mathbf{q}}_i + \frac{\beta - \gamma}{\beta}\dot{\mathbf{q}}_i - \frac{\gamma}{\beta\Delta t}\mathbf{q}_i + \frac{\gamma}{\beta\Delta t}\mathbf{q}_{i+1}.$$

Using Eq. (7.4.160), the three terms in Eq. (7.4.159) are functions of the unknown solution $\mathbf{q}_{i+1}$, and Eq. (7.4.159) can be written as a nonlinear algebraic equation:

$$\mathbf{r}(\mathbf{q}_{i+1}) = \mathbf{M}\ddot{\mathbf{q}}_{i+1-\alpha_m}(\mathbf{q}_{i+1}) + \mathbf{f}_{I,i+1-\alpha_f}(\mathbf{q}_{i+1}) - \mathbf{f}_{E,i+1-\alpha_f}(\mathbf{q}_{i+1}) = 0, \qquad (7.4.161)$$

which is also referred to as a **residual equation**.

The residual equation can be solved using the Newton–Raphson method, a standard root-finding algorithm. The Newton–Raphson method solves a general system of nonlinear equations, $\mathbf{r}(\mathbf{q}) = 0$. Starting from an initial guess $\mathbf{q}^{(0)}$, the method iteratively generates a series of guesses $\mathbf{q}^{(1)}, \mathbf{q}^{(2)}, \ldots, \mathbf{q}^{(n)}$ that eventually converges to the root of the equations $\mathbf{q}^*$.

---

**Algorithm 7.4** Generalized-$\alpha$ method (nonlinear case)

---

**Require:** Mass matrix $\mathbf{M}$, internal force vector $\mathbf{f}_I(\mathbf{q}, \dot{\mathbf{q}}, t)$, external force vector $\mathbf{f}_E(\mathbf{q}, \dot{\mathbf{q}}, t)$, initial conditions $\mathbf{q}_0$ and $\dot{\mathbf{q}}_0$, time step size $\Delta t$, number of steps $N$, parameters $\alpha_f, \alpha_m, \gamma$, and $\beta$, maximum number of Newton–Raphson iterations $M$, and tolerance $\epsilon$.

1: Initialize $i = 0$ and $t_0 = 0$.
2: Compute the initial acceleration

$$\ddot{\mathbf{q}}_0 = \mathbf{M}^{-1}\left[\mathbf{f}_E(\mathbf{q}_0, \dot{\mathbf{q}}_0, 0) - \mathbf{f}_I(\mathbf{q}_0, \dot{\mathbf{q}}_0, 0)\right] \qquad (7.4.167)$$

3: **for** $i = 0, \ldots, N-1$ **do**    ▷ Compute the solution at $t = t_i$
4:    Compute current time $t_i = i\Delta t$
5:    Form the residual equation $\mathbf{r}(\mathbf{q}_{i+1})$ according to Eq. (7.4.161)
6:    Set initial guess for the displacement at $t = t_{i+1}$, $\mathbf{q}_{i+1}^{(0)} = \mathbf{q}_i$
7:    **for** $j = 0, \ldots, M-1$ **do**    ▷ Perform at most $M$ Newton–Raphson iterations
8:       Newton–Raphson step: Solve $\Delta\mathbf{q}$ from

$$\frac{\partial \mathbf{r}}{\partial \mathbf{q}_{i+1}^{(j)}} \Delta\mathbf{q} = -\mathbf{r}(\mathbf{q}_{i+1}^{(j)}) \qquad (7.4.168)$$

9:       Update solution:

$$\mathbf{q}_{i+1}^{(j+1)} = \mathbf{q}_{i+1}^{(j)} + \Delta\mathbf{q} \qquad (7.4.169)$$

10:       **if** $||\mathbf{r}(\mathbf{q}_{i+1}^{(j+1)})|| < \epsilon$ or $j > M$ **then**    ▷ Check convergence
11:          Break
12:       **end if**
13:    **end for**
14:    Let $\mathbf{q}_{i+1}^{(j+1)}$ be the displacement $\mathbf{q}_{i+1}$ at $t = t_{i+1}$.
15:    Compute the new velocity and acceleration $\dot{\mathbf{q}}_{i+1}$ and $\ddot{\mathbf{q}}_{i+1}$ using Eq. (7.4.121)
16: **end for**
17: **return** The solutions $\mathbf{q}, \dot{\mathbf{q}}, \ddot{\mathbf{q}}$ at the time stations $t_0, t_1, \ldots, t_N$.

---

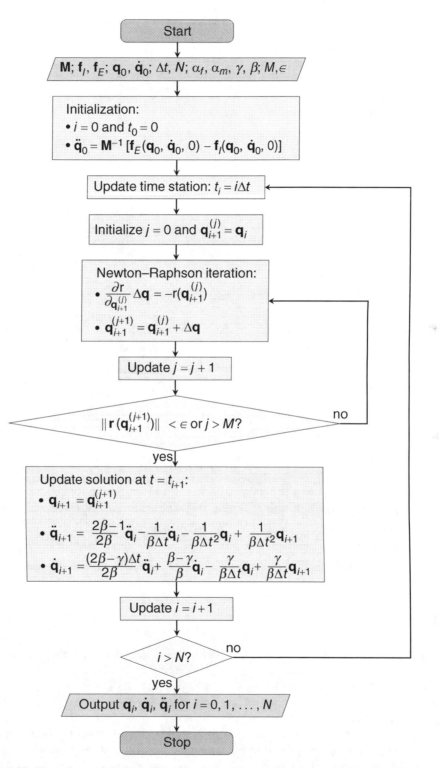

**Figure 7.23** Flow chart of the generalized-$\alpha$ method for nonlinear structural dynamics

At the $j$th iteration, the Newton–Raphson method attempts to find a correction $\Delta\mathbf{q}$ to the previous guess $\mathbf{q}^{(j-1)}$ such that

$$\mathbf{r}(\mathbf{q}^{(j-1)} + \Delta\mathbf{q}) = 0. \tag{7.4.162}$$

The nonlinear equation in Eq. (7.4.162) can be approximated as a linear system using the Taylor series expansion up to the first order:

$$\begin{aligned}
0 = \mathbf{r}(\mathbf{q}^{(j-1)} + \Delta\mathbf{q}) &= \mathbf{r}(\mathbf{q}^{(j-1)}) + (\nabla\mathbf{r})\Delta\mathbf{q} + O(\|\Delta\mathbf{q}\|^2) \\
&\approx \mathbf{r}(\mathbf{q}^{(j-1)}) + (\nabla\mathbf{r})\Delta\mathbf{q},
\end{aligned} \tag{7.4.163}$$

where $\nabla\mathbf{r}$ is the Jacobian matrix of the nonlinear function $\mathbf{r}$. More details on Jacobians are provided in Appendix A.5.2. Using Eq. (7.4.163) the correction $\Delta\mathbf{q}$ is

$$\Delta\mathbf{q} \approx -(\nabla\mathbf{r})^{-1}\mathbf{r}(\mathbf{q}^{(j-1)}) \tag{7.4.164}$$

and the improved guess of root is

$$\mathbf{q}^{(j)} = \mathbf{q}^{(j-1)} + \Delta\mathbf{q}. \tag{7.4.165}$$

The application of the Newton–Raphson method to solve the residual equation, Eq. (7.4.161), requires the Jacobian of $\mathbf{r}$, which is derived as follows:

$$\begin{aligned}
\frac{\partial\mathbf{r}}{\partial\mathbf{q}_{i+1}} &= \mathbf{M}\frac{\partial\ddot{\mathbf{q}}_{i+1-\alpha_m}}{\partial\mathbf{q}_{i+1}} + \frac{\partial\mathbf{f}_{I,i+1-\alpha_f}}{\partial\mathbf{q}_{i+1}} - \frac{\partial\mathbf{f}_{E,i+1-\alpha_f}}{\partial\mathbf{q}_{i+1}} \\
&= \frac{1-\alpha_m}{\beta\Delta t^2}\mathbf{M} + (1-\alpha_f)\left(\frac{\partial\mathbf{f}_I}{\partial\mathbf{q}_{i+1}} + \frac{\gamma}{\beta\Delta t}\frac{\partial\mathbf{f}_I}{\partial\dot{\mathbf{q}}_{i+1}}\right) \\
&\quad - (1-\alpha_f)\left(\frac{\partial\mathbf{f}_E}{\partial\mathbf{q}_{i+1}} + \frac{\gamma}{\beta\Delta t}\frac{\partial\mathbf{f}_E}{\partial\dot{\mathbf{q}}_{i+1}}\right)
\end{aligned} \tag{7.4.166}$$

The reader is referred to Press et al. (2007) for additional details on the general Newton–Raphson method and Belytschko et al. (2014) for a more comprehensive discussion on the numerical solution of nonlinear structural dynamics problems.

The algorithm of the generalized-$\alpha$ method for nonlinear structural dynamic problems is summarized in Algorithm 7.4 and illustrated as a flow chart in Figure 7.23.

## BIBLIOGRAPHY

Ascher, U. M. and Petzold, L. (1998). *Computer Methods for Ordinary Differential Equations and Differential-Algebraic Equations*, pages 73–160. SIAM.

Bathe, K.-J. (2006). *Finite Element Procedures*, pages 768–837. Prentice Hall.

Belytschko, T., Liu, W.-K., Moran, B., and Elkhodary, K. (2014). *Nonlinear Finite Elements for Continua and Structures*, pages 329–416. John Wiley & Sons, 2nd edition.

Chung, J. and Hulbert, G. M. (1993). A time integration algorithm for structural dynamics with improved numerical dissipation: The generalized-$\alpha$ method. *Journal of Applied Mechanics*, 60(2): 371–375. doi: 10.1115/1.2900803.

Felippa, C. A., Park, K.-C., and Farhat, C. (2001). Partitioned analysis of coupled mechanical systems. *Computer Methods in Applied Mechanics and Engineering*, 190(24–25): 3247–3270.

Géradin, M. and Rixen, D. J. (2014). *Mechanical Vibrations: Theory and Application to Structural Dynamics*, pages 539–556. John Wiley & Sons.

Hilber, H. M., Hughes, T. J. R., and Taylor, R. L. (1977). Improved numerical dissipation for time integration algorithms in structural dynamics. *Earthquake Engineering & Structural Dynamics*, 5 (3): 283–292. doi: 10.1002/eqe.4290050306.

Houbolt, J. C. (1950). A recurrence matrix solution for the dynamic response of elastic aircraft. *Journal of the Aeronautical Sciences*, 17(9): 540–550.

Hughes, T. J. R. (2000). *The Finite Element Method: Linear Static and Dynamic Finite Element Analysis*, pages 459–478. Dover Publications.

Nashif, A. D., Jones, D. I. G., and Henderson, J. P. (1985). *Vibration Damping*. John Wiley & Sons

Newmark, N. M. (1959). A method of computation for structural dynamics. *Journal of the Engineering Mechanics Division*, 85(3): 67–94.

Press, W. H., Vetterling, W. T., Teukolsky, S. A., and Flannery, B. P. (2007). *Numerical Recipes: The Art of Scientific Computing*, pages 442–486. Cambridge University Press, 3rd edition.

Wilson, E. L. and Clough, R. W. (1962). Dynamic response by step-by-step matrix analysis. Technical report, Labortorio Nacional de Engenharia Civil, Lisbon, Portugal.

Zienkiewicz, O. C. and Taylor, R. L. (2000). *The Finite Element Method: The Basis*, pages 542–575. Butterworth-Heinemann, 5th edition.

## PROBLEMS

1. In this problem, the trapezoidal rule and the midpoint rule are applied to solve the SDOF oscillator problem presented in section "Numerical Examples for Euler Methods," pp. 384–389:

$$\dot{\mathbf{x}} = \mathbf{f}(\mathbf{x}, t) \equiv \begin{bmatrix} 0 & 1 \\ -1 & 0 \end{bmatrix} \mathbf{x}, \quad \mathbf{x}(0) = \begin{bmatrix} 1 \\ 0 \end{bmatrix}. \tag{P7.29}$$

   (a) Develop the iterative formula for the numerical solution of Eq. (P7.29) using the trapezoidal rule. Compute the numerical solution for five time steps. Evaluate the numerical errors at each time step and compare with the errors of the improved Euler method listed in Table 7.1.

   (b) Develop the iterative formula using the midpoint rule, and show that the iterative formula is identical to the one developed using the trapezoidal rule.

   (c) Next, consider a modified nonlinear problem, where $\mathbf{x} = [x_1, x_2]^T$

$$\dot{\mathbf{x}} = \begin{bmatrix} x_2 + \sin(x_1 x_2) \\ -x_1 + x_2^2 \end{bmatrix}, \quad \mathbf{x}(0) = \begin{bmatrix} 1 \\ 0 \end{bmatrix}. \tag{P7.30}$$

   Develop the iterative formulas using the trapezoidal rule and the midpoint rule, respectively. Identify the differences between the two iterative formulas.

2. The best approach to understanding a numerical algorithm is to implement it on a computer. In this problem, the programming procedures are provided such that one can

develop computer programs for dynamic analysis of a structure using the generalized-$\alpha$ method, as outlined in Algorithm 7.3. It is assumed that only elementary building blocks, such as matrix multiplication and Cholesky decomposition, are available in the programming language (e.g. Python or MATLAB).

(a) Develop a function

$$\alpha_f, \alpha_m, \gamma, \beta = \texttt{setParameters}(\rho_\infty)$$

that evaluates the optimal coefficients $\alpha_f$, $\alpha_m$, $\gamma$, and $\beta$ given the asymptotic spectral radius $\rho_\infty$ using Eqs. (7.4.155) and (7.4.156).

(b) Develop a function

$$\mathbf{L}_1, \mathbf{P}_2^*, \mathbf{P}_3^*, \mathbf{P}_4^* = \texttt{initialize}\,(\mathbf{M}, \mathbf{C}, \mathbf{K}, \alpha_f, \alpha_m, \gamma, \beta, \Delta t)$$

that evaluates the coefficient matrices $\mathbf{P}_1^*, \ldots, \mathbf{P}_4^*$ as defined in Eq. (7.4.147), given the $\mathbf{M}$, $\mathbf{C}$, and $\mathbf{K}$ matrices, coefficients $\alpha_f$, $\alpha_m$, $\gamma$, and $\beta$, and the time step size $\Delta t$. Note that instead of $\mathbf{P}_1^*$ the function should return a lower triangular matrix $\mathbf{L}_1$ such that $\mathbf{P}_1^* = \mathbf{L}_1^T \mathbf{L}_1$.

(c) For conciseness, it is assumed that this and the following functions have access to the system matrices $\mathbf{M}, \mathbf{C}, \mathbf{K}$, and the coefficients $\alpha_f, \alpha_m, \gamma$, and $\beta$. Develop a function

$$\ddot{\mathbf{q}}_0 = \texttt{starting}(\mathbf{q}_0, \dot{\mathbf{q}}_0, \mathbf{f}_0)$$

that evaluates the initial acceleration given the system matrices, the initial conditions, and the initial load vector using Eq. (7.4.119).

(d) Develop a function

$$\mathbf{q}_{i+1}, \dot{\mathbf{q}}_{i+1}, \ddot{\mathbf{q}}_{i+1} = \texttt{timeStepping}(\mathbf{L}_1, \mathbf{P}_2^*, \mathbf{P}_3^*, \mathbf{P}_4^*, \mathbf{q}_i, \dot{\mathbf{q}}_i, \ddot{\mathbf{q}}_i, \mathbf{f}_{i+1})$$

that performs one step of simulation using Eqs. (7.4.121a), (7.4.121b), and (7.4.146). Its inputs include the system matrices, the coefficient matrices, the solutions at the current time step, and the load vector at the next step; and its outputs include the solutions at the next time step. For the sake of computational efficiency, when solving the linear system in Eq. (7.4.146) for $\mathbf{q}_{i+1}$, one should use the lower triangular matrix $\mathbf{L}_1$ for back-substitution.

(e) Develop a program that consecutively executes the functions `setParameters`, `initialize` and `starting`, and subsequently performs the function `timeStepping` in a loop of $N$ iterations that generates the numerical solution of $N$ time steps.

(f) Apply this program to solve the 2-DOF oscillator problem considered in section "Numerical Example for the Newmark-$\beta$ Method," pp. 418–421:

$$\mathbf{M}\ddot{\mathbf{q}} + \mathbf{K}\mathbf{q} = \mathbf{f}_0, \quad \mathbf{q}(0) = [0, 0]^T, \quad \dot{\mathbf{q}}(0) = [0, 0]^T, \tag{P7.31}$$

where

$$\mathbf{f}_0 = \begin{bmatrix} 0.0160 \\ 1.0000 \end{bmatrix}, \quad \mathbf{M} = \begin{bmatrix} 1 & 0 \\ 0 & 4 \end{bmatrix}, \quad \mathbf{K} = \begin{bmatrix} 16 & 0 \\ 0 & 1 \end{bmatrix}.$$

- Solve the problem using $\Delta t = 0.1$ and $\rho_\infty = 0.9$. Compute the numerical solution for five time steps. Evaluate the numerical errors at each time step and compare with the errors of the Newmark-$\beta$ method listed in Table 7.5.
- Solve the problem for $0 \leq t \leq 60$ using (1) $\Delta t = 0.1$ and $\rho_\infty = 0.9$, (2) $\Delta t = 0.1$ and $\rho_\infty = 0.0$, (3) $\Delta t = 0.5$ and $\rho_\infty = 0.9$, and (4) $\Delta t = 0.5$ and $\rho_\infty = 0.0$. Plot the numerical solution against the exact solution. Observe the numerical dissipation in the low- and high-frequency responses in each case. If one would like eliminate the high-frequency response while retaining the low-frequency response, how should one adjust the simulation parameters, that is, $\Delta t$ and $\rho_\infty$?

Note: One can replace the function `setParameters` with an alternative version that computes the coefficients $\gamma$ and $\beta$ using Eq. (7.4.141) and always returns $a_f - a_m - 0$, so that the computer program can generate numerical solutions using the Newmark-$\beta$ methods.

3. The accuracy of a time integration method can be verified numerically by performing a series of simulations associated with decreasing time step sizes. For example, consider again the 2-DOF oscillator problem, Eq. (P7.31), and the generalized-$\alpha$ method with $\rho_\infty = 0.9$. Define the reference time step size $\Delta t_0 = 0.1$ and a reference number of time steps $N_0 = 100$, which correspond to a final time of $t = 10$.

(a) Using a step size $\Delta t_i = \frac{1}{2^i} \Delta t_0$ and number of steps $N_i = 2^i N_0$, generate a numerical solution of Eq. (P7.31), record the solution at the last time step, and denote it as $\mathbf{q}_i$.

(b) Perform the above step for $i = 0, 1, 2, 3, 4$, and $i = 8$, so that one obtains a series of solutions $\{\mathbf{q}_0, \mathbf{q}_1, \mathbf{q}_2, \mathbf{q}_3, \mathbf{q}_4\}$ and a reference solution $\mathbf{q}_8$. The series of solutions should converge to the exact solution at $t = 10$. Usually, the exact solution is unknown, so $\mathbf{q}_8$ is employed to approximate the exact solution.

(c) Compute the approximate numerical error

$$\epsilon_i = \|\mathbf{q}_i - \mathbf{q}_8\|, \quad i = 0, 1, \dots, 4 \tag{P7.32}$$

and plot $\epsilon_i$ against $\Delta t_i$ on a log-log scale. For a $p$th-order method, the convergence of error should satisfy

$$\epsilon = C \Delta t^p, \quad \text{or} \quad \log \epsilon = p \log \Delta t + \log C, \tag{P7.33}$$

where $C$ is a problem-dependent constant. Equation (P7.33) indicates that, if convergence is achieved, the error curve associated with a $p$th-order method should be a straight line with a slope $p$. Verify that in the current problem $p = 2$ for the generalized-$\alpha$ method with $\rho_\infty = 0.9$.

(d) Plot more error curves and make comparisons as specified below:
- Plot the error curves for the generalized-$\alpha$ method with $\rho_\infty = 0.5$ and $\rho_\infty = 0.0$. Do the slopes of the curves change?
- Plot the error curves for the Newmark-$\beta$ method, that is, generalized-$\alpha$ method with $\rho_\infty = 1.0$. Verify that the Newmark-$\beta$ method is also second-order time accurate.

- Plot the error curves for the average acceleration method with damping, using $\rho_\infty = 0.9$. Verify that this method is only first-order time accurate. Note that one can employ the alternative version of program in Problem 2 to generate the numerical solutions.

4. This problem outlines the Wilson-$\theta$ method, which is essentially an extension of the linear acceleration method introduced in Eq. (7.4.108). In the Wilson-$\theta$ method, the acceleration is assumed to be linear from time $t$ to time $t + \theta \Delta t$, where $\theta \geq 0$. When $\theta = 1$, the method reduces to the linear acceleration method. Denote by $\tau$ the increase in time, that is, $0 \leq \tau \leq \theta \Delta t$, it is assumed that

$$\ddot{\mathbf{q}}_{t+\tau} = \ddot{\mathbf{q}}_t + \frac{\tau}{\theta \Delta t}(\ddot{\mathbf{q}}_{t+\theta \Delta t} - \ddot{\mathbf{q}}_t). \tag{P7.34}$$

(a) The Wilson-$\theta$ method is derived in the following procedure:

(i) Integrate Eq. (P7.34) over time, and find the expressions for $\dot{\mathbf{q}}_{t+\tau}$ and $\mathbf{q}_{t+\tau}$.

$$\dot{\mathbf{q}}_{t+\tau} = \dot{\mathbf{q}}_t + \tau \ddot{\mathbf{q}}_t + \frac{\tau^2}{2\theta \Delta t}(\ddot{\mathbf{q}}_{t+\theta \Delta t} - \ddot{\mathbf{q}}_t) \tag{P7.35}$$

$$\mathbf{q}_{t+\tau} = \mathbf{q}_t + \tau \dot{\mathbf{q}}_t + \frac{\tau^2}{2}\ddot{\mathbf{q}}_t + \frac{\tau^3}{6\theta \Delta t}(\ddot{\mathbf{q}}_{t+\theta \Delta t} - \ddot{\mathbf{q}}_t). \tag{P7.36}$$

(ii) Using Eqs. (P7.34) through (P7.36) and $\tau = \theta \Delta t$, express $\ddot{\mathbf{q}}_{t+\theta \Delta t}$ and $\dot{\mathbf{q}}_{t+\theta \Delta t}$ in terms of $\mathbf{q}_{t+\theta \Delta t}$, $\mathbf{q}_t$, $\dot{\mathbf{q}}_t$, and $\ddot{\mathbf{q}}_t$.

$$\ddot{\mathbf{q}}_{t+\theta \Delta t} = \frac{6}{(\theta \Delta t)^2}(\mathbf{q}_{t+\theta \Delta t} - \mathbf{q}_t) - \frac{6}{\theta \Delta t}\dot{\mathbf{q}}_t - 2\ddot{\mathbf{q}}_t \tag{P7.37}$$

$$\dot{\mathbf{q}}_{t+\theta \Delta t} = \frac{3}{\theta \Delta t}(\mathbf{q}_{t+\theta \Delta t} - \mathbf{q}_t) - 2\dot{\mathbf{q}}_t - \frac{\theta \Delta t}{2}\ddot{\mathbf{q}}_t. \tag{P7.38}$$

(iii) Define $\mathbf{f}_\theta = \mathbf{f}_t + \theta(\mathbf{f}_{t+\theta \Delta t} - \mathbf{f}_t)$ and assume that the dynamic equilibrium equation is satisfied at time $t + \theta \Delta t$:

$$\mathbf{M}\ddot{\mathbf{q}}_{t+\theta \Delta t} + \mathbf{C}\dot{\mathbf{q}}_{t+\theta \Delta t} + \mathbf{K}\mathbf{q}_{t+\theta \Delta t} = \mathbf{f}_\theta \tag{P7.39}$$

Using Eqs. (P7.37) and (P7.38), develop a linear system of equations for $\mathbf{q}_{t+\theta \Delta t}$:

$$\mathbf{W}_1 \mathbf{q}_{t+\theta \Delta t} = \mathbf{f}_\theta + \mathbf{W}_2 \mathbf{q}_t + \mathbf{W}_3 \dot{\mathbf{q}}_t + \mathbf{W}_4 \ddot{\mathbf{q}}_t, \tag{P7.40}$$

where $\mathbf{W}_1 = \frac{6}{(\theta \Delta t)^2}\mathbf{M} + \frac{3}{\theta \Delta t}\mathbf{C} + \mathbf{K}$, $\mathbf{W}_2 = \frac{6}{(\theta \Delta t)^2}\mathbf{M} + \frac{3}{\theta \Delta t}\mathbf{C}$, $\mathbf{W}_3 = \frac{6}{\theta \Delta t}\mathbf{M} + 2\mathbf{C}$, and $\mathbf{W}_4 = 2\mathbf{M} + \frac{\theta \Delta t}{2}\mathbf{C}$.

(iv) Once $\mathbf{q}_{t+\theta \Delta t}$ is obtained, one can find $\ddot{\mathbf{q}}_{t+\theta \Delta t}$ from Eq. (P7.37), and subsequently find the solutions at time $t + \Delta t$ from Eqs. (P7.34) through (P7.36) with $\tau = \Delta t$,

$$\ddot{\mathbf{q}}_{t+\Delta t} = \frac{6}{\Delta t^2 \theta^3}(\mathbf{q}_{t+\theta \Delta t} - \mathbf{q}_t) - \frac{6}{\Delta t \theta^2}\dot{\mathbf{q}}_t + \frac{\theta - 3}{\theta}\ddot{\mathbf{q}}_t \tag{P7.41}$$

$$\dot{\mathbf{q}}_{t+\Delta t} = \dot{\mathbf{q}}_t + \frac{\Delta t}{2}(\ddot{\mathbf{q}}_{t+\Delta t} + \ddot{\mathbf{q}}_t) \tag{P7.42}$$

$$\mathbf{q}_{t+\Delta t} = \mathbf{q}_t + \Delta t \dot{\mathbf{q}}_t + \frac{\Delta t^2}{6}(\ddot{\mathbf{q}}_{t+\Delta t} + 2\ddot{\mathbf{q}}_t). \tag{P7.43}$$

(b) Similar to the procedure outlined in Problem 2, develop a computer program for the Wilson-$\theta$ method.

(c) Apply the Wilson-$\theta$ method to solve the 2-DOF oscillator problem. Specifically,
- Solve the problem with $\Delta t = 0.1$ and $\theta = 1.4$. Identify the periodicity error and the amplitude error for both the low- and high-frequency responses. Compare the Wilson-$\theta$ solution with the Newmark-$\beta$ solution in section "Numerical Example for the Newmark-$\beta$ Method," pp. 418–421. What are the disadvantages of the Wilson-$\theta$ method?
- Solve the problem with $\Delta t = 0.1$ for a series of values: $\theta = 1.1, 1.3, 1.5, 1.7$. Are the Wilson-$\theta$ solutions numerically stable for all values of $\theta$?

Note: It is possible to characterize the numerical stability of the Wilson-$\theta$ method using the spectral stability analysis that is presented in this chapter. The interested readers are referred to Bathe (2006). Only a concise discussion of the Wilson-$\theta$ method is provided here. The method is unconditionally stable for $\theta \geq \theta_0$, where $\theta_0 \approx 1.366$, and typically $\theta = 1.4$ is used. The figure below illustrates the spectral radius plot for the Wilson-$\theta$ method with $\theta = 1.4$, where too much numerical dissipation is introduced in the low-frequency range.

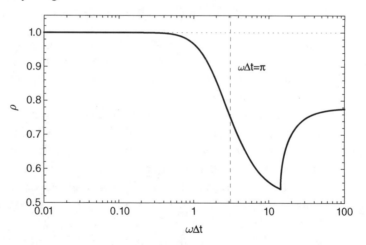

5. The von Kármán plate theory is frequently used to model and analyze the nonlinear response of the skin panel of an aircraft excited by aerodynamic forces. In this problem, the nonlinear vibration of a von Kármán plate is solved numerically using the generalized-$\alpha$ method. The plate is simply supported, semi-infinite, and of unit length, as illustrated in the following figure. Using the Rayleigh–Ritz method, the response of the von Kármán plate is approximated using 2-DOFs $\mathbf{q} = [q_1, q_2]^T$:

$$w(x, t) = q_1(t) \sin (\pi x) + q_2(t) \sin (2\pi x) \qquad (P7.44)$$

From Section 3.8, the equations of motion are

$$\ddot{\mathbf{q}} + [\mathbf{K}_L + \mathbf{K}_N(\mathbf{q})]\, \mathbf{q} = \mathbf{f}_E, \qquad (P7.45)$$

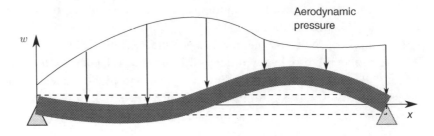

where

$$\mathbf{K}_L = \pi^4 \begin{bmatrix} 1 & 0 \\ 0 & 16 \end{bmatrix}, \quad \mathbf{K}_N(\mathbf{q}) = \pi^4 \begin{bmatrix} \frac{5}{2}q_1^2 + 10q_2^2 & 0 \\ 0 & 10q_1^2 + 40q_2^2 \end{bmatrix} \tag{P7.46}$$

The nonlinear displacement-dependent stiffness matrix term $\mathbf{K}_N$ is due to the geometric nonlinearity in von Kármán plate.

(1) Modify the computer program developed in Problem 2 so that it can solve nonlinear problems using Algorithm 7.4

    (a) The timeStepping function is now defined as

$$\mathbf{q}_{i+1}, \dot{\mathbf{q}}_{i+1}, \ddot{\mathbf{q}}_{i+1} = \text{timeStepping}(\mathbf{q}_i, \dot{\mathbf{q}}_i, \ddot{\mathbf{q}}_i, \mathbf{f}_E, \mathbf{f}_I),$$

    where $\mathbf{f}_E$ and $\mathbf{f}_I$ are nonlinear functions of $\mathbf{q}, \dot{\mathbf{q}}, t$. Note that the Newton-Raphson iteration needs to be implemented in the new timeStepping function.

    (b) Develop a modified generalized-$\alpha$ program that consecutively executes the functions setParameters and starting, and subsequently performs the function timeStepping to generate the numerical solution of $N$ time steps. Note that the initialize function in Problem 2 is no longer necessary.

(2) Assume there is no excitation force, $\mathbf{f}_E = 0$, and the initial conditions are $\mathbf{q}_0 = [0.0, 0.0]^T$ and $\dot{\mathbf{q}}_0 = [1.0, 0.0]^T$. Use a step size $\Delta t = 0.01$ and 2000 steps.

    (a) Solve Eq. (P7.45) using the modified generalized-$\alpha$ program.

    (b) Make Eq. (P7.45) linear by setting $\mathbf{K}_N = 0$, and solve the equation again.

    (c) Compare the nonlinear and linear responses: Do the two responses have the same oscillation frequencies and amplitudes? What are the possible explanations for the differences?

    (d) Repeat the above three steps with a set of new initial conditions $\mathbf{q}_0 = [0.0, 0.0]^T$ and $\dot{\mathbf{q}}_0 = [10.0, 0.0]^T$.

(3) Suppose the panel is subject to an external excitation due to supersonic aerodynamic loads. The external force vector is modeled as

$$\mathbf{f}_E(\mathbf{q}, \dot{\mathbf{q}}) = \mathbf{C}_A \dot{\mathbf{q}} + \mathbf{K}_A \mathbf{q}, \tag{P7.47}$$

where the so-called aerodynamic damping matrix $\mathbf{C}_A$ and aerodynamic stiffness matrix $\mathbf{K}_A$ are, respectively,

$$\mathbf{C}_A = -\mu \begin{bmatrix} 1 & 0 \\ 0 & 1 \end{bmatrix}, \quad \mathbf{K}_A = \lambda \begin{bmatrix} 0 & 8/3 \\ -8/3 & 0 \end{bmatrix} \tag{P7.48}$$

and $\lambda$ is the nondimensional aerodynamic pressure, $\mu = \sqrt{0.001\lambda}$. Using a step size $\Delta t = 0.01$ and initial conditions $\mathbf{q}_0 = [0.01, 0.0]^T$ and $\dot{\mathbf{q}}_0 = [0.0, 0.0]^T$.

(a) Generate the structural responses for (i) $\lambda = 270$, (ii) $\lambda = 275$, and (iii) $\lambda = 280$, and plot the time histories of $q_1$.

(b) Compare the structural responses associated with the three values of $\lambda$. Do all responses decay to zero due to the aerodynamic damping? What are the possible explanations for the differences?

# 8 Structural Dynamics of Rotating Systems

## 8.1 Introduction

Vibrations associated with rotating systems are important in many applications in aerospace and mechanical engineering. Some typical examples are the following: (1) vibrations of shafts to which masses are attached; (2) vibrations of turbine and compressor blades which can be found in jet engines, power-generating equipment and, more recently, wind turbines used for wind–energy conversion; (3) jet engine fan blade vibration problems; (4) propeller blade vibration problems; (5) helicopter rotor-blade vibration problems; and (6) vibrations of flexible appendages attached to spinning spacecraft.

It is also important to note that in the vibrations of several of the systems listed earlier, such as helicopter rotor blades, advanced propeller and fan blades, as well as wind turbine blades, the assumption of small blade deflections is not acceptable anymore. Blades of such systems have often moderate deflections, which require the treatment of geometrically nonlinear effects. It should be noted that the classical linear equations for rotor blades were derived in Houbolt and Brooks (1958). Subsequently, when the importance of moderate deflections was recognized in the aeroelastic behavior of helicopter rotor blades, several formulations accounting for moderate deflections were developed (Hodges and Dowell, 1974; Rosen and Friedmann, 1977). These formulations relied on ordering schemes to neglect higher order terms (H.O.T.s), while retaining terms up to the second order. Currently, there are several formulations which account for large deflections and small strains, as well as blades constructed from composite materials, which are in widespread use in rotary wing vehicles. This is a fascinating topic that has generated a substantial amount of research in the last 50 years; however, the topic is beyond the scope of this book.

The treatment of rotating systems provided in this chapter is limited to the free vibration problems of such systems, that is, determination of frequencies and mode shapes. Furthermore, the problems considered are limited to small deflections or linear problems. However, the free vibration problem only provides the mode shapes and frequencies used subsequently in calculating the forced response or the stability of such rotating systems.

## 8.2 Rotational Motion of a Rigid Body about a Fixed Point

Rotational motion relative to a given system of coordinates does not depend on the choice of the base point. Consider a base point of a rigid body fixed at the origin of Cartesian system X,Y,Z (shown in Figure 8.1). The point P is fixed in the body, and assume that the

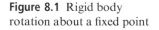

**Figure 8.1** Rigid body rotation about a fixed point

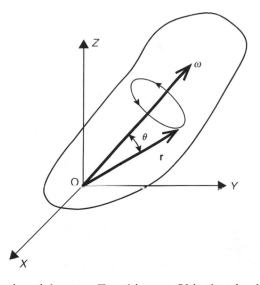

system X,Y,Z is fixed in inertial space. For this case, **V** is the absolute velocity of point P and $\boldsymbol{\omega}$ is the absolute angular velocity of the body. Rotation at any instance occurs about an axis passing through the fixed point O, known as the instantaneous axis of rotation. When the axis is fixed in the X,Y,Z system, then $\boldsymbol{\omega}$ has a fixed direction and point P describes a circular path (as shown in Figure 8.1) with a radius $r \sin \theta$, where $\theta$ is the angle between $\boldsymbol{\omega}$ and **r**. The speed of point P has a magnitude $v$ and is tangential to the circle given by

$$v = r\omega \sin \theta. \tag{8.2.1}$$

The velocity vector can also be expressed as a vector cross product

$$\mathbf{v} = \boldsymbol{\omega} \times \mathbf{r}. \tag{8.2.2}$$

Equation (8.2.1) represents the magnitude of the cross product given by Eq. (8.2.2), and it is oriented in a plane perpendicular to **r** and $\boldsymbol{\omega}$, which is tangential to the circle drawn in Figure 8.1. In a general case, $\omega$ and $\theta$ may vary with time, and Eq. (8.2.1) can be rewritten as

$$\frac{\Delta s}{\Delta t} = r\omega \sin \theta, \tag{8.2.3}$$

where $s$ is the displacement of point P along the circle shown in Figure 8.1. It is assumed that $\Delta \omega$ and $\Delta \theta$ approach zero as $\Delta t \to 0$, which implies that $\dot{\omega}$ is finite. However, $\Delta r$ is zero since point P and the base point O are fixed in the body. Thus, Eq. (8.2.3) is valid for the general case of rigid body rotation about a fixed point. Note also that

$$\mathbf{v} = \dot{\mathbf{r}} = \boldsymbol{\omega} \times \mathbf{r}. \tag{8.2.4}$$

Differentiating the last equation with respect to time again yields

$$\mathbf{a} = \ddot{\mathbf{r}} = \boldsymbol{\omega} \times \dot{\mathbf{r}} + \dot{\boldsymbol{\omega}} \times \mathbf{r}. \tag{8.2.5}$$

Combining Eqs. (8.2.4) and (8.2.5) yields

$$\mathbf{a} = \ddot{\mathbf{r}} = \boldsymbol{\omega} \times (\boldsymbol{\omega} \times \mathbf{r}) + \dot{\boldsymbol{\omega}} \times \mathbf{r}, \qquad (8.2.6)$$

where all the vectors are relative to the X,Y,Z system. The vector $\boldsymbol{\omega} \times (\boldsymbol{\omega} \times \mathbf{r})$ in Eq. (8.2.6) is directed radially inward from point P toward the axis of rotation and perpendicular to it. It is called centripetal acceleration. The force associated with it based on D'Alembert's principle is the centrifugal force.

## 8.2.1  Time Derivative of a Unit Vector

The derivatives of unit vectors, or their rate of change, is similar to the derivative of a position vector $\mathbf{r}$ in a rigid body rotating about a fixed point (Figure 8.1). This is due to the fact that unit vectors have a fixed length. Time derivative of a vector is interpreted as the velocity of the tip of the vector when the other end is fixed. Consider the orthogonal triad of unit vectors $\mathbf{e}_1, \mathbf{e}_2, \mathbf{e}_3$ drawn from the origin of a fixed system X,Y,Z, rotating together as a rigid body with absolute angular velocity $\omega$ (shown in Figure 8.2). Based on Eq. (8.2.4), the time derivative of the unit vectors is

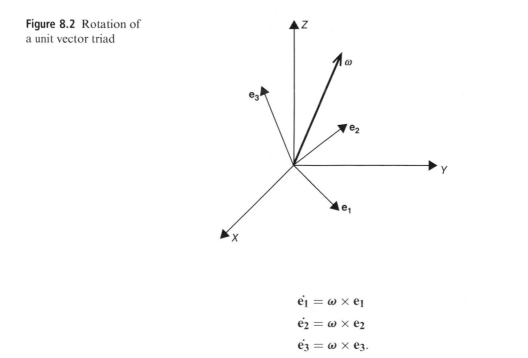

**Figure 8.2** Rotation of a unit vector triad

$$\dot{\mathbf{e}_1} = \boldsymbol{\omega} \times \mathbf{e}_1$$
$$\dot{\mathbf{e}_2} = \boldsymbol{\omega} \times \mathbf{e}_2$$
$$\dot{\mathbf{e}_3} = \boldsymbol{\omega} \times \mathbf{e}_3. \qquad (8.2.7)$$

The triad $\mathbf{e}_1, \mathbf{e}_2, \mathbf{e}_3$ is orthogonal, that is, $\mathbf{e}_1 \times \mathbf{e}_2 = \mathbf{e}_3$, and $\omega$ in general can be written as

$$\boldsymbol{\omega} = \omega_1 \mathbf{e}_1 + \omega_2 \mathbf{e}_2 + \omega_3 \mathbf{e}_3. \qquad (8.2.8)$$

The terms on the l.h.s. of Eq. (8.2.7) can be expanded as a vector cross product to yield

$$\dot{\mathbf{e}}_1 = \omega_3\mathbf{e}_2 - \omega_2\mathbf{e}_3$$
$$\dot{\mathbf{e}}_2 = \omega_1\mathbf{e}_3 - \omega_3\mathbf{e}_1$$
$$\dot{\mathbf{e}}_3 = \omega_2\mathbf{e}_1 - \omega_1\mathbf{e}_2. \tag{8.2.9}$$

If the rotating triad of unit vectors is a Cartesian triad of unit vectors $\mathbf{i}$, $\mathbf{j}$, $\mathbf{k}$ and angular velocity vector $\boldsymbol{\omega}$ is given by

$$\boldsymbol{\omega} = \omega_x\mathbf{i} + \omega_y\mathbf{j} + \omega_z\mathbf{k}, \tag{8.2.10}$$

equations similar to Eqs. (8.2.7) and (8.2.9) can be written by appropriate substitutions. Clearly, the time derivative of a unit vector lies in a plane perpendicular to the vector, as required by the definition of a cross product. At this stage, it is also important to define the terminology used. The terms *relative to* or *with respect to* a given system imply that the vector is viewed by an observer fixed in that system and moving with it. On the other hand, the term *referred to* a certain system implies that the vector is expressed in terms of unit vectors in that system.

## 8.2.2  Relative Motion and Rotating Frames

Newton's laws for motion of a particle require the acceleration to be *absolute*, which implies that it is measured in an inertial reference frame. Furthermore, the acceleration is invariant as long as it is measured relative to *any* inertial frame. For many applications, the motion of a particle is known relative to a rotating frame. A typical example is an element of mass of a rotating helicopter rotor blade, since the blade is part of the rotor which is attached to a moving helicopter. Such a situation is more complicated since one frame is moving while the other frame is rotating. The absolute velocity of a particle P, under such conditions, is given by

$$\mathbf{V}_P = \mathbf{V}_A + \mathbf{V}_{P/A}, \tag{8.2.11}$$

where $\mathbf{V}_A$ is the absolute velocity of a point on frame $A$ and $\mathbf{V}_{P/A}$ is the velocity of particle $P$ relative to $A$, as viewed by an observer in frame $A$ and moving with it. Similarly, the absolute acceleration

$$\mathbf{a}_P = \mathbf{a}_A + \mathbf{a}_{P/A}, \tag{8.2.12}$$

where frame $A$ is moving in pure translation. We are interested in the general case when the moving frame x,y,z is translating and rotating arbitrarily. Our objective is to find the velocity and acceleration of the particle $P$ relative to the inertial frame X,Y,Z in terms of its motion with respect to the non-inertial x,y,z frame. The origin of the non-inertial frame x,y,z is located at $O'$ is defined by a position vector $\mathbf{R}$ relative to the origin $O$ of the X,Y,Z frame, as shown in Figure 8.3. The position of the particle relative to $O'$ is given by $\boldsymbol{\rho}$, so the position of $P$ relative to the X,Y,Z frame is

**Figure 8.3** Moving and
rotating reference frames

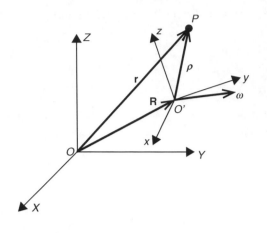

$$\mathbf{r} = \mathbf{R} + \boldsymbol{\rho}. \tag{8.2.13}$$

The corresponding velocity is

$$\mathbf{v} = \dot{\mathbf{r}} = \dot{\mathbf{R}} + \dot{\boldsymbol{\rho}}. \tag{8.2.14}$$

Using Eq. (8.2.4) to express $\dot{\boldsymbol{\rho}}$ in terms of motion relative to a moving frame,

$$\dot{\boldsymbol{\rho}} = \dot{\boldsymbol{\rho}}_r + \boldsymbol{\omega} \times \boldsymbol{\rho}. \tag{8.2.15}$$

The subscript $r$ denotes that the vector is in the moving coordinate system, and $\boldsymbol{\omega}$ is the angular velocity of the x,y,z frame. Writing the expressions in detail,

$$\boldsymbol{\rho} = x\mathbf{i} + y\mathbf{j} + z\mathbf{k} \tag{8.2.16}$$

$$\dot{\boldsymbol{\rho}}_r = \dot{x}\mathbf{i} + \dot{y}\mathbf{j} + \dot{z}\mathbf{k}, \tag{8.2.17}$$

where $\mathbf{i}, \mathbf{j}, \mathbf{k}$ are unit vectors fixed in the x,y,z frame and rotating with it. The absolute velocity of particle $P$ is

$$\mathbf{v} = \dot{\mathbf{r}} = \dot{\mathbf{R}} + \dot{\boldsymbol{\rho}}_r + \boldsymbol{\omega} \times \boldsymbol{\rho}. \tag{8.2.18}$$

The expression for inertial acceleration of particle $P$ is found by taking the time derivative of Eq. (8.2.18); for clarity, the derivatives of the second and third term in Eq. (8.2.18) are taken separately. Taking the derivative of the second term yields

$$\frac{d(\dot{\boldsymbol{\rho}}_r)}{dt} = \ddot{\boldsymbol{\rho}}_r + \boldsymbol{\omega} \times (\dot{\boldsymbol{\rho}}_r). \tag{8.2.19}$$

Taking the derivative of the third term in Eq. (8.2.18) yields

$$\frac{d(\boldsymbol{\omega} \times \boldsymbol{\rho})}{dt} = \dot{\boldsymbol{\omega}} \times \boldsymbol{\rho} + \boldsymbol{\omega} \times (\dot{\boldsymbol{\rho}}_r + \boldsymbol{\omega} \times \boldsymbol{\rho}). \tag{8.2.20}$$

Using Eqs. (8.2.19), (8.2.20), and Eq. (8.2.18) yields

$$\mathbf{a} = \dot{\mathbf{v}} = \ddot{\mathbf{R}} + \ddot{\boldsymbol{\rho}}_r + 2(\boldsymbol{\omega} \times \dot{\boldsymbol{\rho}}_r) + \boldsymbol{\omega} \times (\boldsymbol{\omega} \times \boldsymbol{\rho}) + \dot{\boldsymbol{\omega}} \times \boldsymbol{\rho}_r. \tag{8.2.21}$$

Next, it should be noted that $\rho_r \equiv \rho$, and the subscript $r$ is only used to emphasize operations in the rotating system; therefore, Eq (8.2.21) can be rewritten, using only $\rho$. Also for many applications, the origin of the rotating system coincides with the origin of the inertial system, thus $\mathbf{R} = 0$. Simplify Eq. (8.2.21) to yield

$$\mathbf{a} = \dot{\mathbf{v}} = \ddot{\rho} + 2(\boldsymbol{\omega} \times \dot{\rho}) + \boldsymbol{\omega} \times (\boldsymbol{\omega} \times \rho) + \dot{\boldsymbol{\omega}} \times \rho. \tag{8.2.22}$$

Equation (8.2.22) is important and frequently used in the treatment of dynamics of rotating systems. The terms in the equation have commonly used names that have significant physical meaning. The second term on the r.h.s represents the Coriolis acceleration, associated with the motion of the particle in the rotating system. The third term is the centripetal acceleration term, and the last term can be considered as changing tangential acceleration, as long as the orientation, or axis of rotation, is fixed. However, when the orientation or the axis of rotation changes, such as the rotor shaft of a maneuvering helicopter, the physical interpretation of this term becomes more complicated.

## 8.3  Equations of Motion for Vibrations of Rotating Beams

Consider the vibrations of a rotating beam (shown in Figure 8.4) which is initially straight. It has an elastic axis, coincident with the mass axis and coincident with the centroidal axis of the cross section. In its undeformed state, the elastic axis of the beam is assumed to be coincident with the $x$-axis. The axis of rotation is the $z$-axis (shown in Figure 8.4). It is assumed that the beam is rotating with constant angular speed $\Omega$. Furthermore, in this simplified treatment, only bending in and out of the plane of rotation is considered, and the beam is assumed to be rigid in torsion.

   With these assumptions, the initial position of the point on the undeformed elastic axis of the beam is written as

$$\mathbf{r}_0 = \mathbf{e}_x x, \tag{8.3.1}$$

**Figure 8.4** Geometry of a rotating beam undergoing deflections in and out of the plane of rotation

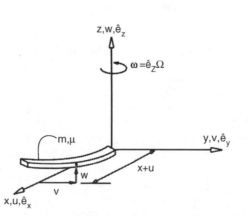

where $\mathbf{e}_x, \mathbf{e}_y, \mathbf{e}_z$ are unit vectors in the $x, y$, and $z$ directions, respectively, rotating with the beam. Also,

$$\boldsymbol{\omega} = \mathbf{e}_z \Omega. \tag{8.3.2}$$

An arbitrary point in the beam is given by

$$\mathbf{r} = \mathbf{e}_x x + \mathbf{e}_y y + \mathbf{e}_z z. \tag{8.3.3}$$

After the deformation, the position vector of a material point in the beam is given by

$$\mathbf{R} = \mathbf{e}_x (x + u) + \mathbf{e}_y (y + v) + \mathbf{e}_z (z + w), \tag{8.3.4}$$

where $u$, $v$, and $w$ are, respectively, the components of the displacement of the material point on the elastic axis of the bar.

From the previous section, the velocity vector (Eq. (8.3.5))

$$\mathbf{V} = \frac{d\mathbf{R}}{dt} + \boldsymbol{\omega} \times \mathbf{R} \tag{8.3.5}$$

and for completeness, the acceleration vector

$$\mathbf{a} = \ddot{\mathbf{R}} + 2\boldsymbol{\omega} \times \dot{\mathbf{R}} + \boldsymbol{\omega} \times (\boldsymbol{\omega} \times \mathbf{R}) + \dot{\boldsymbol{\omega}} \times \mathbf{R}. \tag{8.3.6}$$

Combining Eqs. (8.3.2), (8.3.4), and (8.3.5),

$$\mathbf{V} = \mathbf{e}_x \dot{u} + \mathbf{e}_y \dot{v} + \mathbf{e}_z \dot{w}$$
$$+ \Omega \mathbf{e}_z \times [\mathbf{e}_x (x + u) + \mathbf{e}_y (y + v) + \mathbf{e}_z (z + w)]. \tag{8.3.7}$$

$$= \mathbf{e}_x \dot{u} + \mathbf{e}_y \dot{v} + \mathbf{e}_z \dot{w} + \begin{vmatrix} \mathbf{e}_x & \mathbf{e}_y & \mathbf{e}_z \\ 0 & 0 & \Omega \\ (x+u) & (y+v) & (z+w) \end{vmatrix}$$

$$\mathbf{V} = \mathbf{e}_x [\dot{u} - (y+v)\Omega] + \mathbf{e}_y [\dot{v} + (x+u)\Omega] + \mathbf{e}_z \dot{w}$$

$$\mathbf{V} \cdot \mathbf{V} = \dot{\mathbf{R}} \cdot \dot{\mathbf{R}} = [\dot{u}^2 + \dot{v}^2 + \dot{w}^2 + (y+v)^2 \Omega^2 + (x+u)^2 \Omega^2$$
$$- 2\dot{u}(y+v)\Omega + 2(x+u)\Omega\dot{v}]. \tag{8.3.8}$$

Thus, the kinetic energy of the beam is given by

$$T = \frac{1}{2} \int_v \mu \mathbf{V} \cdot \mathbf{V} dx dy dz, \tag{8.3.9}$$

where $\mu$ is the mass per unit volume of the beam.

The equation of equilibrium can be obtained using two different methods: the direct Newtonian method and Hamilton's principle. Recall Hamilton's principle can be written as

$$\delta \int_{t_1}^{t_2} (T - U - V_E) dt = 0,$$

(8.3.10)

where $U$ is the strain energy and $V_E$ is the potential of externally applied loads.

Next, consider the strain energy $U$. Assume that the structural properties of the rotating beam are adequately described by the Bernoulli–Euler beam theory which is based on the following assumptions: (a) cross sections plane and normal to the undeformed axis of the beam remain plane and normal after the deformation, and (b) shear strains are negligible. With these assumptions,

$$\epsilon_{xy} = \epsilon_{yz} = \epsilon_{xz} = \epsilon_{yy} = \epsilon_{zz} = 0$$

(8.3.11)

and the only nonzero strain component is the axial strain $\epsilon_{xx}$.

An important feature of the rotating beam problem is that even when dealing with small strain vibrations of a rotating elastic beam, one must use the nonlinear strain displacement relations in order to obtain the correct formulation of the problem.

With Eq. (8.3.11), the strain energy is given by

$$U = \frac{E}{2} \int_V \epsilon_{xx}^2 \, dx \, dy \, dz.$$

(8.3.12)

The nonlinear strain in the $x$-direction is defined as

$$\epsilon_{xx} = \frac{\partial u}{\partial x} + \frac{1}{2}\left[\left(\frac{\partial u}{\partial x}\right)^2 + \left(\frac{\partial v}{\partial x}\right)^2 + \left(\frac{\partial w}{\partial x}\right)^2\right].$$

(8.3.13)

*It is important to note that after the introduction of the Euler–Bernoulli assumptions, v and w represent the displacements of a point on the elastic axis of the deformed beam out of the plane of rotation and in the plane of rotation.*

Assuming that the rotating beam experiences a steady-state value of axial tension (due to rotation) leading to an axial displacement $u_0$, from the Euler–Bernoulli assumption, one has

$$u = u_0 - z\frac{\partial w}{\partial x} - y\frac{\partial v}{\partial x}.$$

(8.3.14)

$$\frac{\partial u}{\partial x} = \frac{\partial u_0}{\partial x} - z\frac{\partial^2 w}{\partial x^2} - y\frac{\partial^2 v}{\partial x^2}.$$

(8.3.15)

Substituting Eq. (8.3.15) into Eq. (8.3.13), one has

$$\epsilon_{xx} = \frac{\partial u_0}{\partial x} - z\frac{\partial^2 w}{\partial x^2} - y\frac{\partial^2 v}{\partial x^2} + \frac{1}{2}\left[\left(\frac{\partial u_0}{\partial x} - z\frac{\partial^2 w}{\partial x^2} - y\frac{\partial^2 v}{\partial x^2}\right)^2 + \right.$$
$$\left. + \left(\frac{\partial v}{\partial x}\right)^2 + \left(\frac{\partial w}{\partial x}\right)^2\right].$$

(8.3.16)

Furthermore, the expression for the strain can be examined using a simple ordering scheme where the order of magnitude of the slope of the beam in bending is assumed to be of the order $\epsilon$, which implies $\frac{\partial w}{\partial x} = 0(\epsilon)$.

$$\epsilon_{xx}^2 \cong \underbrace{\left(\frac{\partial u_0}{\partial x}\right)^2}_{\epsilon^6} + z^2 \underbrace{\left(\frac{\partial^2 w}{\partial x^2}\right)^2}_{\epsilon^2} + y^2 \underbrace{\left(\frac{\partial^2 v}{\partial x^2}\right)^2}_{\epsilon^2} - 2z \frac{\partial u_0}{\partial x} \frac{\partial^2 w}{\partial x^2}$$

$$- 2y \underbrace{\frac{\partial u_0}{\partial x} \frac{\partial^2 v}{\partial x^2}}_{\epsilon^4} + 2zy \underbrace{\frac{\partial^2 w}{\partial x^2} \frac{\partial^2 u}{\partial x^2}}_{\epsilon^2} + \underbrace{\frac{\partial u_0}{\partial x} \left(\frac{\partial w}{\partial x}\right)^2}_{\epsilon^5}$$

$$+ \underbrace{\frac{\partial u_0}{\partial x} \left(\frac{\partial v}{\partial x}\right)^2}_{\epsilon^5} \tag{8.3.17}$$

+ H.O.T.s which drop out due to integration.

Regarding the H.O.T.s, note that the following assumptions are made. Elastic strains are small and are of the order $\epsilon^3$ where $\epsilon \cong 0.10$; thus, $\epsilon^2$ is negligible compared to one. Thus, $\frac{\partial u_0}{\partial x} = 0(\epsilon^3)$; $\left(\frac{\partial u_0}{\partial x}\right)^2 = 0(\epsilon^6)$ and slopes are larger, that is, $\frac{\partial w}{\partial x} = 0(\epsilon)$; thus, in Eq. (8.3.17), terms of $O(\epsilon^6)$ and higher are neglected.

Furthermore, when Eq. (8.3.17) is substituted into Eq. (8.3.12) and use is made of the various area integrals are used, one has

$$U = \frac{1}{2}E \int_V \left[ +z^2 \left(\frac{\partial^2 w}{\partial x^2}\right)^2 + y^2 \left(\frac{\partial^2 v}{\partial x^2}\right)^2 - 2z \frac{\partial u_0}{\partial x} \frac{\partial^2 w}{\partial x^2} \right. \tag{8.3.18}$$

$$\left. - 2y \frac{\partial u_0}{\partial x} \frac{\partial^2 v}{\partial x^2} + 2zy \frac{\partial^2 w}{\partial x^2} \frac{\partial^2 v}{\partial x^2} + \frac{\partial u_0}{\partial x} \left(\frac{\partial w}{\partial x}\right)^2 + \frac{\partial u_0}{\partial x} \left(\frac{\partial v}{\partial x}\right)^2 \right] dx\,dy\,dz$$

$$\left. \begin{array}{l} \int\int dy\,dz = A \;;\; \int\int y\,dy\,dz = 0 \;;\; \int\int z\,dz\,dy = 0 \\ \int\int zy\,dy\,dz = 0 \\ \int\int z^2\,dy\,dz = I_y \;;\; \int\int y^2\,dy\,dz = I_z \end{array} \right\} \tag{8.3.19}$$

for the coordinate system coinciding with the area centroid, and a symmetric cross section.

$$U = \int_0^l \frac{EI_y}{2} \left(\frac{\partial^2 w}{\partial x^2}\right)^2 dx + \int_0^L \frac{EI_z}{2} \left(\frac{\partial^2 v}{\partial x^2}\right)^2 dx +$$

$$+ \int_0^L \frac{EA}{2} \left(\frac{\partial u_0}{\partial x}\right) \left(\frac{\partial w}{\partial x}\right)^2 dx + \int_0^L \frac{EA}{2} \left(\frac{\partial u_0}{\partial x}\right) \left(\frac{\partial v}{\partial x}\right)^2 dx. \tag{8.3.20}$$

At this point, one can replace the steady-state axial strain, to the first approximation, by the steady-state axial force $P(x)$, where

$$P(x) = EA \frac{\partial u_0}{\partial x}. \tag{8.3.21}$$

Thus,

$$U = \int_0^L \left[ \frac{EI_y}{2} \left(\frac{\partial^2 w}{\partial x^2}\right)^2 + \frac{EI_z}{2} \left(\frac{\partial^2 v}{\partial x^2}\right)^2 + \frac{P(x)}{2} \left(\frac{\partial w}{\partial x}\right)^2 + \frac{P(x)}{2} \left(\frac{\partial v}{\partial x}\right)^2 \right] dx, \quad (8.3.22)$$

where the axial load is a function of the spanwise location

$$P(x) = \int_x^L m\Omega^2 \eta \, d\eta \quad (8.3.23)$$

and

$$m = \mu dy dz = \text{mass per unit length.}$$

Consider certain simplifications in the expression for the kinetic energy:

$$T = \frac{1}{2} \int_V \mu \left[ \dot{u}^2 + \dot{v}^2 + \dot{w}^2 + (y+v)^2\Omega^2 + (x+u)^2\Omega^2 - \right.$$

$$\left. - 2\dot{u}(y+v)\Omega + 2(x+u)\Omega\dot{v} \right] dx dy dz. \quad (8.3.24)$$

Rotatory inertia is neglected, that is, $\int y^2 dy dz$– terms in the kinetic energy are neglected, an assumption also made in Euler–Bernoulli beam theory. Furthermore, we also use $\int y dy dz = 0$.

We are not interested in the radial (axial) vibration problem; thus, the $\dot{u}^2$ term in Eq. (8.3.24) can be neglected. Also, $u << x$, thus $x + u \cong x$; with these assumptions,

$$T = \frac{1}{2} \int_0^L m[\dot{v}^2 + \dot{w}^2 + v^2\Omega^2 + x^2\Omega^2 - \underline{2\Omega\dot{u}v} + 2x\Omega\dot{v} + \underline{2\Omega\dot{v}u}]dx. \quad (8.3.25)$$

The underlined terms in Eq. (8.3.25) are nonlinear H.O.T.s, due to Coriolis acceleration, and can be neglected in the free vibration problem. However, such terms can be important in stability and response problems. Thus,

$$\delta \int_{t_1}^{t_2} T dt = \int_{t_1}^{t_2} \int_0^L m[\dot{v}\delta\dot{v} + \dot{w}\delta\dot{w} + v\delta v\Omega^2 + x\Omega\delta\dot{v}]dx dt.$$

Integrating by parts,

$$\delta \int_{t_1}^{t_2} T dt = \int_0^L m \left[ \dot{v}\delta v \Big|_{t_1}^{t_2} - \int_{t_1}^{t_2} \ddot{v}\delta v dt + \right.$$

$$+ \dot{w}\delta w \Big|_{t_1}^{t_2} - \int_{t_1}^{t_2} \ddot{w}\delta w dt + \int_{t_1}^{t_2} \Omega^2 v\delta v dt -$$

$$\left. + x\Omega\delta v \Big|_{t_1}^{t_2} - \int_{t_1}^{t_2} \delta v \frac{d}{dt}(\Omega x) dt \right] dx. \quad (8.3.26)$$

Recall that according to Hamilton's principle,

$$\delta v\Big|_{t_1}^{t_2} = \delta w\Big|_{t_1}^{t_2} = 0,$$

thus

$$\delta \int_{t_1}^{t_2} T dt = \int_0^L \int_{t_1}^{t_2} [-m\ddot{v}\delta v - m\ddot{w}\delta w + m\Omega^2 v\delta v]dxdt. \tag{8.3.27}$$

In addition to the kinetic and strain energy, one can introduce external loads acting on the rotating beam given by

$$\mathbf{P} = p_y \mathbf{e}_y + p_z \mathbf{e}_z, \tag{8.3.28}$$

then

$$V_E = -\int_0^L (p_y v + p_z w)dx. \tag{8.3.29}$$

$$\delta V_E = -\int_0^L (p_y \delta v + p_z \delta w)dx. \tag{8.3.30}$$

Using Eq. (8.3.22), one has

$$\delta U = \int_0^L \left[ EI_y \frac{\partial^2 w}{\partial x^2}\frac{\partial^2 \delta w}{\partial x^2} + EI_z \frac{\partial^2 v}{\partial x^2}\frac{\partial^2 \delta v}{\partial x^2} + \right. \\ \left. + P(x)\frac{\partial w}{\partial x}\frac{\partial \delta w}{\partial x} + P(x)\frac{\partial v}{\partial x}\frac{\partial \delta v}{\partial x} \right] dx. \tag{8.3.31}$$

Integrating by parts,

$$\delta U = EI_y \frac{\partial^2 w}{\partial w^2}\frac{\partial \delta w}{\partial x}\Big|_0^L - \frac{\partial}{\partial x}\left( EI_y \frac{\partial^2 w}{\partial x^2} \right)\delta w\Big|_0^L + \int_0^L \frac{\partial^2}{\partial x^2}\left( EI_y \frac{\partial^2 w}{\partial x^2} \right)dx\delta w$$

$$+ EI_z \frac{\partial^2 v}{\partial x^2}\frac{\partial \delta v}{\partial x}\Big|_0^L - \frac{\partial}{\partial x}\left( EI_z \frac{\partial^2 v}{\partial x^2} \right)\delta v\Big|_0^L + \int_0^L \frac{\partial^2}{\partial x^2}\left( EI_z \frac{\partial^2 v}{\partial x^2} \right)dx\delta v$$

$$+ P(x)\frac{\partial w}{\partial x}\delta w\Big|_0^L - \int_0^L \frac{\partial}{\partial x}\left[ P(x)\frac{\partial w}{\partial x} \right]\delta w dx$$

$$+ P(x)\frac{\partial v}{\partial x}\delta v\Big|_0^L - \int_0^L \frac{\partial}{\partial x}\left[ P(x)\frac{\partial v}{\partial x} \right]\delta v dx. \tag{8.3.32}$$

Finally, when Eqs. (8.3.30), (8.3.32), (8.3.27), and (8.3.10) are combined, one obtains the final equation

$$\delta \int_{t_2}^{t_2} (T - U - V_E)dt = 0 = \int_0^L \int_{t_2}^{t_2} [-m\ddot{v}\delta v - m\ddot{w}\delta w \tag{8.3.33}$$

$$+m\Omega^2 v\delta v - \frac{\partial^2}{\partial x^2}\left(EI_y \frac{\partial^2 w}{\partial x^2}\right)\delta w + \frac{\partial}{\partial x}\left[P(x)\frac{\partial w}{\partial x}\right]\delta w$$

$$- \left[\frac{\partial^2}{\partial x^2}\left(EI_z \frac{\partial^2 v}{\partial x^2}\right)\delta v + \frac{\partial}{\partial x}\left[P(x)\frac{\partial v}{\partial x}\right]\delta v\right.$$

$$+ P_y\delta v + P_z\delta w]]\,dxdt + \int_{t_2}^{t_2}\left\{EI_y \frac{\partial^2 w}{\partial x^2}\frac{\partial \delta w}{\partial x}\Big|_0^L\right.$$

$$+ \left[-\frac{\partial}{\partial x}\left(EI_y \frac{\partial^2 w}{\partial x^2}\right) + P(x)\frac{\partial w}{\partial x}\right]\delta w\Big|_0^L$$

$$+ EI_z \frac{\partial^2 v}{\partial x^2}\frac{\partial \delta v}{\partial x}\Big|_0^L + \left[-\frac{\partial}{\partial x}\left(EI_z \frac{\partial^2 v}{\partial x^2}\right) + P(x)\frac{\partial v}{\partial x}\right]\delta v\Big|_0^L\right\}\,dt = 0$$

from which one obtains the final equations of motion and corresponding boundary conditions:

$$\frac{\partial^2}{\partial x^2}\left[EI_y \frac{\partial^2 w}{\partial x^2}\right] - \frac{\partial}{\partial x}\left[P\frac{\partial w}{\partial x}\right] = -m\ddot{w} + p_z(x,t). \tag{8.3.34}$$

$$\frac{\partial^2}{\partial x^2}\left(EI_z \frac{\partial^2 v}{\partial x^2}\right) - \frac{\partial}{\partial x}\left[P(x)\frac{\partial v}{\partial x}\right] = -m\ddot{v} + m\Omega^2 v + p_y(x,t). \tag{8.3.35}$$

Furthermore, for $x = L$, $P(x = L) = 0$; thus, for a cantilevered rotating beam, the boundary conditions are

$$\left.\begin{array}{ll} EI_y \frac{\partial^2 w}{\partial x^2} = 0 \quad \text{and} \quad \frac{\partial}{\partial x}\left(EI_y \frac{\partial^2 w}{\partial x^2}\right) = 0 \quad \text{at } x = L \\ EI_z \frac{\partial^2 v}{\partial x^2} = 0 \quad \text{and} \quad \frac{\partial}{\partial x}\left(EI_z \frac{\partial^2 v}{\partial x^2}\right) = 0 \quad \text{at } x = L \\ \text{also} \qquad v = w = 0 \qquad \text{at } x = 0 \\ \text{and} \qquad \frac{\partial v}{\partial x} = \frac{\partial \omega}{\partial x} = 0 \quad \text{at } x = 0 \end{array}\right\}. \tag{8.3.36}$$

Equations (8.3.34) and (8.3.35) describe, respectively, the dynamics of a rotating beam, performing small amplitude oscillations out of the plane of rotation and in the plane of rotation. When the linear case only is considered, and there is no pitch angle between these two degrees of freedom (DOF), the vibrations in these two mutually perpendicular planes are uncoupled. The vibrations of a beam perpendicular to the plane of rotation are sometimes denoted as flapwise vibrations, while the vibrations in the plane of rotation are denoted as lead-lag vibrations. This terminology is used for rotary wing vehicles (i.e., helicopters).

In comparing Eqs. (8.3.34) and (8.3.35), it is clear that while the equations are similar, considerable differences also exist. The main difference between the two equations is associated with the second term on the right-hand side of Eq. (8.3.35), $m\Omega^2 v$; this term reduces the stiffening due to centrifugal effects, when dealing with vibrations in the plane of rotation.

## 8.4    Alternative Derivation for Vibrations of Rotating Beams

From the preceding discussion of vibrations of axially loaded structures, Eq. (2.9.20), the equations of equilibrium for bending out of the plane of rotation and in the plane of rotation as shown in Figure 8.5 are given by

$$\left.\begin{array}{c} \frac{\partial^2}{\partial x^2}\left(EI_y \frac{\partial^2 w}{\partial x^2}\right) - \frac{\partial}{\partial x}\left(P\frac{\partial w}{\partial x}\right) = p_z(x_1 t) \\ \frac{\partial^2}{\partial x^2}\left(EI_z \frac{\partial^2 v}{\partial x^2}\right) - \frac{\partial}{\partial x}\left(P\frac{\partial v}{\partial x}\right) = p_y(x_1 t) \end{array}\right\}. \tag{8.4.1}$$

Inertia forces, due to accelerations, can be introduced through the loading terms $p_z(x, t)$ and $p_y(x, t)$ in Eq. (8.4.1) using D'Alembert's principle and $P$ is the axial load in tension.

**Figure 8.5** Vibration of a rotating beam: (a) out of the plane of rotation; (b) in the plane of rotation

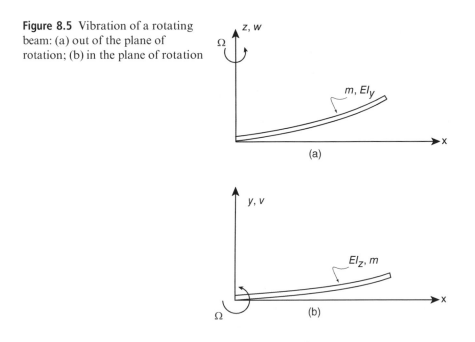

The position vector of a point on the deformed elastic axis of a beam is given by

$$\mathbf{r}_E = (x + u)\mathbf{e_x} + v\mathbf{e_y} + w\mathbf{e_z}, \tag{8.4.2}$$

where it is assumed that the elastic axis and the center of mass in the beam cross section coincide. Using Eq. (8.3.6) with Eq. (8.3.2), one has for $\dot{\Omega} = 0$

$$\mathbf{a} = \ddot{\mathbf{r}}_E + 2\boldsymbol{\omega} \times \dot{\mathbf{R}}_E + \boldsymbol{\omega} \times (\boldsymbol{\omega} \times \mathbf{r}_E) \tag{8.4.3}$$

$$\boldsymbol{\omega} \times \mathbf{r}_E = \begin{vmatrix} \mathbf{e_x} & \mathbf{e_y} & \mathbf{e_z} \\ 0 & 0 & \Omega \\ x+u & v & w \end{vmatrix} = \mathbf{e_x}(-v\Omega) - \mathbf{e_y}[-\Omega(x+u)]$$

$$\boldsymbol{\omega} \times \mathbf{r}_E = -v\Omega\mathbf{e_x} + \Omega(x+u)\mathbf{e_y}$$

$$2\boldsymbol{\omega} \times \dot{\mathbf{R}}_E = -2\dot{v}\Omega\mathbf{e_x} + 2\Omega\dot{u}\mathbf{e_y}$$

$$\boldsymbol{\omega} \times (\boldsymbol{\omega} \times \mathbf{r}_E) = \begin{vmatrix} \mathbf{e_x} & \mathbf{e_y} & \mathbf{e_z} \\ 0 & 0 & \Omega \\ -v\Omega & \Omega(x+u) & 0 \end{vmatrix} = -\mathbf{e_x}\Omega^2(x+u) - \mathbf{e_y}v\Omega^2.$$

$$\begin{aligned} \mathbf{a} &= \ddot{u}\mathbf{e_x} + \ddot{v}\mathbf{e_y} + \ddot{w}\mathbf{e_z} - 2\dot{v}r\mathbf{e_x} + 2\Omega\dot{u}\mathbf{e_y} - \mathbf{e_y}\Omega^2(x+u) - \mathbf{e_y}v\Omega^2 \\ &= \mathbf{e_x}[\ddot{u} - 2\Omega\dot{v} - \Omega^2(x+u)] + \mathbf{e_y}[\ddot{v} - \Omega^2 v + 2\Omega\dot{u}] \\ &\quad + \mathbf{e_z}\ddot{w} = a_x\mathbf{e_x} + a_y\mathbf{e_y} + a_z\mathbf{e_z}. \end{aligned} \tag{8.4.4}$$

From Eq. (8.4.4),

$$\begin{aligned} p_z &= -ma_z = -m\ddot{w}. \\ p_y &= -ma_y = -m[\ddot{v} - \Omega^2 v + 2\Omega\dot{u}]. \end{aligned} \tag{8.4.5}$$

The last term in Eq. (8.4.5) is a Coriolis term which is neglected. Combining Eqs. (8.4.5) and (8.4.1) yields

$$\left. \begin{aligned} \frac{\partial^2}{\partial x^2}\left(EI_y\frac{\partial^2 w}{\partial x^2}\right) - \frac{\partial}{\partial x}\left[P(x)\frac{\partial w}{\partial x}\right] &= -m\ddot{w} \\ \frac{\partial^2}{\partial x^2}\left(EI_z\frac{\partial^2 v}{\partial x^2}\right) - \frac{\partial}{\partial x}\left[P(x)\frac{\partial v}{\partial x}\right] &= -m\ddot{v} + m\Omega^2 v \end{aligned} \right\}, \tag{8.4.6}$$

which are basically identical to Eqs. (8.3.34) and (8.3.35) with

$$P(x) = \int_x^L \Omega^2 m x_1 \, dx_1, \tag{8.4.7}$$

where $x_1$ is a dummy variable and $m$ is the mass distribution per unit length.

### 8.4.1 Example Illustrating the Difference between Vibrations In and Out of the Plane of Rotation

Consider a rotating beam and use Rayleigh's quotient (RQ) to approximate the effect of rotation on the natural frequencies in and out of the plane of rotation (Eqs. (8.3.34) and (8.3.35)).

$$U = \int_0^L \left[ \frac{EI_y}{2} \left( \frac{\partial^2 w}{\partial x^2} \right)^2 + \frac{EI_z}{2} \left( \frac{\partial^2 v}{\partial x^2} \right)^2 + \frac{P(x)}{2} \left( \frac{\partial w}{\partial x} \right)^2 + \frac{P(x)}{2} \left( \frac{\partial v}{\partial x} \right)^2 \right] dx$$

$$P(x) = \int_x^L m\Omega^2 \eta d\eta$$

$$T = \frac{1}{2} \int_0^L m \left[ \dot{v}^2 + \dot{w}^2 + v^2 \Omega^2 \right] dx.$$

Vibration out of plane of rotation with $\omega_{AF}$, the approximate frequency

$$\omega_{AF}^2 = \frac{\int_0^L EI_y w_{,xx}^2 \, dx + \int_0^L P(x) w_{,x}^2 \, dx}{\int_0^L mw^2 dx}.$$

Assume simple function $w(x) = ax^2$

$$\omega_{AF}^2 = \frac{\int_0^L EI(2a)^2 dx + \int_0^L P(x)(2ax)^2 dx}{\int_0^L ma^2 x^4 dx} = \frac{4EIL + \int_0^L P(x)4x^2 dx}{\int_0^L mx^4 dx}$$

$$P(x) = \int_x^L m\Omega^2 \eta d\eta$$

for a uniform beam

$$P(x) = \int_x^L m\Omega^2 \frac{\eta^2}{2}|_x^L = m\Omega^2 \frac{1}{2} \left( L^2 - x^2 \right)$$

$$\int_0^L P(x)(2ax)^2 dx = \int_0^L m\Omega^2 \frac{1}{2} \left( L^2 - x^2 \right) 4x^2 dx = \frac{4m\Omega^2 L^5}{15}$$

$$\omega_{AF}^2 = \frac{4EIL + \frac{4}{15}m\Omega^2 L^5}{\frac{mL^5}{5}}.$$

Next, consider the vibration in the plane of rotation, given by Eq. (8.3.35), rewritten as

$$EI_z \frac{\partial^4 v}{\partial x^4} - \frac{\partial}{\partial x} \left[ P(x) \frac{\partial v}{\partial x} \right] = -m\ddot{v} + m\Omega^2 v$$

$$EI_z \frac{\partial^4 v}{\partial x^4} - \frac{\partial}{\partial x} \left[ P(x) \frac{\partial v}{\partial x} \right] - m\Omega^2 v = -m\ddot{v}$$

for free vibrations $v = v(x)e^{i\omega t}$. For vibrations in the plane of rotation, RQ has to be used in the operator notation form (3.5.4) and the $L[..]$ operator is identified from the last equation, so the approximate frequency for in-plane vibrations $\omega_{AL}^2$

$$L[v] = EI_z\frac{\partial^4 v}{\partial x^4} - \frac{\partial}{\partial x}\left[P(x)\frac{\partial v}{\partial x}\right] - m\Omega^2 v$$

$$\omega_{AL}^2 = \frac{\int_0^L vEI_z\frac{\partial^4 v}{\partial x^4}dx + \int_0^L -\frac{\partial}{\partial x}\left[P(x)\frac{\partial v}{\partial x}\right]vdx - \int_0^L m\Omega^2 v^2 dx}{\int_0^L mv^2 dx}.$$

Integrating by parts the first two terms in the numerator of the last expression,

$$\int_0^L vEI_z\frac{\partial^4 v}{\partial x^4}dx = EI_z v\frac{\partial^3 v}{\partial x^3}\Big|_0^L - EI_z\frac{\partial v}{\partial x}\frac{\partial^2 v}{\partial x^2}\Big|_0^L + \int_0^L EI_z\,(v_{,xx})^2\,dx$$

$$\int_0^L -v\frac{\partial}{\partial x}\left[P(x)\frac{\partial v}{\partial x}\right]dx = vP(x)\frac{\partial v}{\partial x}\Big|_0^L - \int_0^L P(x)v_{,x}^2\,dx.$$

Combining the terms in the approximate frequency equation,

$$\omega_{AL}^2 = \frac{\int_0^L EI_z v_{,xx}^2\,dx + \int_0^L P(x)v_{,x}^2\,dx - \int_0^L m\Omega^2 v^2 dx}{\int_0^L mv^2 dx}.$$

Using an identical mode shape to that used for out-of-plane vibrations $v(x) = ax^2$, for a uniform beam

$$\omega_{AL}^2 = \frac{\int_0^L EI(2a)^2 dx + \int_0^L P(x)(2ax)^2 dx - \int_0^L m\Omega^2 a^2 x^4 dx}{\int_0^L ma^2 x^4 dx}$$

$$= \frac{4EIL + \frac{4}{15}m\Omega^2 L^5 - \frac{1}{5}m\Omega^2 L^5}{\frac{mL^5}{5}} = \frac{4EIL + \frac{1}{15}m\Omega^2 L^5}{\frac{mL^5}{5}}.$$

This illustrates that the effect of rotation is different for out-of-plane and in-plane frequencies. The centrifugal stiffening due to rotation is much stronger for vibrations out of the plane of rotation than that due to inplane of rotation vibrations. The reduction in plane of rotation stiffening due to centrifugal effects is a result of the term $-m\Omega^2 v^2$.

## 8.5    Galerkin Solution of Rotating Beam Vibration

The frequencies and mode shapes for in plane and out of plane of rotation vibrations of beams with various boundary conditions can be obtained conveniently by using a variety of approximate methods. It is important to note that there are no closed form solutions for the rotating beam problem even for the simple case of a beam having constant mass and stiffness distributions along its span.

A convenient method for obtaining approximate solutions to this problem is Galerkin's method. In this case, it is convenient to assume that the rotating mode shapes for a beam having arbitrary mass and stiffness distribution can be expressed as a series in terms of the nonrotating modes, which satisfy all the boundary conditions of the problem. Clearly, mode shapes for nonrotating uniform beams can be generated conveniently. Thus, for vibrations, out-of-plane of rotation or the $xz$ plane one has

$$w(x) = \sum_{q=1}^{n} A_q \phi_q(x), \tag{8.5.1}$$

where the functions $\phi_q(x)$ satisfy all the boundary conditions of the problem. When dealing with a clamped, rotating beam of length $L$, these boundary conditions are

$$\left. \begin{array}{l} \phi_q(x=0) = \phi_q'(x=0) = 0 \\ EI_y \dfrac{\partial^2 \phi_q}{\partial x^2} = \dfrac{\partial}{\partial x}\left(EI_y \dfrac{\partial^2 \phi_q}{\partial x^2}\right) = 0 \quad \text{at } x = L \end{array} \right\}. \tag{8.5.2}$$

For a hinged rotating beam, these would be

$$\phi_q(x=0) = 0 \quad \text{and} \quad EI_y \frac{\partial^2 \phi_q}{\partial x^2} = 0. \tag{8.5.3}$$

The boundary conditions at the free end are the same. Furthermore, for convenience, the equations of motion are nondimensionalized by defining the following quantities:

$$\bar{x} = x/L : \bar{m} = m/mo ; \overline{EI} = EI/EI_0,$$

where $m_0$ and $EI_0$ are, respectively, the mass per unit length and the bending stiffness at the root ($x = 0$) of the beam. Furthermore, introducing the assumption of simple harmonic motion, Eqs. (8.3.34) and (8.3.35) become

$$\frac{EI_{0y}}{L^4} \frac{\partial^2}{\partial \bar{x}^2}\left(\overline{EI}_y \frac{\partial^2 w}{\partial \bar{x}^2}\right) = \frac{\Omega^2 m_0}{L^2} \frac{\partial}{\partial \bar{x}}\left(P^* \frac{\partial w}{\partial \bar{x}}\right) + m_o \omega^2 \frac{mw}{m_o},$$

where

$$P^* = L^2 \int_{\bar{x}}^{1} m(\bar{x}_1)\bar{x}_1 \, d\bar{x}_1.$$

Rearranging the last equation, one has

$$\frac{\partial^2}{\partial \bar{x}^2}\left(E\bar{I}_y\frac{\partial^2 w}{\partial \bar{x}^2}\right) = \frac{\Omega^2 m_0 L^4}{EI_{0y}}\frac{\partial}{\partial \bar{x}}\left(\bar{P}\frac{\partial w}{\partial \bar{x}}\right) + \frac{m_0 L^4}{EI_{0y}}\omega^2 \bar{m}w, \tag{8.5.4}$$

where

$$\bar{P} = \int_{\bar{x}}^{1} \bar{m}(\bar{x}_1)\bar{x}_1 d\bar{x}_1. \tag{8.5.5a}$$

In an analogous manner, for vibrations in the plane of rotation, one has

$$\frac{\partial}{\partial \bar{x}^2}\left(\bar{EI}_z\frac{\partial^2 v}{\partial \bar{x}^2}\right) = \frac{\Omega^2 m_0 L^4}{EI_{0z}}\frac{\partial}{\partial \bar{x}}\left(\bar{P}\frac{\partial v}{\partial \bar{x}}\right) + (\omega^2 + \Omega^2)\frac{m_0 L^4}{EI_{0z}}\bar{m}v. \tag{8.5.5b}$$

Combining Eqs. (8.5.5a), (8.5.4), and (8.5.1), one has

$$\frac{\partial^2}{\partial \bar{x}^2}\left[\bar{EI}_z\frac{\partial^2}{\partial x^2}\left(\sum_{q=1}^{n} A_q\phi_q\right)\right] - \frac{\Omega^2 m_0 L^4}{EI_{0y}}\frac{\partial}{\partial \bar{x}}\left[\bar{P}\frac{\partial}{\partial \bar{x}}\left(\sum_{q=1}^{n} A_q\phi_q\right)\right]$$

$$-\frac{\omega^2 m_0 L^4}{EI_{oy}}\bar{m}\sum_{q=1}^{n} A_q\phi_q = \varepsilon(w). \tag{8.5.6}$$

Clearly, by writing all the terms of Eq. (8.5.6) on the left-hand side, one obtains the error function commonly used in Galerkin's method, Eq. (3.5.3). The undetermined coefficients $\phi_q(\bar{x})$ are obtained from the requirement

$$\int_{0}^{1} \varepsilon(w)\phi_p d\bar{x} = 0 \qquad \text{for} \qquad p = 1, 2, \ldots, n. \tag{8.5.7}$$

Combining Eqs. (8.5.6) and (8.5.7),

$$\int_{0}^{1} \phi_p\frac{\partial^2}{\partial \bar{x}^2}\left[\bar{EI}_y\frac{\partial^2}{\partial \bar{x}^2}\left(\sum_{q=1}^{n} A_q\phi_q\right)\right]d\bar{x}$$

$$= \int_{0}^{1} \phi_p\frac{\Omega^2 m_0 L^4}{EI_{0y}}\frac{\partial}{\partial \bar{x}}\left[\bar{P}\frac{\partial}{\partial \bar{x}}\left(\sum_{q=1}^{n} A_q\phi_q\right)\right]d\bar{x} +$$

$$+ \int_{0}^{1} \phi_p\frac{\omega^2 m_0 L^4}{EI_{0y}}\bar{m}\sum_{q=1}^{n} A_q\phi_q d\bar{x}. \tag{8.5.8}$$

In Eq. (8.5.8), the first term is integrated by parts twice and the second term is integrated by parts once, and making use of the boundary conditions, Eq. (8.5.2), for a cantilevered beam, one has

$$\int_0^1 \phi_p'' \overline{EI}_y \sum_{q=1}^n A_q \phi_q'' d\bar{x} = - \int_0^1 \frac{\Omega m_0 L^4}{EI_{0y}} \bar{P} \phi_p' \sum_{q=1}^n A_q \phi_q' d\bar{x}$$

$$+ \int_0^1 \frac{\omega^2 m_0 L^4}{EI_{0y}} \bar{m} \phi_p \sum_{q=1}^n A_q \phi_q d\bar{x}. \qquad (8.5.9)$$

Interchanging the order of summation and integration, Eq. (8.5.9) can be written as

$$\sum_{q=1}^n A_q \left( \int_0^1 \overline{EI}_y \phi_p'' \phi_q'' d\bar{x} + \frac{\Omega^2 m_0 L^4}{EI_{0y}} \int_0^1 \bar{P} \phi_p' \phi_q' - Q d\bar{x} \right.$$

$$\left. - \omega^2 \frac{m_0 L^4}{EI_{0y}} \int_0^1 \bar{m} \phi_p \phi_q d\bar{x} \right) = 0. \qquad (8.5.10)$$

An analogous derivation can be made for the vibrations of the beam in the plane of rotation. For this case, the in-plane deflection of the elastic axis is expressed as

$$v(x) = \sum_{q=1}^n B_q \psi_q(x). \qquad (8.5.11)$$

The boundary conditions for a cantilevered beam are

$$\left. \begin{array}{l} \psi_q(x=0) = \psi'_q(x=0) = 0 \\ EI_z \frac{\partial^2 \psi_q}{\partial x^2} = \frac{\partial}{\partial x}\left( EI_z \frac{\partial^2 \psi_q}{\partial x^2} \right) = 0 \quad \text{at} \quad x = L \end{array} \right\}. \qquad (8.5.12)$$

Performing a set of operations similar to those performed for the vibrations out of plane of rotation (i.e., combining Eqs. (8.5.11) and (8.5.5a)) yields

$$\sum_{q=1}^n B_q \left( \int_0^1 \overline{EI}_z \psi_q'' \psi_r'' d\bar{x} + \frac{\Omega^2 m_0 L^4}{EI_{0z}} \int_0^1 \bar{P} \psi_q' \psi_r' d\bar{x} \right.$$

$$\left. -(\omega^2 + \Omega^2)\frac{m_0 L^4}{EI_{0z}} \int_0^1 \bar{m} \psi_q \psi_r d\bar{x} \right) = 0 \quad r = 1, 2, \ldots, n. \qquad (8.5.13)$$

Equations (8.5.10) and (8.5.13) hold, if the relations inside the parentheses are zero. Thus, for vibrations out of the plane of rotation, one has

$$A_q \left[ \int_0^1 \overline{EI}_y \phi_p'' \phi_p'' d\bar{x} + \frac{\Omega^2 m_0 L^4}{EI_{0y}} \int_0^1 \bar{P} \phi_p' \phi_q' d\bar{x} - \frac{\omega^2 m_0 L^4}{EI_{0y}} \int_0^1 m \phi_p \phi_q d\bar{x} \right] = 0 \quad p = 1, 2, \ldots, n \qquad (8.5.14)$$

and for vibrations in-the-plane of rotation

$$B_q \left[ \int_0^1 EI_z \psi_r'' \psi_q'' d\bar{x} + \frac{\Omega^2 m_0 L^4}{EI_{0z}} \int_0^1 \bar{P} \psi_r' \psi_q' d\bar{x} - \right.$$

$$\left. -(\omega^2 + \Omega^2)\frac{m_0 L^4}{EI_{0z}} \int_0^1 \bar{m} \psi_r \psi_q d\bar{x} \right] = 0 \qquad r = 1, 2 \ldots, n. \qquad (8.5.15)$$

## 8.6    Numerical Solutions

Numerical solutions to Eqs. (8.5.14) and (8.5.15) are conveniently obtained when the same nonrotating mode shapes are used to generate the mode shapes and frequencies which include the effects of rotation. For convenience, one can define the following quantities:

$$
\left.
\begin{aligned}
I_{yqp} &= \int_0^1 \overline{EI}_y \phi_p'' \phi_q'' d\bar{x} \\
I_{zqr} &= \int_0^1 \overline{EI}_z \psi_r'' \psi_q'' d\bar{x} \\
M_{qp} &= \int_0^1 \bar{m} \phi_q \phi_p d\bar{x} = M_{qr} = \int_0^1 \bar{m} \psi_q \psi_r d\bar{x} \\
S_{qp} &= \int_0^1 \bar{P} \phi_q' \phi_p' d\bar{x} = S_{qr} = \int_0^1 \bar{P} \psi_q^1 \psi_r' d\bar{x}
\end{aligned}
\right\} .
\tag{8.6.1}
$$

With these definitions, Eqs. (8.5.14) and (8.5.15) can be written in their final, matrix form:

$$
[K_{ypq}]\{A_q\} = \omega^2 [M_{pq}]\{A_q\} \qquad \text{out of plane}
\tag{8.6.2}
$$

$$
[K_{zrq}]\{B_q\} = \omega^2 [M_{rq}]\{B_q\} \qquad \text{in plane,}
\tag{8.6.3}
$$

where

$$
[K_{ypq}] = \frac{EI_{0y}}{m_0 L^4}[I_{ypq}] + \Omega^2 [S_{pq}].
\tag{8.6.4}
$$

$$
[K_{zrq}] = \frac{EI_{0z}}{m_0 L^4}[I_{zrq}] + \Omega^2 [S_{rq}] - \Omega^2 [M_{rq}].
\tag{8.6.5}
$$

Equations (8.6.2) and (8.6.3) represent an approximate structural dynamics eigenvalue problem. Clearly, the $[K_{yqp}], [K_{zrq}]$ matrices play the role of the stiffness matrix. The effects of rotation are shown by Eqs. (8.6.4) and (8.6.5). For out of plane of rotation vibrations, the centrifugal force provides a stiffening effect which, under certain conditions, exceeds the structural stiffness contribution. However, Eq. (8.6.5) shows that this stiffening effect is reduced considerably for the case of in plane of rotation vibrations as a result of the last term in the equation.

For certain applications, such as vibrations of helicopter blades, it is convenient to rearrange the last equations so as to express the stiffness in terms of the nonrotating fundamental frequencies.

From Section (3.3), the fundamental frequencies of a nonrotating beam, with constant mass and stiffness distribution, can be written as

$$
\omega_{NRn} = (\beta_n l)^2 \sqrt{\frac{EI_0}{mL^4}}
\tag{8.6.6}
$$

for $n = 1$ and *cantilevered end* conditions.

$$
\beta_1 L = 1.875104 \qquad \text{first frequency}
$$

Thus, instead of using the stiffness, $EI$, one can write

$$
\frac{EI_0}{mL^4} = \frac{\omega_{NR1}^2}{(\beta_1 L)^4}.
$$

Thus, the eigenvalue problem represented by Eqs. (8.6.2) and (8.6.3) can be rewritten as

$$[\bar{K}_{ypq}]\{A\} = \bar{\omega}^2[M_{pg}]\{A\} \qquad (8.6.2a)$$

$$[\bar{K}_{zrq}]\{B\} = \bar{\omega}^2[M_{rq}\{B\} \qquad (8.6.3a)$$

with

$$[\bar{K}_{ypq}] = \frac{\bar{\omega}_{yNR1}^2}{(\beta_1 L)^4}[I_{ypq}] + [S_{pq}] \qquad (8.6.4a)$$

$$[\bar{K}_{zrq}] = \frac{\bar{\omega}_{zNR1}^2}{(\beta_1 L)^4}[I_{zrq}] + [S_{rq}] - [M_{rq}], \qquad (8.6.5a)$$

where

$$\bar{\omega}_{NR1} = \omega_{NR1}/\Omega.$$

The solution of the eigenvalue problems with the modified Eqs. (8.6.4a) and (8.6.5a) will yield the frequencies of the rotating beam, sometimes called rotating frequencies. The eigenvectors corresponding to these eigenvalues will yield the coefficients $A_q$ and $B_q$ indicating the participation of the nonrotating mode shapes, in the appropriate rotating mode shape. Thus, in general, the $s$th mode is given by

$$w_s = \sum_{q=1}^{n} A_q^{(s)}\phi_q.$$

When actual problems are solved, the effect of rotation on the frequency usually turns out to be more important than the effect of rotation in the mode shape.

Some comments on the actual solution of the problem are relevant. First, it should be noted that Yntema (1955) contains closed form expression for $I_{yqp}$, $I_{zqp}$, $M_{qp}$, and $S_{qp}$, in terms of exact characteristic functions of uniform beams. For example, $S_{qp}$, given below (see Yntema, 1955, 35):

$$S_{qp} = \frac{\beta_q^2\beta_p^2}{\beta_q^4 - \beta_p^4}\left[\frac{1}{2}(\alpha_p\beta_p - \alpha_q\beta_q) - \frac{4\beta_q^2\beta_p^2(-1)^{q+p}}{\beta_q^4 - \beta_p^4}\right],$$

which is incorrect; therefore, the use of this reference should be avoided.

In practical situations, integrals arising from the application of Galerkin's method can be easily evaluated using Gaussian quadrature (Appendix C). Numerical integration gives good accuracy when using seven points and single precision. For higher modes, more accuracy may be required; in these cases, quadrature based on double precision arithmetic and 15 Gaussian points gives good results.

Typical results obtained from an analysis identical to the one described in this section based on five nonrotating, uniform, cantilever modes to generate the first three modes of a rotating uniform cantilever beam are shown in Figures 8.6 through 8.8. Figure 8.6 shows the

**Figure 8.6** Frequencies of fundamental rotating modes for vibrations out of the plane of rotation (flap) and in plane of rotations (lag)

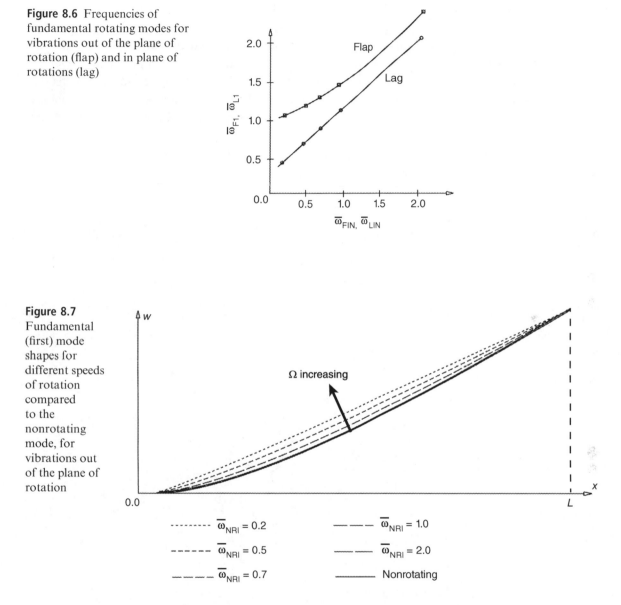

**Figure 8.7** Fundamental (first) mode shapes for different speeds of rotation compared to the nonrotating mode, for vibrations out of the plane of rotation

relations between the fundamental rotating and nonrotating frequencies in the flap (out of the plane of rotation bending) and lag (in the plane of rotation bending) directions, respectively. The horizontal axis shows the nonrotating frequencies and the vertical axis shows the rotating frequencies. The figure shows that for parameters representative of a hingeless rotor blade, the effect of rotation, producing centrifugal stiffening, is much stronger for out of plane of rotation vibrations than for in plane of rotation vibrations. Typical speed of rotations of helicopter rotor blades is 300 RPM, and the speed of rotation is usually kept constant through the entire flight regime.

Figures 8.7 and 8.8 show the effect of rotation on the first two mode shapes: for simplicity, both out of the plane of rotation and in the plane of rotation problems are solved using the same constant mass and stiffness distribution, resulting in the same eigenvectors or mode shapes. The results clearly show that for the cases considered, that are representative of rotor blades, the effect of rotation on the mode shapes can be considered to be a relatively minor correction. However, the effect of rotation on the frequency is quite important. However, it should be mentioned that neglecting the effect of rotation on the modes shapes in aeroelastic stability or response calculations can produce significant errors.

**Figure 8.8**
Second mode shapes for different speeds of rotation compared to the nonrotating mode shape, for vibrations out of the plane of rotation

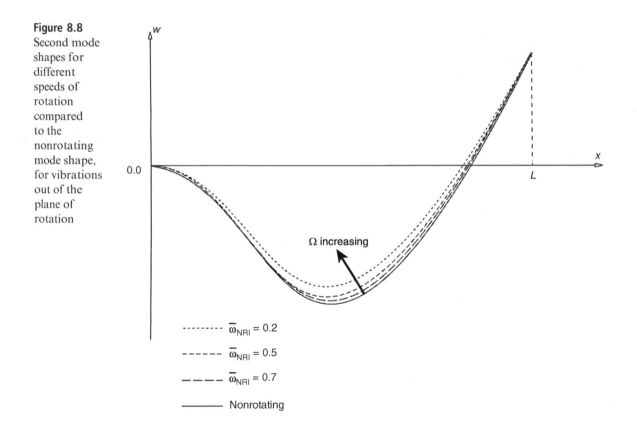

$\bar{\omega}_{NRI} = 0.2$

$\bar{\omega}_{NRI} = 0.5$

$\bar{\omega}_{NRI} = 0.7$

Nonrotating

## 8.7    Finite Element Treatment of Rotating Beam

The axial force in rotating systems is a consequence of the centrifugal effect. As pointed out before, axial forces can be easily introduced using the geometric stiffness matrix; furthermore, it is shown in Section 4.6 that cubic interpolation can be used to construct the conventional parts of the mass and stiffness matrices.

Figure 8.9 illustrates a typical beam type element representing a rotating beam problem, for vibrations out of the plane of rotation. For this case, one must distinguish between the local coordinate $x$ and the global coordinate $r_0$. Furthermore, it is important to note that the element is exposed to a constant axial tension, $P_\ell$ at its outer (right side) edge, and this tension changes within the element, as a result of the centrifugal load.

**Figure 8.9**
Geometry for the finite element formulation for vibrations out of the plane of rotation

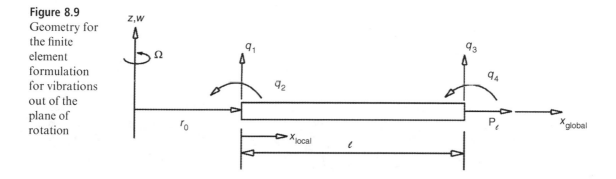

Recall from our previous treatment of an axially loaded structure (and also from Eq. (8.7.1)) that part of the strain energy in bending corresponding to the axial loads is given by

$$U_a = \frac{1}{2} \int_0^\ell P(x) \left[ \left( \frac{\partial w}{\partial x} \right)^2 + \left( \frac{\partial v}{\partial x} \right)^2 \right] dx. \tag{8.7.1}$$

$$v = w = \{\Phi\}^T \{\alpha\}. \tag{8.7.2}$$

Using the same interpolation as before (Eq. (4.6.14))

$$\{\alpha\} = [B]\{q\}. \tag{8.7.3}$$

For conciseness out of plane of rotation, vibrations are denoted as "flap" and the corresponding DOF vector is identified by the subscript $F$. Similarly, in plane of rotation vibrations are denoted as "lag" and the corresponding DOF vector is identified by the subscript $L$.

$$\left. \begin{array}{ll} \text{thus for flapwise motion} & \{q_F\}^T = \lfloor q_1 \quad q_2 \quad q_3 \quad q_4 \rfloor \\ \text{while for lagwise motion} & \{q_L\}^T = \lfloor q_5 \quad q_6 \quad q_7 \quad q_8 \rfloor \end{array} \right\}. \tag{8.7.4}$$

Next, note that

$$P(x) = P_\ell + \Delta P(x), \tag{8.7.5}$$

where

$$P_\ell = P(r_0 + \ell) = \Omega^2 \int_{r_0+\ell}^L m(x_1) x_1 \, dx_1 \tag{8.7.6}$$

and

$$\Delta P(x) = \Omega^2 \int_x^\ell m(\eta)(r_0 + \eta)d\eta. \tag{8.7.7}$$

Combining Eqs. (8.7.1), (8.7.5), (8.7.6), and (8.7.7), one has

$$
\begin{aligned}
U_a &= \frac{1}{2} \int_0^\ell [P_\ell + \Delta P(x)] \left[ \left( \frac{\partial w}{\partial x} \right)^2 + \left( \frac{\partial v}{\partial x} \right)^2 \right] dx = \\
&= \frac{1}{2} \int_0^\ell \left( \Omega^2 \int_{r_0+\ell}^L m(x_1)dx_1 \right) \left[ \left( \frac{\partial w}{\partial x} \right)^2 + \left( \frac{\partial v}{\partial x} \right)^2 \right] dx \\
&\quad + \frac{1}{2} \int_0^\ell \left( \Omega^2 \int_x^\ell m(\eta)(r_0 + \eta)d\eta \right) \left[ \left( \frac{\partial w}{\partial x} \right)^2 + \left( \frac{\partial v}{\partial x} \right)^2 \right] dx \\
&= \frac{1}{2} P_\ell \int_0^\ell \left[ \left( \frac{\partial w}{\partial x} \right)^2 + \left( \frac{\partial v}{\partial x} \right)^2 \right] dx + \frac{1}{2} \int_0^\ell \Delta P(x) \left[ \left( \frac{\partial w}{\partial x} \right)^2 + \left( \frac{\partial v}{\partial x} \right)^2 \right] dx.
\end{aligned}
\tag{8.7.8}
$$

Various assumptions regarding the mass distribution inside the element can be introduced at this stage. One can have a linear, or a constant, mass distribution (more sophisticated distributions are not required). For convenience, it will be assumed that the mass distribution *inside the element is constant*. Then,

$$
\begin{aligned}
\Delta P(x) &= \Omega^2 \int_x^\ell m(r_0 + \eta)d\eta = \Omega^2 m \left( r_0 \eta + \frac{\eta^2}{z} \right) \Big|_x^\ell \\
&= \Omega^2 m \left[ r_0(\ell - x) + \frac{1}{2}(\ell^2 - x^2) \right].
\end{aligned}
\tag{8.7.9}
$$

Thus, one can define a new type of geometric stiffness matrix associated with the effect of rotation given by

$$U_a = \frac{1}{2} \{q_F\}^T [k_{GF}] \{q_F\} + \frac{1}{2} \{q_L\}^T [k_{GL}] \{q_L\}, \tag{8.7.10}$$

where $k_{GF}$ is the new geometric stiffness matrix associated with the flap DOF, and $k_{GL}$ is the geometric stiffness matrix associated with the lag DOF. Consider first, flapping or out of plane of rotation vibrations. Recall from the treatment of structural dynamics of axially loaded structures (Section 4.6) that

$$w = \{\phi\}^T [B] \{q_F\}, \tag{8.7.11}$$

where

$$[B] = \begin{bmatrix} 1 & 0 & 0 & 0 \\ 0 & 1 & 0 & 0 \\ -3/\ell^2 & -2/\ell & 3/\ell^2 & -1/\ell \\ 2/\ell^3 & 1/\ell^2 & -2/\ell^3 & 1/\ell^2 \end{bmatrix}. \tag{8.7.12}$$

Comparing Eqs. (8.7.8) and (8.7.10), it is clear that

$$[k_{GF}] = P_\ell \int_0^\ell [B]^T \{\phi'\}\{\phi'\}^T [B]dx$$

$$+ \int_0^\ell \Delta P(x)[B]^T \{\phi'\}\{\phi'\}^T [B]dx \qquad (8.7.13)$$

$$= P_\ell \int_0^\ell [B]^T \{\phi'\}\{\phi'\}^T [B]dx$$

$$+ \int_0^\ell \Omega^2 m[r_0(\ell - x) + \frac{1}{2}(\ell^2 - x^2)][B]^T \{\phi'\}\{\phi'\}^T [B]dx.$$

It is quite interesting to compare the first term in Eq. (8.7.13) with expression that has been obtained previously for the geometric stiffness matrix. The first term on the right-hand side of Eq. (8.7.13) is quite similar to the geometric stiffness matrix considered previously. There is one important difference; the load $P_\ell$ given by Eq. (8.7.6) depends upon the global coordinate $r_0$ and the mass distribution $m(x_1)$ in addition to being multiplied by $\Omega^2$. The second term contains terms identical to those used in deriving the geometric stiffness matrix. These terms, however, are multiplied by a function of x; thus, evaluation of this integral will yield terms different from those obtained for the conventional geometric stiffness matrix. The second term also contains the global coordinate, $r_0$; thus, the element stiffness properties will also depend on the location of the element. Define the following expressions:

$$\bar{P}_\ell = \frac{P_\ell}{\Omega^2} = \int_{r_0+\ell}^{L} x_1 m(x_1)dx_1 \qquad (8.7.14)$$

and

$$P_\ell \int_0^\ell [B]^T \{\phi'\}\{\phi'\}^T [B]dx = \Omega^2[\bar{k}_G(r_0)], \qquad (8.7.15)$$

where

$$[\bar{k}_G(r_0)] = \frac{\bar{P}_\ell}{\ell} \begin{bmatrix} \frac{6}{5} & & & \\ \frac{\ell}{10} & \frac{2\ell^2}{15} & \text{symmetric} & \\ -\frac{6}{5} & -\frac{\ell}{10} & \frac{6}{5} & \\ \frac{\ell}{10} & -\frac{\ell^2}{30} & -\frac{\ell}{10} & \frac{2\ell^2}{15} \end{bmatrix} \qquad (8.7.16)$$

and

$$\Omega^2[\Delta\bar{k}_G(r_0)] = \int_0^\ell \Omega^2 m[r_0(\ell - x) + \frac{1}{2}(\ell^2 - x^2)][B]^T \{\phi'\}\{\phi'\}^T [B]dx. \qquad (8.7.17)$$

Combining Eqs. (8.7.13) through (8.7.17) yields

$$[k_{GF}] = \Omega^2 \left( \left[\bar{k}_G(r_0)\right] + \left[\Delta \bar{k}_G(r_0)\right] \right) = \Omega^2 [\bar{k}_{GF}].$$

The contributions of the strain energy due to bending can be treated using the conventional bending stiffness properties of the element, which have been previously derived in Section 4.2,

$$[k_{BF}] = \frac{EI_y}{\ell^3} \begin{bmatrix} 12 & 6\ell & -12 & 6\ell \\ & 4\ell^2 & -6\ell & 2\ell^2 \\ \text{sym.} & & 12 & -6\ell \\ & & & 4\ell^2 \end{bmatrix} \tag{8.7.18}$$

and the total stiffness matrix for the element

$$\begin{aligned} [k_F(R_0)] &= [k_{BF}] + \Omega^2([\bar{k}_G(r_0)] + [\Delta \bar{k}_G(r_0)]) \\ &= [k_{BF}] + \Omega^2[\bar{k}_{GF}]. \end{aligned} \tag{8.7.19}$$

The mass or inertia properties of the beam can be taken into account using the regular mass matrix which has been derived in Section 4.2.

$$\begin{aligned} [m] &= \int_0^\ell m[B]^T \{\phi\}\{\phi\}^T [B] dx \\ &= \frac{m\ell}{420} \begin{bmatrix} 156 & 22\ell & 54 & -13\ell \\ & 4\ell^2 & 13\ell & -3\ell^2 \\ & & 156 & -22\ell \\ \textit{symmetric} & & & 4\ell^2 \end{bmatrix}. \end{aligned} \tag{8.7.20}$$

When assembling the element into a master stiffness matrix, one can write symbolically, where the $\mathbf{J}$ is a symbolic matrix that performs the assembly.

$$[M] = \mathbf{J}^T \lceil \mathbf{m}_1, \mathbf{m}_2, \ldots, \mathbf{m}_N \rfloor \mathbf{J} \tag{8.7.21}$$

$$[K_{BF}] = \mathbf{J}^T \lceil \mathbf{k}_{BF1}, \mathbf{k}_{BF2}, \ldots, \mathbf{k}_{BFN} \rfloor \mathbf{J} \tag{8.7.22}$$

and

$$\Omega^2[\bar{K}_{GF}(r_0)] = \mathbf{J}^T \lceil \mathbf{k}_{GF1}, \mathbf{k}_{GF2}, \ldots, \mathbf{k}_{GFN} \rfloor \mathbf{J}. \tag{8.7.23}$$

The complete eigenvalue problem can be written in assembled form, where $\{q_F^*\}$ represents the DOF vector with the constrained DOF eliminated.

$$\omega^2[M]\{q_F^*\} = ([K_{BF}] + \Omega^2[\bar{K}_{GF}(r_0)])\{q_F^*\}. \tag{8.7.24}$$

Consider next the free vibration problem in the plane of rotation. Due to the fact that the same interpolation function is used, one may write, by analogy

$$[k_{BL}] = \frac{EI_z}{\ell^3} \begin{bmatrix} 12 & 6\ell & -12 & 6\ell \\ & 4\ell^2 & -6\ell & 2\ell^2 \\ & & 12 & -6\ell \\ & & & 4\ell^2 \end{bmatrix}. \tag{8.7.25}$$

Similarly,

$$[k_{GL}] = \Omega^2 [\bar{k}_{GF}]. \tag{8.7.26}$$

Finally, the term represented by $m\Omega^2 v$ in Eq. (8.3.35) has to be accounted for. This can be obtained from the kinetic energy (Eq. (8.3.25))

$$T_1 = \frac{1}{2} \int_0^\ell m\Omega^2 v^2 dx. \tag{8.7.27}$$

Using Eq. (8.7.27) it is clear that

$$T_1 = \frac{1}{2} \{q_L\}^T \Omega^2 \int_0^\ell m[B]^T \{\phi\}\{\phi\}^T [B] dx \{q_L\}. \tag{8.7.28}$$

The integrand in Eq. (8.7.28) is again the mass matrix. Thus,

$$T_1 = \frac{1}{2} \{q_L\}^T \Omega^2 [m] \{q_L\}. \tag{8.7.29}$$

When the additional term is included in the formulation of the structural dynamics eigenvalue problem, one has

$$[k_L(r_0)] = [k_{BL}] + \Omega^2 \left([\bar{k}_{GF}] - [m]\right). \tag{8.7.30}$$

Consequently, in a manner analogous to Eqs. (8.7.21) through (8.7.24), the structural dynamics eigenvalue for vibrations, out of the plane of rotation can be written as

$$\omega^2 [M]\{q_L^*\} = ([K_{BL}] + \Omega^2([\bar{K}_{GF}] - [M]))\{q_L^*\}. \tag{8.7.31}$$

## 8.8    Finite Element Treatment of Coupled Vibrations

Rotating beam vibration problems are important in the design of fan, propeller, helicopter rotor, and power-generating turbine blades. In these applications, the blades operate in a fluid (water or air), and their primary purpose is to generate fluid dynamic forces. To achieve this objective, the blades usually have cross sections oriented at some angle $\beta$ with respect to the plane of rotation as shown in Figure 8.10.

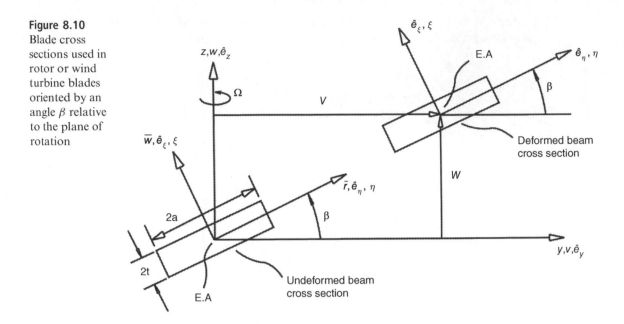

**Figure 8.10**
Blade cross sections used in rotor or wind turbine blades oriented by an angle $\beta$ relative to the plane of rotation

From the geometry of the figure, the following relations between the various unit vectors are evident. Unit vectors are denoted by the quantities: $\hat{e}_x$, $\hat{e}_y$, $\hat{e}_z$, $\hat{e}_\eta$, and $\hat{e}_\zeta$.

$$
\left.
\begin{aligned}
\hat{e}_z &= \hat{e}_\zeta \cos\beta + \hat{e}_\eta \sin\beta \\
\hat{e}_y &= \hat{e}_\eta \cos\beta - \hat{e}_\zeta \sin\beta
\end{aligned}
\right\}. \tag{8.8.1}
$$

Also

$$
\left.
\begin{aligned}
\hat{e}_\zeta &= -\hat{e}_y \sin\beta + \hat{e}_z \cos\beta \\
\hat{e}_\eta &= \hat{e}_y \cos\beta + \hat{e}_z \sin\beta
\end{aligned}
\right\}. \tag{8.8.2}
$$

It was shown previously that the position vector of a point on the deformed elastic axis of the beam can be written as

$$
\begin{aligned}
\mathbf{r}_E &= (x+u)\hat{e}_x + v\hat{e}_y + w\hat{e}_z \\
&= (x+u)\hat{e}_x + v[\hat{e}_\eta \cos\beta - \hat{e}_\zeta \sin\beta] + w[\hat{e}_\zeta \cos\beta + \hat{e}_\eta \sin\beta] \\
&= (x+u)\hat{e}_x + \hat{e}_\eta[v\cos\beta + w\sin\beta] + \hat{e}_\zeta[w\cos\beta - v\sin\beta]. \tag{8.8.3}
\end{aligned}
$$

Furthermore,

$$
\mathbf{r}_E = (x+u)\hat{e}_x + \bar{v}\hat{e}_\eta + \bar{w}\hat{e}_\zeta, \tag{8.8.4}
$$

where $\bar{v}$ and $\bar{w}$ are, respectively, deformation of a point on the elastic axis of the beam in the direction of the major (or principal) axis of the cross section and normal to the principal axis of the cross section (as shown in Figure 8.10). From Eqs. (8.8.3) and (8.8.4),

$$\left.\begin{array}{l} \bar{v} = v\cos\beta + w\sin\beta \\ \bar{w} = w\cos\beta - v\sin\beta \end{array}\right\}.$$

(8.8.5)

Next, consider the axial strain. In dealing with axial strain, we shall concentrate only on the terms due to Euler–Bernoulli bending. The various other terms result only in additional terms associated with axial, centrifugal loading and these have been treated previously. Thus,

$$\varepsilon = u_{0,x} - zw_{,xx} - yv_{,xx} + HOT.$$

(8.8.6)

Using Eq. (8.8.1), the local coordinates of the cross section $\eta$ and $\zeta$ can be related to the $z$- and $y$-coordinates, thus

$$z = \zeta\cos\beta + \eta\sin\beta.$$
$$y = \eta\cos\beta - \zeta\sin\beta.$$

(8.8.7)

Substituting these into Eq. (8.8.6) and concentrating on the strain due to the Euler–Bernoulli term yields

$$\varepsilon_{EB} = -zw_{,xx} - yv_{,xx} = -w_{,xx}(\zeta\cos\beta + \eta\sin\beta) - (\eta\cos\beta - \zeta\sin\beta)v_{,xx}$$
$$= -\eta(w_{,xx}\sin\beta + v_{,xx}\cos\beta) - \zeta(w_{,xx}\cos\beta \quad v_{,xx}\sin\beta).$$

(8.8.8)

The moment components due to Euler–Bernoulli bending in the directions of the principal axes $\eta$, $\zeta$ are (according to the sign convention shown in Figure 8.11):

**Figure 8.11** Bending moment convention in the rotating beam cross section

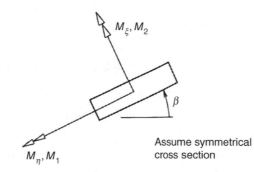

$M_\zeta, M_2$

$\beta$

$M_\eta, M_1$

Assume symmetrical cross section

$$M_1 = M_\eta = -\int_{-a}^{a}\int_{-t}^{t}\sigma\zeta d\zeta d\eta = -\int_{-a}^{a}\int_{-t}^{t}\zeta E\varepsilon_{EB}d\eta d\zeta$$
$$= \int_{-a}^{a}\int_{-t}^{t} d\eta d\zeta\, E[\eta\zeta(w_{,xx}\sin\beta + v_{,xx}\cos\beta) + \zeta^2(w_{,xx}\cos\beta - v_{,xx}\sin\beta)]$$
$$= EI_1(-v_{,xx}\sin\beta + w_{,xx}\cos\beta).$$

(8.8.9)

In Eq. (8.8.9), the product of inertia term is zero, because the cross section is assumed symmetric about the $\eta$-axis. The cross-sectional properties are given by $I_1 = \int_A \zeta^2 d\eta d\zeta$ and $I_2 = \int_A \eta^2 d\eta d\zeta$, and $A$ is the cross-sectional area.

$$
\begin{aligned}
M_2 = M_\zeta &= -\int_{-a}^{a}\int_{-t}^{t} \sigma \eta d\eta d\zeta = -\int_{-a}^{a}\int_{-t}^{t} E\varepsilon_{EB}\eta d\eta d\zeta \\
&= \int_{-a}^{a}\int_{-t}^{t} E[\eta^2(w_{,xx}\sin\beta + v_{,xx}\cos\beta) + \eta\zeta(w_{,xx}\cos\beta - v_{,xx}\sin\beta]d\eta d\zeta \\
&= EI_z(w_{,xx}\sin\beta + v_{,xx}\cos\beta).
\end{aligned}
\tag{8.8.10}
$$

Using these moment components, the strain energy due to Euler–Bernoulli bending can be written as (recall strain energy for a beam $U = \frac{1}{2EI}\int_o^\ell M^2 dx$)

$$
\begin{aligned}
U_{EB} &= \frac{1}{2}\left[\frac{1}{EI_1}\int_0^\ell M_1^2 dx + \frac{1}{EI_2}\int_0^\ell M_2^2 dx\right] \tag{8.8.11} \\
&= \frac{1}{2}\left[\int_0^\ell [EI_1(-v_{,xx}\sin\beta + w_{,xx}\cos\beta)^2 + EI_2(w_{,xx}\sin\beta + v_{,xx}\cos\beta)^2\right]dx \\
&= \frac{1}{2}\int_0^\ell [EI_1 v_{,xx}^2 \sin^2\beta + EI_1 w_{,xx}^2 \cos^2\beta - 2EI_1 v_{,xx}w_{,xx}\sin\beta\cos\beta + \\
&\quad EI_2 w_{xx}^2 \sin^2\beta + EI_2 \cos^2\beta + 2EI_2, v_{xx}w_{xx}\sin\beta\cos\beta dx
\end{aligned}
$$

or for an element, $x$ now is the local coordinate.

$$
\begin{aligned}
U &= \frac{1}{2}\int_0^\ell E[I_1 \cos^2\beta + I_2 \sin^2\beta](w'')^2 dx \\
&\quad + \frac{1}{2}\int_0^\ell E[I_1 \sin^2\beta + I_2 \cos^2\beta](v'')^2 dx \\
&\quad + \frac{1}{2}\int_0^\ell E(I_2 - I_1)2\sin\beta\cos\beta w''v'' dx.
\end{aligned}
\tag{8.8.12}
$$

Equation (8.8.12) can be used to derive a coupled elastic stiffness matrix for the beam vibrating in and out of the plane of rotation

$$
U = \frac{1}{2}\{q\}^T[k_B]\{q^T\},
\tag{8.8.13}
$$

where

$$
\{q\}^T = \underbrace{\lfloor q_1 \ q_2 \ q_3 \ q_4}_{\text{flap}} \ \underbrace{q_5 \ q_6 \ q_7 \ q_8 \rfloor}_{\text{lag}}.
\tag{8.8.14}
$$

Using Eqs. (8.8.12) and (8.7.2), one has

$$
U = \frac{1}{2}\int_0^\ell E[I_1\cos^2\beta + I_2\sin^2\beta]\{q_F\}^T[B]^T\{\phi''\}\{\phi''\}^T[B]dx\{q_F\}
$$
$$
+\frac{1}{2}\int_0^\ell E[I_1\sin^2\beta + I_2\cos^2\beta]\{q_L\}^T[B]^T\{\phi''\}\{\phi''\}^T[B]dx\{q_L\}
$$
$$
+\frac{1}{2}\int_0^\ell E(I_2 - I_1)\sin\beta\cos\beta\{q_L\}^T[B]^T\{\phi''\}\{\phi''\}^T[B]dx\{q_F\}
$$
$$
+\frac{1}{2}\int_0^\ell E(I_2 - I_1)\sin\beta\cos\beta\{q_F\}^T[B]^T\{\phi''\}\{\phi''\}^T[B]dx\{q_L\}.
$$

Next, using the last expression, define the following terms:

$$
[k_{FF}] = \frac{1}{2}\int_0^\ell E(I_1\cos^2\beta + I_2\sin^2\beta)[B]^T\{\phi''\}\{\phi''\}^T[B]dx. \tag{8.8.15}
$$

$$
[k_{LL}] = \int_0^\ell E(I_1\sin^2\beta + I_2\cos^2\beta)[B]^T\{\phi''\}\{\phi''\}^T[B]dx. \tag{8.8.16}
$$

$$
[k_{FL}] = [k_{LF}] = \int_0^\ell E(I_2 - I_1)\sin\beta\cos\beta[B]^T\{\phi''\}\{\phi''\}^T[B]dx. \tag{8.8.17}
$$

Thus, the element stiffness matrix for the coupled structure is given by

$$
[k_B] = \begin{bmatrix} [k_{FF}] & [k_{FL}] \\ [k_{LF}] & [k_{LL}] \end{bmatrix}. \tag{8.8.18}
$$

It should be noted that due to the angle $\beta$ vibrations, out of the plane of rotation and in the plane of rotation are coupled due to the stiffness matrix. This coupling is denoted as the "elastic coupling effect" in helicopter rotor blades.

Performing a symbolic assembly process,

$$
[K_B] = \mathbf{J}^T\lceil \mathbf{k}_{B1}, \mathbf{k}_{B2}, \ldots, \mathbf{k}_{BN} \rfloor \mathbf{J}. \tag{8.8.19}
$$

Combining Eqs. (8.7.24), (8.7.26), and (8.7.20), the coupled eigenvalue problem becomes

$$
\omega^2\begin{bmatrix} [M] & 0 \\ 0 & [M] \end{bmatrix}\begin{Bmatrix} \mathbf{q}_F^* \\ \mathbf{q}_L^* \end{Bmatrix} = \begin{bmatrix} [K_{FF}] & [K_{FL}] \\ [K_{LF}] & [k_{LL}] \end{bmatrix} +
$$

$$
\Omega^2\begin{bmatrix} [\bar{K}_{GF}(r_0)] & 0 \\ 0 & [\bar{K}_{GF}(r_0)] - [M] \end{bmatrix}\begin{Bmatrix} \mathbf{q}_F^* \\ \mathbf{q}_L^* \end{Bmatrix} = \begin{Bmatrix} 0 \\ 0 \end{Bmatrix}. \tag{8.8.20}
$$

Solution of this eigenvalue problem will yield coupled frequencies and mode shapes.

Some typical results using the method described earlier are briefly discussed as follows:

(1) For convergence, four to six elements are sufficient, for generating the first three modes of a rotating beam.
(2) Coupled results are provided in Table 8.1.

**Table 8.1 Effect of elastic coupling on the modes. Note in coupled modes flap and lag motions are both present**

| $\beta$ | FLAP | | LAG | | Comment |
|---|---|---|---|---|---|
| deg | $\omega_{\bar{N}R1}$ | $\omega_{\bar{R}1}$ | $\omega_{\bar{N}R1}$ | $\omega_{\bar{R}1}$ | |
| 0 | 0.3662 | 1.127 | 1.1582 | 1.209 | |
| 5 | 0.3662 | 1.108 | 1.1582 | 1.226 | |
| 10* | 0.3662 | 1.072 | 1.1582 | 1.258 | *For these angles, flap and |
| 15* | 0.3662 | 1.030 | 1.1582 | 1.293 | lag motions are nearly |
| 20* | 0.3662 | 0.986 | 1.1582 | 1.327 | equally represented. |

(3) Note that the elastic coupling effect reduces the frequencies of the out-of-plane vibration modes (flap modes) and increases the frequencies of the in plane of rotation modes (lag modes).

## 8.9    Rotating Shafts

An important problem that belongs to rotating systems class of problems is the rotating shaft. This is a complex problem that has several aspects. It can be considered to be a free vibration problem under certain conditions. However, for most practical applications, it belongs to the structural dynamics response problem. Furthermore, when the problem is considered carefully, it is evident that it is a nonconservative problem of elastic systems that under certain conditions can become unstable.

The simplest problem is an elastic rotating shaft on which a rotating flywheel is rigidly attached as shown in Figure 8.12. The flywheel is relatively heavy compared to the total mass of the shaft.

**Figure 8.12** Typical flexible shaft flywheel configuration

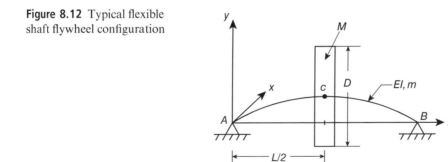

The mass of the flywheel is $M$, its diameter is $D$, and its dimension in the lateral direction is $h$. The center of mass of the flywheel is at $c$. In addition to its mass, two additional important properties are its polar moment of inertia $I_p$ and its diametral moment of inertia $I_d$. The diametral moment of inertia corresponds to rotation about an axis parallel to the *x*-axis, through the center of mass *c*. In Figure 8.12, it is assumed that the shaft is supported by self-centering bearings, at the supports *A* and $B_f$, which act as simple supports. The shaft response problems is more complicated than the free vibration problems treated earlier in this chapter. Furthermore, many different configurations consisting of several flywheels attached to the shaft as well as different types of bearings can be employed in practice. Therefore, it is not surprising that entire books are devoted to this topic (Frisswell et al., 2010; Lalanne and Ferraris, 1998). The problem in various forms is also addressed by special sections and chapters in several books (Bolotin, 1963, 1964).

One of the earliest problems considered is the **critical speed problem**. At the critical speed, the shaft experiences transverse vibrations induced by eccentricity and the amplitudes can reach high values. Such critical speeds coincide with the natural frequencies of the nonrotating shaft. A typical critical speed can be written as

$$\Omega_0 = \sqrt{\frac{k}{M}}.$$

where *M* is the mass of the disk and *k* is a coefficient associated with the shaft stiffness *EI* in bending. By increasing the speed of rotation, one transitions through the critical value, vibrations decrease, and the disk approaches a self-centering state. Actual rotors have a spectrum of critical velocities, and further increase in the speed of rotation $\Omega$ causes encounters with higher critical speeds. This problem is similar to the response of a structural dynamic system to external periodic exciting forces.

Experimental evidence on flexible shafts has shown that self-induced transverse vibrations can occur at speeds different from critical speeds. Such self-induced vibrations have been shown to be connected to internal friction within the shaft. The source of energy for self-excitation is the engine that drives the shaft at constant angular velocity independent of its state. This represents an external energy source that transmits additional power to the structural dynamic system that feeds power driving the vibrations. The power is transmitted to the undisturbed rotational motion through the internal friction associated with the bearings. There are several types of bearings that can contribute to the vibrations in different ways. Oil film bearings can introduce hydrodynamic effects. Magnetic bearings can introduce electromagnetic effects that affect the stability. **Asymmetric bearings** can cause further complications, causing the associated model to become a periodic system (i.e., a system of equations governed by periodic coefficients) that requires special treatment. Periodic systems are discussed in detail in the next chapter (Chapter 9) of this book.

The basic problem, shown in Figure 8.12, is complicated by two features. The load introduced by the rotating flywheel is a point load along the shaft which is a rotating beam. The flywheel represented by the disk has two moments of inertia, $I_p$ and $I_d$, and the bending of the shaft introduces gyroscopic effects due to the precession of the rotating flywheel. Furthermore, the problem becomes more challenging when several flywheels are attached to the rotating shaft.

Due to the complexity and practical importance of the problem, it is not surprising that a vast amount of literature exists on this topic and several books have been completely devoted to this topic (Frisswell et al., 2010; Lalanne and Ferraris, 1998), while others have been partially devoted to the topic (Bolotin, 1963, 1964).

The treatment of the problem in the literature is interesting due to the assumptions made to simplify the problem. For example, the point loading at various locations has resulted in the replacement of the stiffness $EI$ by an equivalent system of discrete springs. Another effect that is important, but sometimes neglected, is the gyroscopic effect. While this problem is best treated by the finite element method, such a treatment has been presented only in Frisswell et al. (2010) and Lalanne and Ferraris (1998). A comprehensive and accurate formulation of the problem is also available in Baruch (1999, 659–663). Therefore, in a book exclusively devoted to the rotating shaft class of problems, the authors introduce simplifying assumptions so as to obtain solutions that capture the physics of the problem. Such an approach is also followed in this section.

### 8.9.1    Flexible Shaft Carrying a Disk at Mid-Span

The rotor is a long and flexible shaft, with circular cross section, and it is supported at the ends by two self-aligning rigid bearings as shown in Figure 8.13. It is assumed that no deflection can occur at the bearing. Furthermore, it is assumed that the mass of the shaft is small compared to the mass of the disk, and therefore the mass of the shaft is neglected. The effect of gravity on the system is also neglected. The dynamic behavior is analyzed by considering the lateral displacement along the $ox$ and $oy$ axes that are perpendicular to the $oz$ axis, as well as rotations about these axes. Additional details about the system are provided in Figures 8.14 and 8.15.

**Figure 8.13** Flexible shaft, with a single disk, on rigid bearings

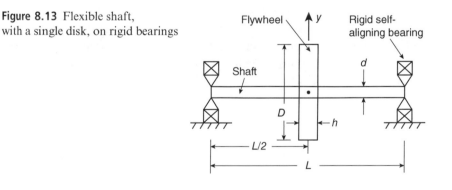

Properties of the system shown in Figure 8.13 are mass of disk $M$, its diameter $D$, horizontal dimension of flywheel $h$, length of shaft $L$, and disk placed at mid-span $L/2$. Bearings rigid and self-aligning. The displacement and rotational DOF are illustrated in Figures 8.14 and 8.15.

The coordinate system used in Figure 8.14 is a right-handed coordinate system and rotations are positive following the right-hand rule (counterclockwise). For this class of problems, it is important to incorporate the gyroscopic effect. Gyroscopic effects are present on masses rotating with constant angular speed about an axis. Gyroscopic moments are a

**Figure 8.14** Right-handed coordinate system used for rotor model

**Figure 8.15** Coordinate system used in the analysis of the shaft

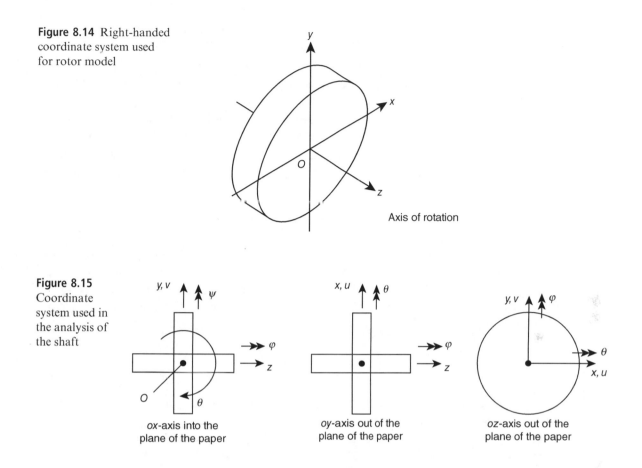

result of conservation of angular momentum (Baruch, 1999). The angular momentum of the rotor about the $oz$ axis is $I_p\Omega$. Rotation of the angular momentum vector by a small angle $\delta\psi$ over a time $\delta t$ produces a precession moment

$$M_x = I_p\Omega\frac{d\phi}{dt} = I_p\Omega\dot{\psi}. \tag{8.9.1}$$

The disk can also be rotated with constant angular rate $\dot{\theta}$ about the $ox$ axis producing a moment

$$M_p = -I_p\Omega\dot{\theta} \tag{8.9.2}$$

since the disk can rotate simultaneously about the $x$ and $y$ axes. The total rate of change of the angular momentum is

$$I_d\ddot{\theta} + I_p\Omega\dot{\psi} = M_x. \tag{8.9.3}$$
$$I_d\ddot{\psi} - I_p\Omega\dot{\theta} = M_y. \tag{8.9.4}$$

The dynamics of the system is described by four coordinates: $u$, $v$, $\theta$, and $\psi$.

This model developed in Frisswell et al. (2010) is an approximation. If one considers a flexible shaft represented by distributed bending stiffness $EI_y$ and $EI_x$, the lateral deflections $u$ and $v$ would be connected to $\theta$ and $\psi$ since these represent the slopes of the beam. The inconsistency in the model used is avoided by representing the stiffness of the shaft by equivalent spring stiffnesses.

The dynamic behavior of the shaft is examined by considering the displacements of the disk in the $ox$- and $oy$-axis directions as well as moments about these axes. Assuming small static displacements of the shaft, linear relations between the applied forces and moments can be used. For **a specific point** on the shaft, corresponding to the location of the disk, one can write

$$f_x = k_{uu}u + k_{u\psi}\psi \tag{8.9.5}$$

$$M_y = k_{\psi\psi}\psi + k_{\psi u}u, \tag{8.9.6}$$

where $f_x$ is the force applied to the shaft in the $ox$ direction and $M_y$ is the moment applied about the $oy$ axis. $k_{uu}$, $k_{\psi u}$, $k_{u\psi}$, and $k_{\psi\psi}$ are stiffness coefficients at a particular location on the shaft. Similar relations are established between the forces and moment associated with the coordinates $v$ and $\theta$.

$$f_y = k_{vv}v + k_{v\theta}\theta \tag{8.9.7}$$

$$M_x = k_{\theta\theta}\theta + k_{\theta v}v \tag{8.9.8}$$

for a conservative system $k_{v\theta} = k_{\theta v}$ and $k_{u\psi} = k_{\psi u}$, producing a symmetric stiffness matrix. These stiffness coefficients are calculated from Euler–Bernoulli beam theory.

The equations of motion are obtained from application of Newton's second law, or using d'Alembert's principle

$$f_x = -M\ddot{u} \tag{8.9.9}$$

$$f_y = -M\ddot{v} \tag{8.9.10}$$

and the applied moments are

$$M_x = -I_d\ddot{\theta} - I_p\Omega\dot{\psi} \tag{8.9.11}$$

$$M_y = -I_d\ddot{\psi} + I_p\Omega\dot{\theta}. \tag{8.9.12}$$

Assuming a circular shaft, the stiffness properties are the same in each direction, thus

$$k_T = k_{uu} = k_{vv} \text{ and } k_{\theta\theta} = k_{\psi\psi} = k_R. \tag{8.9.13}$$

However, due to the sign convention in Figure 8.15, $k_c = k_{v\psi} = -k_{v\theta}$. Combining Eq. (8.9.6) through Eq. (8.9.13) yields the final equations of motion

$$\left.\begin{aligned}
M\ddot{u} + k_T u + k_c\psi &= 0 \\
M\ddot{v} + k_T v - k_c\theta &= 0 \\
I_d\ddot{\theta} + I_p\Omega\dot{\psi} - k_c v + k_R\theta &= 0 \\
I_d\ddot{\psi} - I_p\Omega\dot{\theta} + k_c u + k_R\psi &= 0
\end{aligned}\right\}. \tag{8.9.14}$$

The spring constants are obtained from Euler–Bernoulli beam theory

$$\left.\begin{array}{l} k_T = k_{uu} + k_{vv} - \frac{48EI}{L^3} \\ k_c = k_{u\psi} = -k_{v\theta} = 0 \\ k_R = k_{\psi\psi} = k_{\theta\theta} = \frac{12EI}{L} \end{array}\right\},$$
(8.9.15)

where $EI$ is the bending stiffness of the circular shaft, $L$ is it length, and $M$ is the mass of the disk. The properties of the disk can be found in Likins (1973). $M = \rho\left(\frac{\pi D^2}{4}\right)h$, where $\rho$ is the density of the disk. $I_p$ = polar moment of inertia = $\frac{MD^2}{8}$ for an axis through the center of mass. The diametral moment of inertia about an axis through the center of mass $(I_d)_c = \frac{1}{2}I_p + \frac{1}{12}Mh^2 = M\left(\frac{D^2}{16} + \frac{h^2}{12}\right)$

In most cases associated with rotating shafts, there is limited justification for neglecting gyroscopic effects. Furthermore, when neglecting gyroscopic effects, the solution is independent of $\Omega$, therefore the gyroscopic effect will be included in the solution. Using the second equation of Eq. (8.9.15) $k_c = 0$ and Eq. (8.9.14) simplifies

$$\left.\begin{array}{l} M\ddot{u} + k_T u = 0 \\ M\ddot{v} + k_T v = 0 \\ I_d\ddot{\theta} + I_p\Omega\dot{\psi} + k_R\theta = 0 \\ I_d\ddot{\psi} - I_p\Omega\dot{\theta} + k_R\psi = 0 \end{array}\right\}.$$
(8.9.16)

The first two equations in Eq. (8.9.16) are uncoupled. Assuming a solution in the form

$$u(t) = u_0 e^{st} \; ; \; v(t) = v_0 e^{st} \; ; \; \theta(t) = \theta_0 e^{st} \; ; \; \text{and } \psi(t) = \psi_0 e^{st},$$

where $u_0$, $v_0$, $\theta_0$, and $\psi_0$ are complex constants and substitution of these expressions in the first two equations of Eq. (8.9.16) yields

$$\begin{array}{l} (Ms^2 + k_T)u_0 = 0 \\ (Ms^2 + k_T)v_0 = 0. \end{array}$$
(8.9.17)

Equation (8.9.17), $s^2 = -\frac{k_T}{M}$ ; $s_1 = s_2 = i\sqrt{\frac{k_T}{M}}$ ; $s_3 = s_4 = -i\sqrt{\frac{k_T}{M}}$. These roots are complex conjugate pairs, hence $\omega_1 = \omega_2 = \sqrt{\frac{k_T}{M}}$. The second pair of equations in Eqs. (8.9.16) yields

$$\begin{array}{l} (I_d s^2 + k_R)\theta_0 + I_p\Omega s\psi_0 = 0 \\ -I_p\Omega s\theta_0 + (I_d s^2 + k_R)\psi_0 = 0. \end{array}$$
(8.9.18)

Rewriting Eq. (8.9.18) in matrix form

$$\begin{bmatrix} I_d s^2 + k_R & I_p\Omega s \\ -I_p\Omega s & I_d s^2 + k_R \end{bmatrix} \left\{\begin{array}{c} \theta_0 \\ \psi_0 \end{array}\right\} = 0.$$
(8.9.19)

For a nontrivial solution, the determinant of the matrix in Eq. (8.9.19) is zero, which yields

$$(I_d s^2 + k_R)^2 + (I_p\Omega s)^2 = 0$$
(8.9.20)

$$I_d s^2 + k_R = \pm i I_p\Omega s.$$
(8.9.21)

The last equation is a quadratic equation that yields four solutions:

$$s^2 + \frac{k_R}{I_d} \pm i\frac{I_p\Omega s}{I_d} = 0 \tag{8.9.22}$$

$$s_{3,4} = i\left[\pm\frac{I_p\Omega}{2I_d} + \sqrt{\left(\frac{I_p\Omega}{2I_d}\right)^2 - \frac{k_R}{I_d}}\right] \tag{8.9.23}$$

$$s_{7,8} = -s_{3,4}.$$

Because $s_i = i\omega_i$, $s_{i+4} = i\omega_i$ for $i = 3, 4$, the natural frequencies are

$$\omega_3 = -\frac{I_p\Omega}{2I_d} + \sqrt{\left(\frac{I_p\Omega}{2I_d}\right)^2 - \frac{k_R}{I_d}} \ ; \ \ \omega_4 = \frac{I_p\Omega}{2I_d} + \sqrt{\left(\frac{I_p\Omega}{2I_d}\right)^2 - \frac{k_R}{I_d}}. \tag{8.9.24}$$

These natural frequencies depend on the speed of rotation. As $\Omega \to 0$, the roots approach $\sqrt{\frac{k_R}{I_d}}$. Equation ( 8.9.18) can be rearranged as

$$\left(\frac{\theta_0}{\psi_0}\right)^{(i)} = -\frac{I_p\Omega s_i}{I_d s_i^2 + k_R} = \frac{I_d s_i^2 + k_R}{I_p\Omega s_i}. \tag{8.9.25}$$

The terms on the right-hand side are of the form $-\frac{A}{B} = \frac{B}{A}$ and $\frac{A}{B}$ must equal $\pm i$. Note that $s_i^2 = -\omega_i^2$, where the natural frequency $\omega_i$ is positive, then in Eq. (8.9.25)

$$k_R > w_i^2 I_d \ \text{ and } \ (\frac{\theta_0}{\psi_0})^{(i)} = \begin{cases} -i \text{ if } s_i = i\omega_i \\ i \text{ if } s_i = -i\omega_i \end{cases}$$

$$\text{if } k_R < w_i^2 I_d \ \text{ then } \ (\frac{\theta_0}{\psi_0})^{(i)} = \begin{cases} i \text{ if } s_i = i\omega_i \\ -i \text{ if } s_i = -i\omega_i \end{cases}.$$

The sign of the relationship between $\theta_0$ and $\psi_0$ determines the direction of rotation of the modes. The real response is obtained by adding together the contribution of two complex conjugate modes. Since scaling of the mode shapes is arbitrary, one can assume $\theta_0 = 1$ for all eigenvectors. If $k_R > \omega_i I_d^2$, then the root $s_i = i\omega_i$, $\theta_0 = -\psi_0 i$, hence $\psi_0 = i$. For the complex conjugate, root $\psi_0 = i$, thus the time response

$$\begin{Bmatrix} \theta(t) \\ \psi(t) \end{Bmatrix} = \begin{Bmatrix} 1 \\ i \end{Bmatrix} e^{i\omega_i t} + \begin{Bmatrix} 1 \\ -i \end{Bmatrix} e^{-i\omega_i t} = 2\begin{Bmatrix} \cos\omega_i t \\ -\sin\omega_i t \end{Bmatrix}. \tag{8.9.26}$$

In the $\theta, \psi$ plane (see Figure 8.15), the orbit is a circle. The mode rotates in a clockwise direction. Note that positive spin is defined as counterclockwise, and is called a **backward mode**. Similarly for $k_R < \omega_i^2 I_d$

$$\begin{Bmatrix} \theta(t) \\ \psi(t) \end{Bmatrix} = \begin{Bmatrix} 1 \\ -i \end{Bmatrix} e^{i\omega_i t} + \begin{Bmatrix} 1 \\ i \end{Bmatrix} e^{-i\omega_i t} = 2\begin{Bmatrix} \cos\omega_i t \\ \sin\omega_i t \end{Bmatrix}. \tag{8.9.27}$$

For this case, the mode rotates in the counterclockwise direction, and is called a **forward mode**. A detailed description of the modes is provided in Frisswell et al. (2010).

The purpose of this section is not to give a detailed treatment of the problem. The intent here is to provide a description of the basic rotating shaft problem. It should be emphasized that the current problem is a basic problem, the supports are rigid and symmetric bearings are without friction, and a single flywheel is located at the center of the flexible shaft. It should be emphasized that many interesting effects can occur in rotating shafts. A large number of practical and interesting configurations of rotating shafts are treated in Frisswell et al. (2010).

## BIBLIOGRAPHY

Baruch, H. (1999). *Analytical Dynamics*. McGraw-Hill Book Co.

Bolotin, V. V. (1963). *Nonconservative Problems of the Theory of Elastic Stability*. Pergamon Press.

Bolotin, V. V. (1964). *The Dynamic Stability of Elastic Systems*. Holden Day, Inc.

Frisswell, M. Penny, J. Garvey, S., and Lees, A. (2010). *Dynamics of Rotating Machines*. Cambridge University Press.

Hodges, D. H. and Dowell, E. H. (1974). Nonlinear equations of motion for the elastic bending and torsion of twisted nonuniform rotor blades. Technical Report TN D-7818, NASA.

Houbolt, J. C. and Brooks, G. W. (1958). Differential equations of motion for combined flapwise bending, chordwise bending, and torsion of twisted nonuniform rotor blades. NACA Report 1346.

Hsu, L. and Renbarger, J. (1952). Bending vibration of a rotating beam. In *Proceedings First U. S. National Congress of Applied Mechanics, ASME*.

Lalanne, M. and Ferraris, G. (1998). *Rotordynamics Prediction in Engineering*. Wiley, second edition.

Likins, P. Barbera, F. J., and Baddeley, V. (1973). Mathematical modeling of spinning elastic bodies for modal analysis. *AIAA Journal*, 11(4).

Likins, P. W. (1973). *Elements of Engineering Mechanics*. McGraw-Hill Book Co.

Likins, P. W. (1974). Geometric stiffness characteristics of a rotating elastic appendage. *International Journal of Solids and Structures*, 10: 161–167.

Rao, J. S. (1983). *Rotor Dynamics*. Halsted Press.

Rosen, A. and Friedmann, P. P. (1977). Nonlinear equations of equilibrium for elastic helicopter and wind turbine blades undergoing moderate deformations. Technical Report NASA CR-159478, UCLA.

Yntema, R. T. (1955). Simplified procedures and charts for the rapid estimation of bending frequencies of rotating beams. Technical Note TN3459, NACA.

## PROBLEMS

1.  The figure below describes a rotating beam with a tip mass $M$ at its end. The root of the beam has a simple hinge (i.e., it is simply supported) and the hinge is offset by an amount $e$ from the axis of rotation $z$. The beam is rotating about the $z$-axis with constant angular speed $\Omega$. The length of the beam is $l$. The motion of the beam during its deflection out of the plane of rotation consists of a combination of angular deflection $\beta$ combined with an elastic deformation $w$ as shown in the figure by the dotted line.

You may assume that both the angular deflection and the elastic deflection are small. The distributed mass per unit length of the beam is $m$ (constant) and its stiffness is $EI$ (constant).

(a) Using Hamilton's principle, or a direct Newtonian approach, derive the equations of motion and boundary conditions governing the structural dynamics of the rotating beam vibrating out of the plane of rotation.

(b) Approximate the fundamental frequency of the rotating beam using RQ. To accomplish this objective, you need to select an appropriate shape function. Please justify carefully the selection of the shape function that you have decided to use.

2.  An inventor working on the design of a small unmanned helicopter decided to eliminate the conventional engine-gear configuration. To provide power to the rotor, an innovative tip thruster was developed. A typical blade of this configuration is shown below.

    Each blade consists of uniform mass and stiffness distribution. The mass per unit length is $m$, and the stiffness for bending out of the plane of rotation is $EI_y$ while the stiffness for bending in the plane of rotation is $EI_z$. The length of the blade is $l$ and it is cantilevered at a distance $e$ from the axis of rotation. The tip thruster can provide a constant thrust $T$ in the plane of rotation (the $x - y$ plane) that spins the rotor at a constant angular speed $\Omega$ about the $z$-axis. The blades are cantilevered to the hub. The thruster has a mass $M$ and mass moment of inertia $I_{y'y'}$ and $I_{z'z'}$ about the axes parallel to the $y$ and $z$ axes, respectively. The blade can be assumed to be torsionally rigid.

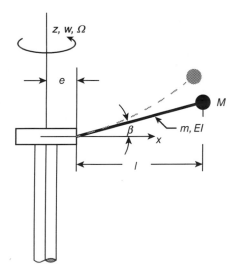

(a) Using Raleigh's quotient, with an appropriate shape function, estimate the natural frequencies of the blade when performing vibrations *in the plane of rotation*, and bending vibrations *perpendicular to the plane of rotation*, when the pitch setting on the blade $\theta = 0$, where the pitch setting (or the pitch angle) is the angle between the $y'$ and $y$ axes, with $y'$ and $z'$ fixed in the blade system.

(b) If the natural frequency of the blade in the plane of rotation is $\omega_{L1}$, what would happen if the thruster malfunctions and instead of producing a constant thrust $T$, it produces a time-varying $T(t) = T \sin \omega_{L1} t$? When discussing this question, address all possible consequences of this event, including the fact that the speed of rotation $\Omega$ can become $\Omega = f(t)$.

(c) How would you change the answer to (a) if the pitch setting on the blade is changed from $\theta = 0$ to $\theta = 10°$? Your answer to this part of the question can be descriptive, but you should provide appropriate mathematical arguments to support your answer.

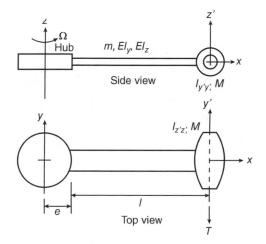

3. A wheel having radius $R$ is rotating at a constant angular speed $\Omega$ about an axis perpendicular to the plane of the paper located at point $O$. Two masses having mass $M$ and polar moments of inertia $I_p$ are attached to the rim of the rotating wheel by **identical** flexible beam segments having distributed mass $m$ per unit length and stiffness $EI$. Assuming that the wheel is rigid, write the equations of dynamic equilibrium and appropriate boundary conditions describing the vibrations of the flexible beam mass system (which has a length of $l$) using Hamilton's principle.

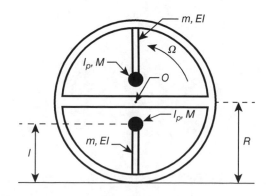

**Hints**

To help you with the solution of the problem, it is important that you keep in mind that the centrifugal force always acts in the radial direction.

You may also assume that $M \gg ml$ and you may neglect the effect of the **centrifugal** force on the distributed mass $m$, but you should include this effect when dealing with the concentrated mass $M$.

Coriolis effect can also be neglected.

4. The structural dynamic system shown below consists of a uniform beam having stiffness $EI$ and mass per unit length $m$. The beam is rotating about the $z$-axis with constant angular velocity $\Omega$. Two masses, $M_1$ and $M_2$, having rotary inertia (about an axis perpendicular to the plane of the paper) $I_{p1}$ and $I_{p2}$, respectively, are attached to the rotating beam at locations $l_1$ and $l_2$, as shown in the figure below.

   (a) Calculate the approximate fundamental frequency for this structural dynamic system using RQ based upon a suitable shape function $w = a\phi(x)$. Use a simple shape function and justify carefully your selection of the function $\phi(x)$.

   (b) How would you modify the answer to (a) given that the angular velocity $\Omega$ is not constant anymore? Instead, it is given by $\Omega(t) = \Omega_0 \sin \omega t$, where $\Omega_0$ is a given constant angular speed.

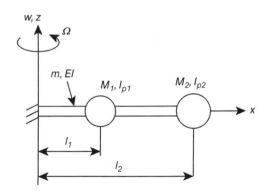

# 9 Stability and Response Problems of Periodic Systems

## 9.1 Introduction

This chapter is devoted to the understanding and practical numerical treatment of dynamic and structural dynamic systems governed by equations with periodic coefficients. Such systems occur frequently in orbital mechanics, applied mechanics where they are frequently denoted by the term parametric excitation problems, nonlinear vibrations, rotating shaft problems, chemical physics applications, electrical engineering, analysis of control systems, and analysis of networks. They also have important aerospace applications since the dynamics and aeromechanics of helicopter rotors in forward flight are governed by such systems. Wind turbine blade dynamics are another application of periodic systems. These problems also appear in rotating satellite dynamics with flexible appendages. Finally, it should be noted that this class of problems has been addressed by books on applied mathematics.

It is not surprising that in several cases researchers working in one area were not aware of the work done in a different area, and similar methods for dealing with the numerical problem were invented and given different names, accompanied by claims that the method was new and original.

## 9.2 Simple Example of a Periodic System

The simply supported beam shown in Figure 9.1 is probably one of the simplest physical systems, where the equation of motion is one with periodic coefficients.

The equation of motion for this case can be obtained from Hamilton's Principle, Section 2.7:

$$EI\frac{\partial^4 v}{\partial x^4} + P\frac{\partial^2 v}{\partial x^2} + m\frac{\partial^2 v}{\partial t^2} = 0. \qquad (9.2.1)$$

$$EI\frac{\partial^4 v}{\partial x^4} + (P_o + P_t \cos\theta t)\frac{\partial^2 v}{\partial x^2} + m\frac{\partial^2 v}{\partial t^2} = 0. \qquad (9.2.2)$$

For such a beam, it has been shown that the mode shapes are given by

$$\phi_k(x) = \sin\frac{k\pi x}{\ell}, \qquad (9.2.3)$$

**Figure 9.1** Simply supported beam with parametric excitation

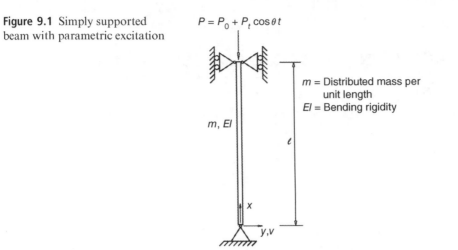

and the natural frequencies are given by

$$\omega_k = \frac{k^2\pi^2}{\ell^2}\sqrt{\frac{EI}{m}}. \tag{9.2.4}$$

Therefore, the solutions are

$$v(x,t) = f_k(t)\phi_k(x). \tag{9.2.5}$$

From Eqs. (9.2.2) through (9.2.5), we have

$$\left[m\frac{d^2f_k}{dt^2} + EI\frac{k^4\pi^4}{\ell^4}f_k - (P_o + P_t\cos\theta t)\frac{k^2\pi^2}{\ell^2}f_k\right]\sin\frac{k\pi x}{\ell} = 0. \tag{9.2.6}$$

Clearly, the solutions $f_k$ must satisfy the relation that the quantity in the brackets should vanish at any value of $t$, giving the differential equation

$$\frac{d^2f_k}{dt^2} + \omega_k^2\left(1 - \frac{P_o + P_t\cos\theta t}{P_k^*}\right)f_k = 0 \qquad k = 1, 2, \tag{9.2.7}$$

where $\omega_k$ is given by (9.2.4) and

$$P_k^* = k^2EI\,\pi^2\Big/\ell^2. \tag{9.2.8}$$

$P^*$ is the $k$th **Euler buckling load**.

Manipulating Eq. (9.2.7), we have

$$\frac{d^2f_k}{dt^2} + \omega_k^2\left(\frac{P_k^* - P_o}{P_k^*} - \frac{P_t\cos\theta t}{P_k^*}\right)f_k =$$

$$\frac{d^2f_j}{dt^2} + \omega_k^2\left(\frac{P_k^* - P_o}{P_k^*}\right)\left[1 - \frac{2P_t\cos\theta t}{1(P_k^* - P_o)}\right]f_k = 0. \tag{9.2.9}$$

Define

$$\Omega_k = \omega_k \sqrt{1 - \frac{P_o}{P_k^*}} \tag{9.2.10}$$

and

$$\mu_k = \frac{P_t}{2(P_k^* - P_o)}, \tag{9.2.11}$$

where $\mu_k$ is the excitation parameter. From Eqs. (9.2.9 to 9.2.11), one has

$$\frac{d^2 f_k}{dt^2} + \Omega_k^2 (1 - 2\mu_k \cos\theta t) f_k = 0 \qquad k = 1, 2, 3, \ldots. \tag{9.2.12}$$

The subscripts $k$ can be omitted because the equation is identical for all modes, thus

$$f'' + \Omega^2(1 - 2\mu \cos\theta t)f = 0. \tag{9.2.13}$$

Equation (9.2.13) is the so-called **Mathieu equation**.

For the more general case, the longitudinal or axial force can be represented by a general periodic function with a period $T$.

$$P(t) = P_o + P_t \phi(t).$$

$$\phi(t + T) = \phi(t).$$

For this case, Eq. (9.2.13) can be rewritten as $[\frac{d}{dt} = (\ )']$

$$f'' + \Omega^2[1 - 2\mu\phi(t)]f = 0. \tag{9.2.14}$$

This equation which is slightly more general is called the **Hill equation**. The most interesting aspect of this equation is that for certain relationships between its coefficients, it has solutions which are unbounded. Equation (9.2.13) can be rewritten as

$$f'' + (\lambda - h^2 \cos 2x)f = 0. \tag{9.2.15}$$

Comparing (9.2.15) and (9.2.13) we have

$$\Omega^2 = \lambda \ ; \ 2\mu\Omega^2 = h.$$
$$\theta = 2 \ ; \ x = t.$$

The solution of Eq. (9.2.15) is well known, and its graphical representation is called the Strut diagram (shown in Figure 9.2).

Figure 9.2 shows the distribution of the regions of instability for the **Mathieu equation**

$$\frac{d^2 f}{dx^2} + (\lambda - h^2 \cos 2x)f = 0.$$

In such a form, the coefficients of the equation depend on the two parameters $\lambda$ and $h^2$, which are plotted as coordinates. **The regions in which the solutions of the equation are unbounded are crosshatched**. As evident from the figure, the regions of instability occupy a

**Figure 9.2**
Ince-Strutt
diagram
identifying
regions of
instability

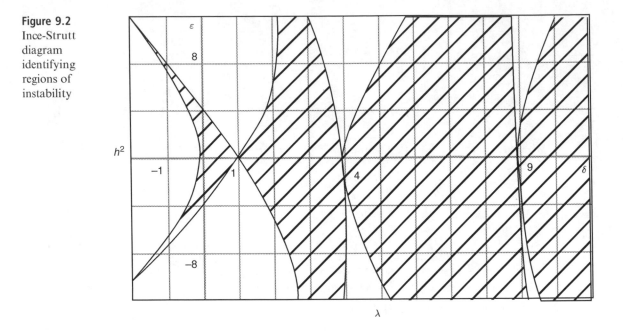

considerable part of the plane of the parameters. Therefore, to answer the question as to whether or not the rod is stable, it is necessary to find the point corresponding to the given ratio of parameters in the $\lambda, h^2$ plane. If a point is located in the non-crosshatched region, it means that the initial straight configuration of the rod is dynamically stable. However, if the same point is in the crosshatched region, then any initial deviation from the straight, undeformed configuration will increase with time in an unbounded manner, that is, the straight form of the rod will be dynamically unstable. The determination of the regions of dynamic instability represents the most important problem that needs to be solved.

The general treatment of equations with periodic coefficients is facilitated by transforming the system into a set of first-order differential equations.
For example, using Eq. (9.2.14), define

$$f' = y_1 \; ; f = y_2, \tag{9.2.16}$$

then

$$y_2' = y_1. \tag{9.2.17}$$

From Eqs. (9.2.16) and (9.2.14), one has

$$y_1' + \Omega^2[1 - 2\mu\phi(t)]y_2 = 0. \tag{9.2.18}$$

Equations (9.2.17) and (9.2.18) can be rewritten in matrix form

$$\mathbf{y}' = \mathbf{A}(t)\mathbf{y}, \tag{9.2.19}$$

where

$$y = \begin{Bmatrix} y_1 \\ y_2 \end{Bmatrix} \; ; A(t) = \begin{bmatrix} 0 & -\Omega^2[1 - 2\,\mu\phi(t)] \\ 1 & 0 \end{bmatrix},$$

$y'$ is called the state variable vector. $\mathbf{A}(t + T) = \mathbf{A}(t)$ is a general periodic matrix.

The **form** of the general solution for a system of first-order periodic, homogeneous equations, such as Eq. (9.2.19), is given by the Floquet theorem.

## 9.3  General Background

As mentioned in the introduction to this chapter, periodic systems have been treated in the past by mathematicians, electrical network researchers, control systems researchers, and applied mechanicians – under the general title of parametric excitation problems. Periodic systems also play a major role in helicopter rotor dynamics.

The general form of these equations can be written as

$$\dot{\mathbf{x}} = \mathbf{A}(\psi)x + \mathbf{f}(\psi), \tag{9.3.1}$$

where $A(\psi + T) = A(\psi)$ is an (n × n) periodic matrix whose elements have a common period $T$ (nondimensional) and $\psi$ is a nondimensional time variable.

Two problems are usually considered:

(1) the response of the system to some forcing represented by the vector $\mathbf{f}(\psi)$;
(2) the stability problem associated with the homogeneous system $\dot{\mathbf{x}} = \mathbf{A}(\psi)\mathbf{x}(\psi)$.

The stability problem will be considered first in this section. To deal with these problems, certain mathematical preliminaries provided next are required.

## 9.4  The Transition Matrix and Its Properties

First, the concept of a **transition matrix** or fundamental matrix is introduced since it is needed for the solution of the homogeneous problem. (Note that the transition matrix is called, sometimes, the matrizant.) **Definition**: The transition matrix is a matrix whose columns contain the linearly independent solutions of the homogeneous equation over the domain of interest. In general, this can be written as

$$\Phi(\psi, \psi_0) = [\boldsymbol{\phi}_1(\psi, \psi_0), \, \boldsymbol{\phi}_2(\psi, \psi_0), \, \ldots, \boldsymbol{\phi}_n(\psi, \psi_0)]. \tag{9.4.1}$$
$$(n \times n) \qquad (n \times 1)$$

Each column, $\boldsymbol{\phi}_1, \boldsymbol{\phi}_2, \ldots, \boldsymbol{\phi}_n$, of the square matrix in Eq. (9.4.1) satisfies the homogeneous differential equation (Eq. (9.3.1)) with the appropriate initial conditions. These initial conditions are given by

$$x_j(\psi_0) = \begin{cases} 1 & j = k \\ 0 & j \neq k \end{cases},$$ 

(9.4.2)

that is, $\Phi(\psi_0, \psi_0) = I$    ($I$ = unit matrix).

Application of each initial condition in Eq. 9.4.2 yields $\phi_k(\psi, \psi_0)$. Clearly, the initial conditions are

$$\mathbf{x}\sim(\psi_0) = \mathbf{x}_0 = \Phi(\psi_0, \psi_0)\mathbf{x}_0, \quad \text{furthermore } \dot{\Phi}(\psi, \psi_0) = \mathbf{0}.$$

(9.4.3)

Some of the important properties of the transition matrix are as follows:

(1)

$$\det|\Phi(\psi, \psi_0)| = \exp\left[\int_{\psi_o}^{\psi} \text{tr } A(\sigma)d\sigma\right],$$

(9.4.4)

where tr $A(\sigma)$ is the trace or sum of the diagonal elements of $A(\psi)$ matrix of the homogeneous system.

(2) If $\Phi$ is a transition matrix, then it satisfies the functional equation

$$\Phi(\psi, \psi_0) = \Phi(\psi, \psi_1)\Phi(\psi_1, \psi_0),$$

(9.4.5)

and this is called the extension property of the transition matrix.

### 9.4.1  Concept of the Matrix Exponential

Consider a matrix $B$ (square matrix), then the matrix exponential

$$e^B = I + B + B^2\Big/2! + \dots B^n\Big/n! \qquad n = 1, 2 \dots \infty,$$

(9.4.6)

where

$$B^2 = BB,$$

$$B^n = BB \dots B \quad n - \text{ times.}$$

## 9.5    The Floquet Theorem

The Floquet theorem was originally stated and proven in a French paper published in 1883 [Floquet, G., Sur le equations differentielles lineaires a coefficients periodiques (On linear differential equations with periodic coefficients) Ann. Ecole Norm Sup. 1883, 12:47–49]. Since the original paper is not readable to the authors, they selected to reference a version of the theorem published and proven in Brockett (1970), pp. 46–48.

**Theorem**: If $A(\psi + T) = A(\psi)$, then the transition matrix associated with the solution of the homogeneous system can be written as

$$\Phi(\psi, \psi_o) = P^{-1}(\psi)e^{R(\psi - \psi_o)}P(\psi_o), \tag{9.5.1}$$

where $R =$ is a constant matrix and $P(\psi + T) = P(\psi)$ is a periodic matrix. **Proof**: Any nonsingular matrix can be expressed as an exponential, define

$$C = e^{RT} = \Phi(T, 0). \tag{9.5.2}$$

Also, define the matrix $P(\psi)$ by

$$P^{-1}(\psi) = \Phi(\psi, 0)e^{-R\psi}. \tag{9.5.3}$$

Proof proceeds in two stages: first it is shown that $P(\psi)$ is periodic and then the proof of Eq. (9.5.1) will be given.
(1) First, it will be shown that if $\Phi(\psi, 0)$ is a transition matrix so is $\Phi(\psi + T, 0)$; if $\Phi(\psi, 0)$ is a transition matrix, it satisfies

$$\dot{\Phi}(\psi, 0) = A(\psi)\Phi(\psi, 0)$$

$$\dot{\Phi}(\psi + T, 0) = A(\psi + T)\Phi(\psi + T, 0), \tag{9.5.4}$$

but

$$A(\psi + T) = A(\psi)$$
$$\dot{\Phi}(\psi + T, 0) = A(\psi)\Phi(\psi + T, 0), \tag{9.5.5}$$

thus $\Phi(\psi + T, 0)$ is also a transition matrix.
**Lemma**: If $\Phi$ is a fundamental matrix solution of $\dot{x} = A(\psi)x$ and if $X(\psi)$ is any matrix solution of this equation, then there is a constant $C$ such that $X(\psi) = \Phi(\psi)C$. To prove this lemma, we only have to show that $\Phi(\psi)^{-1}X(\psi)$ is a constant matrix. This can be shown by taking the derivative

$$\frac{d}{d\psi}\left[\Phi^{-1}(\psi)X(\psi)\right] = \left[\frac{d}{d\psi}\Phi^{-1}(\psi)\right]X(\psi) + $$
$$+\Phi^{-1}(\psi)\frac{d}{d\psi}X(\psi). \tag{9.5.6}$$

Now $\Phi^{-1}(\psi)\Phi(\psi) = I$ and

$$\left[\frac{d}{d\psi}\Phi^{-1}(\psi)\right]\Phi(\psi) + \Phi^{-1}(\psi)\frac{d\Phi(\psi)}{d\psi} = 0,$$

thus

$$\frac{d}{d\psi}\Phi^{-1}(\psi) = -\Phi^{-1}(\psi)\frac{d\Phi(\psi)}{d\psi}\Phi^{-1}(\psi). \tag{9.5.7}$$

Combining Eqs. (9.5.6) and (9.5.5), we have

$$\frac{d}{d\psi}\left[\boldsymbol{\Phi}^{-1}(\psi)\boldsymbol{X}(\psi)\right] = -\boldsymbol{\Phi}^{-1}(\psi)\frac{d\boldsymbol{\Phi}(\psi)}{d\psi}\boldsymbol{\Phi}^{-1}\boldsymbol{X}(\psi) +$$

$$+\boldsymbol{\Phi}^{-1}(\psi)\frac{d}{d\psi}\boldsymbol{X}(\psi). \tag{9.5.8}$$

Using the homogeneous equation and Eq. (9.5.5) and the fact that $\boldsymbol{X}(\psi)$ is a matrix solution of the homogeneous system, we have

$$\frac{d}{d\psi}\left[\boldsymbol{\Phi}^{-1}(\psi)\boldsymbol{X}(\psi)\right] = -\boldsymbol{\Phi}^{-1}\boldsymbol{A}(\psi)\boldsymbol{\Phi}\boldsymbol{\Phi}^{-1}\boldsymbol{X}(\psi) +$$

$$+\boldsymbol{\Phi}^{-1}\boldsymbol{A}(\psi)\boldsymbol{X}(\psi) = -\boldsymbol{\Phi}^{-1}\boldsymbol{A}(\psi)\boldsymbol{X}(\psi) + \boldsymbol{\Phi}^{-1}\boldsymbol{A}(\psi)\boldsymbol{X}(\psi) \tag{9.5.9}$$

$$= 0. \tag{9.5.10}$$

Thus, we have proved the lemma.

Next it is clear that any constant matrix $\boldsymbol{C}$ can be expressed as

$$\boldsymbol{C} = e^{\boldsymbol{R}T}. \tag{9.5.11}$$

Thus, using this lemma

$$\boldsymbol{\Phi}(\psi + T, \psi_0) = \boldsymbol{\Phi}(\psi, \psi_0)\boldsymbol{C}. \tag{9.5.12}$$

Now using the information available we prove that

$$\boldsymbol{P}(\psi + T) = \boldsymbol{P}(\psi) \qquad \text{that is, } \boldsymbol{P} - \text{ is periodic.}$$

Using Eq. (9.5.3)

$$\boldsymbol{P}^{-1}(\psi + T) = \boldsymbol{\Phi}(\psi + T, 0)e^{-\boldsymbol{R}(\psi+0)}. \tag{9.5.13}$$

From Eqs. (9.5.13), (9.4.5), and (9.5.2),

$$\boldsymbol{P}^{-1}(\psi + T) = \boldsymbol{\Phi}(\psi + T, T)\boldsymbol{\Phi}(T, 0)e^{-\boldsymbol{R}(\psi+T)}$$

$$= \boldsymbol{\Phi}(\psi + T, T)e^{\boldsymbol{R}T}e^{-\boldsymbol{R}T}e^{-\boldsymbol{R}\psi} = \boldsymbol{\Phi}(\psi + T, T)e^{-\boldsymbol{R}\psi}. \tag{9.5.14}$$

Next from Eqs. (9.5.12) and (9.4.5),

$$\boldsymbol{\Phi}(\psi + T, T) = \boldsymbol{\Phi}(\psi + T, 0)\boldsymbol{\Phi}(0, T) = \boldsymbol{\Phi}(\psi + T, 0)\boldsymbol{C}^{-1}. \tag{9.5.15}$$

From Eqs. (9.5.14), (9.5.15), and (9.5.12),

$$\boldsymbol{P}^{-1}(\psi + T) = \boldsymbol{\Phi}(\psi + T, T)e^{-\boldsymbol{R}\psi} = \boldsymbol{\Phi}(\psi + T, 0)\boldsymbol{C}^{-1}\boldsymbol{C}^{-1}$$

$$= \boldsymbol{\Phi}(\psi, 0)\boldsymbol{C}\boldsymbol{C}^{-1}\boldsymbol{C}^{-1} = \boldsymbol{\Phi}(\psi, 0)e^{-\boldsymbol{R}\psi_o} = \boldsymbol{P}^{-1}(\psi). \tag{9.5.16}$$

Thus,

$$\boldsymbol{P}^{-1}(\psi + T) = \boldsymbol{P}^{-1}(\psi). \tag{8.4.40a}$$

$\boldsymbol{P}(\psi) = \text{is periodic.}$

(2) Proof of the second part.

From the extension property of the fundamental matrix, Eq. (9.4.5),

$$\Phi(\psi, \psi_0) = \Phi(\psi_0, 0) = \Phi(\psi, 0), \tag{9.5.17}$$

thus

$$\Phi(\psi, \psi_0) = \Phi(\psi, 0)\Phi(\psi_0, 0)^{-1}. \tag{8.4.41a}$$

From Eqs. (8.4.41a) and (9.5.3),

$$\Phi(\psi, \psi_0) = P^{-1}(\psi)e^{R(\psi - \psi_0)}e^{-R\psi_\varrho}\Phi(\psi_0, 0)^{-1} -$$
$$= P^{-1}(\psi)e^{R(\psi - \psi_0)}e^{-R\psi_o}e^{-R\psi_0}P(\psi_0)$$
$$\Phi(\psi, \psi_0) = P^{-1}(\psi)e^{R(\psi - \psi_0)}P(\psi_0).$$

$$\tag{9.5.18}$$

Note: The Floquet theorem does not give us the solution but provides information on the mathematical *form* of the solution.

## 9.6   Relations between the Eigenvalues of $\Phi(T, 0)$ and $R$

Recall that by definition

$$\Phi(T, 0) = e^{RT} = C.$$

Two distinct cases can occur:

(1) The matrix $C$ has $n$-independent eigenvectors associated with $n$-distinct eigenvalues.
(2) The matrix $C$ has repeated eigenvalues and a certain corresponding number of generalized eigenvectors.

*Case (1)*: In this case, it is known from elementary linear algebra (see Noble 1969) that a similarity transformation can be found such that

$$Q^{-1}RQ = \lambda \tag{9.6.1}$$

or

$$R = Q\lambda Q^{-1}, \tag{8.5.42a}$$

where the columns of $Q$ are the $n$-linearly independent eigenvectors of $R$ and $\lambda$ is a diagonal matrix containing the eigenvalues of $R$. From Eqs. (9.5.2) and (8.5.42a) and the definition of the matrix exponential,

$$e^{RT} = e^{Q\lambda Q^{-1}T} = I + Q\lambda TQ^{-1} + 1/2(Q\lambda TQ^{-1})(Q\lambda TQ^{-1}) +$$

$$+ \dots \frac{1}{n!} \underbrace{(Q\lambda TQ^{-1})(Q\lambda TQ^{-1})\dots(Q\lambda TQ^{-1})}_{n - \text{times}} =$$

$$= Q\left[I + \lambda T + \frac{(\lambda T)^2}{2!} + \dots \frac{(\lambda T)^n}{n!}\right]Q^{-1} =$$

$$= Qe^{\lambda T}Q^{-1} = C. \tag{9.6.2}$$

From Eqs. (9.6.2) and (9.5.2),

$$e^{\lambda T} = \Lambda = Q^{-1}\Phi(T,0)Q, \tag{9.6.3}$$

$$e^{\lambda_s T} = \Lambda_s \qquad \Lambda_s - \text{eigenvalues of } C, \tag{9.6.4}$$

where $\lambda_s$ and $\Lambda_s$ are in general complex

$$\left.\begin{array}{c} \lambda_k = \zeta_k + i\omega_k \\ \Lambda_k = \Lambda_{kR} + i\Lambda_{kI} \end{array}\right\}. \tag{9.6.5}$$

From Eqs. (9.6.4) and (9.6.5),

$$\zeta_k = \frac{1}{2T}\ell n\left[\Lambda_{kr}^2 + \Lambda_{kI}^2\right]. \tag{9.6.6}$$

$$\omega_k = \frac{1}{T}\tan^{-1}\left(\frac{\Lambda_{kI}}{\Lambda_{kR}}\right). \tag{9.6.7}$$

It is important to realize that $\omega_k$ can be determined only up to an integer multiple of the common period $T$.

The quantities $\lambda_k$ are denoted *characteristic exponents* and $\Lambda_k$ are called *characteristic multipliers* in Floquet theory.

*Case (2)*: For this case, one must use the Jordan canonical form (see Coddington and Levinson 1955, 80–81).

## 9.7    Consequences of the Floquet Theorem

(1) The stability of the system (Eq. 9.2.19) is determined by the matrix $R$ or the knowledge of the transition matrix at the end of *one* period, $\Phi(T,0)$.

(2) Knowledge of the transition matrix over one period determines the solution to the homogeneous system everywhere throughout the relation

$$\Phi(\psi + sT, 0) = \Phi(\psi, 0)(e^{RT})^s. \tag{9.7.1}$$

Note that Eq. (9.7.1) is a consequence of Eq. (9.4.5), where $0 \leq \psi \leq T$ and $s$ is an integer.

(3) The stability criteria for the system

$$\dot{x} = A(\psi)x(\psi) \tag{9.7.2}$$

is related to the eigenvalues of $R$ or the real part of the characteristic exponent $\zeta_k$. The solutions of the system (9.7.2) approach zeros as $\psi \to \infty$ if

$$\left| \Lambda_{kR}^2 + \Lambda_{kI}^2 \right| < 1 \text{ or } \zeta_k < 0 \quad \text{for } k = 1, 2 \dots n. \tag{9.7.3}$$

## 9.8 Methods for Calculating the Transition Matrix $\Phi(T, 0)$

Clearly, the key to the efficient numerical treatment of equations with periodic coefficients is in the efficient numerical computation of the transition matrix at the end of *one period*. A variety of numerical methods can be used to compute $\Phi(T, 0)$. The difference between the methods is associated with the efficiency, as represented by the computer time required, to compute $\Phi(T, 0)$.

### 9.8.1 Direct Numerical Integration

This straightforward method consists of evaluating the columns of $\Phi(T, 0)$ by integrating the homogeneous system $n$-times. Each integration yields one column $\Phi_k(T, 0)$ of $\Phi(T, 0)$ corresponding to the initial conditions

$$y_j(\psi_0 = 0) = \begin{cases} 1 & j = k \\ 0 & j \neq k \end{cases}. \tag{9.8.1}$$

A fourth-order Runge–Kutta routine for performing this computation is available on the various numerical libraries as well as MATLAB. However, there are many alternatives to this brute force method, which are computationally more efficient.

### 9.8.2 Approximate, Semi-Analytical Methods for Determining $\Phi(T, 0)$

In a series of papers, Hsu has developed various methods for approximating the transition matrix during one period (Hsu 1963; Hsu and Cheng 1973). The most efficient one consists of approximating the periodic matrix $A(\psi)$ by a series of step functions, this method can be considered to be a generalization of the "rectangular ripple" method (Gockel 1972) to multi-dimensional systems. The method consists of evaluating the state transition matrix by dividing a period into a number of equal parts and considering the equations over each interval to be a set of constant coefficient equations.

Each period $T$ is divided into $K$ intervals denoted by

$$\psi_k; k = 0, 1, 2, \dots, k,$$

with

$$0 = \psi_0 < \psi_1 < \ldots < \psi_K = T.$$

The $k$th interval $[\psi_{k-1}, \psi_k]$ is denoted by $\tau_k$ and its size by $\Delta_k = \psi_k - \psi_{k-1}$. In the $k$th interval, the periodic coefficient matrix $A(\psi)$ is replaced by a *constant* matrix $C_k$ defined by

$$C_k = A(\xi_k); \xi_k \epsilon \tau_k.$$
$$C_k = \frac{1}{\Delta_k} \int_{\psi_{k-1}}^{\psi_k} A(r)dr. \tag{9.8.2}$$

Thus, the actual system (Eq. (9.2.19)) is replaced by an approximate system

$$\dot{x}(0,k) = C(\psi,k)x(\psi,k), \tag{9.8.3}$$

where

$$(0,k) = x(0)$$

and

$$C(\psi,k) = \sum_{s=-\infty}^{\infty} \sum_{k=1}^{K} C_k [U(\psi - sT - \psi_{k-1}) - U(\psi - sT - \psi_k)]. \tag{9.8.4}$$

Equation (9.8.4) needs some clarification. The outer sum over $s$ simply covers the whole time ($\psi$) space between $-\infty < \psi < \infty$ and appears in the expression only in order to make it mathematically "fancier." $U(\psi)$ is the Heaviside unit step function, and the difference between two of these step functions at times $\psi_{k-1}$ and $\psi_k$, respectively, results in a rectangular pulse as shown in Figure 9.3.

**Figure 9.3** Heaviside unit step function

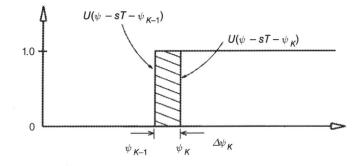

The elementary theory of linear differential equations with *constant coefficients* enables one to write the fundamental matrix (or the transition matrix) of the system with $\Phi_A(0,k) = I$, as

$$\Phi_A(\psi,k) = \exp[(\psi - \psi_{k-1})C_k]\exp(C_{k-1}\Delta_{k-1})\ldots\exp(C_1\Delta_1), \tag{9.8.5}$$

where $\Phi_A$ is an approximate transition matrix which is the solution to the approximate system represented by Eq. (9.8.3). The approximate transition matrix at the end of one period $\psi = T$ is given by

$$H(K) = \Phi_A(T, K) = \exp\left(\Delta_K C_K\right)\exp\left(\Delta_{K-1} C_{K-1}\right)\ldots\exp\left(\Delta_1 C_1\right) =$$
$$= \prod_{i=1}^{K}\exp\left(\Delta_i C_i\right). \tag{9.8.6}$$

With regard to the product sign, it is understood that the order of positioning of the factors is material and that the $k$th factor is to be placed in front of the $(k-1)$th factor. Furthermore, it should be noted that the approximate expression for $\Phi(T,0)$ given by Eq (9.8.6) represents the first level of approximation in Hsu's method.

It can be shown that when the number of division in one period is increased, that is, $K \to \infty$,

$$x(\psi, K) \to x(\psi),$$
$$\Phi_A(\psi, K) \to \Phi(\psi, 0),$$

and

$$H(K) = \Phi_A(T, 0) \to \Phi(T, 0).$$

Thus, the basic problem is the numerically efficient computation of $H(K)$. Using the definition of the matrix exponential,

$$\exp\left(\Delta_i C_i\right) = I + \Delta_i C_i + \frac{(\Delta_i C_i)^2}{2!} + \cdots + \frac{(\Delta_i C_i)^n}{n!}; n \to \infty. \tag{9.8.7}$$

Obviously, the computation of $\exp\left(\Delta_i C_i\right)$ for $n \to \infty$ would be very time-consuming. Fortunately, the time intervals $\Delta_i$ are small and therefore the series in Eq. (9.8.7) converges rapidly, thus the value of the matrix exponential can be accurately approximated by a finite number of terms. Thus,

$$\exp\left(\Delta_i C_i\right) \cong I + \sum_{j=1}^{J}\frac{(\Delta_i C_i)}{j!} = \exp\left(\Delta_i C_i\right) - \sum_{j=J+1}^{\infty}\frac{(\Delta_i C_i)^j}{j!}. \tag{9.8.8}$$

Obviously, the last term in Eq. (9.8.8) represents the error term in this approximation. Equation (9.8.8) is another approximation inherent in this method. Therefore, the approximate value of the transition matrix at the end of one period can be obtained by combining Eqs. (9.8.7) and (9.8.8). Thus,

$$H_A(K) = \Phi_{AA}(T, 0) = \prod_{i=1}^{K}\left\{I + \sum_{j=1}^{J}\frac{(\Delta_i C_i)^j}{j!}\right\}. \tag{9.8.9}$$

Finally, it should be mentioned that error bounds for these approximations have been obtained by Hsu, and it was shown that $\Phi(T, 0) - H_A(K) \sim O(\Delta^2)$ for $J \geq 2$.

Thus, $\mathbf{\Phi}_{AA}(T,0)$ represents the approximate value of the transition matrix at the end of the period after *two* levels of approximation as represented by Eqs. (9.8.6) and (9.8.8).

A number of comments for using this method is provided next. For practical applications, $J = 4$ is sufficient and the number of divisions is $35 < K < 50$.

Typical plots representing computing times for a $4 \times 4$ system and comparison of the numerical accuracy of the method with Hsu's method are shown in Friedmann et al. (1977).

### 9.8.3    Improved Numerical Integration Scheme for Obtaining $\Phi(T,0)$

The direct numerical integration scheme described in Section 9.9.1 has the shortcoming because it requires n-integration passes for an $n \times N$ system; this deficiency is due to the usage of available integration routines in numerical libraries. In actually, by programming one's own Runge–Kutta routine, the integration can be performed in a single pass.

A suitable general Runge–Kutta method is the one using Gill coefficients; these coefficients minimize the local truncation error of the method (see Lapidus and Seinfeld 1971, p. 50).

The Runge–Kutta method provides a solution to the equation

$$\dot{x} = f(\psi, x). \tag{9.8.10}$$

For the fourth order Runge–Kutta method with Gill coefficients, we have

$$x_{i+1} = x_i + \left(\frac{h}{6}\right)[k_1 + 2(1 - \frac{1}{\sqrt{2}})k_1 + 2(1 - \frac{1}{\sqrt{2}})k_3 + \\ + k_4], \tag{9.8.11}$$

where $x_{i+1} = x(\psi_{i+1})$   $x_i = x(\psi_i)$.

$i =$ is the $i$th interval, $h$ is the step size, and the vectors $k_1, k_2, k_3, k_4$ are given by

$$k_1 = f(\psi_i, x_i) = \frac{dx}{d\psi}(\psi_i, x_i) \tag{9.8.12}$$

$$k_2 = f(\psi_i + \frac{1}{2}h, x_i + \frac{1}{2}hk_1) \tag{9.8.13}$$

$$k_3 = f(\psi_1 + \frac{1}{2}h, x_i + (-\frac{1}{2} + \frac{1}{\sqrt{2}})hk_1 + (1 - \frac{1}{\sqrt{2}})hk_2) \tag{9.8.14}$$

$$k_4 = f(\psi_i + h, x_i - \frac{1}{\sqrt{2}}hk_2 + (1 + \frac{1}{\sqrt{2}})hk_3) \tag{9.8.15}$$

for our equations

$$\dot{x} = A(\psi)x,$$

thus

$$k_1 = A(\psi_i)x_i \tag{9.8.16}$$

and

$$k_2 = A(\psi_i + \frac{1}{2}h)x_i + \frac{1}{2}hk_1 =$$
$$= A(\psi_i + \frac{1}{2}h)x_i + \frac{1}{2}hA(\psi_i)x_i$$
$$= [A(\psi_i + \frac{1}{2}h)][I + \frac{1}{2}hA(\psi_i)]x_i \qquad (9.8.17)$$

or

$$k_2 = E(\psi_i)x_i, \qquad (9.8.18)$$

where from Eqs. (9.8.17) and (9.8.18),

$$E(\psi_i) = [A(\psi_i + \frac{1}{2}h)][I + \frac{1}{2}hA(\psi_i)]. \qquad (9.8.19)$$

Similarly, from Eqs. (9.8.16), (9.8.18), and (9.8.14),

$$k_3 = A(\psi_i + \frac{1}{2}h)x_i + (-1\frac{1}{2} + \frac{1}{\sqrt{2}})hA(\psi_i)x_i +$$
$$(1 - \frac{1}{\sqrt{2}})hE(\psi_i)x_i =$$
$$= [A(\psi_i + \frac{1}{2}h)][I + (-\frac{1}{2} + \frac{1}{\sqrt{2}}hA(\psi_i) +$$
$$+(1 - \frac{1}{\sqrt{2}})hE(\psi_i)x_i = F(\psi_i)x_i, \qquad (9.8.20)$$

where

$$F(\psi_i) = A(\psi_i + \frac{1}{2}h)[I + (-\frac{1}{2} + \frac{1}{\sqrt{2}})hA(\psi_i)$$
$$+(1 - \frac{1}{\sqrt{2}})hE(\psi_i)]. \qquad (9.8.21)$$

And from Eqs. (9.8.15), (9.8.16), (9.8.18), and (9.8.20),

$$k_4 = A(\psi_i + h)x_i - \frac{1}{\sqrt{2}}hE(\psi_i)x_i +$$
$$+ (1 - \frac{1}{\sqrt{2}})hF(\psi_i)x_i =$$
$$= A(\psi_i + h)I - \frac{h}{\sqrt{2}}E(\psi_i) + (1 + \frac{1}{\sqrt{2}})hF(\psi_i)x_i$$
$$= G(\psi_i)x_i, \qquad (9.8.22)$$

where

$$G(\psi_i) = [A(\psi_i + h)][I - \frac{h}{\sqrt{2}}E(\psi_i) + (1 + \frac{1}{\sqrt{2}})hF(\psi_i)]. \qquad (9.8.23)$$

Substituting Eqs. (9.8.16), (9.8.18), (9.8.20), and (9.8.22) into Eq. (9.8.11), we have

$$x_{i+1} = x_i + \left(\frac{h}{6}\right)(A(\psi_i)x_i + 2(1 - \frac{1}{\sqrt{2}})E(\psi_i)x_i +$$

$$+2(1 + \frac{1}{\sqrt{2}})F(\psi_i)x_i + G(\psi_i)x_i) =$$

$$x_{i+1} = x_i + \left(\frac{h}{6}\right)[A(\psi_i) + 2(1 - \frac{1}{\sqrt{2}})E(\psi_i) +$$

$$+2(1 - \frac{1}{\sqrt{2}})F(\psi_i) + G(\psi_i)]x_i = K(\psi_i)x_i$$

$$x_{i+1} = K(\psi_i)x_i, \tag{9.8.24}$$

where

$$K(\psi_i) - I + \left(\frac{h}{6}\right)A(\psi_i) + 2(1 - \frac{1}{\sqrt{2}})E(\psi_i)$$

$$+2(1 - \frac{1}{\sqrt{2}})F(\psi_i) + G(\psi_i). \tag{9.8.25}$$

From Eq. (9.8.24), one may write

$$x(\psi_1) = K(\psi_0)x(\psi_0). \tag{9.8.26}$$

$$x(\psi_2) = K(\psi_1)x(\psi_1) = K(\psi_1)K(\psi_0)x(\psi_0). \tag{9.8.27a}$$

$$x(\psi_n) = K(\psi_{n-1})K(\psi_{n-2})\ldots\ldots K(\psi_0)x(\psi_0). \tag{9.8.27b}$$

For the case of a *constant step size* $h = \frac{T}{N}$ ($T$ = nondimensional period, $N$ = number of intervals), and $\psi_0 = 0$ and $\psi_n = T$ from Eq. (9.8.27b),

$$x(T) = K(T - h)\,K(T - 2h)\ldots K(0)x(0)$$

$$= \Phi_A(T,0)x(0) = [\prod_{i=1}^{N} K(T - ih)]x(0) = H(T,0)x(0).$$

Thus,

$$\Phi_A(T,0) = H(T,0). \tag{9.8.28}$$

Thus, the approximate transition matrix at the end of the period is obtained by one integration pass and

$$\Phi_A(T,0) = \prod_{i=1}^{N} K(T - ih), \tag{9.8.29}$$

where it is understood that the position in the product is important and $i$th term is supposed to be **before** the $i + 1$th term.

**Figure 9.4**
Stability
boundary for
region $2\omega_1$

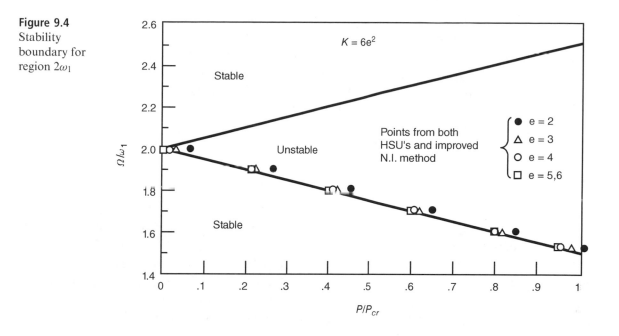

The two methods discussed earlier have been discussed in detail in a paper by Friedmann et al. (1977), where the system described in Figure 9.1, for the parametric excitation of a beam has been represented by conventional beam-type finite elements 4.2, and the effect of the axial time-dependent loading was incorporated using the geometric stiffness matrix Section 4.6. The advantage of this structural dynamic model consists of the feature that by changing the number of elements used in the model, the size of the homogeneous system can be conveniently increased and the computational efficiency of the methods is illustrated. Typical results are shown in Figure 9.4, which shows the stability boundary for the $2\omega_1$ region, shown by the solid lines in the figure. The region inside the lines is unstable and the region outside the lines is stable. The solid lines are obtained from Hsu's method. The points in the lower boundary are obtained by evaluating the transition matrix at the end of the period using both methods described in Sections 9.8.2 and 9.8.3, and evaluating the eigenvalues of the transition matrices at the end of the period. Within the accuracy of the figure the results are identical when the same number of elements is used to model the structural dynamic system. It should also be noted that the various points shown in the figure are obtained using double precision arithmetic and evaluating $\Phi(T, 0)$ by dividing the period $T$ into a number determined by $K = 6e^2$, where $e$ is the number of finite elements used to represent the axially loaded beam. This equation represents an estimate for the step size needed for good convergence, and its derivation is provided in the Appendix of the paper. The step size for the improved numerical method, described in Section 9.8.3, is obtained from numerical experimentation.

Comparison of computer times needed for the implementation of the two methods shows that the improved method is approximately 20% faster than Hsu's method. Results also show that the amount of CPU time required for the computation of $\Phi(T, 0)$ is approximately proportional to $n^3$, where $n$ represents the size of the $A(\psi)$ matrix in the homogeneous problem. A final comment, the excellent agreement between the two methods, is partially accidental due to the fact that the boundary of the $2\omega_1$ region is characterized by almost perfectly straight lines.

### 9.8.4   Additional Comments on Periodic Systems

The treatment of periodic systems is still an ongoing area of research, and to complete the treatment of this topic, additional comments on some more relevant contributions are summarized next.

One area where additional contributions have been made is the efficient computation of the transition matrix $\Phi(T, 0)$ at the end of a period, which is a required quantity for evaluation of the stability of both the homogeneous system and the nonhomogeneous system. For the efficient numerical evaluation of the transition matrix at the end of one period, two studies claim improvements in the required computer time. Sinha and Wu (1991) proposed an approach based on the idea that the periodic matrix can be expanded in terms of Chebyshev polynomials over the common period. The method requires algebraic manipulation of Chebyshev polynomials. This combined analytical and numerical method has been applied to the same example used in Friedmann et al. (1977) and improvements in computational time were obtained. This study was limited to linear homogeneous systems.

Another contribution by Mohan and Gaonkar (2013) focused on presenting and summarizing research on two topics: (1) efficient computation of $\Phi(T, 0)$ using an approach called "fast Floquet," but the topic treated was limited to the rotorcraft problem in forward flight, and (2) the second problem considered was aimed at identifying the eigenvectors or mode shapes that participate in the stability problem, which represents a useful contribution. More recently, the issue of determining and identifying the periodic eigenvectors, representing the coupling between the mode shapes present in a coaxial rotor in hover as well as forward flight treated in Singh and Friedmann (2021), where a novel method for identifying the mode shape analytically was developed and combined with a graphical method for determining the coupling between the modes that produce an instability. It should be noted that the concise summary provided here is not a comprehperidicensive summary of the existing work ongoing in this field. Instead, it illustrates some new developments aimed at resolving some long-standing problems associated with identifying mode shapes in periodic systems and the coupling between them.

Finally, it should be noted that the work done has not been limited to linear periodic systems but has been also extended to the nonlinear response problem of weakly periodic systems, as represented by helicopter rotor problems in forward flight (Friedmann 1990).

## 9.9    Steady-State Response of a Periodic System to Periodic Forcing

Consider a linear dynamical system under a combination of forced excitation, periodic, and parametric excitation. The linear system can be rewritten in first-order state variable form in the following manner:

$$\dot{q}(\psi) = z(\psi) + L(\psi)q(\psi). \tag{9.9.1}$$

The corresponding homogeneous system is given by

$$\dot{q}_H = L(\psi)q_H, \tag{9.9.2}$$

where it is understood that the matrix $[L(\psi)]$ is periodic and has a common period $T$, that is,

$$L(\psi) = L(\psi + T). \tag{9.9.3}$$

The excitation $z(\psi)$ is also periodic, however, in general it can have a different period $T_f$, that is,

$$z(\psi) = z(\psi + T_f). \tag{9.9.4}$$

In such cases, the common period is defined as (Hsu and Cheng, 1973).

$$T_c = m_o T = m_f T_f, \tag{9.9.5}$$

where $m_o$ and $m_f$ are positive integers and $m_o/m_f$ is an irreducible rotational number. For our case, we shall assume for the sake of simplicity that $m_o = m_f = 1$ and $T_c = 2\pi$. This assumption does not introduce any loss of generality in the solution which is discussed next.

Urabe (1965) has shown that for characteristic multipliers of the corresponding homogeneous system (Eq. (9.9.2)) which are different from one, that is, a stable homogeneous system according to Floquet theory, one has one and only one periodic solution of Eq. (9.9.1), given by

$$q = \Phi(\psi) < \int_0^\psi \Phi(s)^{-1}z(s)ds + (I - \Phi(2\pi))^{-1} *$$
$$* \Phi(2\pi) \int_0^{2\pi} \Phi(s)^{-1}z(s)ds. \tag{9.9.6}$$

The general solution for any inhomogeneous equation like Eq. (9.9.1), whether periodic or not, can be mathematically written as Wilson (1971), p. 117.

$$q(\psi) = \Phi(\psi)q(0) + \Phi(\psi) \int_0^\psi \Phi(s)^{-1}z(s)ds, \tag{9.9.7}$$

where in Eqs. (9.9.6) and (9.9.7)

$\Phi(\psi)$ is the transition matrix defined by    $\Phi(\dot{\psi}) = L(\psi)\Phi(\psi)$    and $\Phi(T,0) = \Phi(0) = I$.

Note that the condition for the existence of a unique periodic solution of Eq. (9.9.1) is that the determinant of $(I - \Phi(2\pi))$ be nonzero. This is satisfied if all the multipliers of Eq. (9.9.2), the $\Lambda'_i s$, are different from unity, which is equivalent to the condition that the real parts of all characteristic exponents be nonzero, that is, $\zeta_i \neq 0$. From Floquet theory, if all the $\zeta'_i s < 0$, the homogeneous system is asymptotically stable; if any one $\zeta_i > 0$, the homogeneous system becomes asymptotically unstable. In the former case, Eq. (9.9.6) is the steady-state solution of the linear system, Eq. (9.9.1). When any one $\zeta_i > 0$, the mathematical expression for $q(\psi)$, Eq. (9.9.1), is of no practical significance. Hsu and Cheng (1973) have made some interesting comments in this respect.

Comparing Eqs. (9.9.6) and (9.9.7), it is obvious that Eq. (9.9.6) corresponds to the general solution of any inhomogeneous system, Eq. (9.9.7), with the initial condition given by

$$q(0) = (I - \Phi(2\pi))^{-1} \Phi(2\pi) \int_0^{2\pi} \Phi(s)^{-1} z(s) ds. \tag{9.9.8}$$

With this information in mind, the periodic steady-state solution, or response, $q(\psi)$ is first evaluated using the transition matrix at the end of one period $\psi(2\pi)$. This matrix is evaluated using the approximate, semi-analytical method for determining $[\Phi(T, 0)]$ described in Section 9.8.2 of this book (also Friedmann et al. 1977) using the fourth-order approximation to the matrix exponential, that is,

$$\exp(\Delta_i C_i) = I + \Delta_i C_i + \frac{(\Delta_i C_i)^2}{2} + \frac{(\Delta_i C_i)^3}{3!} + \frac{(\Delta_i C_i)^4}{4!}. \tag{9.9.9}$$

Equation (9.9.8) can be simplified by rewriting it in a more convenient manner:

$$q(0) = (I - \psi(2\psi))^{-1} \int_0^{2\pi} \Phi(2\pi) \Phi(s)^{-1} z(s) ds. \tag{9.9.10}$$

Using the semi-group property of the transition matrix, which establishes a functional equation for the transition matrix,

$$\Phi(\psi, \psi_0) = \Phi(\psi, \psi_1) \Phi(\psi_1, \psi_0),$$

one can express

$$\Phi(2\pi, 0) = \Phi(2\pi, s) \Phi(s, 0) \tag{9.9.11}$$

or

$$\Phi(2\pi) \Phi(s, 0)^{-1} = \Phi(2\pi, s). \tag{9.9.12}$$

Substituting Eq. (9.9.12) into Eq. (9.9.8) yields

$$q(0) = (I - \Phi(2\pi))^{-1} \int_0^{2\pi} \Phi(2\pi, s) z(s) ds. \tag{9.9.13}$$

Equation (9.9.13) is more convenient for the calculation of the initial condition $q(0)$ given by (9.9.8), since it does not require the inversion of $\Phi(s)$. The approximation of the transition matrix given by Eq. (9.9.9) can also be employed to calculate the value of $\Phi(2\pi, s)$ required for the evaluation of $q(0)$. The value of the integral in Eq. (9.9.13) can be evaluated using any conventional integration procedure, such as Simpson's rule. Since normally a revolution, or a period is divided into over 50 intervals, this is an excellent approximation.

Next, by taking this initial condition, the linear system, Eq. (9.9.1), is integrated numerically using a fourth-order Runge–Kutta scheme, with Gill coefficients, which has also been discussed in a previous section dealing with the improved numerical integration scheme for evaluating $\Phi(T,0)$ (Friedmann et al. 1977) performed with a constant step size, which is identical to the step size, used in evaluating the transition matrix at the end of a period $\Phi(T,0)$. This means that the use of Eq. (9.9.6) is actually bypassed.

Convergence of the method is checked by comparing the displacement quantities obtained for the response with the initial conditions and subsequent revolutions or periods, that is, compare

$$q(\psi = 0) \text{ with } q(\psi = 2\pi), q(\psi = 4\pi) \text{ and } q(\psi = 6\pi).$$

Normally, excellent converged solutions are obtained within two or three revolutions.

It should also be mentioned that in Hsu and Cheng (1973) have established a numerical scheme based on Eq. (9.9.1). This method required considerably more computer time than the method described earlier.

Finally, it should be noted that the method described earlier has been very successful in obtaining the periodic response of large horizontal axis wind turbine blades (Kottapalli et al. 1979). The method has also been successfully applied to helicopter aeroelastic problems in forward flight for both linear and nonlinear cases (Friedmann and Kottapalli 1982).

## BIBLIOGRAPHY

Bolotin, V. (1964). *The Dynamic Stability of Elastic Systems*. Holden Day, Inc.

Brockett, R. W., editor (1970). *Finite Dimensional Linear Systems*. John Wiley & Sons.

Coddington, E. and Levinson, N. (1955). *Theory of Ordinary Differential Equations*. McGraw-Hill.

Friedmann, P. P. (1990). Numerical methods for the treatment of periodic systems with applications to structural dynamics and helicopter rotor dynamics. *Computers and Structures*, 35(4): 329–347.

Friedmann, P. P., and Kottapalli, S. B. R. (October 1982). Coupled flap-lag-torsional dynamics of hingeless rotor blades in forward flight. *Journal of the American Helicopter Society*, 28–36.

Friedmann, P. P., Hammond, C. E., and Woo, T. (1977). Efficient numerical treatment of periodic systems with application to stability problems. *International Journal of Numerical Methods in Engineering*, 11: 1117–1136.

Gockel, M. A. (1972). Practical solutions of linear equations with periodic coefficients. *Journal of the American Helicopter Society*, 16: 2–10.

Hsu, C. S. (1963). On parametric excitation of a dynamical system having multiple degrees of freedom. *Journal of Applied Mechanics*, 30: 367–372.

Hsu, C. S. and Cheng, W. H. (1973). Applications of the theory of impulsive parametric excitation and new treatments of general parametric excitation problems. *Journal of Applied Mechanics*, 40: 78–86.

Kottapalli, S. B. R., Friedmann, P. P., and Rosen, A. (1979). Aeroelastic stability and response of horizontal axis wind turbine blades. *AIAA Journal*, 17(12): 1381–1389.

Lapidus, L. and Seinfeld, J. H. (1971). *Numerical Solutions of Ordinary Differential Equations*. Academic Press.

Magnus, W. and Winkler, S. (1966). *Hill's Equation*. Interscience, John Wiley & Sons.

Mohan, R. and Gaonkar, G. H. (2013). A unified assessment of fast floquet, generalized floquet, and periodic eigenvector methods for rotorcraft stability predictions. *Journal of the American Helicopter Society*, 58: 042002–1–042002–12.

Noble, B. (1969). *Applied Linear Algebra*. Prentice Hall.

Richards, J. A. (1983). *Analysis of Periodically Time-Varying Systems*. Springer-Verlag.

Singh, P. and Friedmann, P. P. (2021). Aeromechanics and aeroelastic stability of coaxial rotors. *Journal of Aircraft*, 58(6): 1386–1405.

Sinha, S. C. and Wu, D. H. (1991). An efficient computational scheme for the analysis of periodic systems. *Journal of Sound and Vibration*, 151(1): 91–117.

Urabe, M. (1965). Galerkin's procedure for nonlinear periodic systems. *Archives of Rational Mechanics and Analysis*, 20: 120–152.

Virgin, L. N. (2007). *Vibration of Axially Loaded Structures*. Cambridge University Press.

Wilson, H. K. (1971). *Ordinary Differential Equations*. Addison-Wesley Publishing Co.

## PROBLEMS

1.  Consider the simply supported beam shown in Figure 9.1.

    (a) Model the beam with two beam-type finite elements and solve the resulting version of Eq. (9.2.15) by writing a MATLAB program to calculate the transition matrix at the end of a period using Hsu's method, for the following data $h^2 = 5$ and $\lambda = 10$. Locate the points on the Strut diagram (Figure 9.2). Is the system stable?

    (b) Repeat the calculation performed in (a) by writing a MATLAB code based on the improved numerical integration method for calculating the transition matrix at the end of the period. Compare your results with the results you have obtained in (a). Is the system stable?

    (c) Solve the problem you have solved in (a) using four beam-type finite elements to model the simply supported beam with parametric excitation. Compare your results with the results obtained in (a) and (b), and comment on the differences.

# APPENDIX A

# Linear Algebra and Matrix Methods

Linear structural dynamics problems are usually expressed by systems of ordinary differential equations (ODEs). Solution of linear ODEs requires knowledge of matrix algebra describing operations performed on matrices. Use of matrices in structural dynamics facilitates a systematic representation of the analysis using a sequence of matrix operations. These operations can be easily programmed for computer-based numerical solutions.

Basic concepts in linear algebra and matrix mathematics are presented in this appendix. These are intended to serve as a quick reference guide. Commonly used types of matrices are defined together with fundamental matrix operations. Matrix inversion is an important operation used in the solution of simultaneous linear algebraic equations. Methods for performing matrix inversion are also described so as to provide understanding of solution approaches.

## A.1 Linear Vector Spaces

### A.1.1 Vector Space and Subspace

The concepts of a vector as an $n$-dimensional collection of numbers and linear operators as matrices are familiar. A linear vector space is a collection of elements known as vectors which are not necessarily $n$-dimensional. In order to introduce the idea of a vector space, we define first the concept of a *field*. A field $\mathbf{F}$ is a set of scalars that possess the following properties:

(1) There must exist an element $0 \in \mathbf{F}$ called zero such that for any other element $a \in \mathbf{F}$, $0(a) = 0$ and $0 + a = a$.
(2) There must exist an element $1 \in \mathbf{F}$ called one such that for any other element $a \in \mathbf{F}$, $1(a) = a(1) = a$ and $a/1 = a$.
(3) For every $a \in \mathbf{F}$, there is a unique element $-a \in \mathbf{F}$, called the additive inverse, such that $a + (-a) = 0$.
(4) For every $a \in \mathbf{F}$, there is a unique element $a^{-1} \in \mathbf{F}$, called the multiplicative inverse, such that $a(a)^{-1} = 1$.
(5) If $a \in \mathbf{F}$ and $b \in \mathbf{F}$, then $(a + b) = (b + a) \in \mathbf{F}$. This is known as closure under addition.
(6) If $a \in \mathbf{F}$ and $b \in \mathbf{F}$, then $(ab) = (ba) \in \mathbf{F}$. This is known as closure under multiplication.
(7) If $a \in \mathbf{F}$ and $b \in \mathbf{F}$, and if $b \neq 0$, then $a/b \in \mathbf{F}$.

The set of all real numbers and the set of all complex numbers are both fields. The set of all integers is not a field since the inverse of an integer is not always an integer.

The definition of a vector space is dependent on the field over which it is specified. A *linear vector space* $\mathbf{X}$ is a collection of elements called vectors, defined over a field $\mathbf{F}$, which must satisfy the following properties for any $\mathbf{x}, \mathbf{y}, \mathbf{z} \in \mathbf{X}$:

(1) There exists a unique vector $\mathbf{0} \in \mathbf{X}$ such that $\mathbf{x} + \mathbf{0} = \mathbf{0} + \mathbf{x} = \mathbf{x}$.
(2) For every vector $\mathbf{x} \in \mathbf{X}$, there exists a unique vector $-\mathbf{x}$ such that $\mathbf{x} + (-\mathbf{x}) = (-\mathbf{x}) + \mathbf{x} = \mathbf{0}$.
(3) Closure under addition: $(\mathbf{x} + \mathbf{y}) \in \mathbf{X}$.
(4) Commutativity of addition: $\mathbf{x} + \mathbf{y} = \mathbf{y} + \mathbf{x}$.
(5) Associativity of addition: $(\mathbf{x} + \mathbf{y}) + \mathbf{z} = \mathbf{x} + (\mathbf{y} + \mathbf{z})$.
(6) Closure under scalar multiplication: For any scalars $a, 1, 0 \in \mathbf{F}$, the product $a\mathbf{x} \in \mathbf{X}$. Also, $1\mathbf{x} = \mathbf{x}$ and $0\mathbf{x} = \mathbf{0}$.
(7) Associativity of scalar multiplication: For any scalars $a, b \in \mathbf{F}$, $a(b\mathbf{x}) = (ab)\mathbf{x}$.
(8) Distributivity of scalar multiplication: For any scalars $a, b \in \mathbf{F}$, $(a + b)\mathbf{x} = a\mathbf{x} + b\mathbf{x}$ and $a(\mathbf{x} + \mathbf{y}) = a\mathbf{x} + a\mathbf{y}$.

Let $\mathbf{S}$ be a subset of the vector space $\mathbf{X}$, then $\mathbf{S}$ is a subspace of $\mathbf{X}$ if the following statements are true:

(1) If $\mathbf{x}, \mathbf{y} \in \mathbf{S}$, then $\mathbf{x} + \mathbf{y} \in \mathbf{S}$.
(2) If $\mathbf{x} \in \mathbf{S}$ and $a \in \mathbf{F}$, then $a\mathbf{x} \in \mathbf{S}$.

## A.1.2 Euclidean Space

The *Euclidean space*, denoted as $\mathbb{R}^n$, is the set of all $n$-dimensional vectors with elements from the field of real numbers $\mathbb{R}$. The Euclidean space is an important example of a linear vector space. A vector $\mathbf{x}$ from $\mathbb{R}^n$ is written as

$$\mathbf{x} = \begin{bmatrix} x_1 \\ x_2 \\ \vdots \\ x_n \end{bmatrix}. \tag{A.1.1}$$

Let $\alpha \in \mathbb{R}$, the multiplication between the scalar $\alpha$ and the vector $\mathbf{x}$ is defined as

$$\alpha\mathbf{x} = \begin{bmatrix} \alpha x_1 \\ \alpha x_2 \\ \vdots \\ \alpha x_n \end{bmatrix}. \tag{A.1.2}$$

The addition of two vectors $\mathbf{x}$ and $\mathbf{y}$ is defined as

$$\mathbf{x} + \mathbf{y} = \begin{bmatrix} x_1 + y_1 \\ x_2 + y_2 \\ \vdots \\ x_n + y_n \end{bmatrix}. \tag{A.1.3}$$

Throughout the book, the vectors are defined on Euclidean space, unless otherwise specified.

### A.1.3 Linear Dependence

Consider a set of vectors $\mathbf{x}_1, \mathbf{x}_2, \ldots, \mathbf{x}_n$ in a linear space $\mathbf{X}$ and a set of scalars $a_1, a_2, \ldots, a_n$ in a field $\mathbf{F}$. Then, the vector $\mathbf{x}$ given by

$$\mathbf{x} = a_1\mathbf{x}_1 + a_2\mathbf{x}_2 + \ldots + a_n\mathbf{x}_n \tag{A.1.4}$$

is said to be a *linear combination* of $\mathbf{x}_1, \mathbf{x}_2, \ldots, \mathbf{x}_n$. The vectors $\mathbf{x}_1, \mathbf{x}_2, \ldots, \mathbf{x}_n$ are said to be *linearly independent* if the relation

$$a_1\mathbf{x}_1 + a_2\mathbf{x}_2 + \ldots + a_n\mathbf{x}_n = \mathbf{0} \tag{A.1.5}$$

can be satisfied only when all the coefficients $a_1, a_2, \ldots, a_n$ are identically zero. If Eq. (A.1.5) is satisfied and at least one of the coefficients $a_1, a_2, \ldots, a_n$ is different from zero, then the vectors $\mathbf{x}_1, \mathbf{x}_2, \ldots, \mathbf{x}_n$ are said to be *linearly dependent*, with the implication that one vector is a linear combination of the remaining $n-1$ vectors.

The subspace $\mathbf{S}$ of $\mathbf{X}$ consisting of all the linear combinations of the vectors $\mathbf{x}_1, \mathbf{x}_2, \ldots, \mathbf{x}_n$ is called a subspace spanned by those vectors.

### A.1.4 Basis of a Vector Space

A vector space $\mathbf{X}$ over a field $\mathbf{F}$ is said to be finite-dimensional if there exists a finite set of vectors $\mathbf{x}_1, \mathbf{x}_2, \ldots, \mathbf{x}_n$ that span the space, that is, every vector $\mathbf{y}$ in $\mathbf{X}$ is a linear combination of $\mathbf{x}_1, \mathbf{x}_2, \ldots, \mathbf{x}_n$,

$$\mathbf{y} = a_1\mathbf{x}_1 + a_2\mathbf{x}_2 + \ldots + a_n\mathbf{x}_n. \tag{A.1.6}$$

If $\mathbf{x}_1, \mathbf{x}_2, \ldots, \mathbf{x}_n$ are linearly independent and span $\mathbf{X}$, then they are referred to as the *basis* of $\mathbf{X}$. The coefficients $a_1, a_2, \ldots, a_n$ are the *coordinates* of $\mathbf{y}$, with respect to the basis $\mathbf{x}_1, \mathbf{x}_2, \ldots, \mathbf{x}_n$. The number of vectors in the basis is called the *dimension* of $\mathbf{X}$.

An example of the vector basis is the set of *unit axis vectors* in a Euclidean space $\mathbb{R}^n$,

$$\mathbf{e}_1 = \begin{bmatrix} 1 \\ 0 \\ \vdots \\ 0 \end{bmatrix}, \quad \mathbf{e}_2 = \begin{bmatrix} 0 \\ 1 \\ \vdots \\ 0 \end{bmatrix}, \quad \cdots, \quad \mathbf{e}_n = \begin{bmatrix} 0 \\ 0 \\ \vdots \\ 1 \end{bmatrix}. \tag{A.1.7}$$

A vector $\mathbf{x} \in \mathbb{R}^n$ is expressed using the unit axis vectors as

$$\mathbf{x} = x_1\mathbf{e}_1 + \cdots + x_n\mathbf{e}_n. \tag{A.1.8}$$

## A.1.5    Inner Product and Orthogonal Vector Basis

A vector space $\mathbf{X}$ over a field $\mathbf{F}$ is said to be an *inner product space*, if a unique scalar in $\mathbf{F}$ is assigned to each pair of vectors $\mathbf{x}$ and $\mathbf{y}$ in $\mathbf{X}$. This scalar, called the *inner product*, is denoted as $\langle \mathbf{x}, \mathbf{y} \rangle$ and must satisfy the following properties:

(1) Nonnegativity: For any $\mathbf{x} \in \mathbf{X}$ $\langle \mathbf{x}, \mathbf{x} \rangle \geq 0$ and $\langle \mathbf{x}, \mathbf{x} \rangle = 0$ if and only if $\mathbf{x} = \mathbf{0}$.
(2) Commutativity: $\langle \mathbf{x}, \mathbf{y} \rangle = \langle \mathbf{y}, \mathbf{x} \rangle$.
(3) Distributivity: For any $\mathbf{x}, \mathbf{y}, \mathbf{z} \in \mathbf{F}$, $\langle \mathbf{x}, \mathbf{y} + \mathbf{z} \rangle = \langle \mathbf{x}, \mathbf{y} \rangle + \langle \mathbf{x}, \mathbf{z} \rangle$.
(4) For any $\alpha \in \mathbf{F}$, $\langle \alpha \mathbf{x}, \mathbf{y} \rangle = \langle \mathbf{x}, \alpha \mathbf{y} \rangle = \alpha \langle \mathbf{y}, \mathbf{x} \rangle$.

A common definition of the inner product is

$$\langle \mathbf{x}, \mathbf{y} \rangle = x_1 y_1 + x_2 y_2 + \cdots + x_n y_n. \tag{A.1.9}$$

A measure of the size of a vector $\mathbf{x}$ is called the *norm* and denoted as $\|\mathbf{x}\|$. The norm must satisfy the following properties:

(1) Nonnegativity: For any $\mathbf{x} \in \mathbf{X}$ $\|\mathbf{x}\| \geq 0$ and $\|\mathbf{x}\| = 0$ if and only if $\mathbf{x} = \mathbf{0}$.
(2) Triangular inequality: $\|\mathbf{x} + \mathbf{y}\| \leq \|\mathbf{x}\| + \|\mathbf{y}\|$.
(3) For any $\alpha \in \mathbf{F}$, $\|\alpha \mathbf{x}\| = |\alpha| \|\mathbf{x}\|$.

A commonly used norm is defined using the inner product

$$\|\mathbf{x}\| = \sqrt{\langle \mathbf{x}, \mathbf{x} \rangle} = \sqrt{x_1^2 + \cdots + x_n^2}, \tag{A.1.10}$$

which defines the length of the vector $\mathbf{x}$. The vector of unit length is called a *unit vector*, $\|\mathbf{x}\| = 1$. Any nonzero vector can be normalized to form a unit vector

$$\tilde{\mathbf{x}} = \frac{\mathbf{x}}{\|\mathbf{x}\|}. \tag{A.1.11}$$

Two vectors $\mathbf{x}$ and $\mathbf{y}$ are said to be *orthogonal* if and only if

$$\langle \mathbf{x}, \mathbf{y} \rangle = 0. \tag{A.1.12}$$

A vector basis is said to be *orthogonal* if each pair of vectors in this vector basis are orthogonal. Furthermore, an orthogonal vector basis is said to be *orthonormal* if every vector is of unit length. An example of orthonormal vector basis is the set of unit axis vectors in a Euclidean space, because

$$\|\mathbf{e}_i\| = 1, \quad \langle \mathbf{e}_i, \mathbf{e}_j \rangle = 0 \text{ for } i \neq j. \tag{A.1.13}$$

If a vector space $\mathbf{X}$ has an orthonormal vector basis, the coordinates of a vector $\mathbf{y}$ in $\mathbf{X}$ are found conveniently. The $i$th coordinate of $\mathbf{y}$ is found by the inner product of $\mathbf{y}$ and the $i$th vector basis $\mathbf{x}_i$,

$$\langle \mathbf{y}, \mathbf{x}_i \rangle = \langle a_1\mathbf{x}_1 + \cdots + a_n\mathbf{x}_n, \mathbf{x}_i \rangle$$

$$= a_1 \underbrace{\langle \mathbf{x}_1, \mathbf{x}_i \rangle}_{=0} + \cdots + a_i \underbrace{\langle \mathbf{x}_i, \mathbf{x}_i \rangle}_{=1} + \cdots + a_n \underbrace{\langle \mathbf{x}_n, \mathbf{x}_i \rangle}_{=0}$$

$$= a_i. \tag{A.1.14}$$

## A.1.6  Gram–Schmidt Orthogonalization

In computational studies, it is desirable to work with a vector basis that is orthonormal. The Gram–Schmidt orthogonalization is a procedure that converts an arbitrary vector basis to an orthonormal one. Denote an arbitrary vector basis as $\mathbf{x}_1, \mathbf{x}_2, \ldots, \mathbf{x}_n$, and the desired orthonormal vector basis as $\mathbf{y}_1, \mathbf{y}_2, \ldots, \mathbf{y}_n$.

The Gram–Schmidt process starts by accepting $\mathbf{x}_1$ as the first basis vector $\mathbf{y}_1$. To ensure that $\mathbf{y}_1$ is a unit vector, $\mathbf{x}_1$ needs to be normalized

$$\mathbf{y}_1 = \mathbf{x}_1 / \|\mathbf{x}_1\|. \tag{A.1.15}$$

Next, the second orthonormal basis vector $\mathbf{y}_2$ is constructed from $\mathbf{x}_2$ and needs to be orthogonal to $\mathbf{y}_1$. This is accomplished by

$$\tilde{\mathbf{y}}_2 = \mathbf{x}_2 - \langle \mathbf{x}_2, \mathbf{y}_1 \rangle \mathbf{y}_1, \tag{A.1.16}$$

where the inner product $\langle \mathbf{x}_2, \mathbf{y}_1 \rangle$ finds the coordinate of $\mathbf{x}_2$ in the direction of $\mathbf{y}_1$, and by subtracting $\langle \mathbf{x}_2, \mathbf{y}_1 \rangle \mathbf{y}_1$ from $\mathbf{x}_2$, the component of $\mathbf{x}_2$ that is along the direction of $\mathbf{y}_1$ is removed, leaving only the component of $\mathbf{x}_2$ that is orthogonal to $\mathbf{y}_1$. The new vector $\tilde{\mathbf{y}}_2$ is normalized to produce the second orthonormal basis vector $\mathbf{y}_2$.

$$\mathbf{y}_2 = \tilde{\mathbf{y}}_2 / \|\tilde{\mathbf{y}}_2\|. \tag{A.1.17}$$

One can verify that the vectors $\mathbf{y}_1$ and $\mathbf{y}_2$ are orthogonal by checking

$$\langle \mathbf{y}_1, \mathbf{y}_2 \rangle = \left\langle \mathbf{y}_1, \frac{\tilde{\mathbf{y}}_2}{\|\tilde{\mathbf{y}}_2\|} \right\rangle$$

$$= \frac{1}{\|\tilde{\mathbf{y}}_2\|} \langle \mathbf{y}_1, \mathbf{x}_2 - \langle \mathbf{x}_2, \mathbf{y}_1 \rangle \mathbf{y}_1 \rangle$$

$$= \frac{1}{\|\tilde{\mathbf{y}}_2\|} (\langle \mathbf{y}_1, \mathbf{x}_2 \rangle - \langle \mathbf{x}_2, \mathbf{y}_1 \rangle \langle \mathbf{y}_1, \mathbf{y}_1 \rangle) = 0. \tag{A.1.18}$$

Subsequently, the third orthonormal basis vector $\mathbf{y}_3$ is constructed from $\mathbf{x}_3$ and needs to be orthogonal to all the previous basis vectors, that is, $\mathbf{y}_1$ and $\mathbf{y}_2$. Therefore, the components of $\mathbf{x}_3$ that are along the directions of $\mathbf{y}_1$ and $\mathbf{y}_2$ need to be removed from $\mathbf{x}_3$.

$$\tilde{\mathbf{y}}_3 = \mathbf{x}_3 - \langle \mathbf{x}_3, \mathbf{y}_1 \rangle \mathbf{y}_1 - \langle \mathbf{x}_3, \mathbf{y}_2 \rangle \mathbf{y}_2, \tag{A.1.19}$$

and the third basis vector $\mathbf{y}_3$ is

$$\mathbf{y}_3 = \tilde{\mathbf{y}}_3 / \|\tilde{\mathbf{y}}_3\|. \tag{A.1.20}$$

The rest of the orthonormal basis vectors are computed in a similar manner. The $i$th orthonormal basis vector $\mathbf{y}_i$ is constructed from $\mathbf{x}_i$ by removing its components that are along the directions of the previous basis vectors $\mathbf{y}_1, \mathbf{y}_2, ..., \mathbf{y}_{i-1}$.

$$\tilde{\mathbf{y}}_i = \mathbf{x}_i - \sum_{j=1}^{i-1} \langle \mathbf{x}_i, \mathbf{y}_j \rangle \, \mathbf{y}_j. \tag{A.1.21}$$

$$\mathbf{y}_i = \tilde{\mathbf{y}}_i / \|\tilde{\mathbf{y}}_i\|. \tag{A.1.22}$$

The Gram–Schmidt orthogonalization procedure defined in Eqs. (A.1.15), (A.1.21), and (A.1.22) yields an orthonormal vector basis if the computation is performed in exact arithmetics. In practice, the computations are performed in finite precision, and the procedure, Eqs. (A.1.21) and (A.1.22), may suffer from numerical instability and fail to orthogonalize a vector basis. A description of a numerically stable orthogonalization procedure, called QR decomposition, is provided later in Section A.3.7.

*Example for Gram–Schmidt orthogonalization:* Create a set of orthonormal basis vectors from the following set of vectors:

$$\mathbf{x}_1 = [1, 1, 1, 1]^T, \ \mathbf{x}_2 = [1, -1, 2, 2]^T, \ \mathbf{x}_3 = [0, 1, 0, -1]^T. \tag{A.1.23}$$

First, following Eq. (A.1.15), one normalizes $\mathbf{x}_1$ to generate the first orthonormal basis vector

$$\begin{aligned}
\mathbf{y}_1 &= \frac{1}{\|\mathbf{x}_1\|} \mathbf{x}_1 = \frac{1}{\sqrt{1^2 + 1^2 + 1^2 + 1^2}} [1, 1, 1, 1]^T \\
&= \left[ \frac{1}{2}, \frac{1}{2}, \frac{1}{2}, \frac{1}{2} \right]^T.
\end{aligned} \tag{A.1.24}$$

Next, one may apply Eq. (A.1.21) to remove the component of $\mathbf{x}_2$ that is along the direction of $\mathbf{y}_1$.

$$\begin{aligned}
\tilde{\mathbf{y}}_2 &= \mathbf{x}_2 - \langle \mathbf{x}_2, \mathbf{y}_1 \rangle \, \mathbf{y}_1 \\
&= [1, -1, 2, 2]^T - \left( 1 \times \frac{1}{2} - 1 \times \frac{1}{2} + 2 \times \frac{1}{2} + 2 \times \frac{1}{2} \right) \left[ \frac{1}{2}, \frac{1}{2}, \frac{1}{2}, \frac{1}{2} \right]^T \\
&= [0, -2, 1, 1]^T
\end{aligned} \tag{A.1.25}$$

and normalizing $\tilde{\mathbf{y}}_2$ gives the second orthonormal basis vector

$$\begin{aligned}
\mathbf{y}_2 &= \frac{1}{\|\tilde{\mathbf{y}}_2\|} \tilde{\mathbf{y}}_2 = \frac{1}{\sqrt{0^2 + (-2)^2 + 1^2 + 1^2}} [0, -2, 1, 1]^T \\
&= \left[ 0, -\frac{2}{\sqrt{6}}, \frac{1}{\sqrt{6}}, \frac{1}{\sqrt{6}} \right]^T.
\end{aligned} \tag{A.1.26}$$

Finally, one may apply Eq. (A.1.21) again to remove the component of $\mathbf{x}_3$ that is along the directions of $\mathbf{y}_1$ and $\mathbf{y}_2$.

$$
\begin{aligned}
\tilde{\mathbf{y}}_3 &= \mathbf{x}_3 - \langle \mathbf{x}_3, \mathbf{y}_1 \rangle \mathbf{y}_1 - \langle \mathbf{x}_3, \mathbf{y}_2 \rangle \mathbf{y}_2 \\
&= [0, 1, 0, -1]^T - \left( 0 \times \frac{1}{2} + 1 \times \frac{1}{2} + 0 \times \frac{1}{2} - 1 \times \frac{1}{2} \right) \left[ \frac{1}{2}, \frac{1}{2}, \frac{1}{2}, \frac{1}{2} \right]^T \\
&\quad - \left( 0 \times 0 + 1 \times \frac{-2}{\sqrt{6}} + 0 \times \frac{1}{\sqrt{6}} - 1 \times \frac{1}{\sqrt{6}} \right) \left[ 0, -\frac{2}{\sqrt{6}}, \frac{1}{\sqrt{6}}, \frac{1}{\sqrt{6}} \right]^T \\
&= \left[ 0, 0, \frac{1}{2}, -\frac{1}{2} \right]^T
\end{aligned}
\tag{A.1.27}
$$

and normalizing $\tilde{\mathbf{y}}_3$ gives the third and last orthonormal basis vector

$$
\begin{aligned}
\mathbf{y}_3 &= \frac{1}{\|\tilde{\mathbf{y}}_3\|} \tilde{\mathbf{y}}_3 = \frac{1}{\sqrt{0^2 + 0^2 + (1/2)^2 + (-1/2)^2}} \left[ 0, 0, \frac{1}{2}, -\frac{1}{2} \right]^T \\
&= \left[ 0, 0, \frac{1}{\sqrt{2}}, -\frac{1}{\sqrt{2}} \right]^T .
\end{aligned}
\tag{A.1.28}
$$

One may verify that the basis vectors $\mathbf{y}_1$, $\mathbf{y}_2$, and $\mathbf{y}_3$ are indeed orthogonal to each other.

## A.2  Matrix Definitions

A matrix is defined as a two-dimensional array of scalars and is represented as

$$
\mathbf{P} = \begin{bmatrix} p_{11} & p_{12} & \cdots & p_{1n} \\ p_{21} & p_{22} & \cdots & p_{2n} \\ \cdots & \cdots & \cdots & \cdots \\ p_{m1} & p_{m2} & \cdots & p_{mn} \end{bmatrix},
\tag{A.2.1}
$$

where the scalars $p_{ij}$ ($i = 1, 2, \ldots, m; j = 1, 2, \ldots, n$), called the elements of $P$, belong to a given field $F$. Since the matrix $P$ has $m$ rows and $n$ columns, it is referred to as an $m \times n$ matrix. Note that the first dimension is always the number of rows. The position of the element $p_{ij}$ is in the $i$th row and $j$th column, where $i$ is referred to as the row index and $j$ as the column index.

### A.2.1  Zero or Null Matrix

A matrix with all its elements equal to zero is called a zero or null matrix and is denoted by $\mathbf{0}$. A zero matrix serves the same function in matrix algebra as the number zero does in algebra.

$$
\mathbf{0} = \begin{bmatrix} 0 & 0 & \cdots & 0 \\ 0 & 0 & \cdots & 0 \\ \cdots & \cdots & \cdots & \cdots \\ 0 & 0 & \cdots & 0 \end{bmatrix}.
\tag{A.2.2}
$$

## A.2.2  Square Matrix

If $m = n$, the matrix $\mathbf{P}$ reduces to a square matrix of order $n$.

$$\mathbf{P} = \begin{bmatrix} p_{11} & p_{12} & \cdots & p_{1n} \\ p_{21} & p_{22} & \cdots & p_{2n} \\ \cdots & \cdots & \cdots & \cdots \\ p_{n1} & p_{n2} & \cdots & p_{nn} \end{bmatrix}. \tag{A.2.3}$$

Square matrices are unique and one of the most used in structural dynamics. There are several important types of square matrices which are discussed later.

## A.2.3  Diagonal Matrix

The elements $p_{ii}$ in the square matrix are called the main diagonal elements of $\mathbf{P}$. The remaining elements are referred to as the off-diagonal elements. In the special case where all the off-diagonal elements are zero, $\mathbf{P}$ is referred to as a diagonal matrix. A typical diagonal matrix is given as

$$\mathbf{P} = \begin{bmatrix} p_{11} & 0 & \cdots & 0 \\ 0 & p_{22} & \cdots & 0 \\ \cdots & \cdots & \cdots & \cdots \\ 0 & 0 & \cdots & p_{nn} \end{bmatrix} \tag{A.2.4}$$

or simply

$$\mathbf{P} = \begin{bmatrix} p_{11} & p_{22} \cdots & p_{nn} \end{bmatrix}. \tag{A.2.5}$$

## A.2.4  Unit or Identity Matrix

If $\mathbf{P}$ is a diagonal matrix and all of its diagonal elements are unity, $p_{ii} = 1$, then $\mathbf{P}$ is referred to as a unit or identity matrix. A generic unit matrix is also denoted by $\mathbf{I}$. A unit matrix of order $3 \times 3$ can be written as

$$\mathbf{I} = \begin{bmatrix} 1 & 0 & 0 \\ 0 & 1 & 0 \\ 0 & 0 & 1 \end{bmatrix}. \tag{A.2.6}$$

A unit matrix serves the same function in matrix algebra as unity does in ordinary algebra. The identity matrix can be regarded as a matrix with every element equal to the Kronecker delta, that is, $\mathbf{I} = [\delta_{ij}]$, where

$$\delta_{ij} = \begin{cases} 1 & \text{if} \quad i = j \\ 0 & \text{if} \quad i \neq j \end{cases}.$$

### A.2.5  Triangular Matrix

A triangular matrix is one where all the elements on one side of the main diagonal are zero. A square matrix $\mathbf{P}$ is said to be upper triangular if $p_{ij} = 0$ for $i > j$, that is, all the elements below the main diagonal are zero. Similarly, $\mathbf{P}$ is said to be lower triangular if $p_{ij} = 0$ for $i < j$, that is, all the elements above the main diagonal are zero.

$$\mathbf{P} = \begin{bmatrix} p_{11} & 0 & \dots & 0 \\ p_{21} & p_{22} & \dots & 0 \\ \dots & \dots & \dots & \dots \\ p_{n1} & p_{n2} & \dots & p_{nn} \end{bmatrix}. \tag{A.2.7}$$

If the diagonal elements of an upper or lower triangular matrix are unity, then the matrix is referred to as *unit upper* or *unit lower* triangular matrix, respectively.

### A.2.6  Hessenberg Matrix

A Hessenberg matrix is one where all the elements on one side of the main diagonal, *except* the first subdiagonal, are zero. More specifically, a square matrix $\mathbf{P}$ is said to be upper Hessenberg if $p_{ij} = 0$ for $i > j + 1$, that is, all the elements below the first subdiagonal are zero.

$$\mathbf{P} = \begin{bmatrix} p_{11} & p_{12} & \dots & \dots & p_{1,n-1} & p_{1n} \\ p_{21} & p_{22} & \dots & \dots & p_{2,n-1} & p_{2n} \\ 0 & p_{32} & \dots & \dots & p_{3,n-1} & p_{3n} \\ 0 & 0 & \dots & \dots & \dots & \dots \\ \dots & \dots & \dots & \dots & \dots & \dots \\ 0 & \dots & \dots & 0 & p_{n,n-1} & p_{nn} \end{bmatrix}. \tag{A.2.8}$$

Similarly, a square matrix $\mathbf{P}$ is said to be lower Hessenberg if $p_{ij} = 0$ for $i + 1 < j$, that is, all the elements above the first superdiagonal are zero.

### A.2.7  Banded Matrix and Tridiagonal Matrix

A square matrix $\mathbf{P}$ is called a banded matrix if all its nonzero elements are confined to a band comprising of the main diagonal and one or more adjacent diagonals.

$$\mathbf{P} = \begin{bmatrix} p_{11} & p_{12} & 0 & \dots & 0 \\ p_{21} & p_{22} & p_{23} & \dots & 0 \\ 0 & p_{32} & p_{33} & \dots & 0 \\ \dots & \dots & \dots & \dots & \dots \\ 0 & \dots & p_{(n-1)(n-2)} & p_{(n-1)(n-1)} & p_{(n-1)n} \\ 0 & \dots & 0 & p_{n(n-1)} & p_{nn} \end{bmatrix}. \tag{A.2.9}$$

Specifically, when all its nonzero elements are confined to the main diagonal, subdiagonal, and superdiagonal, the banded matrix is also referred to as a tridiagonal matrix.

### A.2.8   Symmetric and Skew-Symmetric Matrix

If the elements of a square matrix $\mathbf{P}$ are symmetric about the principal diagonal, that is, $p_{ij} = p_{ji}$, then the matrix is said to be symmetric. Otherwise, the matrix is said to be nonsymmetric.

$$\mathbf{P} = \begin{bmatrix} p_{11} & p_{21} & p_{31} & \cdots & p_{n1} \\ p_{21} & p_{22} & p_{32} & \cdots & p_{n2} \\ p_{31} & p_{32} & p_{33} & \cdots & p_{n3} \\ \cdots & \cdots & \cdots & \cdots & \cdots \\ p_{(n-1)1} & \cdots & p_{(n-1)(n-2)} & p_{(n-1)(n-1)} & p_{n(n-1)} \\ p_{n1} & p_{n2} & \cdots & p_{n(n-1)} & p_{nn} \end{bmatrix}. \tag{A.2.10}$$

If the elements of a square matrix $\mathbf{P}$ are such that $p_{ij} = -p_{ji}$ for $i \neq j$, and $p_{ii} = 0$, then the matrix $\mathbf{P}$ is said to be *skew symmetric*.

## A.3   Matrix Operations

### A.3.1   Equality

Two matrices $\mathbf{P}$ and $\mathbf{Q}$ are said to be equal if and only if they have the same dimensions and $p_{ij} = q_{ij}$ for all $i$ and $j$.

### A.3.2   Addition

If $\mathbf{P}$ and $\mathbf{Q}$ are two $m \times n$ matrices, then their sum is defined as a matrix

$$\mathbf{R} = \mathbf{P} + \mathbf{Q}, \tag{A.3.1}$$

with elements

$$r_{ij} = p_{ij} + q_{ij}, \quad i = 1, 2, \ldots, m; j = 1, 2, \ldots, n. \tag{A.3.2}$$

The matrix $\mathbf{R}$ also has dimensions $m \times n$. For example,

$$\begin{bmatrix} 5 & 3 \\ 2 & 0 \\ 1 & 9 \end{bmatrix} + \begin{bmatrix} 1 & -2 \\ -3 & 11 \\ 15 & 1 \end{bmatrix} = \begin{bmatrix} 6 & 1 \\ -1 & 11 \\ 16 & 10 \end{bmatrix}. \tag{A.3.3}$$

Matrix addition is commutative and associative, that is,

$$\mathbf{P} + \mathbf{Q} = \mathbf{Q} + \mathbf{P}. \tag{A.3.4}$$

$$(\mathbf{P} + \mathbf{Q}) + \mathbf{R} = \mathbf{P} + (\mathbf{Q} + \mathbf{R}). \tag{A.3.5}$$

### A.3.3  Multiplication

The product of a matrix $\mathbf{P}$ and a scalar $\alpha$ implies that every element of $\mathbf{P}$ is multiplied by $\alpha$. If $\mathbf{P}$ is an $m \times n$ matrix, then the statement $\mathbf{R} = \alpha\mathbf{P}$ implies

$$r_{ij} = \alpha p_{ij}, \quad i = 1, 2, \ldots, m; j = 1, 2, \ldots, n. \qquad (A.3.6)$$

Two matrices $\mathbf{P}$ and $\mathbf{Q}$ can be multiplied together in order $\mathbf{PQ}$ only when the number of columns in $\mathbf{P}$ is equal to the number of rows in $\mathbf{Q}$. In this case, the matrices $\mathbf{P}$ and $\mathbf{Q}$ are said to be conformable. The matrix product $\mathbf{PQ}$ can be described as $\mathbf{Q}$ premultiplied by $\mathbf{P}$ or $\mathbf{P}$ postmultiplied by $\mathbf{Q}$. If $\mathbf{P}$ is an $m \times n$ matrix and $\mathbf{Q}$ is an $n \times o$ matrix, then the product $\mathbf{R} = \mathbf{PQ}$ is an $m \times o$ matrix with elements

$$r_{ij} = p_{i1}q_{1j} + p_{i2}q_{2j} + \ldots + p_{in}q_{nj} = \sum_{k=1}^{n} p_{ik}q_{kj}. \qquad (A.3.7)$$

For example,

$$\begin{bmatrix} 5 & 3 \\ 2 & 0 \\ 1 & 9 \end{bmatrix} \begin{bmatrix} 1 & -2 & -3 \\ 3 & 5 & 1 \end{bmatrix} \qquad (A.3.8)$$

$$= \begin{bmatrix} 5 \times 1 + 3 \times 3 & 5 \times (-2) + 3 \times 5 & 5 \times (-3) + 3 \times 1 \\ 2 \times 1 + 0 \times 3 & 2 \times (-2) + 0 \times 5 & 2 \times (-3) + 0 \times 1 \\ 1 \times 1 + 9 \times 3 & 1 \times (-2) + 9 \times 5 & 1 \times (-3) + 9 \times 1 \end{bmatrix} \qquad (A.3.9)$$

$$= \begin{bmatrix} 14 & 5 & -12 \\ 2 & -4 & -6 \\ 28 & 43 & 6 \end{bmatrix}. \qquad (A.3.10)$$

Matrix multiplication is in general not commutative.

$$\mathbf{PQ} \neq \mathbf{QP} \qquad (A.3.11)$$

unless one of the matrices is the identity matrix in which case

$$\mathbf{PI} = \mathbf{IP} = \mathbf{P}. \qquad (A.3.12)$$

The matrix product

$$\mathbf{PQ} = 0 \qquad (A.3.13)$$

does not imply that either $\mathbf{P}$ or $\mathbf{Q}$ is a null matrix. For example, consider

$$\begin{bmatrix} 1 & 1 \\ 1 & 1 \end{bmatrix} \begin{bmatrix} 1 & -1 \\ -1 & 1 \end{bmatrix} = \begin{bmatrix} 0 & 0 \\ 0 & 0 \end{bmatrix}. \qquad (A.3.14)$$

The multiplication process described earlier can be extended to products of more than two matrices, provided the adjacent matrices in the product are conformable. Therefore, the order of multiplication is important. Matrix multiplication satisfies the associative and distributive laws, provided that the multiplication sequence is preserved. If $\mathbf{P}$, $\mathbf{Q}$, and $\mathbf{R}$ are $m \times n$, $n \times a$, and $a \times b$ matrices, then

$$\mathbf{S} = (\mathbf{PQ})\mathbf{R} = \mathbf{P}(\mathbf{QR}) = \mathbf{PQR} \qquad (A.3.15)$$

is an $m \times b$ matrix whose elements are given by

$$s_{ij} = \sum_{l=1}^{b} \sum_{k=1}^{n} p_{ik} q_{kl} r_{lj} = \sum_{k=1}^{n} \sum_{l=1}^{b} p_{ik} q_{kl} r_{lj}. \qquad (A.3.16)$$

Matrix product also satisfies distributive laws. If $\mathbf{P}$ and $\mathbf{Q}$ are $m \times n$ matrices, $\mathbf{R}$ is an $a \times m$ matrix, and $\mathbf{S}$ is an $n \times b$, then

$$\mathbf{R}(\mathbf{P} + \mathbf{Q}) = \mathbf{RP} + \mathbf{RQ}. \qquad (A.3.17)$$

$$(\mathbf{P} + \mathbf{Q})\mathbf{S} = \mathbf{PS} + \mathbf{QS}. \qquad (A.3.18)$$

## A.3.4   Transpose

A matrix obtained from interchanging all the rows and columns of a matrix $\mathbf{P}$ is referred to as the transpose of $\mathbf{P}$ and is denoted as $\mathbf{P}^T$. If $\mathbf{P}$ has dimensions $m \times n$, then the dimensions of its transpose $\mathbf{P}^T$ are $n \times m$. Thus, the transpose of a row matrix is a column matrix and vice versa. For example,

$$\begin{bmatrix} 2 & 9 & 3 \\ 1 & 0 & 8 \\ 5 & 7 & 1 \end{bmatrix}^T = \begin{bmatrix} 2 & 1 & 5 \\ 9 & 0 & 7 \\ 3 & 8 & 1 \end{bmatrix}. \qquad (A.3.19)$$

The transpose of a symmetric matrix is itself.

$$\mathbf{P}^T = \mathbf{P} \quad \text{symmetric matrix.} \qquad (A.3.20)$$

$$\mathbf{P}^T = -\mathbf{P} \quad \text{skew-symmetric matrix.} \qquad (A.3.21)$$

If matrices $\mathbf{P}$, $\mathbf{Q}$, and $\mathbf{R}$ are such that $\mathbf{R} = \mathbf{PQ}$, then

$$\mathbf{R}^T = (\mathbf{PQ})^T = \mathbf{Q}^T \mathbf{P}^T. \qquad (A.3.22)$$

Using the transpose, the inner product of two vectors $\mathbf{x}$ and $\mathbf{y}$ is written as

$$\langle \mathbf{x}, \mathbf{y} \rangle = \mathbf{x}^T \mathbf{y} \qquad (A.3.23)$$

and the norm of a vector is thus

$$\|\mathbf{x}\| = \sqrt{\langle \mathbf{x}, \mathbf{x} \rangle} = \sqrt{\mathbf{x}^T \mathbf{x}}. \qquad (A.3.24)$$

## A.3.5  Determinants

The determinant of a square matrix $\mathbf{P}$ is a number denoted by

$$\det \mathbf{P} = |\mathbf{P}| = \begin{vmatrix} p_{11} & p_{12} & \cdots & p_{1n} \\ p_{21} & p_{22} & \cdots & p_{2n} \\ \cdots & \cdots & \cdots & \cdots \\ p_{n1} & p_{n2} & \cdots & p_{nn} \end{vmatrix}. \tag{A.3.25}$$

The value of the determinant can be obtained by expanding in terms of cofactors by the $r$th row as follows:

$$|\mathbf{P}| = \sum_{s=1}^{n} p_{rs}\mathbf{P}_{rs}, \tag{A.3.26}$$

where $\mathbf{P}_{rs}$ is the cofactor of element $p_{rs}$ and is given by

$$\mathbf{P}_{rs} = (-1)^{r+s}|\mathbf{P}_{rs}|, \tag{A.3.27}$$

where $|\mathbf{P}_{rs}|$ is the *minor determinant* corresponding to element $p_{rs}$. The first minor corresponding to the element $p_{rs}$ is defined as the determinant obtained by omitting the $r$th row and the $s$th column of $|\mathbf{P}|$. Therefore, if $|\mathbf{P}|$ is of order $n$, the any first minor is of order $n-1$. This definition can be extended to minors of second, third, and lower orders. The value of the determinant is unique regardless of whether it is calculated by expanding through a row or a column, and regardless of which row or column. As an illustration, let $n = 3$ and expand $\det \mathbf{P}$ by the first row as follows:

$$\det \mathbf{P} = \begin{vmatrix} p_{11} & p_{12} & p_{13} \\ p_{21} & p_{22} & p_{23} \\ p_{31} & p_{32} & p_{33} \end{vmatrix} \tag{A.3.28}$$

$$= p_{11}|\mathbf{P}_{11}| - p_{12}|\mathbf{P}_{12}| + p_{13}|\mathbf{P}_{13}| \tag{A.3.29}$$

$$= p_{11}\begin{vmatrix} p_{22} & p_{23} \\ p_{32} & p_{33} \end{vmatrix} - p_{12}\begin{vmatrix} p_{21} & p_{23} \\ p_{31} & p_{33} \end{vmatrix} + p_{13}\begin{vmatrix} p_{21} & p_{22} \\ p_{31} & p_{32} \end{vmatrix} \tag{A.3.30}$$

$$= p_{11}(p_{22}p_{33} - p_{23}p_{32}) - p_{12}(p_{21}p_{33} - p_{23}p_{31}) + p_{13}(p_{21}p_{32} - p_{22}p_{31}). \tag{A.3.31}$$

Because the value of $|\mathbf{P}|$ is the same, regardless of whether the determinant is expanded by a row or a column, it follows that

$$\det \mathbf{P} = \det \mathbf{P}^{T}. \tag{A.3.32}$$

It is easy to verify that the determinant of a diagonal matrix is equal to the product of the diagonal elements, and the determinant of a triangular matrix is equal to the product of the main diagonal elements. If $|\mathbf{P}| = 0$, then the matrix $\mathbf{P}$ is said to be singular. Some important properties of a determinant are listed below:

(1) If all the elements in a row or a column are zero, then the determinant is zero.
(2) If two rows or columns in a matrix are identical, then the value of the determinant is zero.

(3) Interchanging any two rows or columns changes the sign of the determinant.

(4) If all the elements of one row or column are multiplied by a scalar $\alpha$, then the determinant is multiplied by $\alpha$.

(5) Adding a constant multiple of a row or a column to any other row or column leaves the value of the determinant unchanged. For example,

$$\begin{vmatrix} p_{11} + ap_{31} & p_{12} + ap_{32} & p_{13} + ap_{33} \\ p_{21} & p_{22} & p_{23} \\ p_{31} & p_{32} & p_{33} \end{vmatrix} = \begin{vmatrix} p_{11} & p_{12} & p_{13} \\ p_{21} & p_{22} & p_{23} \\ p_{31} & p_{32} & p_{33}. \end{vmatrix} \tag{A.3.33}$$

## A.3.6  Matrix Inversion

If $\mathbf{P}$ and $\mathbf{Q}$ are two square matrices with dimension $n$ such that

$$\mathbf{PQ} = \mathbf{QP} = \mathbf{I}, \tag{A.3.34}$$

then $\mathbf{Q}$ is said to be the inverse of $\mathbf{P}$ and is denoted by

$$\mathbf{Q} = \mathbf{P}^{-1}. \tag{A.3.35}$$

Also, $\mathbf{P}$ is said to be the inverse of $\mathbf{Q}$ and is denoted by

$$\mathbf{P} = \mathbf{Q}^{-1}. \tag{A.3.36}$$

In order to derive the inverse of a matrix, we first define the adjugate or adjoint matrix as

$$\text{adj } \mathbf{P} = [(-1)^{r+s}|\mathbf{P}_{rs}|]^T, \tag{A.3.37}$$

where $(-1)^{r+s}|\mathbf{P}_{rs}|$ is the cofactor defined in Eq. (A.3.27) and the adjoint matrix is the transpose of the cofactor matrix. It can be shown that

$$\mathbf{P} \text{ adj } \mathbf{P} = |\mathbf{P}|\mathbf{I}, \tag{A.3.38}$$

where $\mathbf{I}$ is the identity matrix of dimension $n$. Therefore,

$$\mathbf{P}^{-1} = \frac{\text{adj } \mathbf{P}}{|\mathbf{P}|}, \tag{A.3.39}$$

which implies that $|\mathbf{P}|$ has to be nonzero for $\mathbf{P}^{-1}$ to exist. To illustrate the inversion process, the inverse of a $2 \times 2$ matrix

$$\mathbf{P} = \begin{bmatrix} p_{11} & p_{12} \\ p_{21} & p_{22} \end{bmatrix} \tag{A.3.40}$$

will be evaluated. Let the inverse matrix be given as

$$\mathbf{P}^{-1} = \begin{bmatrix} q_{11} & q_{12} \\ q_{21} & q_{22} \end{bmatrix}, \tag{A.3.41}$$

then from the definition of matrix inversion $\mathbf{P}^{-1}\mathbf{P} = \mathbf{I}$,

$$\begin{bmatrix} q_{11} & q_{12} \\ q_{21} & q_{22} \end{bmatrix} \begin{bmatrix} p_{11} & p_{12} \\ p_{21} & p_{22} \end{bmatrix} = \begin{bmatrix} q_{11}p_{11} + q_{12}p_{21} & q_{11}p_{12} + q_{12}p_{22} \\ q_{21}p_{11} + q_{22}p_{21} & q_{21}p_{12} + q_{22}p_{22} \end{bmatrix} = \begin{bmatrix} 1 & 0 \\ 0 & 1 \end{bmatrix}. \tag{A.3.42}$$

Equating the corresponding elements and solving the resulting equations for $q_{11}$, $q_{12}$, $q_{21}$, and $q_{22}$ yields

$$\mathbf{P}^{-1} = \frac{1}{p_{11}p_{22} - p_{21}p_{12}} \begin{bmatrix} p_{22} & -p_{12} \\ -p_{21} & p_{11} \end{bmatrix} = \frac{\hat{\mathbf{P}}}{|\mathbf{P}|}, \tag{A.3.43}$$

where $\hat{\mathbf{P}}$ is the cofactor matrix. If the matrix to be inverted is a diagonal matrix

$$\mathbf{P} = [p_{11}\, p_{22}\, \cdots\, p_{nn}], \tag{A.3.44}$$

then

$$\mathbf{P}^{-1} = \left\lfloor \frac{1}{p_{11}}\ \frac{1}{p_{22}}\ \cdots\ \frac{1}{p_{nn}} \right\rfloor. \tag{A.3.45}$$

### A.3.7  Common Matrix Decompositions

This section discusses three types of matrix decompositions that are used in the main body of this text: QR decomposition, LU decomposition, and Cholesky decomposition. Many efficient algorithms exist for these types of decompositions, and such algorithms have been implemented in standard libraries for numerical linear algebra, such as LAPACK and MATLAB. Therefore, this section only introduces the concept of these decompositions without going into detailed implementation.

#### Orthonormal Matrix and QR Decomposition

An $m \times n$ orthonormal matrix $\mathbf{Q}$, $m \geq n$ is said to be orthonormal, if all of its columns are unit vectors and the column vectors are orthogonal to each other. Therefore, the matrix $\mathbf{Q}$ has the following property:

$$\mathbf{Q}^T\mathbf{Q} = \mathbf{I}, \tag{A.3.46}$$

where $\mathbf{I}$ is an $n \times n$ matrix. When $m = n$, Eq. (A.3.46) implies

$$\mathbf{Q}^T = \mathbf{Q}^{-1}. \tag{A.3.47}$$

In other words, the inverse of a square orthonormal matrix is its transpose. Equation (A.3.47) further implies, when $\mathbf{Q}$ is square,

$$\mathbf{Q}^T\mathbf{Q} = \mathbf{Q}\mathbf{Q}^T = \mathbf{I} \tag{A.3.48}$$

and therefore the rows of $\mathbf{Q}$ are unit vectors and the row vectors are also orthogonal to each other.

The *QR decomposition* or factorization factors an arbitrary $m \times n$ matrix $\mathbf{A}$, $m \geq n$, as the product of an $m \times n$ *orthonormal* matrix $\mathbf{Q}$ and an $n \times n$ upper triangular matrix $\mathbf{R}$.

$$\mathbf{P} = \mathbf{QR}. \tag{A.3.49}$$

The QR decomposition is usually employed as a numerically stable approach for carrying out Gram–Schmidt orthogonalization of a set of basis vectors, as discussed in Section A.1.6. Consider an arbitrary set of basis vectors, $\mathbf{p}_1, \mathbf{p}_2, \ldots, \mathbf{p}_m$, of a vector space $\mathbb{R}^m$. The QR decomposition of the matrix

$$\mathbf{P} = [\mathbf{p}_1, \mathbf{p}_2, \ldots, \mathbf{p}_m] = \mathbf{QR} \tag{A.3.50}$$

results in an orthonormal vector basis as the column vectors of $\mathbf{Q}$,

$$\mathbf{Q} = [\mathbf{q}_1, \mathbf{q}_2, \ldots, \mathbf{q}_m]. \tag{A.3.51}$$

## The LU and Cholesky Decomposition

The *LU decomposition* or factorization factors a square matrix $\mathbf{P}$ as the product of a lower triangular matrix $\mathbf{L}$ and an upper triangular matrix $\mathbf{U}$.

$$\mathbf{P} = \mathbf{LU}, \tag{A.3.52}$$

where the diagonal elements of $\mathbf{L}$ are ones. For example, the LU decomposition of a $3 \times 3$ matrix is the following:

$$\begin{bmatrix} p_{11} & p_{12} & p_{13} \\ p_{21} & p_{22} & p_{23} \\ p_{31} & p_{32} & p_{33} \end{bmatrix} = \begin{bmatrix} 1 & 0 & 0 \\ l_{21} & 1 & 0 \\ l_{31} & l_{32} & 1 \end{bmatrix} \begin{bmatrix} u_{11} & u_{12} & u_{13} \\ 0 & u_{22} & u_{23} \\ 0 & 0 & u_{33} \end{bmatrix}. \tag{A.3.53}$$

When $\mathbf{P}$ is symmetric and positive definite, that is,

$$\mathbf{x}^T \mathbf{P} \mathbf{x} > 0, \quad \text{for } \forall \mathbf{x} \neq 0, \tag{A.3.54}$$

the LU decomposition simplifies to the *Cholesky decomposition*,

$$\mathbf{P} = \mathbf{LL}^T, \tag{A.3.55}$$

where the upper triangular matrix becomes the transpose of the lower triangular matrix. For example, the Cholesky decomposition of a $3 \times 3$ symmetric matrix is the following:

$$\begin{bmatrix} p_{11} & p_{21} & p_{31} \\ p_{21} & p_{22} & p_{32} \\ p_{31} & p_{32} & p_{33} \end{bmatrix} = \begin{bmatrix} l_{11} & 0 & 0 \\ l_{21} & l_{22} & 0 \\ l_{31} & l_{32} & l_{33} \end{bmatrix} \begin{bmatrix} l_{11} & l_{21} & l_{31} \\ 0 & l_{22} & l_{32} \\ 0 & 0 & l_{33} \end{bmatrix}. \tag{A.3.56}$$

The LU and Cholesky decompositions are usually employed to solve simultaneous algebraic equations, which are discussed in Section A.3.8.

### A.3.8 Solution of Simultaneous Algebraic Equations

Consider a system of $m$ nonhomogeneous linear algebraic equations in $n$ unknowns $x_1, x_2, \ldots, x_n$.

$$p_{11}x_1 + p_{12}x_2 + \ldots + p_{1n}x_n = c_1$$
$$p_{21}x_1 + p_{22}x_2 + \ldots + p_{2n}x_n = c_2$$
$$\ldots \ldots \quad \ldots$$
$$p_{m1}x_1 + p_{m2}x_2 + \ldots + p_{mn}x_n = c_m,$$

which can be expressed in matrix form as

$$
\begin{bmatrix}
p_{11} & p_{12} & \cdots & p_{1n} \\
p_{21} & p_{22} & \cdots & p_{2n} \\
\cdots & \cdots & \cdots & \cdots \\
p_{m1} & p_{m2} & \cdots & p_{mn}
\end{bmatrix}
\begin{bmatrix}
x_1 \\
x_2 \\
\cdots \\
x_n
\end{bmatrix}
=
\begin{bmatrix}
c_1 \\
c_2 \\
\cdots \\
c_m
\end{bmatrix}
\tag{A.3.57}
$$

or

$$\mathbf{Px} = \mathbf{c}, \tag{A.3.58}$$

where the objective is to find a solution to $\mathbf{x}$. Consider the augmented matrix of the system defined by

$$
\mathbf{Q} = [\mathbf{P}\ \mathbf{c}] =
\begin{bmatrix}
p_{11} & p_{12} & \cdots & p_{1n} & c_1 \\
p_{21} & p_{22} & \cdots & p_{2n} & c_2 \\
\cdots & \cdots & \cdots & \cdots \\
p_{m1} & p_{m2} & \cdots & p_{mn} & c_m
\end{bmatrix}.
\tag{A.3.59}
$$

If a solution to $\mathbf{x}$ exists, then $\mathbf{c}$ is a linear combination of the columns of $\mathbf{P}$. That is, rank $\mathbf{Q}$ = rank $\mathbf{P}$. Therefore, Eq. (A.3.57) has a solution $\mathbf{x}$ if and only if the rank of the augmented matrix $\mathbf{Q}$ is equal to the rank of $\mathbf{P}$. The rank of a matrix $\mathbf{P}$ is equal to the maximum number of linearly independent columns of $\mathbf{P}$, which is also equal to the maximum number of linearly independent rows of $\mathbf{P}$.

In the case where the number of equations is equal to the number of unknowns, the matrix $\mathbf{P}$ is a square matrix of order $n$. A solution $\mathbf{x}$ exists if and only if rank $\mathbf{P}$ = n, and

$$\mathbf{x} = \mathbf{P}^{-1}\mathbf{c}. \tag{A.3.60}$$

The inverse matrix $\mathbf{P}^{-1}$ can be obtained using Eq. (A.3.39), and this method for solving set of simultaneous equations is generally known as Cramer's rule. However, such techniques are not efficient or suitable for higher dimensional systems. Several numerical methods are available as part of software tools or packages, for performing matrix inversion.

The techniques available for solving linear systems of equations can be classified into two categories: direct and iterative methods.

### Direct Methods

The direct methods consist of two steps: (1) triangularization of the system matrix **P** by elimination and (2) back substitution. *Gaussian elimination* is the most commonly used method where a series of row manipulations transform the system matrix into an upper triangular matrix.

*Example for Gaussian elimination:* Consider the set of algebraic equations

$$x_1 + x_2 + 2x_3 = 3$$
$$4x_1 + 3x_2 + 5x_3 = 3$$
$$8x_1 + 3x_2 + 2x_3 = 2$$

which can expressed in a matrix equation

$$\begin{bmatrix} 1 & 1 & 2 \\ 4 & 3 & 5 \\ 8 & 3 & 2 \end{bmatrix} \begin{bmatrix} x_1 \\ x_2 \\ x_3 \end{bmatrix} = \begin{bmatrix} 3 \\ 3 \\ 2 \end{bmatrix}. \tag{A.3.61}$$

Writing the augment matrix

$$\begin{bmatrix} 1 & 1 & 2 & 3 \\ 4 & 3 & 5 & 3 \\ 8 & 3 & 2 & 2 \end{bmatrix}. \tag{A.3.62}$$

Subtracting 4 times first row from second row yields

$$\begin{bmatrix} 1 & 1 & 2 & 3 \\ 0 & -1 & -3 & -9 \\ 8 & 3 & 2 & 2 \end{bmatrix}, \tag{A.3.63}$$

subtracting 8 times first row from the third row yields

$$\begin{bmatrix} 1 & 1 & 2 & 3 \\ 0 & -1 & -3 & -9 \\ 0 & -5 & -14 & -22 \end{bmatrix}, \tag{A.3.64}$$

subtracting 5 times second row from the third row yields

$$\begin{bmatrix} 1 & 1 & 2 & 3 \\ 0 & -1 & -3 & -9 \\ 0 & 0 & 1 & 23 \end{bmatrix}, \tag{A.3.65}$$

which can be solved to obtain

$$x_3 = 23. \tag{A.3.66}$$

Back substitute the value of $x_3$ into the second equation,

$$x_2 = 9 - 3x_3 = -60. \tag{A.3.67}$$

Finally, back substitute the values of $x_2$ and $x_3$ into the first equation,

$$x_1 = 3 - x_2 - 2x_3 = 17. \tag{A.3.68}$$

In practice, Gaussian elimination is implemented via LU decomposition for improved numerical stability in large-scale problems. In this process, the LU decomposition of the system matrix $\mathbf{P}$ is first computed, typically using a reliable computer program,

$$\mathbf{P} = \mathbf{LU}. \tag{A.3.69}$$

Substituting Eq. (A.3.69) into Eq. (A.3.58) yields

$$\mathbf{LUx} = \mathbf{c}. \tag{A.3.70}$$

Let $\mathbf{y} = \mathbf{Ux}$, the solution of the linear system is divided into two steps:

(1) Solve for $\mathbf{y}$ by forward substitution: $\mathbf{Ly} = \mathbf{c}$.
(2) Solve for $\mathbf{x}$ by back substitution: $\mathbf{Ux} = \mathbf{y}$.

When the system matrix is symmetric and positive definite, which is the case for most of the matrices encountered in this book, the Cholesky decomposition can be applied.

$$\mathbf{P} = \mathbf{LL}^T. \tag{A.3.71}$$

The solution of the linear system is modified as follows:

(1) Solve for $\mathbf{y}$ by forward substitution: $\mathbf{Ly} = \mathbf{c}$.
(2) Solve for $\mathbf{x}$ by back substitution: $\mathbf{L}^T\mathbf{x} = \mathbf{y}$.

*Example for LU decomposition:* Consider again the set of algebraic equations in Eq. (A.3.61). The system matrix is decomposed as

$$\begin{bmatrix} 1 & 1 & 2 \\ 4 & 3 & 5 \\ 8 & 3 & 2 \end{bmatrix} = \begin{bmatrix} 1 & 0 & 0 \\ 4 & 1 & 0 \\ 8 & 5 & 1 \end{bmatrix} \begin{bmatrix} 1 & 1 & 2 \\ 0 & -1 & -3 \\ 0 & 0 & 1 \end{bmatrix}. \tag{A.3.72}$$

Note the lower triangular matrix is the same as the one found by Gaussian elimination in Eq. (A.3.65). Next, the lower triangular linear system

$$\begin{bmatrix} 1 & 0 & 0 \\ 4 & 1 & 0 \\ 8 & 5 & 1 \end{bmatrix} \begin{bmatrix} y_1 \\ y_2 \\ y_3 \end{bmatrix} = \begin{bmatrix} 3 \\ 3 \\ 2 \end{bmatrix} \tag{A.3.73}$$

is solved by forward substitution,

$$y_1 = 3 \tag{A.3.74}$$
$$y_2 = 3 - 4y_1 = -9 \tag{A.3.75}$$
$$y_3 = 2 - 8y_1 - 5y_2 = 23, \tag{A.3.76}$$

where note that the solutions $y_1, y_2,$ and $y_3$ are identical to the rightmost column in Eq. ( A.3.65 ). Finally, the upper triangular linear system

$$\begin{bmatrix} 1 & 1 & 2 \\ 0 & -1 & -3 \\ 0 & 0 & 1 \end{bmatrix} \begin{bmatrix} x_1 \\ x_2 \\ x_3 \end{bmatrix} = \begin{bmatrix} 3 \\ -9 \\ 23 \end{bmatrix} \tag{A.3.77}$$

is solved by back substitution, as has been done in Eqs. (A.3.66)–(A.3.68), and yields the solutions $x_1 = 17$, $x_2 = -60$, and $x_3 = 23$.

## Iterative Methods

An iterative method involves repeated application of an algorithm to update an estimate of the solution until it converges. The iterative methods are generally simpler to program and require less memory compared to the direct methods. Classic iterative methods such as the Jacobi and Gauss-Seidel algorithms are referred to as point iterative methods. In these methods, a linear iteration algorithm is defined as

$$\mathbf{x}^k = \mathbf{G}\mathbf{x}^{k-1} + \mathbf{R}, \tag{A.3.78}$$

where $\mathbf{x}^k$ is the approximation at $k$th step and $\mathbf{G}$, $\mathbf{R}$ depend on the system matrices $\mathbf{P}$ and $\mathbf{c}$ in Eq. (A.3.58). In the limiting case when $k \to \infty$, $\mathbf{x}^k$ has to converge to the exact solution

$$\mathbf{x}_k = \mathbf{P}^{-1}\mathbf{c}. \tag{A.3.79}$$

Substituting into Eq. (A.3.78),

$$\mathbf{P}^{-1}\mathbf{c} = \mathbf{G}\mathbf{P}^{-1}\mathbf{c} + \mathbf{R}, \tag{A.3.80}$$

which yields

$$\mathbf{R} = (\mathbf{I} - \mathbf{G})\mathbf{P}^{-1}\mathbf{c} \tag{A.3.81}$$
$$= \mathbf{M}\mathbf{c}, \tag{A.3.82}$$

where $\mathbf{M} = (\mathbf{I} - \mathbf{G})\mathbf{P}^{-1}$. In the point iterative methods, the matrix $\mathbf{P}$ is partitioned as

$$\mathbf{P} = \mathbf{L} + \mathbf{D} + \mathbf{U}, \tag{A.3.83}$$

where $\mathbf{L}$ is a strictly lower triangular matrix, $\mathbf{D}$ is a diagonal matrix, and $\mathbf{U}$ is a strictly upper triangular matrix. The various linear point iterative methods differ in the choice of $\mathbf{G}$ and $\mathbf{M}$.

*Jacobi method:* In this method, the iteration matrices are given by

$$\mathbf{G} = -\mathbf{D}^{-1}(\mathbf{L} + \mathbf{U}). \tag{A.3.84}$$
$$\mathbf{M} = -\mathbf{D}^{-1}. \tag{A.3.85}$$

*Gauss-Seidel method:* In this method, the iteration matrices are given by

$$\mathbf{G} = -(\mathbf{L} + \mathbf{D})^{-1}\mathbf{U}. \tag{A.3.86}$$

$$\mathbf{M} = (\mathbf{L} + \mathbf{D})^{-1}. \tag{A.3.87}$$

There are several other advanced iterative methods available in the literature which are more efficient and stable (Saad, 2003).

## A.3.9  Pseudoinverse

The earlier discussion on matrix inversion is for a square matrix. However, some applications such as the least squares estimation require inversion of a rectangular matrix. For a rectangular matrix, a pseudoinverse can be defined. The pseudoinverse $\mathbf{P}^+$ of an $m \times n$ matrix $\mathbf{P}$ is an $n \times m$ matrix satisfying

$$\mathbf{P}\mathbf{P}^+\mathbf{P} = \mathbf{P}. \tag{A.3.88}$$

$$\mathbf{P}^+\mathbf{P}\mathbf{P}^+ = \mathbf{P}^+. \tag{A.3.89}$$

$$(\mathbf{P}\mathbf{P}^+)^T = \mathbf{P}\mathbf{P}^+. \tag{A.3.90}$$

$$(\mathbf{P}^+\mathbf{P})^T = \mathbf{P}^+\mathbf{P}. \tag{A.3.91}$$

Note that the properties, Eqs. (A.3.90) and (A.3.91), essentially require $\mathbf{P}\mathbf{P}^+$ and $\mathbf{P}^+\mathbf{P}$ to be symmetric matrices.

If $\mathbf{P}$ has linearly independent columns, that is rank($\mathbf{P}$) = $n$, then $\mathbf{P}^T\mathbf{P}$ is invertible and

$$\mathbf{P}^+ = (\mathbf{P}^T\mathbf{P})^{-1}\mathbf{P}^T. \tag{A.3.92}$$

Since $\mathbf{P}^+\mathbf{P} = \mathbf{I}$, $\mathbf{P}^+$ is called the *left inverse*. Similarly, if $\mathbf{P}$ has linearly independent rows, that is rank($\mathbf{P}$) = $m$, then $\mathbf{P}\mathbf{P}^T$ is invertible and

$$\mathbf{P}^+ = \mathbf{P}^T(\mathbf{P}\mathbf{P}^T)^{-1}. \tag{A.3.93}$$

Since $\mathbf{P}\mathbf{P}^+ = \mathbf{I}$, $\mathbf{P}^+$ is called the *right inverse*.

*Example of pseudoinverse:* Consider a rectangular matrix $\mathbf{P}$.

$$\mathbf{P} = \begin{bmatrix} 1 & 2 \\ 0 & 1 \\ 1 & 1 \end{bmatrix}. \tag{A.3.94}$$

The rank of $\mathbf{P}$ is 2 and hence the left inverse of $\mathbf{P}$ exists and can be computed using Eq. (A.3.92).

$$\mathbf{P}^+ = (\mathbf{P}^T\mathbf{P})^{-1}\mathbf{P}^T$$

$$= \left( \begin{bmatrix} 1 & 0 & 1 \\ 2 & 1 & 0 \end{bmatrix} \begin{bmatrix} 1 & 2 \\ 0 & 1 \\ 1 & 1 \end{bmatrix} \right)^{-1} \begin{bmatrix} 1 & 0 & 1 \\ 2 & 1 & 1 \end{bmatrix}$$

$$= \begin{bmatrix} 2 & 3 \\ 3 & 6 \end{bmatrix}^{-1} \begin{bmatrix} 1 & 0 & 1 \\ 2 & 1 & 1 \end{bmatrix} = \begin{bmatrix} 2 & -1 \\ -1 & 2/3 \end{bmatrix} \begin{bmatrix} 1 & 0 & 1 \\ 2 & 1 & 1 \end{bmatrix}$$

$$= \begin{bmatrix} 0 & -1 & 1 \\ 1/3 & 2/3 & -1/3 \end{bmatrix}. \tag{A.3.95}$$

One may verify that $\mathbf{P}^+\mathbf{P}$ is an identity matrix, that is,

$$\mathbf{P}^+\mathbf{P} = \begin{bmatrix} 0 & -1 & 1 \\ 1/3 & 2/3 & -1/3 \end{bmatrix} \begin{bmatrix} 1 & 2 \\ 0 & 1 \\ 1 & 1 \end{bmatrix} = \begin{bmatrix} 1 & 0 \\ 0 & 1 \end{bmatrix} = \mathbf{I}. \tag{A.3.96}$$

Therefore, $\mathbf{P}^+$ satisfies the pseudoinverse property, Eq. (A.3.91), that $\mathbf{P}^+\mathbf{P}$ is symmetric. Next, one can also verify that the pseudoinverse properties, Eqs. (A.3.88) and (A.3.89) are satisfied:

$$\mathbf{P}\mathbf{P}^+\mathbf{P} = \mathbf{P}(\mathbf{P}^+\mathbf{P}) = \mathbf{P}\mathbf{I} = \mathbf{P}. \tag{A.3.97}$$

$$\mathbf{P}^+\mathbf{P}\mathbf{P}^+ = (\mathbf{P}^+\mathbf{P})\mathbf{P}^+ = \mathbf{I}\mathbf{P}^+ = \mathbf{P}^+. \tag{A.3.98}$$

Finally, for property, Eq. (A.3.90), one may verify by the following calculation

$$\mathbf{P}\mathbf{P}^+ = \begin{bmatrix} 1 & 2 \\ 0 & 1 \\ 1 & 1 \end{bmatrix} \begin{bmatrix} 0 & -1 & 1 \\ 1/3 & 2/3 & -1/3 \end{bmatrix} = \begin{bmatrix} 2/3 & 1/3 & 1/3 \\ 1/3 & 2/3 & -1/3 \\ 1/3 & -1/3 & 2/3 \end{bmatrix}. \tag{A.3.99}$$

It is clear that $\mathbf{P}\mathbf{P}^+$ is a symmetric matrix, and hence $(\mathbf{P}\mathbf{P}^+)^T = \mathbf{P}\mathbf{P}^+$.

## A.4    Linear Transformation

### A.4.1    Definitions

Using the unit axis vectors, a vector $\mathbf{x}$ on the vector space $\mathbf{X}$ is expressed as

$$\mathbf{x} = x_1\mathbf{e}_1 + \cdots + x_n\mathbf{e}_n. \tag{A.4.1}$$

The multiplication of an $n \times n$ matrix $\mathbf{A}$ and the vector $\mathbf{x}$ can be regarded as a *linear transformation* on the vector space $\mathbf{X}$ that maps a vector $\mathbf{x}$ to a vector $\tilde{\mathbf{x}}$,

$$\tilde{\mathbf{x}} = \mathbf{A}\mathbf{x}. \tag{A.4.2}$$

In other words, $\mathbf{A}$ represents a linear transformation under the unit axis vectors.

Alternatively, $\mathbf{x}$ can be expressed using an arbitrary vector basis $\mathbf{p}_1, \ldots, \mathbf{p}_n$ of $\mathbf{X}$ as

$$\mathbf{x} = a_1\mathbf{p}_1 + \cdots + a_n\mathbf{p}_n \tag{A.4.3}$$

or more conveniently in a form of matrix-vector multiplication,

$$\mathbf{x} = \mathbf{P}\mathbf{a} \tag{A.4.4}$$

where $\mathbf{P}$ is a matrix with the basis vectors as the columns and $\mathbf{a}$ is the vector of coordinates.

$$\mathbf{P} = [\mathbf{p}_1, \ldots, \mathbf{p}_n], \quad \mathbf{a} = [a_1, \ldots, a_n]^T. \tag{A.4.5}$$

The vector $\mathbf{a}$ is a representation of $\mathbf{x}$ in the vector basis $\mathbf{P}$.

Similar to Eq. (A.4.4), the transformed vector $\tilde{\mathbf{x}}$ is expressed using the same set of vector basis $\mathbf{P}$,

$$\tilde{\mathbf{x}} = \mathbf{P}\tilde{\mathbf{a}}, \tag{A.4.6}$$

where $\tilde{\mathbf{a}}$ is the vector of coordinates of $\tilde{\mathbf{x}}$ and a representation of $\tilde{\mathbf{x}}$ in the vector basis $\mathbf{P}$. The matrix $\mathbf{P}$ is called the *transformation matrix* or the change-of-basis matrix.

It is of interest to find a matrix $\mathbf{B}$ that represents the *same* linear transformation as $\mathbf{A}$, but using the vector basis $\mathbf{P}$. That is, to find $\mathbf{B}$ that transforms $\mathbf{a}$ to $\tilde{\mathbf{a}}$. This matrix $\mathbf{B}$ is found by replacing $\mathbf{x}$ and $\tilde{\mathbf{x}}$ in Eq. (A.4.2) using Eqs. (A.4.4) and (A.4.6), respectively.

$$\mathbf{P}\tilde{\mathbf{a}} = \mathbf{A}(\mathbf{P}\mathbf{a})$$
$$\tilde{\mathbf{a}} = \mathbf{P}^{-1}\mathbf{A}\mathbf{P}\mathbf{a} \equiv \mathbf{B}\mathbf{a}, \tag{A.4.7}$$

where

$$\mathbf{B} = \mathbf{P}^{-1}\mathbf{A}\mathbf{P}. \tag{A.4.8}$$

Equation (A.4.8) is called *similarity transformation*, and the two matrices $\mathbf{A}$ and $\mathbf{B}$ are said to be *similar*.

## A.4.2 Orthonormal Transformation

When the transformation matrix $\mathbf{P}$ is orthonormal, that is, $\mathbf{P}^{-1} = \mathbf{P}^T$, the similarity transformation is called *orthonormal transformation*, and $\mathbf{A}$ and $\mathbf{B}$ are related by

$$\mathbf{B} = \mathbf{P}^T\mathbf{A}\mathbf{P}. \tag{A.4.9}$$

The orthonormal transformation preserves the symmetry. That means, if $\mathbf{A}$ is symmetric, then $\mathbf{B}$ is symmetric too.

$$\mathbf{B}^T = (\mathbf{P}^T\mathbf{A}\mathbf{P})^T = \mathbf{P}^T\mathbf{A}^T\mathbf{P} = \mathbf{P}^T\mathbf{A}\mathbf{P} = \mathbf{B}. \tag{A.4.10}$$

## A.4.3 Matrix Diagonalization

An important vector basis for linear transformation is the one constructed using the eigenvectors of a square matrix. In many cases, the similarity transformation based on the eigenvectors can be used to diagonalize a square matrix.

### Eigenvalues and Eigenvectors

The standard eigenvalue problem of an $n \times n$ real matrix $\mathbf{A}$ is defined as

$$\mathbf{Au} = \lambda\mathbf{u}, \tag{A.4.11}$$

where $\lambda$ is the *eigenvalue* and $\mathbf{u}$ is the *eigenvector*.

The eigenvalues of an $n \times n$ matrix $\mathbf{A}$ are the roots of the *characteristic polynomial $p(\lambda)$*.

$$p(\lambda) \equiv |\mathbf{A} - \lambda\mathbf{I}| = 0. \tag{A.4.12}$$

The characteristic polynomial always have $n$ roots in the complex domain, which correspond to the $n$ eigenvalues of $\mathbf{A}$. The largest magnitude of the eigenvalues of $\mathbf{A}$

$$\rho(\mathbf{A}) = \max_i |\lambda_i|, \quad i = 1, 2, \cdots, n \tag{A.4.13}$$

is defined as the *spectral radius* of $\mathbf{A}$. The spectral radius is a useful quantity to characterize the distribution of eigenvalues.

Some of the eigenvalues may have identical values. Assuming that there are $k$ distinct values of eigenvalues, $\lambda_1, \lambda_2, \cdots, \lambda_k, k \leq n$, and the $i$th eigenvalue $\lambda_i$ is repeated $\mu_i$ times, the characteristic polynomial can be factored into the product of $k$ terms,

$$p(\lambda) = \prod_{i=1}^{k} (\lambda_i - \lambda)^{\mu_i}. \tag{A.4.14}$$

The factor $\mu_i$ is the *algebraic multiplicity* of the eigenvalue $\lambda_i$, and the total algebraic multiplicity equals to the dimension of $\mathbf{A}$, that is,

$$\sum_{i=1}^{k} \mu_k = n \tag{A.4.15}$$

The eigenvectors $\mathbf{u}$ associated with the $i$th eigenvalue $\lambda_i$ are the vectors that satisfy

$$(\mathbf{A} - \lambda_i\mathbf{I})\mathbf{u} = 0. \tag{A.4.16}$$

The eigenvectors $\mathbf{u}$ span a subspace called the *eigenspace*. The eigenvectors are usually represented by an orthonormal vector basis of the eigenspace. The dimension of the eigenspace associated with eigenvalue $\lambda_i$ is the *geometric multiplicity* of $\lambda_i$, denoted by $\gamma_i$. For any $n \times n$ matrix,

$$1 \leq \gamma_i \leq \mu_i \leq n. \tag{A.4.17}$$

The total geometric multiplicity of $\mathbf{A}$ is $\gamma = \sum_{i=1}^{k} \gamma_k$, and

$$k \leq \gamma \leq n, \tag{A.4.18}$$

where $k$ is the number of distinct eigenvalues of $\mathbf{A}$. The total geometric multiplicity equals to the number of linearly independent eigenvectors of $\mathbf{A}$.

A square matrix is *non-defective* if and only if the algebraic and geometric multiplicities of all eigenvalues coincide, that is, $\mu_i = \gamma_i$ for $i = 1, \cdots, k$. Otherwise, the matrix is *defective*.

*Example of non-defective matrix:* Consider a $3 \times 3$ matrix

$$\mathbf{A} = \begin{bmatrix} 1/2 & 0 & 1 \\ 0 & 1 & 0 \\ 0 & 0 & 1 \end{bmatrix}. \tag{A.4.19}$$

The characteristic polynomial of $\mathbf{A}$ is

$$p(\lambda) = |\mathbf{A} - \lambda\mathbf{I}| = (1/2 - \lambda)(1 - \lambda)^2. \tag{A.4.20}$$

The eigenvalues are therefore

$$\lambda_1 = 1/2, \quad \lambda_2 = \lambda_3 = 1. \tag{A.4.21}$$

The eigenvalue of $1/2$ has an algebraic multiplicity of 1, while the eigenvalue of 1 has an algebraic multiplicity of 2.

The eigenvector associated with $\lambda = \lambda_1 = 1/2$ satisfies

$$(\mathbf{A} - \lambda\mathbf{I})\mathbf{u} \equiv \begin{bmatrix} 0 & 0 & 1 \\ 0 & 1/2 & 0 \\ 0 & 0 & 1/2 \end{bmatrix} \begin{bmatrix} u_1 \\ u_2 \\ u_3 \end{bmatrix} = 0, \tag{A.4.22}$$

which simplifies to $u_2 = u_3 = 0$ while $u_1$ is arbitrary. Choosing $u_1 = 1$, the eigenvector associated with $\lambda = 1/2$ is

$$\mathbf{u}_1 = [1, 0, 0]^T. \tag{A.4.23}$$

The eigenvector associated with $\lambda = \lambda_2 = 1$ satisfies

$$(\mathbf{A} - \lambda\mathbf{I})\mathbf{u} \equiv \begin{bmatrix} -1/2 & 0 & 1 \\ 0 & 0 & 0 \\ 0 & 0 & 0 \end{bmatrix} \begin{bmatrix} u_1 \\ u_2 \\ u_3 \end{bmatrix} = 0, \tag{A.4.24}$$

which simplifies to $u_1 = 2u_3$ while $u_2$ is arbitrary. Choosing $u_2 = 1$ and $u_3 = 0$, one eigenvector associated with $\lambda = 1$ is

$$\mathbf{u}_2 = [0, 1, 0]^T. \tag{A.4.25}$$

Choosing $u_2 = 0$ and $u_3 = 1$, another eigenvector associated with $\lambda = 1$ is

$$\mathbf{u}_3 = [2, 0, 1]^T. \tag{A.4.26}$$

The eigenvectors $\mathbf{u}_2$ and $\mathbf{u}_3$ are linearly independent.

There is one eigenvector associated with the eigenvalue of $1/2$ and hence its geometric multiplicity is 1. There are two eigenvectors associated with the eigenvalue of 1 and hence its geometric multiplicity is 2. The algebraic and geometric multiplicities of all the eigenvalues coincide and therefore $\mathbf{A}$ is non-defective.

*Example of defective matrix:* Consider a $3 \times 3$ matrix

$$\mathbf{B} = \begin{bmatrix} 1/2 & 0 & 1 \\ 0 & 1 & 1 \\ 0 & 0 & 1 \end{bmatrix}. \tag{A.4.27}$$

The characteristic polynomial of $\mathbf{B}$ is the same as the one in the previous example, Eq. (A.4.20), and the eigenvalues are therefore $\lambda_1 = 1/2$ and $\lambda_2 = \lambda_3 = 1$. The algebraic multiplicities for $\lambda = 1/2$ and $\lambda = 1$ are 1 and 2, respectively.

The procedure for finding the eigenvector associated with $\lambda = \lambda_1 = 1/2$ is similar to the procedure in Eq. (A.4.22), and the eigenvector is

$$\mathbf{u}_1 = [1, 0, 0]^T. \tag{A.4.28}$$

Therefore, the eigenvalue of $1/2$ has a geometric multiplicity of 1, which coincides with its algebraic multiplicity.

The eigenvector associated with $\lambda = \lambda_2 = 1$ satisfies

$$(\mathbf{A} - \lambda\mathbf{I})\mathbf{u} \equiv \begin{bmatrix} -1/2 & 0 & 1 \\ 0 & 0 & 1 \\ 0 & 0 & 0 \end{bmatrix} \begin{bmatrix} u_1 \\ u_2 \\ u_3 \end{bmatrix} = 0, \tag{A.4.29}$$

which simplifies to $u_1 = u_3 = 0$ while $u_2$ is arbitrary. Choosing $u_2 = 1$, the eigenvector associated with $\lambda = 1$ is

$$\mathbf{u}_2 = [0, 1, 0]^T \tag{A.4.30}$$

and $\mathbf{u}_2$ is the *only* linearly independent eigenvector for the eigenvalue $\lambda = 1$. Therefore, unlike the previous example, the eigenvalue of 1 has a geometric multiplicity of 1, which differs from its algebraic multiplicity. This indicates that $\mathbf{B}$ is a defective matrix.

### Diagonalization of a Non-Defective Matrix

When an $n \times n$ matrix $\mathbf{A}$ is non-defective, it has $n$ linearly independent eigenvectors, and it can be diagonalized using a transformation matrix constructed by its eigenvectors, that is,

$$\mathbf{P} = [\mathbf{u}_1, \mathbf{u}_2, \cdots, \mathbf{u}_n] \tag{A.4.31}$$

and $\mathbf{A}$ is diagonalized as

$$\mathbf{\Lambda} = \mathbf{P}^{-1}\mathbf{A}\mathbf{P} \quad \text{or} \quad \mathbf{A} = \mathbf{P}\mathbf{\Lambda}\mathbf{P}^{-1}, \tag{A.4.32}$$

where $\mathbf{\Lambda}$ is a diagonal matrix whose diagonal elements are $\lambda_1, \lambda_2, \cdots, \lambda_n$.

*Example of matrix diagonalization:* Consider the $3 \times 3$ non-defective matrix $\mathbf{A}$ in Eq. (A.4.19 ). Its eigenvalues and eigenvectors have been found in Eqs. (A.4.21), (A.4.23 ), (A.4.25), and (A.4.26). One may construct a transformation matrix

$$\mathbf{P} = [\mathbf{u}_1, \mathbf{u}_2, \mathbf{u}_3] = \begin{bmatrix} 1 & 0 & 2 \\ 0 & 1 & 0 \\ 0 & 0 & 1 \end{bmatrix} \tag{A.4.33}$$

and diagonalize the matrix $\mathbf{A}$

$$\mathbf{P}^{-1}\mathbf{A}\mathbf{P} = \begin{bmatrix} 1 & 0 & -2 \\ 0 & 1 & 0 \\ 0 & 0 & 1 \end{bmatrix} \begin{bmatrix} 1/2 & 0 & 1 \\ 0 & 1 & 0 \\ 0 & 0 & 1 \end{bmatrix} \begin{bmatrix} 1 & 0 & 2 \\ 0 & 1 & 0 \\ 0 & 0 & 1 \end{bmatrix} = \begin{bmatrix} 1/2 & 0 & 0 \\ 0 & 1 & 0 \\ 0 & 0 & 1 \end{bmatrix}. \tag{A.4.34}$$

### A.4.4  Canonical Form of a Matrix

The defective matrices cannot be diagonalized, but can be transformed to an almost-diagonal form, called the *Jordan normal form* or *Jordan canonical form*. The Jordan normal form, usually denoted by $\mathbf{J}$, has nonzero elements only on the diagonal and the first upper diagonal. The diagonal consists of the eigenvalues of $\mathbf{A}$ while the upper diagonal elements are either 1 or 0.

Suppose an $n \times n$ matrix $\mathbf{A}$ has $\gamma$ linearly independent eigenvectors $\mathbf{u}_1, \cdots, \mathbf{u}_\gamma$, and each eigenvector is associated with an eigenvalue $\lambda_i$ with algebraic multiplicity $\mu_i$ and geometric multiplicity $\gamma_i$. The Jordan normal form of $\mathbf{A}$ is a block-diagonal matrix

$$\mathbf{J} = \begin{bmatrix} \mathbf{J}_1 & 0 & 0 & 0 \\ 0 & \mathbf{J}_2 & 0 & 0 \\ 0 & 0 & \ddots & 0 \\ 0 & 0 & 0 & \mathbf{J}_\gamma \end{bmatrix}, \tag{A.4.35}$$

where each $\mathbf{J}_i$ is called a *Jordan block* that is associated with eigenvalue $\lambda_i$

$$\mathbf{J}_i = \begin{bmatrix} \lambda_i & 1 & 0 & 0 \\ 0 & \lambda_i & \ddots & 0 \\ 0 & 0 & \ddots & 1 \\ 0 & 0 & 0 & \lambda_i \end{bmatrix}. \tag{A.4.36}$$

In the Jordan block $\mathbf{J}_i$, every diagonal element is the $i$th eigenvalue of $\mathbf{A}$ and every super-diagonal element is 1. The size of $\mathbf{J}_i$ equals to $(\mu_i - \gamma_i + 1)$. In other words, if the algebraic and geometric multiplicities of $\lambda_i$ coincide, then the Jordan block is $1 \times 1$, that is, simply a scalar of $\lambda_i$. When a matrix is non-defective, its Jordan normal form simplifies to be a diagonal matrix consisting of the eigenvalues.

The construction of the transformation matrix that transforms a defective matrix to its Jordan normal form is out of the scope of this book. However, the numerical methods for computing the Jordan normal form of a matrix are available in many standard linear algebra software tools or packages. The interested readers are referred to Strang (2017) for more details. In the following, an example of Jordan normal form is provided.

*Example of Jordan normal form:* Consider the $3 \times 3$ defective matrix $\mathbf{B}$ in Eq. (A.4.27). The Jordan normal form of $\mathbf{B}$ is

$$\mathbf{J} = \begin{bmatrix} \mathbf{J}_1 & 0 \\ 0 & \mathbf{J}_2 \end{bmatrix} = \begin{bmatrix} 1/2 & 0 & 0 \\ 0 & 1 & 1 \\ 0 & 0 & 1 \end{bmatrix}. \tag{A.4.37}$$

The first Jordan block $\mathbf{J}_1$ corresponds to the eigenvalue 1 that has algebraic and geometric multiplicities of 1. The second Jordan block $\mathbf{J}_2$ corresponds to the eigenvalue 2 that has an algebraic multiplicity of 2 and a geometric multiplicity of 1.

For this particular example, the transformation matrix $\mathbf{P}$ in Eq. (A.4.33) may be used to transform $\mathbf{B}$ into its Jordan normal form,

$$\mathbf{P}^{-1}\mathbf{B}\mathbf{P} = \begin{bmatrix} 1 & 0 & -2 \\ 0 & 1 & 0 \\ 0 & 0 & 1 \end{bmatrix} \begin{bmatrix} 1/2 & 0 & 1 \\ 0 & 1 & 1 \\ 0 & 0 & 1 \end{bmatrix} \begin{bmatrix} 1 & 0 & 2 \\ 0 & 1 & 0 \\ 0 & 0 & 1 \end{bmatrix} = \begin{bmatrix} 1/2 & 0 & 0 \\ 0 & 1 & 1 \\ 0 & 0 & 1 \end{bmatrix}. \tag{A.4.38}$$

## A.4.5 Matrix Functions

Scalar functions $f(x)$, such as powers, exponentials, and trigonometric functions, may be extended to matrix functions $f(\mathbf{A})$ that maps a square matrix $\mathbf{A}$ to another matrix of the same size. This extension is achieved using the canonical form of a matrix.

### General Matrix Functions

For diagonalizable matrices, $\mathbf{A} = \mathbf{P}\mathbf{\Lambda}\mathbf{P}^{-1}$, the matrix function is defined as

$$f(\mathbf{A}) = \mathbf{P}f(\mathbf{\Lambda})\mathbf{P}^{-1}, \tag{A.4.39}$$

where

$$f(\mathbf{\Lambda}) = \begin{bmatrix} f(\lambda_1) & 0 & 0 \\ 0 & \ddots & 0 \\ 0 & 0 & f(\lambda_n) \end{bmatrix}. \tag{A.4.40}$$

For non-diagonalizable matrices, $\mathbf{A} = \mathbf{P}\mathbf{J}\mathbf{P}^{-1}$, the matrix function is defined as

$$f(\mathbf{A}) = \mathbf{P}f(\mathbf{J})\mathbf{P}^{-1}, \tag{A.4.41}$$

where

$$f(\mathbf{J}) = \begin{bmatrix} f(\mathbf{J}_1) & 0 & 0 \\ 0 & \ddots & 0 \\ 0 & 0 & f(\mathbf{J}_\gamma) \end{bmatrix} \tag{A.4.42}$$

and for a Jordan block of size $m \times m$,

$$f(\mathbf{J}) = \begin{bmatrix} f(\lambda) & f'(\lambda) & \frac{1}{2}f''(\lambda) & \cdots & \frac{1}{(m-1)!}f^{(m-1)}(\lambda) \\ 0 & f(\lambda) & f'(\lambda) & \cdots & \frac{1}{(m-2)!}f^{(m-2)}(\lambda) \\ \vdots & \vdots & \ddots & \ddots & \vdots \\ 0 & 0 & \cdots & f(\lambda) & f'(\lambda) \\ 0 & 0 & \cdots & 0 & f(\lambda) \end{bmatrix}. \tag{A.4.43}$$

## Matrix Power and Matrix Exponential

Two matrix functions of particular interest in this book are the matrix power $f(\mathbf{A}) = \mathbf{A}^k$ and the matrix exponential $f(\mathbf{A}) = \exp(\mathbf{A})$.

For diagonalizable matrices, the power and exponential functions are computed as, respectively

$$\mathbf{A}^k = \mathbf{P}\mathbf{\Lambda}^k\mathbf{P}^{-1} \equiv \mathbf{P} \begin{bmatrix} \lambda_1^k & 0 & 0 \\ 0 & \ddots & 0 \\ 0 & 0 & \lambda_n^k \end{bmatrix} \mathbf{P}^{-1}. \tag{A.4.44}$$

$$\exp(\mathbf{A}) = \mathbf{P}\exp(\mathbf{\Lambda})\mathbf{P}^{-1} \equiv \mathbf{P} \begin{bmatrix} \exp(\lambda_1) & 0 & 0 \\ 0 & \ddots & 0 \\ 0 & 0 & \exp(\lambda_n) \end{bmatrix} \mathbf{P}^{-1}. \tag{A.4.45}$$

For non-diagonalizable matrices, the computation of power and exponential functions depends on the specific form of the Jordan block. Take an example of a $2 \times 2$ matrix $\mathbf{B}$ having an eigenvalue $\lambda$ of algebraic multiplicity 2 and geometric multiplicity 1. The Jordan normal form of $\mathbf{B}$ consists of only one Jordan block, and the canonical form is

$$\mathbf{B} = \mathbf{P}\mathbf{J}\mathbf{P}^{-1} \equiv \mathbf{P} \begin{bmatrix} \lambda & 1 \\ 0 & \lambda \end{bmatrix} \mathbf{P}^{-1}. \tag{A.4.46}$$

Using Eq. (A.4.43), the power and exponential functions are computed as, respectively

$$\mathbf{B}^k = \mathbf{P}\mathbf{J}^k\mathbf{P}^{-1} \equiv \mathbf{P} \begin{bmatrix} \lambda^k & k\lambda^{k-1} \\ 0 & \lambda^k \end{bmatrix} \mathbf{P}^{-1}. \tag{A.4.47}$$

$$\exp(\mathbf{B}) = \mathbf{P}\exp(\mathbf{\Lambda})\mathbf{P}^{-1} \equiv \mathbf{P} \begin{bmatrix} \exp(\lambda) & \exp(\lambda) \\ 0 & \exp(\lambda) \end{bmatrix} \mathbf{P}^{-1}. \tag{A.4.48}$$

## Boundedness of Matrix Powers

It is of interest to examine the boundedness of $\mathbf{A}^k$ as $k \to \infty$. For the diagonalizable case, Eq. (A.4.44), $\mathbf{P}$ is a constant matrix and only the term $\mathbf{\Lambda}^k$ depends on $k$; therefore, the boundedness of $\mathbf{A}^k$ is equivalent to that of $\mathbf{\Lambda}^k$. The term $\mathbf{\Lambda}^k$ is bounded if the absolute value of each diagonal element, that is, the eigenvalue, is less than or equal to 1. The reverse is true as well. Therefore, the matrix power $\mathbf{A}^k$ is bounded when $k \to \infty$ if and only if

$$|\lambda_i| \le 1, \quad i = 1, 2, \ldots, n \tag{A.4.49}$$

or

$$\rho(\mathbf{A}) \le 1. \tag{A.4.50}$$

For the non-diagonalizable case, Eq. (A.4.47), the boundedness of $\mathbf{A}^k$ is equivalent to that of $\mathbf{J}^k$, which requires each element in $\mathbf{J}^k$ to be bounded. In order to have terms such as

$k\lambda^{k-1}$ to be bounded when $k \to \infty$, the absolute value of the eigenvalue needs to be *strictly* less than 1. Therefore, the power of a defective matrix is bounded if and only if

$$\rho(\mathbf{A}) < 1. \tag{A.4.51}$$

### Examples of Matrix Powers

*Example for the diagonalizable case:* Consider the $3 \times 3$ non-defective matrix $\mathbf{A}$ in Eq. (A.4.34).

$$\mathbf{A} = \begin{bmatrix} 1/2 & 0 & 1 \\ 0 & 1 & 0 \\ 0 & 0 & 1 \end{bmatrix}$$

$$= \mathbf{P\Lambda P}^{-1} = \begin{bmatrix} 1 & 0 & 2 \\ 0 & 1 & 0 \\ 0 & 0 & 1 \end{bmatrix} \begin{bmatrix} 1/2 & 0 & 0 \\ 0 & 1 & 0 \\ 0 & 0 & 1 \end{bmatrix} \begin{bmatrix} 1 & 0 & -2 \\ 0 & 1 & 0 \\ 0 & 0 & 1 \end{bmatrix}. \tag{A.4.52}$$

Note that spectral radius of $\mathbf{A}$ is 1.

Using Eq. (A.4.44), the $k$th power of $\mathbf{A}$ is

$$\mathbf{A}^k = \begin{bmatrix} 1 & 0 & 2 \\ 0 & 1 & 0 \\ 0 & 0 & 1 \end{bmatrix} \begin{bmatrix} (1/2)^k & 0 & 0 \\ 0 & 1 & 0 \\ 0 & 0 & 1 \end{bmatrix} \begin{bmatrix} 1 & 0 & -2 \\ 0 & 1 & 0 \\ 0 & 0 & 1 \end{bmatrix}. \tag{A.4.53}$$

Using Eq. (A.4.53), when $k = 4$,

$$\mathbf{A}^4 = \begin{bmatrix} 1/16 & 0 & 15/8 \\ 0 & 1 & 0 \\ 0 & 0 & 1 \end{bmatrix}. \tag{A.4.54}$$

One may verify the correctness of Eq. (A.4.54) by computing $\mathbf{A}^4$ using matrix multiplication.

As $k \to \infty$, $(1/2)^k \to 0$, and

$$\lim_{k \to \infty} \mathbf{A}^k = \begin{bmatrix} 1 & 0 & 2 \\ 0 & 1 & 0 \\ 0 & 0 & 1 \end{bmatrix} \begin{bmatrix} 0 & 0 & 0 \\ 0 & 1 & 0 \\ 0 & 0 & 1 \end{bmatrix} \begin{bmatrix} 1 & 0 & -2 \\ 0 & 1 & 0 \\ 0 & 0 & 1 \end{bmatrix} = \begin{bmatrix} 0 & 0 & 2 \\ 0 & 1 & 0 \\ 0 & 0 & 1 \end{bmatrix} \tag{A.4.55}$$

and therefore the matrix power $\mathbf{A}^k$ is bounded when $k \to \infty$.

*Example for the non-diagonalizable case:* Consider the $3 \times 3$ defective matrix $\mathbf{B}$ in Eq. (A.4.38).

$$\mathbf{B} = \begin{bmatrix} 1/2 & 0 & 1 \\ 0 & 1 & 1 \\ 0 & 0 & 1 \end{bmatrix}$$

$$= \mathbf{PJP}^{-1} = \begin{bmatrix} 1 & 0 & 2 \\ 0 & 1 & 0 \\ 0 & 0 & 1 \end{bmatrix} \begin{bmatrix} 1/2 & 0 & 0 \\ 0 & 1 & 1 \\ 0 & 0 & 1 \end{bmatrix} \begin{bmatrix} 1 & 0 & -2 \\ 0 & 1 & 0 \\ 0 & 0 & 1 \end{bmatrix}. \tag{A.4.56}$$

Note that spectral radius of $\mathbf{B}$ is 1, which is the same as the previous example.

Using Eq. (A.4.47), the $k$th power of $\mathbf{B}$ is

$$\mathbf{B}^k = \begin{bmatrix} 1 & 0 & 2 \\ 0 & 1 & 0 \\ 0 & 0 & 1 \end{bmatrix} \begin{bmatrix} (1/2)^k & 0 & 0 \\ 0 & 1 & k \\ 0 & 0 & 1 \end{bmatrix} \begin{bmatrix} 1 & 0 & -2 \\ 0 & 1 & 0 \\ 0 & 0 & 1 \end{bmatrix} \tag{A.4.57}$$

and

$$\lim_{k \to \infty} \mathbf{B}^k = \begin{bmatrix} 1 & 0 & 2 \\ 0 & 1 & 0 \\ 0 & 0 & 1 \end{bmatrix} \begin{bmatrix} 0 & 0 & 0 \\ 0 & 1 & k \\ 0 & 0 & 1 \end{bmatrix} \begin{bmatrix} 1 & 0 & -2 \\ 0 & 1 & 0 \\ 0 & 0 & 1 \end{bmatrix} = \begin{bmatrix} 0 & 0 & 2 \\ 0 & 1 & k \\ 0 & 0 & 1 \end{bmatrix}. \tag{A.4.58}$$

Therefore, the matrix power $\mathbf{B}^k$ is unbounded when $k \to \infty$, even if the spectral radius of $\mathbf{B}$ is 1.

## A.5    Applications in Multivariate Calculus

The vectors and matrices are used frequently in multivariate calculus to simplify the notation of partial derivatives.

### A.5.1    Gradients and Hessians

Consider a scalar function $f(\mathbf{x})$ with a vector of input $\mathbf{x} = [x_1, x_2, \ldots, x_n]^T$. The gradient vector of $f$ is denoted as $\nabla f$ and defined as

$$\nabla f = \begin{bmatrix} \frac{\partial f}{\partial x_1} \\ \frac{\partial f}{\partial x_2} \\ \vdots \\ \frac{\partial f}{\partial x_n} \end{bmatrix}. \tag{A.5.1}$$

The Hessian matrix of $f$ is denoted as $\nabla^2 f$ and defined as an $n \times n$ matrix,

$$\nabla^2 f = \begin{bmatrix} \frac{\partial^2 f}{\partial x_1^2} & \frac{\partial^2 f}{\partial x_1 \partial x_2} & \cdots & \frac{\partial^2 f}{\partial x_1 \partial x_n} \\ \frac{\partial^2 f}{\partial x_2 \partial x_1} & \frac{\partial^2 f}{\partial x_2^2} & \cdots & \frac{\partial^2 f}{\partial x_2 \partial x_n} \\ \vdots & \vdots & \ddots & \vdots \\ \frac{\partial^2 f}{\partial x_n \partial x_1} & \frac{\partial^2 f}{\partial x_n \partial x_2} & \cdots & \frac{\partial^2 f}{\partial x_n^2} \end{bmatrix}. \tag{A.5.2}$$

The $i$th column of Hessian matrix $\nabla^2 f$ can be viewed as the gradient vector of the $i$th element of $\nabla f$. Due to the property of partial derivatives, $\frac{\partial^2 f}{\partial x_i \partial x_j} = \frac{\partial^2 f}{\partial x_j \partial x_i}$, the Hessian matrix is symmetric.

Using the gradient and Hessian, the Taylor series expansion of a scalar function $f(\mathbf{x})$ is written up to the second order

$$f(\mathbf{x} + \Delta\mathbf{x}) = f(\mathbf{x}) + (\nabla f)^T \Delta\mathbf{x} + \frac{1}{2}(\Delta\mathbf{x})^T (\nabla^2 f)\Delta\mathbf{x} + O(\|\Delta\mathbf{x}\|^3). \tag{A.5.3}$$

A special multivariate function that is frequently encountered in structural dynamics is of the following quadratic form:

$$q(\mathbf{x}) = \frac{1}{2}\mathbf{x}^T \mathbf{A}\mathbf{x} = \frac{1}{2}\sum_{i=1}^{n}\sum_{j=1}^{n} a_{ij}x_i x_j. \tag{A.5.4}$$

Next the gradient vector and Hessian matrix of $q$ are derived.

The $k$th element of $\nabla q$ is

$$\frac{\partial q(\mathbf{x})}{\partial x_k} = \frac{\partial}{\partial x_k}\left(\frac{1}{2}\sum_{i=1}^{n}\sum_{j=1}^{n} a_{ij}x_i x_j\right)$$

$$= \frac{1}{2}\sum_{i=1}^{n}\sum_{j=1}^{n} a_{ij}\left(\frac{\partial x_i}{\partial x_k}x_j + x_i\frac{\partial x_j}{\partial x_k}\right)$$

$$= \frac{1}{2}\sum_{i=1}^{n}\sum_{j=1}^{n} a_{ij}\delta_{ik}x_j + a_{ij}x_i\delta_{jk}$$

$$= \frac{1}{2}\sum_{j=1}^{n} a_{kj}x_j + \frac{1}{2}\sum_{i=1}^{n} a_{ik}x_i, \tag{A.5.5}$$

where the Kronecker delta is used: $\delta_{ij} = 1$ if $i = j$, otherwise $\delta_{ij} = 0$. The gradient vector $\nabla q$ is therefore

$$\nabla q = \frac{1}{2}\left(\mathbf{A} + \mathbf{A}^T\right)\mathbf{x} \tag{A.5.6}$$

The $(k, l)$th element of Hessian matrix of $q$ is the partial derivative of $\frac{\partial q(\mathbf{x})}{\partial x_k}$ with respect to $x_l$. Utilizing Eq. (A.5.5)

$$\frac{\partial^2 q(\mathbf{x})}{\partial x_k \partial x_l} = \frac{\partial}{\partial x_l}\left(\frac{1}{2}\sum_{j=1}^{n} a_{kj}x_j + \frac{1}{2}\sum_{i=1}^{n} a_{ik}x_i\right)$$

$$= \frac{1}{2}\sum_{j=1}^{n} a_{kj}\delta_{jl} + \frac{1}{2}\sum_{i=1}^{n} a_{ik}\delta_{il}$$

$$= \frac{1}{2}(a_{kl} + a_{lk}). \tag{A.5.7}$$

The Hessian matrix $\nabla^2 q$ is therefore

$$\nabla^2 q = \frac{1}{2}\left(\mathbf{A} + \mathbf{A}^T\right). \tag{A.5.8}$$

When $\mathbf{A}$ is symmetric, Eqs. (A.5.6) and (A.5.8) simplify as

$$\nabla q = \nabla \left( \frac{1}{2} \mathbf{x}^T \mathbf{A} \mathbf{x} \right) = \mathbf{A} \mathbf{x}. \tag{A.5.9}$$

$$\nabla^2 q = \nabla^2 \left( \frac{1}{2} \mathbf{x}^T \mathbf{A} \mathbf{x} \right) = \mathbf{A}. \tag{A.5.10}$$

## A.5.2  Jacobians

Now consider a vector function $\mathbf{f}(\mathbf{x})$ with a vector of input $\mathbf{x} = [x_1, x_2, \ldots, x_n]^T$ and a vector of output $\mathbf{y} = [y_1, y_2, \ldots, y_m]^T$. The Jacobian matrix, or in short Jacobian, of $\mathbf{f}$ is denoted as $\nabla \mathbf{f}$ and defined as an $m \times n$ matrix

$$\nabla \mathbf{f} = \begin{bmatrix} \frac{\partial f_1}{\partial x_1} & \frac{\partial f_1}{\partial x_2} & \cdots & \frac{\partial f_1}{\partial x_n} \\ \frac{\partial f_2}{\partial x_1} & \frac{\partial f_2}{\partial x_2} & \cdots & \frac{\partial f_2}{\partial x_n} \\ \vdots & \vdots & \ddots & \vdots \\ \frac{\partial f_m}{\partial x_1} & \frac{\partial f_m}{\partial x_2} & \cdots & \frac{\partial f_m}{\partial x_n} \end{bmatrix}. \tag{A.5.11}$$

The Hessian matrix $\nabla^2 f$ of a scalar function $f$ may be considered as the Jacobian matrix of the gradient vector $\nabla f$.

Using the Jacobian, the Taylor series expansion of a vector function $\mathbf{f}(\mathbf{x})$ is written up to the first order,

$$\mathbf{f}(\mathbf{x} + \Delta \mathbf{x}) = \mathbf{f}(\mathbf{x}) + (\nabla \mathbf{f}) \Delta \mathbf{x} + O(\|\Delta \mathbf{x}\|^2). \tag{A.5.12}$$

## BIBLIOGRAPHY

Bathe, K.-J. and Wilson, E. L. (1976). *Numerical Methods in Finite Element Analysis*. Prentice Hall.

Craig Jr, R. R. and Kurdila, A. J. (2006). *Fundamentals of Structural Dynamics*. John Wiley & Sons.

Meirovitch, L. (1997) *Principles and Techniques of Vibrations*, volume 1. Prentice Hall.

Noble, B. (1987). *Applied Linear Algebra*. Pearson Education International, 3rd edition.

Norrie, D. H. and De Vries, G. (1978). *An Introduction to Finite Element Analysis*. Academic Press.

Saad, Y. (2003). *Iterative Methods for Sparse Linear Systems*. SIAM, 2nd edition.

Strang, G. (2017). *Introduction to Linear Algebra*. Wellesley-Cambridge Press, 5th edition.

# APPENDIX B

# Laplace Transforms

The Laplace transform is a mathematical tool that can be used to simplify the solution of linear ordinary differential equations (ODEs). It converts commonly used functions such as the trigonometric and exponential functions into algebraic functions of a complex variable $s$. Integration and differentiation in the time domain are replaced by algebraic operations in the complex plane. Thus, linear ODEs encountered in structural dynamics can be transformed into algebraic equations in the Laplace domain $s$. Once the algebraic equation is solved for $s$, an inverse Laplace transform can be used to find a solution to the ODE in the time domain. The transform and its inverse constitute a transform pair, and for many transformations in use there are summary tables available.

This appendix describes the Laplace transform method and its application to some typical functions and operations encountered in structural dynamics. A table of typically used Laplace transform pairs is also provided.

## B.1  Laplace Transform

### B.1.1  Complex Variables and Functions

A complex number has a real part and an imaginary part. In the Laplace domain, the complex variable

$$s = \sigma + \iota \omega \tag{B.1.1}$$

is employed, where $\sigma$ is the real part and $\omega$ is the imaginary part. A complex function $F(s)$ is expressed as

$$F(s) = F_r + \iota F_i, \tag{B.1.2}$$

where $F_r$ is the real part and $F_i$ is the imaginary part. The magnitude of $F(s)$ is given as

$$|F(s)| = \sqrt{F_r^2 + F_i^2} \tag{B.1.3}$$

and the phase angle is

$$\phi(s) = \arctan\left(\frac{F_i}{F_r}\right). \tag{B.1.4}$$

The phase angle is measured counterclockwise from the positive real axis.

### B.1.2 Laplace Transform

Consider a function $f(t)$ defined for all $t \geq 0$. Its Laplace transform is defined as

$$\mathscr{L}[f(t)] = F(s) = \int_0^{\infty} e^{-st} f(t) dt. \tag{B.1.5}$$

The reverse process of finding the time function $f(t)$ from the Laplace transform $F(s)$ is called the inverse Laplace transform and is represented as

$$\mathscr{L}^{-1}[F(s)] = f(t). \tag{B.1.6}$$

The Laplace transform exists if the function $f(t)$ is piecewise continuous and the Laplace integral converges. The integral will converge if $f(t)$ is continuous in every finite interval in $t > 0$ and if it is of exponential order as $t$ approaches infinity, that is,

$$|e^{-st} f(t)| < C e^{-(s-a)t}, \quad \text{Re } s > a, \tag{B.1.7}$$

where $C$ is a constant. The inequality in Eq. (B.1.7) implies that $f(t)$ cannot increase faster than $Ce^{at}$ with increasing $t$.

### B.1.3 Euler's Theorem

Euler's theorem is used to derive Laplace transforms of trigonometric functions. Euler's theorem states that

$$\cos\theta + \iota \sin\theta = e^{\iota\theta}. \tag{B.1.8}$$

This can be shown by considering the power series expansions of the trigonometric functions $\cos\theta$ and $\sin\theta$:

$$\cos\theta = 1 - \frac{\theta^2}{2!} + \frac{\theta^4}{4!} - \frac{\theta^6}{6!} + \cdots, \tag{B.1.9}$$

$$\sin\theta = \theta - \frac{\theta^3}{3!} + \frac{\theta^5}{5!} - \frac{\theta^7}{7!} + \cdots. \tag{B.1.10}$$

Therefore,

$$\cos\theta + \iota \sin\theta = 1 + (\iota\theta) + \frac{(\iota\theta)^2}{2!} + \frac{(\iota\theta)^3}{3!} + \frac{(\iota\theta)^4}{4!} + \cdots = e^{\iota\theta}. \tag{B.1.11}$$

Since the complex conjugate

$$\cos\theta - \iota \sin\theta = e^{-\iota\theta}, \tag{B.1.12}$$

it can be shown that

$$\cos\theta = \frac{1}{2}(e^{\iota\theta} + e^{-\iota\theta}), \tag{B.1.13}$$

$$\sin\theta = \frac{1}{2\iota}(e^{\iota\theta} - e^{-\iota\theta}). \tag{B.1.14}$$

## B.1.4    Examples

Laplace transforms of a few commonly encountered functions are derived next.

### Exponential Function

Consider the exponential function

$$f(t) = Ae^{-\alpha t}, \tag{B.1.15}$$

where $A$ and $\alpha$ are constants. The Laplace transform of this exponential function can be obtained as

$$\mathscr{L}[Ae^{-\alpha t}] = \int_0^\infty Ae^{-\alpha t}e^{-st}dt = A\int_0^\infty e^{-(\alpha+s)t}dt = \frac{A}{s+\alpha}. \tag{B.1.16}$$

### Step Function

Consider the step function

$$f(t) = 0 \text{ for } t < 0 \tag{B.1.17}$$

$$= Ae^{-\alpha t} \text{ for } t > 0, \tag{B.1.18}$$

where $A$ is a constant. Note that the step function is a special case of the exponential function $Ae^{-\alpha t}$, with $\alpha = 0$. Therefore, the Laplace transform is given by

$$\mathscr{L}[f(t)] = \int_0^\infty Ae^{-st}dt = \frac{A}{s}. \tag{B.1.19}$$

### Sine Function

Consider the sine function

$$f(t) = A\sin\omega t, \tag{B.1.20}$$

where $A$ and $\omega$ are constants. Using the Euler's theorem in Eq. (B.1.14),

$$\sin\omega t = \frac{1}{2\iota}(e^{\iota\omega t} - e^{\iota\omega t}). \tag{B.1.21}$$

Hence,

$$\mathscr{L}[A\sin\omega t] = \frac{A}{2\iota}\int_0^\infty (e^{\iota\omega t} - e^{\iota\omega t})e^{-st}dt \tag{B.1.22}$$

$$= \frac{A}{2\iota}\frac{1}{s-\iota\omega} - \frac{A}{2\iota}\frac{1}{s+\iota\omega} \tag{B.1.23}$$

$$= \frac{A\omega}{s^2+\omega^2}. \tag{B.1.24}$$

Similarly, it can be shown that

$$\mathcal{L}[A\cos\omega t] = \frac{As}{s^2 + \omega^2}. \tag{B.1.25}$$

## B.2    Properties of Laplace Transformation

Next, a few important properties of the Laplace transform are presented.

If a function $f(t)$ has a Laplace transform, then it can be shown from Eq. (B.1.5) that

$$\mathcal{L}[af(t)] = a\mathcal{L}[f(t)]. \tag{B.2.1}$$

Similarly, for a sum of functions $f_1(t)$ and $f_2(t)$, the Laplace transform is

$$\mathcal{L}[f_1(t) + f_2(t)] = \mathcal{L}[f_1(t)] + \mathcal{L}[f_2(t)]. \tag{B.2.2}$$

### B.2.1    Laplace Transform of a Derivative

Solving ODEs using Laplace transforms requires transformation of derivatives of $f(t)$. Assuming that the Laplace transform of $f(t)$ exists, the Laplace transform of the derivative is given by

$$\mathcal{L}[\frac{df(t)}{dt}] = \int_0^\infty e^{-st}\frac{df(t)}{dt}dt \tag{B.2.3}$$

$$= e^{-st}f(t)|_0^\infty - \int_0^\infty (-se^{-st})f(t)dt \tag{B.2.4}$$

$$= -f(0) + sF(s). \tag{B.2.5}$$

In general, for the $n$th derivative of $f(t)$.

$$\mathcal{L}[\frac{d^n f(t)}{dt^n}] = s^n F(s) - f^{(n-1)}(0) - sf^{(n-1)}(0) - \cdots - s^{(n-1)}f(0), \tag{B.2.6}$$

where

$$f^{(n-j)}(0) = \frac{d^{(n-j)}f(t)}{dt^{(n-j)}}|_{t=0}. \tag{B.2.7}$$

### B.2.2    Final Value Theorem

The final value theorem relates the steady-state behavior of $f(t)$ to the behavior of $sF(s)$ in the neighborhood of $s = 0$. This theorem applies if and only if $\lim_{t\to\infty} f(t)$ exists, that is if $f(t)$ settles down to a definite value. The theorem states that if $f(t)$ and $df(t)/dt$ have Laplace transforms, then

$$\lim_{t\to\infty} f(t) = \lim_{s\to 0} sF(s). \tag{B.2.8}$$

### B.2.3    Laplace Transform of an Integral

If the Laplace transform $F(s) = \mathscr{L}[f(t)]$ exists, then

$$\mathscr{L}[\int f(t)dt] = \frac{F(s)}{s} + \frac{f^{-1}(0)}{s}, \tag{B.2.9}$$

showing that integration in the time domain is converted into division in the $s$ domain.

### B.2.4    Complex Differentiation

If $F(s) = \mathscr{L}[f(t)]$, then, except at the poles of $F(s)$,

$$\mathscr{L}[tf(t)] = -\frac{d}{ds}F(s). \tag{B.2.10}$$

In general,

$$\mathscr{L}[t^n f(t)] = (-1)^n \frac{d^n}{ds^n}F(s). \tag{B.2.11}$$

### B.2.5    Convolution Integral

Consider two functions $f_1(t)$ and $f_2(t)$, with Laplace transforms $F_1(s)$ and $F_2(s)$, respectively. Then, the convolution integral is given as

$$f_1(t) * f_2(t) = \int_0^t f_1(\tau)f_2(t-\tau)d\tau = \int_0^\infty f_1(t-\tau)f_2(\tau)d\tau, \tag{B.2.12}$$

where $*$ represents the convolution operation. The Laplace transform of the convolution integral is given as

$$\mathscr{L}[f_1(t) * f_2(t)] = F_1(s)F_2(s). \tag{B.2.13}$$

Conversely, if the Laplace transform of a function is given by a product of two Laplace transform functions $F_1(s)$ and $F_2(s)$, then the corresponding time function is given by the convolution integral $f_1(t) * f_2(t)$.

### B.2.6    Shifting Theorem

Consider the Laplace transform of $e^{-at}f(t)$, where $a$ is real or complex:

$$\mathscr{L}[e^{-at}f(t)] = \int_0^\infty e^{-at}f(t)e^{-st}dt = F(s+a). \tag{B.2.14}$$

The effect of multiplying $f(t)$ by $e^{at}$ in the time domain is to shift the transform of $F(s)$ by $a$ in the $s$ domain.

### B.2.7   Laplace Transform of ODEs

As an example, consider the differential equation of motion for a mass–spring–damper system

$$m\ddot{x}(t) + c\dot{x}(t) + kx(t) = f(t), \tag{B.2.15}$$

where $m$, $c$, and $k$ are the mass, damping, and spring constants, respectively, $x(t)$ is the displacement of $m$, and $F(t)$ is the external force acting on $m$. Applying the Laplace transform on both sides of Eq. (B.2.15) yields

$$m[s^2 X(s) - sx(0) - \dot{x}(0)] + c[sX(s) - x(0)] + kX(s) = F(s), \tag{B.2.16}$$

where $x(0)$ and $\dot{x}(0)$ are the initial displacement and velocity, respectively. Solving for $X(s)$ results in

$$X(s) = \frac{1}{ms^2 + cs + k}F(s) + \frac{ms + c}{ms^2 + cs + k}x(0) + \frac{m}{ms^2 + cs + k}\dot{x}(0). \tag{B.2.17}$$

An inverse Laplace transform of $X(s)$ will yield the appropriate solution to $x(t)$.

## B.3   Inverse Laplace Transformation

Converting the Laplace transform, expressed in terms of a complex variable, back into the time domain is called an inverse Laplace transform and is represented as

$$\mathcal{L}^{-1}[F(s)] = f(t). \tag{B.3.1}$$

Mathematically, $f(t)$ is found from $F(s)$ by evaluating the following inversion integral

$$f(t) = \frac{1}{2\pi \iota} \int_{c-\iota\infty}^{c+\iota\infty} e^{st} F(s)ds, \tag{B.3.2}$$

where the path of integration is a line parallel to the imaginary axis crossing the real axis at $c$ and extending from $-\infty$ to $+\infty$. A convenient method for obtaining inverse Laplace transforms is by using Laplace transform tables. If a particular transform $F(s)$ cannot be found in a table, then it can be expanded using partial fractions and written in terms of simple functions of $s$ for which the inverse Laplace transforms are already known.

### B.3.1   Method of Partial Fractions

In dynamic systems, the Laplace transform $F(s)$ frequently appears in the form

$$F(s) = \frac{A(s)}{B(s)}, \tag{B.3.3}$$

where $A(s)$ and $B(s)$ are polynomials in $s$. In general, $B(s)$ is a polynomial of higher degree than $A(s)$, and it can be written in a factored form

$$B(s) = (s - b_1)(s - b_2)\ldots(s - b_n), \tag{B.3.4}$$

where $b_1, b_2, \ldots, b_n$ are the roots of the polynomial. In the case where all the roots are distinct, the Laplace transform can be expressed as

$$F(s) = \frac{A(s)}{B(s)} = \frac{C_1}{s - b_1} + \frac{C_2}{s - b_2} + \cdots + \frac{C_n}{s - b_n}, \qquad (B.3.5)$$

where the coefficients $C_k$ are given by

$$C_k = \lim_{s \to b_k} [(s - b_k)F(s)]. \qquad (B.3.6)$$

Using the table of Laplace transforms provided in the next section, and the shifting theorem in Eq. (B.2.14), we have

$$\mathscr{L}^{-1}[\frac{1}{s - b_k}] = e^{b_k t}, \qquad (B.3.7)$$

so that the inverse Laplace transform of Eq. (B.3.5) is

$$f(t) = \mathscr{L}^{-1}[F(s)] = C_1 e^{b_1 t} + C_2 e^{b_2 t} + \cdots + C_n e^{b_n t}. \qquad (B.3.8)$$

## B.4    Table of Laplace Transform Pairs

Several commonly used Laplace transform pairs are given in Table B.1. An example illustrating the application of Laplace transform to solve differential equations is provided below.

### B.4.1    Example

Find the solution $x(t)$ of the differential equation

$$\ddot{x} + 2\dot{x} + 5x = 3, \; x(0) = 0, \; \dot{x}(0) = 0. \qquad (B.4.1)$$

Taking the Laplace transform of the differential equation,

$$s^2 X(s) + 2s X(s) + 5X(s) = \frac{3}{s}. \qquad (B.4.2)$$

Solving for X(s), we find

$$X(s) = \frac{3}{s(s^2 + 2s + 5)} \qquad (B.4.3)$$

$$= \frac{3}{5s} - \frac{3}{5}\frac{s + 2}{s^2 + 2s + 5} \qquad (B.4.4)$$

$$= \frac{3}{5s} - \frac{3}{10}\frac{2}{(s + 1)^2 + 2^2} - \frac{3}{5}\frac{s + 1}{(s + 1)^2 + 2^2}. \qquad (B.4.5)$$

**Table B.1 Table of Laplace transforms**

| $f(t)$ | $F(s)$ |
| --- | --- |
| Unit impulse function $\delta(t)$ | $1$ |
| Unit step function $1(t)$ | $\frac{1}{s}$ |
| $t$ | $\frac{1}{s^2}$ |
| $t^n$ | $\frac{n!}{s^{n+1}}$ |
| $e^{-at}$ | $\frac{1}{s+a}$ |
| $te^{-at}$ | $\frac{1}{(s+a)^2}$ |
| $\cos \omega t$ | $\frac{s}{s^2+\omega^2}$ |
| $\sin \omega t$ | $\frac{\omega}{s^2+\omega^2}$ |
| $\cosh \omega t$ | $\frac{s}{s^2-\omega^2}$ |
| $\sinh \omega t$ | $\frac{\omega}{s^2-\omega^2}$ |
| $e^{-at} \cos \omega t$ | $\frac{s+a}{(s+a)^2+\omega^2}$ |
| $e^{-at} \sin \omega t$ | $\frac{\omega}{(s+a)^2+\omega^2}$ |
| $1 - e^{-at}$ | $\frac{a}{s(s+a)}$ |
| $1 - \cos \omega t$ | $\frac{\omega^2}{s(s^2+\omega^2)}$ |
| $\omega t - \sin \omega t$ | $\frac{\omega^3}{s^2(s^2+\omega^2)}$ |
| $\omega t \cos \omega t$ | $\frac{\omega(s^2-\omega^2)}{(s^2+\omega^2)^2}$ |
| $\omega t \sin \omega t$ | $\frac{2\omega^2 s}{(s^2+\omega^2)^2}$ |
| $t \cosh \omega t$ | $\frac{(s^2+\omega^2)}{(s^2-\omega^2)^2}$ |
| $t \sinh \omega t$ | $\frac{2\omega s}{(s^2-\omega^2)^2}$ |
| $\frac{\omega}{\sqrt{1-\zeta^2}} e^{-\zeta \omega t} \sin(\omega\sqrt{1-\zeta^2}t)$ | $\frac{\omega^2}{s^2+2\zeta\omega s+\omega^2}$ |
| $-\frac{1}{\sqrt{1-\zeta^2}} e^{-\zeta \omega t} \sin(\omega\sqrt{1-\zeta^2}t - \tan^{-1}\frac{\sqrt{1-\zeta^2}}{\zeta})$ | $\frac{s}{s^2+2\zeta\omega s+\omega^2}$ |

Taking the inverse Laplace transform

$$x(t) = \mathscr{L}^{-1}[X(s)] \tag{B.4.6}$$

$$= \frac{3}{5}\mathscr{L}^{-1}[\frac{1}{s}] - \frac{3}{10}\mathscr{L}^{-1}[\frac{2}{(s+1)^2+2^2}] - \frac{3}{5}\mathscr{L}^{-1}[\frac{s+1}{(s+1)^2+2^2}] \tag{B.4.7}$$

$$= \frac{3}{5} - \frac{3}{10}e^{-t}\sin 2t - \frac{3}{5}e^{-t}\cos 2t, \tag{B.4.8}$$

which is the solution for the given differential equation.

## BIBLIOGRAPHY

Meirovitch, L. (1997). *Principles and Techniques of Vibrations*, Prentice Hall Inc.
Ogata, K. (1996). *Modern Control Engineering*, Prentice Hall.
Reid, J. G. (1983). *Linear System Fundamentals: Continuous and Discrete, Classic and Modern*, McGraw-Hill.
Schiff, J. L. (1999). *The Laplace Transform: Theory and Applications*, Springer-Verlag.

# APPENDIX C

# Gaussian Quadrature

The value of the integral

$$\int_a^b f(x)dx = \int_a^b p_n(x)dx + \int_a^b R_n(x)dx \tag{C.1}$$

is approximated with an $n$th degree interpolating polynomial $p_n(x)$, where $R_n(x)$ is the error term of the $n$th degree interpolating polynomial. Since the $x_i$ (or interpolation points) are not yet specified, the Lagrangian form of the interpolating polynomial, which permits arbitrarily spaced points, is used together with its error term

$$f(x) = p_n(x) + R_n(x)$$
$$= \sum_{i=0}^{n} L_i(x)f(x_i) + \left[\prod_{i=0}^{n}(x - x_i)\right]\frac{f^{(n+1)}(\xi)}{(n+1)!}, \tag{C.2}$$
where $a < \xi < b$

$$\text{and } L_i(x) = \prod_{\substack{j=0 \\ j \neq i}}^{n} \left(\frac{x - x_j}{x - x_i}\right). \tag{C.3}$$

To simplify the development somewhat, but without removing any generality of the result, the interval of integration will be changed from $[a, b]$ to $[-1, 1]$ by a suitable transformation of variable. Assume that all the base points are in the interval of integration, that is, $a \leq x_0, x_1, \ldots, x_n \leq b$. Let the new variable $z$ be defined by

$$z = \frac{2x - (a+b)}{(b-a)} \tag{C.4 a}$$

$$\text{also} \quad x = \frac{b-a}{2}z + \frac{a+b}{2}, \tag{C.4 b}$$

obviously $z$ varies between $-1 \leq z \leq 1$.

Also define a new function $F(z)$ such that

$$F(z) = f(x) = f\left[\frac{(b-a)z + (a+b)}{2}\right]. \tag{C.5}$$

With this transformation, Eq. ( C.2 ) becomes

$$F(z) = \sum_{i=0}^{n} L_i(z)F(z) + \left[\prod_{i=0}^{n}(z - z_i)\right] \frac{F^{(n+1)}(\bar{\xi})}{(n+1)!}, \tag{C.6}$$

$$\text{where} \qquad L_i(z) = \prod_{\substack{j=0 \\ j\neq i}}^{n} \frac{(z - z_j)}{(z_i - z_j)} \qquad \text{and} \qquad -1 \leq \bar{\xi} \leq 1. \tag{C.7}$$

With these relations, the weights and the error estimate can be obtained. The derivations can be found in Carnahan et al. (1969).

The error term is given by

$$E_n(z) = \frac{z^{2n+3}[(n+1)!]^4}{(2n+3)[(2n+2)!]^3} F^{(2n+2)}(\bar{\xi}). \tag{C.8}$$

The integral is obtained from

$$\int_{-1}^{1} F(z)dz = \sum_{i=0}^{n} w_i F(z_i), \tag{C.9}$$

where the weights $w_i$ corresponding to the station $z_i$ are given in Table C.1.

It should be noted that provided that the magnitudes of higher order derivatives decrease or do not increase substantially with increasing $n$, the Gauss–Legendre formulas are significantly more accurate than the equal-interval formulas for comparable values of $n$. *The reason being that the stations $z_i$ are selected so as to minimize the error term.*

## Table C.1 Roots of the Legendre Polynomials $P_{n+1}(z)$ and the Weight Factors for the Gauss–Legendre Quadrature

| Roots ($z_i$) | $\int_{-1}^{1} F(z)dz = \sum_{i=0}^{n} w_i F(z_i)$ | Weight Factors ($w_i$) |
|---|---|---|
| | *Two-Point Formula* | |
| | $n = 1$ | |
| ±0.57735 02691 89626 | | 1.00000 00000 00000 |
| | *Three-Point Formula* | |
| | $n = 2$ | |
| 0.00000 00000 00000 | | 0.88888 88888 88889 |
| ±0.77459 66692 41483 | | 0.55555 55555 55556 |
| | *Four-Point Formula* | |
| | $n = 3$ | |
| ±0.33998 10435 84856 | | 0.65214 51548 62546 |
| ±0.86113 63115 94053 | | 0.34785 48451 37454 |
| | *Five-Point Formula* | |
| | $n = 4$ | |
| 0.00000 00000 00000 | | 0.56888 88888 88889 |
| ±0.53846 93101 05683 | | 0.47862 86704 99366 |
| ±0.90617 98459 38664 | | 0.23692 68850 56189 |

## Table C.1 (cont.)

| Roots ($z_i$) | $\int_{-1}^{1} F(z)dz = \sum_{i=0}^{n} w_i F(z_l)$ | Weight Factors ($w_i$) |
|---|---|---|
| | *Six-Point Formula* $n = 5$ | |
| ±0.23861 91860 83197 | | 0.46791 39345 72691 |
| ±0.66120 93864 66265 | | 0.36076 15730 48139 |
| ±0.93246 95142 03152 | | 0.17132 44923 79170 |
| | *Ten-Point Formula* $n = 9$ | |
| ±0.14887 43389 81631 | | 0.29552 42247 14753 |
| ±0.43339 53941 29247 | | 0.26926 67193 09996 |
| ±0.67940 95682 99024 | | 0.21908 63625 15982 |
| ±0.86506 33666 88985 | | 0.14945 13491 50581 |
| ±0.97390 65285 17172 | | 0.06667 13443 08688 |
| | *Fifteen-Point Formula* $n = 14$ | |
| ±0.00000 00000 00000 | | 0.20257 82419 25561 |
| ±0.20119 40939 97435 | | 0.19843 14853 27111 |
| ±0.39415 13470 77563 | | 0.18616 10001 15562 |
| ±0.57097 21726 08539 | | 0.16626 92058 16994 |
| ±0.72441 77313 60170 | | 0.13957 06779 26154 |
| ±0.84820 65834 10427 | | 0.10715 92204 67172 |
| ±0.93727 33924 00706 | | 0.07036 60474 88108 |
| ±0.98799 25180 20485 | | 0.03075 32419 96117 |

# BIBLIOGRAPHY

Carnahan, B., Luther, H., and Wilkes, J. (1969). *Applied Numerical Methods*, pages 101–105. John Wiley & Sons.

# APPENDIX D

# Pseudocode for Describing Algorithms

A computer algorithm is a finite sequence of well-defined operations or *statements*. The sequence always starts from a statement specifying the required inputs to the algorithm and ends with a statement specifying the output produced by the algorithm.

The numerical algorithms in this book, such as Algorithm A.1 on Page 265, are presented in the form of pseudocode. Pseudocode generally does not actually obey the syntax rules of any particular computer language. The convention used in this book is described later.

A minimal example of an algorithm is illustrated in Algorithm A.1. The line that starts with "Require" defines the input, a number $N$; Line 1 performs a multiplication $M = 2N$; Line 2, which starts with "return," specifies the output $M$. In Line 1, the text placed to the right of the triangle is a *comment*. The comment provides an explanation to the statement, but is not involved in the execution of the algorithm. If one supplies an input of $N = 2$ to Algorithm A.1, the algorithm returns $M = 2 \times 2 = 4$.

---
**Algorithm A.1** Minimal example of an algorithm

**Require:** A number $N$.
  1: Compute $M = 2N$.                                            ▷ $M$ is two times of $N$
  2: **return** The computed number $M$.

---

There are three special statements that are used in this book: the *assign* statement ($\leftarrow$), the *if*-statement, and the *for*-loop.

The *assign* statement sets the value to a variable

$$\text{variable} \leftarrow \text{value},$$

where the value can be an expression. An example snippet of pseudocode, consisting of two *assign* statements, is presented in Algorithm A.2. In Line 1, a value of 1 is assigned to a variable $a$. In Line 2, the expression $3a + 1$ is first evaluated to be 4, which is subsequently assigned *back* to $a$. Therefore, in Line 3, a value of 4, instead of 1, is returned.

---
**Algorithm A.2** Example of assign statements

**Require:** A number $a$.
  1: $a \leftarrow 1$                                                        ▷ Set $a = 1$
  2: $a \leftarrow 3a + 1$                                                  ▷ Now $a = 4$
  3: **return** The computed number $a$.

---

The *if*-statement performs different computations depending on whether a user-specified condition evaluates to be true or false. An example is presented in Algorithm A.3. If the input $a$ is an odd number, then compute $3a + 1$ and assign the new value to $a$; otherwise, compute $a/2$ and assign the new value to $a$. For instance, when $a = 3$, Line 2 will be executed and the value $a = 3 \times 3 + 1 = 10$ is returned; when $a = 4$, Line 4 will be executed and the value $a = 4/2 = 2$ is returned.

---
**Algorithm A.3** Example of *if*-statement

---
**Require:** A number $a$.
1: **if** $a$ is odd **then**
2:     $a \leftarrow 3a + 1$
3: **else**
4:     $a \leftarrow a/2$
5: **end if**
6: **return** The computed number $a$.

---

If nothing needs to be done when the user-specified condition is false, the "else" branch can be omitted in the *if*-statement. An example is presented in Algorithm A.4. If the input $a$ is an odd number, then compute $3a + 1$, assign the new value to $a$, and return this new value; otherwise, $a$ remains unchanged and is returned as the output.

---
**Algorithm A.4** Example of *if*-statement without *else*

---
**Require:** A number $a$.
1: **if** $a$ is odd **then**
2:     $a \leftarrow 3a + 1$
3: **end if**
4: **return** The computed number $a$.

---

The *for*-loop specifies iterations, which execute a set of statements repeatedly. An example is presented in Algorithm A.5, where the operation "add 1 to $N$" is repeated four times. For instance, when $N$ is 2 initially, after the execution of the loop, $N$ becomes 6.

---
**Algorithm A.5** Example of *for*-statement

---
**Require:** A number $N$.
1: **for** $i = 1, \ldots, 4$ **do**
2:     $N \leftarrow N + 1$                                    ▷ Add 1 to $N$
3: **end for**
4: **return** The computed number $N$.

---

It is typical to incorporate an *if*-statement in the *for*-loop, which will serve as a *termination* condition. An example is presented in Algorithm A.6, where the operation "add 1 to $N$" is repeated either four times or until $N$ is larger than 8. For instance, when $N$ is 2 initially, the loop will execute four times, and $N$ becomes 6. However, when $N$ is 6 initially, the loop will only execute three times; when $N$ becomes 9, the condition in Line 3 is satisfied, and therefore Line 4 is executed to terminate the loop.

---

**Algorithm A.6** Example containing both *for-* and *if*-statements

---

**Require:** A number $N$.

1: **for** $i = 1, \ldots, 4$ **do**
2:     $N \leftarrow N + 1$                                    ▷ Add 1 to $N$
3:     **if** $N > 8$ **then**
4:         Break                                    ▷ Terminate the loop
5:     **end if**
6: **end for**
7: **return** The computed number $N$.

---

A final complete example of pseudocode is presented in Algorithm A.7, which contains all the three special statements. The algorithm applies a function

$$f(a) = \begin{cases} 3a + 1, & a \text{ is odd} \\ a/2, & \text{otherwise} \end{cases}$$

to an input integer repeatedly for either $N$ times or until the value of $a$ becomes 1. For instance, when $a = 5$, $N = 6$, intermediate values of $a$ computed by the loop are 16, 8, 4, 2, 1, and the loop terminates after the 5th iteration and the algorithm returns 1; when $a = 12$, $N = 6$, intermediate values of $a$ computed by the loop are 6, 3, 10, 5, 16, 8, and the loop completes all the six iterations and the algorithm returns 8.

---

**Algorithm A.7** An example algorithm

---

**Require:** An integer $a$, maximum number of iterations $N$.

1: **for** $i = 1, \ldots, N$ **do**
2:     **if** $a$ is odd **then**
3:         $a \leftarrow 3a + 1$
4:     **else**
5:         $a \leftarrow a/2$
6:     **end if**
7:     **if** $a = 1$ **then**
8:         Break
9:     **end if**
10: **end for**
11: **return** The computed number $a$.

---

# Index